The Superfluid Phases of Helium 3

The Superfluid Phases of Helium 3

DIETER VOLLHARDT

Rheinisch-Westfälische Technische Hochschule Aachen, Federal Republic of Germany

PETER WÖLFLE

Universität Karlsruhe Federal Republic of Germany

Taylor & Francis
London • New York • Philadelphia
1990

UK Taylor and Francis Ltd, 4 John St, London WC1N 2ET

USA Taylor and Francis Inc., 1900 Frost Road, Suite 101, Bristol, PA 19007

British Library Cataloguing in Publication Data

Vollhardt, Dieter
The superfluid phases of helium 3
1. Helium
I. Title II. Wölfle, Peter
546′.751588

ISBN 0-85066-412-8

Library of Congress Cataloging-in-Publication Data

Vollhardt, Dieter.
The superfluid phases of helium 3 / Dieter Vollhardt, Peter Wölfle.
p. cm.
Includes bibliographical references
ISBN 0-85066-412-8
1. Helium—Isotopes. 2. Liquid helium. 3. Superfluidity.
I. Wölfle, Peter. II. Title. III. Title: Superfluid phases of helium three.
QC145.45.H4V65 1990
530.4′2—dc20 90-10762
CIP

Cover design by Jordan and Jordan, Fareham, Hampshire, UK
Typeset in 10/12 point Times by
The Universities Press (Belfast) Ltd.

Printed in Great Britain by Burgess Science Press, Basingstoke on paper which has a specified pH value on final paper manufacture of not less than 7.5 and is therefore ‘acid free’.

Preface

The superfluid phases of ^{3}He were discovered in 1971, i.e. almost 20 years ago. Since then, around 2000 papers have been published on this subject. There is now consensus in the low temperature physics community that the basic physics of the superfluid phases of ^{3}He is well understood. Indeed, there are many examples where the theory has been refined to the point that excellent quantitative agreement with experiments is achieved.

We therefore feel that the time has come to write a book that collects not only the ideas and concepts on superfluid ^{3}He that have been developed in these years, but also the methods of theoretical physics used and the results obtained. We have attempted to arrive at a coherent presentation with little redundancy, while avoiding gaps or leaps in the argumentation. The enormous amount of material (the true extent of which we came to realize only after we had started the project) made it inevitable, however, that we treat some of the more technical theories, which require a longer exposition, as well as some of the more specialized topics, in a different way than we should have liked given infinite book space and working time.

The need for a book on superfluid ^{3}He has not been felt for a long time because several excellent reviews were available, in particular the early articles by Wheatley (1975) and Lee and Richardson (1978) on experiment and by Leggett (1975a) and Anderson and Brinkman (1975, 1978) on theory. The more recent reviews on superfluid ^{3}He have already started to specialize; these include the excellent articles by Brinkman and Cross (1978) on spin and orbital properties, by Volovik (1984b) on the superfluid properties of the A phase, by Hall and Hook (1986) on hydrodynamics, and by Fetter (1986) and Salomaa and Volovik (1987) on rotating superfluid ^{3}He. A general discussion of all aspects of superfluid ^{3}He clearly requires the form of a book.

This book is on the theory of superfluid ^{3}He, although contact with experiment is made in many specific cases. It is intended for readers with a background in quantum mechanics and statistical mechanics, but it does not assume previous knowledge of the theory of superfluid ^{4}He or of superconductivity, or indeed of any such more specialized subject in condensed matter theory. In order to keep the entrance barrier as low as possible, we have decided not to make use of advanced theoretical techniques such as Green's functions or functional-integral methods, since such techniques require the reader to have considerable training in many-body theory. In our view the limitations caused by this self-imposed restriction are still tolerable. Nevertheless, we have also attempted to address those topics requiring more elaborate techniques, so as to provide the reader with the information that is necessary to find his/her own way in the literature.

The first part of the book (Chapters 1–4) is designed to provide an introduction to the subject, while the later chapters are intended to give an exposition of the important new concepts that grew out of the more conventional theories in the course of time. Quite naturally, some of the topics—in particular those where we have ourselves been actively involved—have been covered in more detail than others. In general we have tried to illuminate the basic concepts rather than to stress the details of the actual calculations or the detailed comparison with experiment. We have gone to great pains to do justice to most of the theory papers that have been written on the subject and to at least reference the work if space has not permitted a somewhat longer discussion. The selection of a particular theoretical formulation in cases where several exist in the literature is necessarily subjective. We apologize to those who feel that their work has not been represented adequately, and we ask them to blame this on our ignorance and inability to do better, rather than on evil intention or presumptuousness. In any case, we welcome any feedback, be it positive or negative.

Superfluid ^{3}He is one of the most fascinating condensed matter systems yet discovered. It is a test system for most of the ideas of modern theoretical physics: (i) the simultaneous occurence of several broken symmetries, including non-Abelian gauge symmetries; (ii) the existence of a multitude of topological defects of the order-parameter field; (iii) the possibility of massless and massive collective modes of the order parameter. On the other hand, superfluid ^{3}He is the prototype system for anisotropic superfluidity. It has helped to shape our understanding of neutron stars, anisotropic superconductors such as may be realized in heavy-fermion systems, or even ceramic high-T_c superconductors. We hope that this book can serve as a source of education and information on superfluid ^{3}He, not only for those working in this very field, but also in particular for those who want to carry over the concepts and methods to other systems.

The book is divided into twelve chapters. Chapter 1 introduces the subject with a discussion of the qualitative properties of liquid ^{3}He as compared with liquid ^{4}He and other systems. The historical development that finally led to the discovery of the superfluid phases of ^{3}He is sketched. In addition, we present an extended elementary discussion of the properties of superfluid ^{3}He and the concepts involved. We believe that this will be of help to those readers who are not familiar with the subject at all. Chapter 2 presents the theory of normal liquid ^{3}He in as much as it is needed for the later development of the theory of the superfluid phases. It covers the phenomenological Fermi-liquid theory as well as recent attempts to calculate the effective interaction leading to superfluidity. Chapter 3 deals with the standard weak-coupling pairing theory of p-wave paired superfluids. The order parameter of such a superfluid is introduced. The ground state and single-particle excited states as well as thermodynamic properties are calculated explicitly for two model states, the so-called ABM and BW states. Chapter 4 provides a summary of important experimental observations as well as a sketch of the development of experimental techniques. The experimental data are found to support the identifications of the two observed superfluid phases with the two model states mentioned above, in spite of the fact that the weak-coupling theory of Chapter 3 predicts only one stable state, the BW state. This calls for a discussion of corrections to weak-coupling theory, presented in Chapter 5. The general Landau–Ginzburg representation of the free energy given there also allows for a convenient discussion of the effects of a uniform static magnetic field on the phase diagram and the order-parameter structure. Chapter 6 addresses the symmetry properties of the superfluid phases of ^{3}He. Broken symmetries and remaining symmetries are discussed in detail. A general classification scheme for p-wave order parameters is presented. In the final part of this chapter the effect of internal and external fields in orienting the (degenerate) preferred directions of the two order-parameter structures is considered. This sets the stage for a detailed discussion of spatially varying order-parameter configurations, so-called textures, and superflow in Chapter 7. After the derivation of the gradient free energies and bending free energies, superflow is discussed as a first application. Then the nature of defects in the two phases is discussed, first for an infinite system and later for finite geometries. Symmetry considerations and topological properties play an important role in this chapter. The properties of linear defects, or vortices, are studied in detail, with an application to rotating superfluid ^{3}He, as well as those of planar defects, or solitons. The effect of superflow on the textures and the superfluid state itself is also discussed. The remaining chapters deal mainly with dynamical phenomena. In Chapter 8 the theory of nuclear magnetic resonance in superfluid ^{3}He is presented. A complete solution of the weakly excited spin system is given for uniform as well as nonuniform textures. A few important nonlinear resonance phenomena are discussed in detail. The effect of relaxation processes is also considered. Chapter 9 contains a complete derivation of the phenomenological hydrodynamic equations, based only on symmetry principles. The independent thermodynamic derivatives and transport coefficients in these equations are identified. They are calculated from kinetic theory in Chapter 10, which contains a rather detailed exposition of the kinetic theory of anisotropic Fermi superfluids. Chapter 11 addresses the collective modes of the system, both in the hydrodynamic and collisionless regimes. The theory here is based on the hydrodynamic equations and the kinetic theory, and again symmetry considerations are of value. Finally, in Chapter 12 we take another look at macroscopic quantum coherence in superfluid ^{3}He and discuss some rather subtle consequences of broken parity and time-reversal symmetry in the two superfluid phases of ^{3}He.

We began writing this book in 1983 when both of us were in Munich, namely at the Max-Planck-Insitut für Physik and Astrophysik (DV and PW) and the Technische Universität München (PW). Since then, both of us have moved to different institutions: DV to the Technische Hochschule Aachen, and PW first to the University of Florida, Gainesville, and then to the Universität Karlsruhe. We gratefully acknowledge the support by these institutions for our book project.

In the course of writing we have had numerous discussions with colleagues and friends around the world, which have improved our understanding of the subject. These include many of our colleagues at the institutions with which we were associated during that period of time. In particular, we have profited from discussions with A. Buras, D. Einzel, G. Eska, I. Fomin, G. G. Ihas, P. Kumar, A. J. Leggett, M. Liu, K. Maki, R. Peccei, P. Ramond, H. Saller and L. Stodolsky. Special thanks go to D. Einzel, T. Kopp and M. Liu for critically reading parts of the manuscript and for suggesting explicit improvements. Above all, we are grateful to A. L. Fetter, H. E. Hall, J. R. Hook, A. J. Leggett and D. Rainer, who have been so generous with their time in aiding us with their criticism and comments on the completed manuscript. Our cordial thanks go to Mrs Wiltrud (Willi) Huber at the Max-Planck-Institut für Physik und Astrophysik, Munich, for her selfless effort and wonderful spirit in typing the largest part of the manuscript between 1983 and 1987, and for being so helpful in every respect. We are also extremely grateful to Mrs G. Kramp-Salecker for typing the final parts of the book and for her invaluable assistance in preparing and organizing the completed manuscript. Furthermore, we should like to thank Mrs I. Kuchenbecker at the Technische Universität München, as well as Mr H. Mattke and Mr R. Roeger at the Technische Hochschule Aachen for their careful drawing of the figures. Lastly, we are grateful to Andrea Wölfle and Jutta Muttenhammer-Vollhardt for their help with the references. We also acknowledge the smooth collaboration with the editorial staff of Taylor and Francis, in particular with Julie Lancashire.

Finally, and above all, we thank our wives Jutta and Uschi and our families for their endless support and unselfish sacrifice of many leisurely weekends during the past five years. Without them, this book would have hardly been finished.

D. VOLLHARDT, P. WÖLFLE

Contents

List of Frequently Used Symbols

Symbol	Meaning	Definition
Latin symbols		
A	prefactor of A-phase free energy	below (7.152) and in (7.233)
$\boldsymbol{A}$, $A_{\mu j}$ $(\mu, j = 1, 2, 3)$	normalized order-parameter matrix and components thereof	(5.1a)
C_{N}	specific heat in the normal phase	(2.14)
C_0	coefficient in A-phase supercurrent	(7.21c), (7.39)
$\boldsymbol{C}$, C_{ij} $(i, j = 1, 2, 3)$	tensor coefficient of A-phase bending energy and components thereof	(7.23b), (7.37b)
$C_{\mu m}(r)$ $(\mu, m = 0, +, -)$	normalized order-parameter matrix in basis of spherical harmonics	(7.86)
$\boldsymbol{d}(\boldsymbol{k})$, $d_\mu(\boldsymbol{k})$ $(\mu = 1, 2, 3)$	order-parameter vector in spin space and components thereof	(3.34b)
$\boldsymbol{d}$, $d_{\mu j}$ $(\mu, j = 1, 2, 3)$	order-parameter matrix for p-wave state	(3.64)
$\hat{\boldsymbol{d}}$	"$\boldsymbol{d}$ vector": preferred direction in spin of the A phase	(3.68)
$\hat{\boldsymbol{e}}, \hat{\boldsymbol{f}}$ $(\hat{\boldsymbol{d}} = \boldsymbol{e} \times \hat{\boldsymbol{f}})$	triad of unit vectors in	(5.27) and

	spin space	below (6.15)
$E_{\boldsymbol{k}}$	energy of Bogoliubov quasiparticle in state $\boldsymbol{k}$	(3.32)
$\mathbf{F}_{\boldsymbol{k}}$, $F_{\boldsymbol{k}\alpha\beta}$ $(\alpha, \beta = \uparrow, \downarrow)$	pair-amplitude matrix (off-diagonal distribution-function matrix) in spin space and components thereof	(3.43b)
F_l^{s}, F_l^{a}	Landau parameters of Fermi-liquid interaction	(2.6)
$\boldsymbol{g}$, g_i $(i = 1, 2, 3)$	momentum density and components thereof	(2.25)
$\boldsymbol{g}_{\mathrm{n}}$, $\boldsymbol{g}_{\mathrm{s}}$	momentum densities of normal and superfluid components	(3.85), (7.14b)
$g_{\mathrm{D}}(T)$	dipole coupling constant	(6.101)
$\boldsymbol{H}$	magnetic field	
H^*	characteristic magnetic field for orientation of A-phase order parameter	(7.28c)
$\boldsymbol{j}$	particle number current density	(2.21b)
$\boldsymbol{j}_\varepsilon$	energy current density	(2.23)
$\boldsymbol{j}_a$	heat current density	(9.73)
$\boldsymbol{j}_\mu^\sigma$ $(\mu = 1, 2, 3)$	spin current density for spin component S_μ	(2.24)
k_{F}	Fermi wavenumber	(2.1)
$\hat{\boldsymbol{l}}$	"$\hat{\boldsymbol{l}}$ vector": preferred direction of A-phase order parameter in orbital space	(3.70)
$\hat{\boldsymbol{m}}$, $\hat{\boldsymbol{n}}$ $(\hat{\boldsymbol{l}} = \hat{\boldsymbol{m}} \times \hat{\boldsymbol{n}})$	triad of unit vectors in orbital space	(3.68) and below (6.15)
m^*	effective mass of quasiparticles in the normal phase	(2.3)
m_{s}	"spin" effective mass	(10.44)
$\hat{\boldsymbol{n}}$	"$\hat{\boldsymbol{n}}$ vector": axis of spin–orbit rotation in B phase	(6.106)
$\mathbf{n}_{\boldsymbol{k}}$, $n_{\boldsymbol{k}\alpha\beta}$ $(\alpha, \beta = \uparrow, \downarrow)$	diagonal distribution-function matrix in spin space	(3.43a)
N_{F}	density of quasiparticle states of both spin components in the normal phase at the Fermi energy	(2.4a)

Symbol	Meaning	Reference
$N(0)=\frac{1}{2}N_{\mathrm{F}}$	density of states for one spin component	(3.7)
P	thermodynamic pressure	
$\mathbf{R}$, $R_{\mu j}$ ($\mu, j=1, 2, 3$)	spin–orbit rotation matrix and components thereof	(3.67)
$\boldsymbol{R}_{\mathrm{D}}$	dipole torque	(8.7)
$\boldsymbol{S}$	spin density	(2.9)
s	entropy density	(2.13)
T	temperature	
T_{F}	Fermi temperature	below (2.14)
T_{c}	transition temperature to superfluid state	(3.48)
v_{F}	Fermi velocity	(2.3)
$\boldsymbol{v}_{\mathrm{s}}$	superfluid velocity	(7.5), (7.10)
$\boldsymbol{v}_{\mathrm{n}}$	normal-fluid velocity	above (3.84)
$Y_0(T)$	averaged Yosida function	(3.101c)

Greek symbols

Symbol	Meaning	Reference
β_i ($i=1, \ldots, 5$)	coefficients of fourth-order invariants in free-energy functional	(5.4)
$\boldsymbol{\Delta}_{\boldsymbol{k}}$, $\Delta_{\boldsymbol{k}\alpha\beta}$ ($\alpha, \beta=\uparrow, \downarrow$)	energy-gap matrix in spin space and components thereof	(3.17), (3.34a)
$\Delta(T)$	average energy gap	(3.57)
$\Delta_0(T)$	maximum value of energy gap	(3.68a), (3.69)
$\hat{\boldsymbol{\Delta}}$	complex unit vector in orbital space (A phase)	(7.12a)
$\Delta_{\uparrow\uparrow}$, $\Delta_{\downarrow\downarrow}$	components of order parameter in a magnetic field (A phase)	(5.50)
$\Delta_{\perp}$, $\Delta_{\parallel}$	components of order parameter in a magnetic field (B phase)	below (5.66b)
θ_{L}	Leggett angle	below (6.107)
λ_{D}	dimensionless dipole coupling parameter	(6.98)
λ_l	coupling constant of pair interaction	(2.97)
μ	chemical potential	(2.3)
$\xi(T)$	temperature-dependent coherence length	(7.18c)

ξ_0	zero-temperature coherence length	(7.18b)
$\xi_{\boldsymbol{k}}$	quasiparticle energy in the normal state	(3.4)
ξ_D	dipole coherence length	(7.27a)
ξ_H^A, ξ_H^B	magnetic healing lengths in A and B phases	(7.28a), (7.29a)
ξ_s	surface healing length	(7.31a)
$\boldsymbol{\Pi}$, Π_{ij} $(i, j = 1, 2, 3)$	stress tensor and components thereof	(2.22)
ρ	mass density	below (2.10)
$\boldsymbol{\rho}_n$, $\boldsymbol{\rho}_n^0$	normal-fluid density tensor (with and without Fermi-liquid corrections)	(3.86), (3.87b)
$\boldsymbol{\rho}_\mathrm{s}$, $\boldsymbol{\rho}_\mathrm{s}^0$	superfluid density tensor (with and without Fermi-liquid corrections)	(7.15b), (7.23a)
$\boldsymbol{\sigma}_\mu$	μth Pauli spin matrix	below (2.6)
$\tau_{\boldsymbol{k}}$	lifetime of quasiparticle in state $\boldsymbol{k}$	(2.37)
$\tau_\mathrm{N}^0(T)$	quasiparticle lifetime at the Fermi level in the normal state	(2.38)
χ_N	Pauli spin susceptibility of normal liquid ^{3}He	(2.18)
ω_L	Larmor frequency	below (8.16)
Ω_A, Ω_B	longitudinal NMR frequencies in A and B phases	(8.20b), (8.24)

1

Introduction

1.1 THE HELIUM LIQUIDS

There are two stable isotopes of the chemical element helium: helium 3 and helium 4, conventionally denoted by ^{3}He and ^{4}He respectively. The existence of the heavier one, ^{4}He, had already been established indirectly in 1871 by its characteristic line in the solar spectrum (hence the name "helium"). Then in 1895 Ramsay succeeded in obtaining an actual sample of ^{4}He gas by heating the uranium ore cleveite. It was later found that ^{4}He is also part of the Earth's atmosphere, but only at a fraction of about 2×10^{-5}. Kamerlingh Onnes was the first to liquefy ^{4}He in 1908. For a fascinating account of the history of early helium physics and of low-temperature physics in general, we refer the reader to the books by Keesom (1942) and Mendelssohn (1977).

The discovery and identification of the lighter isotope ^{3}He was only made much later (Oliphant *et al.* 1933). In fact, at that time it was thought that the ^{3}He isotope was unstable and that tritium (^{3}H) was stable. Using a cyclotron as a mass spectrograph, Alvarez and Cornog (1939a,b) then showed that ^{3}He is indeed a *stable* isotopic constituent of "ordinary" helium. However, in the atmosphere ^{3}He constitutes only one part in a million of the total helium content, which is so small anyway. The quantities of ^{3}He needed for low-temperature experiments can only be produced by nuclear reactions such as

$$ {}^6_3\mathrm{Li} + {}^1_0\mathrm{n} \rightarrow {}^3_1\mathrm{H} + {}^4_2\alpha \tag{1.1a} $$

and the subsequent decay of the tritium

$$ {}^3_1\mathrm{H} \rightarrow {}^3_2\mathrm{He} + \mathrm{e}^- \tag{1.1b} $$

Clearly, this was not possible until after the end of the Second World War. The condensation of ^{3}He gas and the first experimental work on the new liquid were achieved by Sydoriak *et al.* (1949a,b). Somewhat larger quantities of ^{3}He became available only in the late 1950s. Since then, macroscopic samples of ^{3}He have been investigated intensively and at lower and lower temperatures.

By the time experimental ^{3}He physics eventually started, the theoretical and experimental investigations of ^{4}He had been going on for such a long time and had yielded so many novel results that "helium" was used synonymously for ^{4}He. While this tradition can be found even in today's scientific literature, one may expect the situation to change in the future. After all, the low-temperature phases (normal, superfluid, solid) of ^{3}He have led to a unique multitude of fundamentally new concepts and have thereby enlarged our knowledge of the possible states of condensed matter more than any other single-component system previously studied.

From a microscopic point of view, helium atoms are structureless spherical particles interacting via a two-body potential that is well understood (see Chapter 2). The attractive part of the potential, arising from weak van der Waals-type dipole (and higher multipole) forces, causes helium gas to condense into a liquid state at temperatures of 3.2 K and 4.2 K for ^{3}He and ^{4}He respectively, at normal pressure. The pressure versus temperature phase diagrams of ^{3}He and ^{4}He are shown in Figs. 1.1 and 1.2. When the temperature is decreased even further one finds that the helium liquids, unlike all other known liquids, do not solidify unless a pressure of around 30 bar is applied. This is the first remarkable indication of macroscopic quantum effects in these systems. The origin of this unusual behaviour lies in the quantum-mechanical uncertainty principle, which requires that a quantum particle can never be completely at rest at given position, but rather performs a zero-point motion about the average position. The smaller the mass of the particle and the weaker the binding force, the stronger these oscillations are. In most solids the zero-point motion is confined to a small volume of only a fraction of the lattice-cell volume. In the case of helium, however, two features combine to prevent the formation of a crystalline solid with a rigid lattice structure: (i) the strong zero-point motion arising from the small atomic mass (helium is the second-lightest element in the periodic table); and (ii) the weakness of the attractive interaction due to the high symmetry of these simple atoms.

It is this very property of helium—of staying liquid—that makes it such a valuable system for observing quantum behaviour on a macroscopic scale.

Quantum effects are also responsible for the strikingly different behaviours of ^{4}He and ^{3}He at even lower temperatures. Whereas ^{4}He undergoes a second-order phase transition (Kamerlingh Onnes 1911b, Kamerlingh Onnes and Boks 1924, Keesom and Wolfke 1928) into a state later shown to be superfluid (Kapitza 1938, Allen and Misener 1938), i.e. where the liquid is capable of flowing through narrow capillaries or tiny pores without friction, no such transition is observed in liquid ^{3}He in the same temperature range (see Figs. 1.1 and 1.2). The properties of liquid ^{3}He below 1 K are nevertheless found to be increasingly different from those of a classical liquid. It is only at a temperature roughly one thousandth of the transition temperature of ^{4}He that ^{3}He also becomes superfluid, and in fact forms *several* superfluid phases, each of which has a much more complex structure than that of superfluid ^{4}He.

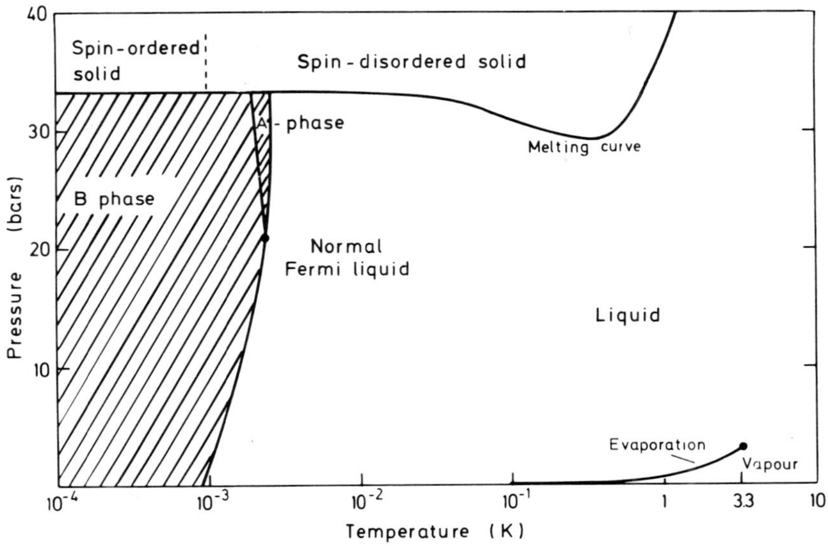

Figure 1.1 Pressure versus temperature phase diagram for ^{3}He; note the logarithmic temperature scale.

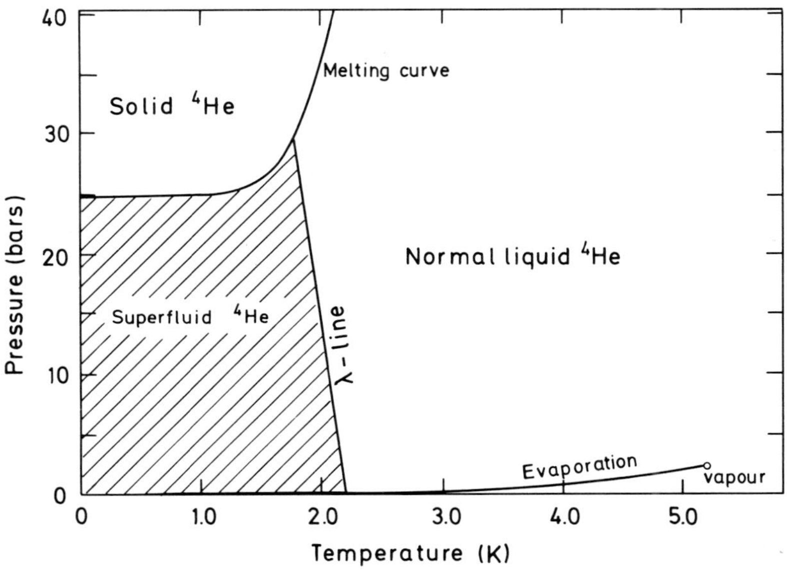

Figure 1.2 Pressure versus temperature phase diagram for ^{4}He; linear temperature scale.

The striking difference in the behaviours of ^{3}He and ^{4}He at low temperatures is a consequence of the laws of quantum theory as applied to systems of identical particles, i.e. the laws of quantum statistics. The ^{4}He atom, being composed of an even number of electrons and nucleons, has spin zero and consequently obeys Bose–Einstein statistics. In contrast, the ^{3}He nucleus consists of *three* nucleons, whose spins add up to give a total nuclear spin of $I = \frac{1}{2}$, making the total spin of the entire ^{3}He atom $\frac{1}{2}$ as well. Consequently liquid ^{3}He obeys Fermi–Dirac statistics. So it is the tiny nuclear spin, buried deep inside the helium atom, that is responsible for all the differences of the macroscopic properties of the two isotopes.

Since in a Bose system single-particle states may be multiply occupied, at low temperatures this system has a tendency to condense into the lowest-energy single-particle state (Bose–Einstein condensation). It is believed that the superfluid transition in ^{4}He is a manifestation of Bose–Einstein condensation. The all-important qualitative feature of the Bose condensate is its phase rigidity, i.e. the fact that it is energetically favourable for the particles to condense into a single-particle state of fixed quantum-mechanical phase, such that the global gauge symmetry is spontaneously broken. As a consequence, macroscopic flow of the condensate is (meta)stable, giving rise to the phenomenon of superfluidity.

In a Fermi system, on the other hand, the Pauli exclusion principle allows only single occupation of fermion states. The ground state of the Fermi gas is therefore the one in which all single-particle states are filled up to a limiting energy, the Fermi energy E_F. As predicted by Landau (1956, 1957, 1958) and later verified experimentally (for a review see Wheatley 1966), the properties of ^{3}He well below its Fermi temperature $T_F = E_F/k_B \approx 1$ K are similar to those of a degenerate Fermi gas. In particular, the formation of a phase-rigid condensate is not possible in this framework. Until the mid-1950s a superfluid phase of liquid ^{3}He was therefore believed to be ruled out. On the other hand, it is most remarkable that the property of superfluidity (see F. London 1950, 1954) was indeed first discovered experimentally in a *Fermi* system, namely that of the "liquid" of conduction electrons in a superconducting metal (Kamerlingh Onnes 1911a). The superfluidity of ^{4}He was only found more than 25 years later.

The key to the theory of superconductivity (Bardeen, Cooper and Schrieffer (BCS) 1957) turned out to be the formation of "Cooper pairs", i.e. pairs of electrons with opposite momentum $\boldsymbol{k}$ and spin projection σ: $(\boldsymbol{k}\uparrow, -\boldsymbol{k}\downarrow)$. These particular Cooper pairs are structureless objects, i.e. the two partners form a spin-singlet state in a relative s-wave orbital state. Cooper pairs may therefore be looked upon in a way as composite bosons, which all have the same pair wave function and are all in the same quantum-mechanical state. Hence in this picture the transition to the superconducting state corresponds to the formation of Cooper pairs that are automatically Bose-condensed, the condensate being characterized by macroscopic quantum coherence. Such a picture requires some qualification (see Chapter 12), but is nevertheless very helpful for an understanding of many basic properties of superconductors.

While in free space an attractive force has to be sufficiently strong to bind two electrons, inside the metal the presence of the filled Fermi sea of conduction electrons blocks the decay of a Cooper pair, so that an *arbitrarily* small attractive interaction leads to the formation of stable Cooper pairs. The attractive interaction between the electrons of a Cooper pair in a conventional superconducting metal is due to the exchange of virtual phonons (electron–phonon interaction). If the phonon-mediated interaction is strong enough to overcome the repulsive Coulomb interactions between the two electrons then a transition into a superconducting state may occur. On the other hand, any other mechanism leading to attraction between electrons at the Fermi surface is equally well suited for producing superconductivity.

1.2 EARLY HISTORY OF SUPERFLUID ^{3}He

Given the success of the BCS theory in the case of superconductivity, it was natural to ask whether a similar mechanism might also work for liquid ^{3}He. Since there is no underlying crystal lattice in the liquid that could mediate the attractive force, the attraction must clearly be an intrinsic property of the one-component ^{3}He liquid itself. The main feature of the interatomic ^{3}He potential is the strong repulsive component at short distances, and the weak van der Waals attraction at medium and long distances. It soon became clear that, in order to avoid the hard repulsive core and thus make optimal use of the attractive part of the potential, the ^{3}He atoms would have to form Cooper pairs in a state of *nonzero* relative angular momentum l. In this case the Cooper-pair wave function vanishes at zero relative distance, thus cutting out the most strongly repulsive part of the potential. In a complementary classical picture one might imagine the partners of a Cooper pair revolving about their centre of gravity, thus being kept away from each other by the centrifugal force.

A first estimate of the transition temperature for Cooper pairs with large relative angular-momentum quantum number, bound by the long-range tail of the van der Waals attraction (which was argued to be essentially unrenormalized by many-body effects), yielded unattainably low values (Pitaevskii 1959). Another line of approach was based on an approximate calculation of the effective pair interaction, taking multiple scattering into account; this yielded an attraction for d-wave pairs (Brueckner *et al.* 1960) but repulsion for s-wave pairs (Cooper *et al.* 1959). Other proposals for anisotropic, i.e. non-s-wave Cooper pairing, were due to Thouless (1960), Emery and Sessler (1960) and Galasiewicz (1960, 1969). However, from a modern point of view, none of these attempts was sophisticated enough. As more experimental data on liquid ^{3}He became available, it was soon realized that this is a strongly interacting system. The entities forming the Cooper pairs are not the bare ^{3}He atoms but are rather the quasiparticles of Landau's theory. These quasiparticles are single-particle excitations, which are sometimes viewed as particles surrounded by a polarization cloud of other particles. In fact, the effective mass of such quasiparticles may be as much as six times the bare atomic mass. Similarly, the interactions between quasiparticles were found to be very strong. It is then not surprising that the bare atomic potential bears little resemblance to the effective quasiparticle potential.

These difficulties notwithstanding, several authors considered the properties of non-s-wave pairing states. Anderson and Morel (1960, 1961) gave an extensive discussion of the thermodynamic properties of general anisotropic states, concentrating on d-wave states, but introducing also the socalled "axial" p-wave state (later named the Anderson–Brinkman–Morel (ABM) state). This state has the peculiar feature that the energy-gap function has nodes (i.e. zero-points) on the Fermi surface, with the orbital-angular-momentum projection pointing along the direction of these nodes. In fact this state turned out to describe one of the superfluid phases, the A phase, which was to be discovered only much later. These authors realized that the antisymmetry of the Cooper-pair wave function under exchange of the two particles requires the spin state to be the triplet state for any odd angular momentum. In contrast, any even-parity Cooper pair (e.g. a d-wave pair) is in the spin-singlet state. Clearly, a general p-wave Cooper pair must therefore have *three* spin substates—not only two, like that discussed by Anderson and Morel (1960, 1961). Indeed, Vdovin (1963) and Balian and Werthamer (1963) showed that, within weak-coupling theory, a state with an equal admixture of all three states is energetically favoured at all temperatures. The energy gap of this state (later called the Balian–Werthamer (BW) state) was found to be isotropic, just as in the case of s-wave pairing, despite the intrinsic anisotropy of the Cooper-pair wave function. Nevertheless, its magnetic properties are as complex and anisotropic as that of the ABM state. The BW state was also shown to sustain order-parameter collective oscillations of a type never encountered before (Vdovin 1963), although this work remained unknown outside the Soviet Union. The BW state turned out to describe the superfluid B phase, observed at temperatures below which the A phase is stable.

Another important early advance in the description of the properties of these hypothetical states was made by Leggett (1965a,b), who showed that the thermodynamic properties would be strongly renormalized (in a temperature-dependent way) by Fermi-liquid interaction effects.

Meanwhile, a better understanding of the physically relevant processes leading to large renormalization effects, and in particular of the important role of spin fluctuations, began to emerge. It was realized that ferromagnetic spin fluctuations favour spin-triplet/odd-l states over spin-singlet/even-l states (Emery 1964). The consequences of spin fluctuations for the thermodynamic and transport properties were worked out and found to be in qualitative agreement with experiment (Berk and Schrieffer 1966, Doniach and Engelsberg 1966, Brenig and Mikeska 1967, Brenig *et al.* 1967, Rice 1967a,b, Brinkman and Engelsberg 1968, Riedel 1968). A detailed calculation of the influence of spin fluctuations on the transition temperature for anisotropic pairing gave clear preference for spin-triplet/odd-l pairing (Layzer and Fay 1968, 1971, Nakajima 1973) and yielded values of T_c that were rather encouraging (Layzer and Fay 1968). A quantitative prediction of T_c, however, was not possible on this basis. For an account of early theoretical work on superfluid ^{3}He see Anderson and Brinkman (1975, 1978).

When the superfluid phases of ^{3}He were finally discovered in 1971 at temperatures of about 2.6 mK and 1.8 mK respectively (Osheroff *et al.* 1972a), in an experiment actually designed to observe a magnetic phase transition in solid ^{3}He, the results came as a great surprise.

1.3 ELEMENTARY DISCUSSION OF SUPERFLUID ^{3}He

Soon after the discovery of the phase transitions by Osheroff, Richardson and Lee (1972a), it was possible to identify altogether *three* distinct stable superfluid phases of bulk ^{3}He (see Chapter 4); these are referred to as the A, B and A_1 phases. In zero magnetic field only the A and B phases are stable. In particular, in zero field the A phase only exists within a finite range of temperatures, above a critical pressure of about 21 bar. Hence its region of stability in the pressure–temperature phase diagram has a roughly triangular shape as shown in Fig. 1.1. The B phase, on the other hand, occupies the largest part of this phase diagram and is found to be stable down to the lowest temperatures attained so far. Application of an external magnetic field has a strong influence on this phase diagram. First of all, the A phase is now stabilized down to zero pressure. Secondly, an entirely new phase, the A_1 phase, appears as a narrow wedge between the normal state and the A and B phases. Since the magnetic field changes the structure of the A and B phases somewhat, they are referred to as A_2 and B_2 phases respectively. Thus the A_2 and B_2 phases are just the old A and B phases in the presence of a magnetic field. This is summarized in Table 1.1, and the full P–H–T phase diagram is shown in Fig. 4.2.

Owing to the theoretical work on anisotropic superfluidity that had been carried out before the actual discovery of superfluid ^{3}He, progress in understanding the detailed nature of the phases was very rapid. This was clearly also due to the excellent contact between experimentalists and theorists, which greatly helped to develop the right ideas at the right time. In particular, it fairly soon became possible to identify the A phase and the B phase as realizations of the states studied previously by Anderson and Morel (1960, 1961) and Balian and Werthamer (1963) respectively. Therefore the A phase is described by the so-called "Anderson–Brinkman–

Table 1.1 Superfluid phases of ^{3}He.

External magnetic field H	Stable phases
$H = 0$	A phase B phase
$H \neq 0$	A_1 phase A_2 phase (≡A phase in a magnetic field) B_2 phase (≡B phase in a magnetic field)

Morel" (ABM) state, while the B phase is described by the "Balian–Werthamer" (BW) state. Consequently, "A phase" and "ABM state" are now used as synonyms; the same is true in the case of "B phase" and "BW state". (The fact that the ABM state describes the A phase and the BW state the B phase is a very fortunate concidence—if it was the other way around, it would be quite confusing!).

Although the three superfluid phases all have very different properties, they have one important thing in common: the Cooper pairs in all three phases are in a state with *parallel* spin ($S = 1$) and relative orbital angular momentum $l = 1$. This kind of pairing is referred to as "spin-triplet p-wave pairing". In contrast, prior to the discovery of the superfluid phases of ^{3}He, Cooper pairing in superconductors was only known to occur in a state with opposite spins ($S = 0$) and $l = 0$, i.e. in a "spin-singlet s-wave state". It should be noted that Cooper pairs in a superconductor and in superfluid ^{3}He are therefore very different entities: in the former case pairs are formed by pointlike, structureless electrons and are spherically symmetric, while in the case of ^{3}He Cooper pairs are made of actual atoms (or rather of quasiparticles involving ^{3}He atoms) and have an internal structure themselves.

1.3.1 The internal structure of Cooper pairs

Quantum-mechanically, a spin-triplet configuration ($S = 1$) of two particles has three substates with different spin projection S_z. They may be represented as $|\uparrow\uparrow\rangle$ with $S_z = +1$, $2^{-1/2}(|\uparrow\downarrow\rangle + |\downarrow\uparrow\rangle)$ with $S_z = 0$ and $|\downarrow\downarrow\rangle$ with $S_z = -1$. The pair wave function Ψ is in general a linear superposition of all three spin substates, i.e.

$$\Psi = \psi_{1,+}(\boldsymbol{k})\,|\uparrow\uparrow\rangle + \psi_{1,0}(\boldsymbol{k})(|\uparrow\downarrow\rangle + |\downarrow\uparrow\rangle) + \psi_{1,-}(\boldsymbol{k})\,|\downarrow\downarrow\rangle, \qquad (1.2)$$

where $\psi_{1,+}(\boldsymbol{k})$, $\psi_{1,0}(\boldsymbol{k})$ and $\psi_{1,-}(\boldsymbol{k})$ are the three complex-valued amplitudes of the respective substates. In the case of a superconductor, where $S = 0$ and $l = 0$, the pair wave function is much simpler, i.e. it is given by only a single component

$$\Psi_{sc} = \psi_0(|\uparrow\downarrow\rangle - |\downarrow\uparrow\rangle), \tag{1.3}$$

with a single amplitude ψ_0.

So far we have only taken into account that, since $S = 1$, there are three substates for the spin. The same is of course true for the relative orbital angular momentum $l = 1$ of the Cooper pair, which also has three substates $l_z = 0, \pm 1$. This fact is important if we want to investigate the amplitudes $\psi_{1,+}(\boldsymbol{k})$ etc. further. They still contain the complete information about the space (or momentum) dependence of Ψ. The pair wave function Ψ is therefore characterized by three spin substates and three orbital substates, i.e. by altogether $3 \times 3 = 9$ substates with respect to the spin and orbital dependence. Each of these nine substates is connected with a complex-valued parameter. Here we see the essential difference between Cooper pairs with $S = l = 0$ (conventional superconductors) and $S = l = 1$ (^{3}He): their pair wave functions are very different. In the former case a single complex-valued parameter is sufficient for its specification, in the latter case of superfluid ^{3}He *nine* such parameters are required. This also expresses the fact that a Cooper pair in superfluid ^{3}He has an internal structure, while that for a conventional superconductor does not: because $l = 1$, it is intrinsically *anisotropic.* This anisotropy may conveniently be described by specifying some direction with respect to a quantization axis both for the spin and the orbital component of the wave function.

In order to understand the novel properties of superfluid ^{3}He, it is therefore important to keep in mind that there are two characteristic directions that specify a Cooper pair. Here lies the substantial difference from a superconductor and the origin of the multitude of unusual phenomena occurring in superfluid ^{3}He: the structure of the Cooper pair is characterized by *internal degrees of freedom.* Nevertheless, in both cases the superfluid/superconducting state can be viewed as the condensation of a macroscopic number of these Cooper pairs into the same quantum-mechanical state, similar to a Bose–Einstein condensation.

1.3.2 Broken symmetry and the order parameter

In the normal liquid state Cooper pairs do not exist. Obviously, in the superfluid a new state of order appears, which spontaneously sets in at the critical temperature T_c. This particular transition from the normal to the superfluid, i.e. into the ordered state, is called "continuous", since the condensate—and hence the state of order—builds up continuously. This fact may be expressed quantitatively by introducing an "order parameter" that is finite for $T < T_c$ and zero for $T \geqslant T_c$. A well-known example of such a transition is that from a paramagnetic to a ferromagnetic state of a metal when the system is cooled below the Curie temperature. In the paramagnetic regime the spins of the particles are disordered such that the average magnetization $\langle \boldsymbol{M} \rangle$ of the system is zero. By contrast, in the ferromagnetic phase the spins are more or less aligned and $\langle \boldsymbol{M} \rangle$ is thus finite. In this case the system exhibits long-range order of the spins. The degree of ordering is quantified by $|\langle \boldsymbol{M} \rangle|$, the magnitude of the magnetization. Hence $\boldsymbol{M}$ is called the "order parameter" of the ferromagnetic state. Clearly, the existence of a preferred direction $\boldsymbol{M}$ of the spins implies that the symmetry of the ferromagnet under spin rotations is reduced ("broken") when compared with the paramagnet: the directions of the spins are no longer isotropically distributed, and the system will therefore no longer be invariant under a spin rotation. This phenomenon is called "spontaneously broken symmetry"; it is of fundamental importance in the theory of phase transitions. It describes the property of a macroscopic system (i.e. a system in the thermodynamic limit) that is in a state that does not have the full symmetry of the microscopic dynamics.

The concept of spontaneously broken symmetry also applies to superconductivity and superfluid ^{3}He. In this case the order parameter measures the existence of Cooper pairs and is given by the probability amplitude for a pair to exist at a given temperature. It follows from the discussion of the possible structure of a Cooper pair in superfluid ^{3}He that the associated order parameter will reflect this structure and the allowed internal degrees of freedom. What then are the spontaneously broken symmetries in superfluid ^{3}He?

As already mentioned, the interparticle forces between the ^{3}He atoms are rotationally invariant in spin and orbital space and, of course, conserve particle number. The latter symmetry gives rise to a somewhat abstract symmetry called "gauge symmetry". Nevertheless, gauge symmetry is spontaneously broken in any superfluid or superconductor (see Chapter 3). In addition, in an odd-parity pairing superfluid as in the case of ^{3}He, where $l = 1$, the pairs are necessarily in a spin-triplet state, implying that rotational symmetry in spin space is broken, just as in a magnet. At the same time, the anisotropy of the Cooper-pair wave function in orbital space calls for a spontaneous breakdown of orbital rotation symmetry, as in liquid crystals. All three symmetries are therefore simultaneously broken in superfluid ^{3}He. This implies that the A phase, for example, may be considered as a "superfluid nematic liquid crystal with (anti)ferromagnetic character". One might think that a study of the abovementioned broken symmetries could be performed much more easily by investigating them separately, i.e. within the isotropic superfluid, the magnet, the liquid crystal etc. itself. However, the combination of several *simultaneously* broken continuous symmetries is more than just the simple sum of the properties of all these known systems. Some of the symmetries broken in superfluid ^{3}He are "relative" symmetries, such as spin–orbit rotation symmetry or gauge–orbit symmetry (Leggett 1972, 1973b, Liu and Cross 1978). Because of this, a rigid connection is established between the corresponding degrees of freedom of the condensate, leading to long-range order only in the combined (and not in the individual) degrees of freedom. This particular kind of broken symmetry, for example the so-called "spontaneously broken spin–orbit symmetry", gives rise to very unusual behaviour, as will be discussed later. The whole concept of broken symmetries in superfluid ^{3}He and its consequences will be discussed in detail in Chapter 6.

It is clear that in principle the internal degrees of freedom of a spin-triplet p-wave state allow for many different Cooper-pair states and hence superfluid phases. (This is again different from ordinary superconductivity with $S=0$, $l=0$ pairing, where only a *single* phase is possible.) Of these different states, the one with the lowest energy for given external parameters will be realized. In fact, Balian and Werthamer (1963) showed, that, within a conventional "weak-coupling" approach, of all possible states there is precisely one state (the BW state) that has the lowest energy at *all* temperatures. This state is the one that describes the B phase of superfluid ^{3}He. The state originally discussed by these authors is one in which the orbital angular momentum $\boldsymbol{l}$ and spin $\boldsymbol{S}$ of a Cooper pair couple to a total angular momentum $\boldsymbol{J}=\boldsymbol{l}+\boldsymbol{S}=0$. This $^3P_0$ state is, however, only a special case of a more general one with the same energy (in the absence of spin–orbit interaction), obtained by an arbitrary rotation of the spin axes relative to the orbital axes of the Cooper-pair wave function. Such a rotation may be described mathematically by specifying a rotation axis $\hat{\boldsymbol{n}}$ and a rotation angle θ. In the BW state all three spin substates in (1.2) occur with equal measure. This state has a rather surprising property: in spite of the intrinsic anisotropy, the state has an *isotropic* energy gap (see Fig. 3.4a). (The energy gap is the amount by which the system lowers its energy in the condensation process, i.e. it is the minimum energy required for the excitation of a single particle out of the condensate.) Therefore the BW phase resembles ordinary superconductors in several ways. On the other hand, even though the energy gap is isotropic, the BW state *is* intrinsically anisotropic. This is clearly seen in dynamic experiments in which the Cooper-pair structure is distorted. For this reason the BW state is sometimes referred to as "pseudo-isotropic". Owing to the quantum coherence of the superfluid state, the rotation axis $\hat{\boldsymbol{n}}$ and angle θ characterizing a Cooper pair in the BW state are macroscopically defined degrees of freedom, whose variation is physically measurable.

Since in weak-coupling theory the BW state always has the lowest energy, an explanation of the existence of the A phase of superfluid ^{3}He obviously requires one to go beyond such an approach and to include "strong-coupling effects", as will be discussed in Chapter 5. In view of the fact that at present microscopic theories are not capable of computing transition temperatures for ^{3}He, it is helpful to single out a particular effect that can explain the stabilization of the A phase over the B phase. As shown by Anderson and Brinkman (1973), there is such a conceptually simple effect, which is based on a feedback mechanism: the pair correlations in the condensed state change the pairing interaction between the ^{3}He quasiparticles, the modification depending on the actual state itself. As a specific mechanism, these authors considered the role of spin fluctuations and showed that a stabilization of the state first considered by Anderson and Morel (1960, 1961) is indeed possible. This only happens at somewhat elevated pressures, since spin fluctuations become more pronounced only at higher pressures. This state, which is now referred to as the "ABM state" (from the initials of these three authors), does indeed describe the A phase. It has the property that, in contrast with ^{3}He-B, its magnetic susceptibility is essentially the same as that of the normal liquid. This is a clear indication that in this phase the spin substate with $S_z = 0$, which is the only one that can be reduced appreciably by an external magnetic field, is absent. Therefore ^{3}He-A is composed only of $|\uparrow\uparrow\rangle$ and $|\downarrow\downarrow\rangle$ Cooper pairs. This implies that the anisotropy axis of the spin part of the Cooper-pair wave function, called $\hat{\boldsymbol{d}}$, has the same, fixed, direction in every pair. (More precisely, $\hat{\boldsymbol{d}}$ is the direction along which the total spin of the Cooper pair vanishes: $\hat{\boldsymbol{d}} \cdot \boldsymbol{S} = 0$.) Likewise, the direction of the relative orbital angular momentum $\hat{\boldsymbol{l}}$ is the same for all Cooper pairs. Therefore in the A phase the anisotropy axes $\hat{\boldsymbol{d}}$ and $\hat{\boldsymbol{l}}$ of the Cooper-pair wave function are long-range-ordered, i.e. are preferred directions in the whole macroscopic sample. This implies a pronounced anisotropy of this phase in all its properties. In particular, the value of the energy gap now explicitly depends on the direction in $\boldsymbol{k}$ space on the Fermi sphere and takes the form

$$\Delta_{\hat{\boldsymbol{k}}}(T) = \Delta_0(T)[1 - (\hat{\boldsymbol{k}} \cdot \hat{\boldsymbol{l}})^2]^{1/2}. \tag{1.4}$$

Hence the gap vanishes at two points on the Fermi sphere, namely along $\pm\hat{\boldsymbol{l}}$ (see Fig. 3.4b). Because of the existence of an axis $\hat{\boldsymbol{l}}$, the ABM state is also called the "axial state". The existence of nodes implies that in general quasiparticle excitations may take place at arbitrarily low temperatures. Therefore, in contrast with ^{3}He-B or ordinary superconductors, there is a finite density of states for excitations with energies below the average gap energy, leading for example to a specific heat proportional to T^3 at low temperatures (see Chapters 3 and 7).

The third experimentally observable superfluid phase of ^{3}He, the A_1 phase, is only stable in the presence of an external magnetic field. In this phase Cooper pairs are all in a single spin substate, the $|\uparrow\uparrow\rangle$ state, corresponding to $S_z = +1$; the components with $|\uparrow\downarrow\rangle + |\downarrow\uparrow\rangle$ and $|\downarrow\downarrow\rangle$ states are missing. It is therefore a *magnetic* superfluid, the first ever observed in nature.

1.3.3 Orientational effects

For a pair-correlated superfluid, the pairing interaction is the most important interaction, since it is responsible for the formation of the condensate itself. Nevertheless, there also exist other, much weaker, interactions, which may not be important for the actual transition to the pair-condensed state, but which do become important if their symmetry differs from the aforementioned. In particular, they may be able to break remaining degeneracies.

The dipole–dipole interaction and other weak effects

The dipole–dipole interaction between the nuclear spins of the ^{3}He atoms leads to a very weak, spatially strongly anisotropic, coupling. The relevant coupling constant $g_D(T)$ is given by

$$g_D(T) \approx \frac{\mu_0^2}{a^3}\left(\frac{\Delta(T)}{E_F}\right)^2 n. \tag{1.5}$$

Here μ_0 is the nuclear magnetic moment, such that μ_0^2/a^3 is the average dipole energy of two particles at relative distance a (the average atomic distance), while the second factor measures the probability for these two particles to form a Cooper pair, and n is the overall particle density. Since μ_0^2/a^3 corresponds to about 10^{-7} K, this energy is extremely small and the resulting interaction of quasiparticles at temperatures of the order of 10^{-3} K might be expected to be completely swamped by thermal fluctuations. This is indeed true in a normal system. However, the dipole–dipole interaction implies a spin–orbit coupling and thereby has a symmetry different from that of the pairing interaction. In the condensate the symmetries with respect to a rotation in spin and orbital space are spontaneously broken, leading to long-range order (for example of $\hat{\boldsymbol{d}}$ and $\hat{\boldsymbol{l}}$ in the case of ^{3}He-A). Nevertheless, the pairing interaction does not fix the *relative* orientation of these preferred directions, leaving a continuous degeneracy. As pointed out by Leggett (1973a,b, 1974a), in this situation the tiny dipole interaction is able to lift the degeneracy, namely by choosing that particular relative orientation for which the dipolar energy is minimal. Thereby this interaction becomes of *macroscopic* importance. One may also view this effect as a macroscopic amplification of a microscopic interaction via the quantum coherence of the pair condensate. This correlation is equivalent to a *permanent* local magnetic field of about 30 G at any point in the superfluid (in a liquid!). In ^{3}He the dipolar interaction is minimized by a parallel orientation of $\hat{\boldsymbol{d}}$ and $\hat{\boldsymbol{l}}$; for details see Section 6.3.

Besides the dipole interaction, there are also other, even weaker, interactions, which—although completely unimportant in a normal system—may be amplified to macroscopic size by the long-range order in the superfluid (Leggett 1977a,b, 1978a). For example, the electromagnetic interaction of the two ^{3}He atoms in a Cooper pair leads to a minute distortion of the electronic shell of the ^{3}He atom, thereby shifting the centres of the positive and negative charges relative to each other. The orbital motion within the Cooper pair then leads to a magnetic moment along $\hat{\boldsymbol{l}}$ in every Cooper pair. Because of the long-range order of $\hat{\boldsymbol{l}}$, these magnetic moments add up to an equivalent magnetic field of about 0.02 G. Although this is very small, it makes the system behave differently upon reversal of the direction of an external magnetic field. Such a change has indeed been detected (Paulson and Wheatley 1978a). Therefore ^{3}He-A is a *liquid* orbital ferromagnet!

Leggett (1977a) also predicted that the macroscopic quantum coherence in superfluid ^{3}He will raise weak-interaction effects in elementary particle physics to a macroscopic level. For example, the interaction between electrons and protons within a ^{3}He atom has a parity-violating part given by the exchange of a neutral Z^0 boson. This effect is expected to induce a finite electric dipole moment along the preferred direction $\hat{\boldsymbol{n}}$ in ^{3}He-B, with an overall magnitude of about $10^{-12} e\,\mathrm{cm}^{-2}$, where e is the electronic charge. This is still a very small effect, but it does not seem to be hopelessly outside experimental reach. If an experiment succeeded in measuring this effect, it would be the first detection of parity violation on a macroscopic scale.

The amplification of these and other effects due to the specific long-range order in superfluid ^{3}He is discussed in Chapter 12. Clearly, of all those interactions the nuclear dipole interaction is by far the most important. Its effect on the relative orientation of the spontaneously preferred directions in spin and orbital space has to be included in any description of low-energy phenomena in superfluid ^{3}He (i.e. where the order-parameter structure itself remains unchanged).

Effect of a magnetic field

An external magnetic field acts on the nuclear spins and thereby leads to an orientation of the preferred direction in spin space. In the case of ^{3}He-A the orientation energy is minimal if $\hat{\boldsymbol{d}}$ is perpendicular to the field $\boldsymbol{H}$, since (taking into account $\hat{\boldsymbol{d}} \cdot \boldsymbol{S} = 0$) this orientation guarantees $\boldsymbol{S} \parallel \boldsymbol{H}$.

Walls

Every experiment is performed in a volume of finite size. Clearly, the walls will have some effect on the liquid inside. In superfluid ^{3}He this effect may readily be understood by using a simple picture. Let us view the Cooper pair as a kind of giant "molecule" of two ^{3}He quasiparticles orbiting around each other. For a pair not to bump into a wall, this rotation will have to take place in a plane parallel to the wall. In the case of ^{3}He-A, where the orbital angular momentum $\hat{\boldsymbol{l}}$ has the same direction in all Cooper pairs (standing perpendicular on the plane of rotation), this means that $\hat{\boldsymbol{l}}$ has to be oriented perpendicular to the wall. So there exists a strict orientation of $\hat{\boldsymbol{l}}$ caused by the walls (Ambegaokar *et al.* 1974). In the B phase, with its (pseudo)isotropic order parameter, the orientational effect is not as pronounced, but there are qualitatively similar boundary conditions.

1.3.4 Textures

From the above discussion, it is clear that the preferred directions $\hat{\boldsymbol{l}}$ and $\hat{\boldsymbol{d}}$ in ^{3}He-A are in general subject to different, often competing, orientational effects (for simplicity, we shall limit our description to ^{3}He-A). At the same time, the condensate will oppose any spatial variation of its long-range order. Any "bending" of the order-parameter field will therefore increase the energy, thus giving an internal stiffness or rigidity to the system. While the orientational effects might want $\hat{\boldsymbol{d}}$ and $\hat{\boldsymbol{l}}$ to adjust on the smallest possible lengthscale, the bending energy wants to keep the configuration as uniform as possible. Altogether, the competition between these two opposing effects will lead to a smooth spatial variation of $\hat{\boldsymbol{d}}$ and $\hat{\boldsymbol{l}}$ throughout the sample, called a "texture". This nomenclature is borrowed from the physics of liquid crystals, where similar orientational effects of the preferred directions occur (de Gennes 1974). The largest part of Chapter 7, which itself is by far the longest chapter in this book, is devoted to the discussion of textures in superfluid ^{3}He.

The bending energy and all quantitatively important orientational energies are invariant under the replacement $\hat{\boldsymbol{d}} \to -\hat{\boldsymbol{d}}$, $\hat{\boldsymbol{l}} \to -\hat{\boldsymbol{l}}$. A state where $\hat{\boldsymbol{d}}$ and $\hat{\boldsymbol{l}}$ are parallel therefore has the same energy as one where $\hat{\boldsymbol{d}}$ and $\hat{\boldsymbol{l}}$ are antiparallel. This leads to two different, degenerate, ground states. There is then the possibility that in one part of the sample the system is in one ground state and in the other in a different ground state. Where the two configurations meet they form a planar "defect" in the texture, called a "domain wall" (Maki 1977a) (see Fig. 7.22). This is in close analogy to the situation in a ferromagnet composed of domains with a different orientations of the magnetization. Domain walls are spatially localized and are quite stable against external perturbations. In fact, their stability is guaranteed by the specific nature of the order-parameter structure of ^{3}He-A. Mathematically, this structure may be analysed according to its topological properties. The stability of a domain wall can then be traced back to the existence of a conserved "topological charge". Using the same mathematical approach, one can show that the order-parameter fields of the superfluid phases of ^{3}He not only allow for planar defects but also for point and line defects, called "monopoles" and "vortices" respectively. Defects can be "nonsingular" or "singular", depending on whether the core of the defect remains superfluid or whether it is forced to become normal liquid. The concept of vortices is of course well known from superfluid ^{4}He. However, since the order-parameter structure of superfluid ^{3}He is so much richer than that of superfluid ^{4}He, there exist a wide variety of different vortices in these phases. Their detailed structure has been the subject of intensive investigation, in particular in the context of experiments on rotating superfluid ^{3}He, where they play a central role (Hakonen and Lounasmaa 1987); see Chapter 7.

1.3.5 Superfluid mass currents in ^{3}He-A

Superfluids owe their name to their ability to flow through tiny pores and narrow slabs, apparently without friction. This extraordinary property, which is in sharp contrast with our usual experience with liquids, is certainly the most impressive manifestation of the macroscopic quantum coherence of the pair condensate in a Fermi system. Mathematically, superfluidity finds its expression in a complex-valued order parameter, i.e. the existence of *macroscopic* quantum-mechanical phase variables.

To understand the flow properties of a superfluid, it is helpful to consider it as a system of two independent, interpenetrating components: a "superfluid" and a "normal" component, with densities ρ_s and ρ_n respectively, such that $\rho_s + \rho_n = \rho$, with ρ the total density. At $T \geqslant T_c$ one has $\rho_s = 0$ and $\rho_n = \rho$, while at $T = 0$ one has $\rho_s = \rho$ and $\rho_n = 0$. This "two-fluid model" is useful in many ways (but, of course, it is only a *model*, since in reality the liquid is not composed of distinct superfluid and normal particles). In this model a mass current $\boldsymbol{g}$ is the sum of a superfluid and a normal part $\boldsymbol{g}_s = \rho_s \boldsymbol{v}_s$ and $\boldsymbol{g}_n = \rho_n \boldsymbol{v}_n$ respectively. Here $\boldsymbol{v}_s$ and $\boldsymbol{v}_n$ are the velocities of the two components. In a superfluid with isotropic order parameter, $\Psi = \psi_0 e^{i\phi}$, as in ^{4}He, the superfluid velocity $\boldsymbol{v}_s$ is simply given by the gradient of the macroscopic, i.e. long-range-ordered, phase of the order parameter:

$$\boldsymbol{v}_s = \frac{\hbar}{m_4} \boldsymbol{\nabla} \phi, \tag{1.6}$$

where m_4 is the mass of a ^{4}He atom. This simple dependence implies

$$\boldsymbol{\nabla} \times \boldsymbol{v}_s = 0, \tag{1.7}$$

i.e. the flow is "curl-free"—there is no rotational motion (we do not consider the rotation of the system as a whole). For any closed contour $\mathscr{C}$ in the liquid, we may define the "circulation" κ as the line integral over the velocity:

$$\kappa = \oint_{\mathscr{C}} d\boldsymbol{s} \cdot \boldsymbol{v}_s = \int_{\text{area}} d\boldsymbol{f} \cdot (\boldsymbol{\nabla} \times \boldsymbol{v}_s). \tag{1.8}$$

Using (1.7), it follows that $\kappa = 0$. On the other hand, in a container with a hole (e.g. a vessel with the shape of a doughnut—a "torus") the situation is quite different. The circulation κ along a contour $\mathscr{C}$ enclosing the hole will clearly be finite as long as $\boldsymbol{v}_s \neq 0$. Since the phase of the order parameter is a macroscopic quantity, it is well defined along $\mathscr{C}$ and can only change by multiplies of 2π when $\mathscr{C}$ is traversed. In this case κ is *quantized* and (1.8) takes the well-known form of the Bohr–Sommerfeld quantization condition for electronic orbits in an atom:

$$\kappa = \frac{h}{m_4} N. \tag{1.9}$$

Here $N = 0, \pm 1, \pm 2, \ldots$, and h/m_4 is the so-called "flux quantum". This means that supercurrents in a toroidal geometry are quantized and hence cannot decay continuously: they are (meta)stable, their lifetime being of the order of cosmological times.

In an anisotropic superfluid such as ^{3}He-A the situation is very different. Here the order parameter not only has a phase but also has an *orientation*, namely the preferred direction $\hat{\boldsymbol{l}}$ (the spin structure is unimportant in this discussion). The anisotropy implies that the system can now distinguish directions relative to $\hat{\boldsymbol{l}}$. In this case the superfluid velocity will depend not only on the spatial change of the phase ϕ but also on that of $\hat{\boldsymbol{l}}$. Parametrizing $\hat{\boldsymbol{l}}$ by azimuthal (β) and polar (α) angles, $\boldsymbol{v}_s$ now takes the form

$$\boldsymbol{v}_s = \frac{\hbar}{2m_3}(\nabla\phi - \cos\beta\,\nabla\alpha) \tag{1.10}$$

(note that the mass of a Cooper pair, $2m_3$, appears in (1.10)). The fact that $\boldsymbol{v}_s$ is no longer given by a phase gradient has an immediate, drastic, consequence: $\nabla \times \boldsymbol{v}_s \neq 0$. This means that $\nabla \times \boldsymbol{v}_s$, and thereby κ in (1.7), can in general take *any* value, depending on how $\hat{\boldsymbol{l}}$ changes. There is no longer any flux quantization! Consequently, a superfluid mass current in ^{3}He-A is unstable (Mermin and Ho 1976). Therefore ^{3}He-A does not seem to be a "superfluid" at all. However, we have so far neglected the surface of the container. Since $\hat{\boldsymbol{l}}$ has to be perpendicular to the wall, i.e. it has a fixed orientation there, the circulation at the wall is quantized and the stability of a supercurrent is guaranteed at least very near to the surface. These and related questions are discussed in detail in Chapter 7.

1.3.6 Dynamic properties

From the discussion presented so far, we have already seen that the static properties of an anisotropic superfluid are very unusual. Clearly, the dynamic properties can be expected to be at least as new and diverse. Indeed, the fact that in superfluid ^{3}He Cooper pairs have an internal structure can only be investigated in detail by studying the dynamics, i.e. the frequency and momentum dependence, of the condensate. One may roughly distinguish between magnetic and nonmagnetic dynamic properties, depending on whether the magnetization of the system is probed or whether properties such as mass transport (see Chapter 10) or the propagation of sound are studied.

For the investigation of dynamical effects, it is instructive to have an idea of the typical frequencies inherent to the superfluid condensate. For this, we again employ the two-fluid model. Both the normal and the superfluid components are essentially characterized by a single timescale each: for the normal component this is the quasiparticle lifetime τ, and for the superfluid component it is $\hbar/\Delta(T)$, where $\Delta(T)$ is the average of the temperature-dependent energy gap. The orders of magnitude of the equivalent frequencies are given by $\tau^{-1} \approx 10\,\text{MHz}$ and $\Delta(T)/\hbar \approx 10^3 (1 - T/T_c)^{1/2}\,\text{MHz}$, i.e. usually one has $\tau^{-1} \ll \Delta(T)/\hbar$. For frequencies ω much smaller than either of these characteristic values, the liquid is always in local thermodynamic equilibrium, since the system always has sufficient time to adjust to any change induced on the timescale ω^{-1}. This is called the "hydrodynamic regime", which is important for a couple of reasons: (i) in this regime knowledge of the conserved quantities and of those describing the broken symmetries is sufficient to describe the properties of the system (see Chapter 9); and (ii) this regime is experimentally well accessible. The multitude of broken symmetries in superfluid ^{3}He consequently leads to very rich hydrodynamics, which describes the various low-frequency collective excitations of the system. Here the word "collective" (as opposed to "single-particle") means that a macroscopic number of particles is involved in a coherent fashion.

Spin dynamics

Investigations of the collective magnetic (i.e. spin-dependent) properties of the superfluid phases of ^{3}He by nuclear magnetic resonance (NMR) were particularly useful in identifying the explicit order-parameter structure of these phases. In usual NMR experiments the system under investigation is brought into a strong constant external magnetic field $\boldsymbol{H}_0 = H_0\hat{\boldsymbol{z}}$, which forces the (nuclear) spin $\boldsymbol{S}$ to precess about $\boldsymbol{H}_0$. By applying a weak high-frequency magnetic field $\boldsymbol{H}_{\text{rf}}$ perpendicular to $\boldsymbol{H}_0$, one is able to induce transitions in S_z, the component along $\boldsymbol{H}_0$, of magnitude $\pm\hbar$. This effect is observed as an energy absorption from the magnetic field. In the case of noninteracting spins these transitions occur *exactly* at the energy $\gamma\hbar H_0$, i.e. at the Larmor frequency $\omega_L = \gamma H_0$, where γ is the gyromagnetic ratio of the nucleus. How does this change in the presence of interactions? For a spin of magnitude $\frac{1}{2}\hbar$, as in the case of the ^{3}He nucleus, a very general statement is possible (Leggett 1972, 1973b): as long as the interactions are spin-*conserving*, there is no change at all—the resonance remains at ω_L. On the other hand, for spin-nonconserving interactions, such as the spin–orbit interaction caused by the dipole coupling of the nuclear spins, a frequency shift may indeed occur. However, such a "nonsecular" shift will usually be very small, namely at most of the order of the linewidth. The experimental data obtained by Osheroff *et al.* (1972b) in connection with their discovery of the superfluid phases therefore came as a great surprise—they found that the resonance, although still very sharp, occurred at frequencies substantially higher than ω_L. The origin of this large shift was especially mysterious, since it obviously corresponded to a *constant* local magnetic field of order 30 G surrounding the nuclear spins in the liquid.

The solution to this puzzle was found by Leggett (1972, 1973b, 1974a), who showed that the NMR shifts are a consequence of the broken symmetries of the spin-triplet p-wave condensate, which he named "spontaneously broken spin–orbit symmetry". As explained earlier, the meaning of this concept is that the preferred directions in spin and orbital space are long-range-ordered (individually so, or in a combined way) and the tiny dipole interaction may take advantage of this situation by lifting the remaining degeneracy. The macroscopic quantum coherence of the condensate therefore raises the dipole coupling to macroscopic importance. In this way, Leggett (1974a) was able to calculate the general NMR response of a spin-triplet p-wave condensate. In particular, in the A phase the transverse NMR frequency ω_t is given by

$$\omega_t^2 = \omega_L^2 + \Omega_A^2(T), \tag{1.11}$$

where $\Omega_A^2(T)$ is proportional to the dipole coupling constant (see (1.5)). It should be noted that the field and temperature dependences of ω_t are neatly separated in a "Pythagorean" form: ω_L only depends on H_0 and Ω_A only on T! In fact, Leggett (1974a) worked out a complete theory of spin dynamics, whose predictions were experimentally confirmed in every detail; this will be discussed in Chapter 8. For example, the equation of motion of the total spin $\boldsymbol{S}$ is given by

$$\dot{\boldsymbol{S}} = \gamma \boldsymbol{S} \times \boldsymbol{H} + \boldsymbol{R}_D, \tag{1.12}$$

where $\boldsymbol{H} = \boldsymbol{H}_0 + \boldsymbol{H}_{rf}$ is the total external magnetic field and a dot over a symbol indicates the time derivative. Here $\boldsymbol{R}_D$ is the anisotropic "dipole torque", which itself depends on the change of the dipole energy under a reorientation of the order parameter. In the normal phase $\boldsymbol{R}_D$ is always zero. In the superfluid one has $\boldsymbol{R}_D \neq 0$, except for static situations. If the system is displaced from static equilibrium (for example by applying $\boldsymbol{H}_{rf}$), $\boldsymbol{R}_D$ acts as a restoring force. For example, in the A phase a periodic oscillation of $\boldsymbol{S}$ will lead to an oscillation of $\hat{\boldsymbol{d}}$, the preferred direction in spin space, around the orbital degree of freedom $\hat{\boldsymbol{l}}$ (which may be assumed to remain fixed because it cannot move very quickly). Equation (1.12) led Leggett to a spectacular prediction: even if the high-frequency field $\boldsymbol{H}_{rf}$ is oriented *parallel* to $\boldsymbol{H}_0$, there is a resonance, i.e. there exists a longitudinal spin resonance! Since in this case $(\boldsymbol{S} \times \boldsymbol{H})_z = 0$, (1.12) yields $dS_z/dt = R_{Dz}$. In a normal system there can be no resonance since there is no restoring force: the z component of the magnetization will simply relax exponentially but will not oscillate. How then can we understand the nature of the longitudinal oscillation in the case of superfluid ^{3}He-A? The A phase only consists of the two spin substates $|\uparrow\uparrow\rangle$ and $|\downarrow\downarrow\rangle$. They may be viewed as essentially independent interpenetrating superfluids, which are only very weakly coupled by the spin-nonconserving dipole coupling. This coupling allows for a transition of $|\uparrow\uparrow\rangle$ pairs into $|\downarrow\downarrow\rangle$ pairs, and vice versa. (The situation is quite similar to a pair of weakly coupled superconductors, where Cooper pairs can tunnel from one superconductor to the other (the

"Josephson effect"); the difference here is that the two subsystems fill the same volume, i.e. they are not spatially separated). Applying a high-frequency magnetic field parallel to the static field $\boldsymbol{H}_0$ leads to oscillatory nonequilibrium between the two spin subsystems, with the dipole interaction acting as a restoring force. The resonant frequency of this longitudinal oscillation occurs at

$$\omega_\ell = \Omega_{\mathrm{A}}(T), \tag{1.13}$$

where Ω_{A} is the frequency that has already appeared in the expression for the transverse frequency (1.11).

Any texture formed by the order-parameter field changes the dipole torque $\boldsymbol{R}_{\mathrm{D}}$ in a very specific way. Therefore the measurement of NMR shifts, in combination with the corresponding theory, provides the most versatile, and at the same time sensitive, tool for the investigation of order-parameter textures.

NMR frequencies are generally considerably smaller than the characteristic frequencies τ^{-1} and $\Delta(T)/\hbar$ of the normal and superfluid components. Hence such experiments take place in the hydrodynamic regime. At such low frequencies, i.e. energies, the magnitude of the order parameter $\Delta(T)$ does not change at all—only the orientation of its spin part varies. Hence the *structure* of the order parameter is left intact—the dynamics is due to a "rigid" excitation of the order parameter. At higher frequencies, $\omega \approx \Delta(T)$, this changes dramatically. To understand the consequences of this, it is again helpful to view a Cooper pair as some kind of diatomic molecule. As in the case of a molecule, an energy of the order of the binding energy will lead to internal excitations such as rotational and vibrational states.

Ultrasound excitations

Such a situation occurs in experiments measuring the attenuation of ultrasound at sound frequencies close to $\Delta(T)/\hbar$. Quite unexpectedly, one finds that the sound attenuation of the superfluid has a sharp maximum directly below the transition temperature T_{c}; this maximum depends strongly on the frequency ω. These and other phenomena are explained by collective excitations of the order-parameter structure of the condensate. They owe their existence to pair correlations in a state with nonzero relative orbital angular momentum, which imply an internal structure of the Cooper pair (Wölfle 1973a, 1978a). This structure allows for the excitation of high-frequency ($\omega \approx \Delta(t)/\hbar$) collective oscillations (pair-vibration modes). Besides this, there is also the possibility of a break-up of the Cooper pair. Pair breaking is only possible if the energy $\hbar\omega$ of the sound wave is larger than the minimum energy for breaking a pair, $2\Delta_{\hat{\boldsymbol{k}}}(T)$. Here $\Delta_{\hat{\boldsymbol{k}}}(T)$ is the energy gap, which in general depends on $\hat{\boldsymbol{k}}$, the position on the Fermi sphere. For smaller energies, only vibrations can be excited. A detailed theory of sound absorption, including damping effects etc., has been developed (see Chapter 11) and is in good agreement with experiments.

In particular, the existence of isotropic and anisotropic energy gaps in the B and A phase respectively led to early identification of these two phases. Indeed, in the B phase sound attenuation is independent of the direction of the sound entering the probe. By contrast, in the A phase it strongly depends on the relative orientation of the sound wave to the anisotropy axis $\hat{\boldsymbol{l}}$. This orientation dependence is very remarkable: by coupling to the nuclear spins, a weak external magnetic field of the order of 30 G is able to change the direction of $\hat{\boldsymbol{l}}$ and thereby to modify the sound absorption. It is the coherent ordering of *nuclear* spins that is ultimately responsible for the anisotropy of sound absorption!

This concludes our elementary discussion of the properties of superfluid ^{3}He. There are many other astonishing aspects of these anisotropic phases, which will also be discussed in this book.

1.4 RELATION TO OTHER FIELDS

Why spend so much effort on sorting out and explaining the strange behaviour of states of matter that are not even found in nature, at temperatures well outside the reach of even a well-equipped low-temperature laboratory? Partly, of course, "because it's there", and because—like any other system—superfluid ^{3}He deserves to be studied in its own right. However, what is even more important is that superfluid ^{3}He is a model system that exemplifies many of the concepts of modern theoretical physics and, as such, has given us, and will further provide us, with new insights into the functioning of quantum-mechanical many-body systems close to their ground state.

As discussed in Section 1.3, the key to understanding superfluid ^{3}He is "spontaneously broken symmetry". In this respect there are also very fundamental connections with particle physics, deriving from the interpretation of the order-parameter field as a quantum field with a rich group structure. The collective modes of the order parameter as well as the localized topological defects in a given ground-state configuration are the particles of this quantum field theory. Various anomalies known from particle physics can be identified in the ^{3}He model system, and one may hope that insights gained from the study of superfluid ^{3}He will turn out to be useful in elementary particle theory (Volovik 1987).

There are several other physical systems for which the ideas developed in the context of superfluid ^{3}He are relevant or may be relevant in the future. Dilute solutions of ^{3}He in ^{4}He constitute a system of fermions moving in a background of superfluid ^{4}He. It is expected that a transition into a pair-correlated state should take place, which would make this the first system of two interpenetrating superfluids (Bashkin and Meyerovich 1981). Unfortunately, the transition temperature is estimated to be rather low, around 5 μK or even much lower. The symmetry of the pairing is predicted to be s-wave at low concentrations (<3%) of ^{3}He, with the possibility of p-wave formation at higher concentrations (3–6%) (Pfitzner 1984, Hsu 1984, Hsu and Pines 1985). The transition has not yet been observed, despite considerable effort (Owers-Bradley *et al.* 1983).

Another anisotropic superfluid system that does already exist in nature is not accessible for laboratory experiments: this is the nuclear matter forming the cores of neutron stars. There the pairing of neutrons has been calculated to be of p-wave symmetry. Because of the strong spin–orbit nuclear force, the total angular momentum of the Cooper pairs is $J=2$ (Hoffberg *et al.* 1970, Muzikar *et al.* 1980, Sauls *et al.* 1982, Pines and Alpar 1985).

Above all, an anisotropic superconducting state would be most exciting. There are now strong indications that superconductivity in the so-called "heavy-fermion" systems, first discovered by Steglich *et al.* (1979), is, at least in some cases, due to the formation of anisotropic pairs with d-wave or possibly p-wave symmetry (for reviews see Stewart 1984, Lee *et al.* 1986, Gorkov 1987, Ott 1987, Fulde *et al.* 1988). Many of the concepts and ideas developed for superfluid ^{3}He have been adapted to these systems. One must keep in mind, however, that the charge of the electrons and the presence of the underlying crystal lattice may lead to qualitatively new behaviour (Leggett 1987). Other systems where relations to superfluid ^{3}He may eventually appear are the recently discovered high-T_c oxide superconductors (Bednorz and Müller 1986, Wu *et al.* 1987), which are now being investigated with an intensity unparalleled in the history of condensed-matter physics.

The above discussion shows that superfluid ^{3}He is a field of continuing interest. We hope that this book will contribute to the future development and expansion of our understanding of anisotropic superfluid systems by generating the interest of the beginner, educating the serious student of the subject and providing reference material for the researcher in this field.

1.5 REVIEWS AND INTRODUCTORY ARTICLES ON SUPERFLUID ^{3}He

1.5.1 General reviews

Theory

Leggett A J 1975 A theoretical description of the new phases of liquid ^{3}He. *Rev. Mod. Phys.* **47** 331

Anderson P W and Brinkman W F 1975 Theory of anisotropic superfluidity in ^{3}He. In *The Helium Liquids* (*Proceedings of the 15th Scottish Universities Summer School, 1974*), ed. J G M Armitage and I E Farquhar (Academic Press, London), p. 315

Anderson P W and Brinkman W F 1978 Theory of anisotropic superfluidity in ^3He (updated version of the above article). In *The Physics of Liquid and Solid Helium,* Part II, ed. K H Bennemann and J B Ketterson (Wiley, New York), p. 177

Wölfle P 1979 Low temperature properties of liquid ^3He. *Rep. Prog. Phys.* **42** 269

Mineev V P 1983 Superfluid ^3He: introduction to the subject. *Usp. Fiz. Nauk* **139** 303 [*Sov. Phys. Usp.* **26** 160 (1983)]

Experiment

Wheatley J C 1975 Experimental properties of superfluid ^3He. *Rev. Mod. Phys.* **47** 415

Lee D M and Richardson R C 1978 Superfluid ^3He. In *The Physics of Liquid and Solid Helium,* Part II, ed. K H Bennemann and J B Ketterson (Wiley, New York), p. 287

1.5.2 Reviews of specific topics

Wheatley J C 1978 Further experimental properties of superfluid ^3He. In *Progress in Low Temperature Physics,* Vol. VIIa, ed. D F Brewer (North-Holland, Amsterdam), p. 1

Brinkman W F and Cross M C 1978 Spin and orbital dynamics of superfluid ^3He. In *Progress in Low Temperature Physics,* Vol. VIIa, ed. D F Brewer (North-Holland, Amsterdam), p. 105

Osheroff D D 1978 Recent experiments in superfluid ^3He. *J. Physique* **39** Colloq. C-6, Vol. III, p. 1270 (Proceedings of the 15th International Conference on Low Temperature Physics, LT-15)

Wölfle P 1978 Sound propagation and kinetic coefficients in superfluid ^3He. In *Progress in Low Temperature Physics,* Vol. VIIa, ed. D F Brewer (North-Holland, Amsterdam), p. 191

Volovik G E 1979 Superfluid ^3He. Hydrodynamics and inhomogeneous states. In *Soviet Scientific Reviews,* Section A, *Physics Reviews,* Vol. 1 (Harwood Academic Publishers, Chur), p. 23

Fomin I A 1981 Perturbation method in nonlinear spin dynamics of the superfluid phase of ^3He. In *Soviet Scientific Reviews,* Section A, *Physics Reviews,* Vol. 3 (Harwood Academic Publishers, Chur), p. 275

Serene J W and Rainer D 1983 The quasiclassical approach to superfluid ^3He. *Phys. Rep.* **101** 221

Volovik G E 1984 Superfluid properties of ^3He-A. *Usp. Fiz. Nauk* **143** 73 [*Sov. Phys. Usp.* **27** 363 (1984)]

Hall H E and Hook J R 1986 The hydrodynamics of superfluid ^3He. In *Progress in Low Temperature Physics,* Vol. IX, ed. D F Brewer (North-Holland, Amsterdam), p. 143

Fetter A L 1986 Vortices in rotating superfluid ^3He. In *Progress in Low Temperature Physics,* Vol. X, ed. D F Brewer (North-Holland, Amsterdam), p. 1

Maki K 1986 Solitons in superfluid ^3He. In *Solitons,* ed. S E Trullinger, V E Zakharov and V L Pokrovskii (North-Holland, Amsterdam), p. 435

Salomaa M M and Volovik G E 1987 Quantized vortices in superfluid ^3He. *Rev. Mod. Phys.* **59** 533

1.5.3 Discussions in books on related subjects

Abragam A and Goldman M 1982 *Nuclear Magnetism: Order and Disorder* (Clarendon Press, Oxford), Chap. 4.
Tilley D R and Tilley J 1986 *Superfluidity and Superconductivity,* 2nd edn (Adam Hilger, Bristol), Chap. 9
Wilks J and Betts D S 1987 *An Introduction to Liquid Helium,* 2nd edn (Clarendon Press, Oxford), Chap. 9.

1.5.4 Introductory articles

Lounasmaa O V 1974 The superfluid phases of liquid ^{3}He. *Contemp. Phys.* **15** 353
Wheatley J C 1976 Superfluid phases of helium three. *Phys. Today* **29** 32
Leggett A J 1976 The new phase of ^{3}He. *Endeavour* **35** 83
Mermin N D and Lee D M 1976 Superfluid helium 3. *Scientific American,* **235** (December) 56
Cross M C 1977 New phases of helium 3. *Sci. Prog.* **64** 157
Vollhardt D 1982 Suprafluides Helium-3: Die Superflüssigkeit. *Phys. Blätter* **39** 41, **39** 120, **39** 151
Dobbs E R 1983 Superfluid helium three. *Contemp. Phys.* **24** 389
Mineev V P, Salomaa M M and Lounasmaa O V 1986 Superfluid ^{3}He in rotation. *Nature* **324** 333
Hakonen P and Lounasmaa O V 1987 Vortices in rotating superfluid ^{3}He. *Phys. Today* **40** 70

2

Theory of Normal Fermi Liquids

In the temperature range from well below the Fermi temperature T_F (≈ 1 K) down to the transition temperature T_c of the superfluid phases the properties of liquid ^{3}He can be accounted for by the so-called Landau Fermi-liquid model. Starting from the perfect Fermi gas, this model introduces the effects of interactions between the atoms in a phenomenological way (Landau 1956, 1957). It is based on the concept of elementary excitations, according to which the low-energy properties of an interacting many-body system can be described in terms of a rarefied gas of elementary excitations or "quasiparticles". This might appear somewhat surprising. After all, in liquid ^{3}He the interparticle distance is of the same order as the range of the interatomic potential, and therefore interaction effects between the hard-sphere atoms can be expected to be very important. However, it turns out that at temperatures $T \ll T_F$ the relevant degrees of freedom are severely reduced by the Fermi statistics. The majority of particles are frozen into states deep in the Fermi sea. Only a fraction T/T_F of particles participate actively in interaction processes.

2.1 THE QUASIPARTICLE CONCEPT

The energy eigenstates of the noninteracting (spin-$\frac{1}{2}$) Fermi gas are specified by the number of particles $N_{\boldsymbol{k}\alpha}$ (=0, 1) in each of the single-particle states with momentum $\hbar\boldsymbol{k}$ and spin projection α. For the purpose of describing macroscopic properties, it is sufficient to introduce a smoothed distribution function $n_{\boldsymbol{k}\alpha}$ ($0 \leqslant n_{\boldsymbol{k}\alpha} \leqslant 1$) in place of the highly discontinuous occupation numbers $N_{\boldsymbol{k}\alpha}$ by averaging $N_{\boldsymbol{k}\alpha}$ over a group of neighbouring states. In the ground state all single-particle states with momentum less than the Fermi momentum $\hbar k_F$ are occupied ($n_{\boldsymbol{k}\alpha} = 1$) and all other states are empty ($n_{\boldsymbol{k}\alpha} = 0$). Employing the usual periodic boundary conditions in unit volume,

k_F is related to the particle density n by

$$n = \sum_{k\alpha} n_{k\alpha} = \frac{k_F^3}{3\pi^2}. \tag{2.1}$$

Let us now imagine the interaction between the particles to be turned on adiabatically. If the single-particle energy spectrum of the interacting system is in one-to-one correspondence with the Fermi-gas spectrum and if the ground state has the full symmetry of the Hamiltonian then the system is referred to as "normal". In contrast, in a superfluid state this one-to-one correspondence does not exist owing to a condensation of degrees of freedom into a macroscopic quantum state. Obviously the state of a normal Fermi liquid is then still completely characterized by the distribution function $n_{k\alpha}$ of the corresponding noninteracting system. This is true at least as long as any collective excitations that are introduced by the interaction are negligible. The elementary excitations of a normal Fermi liquid correspond to the particle and hole excitations of the perfect Fermi gas and are referred to as "quasiparticles" and "quasiholes"; for detailed discussions see Nozières (1964), Pines and Nozières (1966) and Baym and Pethick (1978).

The energy of a quasiparticle, $\epsilon_{k\alpha}$, is defined as the amount of energy by which the total energy of the system E increases, if a quasiparticle is added to the unoccupied state $\boldsymbol{k}, \alpha$, or

$$\delta E = \sum_{k\alpha} \epsilon_{k\alpha}\, \delta n_{k\alpha\alpha}, \tag{2.2}$$

where $\delta n_{k\alpha\alpha}$ is the corresponding change in the distribution function. The isotropy of the system requires $\epsilon_{k\alpha}$ to be a function of $|\boldsymbol{k}|$ only.

Note that while $n_{k\alpha}$ in (2.1) is characterized by a single spin index, the quantity $\delta n_{k\alpha\beta}$ carries *two* spin indices. The $n_{k\alpha}$ ($\alpha = \uparrow, \downarrow$) are actually the eigenvalues of the more general quantity $n_{k\alpha\beta}$, which is the spin density matrix of quasiparticles with momentum $\hbar\boldsymbol{k}$. Changing the distribution function will therefore in general give rise to off-diagonal contributions $\delta n_{k\alpha-\alpha}$, too. In the same way, $\epsilon_{k\alpha}$ represents the eigenvalues of the energy matrix $\epsilon_{k\alpha\beta}$ for quasiparticles, which will be discussed below.

The quasiparticle energy may be parametrized in the vicinity of the Fermi energy by

$$\left.\begin{aligned} \epsilon_{k\alpha} &= \mu + \hbar v_F(k - k_F) \quad \text{for} \quad |\epsilon_{k\alpha} - \mu| \ll \mu, \\ v_F &\equiv \frac{1}{\hbar}\frac{\partial \epsilon_{k\alpha}}{\partial k}\bigg|_{k_F} = \hbar\frac{k_F}{m^*}, \end{aligned}\right\} \tag{2.3}$$

where μ is the chemical potential (equal to the Fermi energy ϵ_F at $T = 0$), v_F is the Fermi velocity and m^* is the effective mass of the quasiparticles. For liquid ^{3}He, the ratio of effective mass to bare mass m^*/m varies from about 3 at low pressure to about 6 at melting pressure.

The effective mass determines the density of states at the Fermi level, N_F, defined by

$$N_{\mathrm{F}} \equiv \sum_{\boldsymbol{k}\alpha} \delta(\epsilon_{\boldsymbol{k}\alpha} - \mu) = \sum_{\alpha} \int \frac{\mathrm{d}^3 k}{(2\pi)^3} \delta(\epsilon_{\boldsymbol{k}\alpha} - \mu) \tag{2.4a}$$

$$= \frac{m^* k_{\mathrm{F}}}{\pi^2 \hbar^2}. \tag{2.4b}$$

It should be emphasized that the quasiparticle states are not energy eigenstates. As will be shown in Section 2.3, the decay rate of a quasiparticle in the state $\boldsymbol{k}, \alpha$ is proportional to $(\epsilon_k - \mu)^2$, or, equivalently, in equilibrium at temperature T, to $(k_{\mathrm{B}}T)^2$.

As a further result of the quasiparticle interaction, the quasiparticle energy depends on the configuration of the surrounding quasiparticles. A local change of the quasiparticle distribution $\delta n_{\boldsymbol{k}\alpha\beta}(\boldsymbol{r}, t)$ (equivalent to a local polarization of the quasiparticle system) causes the local quasiparticle energy to change by

$$\delta\epsilon_{\boldsymbol{k}\alpha\beta}(\boldsymbol{r}, t) = \sum_{\boldsymbol{k}'\alpha'\beta'} f_{\boldsymbol{k}\alpha\beta\boldsymbol{k}'\alpha'\beta'} \, \delta n_{\boldsymbol{k}'\alpha'\beta'}(\boldsymbol{r}, t). \tag{2.5}$$

Here $f_{\boldsymbol{k}\alpha\beta\boldsymbol{k}'\alpha'\beta'}$ is the Fermi-liquid interaction introduced by Landau (1956, 1957). Taking into account the short-range character of the interatomic potential and the isotropy of the system in position space and spin space, the Fermi-liquid interaction may be parametrized as

$$f_{\boldsymbol{k}\alpha\beta\boldsymbol{k}'\alpha'\beta'} = N_{\mathrm{F}}^{-1} \sum_{l=0}^{\infty} P_l(\hat{\boldsymbol{k}} \cdot \hat{\boldsymbol{k}}') \left[F_l^{\mathrm{s}} \, \delta_{\alpha\beta} \, \delta_{\alpha'\beta'} + F_l^{\mathrm{a}} \sum_{\mu=1}^{3} (\sigma_\mu)_{\alpha\beta} (\sigma_\mu)_{\alpha'\beta'} \right], \tag{2.6}$$

where the $\boldsymbol{\sigma}_\mu$ are the three Pauli matrices, $P_l(\cos\theta)$ are the Legendre polynomials and $\hat{\boldsymbol{k}} = \boldsymbol{k}/|\boldsymbol{k}|$. F_l^{s} and F_l^{a} are the spin-symmetric and spin-antisymmetric dimensionless Landau parameters. For many purposes, it is sufficient to keep the Landau parameters for $l = 0$ and 1 only.

The first few terms in $\delta\epsilon_{\boldsymbol{k}\alpha\beta}$ may be expressed in terms of mean fields by

$$\delta\epsilon_{\boldsymbol{k}\alpha\beta} = N_{\mathrm{F}}^{-1} \left[F_0^{\mathrm{s}} \, \delta n \, \delta_{\alpha\beta} + F_0^{\mathrm{a}} \frac{2}{\hbar} \sum_{\mu} (\sigma_\mu)_{\alpha\beta} S_\mu \right] + \frac{1}{3\rho} F_1^{\mathrm{s}} \frac{m}{m^*} \hbar \boldsymbol{k} \cdot \boldsymbol{g} \delta_{\alpha\beta}, \tag{2.7}$$

where

$$\delta n = \sum_{\boldsymbol{k}\alpha} \delta n_{\boldsymbol{k}\alpha\alpha}, \tag{2.8}$$

$$S_\mu = \frac{\hbar}{2} \sum_{\boldsymbol{k}\alpha\beta} (\sigma_\mu)_{\beta\alpha} \, \delta n_{\boldsymbol{k}\alpha\beta}, \tag{2.9}$$

$$\boldsymbol{g} = \sum_{\boldsymbol{k}\alpha} \hbar \boldsymbol{k} \, \delta n_{\boldsymbol{k}\alpha\alpha} \tag{2.10}$$

are the respective changes in particle number density, spin density and momentum density. Here $\rho = mn$ is the mass density in equilibrium and the symbol $\boldsymbol{g}$ is used to distinguish the momentum density $\boldsymbol{g}$ from the particle number current, conventionally denoted by $\boldsymbol{j}$; in the present case, where Galilean invariance is valid, one has $\boldsymbol{g} = m\boldsymbol{j}$. Note that the mean fields entering in (2.7) are the only conserved densities that may appear here.

2.2 THERMODYNAMIC PROPERTIES

In a state of thermodynamic equilibrium characterized by the temperature T and the chemical potential μ the quasiparticle distribution is given by the Fermi function

$$n_{\boldsymbol{k}\alpha} = \left[\exp\left(\frac{\epsilon_{\boldsymbol{k}\alpha} - \mu}{k_{\mathrm{B}}T}\right) + 1\right]^{-1}. \tag{2.11}$$

This follows from the one-to-one correspondence of the quasiparticle states with the states of the noninteracting system by the fact that the Fermi distribution is obtained from purely combinational considerations.

2.2.1 Entropy and specific heat

The entropy S of the quasiparticle system is given by the expression for the ideal Fermi gas, with, however, the free-particle energy replaced by the quasiparticle energy:

$$S = -k_{\mathrm{B}} \sum_{\boldsymbol{k}\alpha} [n_{\boldsymbol{k}\alpha} \ln n_{\boldsymbol{k}\alpha} + (1 - n_{\boldsymbol{k}\alpha}) \ln (1 - n_{\boldsymbol{k}\alpha})]. \tag{2.12}$$

Substituting the equilibrium quasiparticle distribution (2.11) and replacing the sum over momenta by an integration over energy, one finds the entropy density

$$s = \tfrac{1}{3}\pi^2 N_{\mathrm{F}} k_{\mathrm{B}}^2 T. \tag{2.13}$$

The specific heat at constant volume (equal to that at constant pressure up to terms of order $(T/T_{\mathrm{F}})^3$) is then

$$C_{\mathrm{N}} = T\left(\frac{\partial s}{\partial T}\right)_V \equiv s = \frac{m^* k_{\mathrm{F}}}{3\hbar^2} k_{\mathrm{B}}^2 T, \tag{2.14}$$

inserting the expression for N_{F}, (2.4). The specific heat is linear in T and proportional to the effective mass. Introducing the Fermi temperature $T_{\mathrm{F}} = \hbar^2 k_{\mathrm{F}}^2 / 2m^* k_{\mathrm{B}}$ in (2.14), it is seen that C_N is roughly T/T_{F} times the classical specific heat, as anticipated. The specific heat of liquid ^{3}He (see Fig. 2.1) is observed to follow a linear law below 20 mK (Abel *et al.* 1966, Greywall 1983, 1986).

The effective mass ratio m^*/m may be related to the Landau parameter F_1^s by a symmetry argument (Landau 1956, Pines and Nozières 1966). Consider a Galilean transformation to a system moving with constant

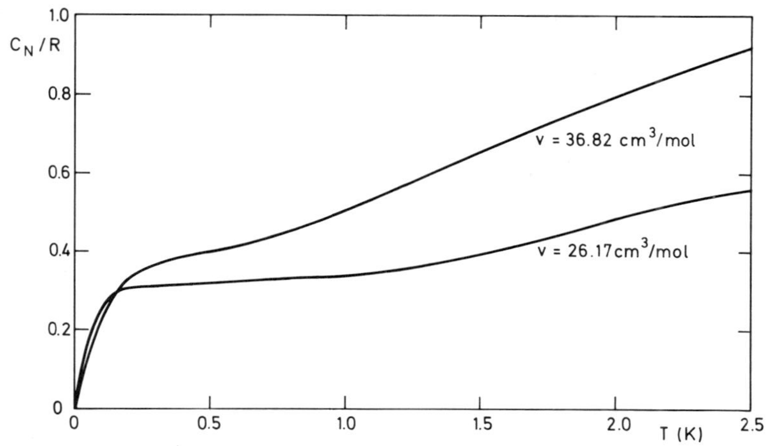

Figure 2.1 Specific heat of normal liquid ^{3}He as a function of temperature for the two molar volumes $v = 36.8\ \text{cm}^3/\text{mol}$ and $v = 26.2\ \text{cm}^3/\text{mol}$ corresponding to pressures of 0 bar and 30 bar in the limit $T \to 0$. (After Greywall (1983).)

velocity $-\boldsymbol{u}$. In the moving frame, the momentum of a quasiparticle is changed by $\Delta \boldsymbol{p} = m\boldsymbol{u} \equiv \hbar\, \Delta \boldsymbol{k}$. Therefore the quasiparticle distribution function in the moving frame is that of a Fermi sphere shifted by $\Delta \boldsymbol{k}$. This implies a change in the quasiparticle energy relative to that in the rest frame of

$$\begin{aligned} \delta\epsilon_{\boldsymbol{k}\alpha\alpha} &= \Delta \boldsymbol{k} \cdot \left(\boldsymbol{\nabla}_{\boldsymbol{k}} \epsilon_{\boldsymbol{k}\alpha\alpha} - \sum_{\boldsymbol{k}'\beta} f_{\boldsymbol{k}\alpha\alpha\boldsymbol{k}'\beta\beta} \boldsymbol{\nabla}_{\boldsymbol{k}'} n_{\boldsymbol{k}'\beta} \right) \\ &= \hbar \boldsymbol{k} \cdot \boldsymbol{u} \frac{m}{m^*} (1 + \tfrac{1}{3} F_1^s), \end{aligned} \tag{2.15a}$$

where we have taken into account that the particular change in the distribution function singles out the $l = 1$ spin-symmetric component of the Fermi-liquid interaction. We have also used the relation

$$-\sum_{k\alpha} \frac{\partial n_{\boldsymbol{k}}}{\partial \epsilon_{\boldsymbol{k}}} \dots = N_F \int \frac{d\Omega_{\boldsymbol{k}}}{4\pi} \dots \quad ,$$

which holds to order T/T_F.

On the other hand, the change in quasiparticle energy may be expressed directly in terms of the quasiparticle momentum by invoking the principle of Galilean relativity as

$$\delta\epsilon_{\boldsymbol{k}\alpha\alpha} = \hbar \boldsymbol{k} \cdot \boldsymbol{u}. \tag{2.15b}$$

Equating (2.15a) and (2.15b), one finds the so-called effective-mass relation

$$\frac{m^*}{m} = 1 + \tfrac{1}{3} F_1^s. \tag{2.16}$$

The values for F_1^s obtained from m^*/m via (2.16) range from about 6 at saturated vapour pressure to about 14 at melting pressure. The fact that F_1^s is large shows that liquid ^{3}He is indeed a strongly correlated Fermi liquid.

2.2.2 Spin susceptibility

In the presence of a magnetic field $\boldsymbol{H}$ applied in the z direction the quasiparticle energies are shifted by the Zeeman energy $-\mu_0 H\alpha$, where $\mu_0 = \frac{1}{2}\gamma\hbar$ is the magnetic moment of the ^{3}He nucleus, with γ the gyromagnetic ratio, and $\alpha = \pm 1$. It should be borne in mind that the g factor of the ^{3}He nucleus is negative. Consequently, spin and magnetic moment point into opposite directions. It is therefore convenient to let the spin quantization axis be directed *opposite* to an external magnetic field too. This choice implies that in a magnetic field the number of up-spin particles is enhanced over that of the down-spin particles, in accordance with the intuitive notion. In addition, the quasiparticle energies change owing to the change in the distribution function. The spin polarization induced by the magnetic field is given by

$$\begin{aligned}\delta n_z &= \sum_{\boldsymbol{k}\alpha} \alpha\, \delta n_{\boldsymbol{k}\alpha\alpha} \\ &= \sum_{\boldsymbol{k}\alpha} \alpha \frac{\partial n_{\boldsymbol{k}\alpha}}{\partial \epsilon_{\boldsymbol{k}\alpha}} (\delta\epsilon_{\boldsymbol{k}\alpha\alpha} - \mu_0 \alpha H).\end{aligned} \tag{2.17}$$

Since $\alpha\, \partial n/\partial\epsilon$ in (2.17) is isotropic and odd in α, only the $l=0$ spin-antisymmetric part of $\delta\epsilon_{\boldsymbol{k}\alpha\alpha}$ given by $N_F^{-1} F_0^a\, \delta n_z$ contributes.

Performing the sums on momenta and spins, the spin susceptibility is found as

$$\chi_N = \chi_N^0/(1 + F_0^a), \tag{2.18}$$

where $\chi_N^0 = \mu_0^2 N_F$ is the Pauli spin susceptibility of a Fermi gas with effective mass m^* and density of states N_F. χ_N is seen to be independent of temperature, roughly T/T_F times the Curie susceptibility of a nondegenerate spin gas. In ^{3}He the spin susceptibility is observed to be independent of temperature below about 40 mK (Wheatley 1966). χ is enhanced over the value of a free Fermi gas with effective mass m^* by roughly a factor 4, implying that $F_0^a \approx -0.75$. The Landau parameter F_0^a is only weakly pressure-dependent.

2.2.3 Compressibility

The response of the number density to a change in chemical potential is obtained in complete analogy with the spin susceptibility as

$$\frac{\partial n}{\partial \mu} = \frac{N_F}{1 + F_0^s}. \tag{2.19a}$$

The compressibility $\kappa_N = (1/n^2)\,\partial n/\partial\mu$ (isothermal and adiabatic compressibility are equal up to order $(T/T_F)^2$), which is

$$\kappa_N = \frac{1}{n^2}\frac{N_F}{1+F_0^s}, \tag{2.19b}$$

determines the velocity of sound c_1 via

$$c_1^2 = (\rho\kappa_N)^{-1} = \tfrac{1}{3}(1+F_0^s)(1+\tfrac{1}{3}F_1^s)v_F^2. \tag{2.20}$$

Thus the Landau parameter F_0^s may be inferred from the speed of sound. The values for F_0^s found in this way (Wheatley 1966, 1970) range from about 10 at s.v.p. to about 100 at melting pressure. Liquid ^{3}He is seen to be much less compressible than a noninteracting Fermi gas of the same density.

The values of the Landau parameters F_0^a, F_0^s and F_1^s, related to thermodynamic quantities as discussed above, are listed in Table 2.1 for pressures between $P=0$ and melting pressure. They are based on high-precission measurements of the low-temperature specific heat C_N, i.e. of the effective mass m^*/m, by Greywall (1986), with F_0^s and F_0^a taken from Wheatley (1975) but corrected for the newly determined m^*/m. Higher Landau parameters, such as F_1^a and F_2^s, are much harder to obtain experimentally. The parameter F_1^a may be extracted from measurements of the normal-state spin diffusion in a high magnetic field (Corruccini *et al.* 1971), from finite-temperature measurements of the specific heat (Greywall

Table 2.1 Values of the Landau parameters F_0^a, F_0^s and F_1^s in ^{3}He together with the molar volume and the effective mass ratio m^*/m for pressures between $P=0$ and melting pressure. The values of V, m^*/m and F_1^s are taken from Greywall (1986), whereas F_0^a, F_0^s are from Wheatley (1975) but corrected for the newly determined m^*/m. At the highest pressure ($P=34.39$ bar) this recalculation was done using Wheatley's values at $P=34.36$ bar.

P (bar)	V (cm^3)	m^*/m	F_1^s	F_0^s	F_0^a
0	36.84	2.80	5.39	9.30	−0.695
3	33.95	3.16	6.49	15.99	−0.723
6	32.03	3.48	7.45	22.49	−0.733
9	30.71	3.77	8.31	29.00	−0.742
12	29.71	4.03	9.09	35.42	−0.747
15	28.89	4.28	9.85	41.73	−0.753
18	28.18	4.53	10.60	48.46	−0.757
21	27.55	4.78	11.34	55.20	−0.755
24	27.01	5.02	12.07	62.16	−0.756
27	26.56	5.26	12.79	69.43	−0.755
30	26.17	5.50	13.50	77.02	−0.754
33	25.75	5.74	14.21	84.79	−0.755
34.39	25.50	5.85	14.56	88.47	−0.753

1983) and from the spin-wave spectrum in the superfluid B phase (Osheroff *et al.* 1977). Thereby F_1^a was found to decrease from -0.5 to -1.0 with increasing pressure. Hence, as in the case of F_0^a, it is small and negative. The parameter F_2^s may be obtained from the difference in the velocity of zero and first sound (see (2.77a)). In this way Engel and Ihas (1985) found $F_2^s \approx -0.8$ at $P = 0$, increasing to about $+0.8$ at melting pressure.

2.3 TRANSPORT PROPERTIES

Let us now turn to dynamical properties. At frequencies low enough for collisions between quasiparticles to establish local equilibrium within one cycle, the state of the system is described by the local values of the particle number density $n(\boldsymbol{r}, t)$, spin population density $n_z(\boldsymbol{r}, t)$, momentum density $\boldsymbol{g}(\boldsymbol{r}, t)$ and energy density $\epsilon(\boldsymbol{r}, t)$.

2.3.1 Hydrodynamic equations

The conserved quantities obey the set of hydrodynamic equations for an ordinary paramagnetic fluid (Landau and Lifshitz 1959b), namely the continuity equation resulting from mass conservation

$$\partial_t \rho + \nabla \cdot \boldsymbol{g} = 0, \tag{2.21a}$$

or, equivalently, from particle number conservation ($n = \rho/m$, $\boldsymbol{j} = \boldsymbol{g}/m$)

$$\partial_t n + \nabla \cdot \boldsymbol{j} = 0, \tag{2.21b}$$

the momentum-conservation equation

$$\partial_t \boldsymbol{g} + \nabla \boldsymbol{\Pi} = 0, \tag{2.22}$$

and the energy-conservation equation

$$\partial_t \epsilon + \nabla \cdot \boldsymbol{j}_\epsilon = 0. \tag{2.23}$$

Conservation of spin gives rise to the continuity equation for the spin density S_z (z is the axis of quantization):

$$\partial_t S_\mu + \nabla \cdot \boldsymbol{j}_\mu^\sigma = 0. \tag{2.24}$$

The currents appearing in these equations can be expressed in terms of the thermodynamically conjugate fields. The mass current, which is identical with the momentum density, is given by the product of mass density ρ and fluid velocity $\boldsymbol{v}$:

$$\boldsymbol{g} = \rho \boldsymbol{v}. \tag{2.25}$$

The stress tensor $\boldsymbol{\Pi}$ is not a conserved quantity. This implies that in addition to the reactive part of Π_{ij}, i.e. the pressure P, there is a dissipative part Π_{ij}^{D}

proportional to gradients of the velocity field, such that in the hydrodynamic regime

$$\Pi_{ij} = P\delta_{ij} + \rho v_i v_j + \Pi_{ij}^{\mathrm{D}}, \tag{2.26a}$$

where

$$\Pi_{ij}^{\mathrm{D}} = -\eta(\nabla_i v_j + \nabla_j v_i - \tfrac{2}{3}\delta_{ij}\nabla \cdot \boldsymbol{v}) - \zeta\delta_{ij}\nabla \cdot \boldsymbol{v}. \tag{2.26b}$$

Here η and ζ are the coefficients of shear viscosity and bulk viscosity respectively (Landau and Lifshitz 1959b).

Similarly, the energy current is given to linear order in $\boldsymbol{v}$ by

$$\boldsymbol{j}_\epsilon = \frac{\mu}{m}\boldsymbol{g} + Ts\boldsymbol{v} - \kappa\nabla T, \tag{2.27}$$

where s is the entropy per volume and κ is the thermal conductivity.

The spin current, finally, is also not conserved and is given by the sum of a convective and a dissipative term:

$$\boldsymbol{j}_\mu^\sigma = S_\mu \boldsymbol{v} + D_\sigma \nabla S_\mu, \tag{2.28}$$

where the spin diffusion coefficient D_σ completes the list of transport coefficients.

2.3.2 Kinetic equation

The transport coefficients may be calculated in Landau's theory by generalizing the quasiparticle description to phenomena varying in space and time. The time evolution of the local quasiparticle distribution function $n_{\boldsymbol{k}\alpha}(\boldsymbol{r}, t)$ is described by the kinetic equation (Landau 1957)

$$\partial_t n_{\boldsymbol{k}\alpha\beta} + \sum_{\alpha'} \nabla_{\boldsymbol{k}}\epsilon_{\boldsymbol{k}\alpha\alpha'} \cdot \nabla_r n_{\boldsymbol{k}\alpha'\beta} - \sum_{\alpha'} \nabla_r \epsilon_{\boldsymbol{k}\alpha\alpha'} \cdot \nabla_{\boldsymbol{k}} n_{\boldsymbol{k}\alpha'\beta} = I\{n_{\boldsymbol{k}\alpha\beta}\}. \tag{2.29}$$

The left-hand side of (2.29) accounts for the continuous change of the distribution due to the streaming of quasiparticles in phase space. The collision integral I describes discontinuous changes brought about by collisions.

It is now necessary to express the hydrodynamic densities and currents in terms of the quasiparticle distribution function. This has been done in part in (2.8)–(2.10). The remaining quantities are obtained as follows (Nozières 1964). Multiplying the kinetic equation (2.29) by $\hbar\boldsymbol{k}$ and summing over $\boldsymbol{k}$, one generates the momentum conservation law (2.22). Since momentum is conserved in each quasiparticle collision, the collision integral drops out. By comparison with (2.22), the stress tensor is identified as

$$\Pi_{ij} = \sum_{\boldsymbol{k}\alpha\beta} \hbar k_i v_{\boldsymbol{k}j\alpha\beta} n_{\boldsymbol{k}\beta\alpha}, \tag{2.30}$$

where we have introduced the quasiparticle velocity

$$\boldsymbol{v}_{\boldsymbol{k}\alpha\beta} = \boldsymbol{\nabla}_{\boldsymbol{k}} \epsilon_{\boldsymbol{k}\alpha\beta}. \tag{2.31}$$

Similarly, the expression for the energy current $\boldsymbol{j}_\epsilon(\boldsymbol{r}, t)$ for the spin-symmetric case is derived, using the definition of the energy-density change

$$\delta\epsilon = \sum_{\boldsymbol{k}\alpha} \epsilon_{\boldsymbol{k}\alpha\alpha} n_{\boldsymbol{k}\alpha\alpha}, \tag{2.32}$$

by multiplying with $\epsilon_{\boldsymbol{k}\alpha\alpha}$ and summing over $\boldsymbol{k}, \alpha$. The result is

$$\boldsymbol{j}_\epsilon = \sum_{\boldsymbol{k}\alpha} \epsilon_{\boldsymbol{k}\alpha\alpha} \boldsymbol{v}_{\boldsymbol{k}\alpha\alpha} n_{\boldsymbol{k}\alpha\alpha}. \tag{2.33}$$

Finally, the spin current $\boldsymbol{j}_\mu$ is obtained as

$$\boldsymbol{j}_\mu^\sigma = \frac{\hbar}{2} \sum_{\boldsymbol{k}\alpha\alpha'\beta'} (\sigma_\mu)_{\alpha\alpha'} v_{\boldsymbol{k}\alpha'\beta'} n_{\boldsymbol{k}\beta'\alpha}. \tag{2.34}$$

2.3.3 Quasiparticle lifetime

Before we turn to a detailed discussion of the collision integral, let us first consider the lifetime of a quasiparticle, $\tau_{\boldsymbol{k}\alpha}$. At low temperatures, when binary collisions dominate, this is given by the Golden-Rule expression

$$\frac{1}{\tau_{\boldsymbol{k}_1\alpha_1}} = \frac{2\pi}{\hbar} \sum_{234} |a(1, 2; 3, 4)|^2 \, n_2(1 - n_3)(1 - n_4)\delta_{\boldsymbol{k}_1+\boldsymbol{k}_2, \boldsymbol{k}_3+\boldsymbol{k}_4} \times \delta_{\alpha_1+\alpha_2, \alpha_3+\alpha_4} \, \delta(\epsilon_1 + \epsilon_2 - \epsilon_3 - \epsilon_4). \tag{2.35}$$

Here the delta functions express conservation of momentum, spin and energy in the scattering process, and the Fermi factors represent the probability for state 2 $(=\boldsymbol{k}_2, \alpha_2)$ to be occupied and 3, 4 to be empty. The transition probability is given by the square of the quasiparticle scattering amplitude $a(1, 2; 3, 4)$ for the binary collision process $(1, 2) \to (3, 4)$.

Spin-rotation invariance implies that there are only two independent scattering amplitudes because both the initial and final states have to be either spin-singlet or spin-triplet. Also, in the limit of low temperatures $(T \ll T_F)$ only quasiparticles in the immediate vicinity of the Fermi energy are excited, so that their momenta are approximately equal to the Fermi momentum. Taking into account momentum conservation and rotation invariance in momentum space, the quasiparticle scattering amplitude may be taken to depend only on two angular variables. They are customarily chosen as the angle θ between the momenta of the incoming particles, i.e. $\cos\theta = \hat{\boldsymbol{p}}_1 \cdot \hat{\boldsymbol{p}}_2$, and the angle ϕ between the planes spanned by the momentum vectors of the incoming particles and the outgoing particles, i.e. $\cos\phi = (\hat{\boldsymbol{p}}_1 \times \hat{\boldsymbol{p}}_2) \cdot (\hat{\boldsymbol{p}}_3 \times \hat{\boldsymbol{p}}_4)/|\hat{\boldsymbol{p}}_1 \times \hat{\boldsymbol{p}}_2| \, |\hat{\boldsymbol{p}}_3 \times \hat{\boldsymbol{p}}_4|$, as shown in Fig. 2.2 (Abrikosov and Khalatnikov 1959). Introducing dimensionless scattering

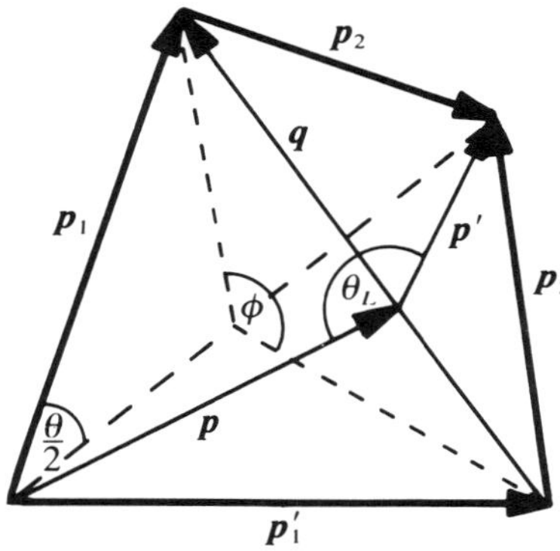

Figure 2.2 Pictorial representation of the variables of the quasiparticle scattering amplitude in terms of the momentum vectors $\boldsymbol{p}_1$, $\boldsymbol{p}_2$ and $\boldsymbol{p}'_1$, $\boldsymbol{p}'_2$ of incoming and outgoing quasiparticles respectively.

amplitudes $A^{0,1}(\theta, \phi)$ for the spin singlet and triplet states respectively, we may write

$$a(1\,2; 3\,4) = \frac{1}{4N_F}\left[(3A^1 + A^0)\delta_{\alpha_1\alpha_3}\,\delta_{\alpha_2\alpha_4} + (A^1 - A^0)\sum_{\mu}(\sigma_\mu)_{\alpha_1\alpha_3}(\sigma_\mu)_{\alpha_2\alpha_4}\right], \tag{2.36}$$

where $(\sigma_\mu)_{\alpha\beta}$ is the α, β component of the Pauli matrix $\boldsymbol{\sigma}_\mu$.

Transforming the sums over momenta in (2.35) into integrals over energy and the angles θ, ϕ, one obtains for the quasiparticle relaxation rate

$$\frac{1}{\tau_k} = \left[1 + \left(\frac{\epsilon_k - \mu}{\pi k_B T}\right)^2\right]\frac{1}{\tau_N^0}, \tag{2.37}$$

with the quasiparticle lifetime on the Fermi surface given by

$$\tau_N^0 = \frac{64\hbar}{\pi^3}\frac{\epsilon_F}{(k_B T)^2}\langle W\rangle_a^{-1}. \tag{2.38}$$

Here we have defined the angular average over the transition probability

$$W(\theta, \phi) = |A^0|^2 + 3\,|A^1|^2 \tag{2.39}$$

as

$$\langle W\rangle_a = \int_0^1 d(\cos\tfrac{1}{2}\theta)\int_0^{2\pi}\frac{d\phi}{2\pi}W(\theta, \phi),$$

such that $\langle 1\rangle_a = 1$. It follows from symmetry that $\tau_{k\alpha}$ is isotropic and independent of spin.

The quasiparticle relaxation rate is seen to be proportional to T^2 or $(\epsilon_k - \mu)^2$, whichever is larger, and hence becomes much smaller than the excitation energy itself for sufficiently low temperatures. This property forms the basis of the quasiparticle picture, because it guarantees that the quasiparticles are sufficiently long-lived to be treated as well-defined entities. Note that this is a consequence of the Fermi statistics and of energy conservation, leading to the restriction $\epsilon_i - \mu < k_B T$ in the two independent energy integrals in (2.35).

It is difficult to measure the quasiparticle lifetime τ_N^0 in the normal state. However, one can determine τ_N^0 from experiments probing relaxation effects in the superfluid state close to the transition. In this way one finds a value $T^2\tau_N^0 \approx 0.3\ \mu\text{s}\ (\text{mK})^2$ (Wheatley 1978) at intermediate pressure. Using this number, we may estimate the temperature T_Q at which the quasiparticle energy width $\hbar/\tau_N^0$ is equal to the thermal energy $k_B T$ as approximately 40 mK. Strictly speaking, the quasiparticle description of liquid ^{3}He is only valid in the temperature regime below T_Q.

Having calculated the quasiparticle lifetime, we are already in a position to derive the low-temperature limiting behaviour of the transport coefficients (Abrikosov and Khalatnikov 1959). All we need for a qualitative discussion is to recall the results of the kinetic theory of gases as applied to the quasiparticle gas. The gas-kinetic expressions for the shear viscosity η_0, the thermal conductivity κ_0 and the (spin) diffusion coefficient D_σ^0 read

$$\left.\begin{aligned} \eta_0 &= \tfrac{1}{5}\rho v_F^2 \tau_N^0 \propto T^{-2}, \\ \kappa_0 &= \tfrac{1}{3}C_N v_F^2 \tau_N^0 \propto T^{-1}, \\ D_\sigma^0 &= \tfrac{1}{3} v_F^2 \tau_N^0 \propto T^{-2}. \end{aligned}\right\} \tag{2.40}$$

Here we have made use of the fact that the quasiparticle velocity is given by the Fermi velocity v_F and that the collision time is of the order of τ_N^0. All the transport coefficients diverge as $T \to 0$ because the mean free path $l_0 = v_F \tau_N^0$ of the excitations becomes infinitely long. While η and D_σ have a T^{-2} dependence, which is due solely to τ_N^0, the thermal conductivity diverges only as T^{-1} owing to the linear T dependence of the specific heat C_N.

This behaviour has been observed experimentally, although only at temperatures below about 30 mK, i.e. far below the Fermi temperature (Wheatley 1966, 1970, 1975, Greywall 1984). The reason for the substantial deviations from Landau theory at intermediate temperatures (see e.g. Fig. 2.1 for the specific heat) lies in the excitation of low-lying collective modes, in particular spin fluctuations (Berk and Schrieffer 1966, Doniach and Engelsberg 1966, Brenig *et al.* 1967, Brinkman and Engelsberg 1968, Riedel 1968, Amit *et al.* 1968a,b, Baym and Pethick 1978, Mishra and Ramakrishnan 1985). These effects are, strictly speaking, beyond the regime of applicability of Fermi-liquid theory. The temperature dependence of the thermodynamic quantities at finite temperatures, i.e. above the low-temperature regime where Landau's theory is valid, has been considered in several model calculations. These include (i) a systematic extension of Landau theory (Pethick and Carneiro 1973, Carneiro and Pethick 1977,

Coffey and Pethick 1988), (ii) the spin-fluctuation model (Levin and Valls 1983), and (iii) the Gutzwiller–Hubbard lattice-gas description, discussed below in more detail (Seiler *et al.* 1986).

While the simple expressions given in (2.40) certainly give the correct temperature dependence, their prefactor only gives the correct order of magnitude. In the following we shall show that (and how) these prefactors can be calculated exactly. This, however, requires a more widely ranging treatment.

2.3.4 Local equilibrium

The collision integral in the kinetic equation (2.29) is the rate of change of the distribution function for a nonequilibrium state by collisions. It is precisely given by the relaxation rate (2.35) of a nonequilibrium state times the distribution function $n_{\mathbf{k}_1\alpha_1}$, i.e.

$$I_{\mathbf{k}_1\alpha_1}\{n_{\mathbf{k}\alpha}\} = -\frac{2\pi}{\hbar}\sum_{234} |a(1\,2;3\,4)|^2\, n_1 n_2(1-n_3)(1-n_4) \times \delta_{\mathbf{k}_1+\mathbf{k}_2,\mathbf{k}_3+\mathbf{k}_4}\delta_{\alpha_1+\alpha_2,\alpha_3+\alpha_4}\delta(\epsilon_1+\epsilon_2-\epsilon_3-\epsilon_4). \tag{2.41}$$

In (2.41) both ϵ_i and n_i have to be interpreted as nonequilibrium quantities, i.e. $\epsilon_i = \epsilon_{\mathbf{k}_i\alpha_i}(\mathbf{r}, t)$ etc.

For simplicity, we restrict our discussion here to distribution functions that are diagonal in spin space, i.e. we do not consider the case of a spin polarization perpendicular to an applied magnetic field.

The delta functions in (2.41), which are again a consequence of the conservation laws, guarantee that the collision integral is zero for the local-equilibrium distribution:

$$n^{\text{l.e.}}_{\mathbf{k}\alpha}(\mathbf{r}, t) = \left\{\exp\left[\frac{\epsilon^{\text{l.e.}}_{\mathbf{k}\alpha}(\mathbf{r}, t) - \mu_\alpha(\mathbf{r}, t) - \hbar\mathbf{k}\cdot\mathbf{v}(\mathbf{r}, t)}{k_B T(\mathbf{r}, t)}\right] + 1\right\}^{-1}, \tag{2.42}$$

which is determined by the local chemical potentials μ_α for both spin species, the fluid velocity $\mathbf{v}$, the temperature T and the local quasiparticle energy $\epsilon^{\text{l.e.}}_{\mathbf{k}\alpha}(\mathbf{r}, t)$. Note that the only changes of the quasiparticle energy that may occur in local equilibrium are those related to the conserved quantities. The changes are precisely given by (2.7). The role of the collisions is to drive $n_{\mathbf{k}\alpha}$ towards the local-equilibrium solution $n^{\text{l.e.}}_{\mathbf{k}\alpha}$.

In a usual transport problem two features allow for a considerable simplification: (i) the deviation from local equilibrium is small, and (ii) the variations of $n_{\mathbf{k}\alpha}(\mathbf{r}, t)$ are slow in space and time. One may then linearize the collision integral in the deviation from local equilibrium and approximate the distribution function on the left-hand side of the kinetic equation, where only derivatives of $n_{\mathbf{k}\alpha}$ appear, by the local-equilibrium solution. This procedure results in the following linearized Boltzmann equation

$$\frac{\partial n_{\boldsymbol{k}\alpha}^{\text{l.e.}}}{\partial \epsilon_{\boldsymbol{k}\alpha}}\left[\frac{1}{2\eta}\hbar\boldsymbol{k}\boldsymbol{\Pi}^{\mathrm{D}}\boldsymbol{v}_{\boldsymbol{k}\alpha}+\frac{\epsilon_{\boldsymbol{k}\alpha}-\mu}{k_{\mathrm{B}}T}\frac{1}{\kappa}\boldsymbol{v}_{\boldsymbol{k}\alpha}\cdot\boldsymbol{j}_{\epsilon}^{\mathrm{D}}\right.$$
$$\left.+\frac{1}{9\zeta}\hbar(\boldsymbol{k}\cdot\boldsymbol{v}_{k}-k_{\mathrm{F}}v_{\mathrm{F}})\operatorname{tr}\boldsymbol{\Pi}^{\mathrm{D}}+\alpha\frac{\mu_0^2}{\chi D_\sigma}\boldsymbol{v}_{\boldsymbol{k}\alpha}\cdot\boldsymbol{j}_z^{\mathrm{D}}\right]=\delta I_{\boldsymbol{k}\alpha}, \quad (2.43)$$

where $\boldsymbol{\Pi}^{\mathrm{D}}$, $\boldsymbol{j}_{\epsilon}^{\mathrm{D}}$ and $\boldsymbol{j}_z^{\mathrm{D}}$ are the dissipative parts of the stress tensor, the energy current (or equivalently the heat current) and the spin current, as defined in (2.30)–(2.34), and η, ζ, κ and D_σ are the transport coefficients. In deriving (2.43), we have made use of the quasiparticle energy (2.7) and the conservation laws (2.21)–(2.24) as well as various thermodynamic relations. Terms quadratic in the gradients have been neglected. The fact that the entropy is a factor T/T_{F} smaller than its classical value gives rise to an effective decoupling of thermal and mechanical degrees of freedom. In (2.43) we have omitted several of these cross-terms, which are of order T/T_{F} relative to the terms kept. It is seen that the term in (2.43) involving the bulk viscosity is of this kind, since the average of $\boldsymbol{k}\cdot\boldsymbol{v}_k - k_{\mathrm{F}}v_{\mathrm{F}}$ is of order T/T_{F}.

2.3.5 The linearized collision integral

We are now ready to calculate the transport coefficients in the limit of low temperatures. For this purpose, we shall first give a detailed discussion of the linearized collision integral, separating energy dependence from angular dependence. The calculation of any particular transport coefficient may then be reduced to the solution of a one-dimensional integral equation in the energy variable. This equation may be solved exactly. Consequently, any transport coefficient can be expressed by a rapidly convergent series in the eigenvalues of the respective integral operator and an angular average of the quasiparticle scattering cross-section (Højgaard Jensen *et al.* 1968, 1969, Brooker and Sykes 1968, Sykes and Brooker 1970).

The linearized collision integral $\delta I_{\boldsymbol{k}\alpha}$ is obtained from (2.41) by expanding n_i and ϵ_i to lowest order in the deviation from local equilibrium:

$$\left.\begin{aligned}\delta n_{\boldsymbol{k}\alpha}&=n_{\boldsymbol{k}\alpha}-n_{\boldsymbol{k}\alpha}^{\text{l.e.}},\\ \delta\epsilon_{\boldsymbol{k}\alpha}&=\epsilon_{\boldsymbol{k}\alpha}-\epsilon_{\boldsymbol{k}\alpha}^{\text{l.e.}}.\end{aligned}\right\} \quad (2.44)$$

The two contributions from expanding n_i and ϵ_i (in the argument of the energy delta function) may be combined into the function

$$\delta n'_{\boldsymbol{k}\alpha}=\delta n_{\boldsymbol{k}\alpha}-\frac{\partial n_{\boldsymbol{k}\alpha}^{\text{l.e.}}}{\partial\epsilon_{\boldsymbol{k}\alpha}}\delta\epsilon_{\boldsymbol{k}\alpha}. \quad (2.45)$$

Separating spin, energy and angular dependence in a fashion analogous to the treatment of the quasiparticle collision rate (2.35), the collision integral may be written in the form

$$\delta I_{k\alpha} = -\frac{1}{\tau_k}\delta n'_{k\alpha} + (\pi^2 k_B T N_F \tau_N^0)^{-1} \sum_{k'\beta} C_{k\alpha k'\beta}\, \delta n'_{k'\beta}, \tag{2.46}$$

where $C_{k\alpha k'\beta}$ is the linear operator

$$C_{k\alpha k'\beta} = K(x, x') \sum_{l=0}^{\infty} (2l+1) P_l(\hat{\boldsymbol{k}} \cdot \hat{\boldsymbol{k}}') \sum_{t=\pm} [\lambda_{lt}^{s} + \alpha\beta\lambda_{lt}^{a}]\tilde{\Pi}^t, \tag{2.47}$$

where $K(x, x')$ is defined in (2.55). The energy variables x, x' are defined by $x = (\epsilon_k - \mu)/k_B T$. The particle–hole symmetry operator $\tilde{\Pi}^t$ projects onto functions even $(t = +)$ and odd $(t = -)$ in the energy variable x'. The angular dependence of the collision process is contained in the parameters λ_{lt}^s, λ_{lt}^a, which are given by the angular averages of the singlet and triplet scattering amplitudes $A^{0,1}$ as

$$\left.\begin{aligned} \lambda_{lt}^{s} &= \frac{\langle W[-tP_l(\hat{\boldsymbol{k}}_1 \cdot \hat{\boldsymbol{k}}_2) + P_l(\boldsymbol{k}_1 \cdot \boldsymbol{k}_3) + P_l(\hat{\boldsymbol{k}}_1 \cdot \hat{\boldsymbol{k}}_4)]\rangle_a}{\langle W \rangle_a}, \\ \lambda_{lt}^{a} &= \frac{\langle -W_2 t P_l(\hat{\boldsymbol{k}}_1 \cdot \hat{\boldsymbol{k}}_2) + W_3 P_l(\hat{\boldsymbol{k}}_1 \cdot \hat{\boldsymbol{k}}_3) + W_4 P_l(\hat{\boldsymbol{k}}_1 \cdot \hat{\boldsymbol{k}}_4)\rangle_a}{\langle W \rangle_a}, \end{aligned}\right\} \tag{2.48}$$

where W has been defined in (2.39) and

$$\left.\begin{aligned} W_2 &= (A^1)^2 - (A^0)^2, \\ W_{3,4} &= (A^1 \pm A^0)A^1. \end{aligned}\right\} \tag{2.49}$$

The arguments of the Legendre polynomials are given in terms of the scattering angles θ and ϕ by

$$\left.\begin{aligned} \hat{\boldsymbol{k}}_1 \cdot \hat{\boldsymbol{k}}_2 &= \cos\theta, \\ \hat{\boldsymbol{k}}_1 \cdot \hat{\boldsymbol{k}}_{3,4} &= \cos^2 \tfrac{1}{2}\theta \pm \sin^2 \tfrac{1}{2}\theta \cos\phi. \end{aligned}\right\} \tag{2.50}$$

Note that, as a consequence of particle-number, momentum, energy and spin conservation, one has $\lambda_{0+}^s = 1$, $\lambda_{1+}^s = 1$, $\lambda_{0-}^s = 3$ and $\lambda_{0+}^a = 1$.

For each transport process the symmetry of the driving force on the left-hand side of (2.43) singles out exactly one component of the collision integral. Thus in the case of the viscosity coefficient the driving term has $l = 2$ orbital symmetry and is spin-symmetric and particle–hole-symmetric. The relevant parameter in the collision integral, λ_{2+}^s, may be written more explicitly as

$$\lambda_{2+}^s = 1 - 3\frac{\langle W \sin^4 \frac{1}{2}\theta \sin^2\phi \rangle_a}{\langle W \rangle_a}. \tag{2.51}$$

Similarly, the parameters relevant for heat transport and spin diffusion are

λ^{s}_{1-} and λ^{a}_{1+}. They may be written as

$$\lambda^{s}_{1-} = 1 + 2\frac{\langle W \cos\theta\rangle_{a}}{\langle W\rangle_{a}}, \tag{2.52}$$

$$\lambda^{a}_{1+} = 1 - 2\frac{\langle(|A^{0}|^{2} + |A^{1}|^{2})\sin^{2}\tfrac{1}{2}\theta\,(1-\cos\phi)\rangle_{a}}{\langle W\rangle_{a}}. \tag{2.53}$$

2.3.6 Calculation of the transport coefficients

The problem of calculating the kinetic coefficients of a Fermi liquid is seen to be reduced to solving a one-dimensional integral equation in the energy variable x, of the form

$$(\pi^{2} + x^{2})\,\delta n'(x) - \lambda\int_{-\infty}^{\infty} dx'\,K(x, x')\,\delta n'(x') = \frac{F(x)}{\cosh^{2}\tfrac{1}{2}x}. \tag{2.54}$$

Here $\delta n'(x)$ is the distribution function $\delta n'_{\mathbf{k}}$, where the angular dependence has been separated off and $F(x)$ is constant or proportional to x, depending on the symmetry of the driving force. The integral kernel is defined by

$$K(x, x') = \frac{\cosh\tfrac{1}{2}x'}{\cosh\tfrac{1}{2}x}\,\frac{x - x'}{\sinh[\tfrac{1}{2}(x - x')]}. \tag{2.55}$$

In the following we discuss the solution of the integral equation (2.54) (Højgaard Jensen *et al.* 1968, Brooker and Sykes 1968). It is seen that, by factoring out the cosh $\tfrac{1}{2}x$ dependence, one is left with a positive symmetric kernel of the convolution type. This allows a transformation of the integral equation by Fourier analysis to an ordinary linear second-order differential equation, which may be solved in terms of tabulated functions. In this way it is found that the eigenvalue problem

$$(\pi^{2} + x^{2})\psi_{n}(x) - k_{n}\int_{-\infty}^{\infty} dx'\,K(x, x')\psi_{n}(x') = 0 \tag{2.56}$$

has the solution

$$\psi_{n}(x) = \frac{1}{\cosh\tfrac{1}{2}x}\int_{-\infty}^{\infty} dk\,e^{ikx}P^{1}_{n}(\tanh\pi k), \tag{2.57}$$

where the eigenvalues are given by

$$k_{n} = \tfrac{1}{2}n(n+1), \quad n = 1, 2, \ldots. \tag{2.58}$$

The functions $P^{1}_{m}(z)$ are associated Legendre polynomials. The eigenfunctions $\psi_{n}(x)$ are orthogonal with respect to the weight function

$$p(x) = (\pi^{2} + x^{2})\cosh^{2}\tfrac{1}{2}x \tag{2.59}$$

and may be normalized as

$$\int dx\ \psi_n(x)p(x)\psi_m(x) = \delta_{n,m}. \tag{2.60}$$

Also, the eigenfunctions $\psi_n(x)$ have definite parity $(-1)^{n+1}$.

The operator $K(x, x')$ has the following representation in terms of eigenfunctions:

$$K(x, x') = \frac{1}{\cosh^2 \frac{1}{2}x} \sum_{n=1}^{\infty} p(x)\psi_n(x)\frac{1}{k_n}\psi_n(x')p(x'), \tag{2.61}$$

One may now expand the distribution function $\delta n'(x)$ for a particular transport problem in terms of eigenfunctions of the correct parity as

$$\delta n'(x) = \sum_{\substack{n \\ \text{odd/even}}} a_n\psi_n(x) \tag{2.62}$$

and substitute this into the transport equation (2.54). With the aid of (2.61) and the orthonormality of the ψ_n, the coefficients a_n are found as

$$a_n = \frac{1}{1-\lambda/k_n}\int dx\ F(x)\psi_n(x). \tag{2.63}$$

For example, in the case of the shear viscosity η we consider a shear-flow velocity field $\boldsymbol{v}$, where $\nabla_x v_y \neq 0$. It may then be inferred from the driving force (2.43) that the distribution function takes the form

$$\delta n'_{\boldsymbol{k}\alpha} = \hat{k}_x\hat{k}_y \sum_{n\ \text{odd}} a_n\psi_n(x). \tag{2.64}$$

The driving force is $F(x) = \frac{1}{4}\pi^2\tau_N^0(\hbar k_F v_F/k_B T)\nabla_x v_y$, where the angle-dependent factor $\hat{k}_x\hat{k}_y$ is not included by definition. By comparing the hydrodynamic expression (2.26) for the dissipative part of the stress tensor, and the kinetic expression (2.30), linearized in δn and $\delta\epsilon$, one sees that the shear-viscosity coefficient is given by

$$\eta = \left[\hbar k_F v_F \sum_{\boldsymbol{k}\alpha} \hat{k}_x\hat{k}_y\ \delta n'_{\boldsymbol{k}\alpha}\right] \Big/ (\nabla_x v_y), \tag{2.65}$$

Substituting the eigenfunctions expansion (2.64) into (2.65) and using (2.63), one finds

$$\eta = \eta_0 \frac{m^*}{m} f_\eta(\lambda_{2+}^s), \tag{2.66}$$

where η_0, the usual kinetic expression for the viscosity, has been introduced in (2.40).

Incidentally, the result $\eta = \eta_0$ would have been obtained had we approximated the collision integral by $\delta I_{\boldsymbol{k}\alpha} = -\delta n'_{\boldsymbol{k}\alpha}/\tau_N^0$. The correction factor $f_\eta(\lambda_{2+}^s)$ depends on the details of the collision integral through the parameter λ_{2+}^{s}. It is given exactly by

$$f_\eta(\lambda) = \sum_{n\ \mathrm{odd}} \frac{1}{1-\lambda/k_n} \left| \tfrac{1}{2}\pi \int_{-\infty}^{\infty} \mathrm{d}x\, \psi_n(x) \right|^2. \tag{2.67}$$

Performing the integral and rewriting the sum over n, one gets

$$f_\eta(\lambda) = \frac{\pi^2}{12} + \frac{\lambda}{4} \sum_{n=0}^{\infty} \frac{4n+3}{[(2n+1)(n+1)]^2[(2n+1)(n+1)-\lambda]}. \tag{2.68}$$

The sum over n is rapidly convergent. Given values of λ_{2+}^{s} in the range 0.6–0.7 as determined from calculations of the scattering amplitude, the first term of the sum already provides an accuracy of better than one per cent.

The coefficient of the bulk viscosity ζ turns out to be smaller than the shear viscosity η by a factor $(T/T_\mathrm{F})^2$, and hence can be neglected. Here one of the factors T/T_F comes from the driving force (2.43), while the other one is due to the term $\boldsymbol{k} \cdot \boldsymbol{v}_k - k_\mathrm{F} v_\mathrm{F}$ in the expression for the diagonal part of the dissipative stress tensor.

The calculation of the thermal conductivity proceeds along the same lines, except for the fact that the distribution function has odd particle–hole symmetry. The result is

$$\kappa = \kappa_0 f_\kappa(\lambda_{1-}^{s}), \tag{2.69}$$

where κ_0 is defined in (2.40) and

$$f_\kappa(\lambda) = 3 - \frac{\pi^2}{4} + \frac{3\lambda}{4} \sum_{n=0}^{\infty} \frac{4n+5}{[(2n+3)(n+1)]^2[(2n+3)(n+1)-\lambda]}. \tag{2.70}$$

The calculation of the spin-diffusion coefficient, finally, leads to the same integral equation that appeared in the shear-viscosity problem. The result is

$$D_\sigma = D_\sigma^0(1+F_0^a) f_\eta(\lambda_{1+}^{a}), \tag{2.71}$$

where D_σ^0 has been defined in (2.40).

The preceding discussion has shown how the transport parameters may be expressed rigorously in terms of the scattering properties of quasiparticles on the Fermi surface. On the other hand, a calculation of the scattering amplitude cannot be performed within the framework of Landau theory since it involves interaction processes with momentum exchange of the order of the Fermi momentum. In Section 2.5 we shall address some of the recent attempts to calculate the scattering amplitude.

2.4 COLLECTIVE MODES

In the hydrodynamic regime ($\omega\tau \ll 1$), the only well defined collective mode is normal longitudinal sound ("first sound"), with velocity given by (2.20). The sound attenuation α_1 is dominated by viscous processes and is given by the hydrodynamic expression (Landau and Lifshitz 1959b)

$$\alpha_1 = \frac{2\omega^2}{3c_1^3\rho}\eta, \tag{2.72}$$

where $\rho = nm$ is the mass density (see also below (2.76)).

As the frequency of sound is increased, the damping grows as ω^2, until the sound wave becomes overdamped. However, at low temperatures such that $\hbar/\tau_N^0 \ll k_B T$ there exists a regime in which $\omega\tau \gg 1$ owing to the fact that the quasiparticle relaxation rate $1/\tau_N^0$ is independent of sound frequency as long as $\hbar\omega \ll 2\pi k_B T$ (Landau 1957). In this "collisionless" regime the system oscillates many times between successive collisions of a quasiparticle such that the oscillations are only weakly damped. The existence of well-defined collective modes with linear dispersion in the collisionless regime was predicted by Landau (1957) on the basis of the transport equation (2.29). Landau called these excitations "zero sound". Generally speaking, zero sound is a wave-like excitation of the mean fields in a Fermi liquid, which appear in the expression for the quasiparticle energy (2.7).

Linearizing the kinetic equation (2.29) in the deviations from equilibrium δn and $\delta\epsilon$ and taking the Fourier transform, one obtains

$$(\omega - \boldsymbol{v}_k \cdot \boldsymbol{q})\,\delta n_{\boldsymbol{k}\alpha} + \boldsymbol{v}_k \cdot \boldsymbol{q}\frac{\partial n_k}{\partial \epsilon_k}\delta\epsilon_{\boldsymbol{k}\alpha} = \mathrm{i}I\{\delta n_{\boldsymbol{k}\alpha}\}. \tag{2.73}$$

The zero-sound modes are found as eigensolutions of (2.73), putting the collision integral equal to zero. In principle, each of the spherical harmonic components of (2.7) is capable of oscillations. However, in ^{3}He only longitudinal isotropic zero sound is well defined. Other modes either do not exist for lack of a restoring force (spin zero sound except in high magnetic fields) or are strongly damped by collisions (transverse zero sound) in the temperature regime of interest. The reasons for the remarkable stability of longitudinal zero sound in ^{3}He are twofold. (1) Owing to the large repulsive interaction constants F_0^s and F_1^s, the sound velocity is much larger than the Fermi velocity (see (2.20)), so that the direct decay of sound quanta into particle–hole pairs ("Landau damping") is prohibited by energy and momentum conservation. For the same reason ($c_1 \gg v_F$) the deformation of the Fermi sphere caused by zero sound is nearly spherical because the coupling to higher-l components is proportional to $(v_F/c_1)^l$. Also, the zero-sound velocity is nearly equal to the velocity of first sound (see below). (2) The damping of a nearly spherically symmetric oscillation of the Fermi sphere is strongly reduced. This is because the $l = 0$ and $l = 1$ components of δn_k are not affected by collisions—a consequence of conservation of particle number ($l = 0$) and momentum ($l = 1$).

The first property allows a simple description of longitudinal sound propagation in liquid ^{3}He (Wölfle 1979). Let us start with the conservation laws for particle number and momentum (2.21) and (2.22).

Outside the hydrodynamic regime the stress tensor is not simply given by (2.26). Rather, we expect the stress tensor to relax towards the local-equilibrium value (2.26). Considering the rotational invariance of the system, there exist only two independent relaxation rates, one for the trace of $\mathbf{\Pi}$ and the other for the traceless part $\Pi_{ij}^{\mathrm{c}} = \Pi_{ij} - \frac{1}{3}\delta_{ij}\,\mathrm{tr}\,\mathbf{\Pi}$. It follows from (2.26a) that in the hydrodynamic regime the trace of $\mathbf{\Pi}$ is given to a very good approximation by the local-equilibrium value, the pressure $\frac{1}{3}\,\mathrm{tr}\,\mathbf{\Pi} = P$. This is so because the bulk-viscosity coefficient turned out to be negligibly small. Assuming that the relaxation times for nonconserved currents are proportional to the respective transport coefficients, we are led to the conclusion that $\frac{1}{3}\,\mathrm{tr}\,\mathbf{\Pi}$ is equal to the instantaneous value of the pressure. Combining the two conservation laws (2.21) and (2.22) and using the relation $\nabla P = (\partial P/\partial\rho)\nabla\rho = c_1^2\nabla\rho$, one finds the following wave equation for the mass density:

$$\partial_t^2\rho - c_1^2\nabla^2\rho = \sum_{i,j}\nabla_i\nabla_j\Pi_{ij}^{\mathrm{c}}. \tag{2.74}$$

The traceless part of the stress tensor, on the other hand, approximately satisfies the equation

$$\partial_t\Pi_{ij}^{\mathrm{c}} = \frac{1}{\tau_z}(\Pi_{ij}^{\mathrm{c}} - \Pi_{ij}^{\mathrm{D}}), \tag{2.75}$$

which describes the relaxation of Π_{ij}^{c} towards the hydrodynamic value Π_{ij}^{D} given in (2.26b). The relaxation time τ_z turns out to be given in terms of the viscous relaxation time $\tau_\eta \equiv \tau_{\mathrm{N}}^0 f_\eta(\lambda_{2+}^{\mathrm{s}})$ as $\tau_z = \tau_\eta/(1+\frac{1}{5}F_2^{\mathrm{s}})$. In writing (2.75) we have neglected terms like the divergence of a current of Π_{ij}^{c}, because they only give rise to small contributions of order v_{F}/c_1.

With the aid of (2.21), (2.26b), (2.74) and (2.75), the dispersion relation of sound for arbitrary $\omega\tau$ is found as

$$\omega^2 = c_1^2 q^2\left(1 - \frac{4}{3}\frac{\eta}{\rho c_1^2}\frac{\mathrm{i}\omega}{1-\mathrm{i}\omega\tau_z}\right). \tag{2.76}$$

Solving (2.76) for the momentum q, the sound velocity c and sound attenuation α are obtained from $c = \omega/\mathrm{Re}\,(q)$ and $\alpha = -\mathrm{Im}\,(q)$ respectively. In the hydrodynamic regime ($\omega\tau \ll 1$) one thereby recovers the expression (2.72). On the other hand, in the collisionless limit ($\omega\tau \gg 1$) one finds

$$c_0 = c_1\left\{1 + \frac{2}{15}\frac{m^*}{m}(1+\tfrac{1}{5}F_2^{\mathrm{s}})\left(\frac{v_{\mathrm{F}}}{c_1}\right)^2 + O\left[\left(\frac{v_{\mathrm{F}}}{c_1}\right)^4\right]\right\}, \tag{2.77a}$$

$$\alpha_0 = \frac{2}{15}\frac{m^*}{m}\frac{1+\frac{1}{5}F_2^{\mathrm{s}}}{v_{\mathrm{F}}\tau_z}\left(\frac{v_{\mathrm{F}}}{c_1}\right)^3\left\{1 + O\left[\left(\frac{v_{\mathrm{F}}}{c_1}\right)^2\right]\right\}. \tag{2.77b}$$

As mentioned above, the deviation of the zero-sound velocity c_0 from the hydrodynamic sound velocity c_1 is only of order $(v_{\mathrm{F}}/c_1)^2$ small. The attenuation α_0, however, is a factor of $(v_F/c_1)^3$ smaller than what one might expect on dimensional grounds.

The transition from first to zero sound may be achieved by working at fixed frequency and lowering the temperature, whereby $\tau \propto T^{-2}$ increases (Keen *et al.* 1963; Abel *et al.* 1965). One finds an increase in sound velocity by a fraction of a per cent and a maximum in the sound attenuation at $\omega\tau \approx 1$, in agreement with (2.77a,b) (see Fig. 2.3). There are other forms of zero sound that have been found experimentally, too, although they are much less well defined, e.g. *transverse* zero sound (Roach and Ketterson 1976a,b, Lea *et al.* 1977) and *spin* zero sound (Masuhara *et al.* 1984). The coupling of density and spin zero sounds induced by a small particle–hole asymmetry has also been discussed (Ketterson 1983, Fishman and Sauls 1985).

The density response of liquid ^{3}He at frequencies several times the Fermi frequency has been studied by means of neutron scattering, (Sköld *et al.* 1976, Sköld and Pelizzari 1976, 1980, Scherm *et al.* 1974, 1987, Stirling

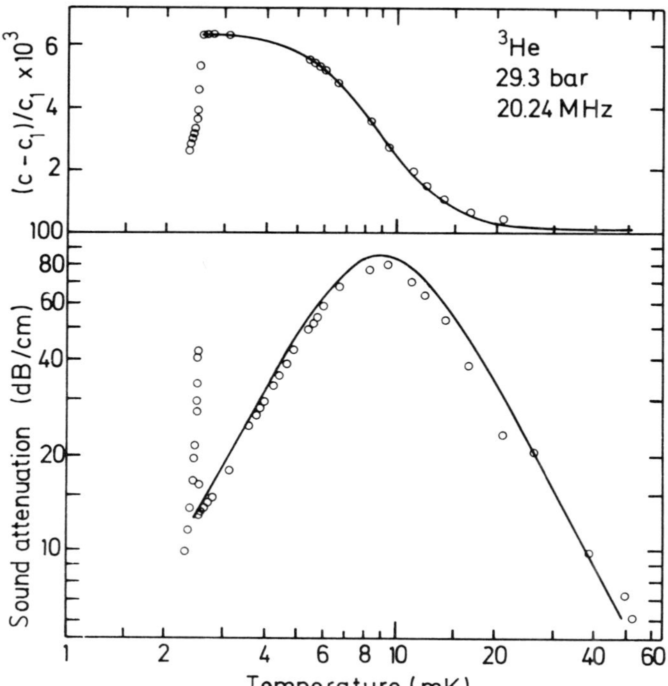

Figure 2.3 Sound attenuation and velocity change in liquid ^{3}He versus temperature as measured by Ketterson *et al.* (1975). The solid lines are the result of a fit to (2.76). The onset of the superfluid state is signalled by the sharp features at the low-temperature end. (After Wölfle (1978a).)

et al. 1976, Stirling 1978). The scattering is dominated by excitation of particle–hole pairs generating a broad continuum for energy transfer $\hbar\omega$ and momentum transfer q satisfying $|\hbar\omega - q^2/2m^*| < qv_F$. Sköld and Pelizzari (1976, 1980) observed what seems to be a continuation of the zero-sound mode into this high-frequency regime. The existence of such a mode was anticipated by Pines (1966) and Aldrich and Pines (1978) on the basis of the so-called "polarization potential theory" to be discussed below (see also Aldrich *et al.* 1976).

2.5 THEORIES BEYOND THE FERMI-LIQUID MODEL

Landau's theory of Fermi liquids is phenomenological in the sense that the observable properties are expressed in terms of an effective interaction (something like a set of pseudopotentials), namely the interaction function $f_{\mathbf{k}\alpha\mathbf{k}'\beta}$ or, more generally, the scattering amplitudes $A^{0,1}(\theta, \phi)$. It is then argued that these pseudopotentials are well described by a small number of parameters, which depend on pressure but not on temperature and magnetic field provided that $T \ll T_F$ and $\gamma\hbar H \ll E_F$. In the following we shall discuss the different approaches towards a calculation of these interaction functions.

2.5.1 Microscopic theories

A microscopic theory starting from atoms with a bare mass m interacting via the molecular ^{3}He–^{3}He potential has been a challenge for more than two decades. In fact, it is only since the early 1970s that we have slowly begun to understand better the gross features of the microscopic physics of liquid helium. This development is based on several different approaches.

There are now powerful Green's-function Monte Carlo techniques available for integrating the many-particle Schrödinger equation. They have mainly been used to calculate ground-state properties of the quantum liquids, i.e. of ^{4}He (Ceperley and Kalos 1979) and, in particular, of ^{3}He (Lee *et al.* 1981, Schmidt and Kalos 1984, Panoff and Carlson, 1989). There is hope that it will eventually be feasible to calculate finite-temperature equilibrium properties and even transport properties using these methods.

Early theories attempted to cope with the large repulsive core of the potential (see Fig. 2.4), which causes multiple-scattering processes to be very important, by concentrating on the calculation of what is essentially a two-particle scattering matrix (Brueckner and Gammel 1958, Burckhardt 1968, Østgaard 1968a,b,c, 1969a,b, Bertsch 1969, Glyde and Khanna 1977, 1980, Glyde and Hernadi 1983, 1984a,b). Such an approach, however, does not account for the phenomenon of steric hindrance: the interaction of two particular atoms is strongly influenced by the blocking effect in real space of

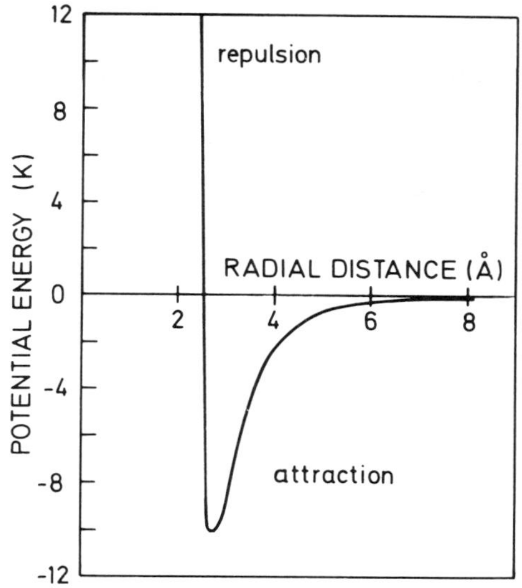

Figure 2.4 Interaction potential of two ^{3}He atoms as a function of separation.

other nearby atoms. These short-range correlation effects are best described within a variational approach (Feenberg 1969, Woo 1976, Pandharipande and Wiringa 1979, Campbell 1978, Kurten and Campbell 1981, Owen and Ripka 1983, Manousakis *et al.* 1983). There a variational wave function in real space is constructed by multiplying the Slater determinant by products of correlation factors (Jastrow factors) for each pair. This reduces the probability for atoms to be in a spatially close configuration. The latter approach has been refined using elements from many-body diagram theory (hypernetted-chain summation) as well as the theory of classical liquids (see Krotscheck 1983).

The results of a generalized Jastrow-type variational calculation for the ground-state energy of both ^{4}He and ^{3}He is in reasonable agreement with the exact Green's-function Monte Carlo calculations (Pandharipande and Wiringa 1979). A systematic improvement of the variational method is attempted within the correlated basis function (CBF) approach (Feenberg 1969). There the correlation operators determined in a variational calculation are used to construct a set of nonorthogonal basis states, on which a perturbation theory is built. This has the advantage of avoiding the complications stemming from the hard-core part of the potential. The method provides good results for the ground-state energies (Pandharipande and Wiringa 1969, Krotscheck 1983). It holds promise for a calculation of excited-state properties (Krotscheck, 1983). However, those properties involving details of the excitation spectra, in particular of ^{3}He near the Fermi surface, remain difficult to calculate.

At the present stage it is useful to consider also theories on a level between the fully microscopic and the semiclassical quasiparticle regimes. Theories of this type may serve to bring out the salient properties of a many-body system on the microscopic scale, at the cost of introducing certain model parameters like an effective mass or an effective interaction function. The latter quantities may be more amenable to microscopic calculation than the directly observable quantities. Let us consider such a theory in more detail.

2.5.2 Polarization potential theory

Any microscopic theory of liquid helium has to face two basic difficulties. First, the radius of the repulsive hard core in the interatomic potential is large—up to 70% of the average interparticle distance. The repulsion leads to a spatial correlation between the atoms, causing them to avoid each other. Consequently, there is a large steric-hindrance effect, which may be expected to increase the radius of the effective interation, while at the same time the spatial correlations transform the hard-core into an effective soft-core potential. Secondly, the strong short-range repulsion gives rise to large screening effects, because the system is more rigid. In other words, there also exist strong long-range correlations.

One may expect these properties to be relatively insensitive to the quantum statistics of the particles involved and therefore to be shared by both liquid ^{3}He and ^{4}He. The importance of polarization effects led Pines and coworkers (Pines 1966, Aldrich *et al.* 1976, Aldrich and Pines 1978) to the formulation of the so-called "polarization potential theory". It is well known that the density response of a many-body system is screened by the effect of interactions giving rise to density fluctuations. In the helium liquids additional screening is caused by longitudinal current fluctuations (the so-called "backflow"), which are directly connected to density fluctuations via the continuity equation. Fluctuations of a more complex structure are more heavily damped than those of the conserved variables and contribute much less. Employing the concept of polarization potentials, the density response function $\chi(\boldsymbol{q}, \omega)$ may be represented in terms of a simpler "screened" response function $\chi^{sc}(\boldsymbol{q}, \omega)$ and two effective interaction functions $f^s(\boldsymbol{q})$ and $g^s(\boldsymbol{q})$, which provide the coupling to density and current fluctuations respectively (Aldrich and Pines 1978):

$$\chi(\boldsymbol{q}, \omega) = \frac{\chi^{sc}(\boldsymbol{q}, \omega)}{1 - [f^s(\boldsymbol{q}) + (\omega^2/q^2)g^s(\boldsymbol{q})]\chi^{sc}(\boldsymbol{q}, \omega)}. \tag{2.78a}$$

An analogous expression holds for the spin-density response function of the Fermi liquid ^{3}He. It involves two further functions $f^a(\boldsymbol{q})$ and $g^a(\boldsymbol{q})$. The

$q=0$ values of $f^{s,a}(\boldsymbol{q})$ and $g^{s,a}(\boldsymbol{q})$ are related to the Landau parameters by

$$\left.\begin{aligned} f^{s,a}(0) &= N_F^{-1} F_0^{s,a}, \\ g^{s,a}(0) &= \frac{m}{3n} F_1^{s,a}. \end{aligned}\right\} \tag{2.78b}$$

One may argue that the long-range part of the spin-symmetric density interaction function $f^s(\boldsymbol{q})$ must essentially be given by the van der Waals attractive part of the bare potential. The short-range part may be modelled by a soft core of radius roughly equal to the hard-core radius and strength adjusted to the experimental compressibility $\partial n/\partial\mu = \chi(\omega=0,\ \boldsymbol{q}\to 0) \approx 1/f^s(\boldsymbol{q}=0)$. It is an experimental fact that $\partial n/\partial\mu$ is approximately the same for ^{4}He and ^{3}He, except for pressures well below 10 bar, where the value for ^{4}He is somewhat larger. This may be taken as an indication that $f^s(\boldsymbol{q})$ is nearly the same for both liquids. In other words, the influence of the statistics on $f^s(\boldsymbol{q})$ is clearly dominated by the influence of the short-range repulsion.

The spin-symmetric current-density interaction is much less well known. A reasonable description of the neutron-scattering data on both ^{4}He and ^{3}He can be obtained (Aldrich and Pines 1978) by again assuming $g^s(\boldsymbol{q})$ to be essentially the same for both systems. Finally, the screened density response function $\chi^{sc}(\boldsymbol{q},\omega)$ is modelled by the well-known expression for a free Bose or Fermi gas of particles with an effective mass m^*, plus a structureless multiparticle contribution.

The polarization potential theory appears to be capable of describing both liquid ^{4}He and ^{3}He in a unified way.

2.5.3 Extended Fermi-liquid theory

There arises the natural question of how to calculate the effective interaction functions. For the Fermi liquid ^{3}He this may be attempted in the framework of an extended Landau quasiparticle theory that consists of a quantum kinetic equation for the distribution function $n_{\boldsymbol{p}}(\boldsymbol{q},\omega)$ for $|\boldsymbol{q}| < 2k_F$. Such a theory was first considered by Babu and Brown (1973), who pointed out the importance of many-body polarization effects for the calculation of the Landau parameters as well as for the scattering amplitude. This idea was taken up by Ainsworth *et al.* (1983), Bedell and Ainsworth (1984) and Quader and Bedell (1985), who calculated the polarization-induced part of the interaction, starting from a so-called "direct interaction" as input. A somewhat different way of including many-body polarization effects into the Landau effective interaction has been proposed by Pfitzner and Wölfle (1984, 1986, 1987). This approach is based on an approximation in which the integral equation for the dimensionless scattering amplitude

$A = N_F \ a(1\,2;1'\,2')$ proposed by Landau (1958) is generalized to finite momentum transfer q:

$$A^{\lambda}_{kk'}(\boldsymbol{q}) = F^{\lambda}_{kk'}(\boldsymbol{q}) + \sum_{k''} F^{\lambda}_{kk''}(\boldsymbol{q})\chi_{k''}(\boldsymbol{q})A^{\lambda}_{k''k'}(\boldsymbol{q}), \tag{2.79a}$$

where

$$\chi_k(\boldsymbol{q}) = N_F^{-1}\frac{n_{k+q/2} - n_{k-q/2}}{\epsilon_{k+q/2} - \epsilon_{k-q/2}} \tag{2.79b}$$

is the particle–hole propagator describing the propagation of a quasiparticle with momentum $\boldsymbol{p} = \hbar(\boldsymbol{k} + \frac{1}{2}\boldsymbol{q})$ and a quasihole with momentum $\hbar(\boldsymbol{k} - \frac{1}{2}\boldsymbol{q})$. The superscript $\lambda = \mathrm{s, a}$ denotes spin-symmetric and spin-antisymmetric amplitudes respectively. In this notation the momenta of the ingoing and outgoing quasiparticles are given by $(\boldsymbol{p}_1, \boldsymbol{p}_2)$ and $(\boldsymbol{p}'_1, \boldsymbol{p}'_2)$ respectively, as shown in Fig. 2.2, where $\boldsymbol{p}_1 = \hbar(\boldsymbol{k} + \frac{1}{2}\boldsymbol{q})$, $\boldsymbol{p}_2 = \hbar(\boldsymbol{k}' - \frac{1}{2}\boldsymbol{q})$, $\boldsymbol{p}'_1 = \hbar(\boldsymbol{k}' + \frac{1}{2}\boldsymbol{q})$ and $\boldsymbol{p}'_2 = \hbar(\boldsymbol{k} - \frac{1}{2}\boldsymbol{q})$.

The spin-symmetric (spin-antisymmetric) scattering amplitudes $A^{\mathrm{s,a}}$ are related to the spin-singlet and spin-triplet amplitudes $A^{0,1}$ introduced earlier by

$$\left.\begin{aligned} A^{\mathrm{s}} &= \tfrac{1}{4}(3A^1 + A^0), \\ A^{\mathrm{a}} &= \tfrac{1}{4}(A^1 - A^0). \end{aligned}\right\} \tag{2.80}$$

In the Landau limit, $q \to 0$, the particle–hole propagator is equal to $N_F^{-1}\,\partial n/\partial\epsilon$. Performing a Legendre expansion, the integral equation (2.79a) may be solved in the approximation where the dependence of A and F on the absolute magnitude of the momenta $|\boldsymbol{k}''|$ is neglected. This yields the well-known result (Landau 1958)

$$A^{\mathrm{s,a}}_{kk'}(q = 0) = A^{\mathrm{s,a}}(\theta, \phi = 0) = \sum_{l=0}^{\infty} A^{\mathrm{s,a}}_l P_l(\hat{\boldsymbol{k}} \cdot \hat{\boldsymbol{k}}'), \tag{2.81a}$$

where

$$A^{\mathrm{s,a}}_l = \frac{F^{\mathrm{s,a}}_l}{1 + F^{\mathrm{s,a}}_l/(2l+1)}. \tag{2.81b}$$

The Pauli principle requires the scattering amplitude to be antisymmetric under exchange of either the two incoming or the two outgoing quasiparticles (exchange symmetry). In terms of the singlet/triplet amplitudes, this property can be expressed as

$$A^{0,1}(\theta, \phi) = \pm A^{0,1}(\theta, \phi + \pi). \tag{2.82}$$

In the forward-scattering limit $\theta \to 0$ the triplet scattering amplitude has to vanish, i.e.

$$A^1(0, \phi) = \sum_{l=0}^{\infty} (A^{\mathrm{s}}_l + A^{\mathrm{a}}_l) = 0, \tag{2.83}$$

leaving us with a useful sum rule for the coefficients $A^{\mathrm{s,a}}$.

The requirement of exchange symmetry makes it difficult to calculate A approximately from (2.79a), since, for any given Landau interaction function $F_{\boldsymbol{kk}'}(\boldsymbol{q})$, the solution A of (2.79b) will in general not be exchange-symmetric. However, the fact that the dependences on $(\boldsymbol{k}, \boldsymbol{k}')$ and $\boldsymbol{q}$ are highly correlated may be turned to advantage: for given q dependence of the first few Landau interaction parameters $F_l(\boldsymbol{q})$ $(l = 0, 1)$, a set of higher Landau parameters may be generated such that the resulting scattering amplitude satisfies exchange symmetry (Pfitzner and Wölfle 1983a,b). Unfortunately, the q dependence of the Landau parameters is not yet known very well, although some insight, especially on $F_0^s(\boldsymbol{q})$, may be gained from the polarization potential approach of Aldrich and Pines (1978) discussed above. An earlier calculation of the scattering amplitude by Bedell and Pines (1980a) along these lines, based on the polarization potential theory, in which exchange symmetry is imposed via a simple approximation, already provides good agreement with experiment. It appears that the effective interaction $F_{\boldsymbol{kk}'}(\boldsymbol{q})$ is more sensitive to small changes in the system parameters than the scattering amplitude $A_{\boldsymbol{k}\bar{\boldsymbol{k}}'}(\boldsymbol{q})$, a fact also apparent from the large values of $F_0^s(\boldsymbol{q})$ and $F_1^s(\boldsymbol{q})$ compared with $A_l(\boldsymbol{q})$, which is at most of order unity. It is therefore more promising to concentrate on the calculation of F rather than A.

The effective interaction function $F_{\boldsymbol{kk}'}(\boldsymbol{q})$, defined as the particle–hole irreducible vertex function, can be shown to satisfy an integral equation in the "crossed" particle–hole channel, describing the propagation of the quasiparticle with momentum $\boldsymbol{k} + \frac{1}{2}\boldsymbol{q}$ and the quasiparticle with $\boldsymbol{k}' + \frac{1}{2}\boldsymbol{q}$ rather than $\boldsymbol{k}' - \frac{1}{2}\boldsymbol{q}$ (Nozières 1964),

$$\bar{F}^{\lambda}_{\boldsymbol{kk}'}(\boldsymbol{q}) = I^{\lambda}_{\boldsymbol{kk}'}(\boldsymbol{q}) + \sum_{\boldsymbol{k}''} F^{\lambda}_{\boldsymbol{kk}''}(\boldsymbol{q})\chi_{\boldsymbol{k}''}(\boldsymbol{q})A^{\lambda}_{\boldsymbol{k}''\boldsymbol{k}'}(\boldsymbol{q}), \tag{2.84}$$

where $\lambda = \mathrm{s, a}$ and the bar denotes the exchange operation, i.e.

$$\bar{F}^{\lambda}_{\boldsymbol{kk}'}(\boldsymbol{q}) = -\tfrac{1}{2}[F^{\mathrm{s}}_{\bar{\boldsymbol{k}}\bar{\boldsymbol{k}}'}(\bar{\boldsymbol{q}}) + m_{\lambda}F^{\mathrm{a}}_{\bar{\boldsymbol{k}}\bar{\boldsymbol{k}}'}(\bar{\boldsymbol{q}})],$$

where $m_{\mathrm{s}} = 3$, $m_{\mathrm{a}} = -1$ and $\bar{\boldsymbol{k}} = \frac{1}{2}(\boldsymbol{k} + \boldsymbol{k}' + \boldsymbol{q})$, $\bar{\boldsymbol{k}}' = \frac{1}{2}(\boldsymbol{k} + \boldsymbol{k}' - \boldsymbol{q})$, $\bar{\boldsymbol{q}} = \boldsymbol{k} - \boldsymbol{k}'$. Here $I_{\boldsymbol{kk}'}(\boldsymbol{q})$ is the totally irreducible vertex part (i.e. it does not contain polarization contributions in either particle–hole channel), and will be called "direct interaction". The direct interaction I is exchange-symmetric, although F is not.

The scattering amplitude may be expressed completely in terms of I and F:

$$A^{\lambda}_{\boldsymbol{kk}'}(\boldsymbol{q}) = -I^{\lambda}_{\boldsymbol{kk}'}(\boldsymbol{q}) + F^{\lambda}_{\boldsymbol{kk}'}(\boldsymbol{q}) + \bar{F}^{\lambda}_{\boldsymbol{kk}'}(\boldsymbol{q}), \tag{2.85}$$

If one now replaces A in (2.84) or equivalently (2.79a) by I and F via (2.85), one obtains a nonlinear integral equation for F in terms of I. This integral equation may be solved numerically by making use of the eigenfunctions of the exchange operator, described below. Indeed, an essentially parameter-free calculation of the Landau parameters using a realistic single-particle spectrum and a direct interaction derived from the CBF approach (Krotscheck 1983) yields good agreement with the experimental values (Pfitzner and Wölfle 1984, 1986, 1987). Also, the pair interaction turns out to be most attractive in the $l = 1$ state (see below).

It is worth mentioning that the solution of the coupled equations (2.79a) and (2.85) for A and F has certain universal features. For large repulsive direct interaction I, one finds that the spin-symmetric Landau parameter F_0^s scales with I to large positive values, whereas F_0^a saturates at a negative value of about $(2\chi_0)^{-1}$, where χ_0 is an average of the particle–hole propagator $\chi_{\boldsymbol{k}}(q)$. This may be verified at once in the approximation where only the s-wave component of A, F and I is kept. For large negative I, one finds a similar result, with the spin-symmetric and spin-antisymmetric components interchanged (Pfitzner and Wölfle 1986, 1987).

2.5.4 Lattice-gas model

The above results are very similar to those obtained within the context of a lattice-gas description of ^{3}He using Gutzwiller's solution of the Hubbard model (Anderson and Brinkman 1975, Vollhardt 1984a, Vollhardt *et al.* 1987). This approach stresses the importance of the fact that ^{3}He atoms want to stay well separated so as to avoid the mutual and short-range interaction between the hard cores. Within such a picture the large effective mass of the ^{3}He quasiparticles and the small compressibility of the liquid, i.e. the largeness of F_0^s and F_1^s, can be understood in a natural way.

The hard-core repulsion is described by an interaction that only acts on the same lattice site and if the particles have opposite spin. The effective Hamiltonian has the form

$$H = \sum_{\langle i,j \rangle} \sum_{\sigma} t_{ij} c_{i\sigma}^{+} c_{j\sigma} + U \sum_{i} n_{i\uparrow} n_{i\downarrow}. \tag{2.86}$$

Here the first term represents the kinetic energy, where t_{ij} is a (nearest-neighbour) hopping constant and $c_{i\sigma}^{+}$, $c_{i\sigma}$ are the usual fermion creation and destruction operators for particles with spin σ at site i. This term is quantum-mechanical in origin. The second term describes the interaction, where U is an effective interaction constant and $n_{i\sigma} = c_{i\sigma}^{+} c_{i\sigma}$ are density operators. This lattice model, known as the Hubbard model (Gutzwiller 1963, Hubbard 1963, Kanamori 1963), was originally introduced to treat electronic correlations and magnetic ordering in narrowband electron systems where the Coulomb interaction could be assumed to be strongly screened. Note that in the case of a liquid like ^{3}He the existence of a lattice (whose actual coordination should not, of course, influence the results for the physical properties of the liquid very strongly) serves two purposes: (i) it describes the well-known short-range solid-like correlations of the liquid; and (ii) it provides definite sites, i.e. a finite probability, for interactions. In this way, the finite size of the ^{3}He atoms is simulated. Furthermore, the lattice prevents the system from flying apart, i.e. it mimics an attractive interaction not present in (2.86) itself.

In fact, the hard cores of the ^{3}He atoms do not allow two atoms to be strictly at the same lattice site, in contrast with the case of pointlike electrons. One should therefore consider the atoms to be distributed in a regular array of *cells,* which in the case of a solid are singly occupied. There will be a finite probability for double occupancy of a given cell. This corresponds to an interstitial in the lattice, which in turn is associated with the existence of a vacancy. It is then natural to identify the energy of formation of an interstitial with the on-site repulsion U in the model. In the following we shall nevertheless use the expression "doubly occupied site" when we actually mean "doubly occupied cell".

A model with the same kind of interaction as in (2.86) but *without* an underlying lattice, i.e. a continuum model of fermions interacting via a delta-function potential, is used as starting point for the so-called "paramagnon theory" (Berk and Schrieffer 1966, Doniach and Engelsberg 1966, Béal-Monod *et al.* 1968; see also Levin and Valls 1983). It should be noted, however, that, strictly speaking, a zero-range repulsive interaction does not influence, i.e. correlate, freely moving fermions in dimensions $d \geq 3$ at all (Herring 1966, Stamp 1985). The paramagnon theory assumes that long-lived and long-ranged spin fluctuations ("paramagnons") determine the low-temperature properties of liquid ^{3}He. Within that particular theory the strong enhancement of the spin susceptibility is interpreted as a consequence of an incipient ferromagnetic instability of the liquid. This is why liquid ^{3}He has long been called "almost-ferromagnetic".

A variational solution for the lattice model (2.86) was proposed by Gutzwiller (1963, 1965). He constructed a variational wave function that restricts local density fluctuations in $|\psi_0\rangle$, the wave function of the noninteracting system, i.e. one that reduces the weight of those spin configurations in $|\psi_0\rangle$ that allow for too many on-site interactions. The approach is therefore similar in spirit to that of the correlated basis function (CBF) approach (Feenberg 1969, Woo 1976). To calculate the expectation value of (2.86), Gutzwiller used an approximation that is equivalent to calculating the *classical* statistical weights of different configurations of particles (for a discussion see Vollhardt 1984a) and that becomes exact in $d=\infty$ dimensions (Metzner and Vollhardt 1988, 1989). If an equal number of particles and lattice sites is assumed (half-filled band) then the results of this approximation describe a transition to a localized state with single occupancy of lattice sites at a finite interaction strength $U=U_c$ (Brinkman and Rice 1970). Although the transition itself is an artefact of the approximation, the results for $U \lesssim U_c$ for the effective mass, the spin susceptibility and the compressibility closely resemble the experimentally measured behaviours of these quantities for normal liquid ^{3}He at $T \ll T_F$ (Anderson and Brinkman 1975, Vollhardt 1984a). These results may be understood within the concept of a simplified Landau theory. The effective mass m^* and the Landau parameters F_0^a and F_0^s are thus obtained as

$$\frac{m^*}{m}=\frac{1}{1-\bar{U}^2}, \tag{2.87}$$

$$F_0^a=-p\left[1-\frac{1}{(1+\bar{U})^2}\right], \tag{2.88}$$

$$F_0^s=p\left[\frac{1}{(1-\bar{U})^2}-1\right], \tag{2.89}$$

where $\bar{U}=U/U_c$ and $p\approx 1$. Note that $F_0^a(\bar{U})=F_0^s(-\bar{U})$, i.e. changing the interaction from repulsive to attractive leads to an interchange of the spin-symmetric and spin-antisymmetric components. For $U\rightarrow U_c$, m^* and F_0^s diverge, while F_0^a saturates at a finite negative value ($F_0^a\approx-\frac{3}{4}$). If one eliminates the unknown interaction parameter $\bar{U}$ in favour of the pressure-dependent effective mass (2.87), one may calculate the pressure dependence of F_0^a and F_0^s, i.e. of the spin susceptibility χ_N, (2.18), and of the compressibility κ_N, (2.19b).

One finds that for increasing pressure, F_0^a is almost pressure-independent. This implies that the pressure dependence of χ_N comes almost entirely from that of the effective mass. Consequently the ratio $\chi_N/\chi_N^0=(1+F_0^a)^{-1}\approx 4$ is also essentially pressure-independent, as is observed experimentally (Wheatley 1975, Greywall 1983). Hence liquid ^{3}He is not found to be close to a ferromagnetic transition, as concluded within the paramagnon model.

The compressibility κ_N is found to be strongly reduced and to decrease further for increasing pressure. This is also observed in experiments. The result that in the limit $m^*\rightarrow\infty$ (which is not actually relevant for ^{3}He) F_0^s diverges as $F_0^s\propto(m^*/m)^2$, leading to vanishing compressibility, may be traced back to the assumption of a half-filled band (see below).

The remarkable agreement between the results of this simple model, which are determined by the vicinity of a localization transition of particles, with those found experimentally for normal liquid ^{3}He is the reason why ^{3}He has been called an "almost-localized" Fermi liquid (Anderson and Brinkman 1975, Vollhardt 1984a; see also Castaing and Nozières (1979), who conclude that liquid ^{3}He is solid-like or "almost-solid"). The static properties of normal liquid ^{3}He may then be understood to be caused mainly by the tendency of the hard cores of the particles to keep well separated.

The model described above, which assumed an equal number of particles and lattice sites, may be generalized to a lattice-gas model with variable density (Vollhardt *et al.* 1987). In the case that the underlying lattice is taken as incompressible, it is found that the particle density actually tends towards half-band filling as the localization transition is approached. In this limit the compressibility now stays finite. If the lattice is given a finite compressibility itself (after all, even the solid is compressible) then the spurious localization *transition* disappears. Thereby the (unphysical) singular features of the results of the model are removed while the essential dependences of m^*/m and χ_N on pressure, which are in good agreement with experiment, remain. These properties are nevertheless still determined by the "almost-localized" character of the system. This model also allows one to deduce a connection with microscopic models by relating the short-range interaction parameter U to the soft-core part of the polarization potential of Aldrich and Pines (1978).

An extension of the lattice-gas model for ^3He to finite temperatures was discussed by Seiler *et al.* (1986). Within the model considered, three different temperature regimes may be distinguished: (i) for $T \ll T_F$ the system is a Fermi liquid; (ii) for $T_F \lessapprox T \lessapprox U$ it shows classical behaviour, but is still strongly correlated; while (iii) for $T \gg U$ the system behaves in a free-particle-like manner. In the second temperature regime the entropy of the almost-localized fermions is bounded by $k_B \ln 2$ per ^3He atom, because in this temperature range essentially only singly occupied lattice sites exist. Such a bound had already been assumed much earlier by Goldstein (1954). Assuming a particular functional form for the temperature dependence of the entropy to fit the experimental data of Greywall (1983), Dyugaev (1984) obtained a similar but somewhat higher bound. Within this framework, the crossover from Fermi-liquid to classical behaviour at $T \approx T_F$ has been described, which is otherwise outside the scope of Fermi-liquid theory itself. Good qualitative agreement with the measured specific heat, spin susceptibility and thermal expansion coefficient is obtained. In particular, it is found that the anomalous behaviour of the specific heat (Fig. 2.1), i.e. the sharp kink at about 200 mK and the plateau-like structure above this temperature, is caused by the bound of $R \ln 2$ on the entropy.

2.5.5 Phenomenological approaches

Although the various approaches to calculating the Landau interaction function and the scattering amplitude from microscopic theory look promising, a complete and accurate theory is not yet available. In this situation it is useful to take a certain amount of experimental information as input, for example the Landau parameters $F_0^{s,a}$, F_1^s and some of the transport parameters, in order to determine a model form of the scattering amplitude. The scattering amplitude obtained in this way may then be used to calculate other experimental quantities like higher-order Landau parameters, the pair-interaction parameters essential for a description of pair correlations, and various transport and relaxation parameters in the normal and superfluid phases.

A convenient representation of the scattering amplitude that conserves exchange symmetry is the so-called effective-potential approximation. It can be derived by assuming the scattering amplitude to be given in the Born approximation in terms of effective potentials $W^{0,1}(\boldsymbol{k}-\boldsymbol{k}')$ for the singlet and triplet channels respectively (Sauls and Serene 1981a, Levin and Valls 1979a,b, Wölfle 1979):

$$A^s(\theta, \phi) = W^s(\boldsymbol{k}_1 - \boldsymbol{k}_3) + (-1)^s W^s(\boldsymbol{k}_1 - \boldsymbol{k}_4), \tag{2.90}$$

where $s = 0, 1$. The coefficients of the Legendre expansion of W^s,

$$W^s(\boldsymbol{k}_1 - \boldsymbol{k}_3) = \sum_{l=0}^{\infty} W_l^s P_l(\hat{\boldsymbol{k}}_1 \cdot \hat{\boldsymbol{k}}_3), \tag{2.91}$$

are completely determined by the coefficients $A^{\mathrm{s,a}}$ of the scattering amplitude in the Landau limit, and hence by the Landau parameters (via (2.81)). These relations are

$$\left.\begin{aligned} W_l^0 &= A_l^{\mathrm{s}} - 3A_l^{\mathrm{a}}, \\ W_l^1 &= A_l^{\mathrm{s}} + A_l^{\mathrm{a}}, \end{aligned}\right\} \quad l > 0 \tag{2.92}$$

and

$$W_0^0 = \frac{1}{2}\left(A_0^{\mathrm{s}} - 3A_0^{\mathrm{a}} - \sum_{l=1}^{\infty} W_l^0\right).$$

Keeping only components with $l = 0, 1$, one arrives at the so-called s–p approximation (Dy and Pethick 1969), which has been widely used. Although the scattering amplitude is not expected to be a strongly anisotropic function (apart from the weak singularity at $\theta = \pi$ associated with pair formation), the s–p approximation is not sufficient. Using terms up to $l = 3$, satisfactory agreement with the transport parameters may be obtained (Sauls and Serene 1981a).

It should be kept in mind, however, that (2.90) is based on an unjustified assumption: the scattering of two quasiparticles is treated as an isolated process, whereas one should expect three-particle and higher-order interaction effects to be present as well. These limitations may be avoided by expanding the scattering amplitude in terms of a complete orthonormal set of exchange-symmetric functions: the eigenfunctions of the exchange operator. These functions are given by (Pfitzner 1985a)

$$X_{lk}(\theta, \phi) = [(k+1)(2l+1)]^{1/2}(-1)^l P_l(\cos\phi)(\sin\tfrac{1}{2}\theta)^{2l} P_{k-l}^{(2l+1,0)}(\cos\theta), \tag{2.93}$$

for $l \leq k$, and $l, k = 0, 1, \ldots$. Here $P_l(x)$ are Legendre polynomials and $P_{k-l}^{(2l+1,0)}(x)$ are Jacobi polynomials of index $(2l+1, 0)$ and degree $k-l$, defined by

$$P_{k-l}^{(2l+1,0)}(x) = 2^{l-k} \sum_{m=0}^{k-l} \binom{k+l+1}{m} \binom{k-l}{k-l-m} (x-1)^{k-l-m}(x+1)^m. \tag{2.94}$$

The parity of X_{lk} under the exchange operation defined by (2.82) is $(-1)^l$, as determined by the $P_l(\cos\phi)$ factor. The X_{lk} obey the orthonormality relations

$$\frac{1}{2}\int_{-1}^{1} d(\cos\theta)\int_{-1}^{1} d(\cos\phi)\sin^2\tfrac{1}{2}\theta\, X_{lk}(\theta,\phi)X_{l'k'}(\theta,\phi) = \delta_{ll'}\delta_{kk'}. \quad (2.95)$$

The number of eigenfunctions X_{lk} of degree not higher than n in $\cos\theta$ and $\cos\phi$ is $\frac{1}{2}(n+1)(n+2)$. For $n=3$, there are 10 eigenfunctions, 6 of even and 4 of odd symmetry, as compared with the 7 terms in the equivalent effective-potential approximation. The 10 parameters a_{lk} in the expansion

$$\left.\begin{aligned} A^0(\theta,\phi) &= \sum_{k=0}^{3}\sum_{l\text{ even}}^{k} a_{lk}X_{lk}(\theta,\phi), \\ A^1(\theta,\phi) &= \sum_{k=1}^{3}\sum_{l\text{ odd}}^{k} a_{lk}X_{lk}(\theta,\phi) \end{aligned}\right\} \quad (2.96)$$

may serve to characterize the quasiparticle scattering amplitude with an accuracy of about 10%.

A fit of this form (employing the six lowest eigenfunctions) to the presently available data on the Landau parameters $F_0^{s,a}$, F_1^s and the transport coefficients η, κ and D yields the set of coefficients shown in Table 2.2 (Pfitzner 1985a). The fit does not improve noticeably if 10 eigenfunctions and additional experimental information are used. This appears to be due to the internal inconsistency of the various experimental data. It is of interest to calculate the q dependence of the Landau parameters from the phenomenologically determined scattering amplitude by solving (2.78) for $F_{kk'}^{s,a}(q)$ defined by the analogue of (2.81a). The result is shown in Fig. 2.5. It is seen that the interaction functions in the density (and spin-density) channel increase in strength at finite q values, before they become small at $q>2k_F$. Similar forms have been deduced from real-space models of the polarization potentials employed by Aldrich and Pines (1978) and from microscopic CBF calculations by Krotscheck and collaborators (Krotschek 1983).

2.5.6 Pair interaction

Finally, we consider the pair interaction, i.e. the interaction responsible for the superfluid phase transition in liquid ^{3}He. As will be discussed in detail in

Table 2.2 Parameters a_{lk} characterizing the quasiparticle scattering amplitude (2.96) and pair-interaction coupling constants λ_l according to (2.98).

Pressure (bar)	a_0	a_{01}	a_{02}	a_{22}	a_{11}	a_{12}	λ_0	λ_1	λ_2
0	4.18	2.39	−0.12	1.62	1.35	−0.22	0.15	−0.33	0.31
12	5.30	3.50	−0.07	1.74	1.54	0.14	0.06	−0.28	0.34
21	5.54	3.62	−0.09	1.82	1.65	0.14	0.07	−0.30	0.35
34	5.13	3.23	0.34	2.10	1.75	−0.02	0.29	−0.36	0.40

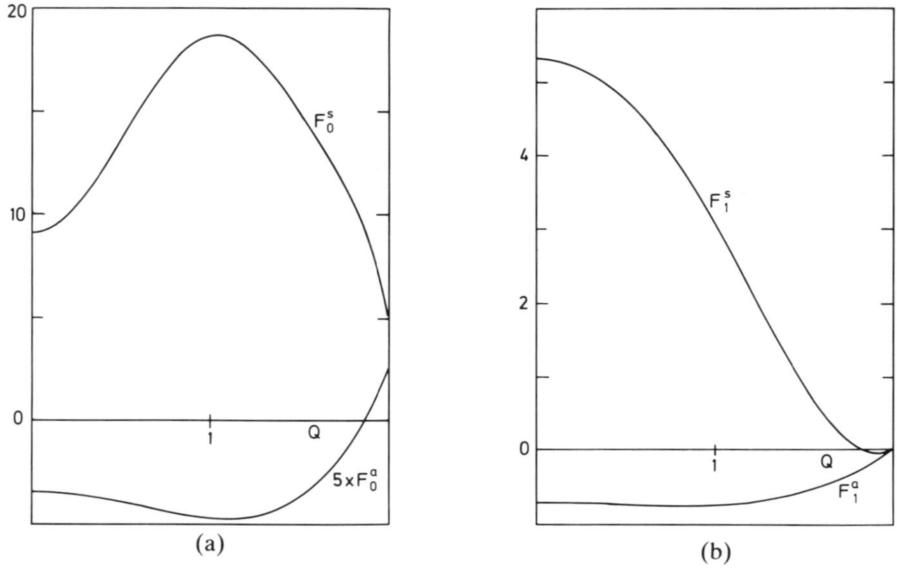

Figure 2.5 Momentum-dependent Landau parameters as functions of the reduced particle–hole momentum $Q = q/k_F$. (After Pfitzner and Wölfle (1987).)

Chapter 3, this is the interaction between two quasiparticles with momenta ($\boldsymbol{k}$, $-\boldsymbol{k}$). The matrix element $V_{\boldsymbol{k}-\boldsymbol{k}'}$ of this interaction describes the process where two quasiparticles with momenta ($\boldsymbol{k}$, $-\boldsymbol{k}$) are scattered into states ($\boldsymbol{k}'$, $-\boldsymbol{k}'$). There are arguments (Patton and Zaringhalam 1975) that the pair interaction is given to a good approximation by the scattering amplitude $A(\theta, \phi)$ for $\theta = \pi$, and $\cos\phi = \hat{\boldsymbol{k}} \cdot \hat{\boldsymbol{k}}'$. More precisely, the exchange-symmetrized interaction V is given by $a(\theta, \phi)$ averaged over a small region around θ, such that the weak logarithmic singularity in $A(\theta, \phi)$ for $\theta = \pi$ is removed. The pair-interaction constants V_l, obtained from a Legendre expansion of $V_{\boldsymbol{k}-\boldsymbol{k}'}$ (see (3.5a)), may then be calculated from

$$\lambda_l \equiv N(0)V_l = \frac{1}{4}\int_{-1}^{1} \mathrm{d}(\cos\phi)\, P_l(\cos\phi) A^s(\pi, \phi), \tag{2.97}$$

with $s = 0$ or 1 for l even or odd. Substituting the expression (2.96) and using the property of the Jacobi polynomials $P_n^{(a,0)}(-1) = (-1)^n$, one obtains

$$\lambda_l = \tfrac{1}{4}(2l+1)^{-1/2} \sum_{k=l}^{\infty} (k+1)^{1/2}(-1)^k a_{lk}. \tag{2.98}$$

For the values of a_{lk} given in Table 2.2, one finds λ_0, λ_1, λ_2 as shown in the same table.

As can be seen from Table 2.2, the pair-interaction constants λ_l are positive (repulsive interaction) for $l = 0$, 2 and negative for the $l = 1$ channel (see also Bedell and Pines 1980b). An attractive interaction in the p-wave channel mediated by spin fluctuations was first proposed by Layzer and Fay (1968, 1971); see also Østgaard (1973, 1974, 1975). In the next chapter it will be shown that the attractive p-wave attraction leads to a phase transition characterized by the formation of pairs in a relative p-wave state.

This concludes our review of the properties of normal liquid ^{3}He. We have focused mainly on the limiting low-temperature behaviour, which provides the starting point for the theory of the superfluid phases, our main subject. It appears that the semiphenomenological Landau theory of Fermi liquids provides an accurate description of the low-temperature properties of normal liquid ^{3}He. A reliable microscopic theory of liquid ^{3}He is still lacking, although important progress has been made in recent years.

Much interest has recently been devoted to the behaviour of liquid ^{3}He at intermediate temperatures up to the Fermi temperature, where precise measurements (Greywall 1983) for example of the specific heat show distinct structure. This behaviour appears to reflect a single-particle spectrum with a rapidly decreasing effective mass, as one moves away from the Fermi edge, both in momentum and energy (Brown *et al.* 1982, Krotscheck 1983). Naturally, the structure of the excitation spectrum for higher energies and momenta also enters a microscopic calculation at zero temperature. A complete theory of normal liquid ^{3}He must account for the quasiparticle energy and interaction in a consistent way.

FURTHER READING

Abrikosov A A and Khalatnikov I M 1959 *Rep. Prog. Phys.* **22** 329

Baym G and Pethick C J 1978 in *The Physics of Liquid and Solid Helium,* Part II, ed. K H Bennemann and J B Ketterson (Wiley, New York), p. 1

Campbell C E 1978 in *Progress in Liquid Physics,* ed. C A Croxton (Wiley, New York), p. 213

Feenberg E 1969 *Theory of Quantum Fluids* (Academic Press, New York)

Keller W E 1969, *Helium-3 and Helium-4* (Plenum, New York)

Landau L D 1956 *Zh. Eksp. Teor. Fiz.* **30** 1058 [*Sov. Phys. JETP* **3** 920 (1957)]

Landau L D 1957 *Zh. Eksp. Teor. Fiz.* **32** 59 [*Sov. Phys. JETP* **5** 101 (1957)]

Landau L D 1958 *Zh. Eksp. Teor. Fiz.* **35** 97 [*Sov. Phys. JETP* **8** 70 (1959)]

Leggett A J 1975 *Rev. Mod. Phys.* **47** 331

Levin K and Valls O T 1983 *Phys. Rep.* **98** 1

Nozières P 1974 *Theory of Interacting Fermi Systems* (Benjamin, New York)

Pines D 1985 in *Highlights of Condensed Matter Theory*: *Course 89 of the Varenna Summer School* (Societa Italiana di Fisica, Bologna, Italy), p. 580

Pines D and Nozieres P 1966 *The Theory of Quantum Liquids* (Benjamin, New York)

Vollhardt D 1984 *Rev. Mod. Phys.* **56** 99

Wheatley J C 1970 in *Progress in Low Temperature Physics,* Vol. VI, ed. J C Gorter (North-Holland, Amsterdam), p. 77

Wilks J 1967 *The Properties of Liquid and Solid Helium* (Clarendon Press, Oxford)

Wölfle P 1979 *Rep. Prog. Phys.* **42** 269

3

Pair Correlations in the Weak-Coupling Limit

The concept of pair correlations in interacting Fermi systems was first introduced in 1957 by Bardeen, Cooper and Schrieffer (1957) (BCS) to provide a microscopic understanding of superconductivity. This phenomenon had resisted theoretical explanation for more than 40 years. According to modern interpretation, the salient feature of this theory is the concept of spontaneously broken gauge symmetry, or, in more elementary language, the fact that the pair-correlated state is described by a complex rather than a real-valued order parameter. This fact is usually related to the appearance of a gap in the single-particle excitation spectrum. The phase rigidity of the wave function (or order parameter) of the pair-correlated state is responsible for the property of superfluidity or superconductivity, which is probably the most spectacular property of these systems. We shall see later to what extent the new phases of ^{3}He deserve the name superfluid. In this chapter we present the so-called weak-coupling theory of pair correlations in Fermi systems. As the interaction giving rise to pair correlations becomes weaker, we should expect both the transition temperature T_c to the pair-correlated state to decrease and the characteristic length ξ_0 of pair correlations to increase. The small parameter of weak-coupling theory is the ratio T_c/T_F (or equivalently $(k_F\xi_0)^{-1}$). For liquid ^{3}He, this ratio is of order 10^{-3}, suggesting that weak-coupling theory, which becomes exact in the limit $T_c/T_F \rightarrow 0$, should be a reasonable starting point. Why weak-coupling theory is not quite as good as one would expect from the smallness of the parameter T_c/T_F will be discussed in Chapter 5.

3.1 COOPER INSTABILITY

Let us first consider a simple model problem introduced by Cooper (1956) in the preliminary stages of the development of BCS pairing theory, which illuminates best the basic idea of pairing theory. Consider a system of N

non-interacting identical fermions of mass m and spin $\frac{1}{2}$ in the ground state (the "Fermi sea"). If we add two more particles, the ground state of the $N+2$ particle system is of course obtained by putting the two particles in the lowest available states at the Fermi energy. However, what happens if we now switch on an attractive interaction $V(\boldsymbol{r}_1 - \boldsymbol{r}_2)$ just between the two added particles?

Quite generally, the wave function of our model system is given by the antisymmetrized product of a correlated pair wave function $\Phi(\boldsymbol{r}_1, \boldsymbol{r}_2; \alpha, \beta)$ and an N-particle Slater determinant describing the Fermi sea. The pair wave function in turn is a product of the centre-of-mass plane wave, the wave function describing the relative motion $\psi(\boldsymbol{r}_1 - \boldsymbol{r}_2)$, and the spin functions $\chi(\alpha, \beta)$:

$$\Phi(\boldsymbol{r}_1, \boldsymbol{r}_2; \alpha, \beta) = \exp\left[\tfrac{1}{2}\mathrm{i}\boldsymbol{P} \cdot (\boldsymbol{r}_1 + \boldsymbol{r}_2)\right] \psi(\boldsymbol{r}_1 - \boldsymbol{r}_2)\chi(\alpha, \beta). \tag{3.1}$$

Since the particles are identical by assumption, the pair wave function must be antisymmetric upon interchanging particles 1 and 2, implying that either $\psi(\boldsymbol{r})$ is odd and $\chi(\alpha, \beta)$ is even (spin-triplet state) or vice versa (singlet state).

The important effect of the particles in the Fermi sea is to block the single-particle states below the Fermi energy. This is most simply taken into account by working in the momentum representation. Defining Fourier components by

$$\psi(\boldsymbol{r}) = \sum_{\boldsymbol{k}} \mathrm{e}^{\mathrm{i}\boldsymbol{k}\cdot\boldsymbol{r}}\psi_{\boldsymbol{k}} = (2\pi)^{-3}\int \mathrm{d}^3k\, \mathrm{e}^{\mathrm{i}\boldsymbol{k}\cdot\boldsymbol{r}}\psi_{\boldsymbol{k}}, \tag{3.2a}$$

$$V_{\boldsymbol{k}} = \int \mathrm{d}^3r\, \mathrm{e}^{-\mathrm{i}\boldsymbol{k}\cdot\boldsymbol{r}}V(\boldsymbol{r}), \tag{3.2b}$$

the pair Schrödinger equation takes the form

$$(\xi_{\boldsymbol{k}+\boldsymbol{P}/2} + \xi_{\boldsymbol{k}-\boldsymbol{P}/2} - E)\psi_{\boldsymbol{k}} = -(2\pi)^{-3}\int_{k'>k_{\mathrm{F}}} \mathrm{d}^3k'\, V_{\boldsymbol{k}-\boldsymbol{k}'}\psi_{\boldsymbol{k}'}. \tag{3.3}$$

Here the blocking effect must be taken into account in the sum over intermediate states $\boldsymbol{k}'$. For the following, it is useful to measure single-particle energies from the Fermi level, defining the quasiparticle energy

$$\xi_{\boldsymbol{k}} = \frac{\hbar^2k^2}{2m} - \mu, \tag{3.4}$$

where μ is the chemical potential (equal to the Fermi energy ϵ_{F} at zero temperature).

It is clear from (3.3) that the lowest energy E is obtained when the two particles have equal and oppositely directed momenta. We put $\boldsymbol{P} = 0$ in the following, because finite $\boldsymbol{P}$ corresponds to a trivial centre-of-mass motion of the pair with momentum $\hbar\boldsymbol{P}$.

Equation (3.3) may be separated into angular-momentum components owing to the assumed spherical symmetry of the interaction potential. By expanding $V_{\boldsymbol{k}-\boldsymbol{k}'}$ and $\psi_{\boldsymbol{k}}$ in terms of Legendre polynomials $P_l(\hat{\boldsymbol{k}} \cdot \hat{\boldsymbol{k}}')$ and spherical harmonics $Y_{lm}(\hat{\boldsymbol{k}})$, one has

$$V_{\boldsymbol{k}-\boldsymbol{k}'} = \sum_{l=0}^{\infty} (2l+1) V_l(k, k') P_l(\hat{\boldsymbol{k}} \cdot \hat{\boldsymbol{k}}'), \tag{3.5a}$$

$$\psi_{\boldsymbol{k}} = \sum_{lm} a_{lm} \psi_l(k) Y_{lm}(\hat{\boldsymbol{k}}). \tag{3.5b}$$

In order to allow for an explicit analytic solution of (3.3), we now introduce a further simplifying assumption, by putting $V_l(k, k')$ equal to a constant within a thin shell around the Fermi surface and zero elsewhere:

$$V_l(k, k') = \begin{cases} V_l & (|\xi_{\boldsymbol{k}}|, |\xi_{\boldsymbol{k}'}| \leqslant \epsilon_c \ll \epsilon_F), \\ 0 & \text{otherwise}. \end{cases} \tag{3.6}$$

Equation (3.3) then reduces to the separable integral equation

$$(2\xi_{\boldsymbol{k}} - E)\psi_l(k) = -V_l N(0) \int_0^{\epsilon_c} \psi_l(k') \, \mathrm{d}\xi_{\boldsymbol{k}'}, \tag{3.7}$$

where the density of states (for one spin species) $N(0)$ has been taken out of the integral, $N(\xi)$ being a slowly varying function on the scale of ϵ_c. It is worth noting that (3.7) has essentially the same form as the Schrödinger equation for just two particles in two spatial dimensions. This is due to the fact that the effective density of states in the Cooper problem is finite (and constant) above the Fermi energy and zero below, exactly as in the case of the scattering of two particles in two dimensions.

By inspection of (3.7), one can see immediately that for attractive interaction ($V_l < 0$) the energy eigenvalue E is necessarily negative. So, in the presence of the Fermi sea, the two test particles form a bound state for arbitrarily weak attractive interaction. The bound state is formed with relative orbital angular momentum L corresponding to the strongest attractive interaction parameter V_L. The correlated pair is called a "Cooper pair".

Cooper's problem is in marked contrast with the two-particle problem without blocking effect, where a bound state is formed only if the potential is sufficiently attractive. An accurate necessary condition for the existence of a bound state in three dimensions is (Rosen 1982)

$$\int |V(\boldsymbol{r})|^{3/2} \, \mathrm{d}^3 r > \frac{\pi^2}{4} \left(\frac{3\hbar^2}{2m}\right)^{3/2}, \tag{3.8}$$

and the lowest bound state is always an s-state.

In the so-called "weak-coupling" limit, i.e. for $N(0)|V_L| \ll 1$, the solution of (3.7) is given by

$$E = -2\epsilon_c \exp\left(-\frac{2}{N(0)\,|V_L|}\right), \quad V_L = \min_l \{V_l\} < 0, \tag{3.9a}$$

$$\psi_{\boldsymbol{k}} = \sum_m a_{Lm} \frac{\theta(\epsilon_c - |\xi_{\boldsymbol{k}}|)}{2\xi_{\boldsymbol{k}} - E} Y_{Lm}(\hat{\boldsymbol{k}}), \tag{3.9b}$$

where $\theta(x)$ is the unit step function.

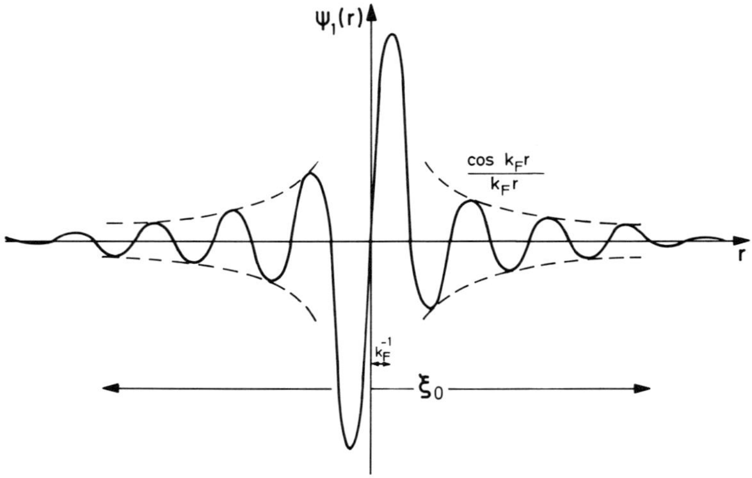

Figure 3.1 A Cooper-pair wave function with relative orbital angular momentum $L = 1$, as given by (3.10b), is shown schematically. Note that the lengths k_F^{-1} and ξ_0 are not to scale.

The wave function of the pair in position space, obtained by Fourier-inverting (3.9b), is found as

$$\psi(\boldsymbol{r}) = \sum_m a_{Lm} Y_{Lm}(\hat{\boldsymbol{r}}) \psi_L(r), \tag{3.10a}$$

where

$$\psi_L(r) = \int_0^{\epsilon_c} \mathrm{d}\xi_{\boldsymbol{k}} \frac{N(0)}{2\xi_{\boldsymbol{k}} + |E|} j_L(kr). \tag{3.10b}$$

The spherical Bessel functions $j_L(x)$ vanish as x^L for $x \to 0$, possess a first maximum at $x \approx L$ and approach the asymptotic behaviour $\sin(x - \frac{1}{2}L\pi)/x$ for $x \gg 1$. However, the $\boldsymbol{k}$ components of the wave function with $\xi_{\boldsymbol{k}} < |E|$ interfere destructively, causing the wave function to vanish more rapidly than $1/r$ for $r > \hbar v_F/|E|$. Thus the extension of the pair wave function, the so-called correlation length or coherence length, is given by $\xi_0 \approx \hbar v_F/|E|$.

A schematic picture of $\psi_1(r)$, the wave function for $L = 1$, is shown in Fig. 3.1.

3.2 GENERALIZED PAIRING THEORY

If the formation of a bound pair is energetically advantageous in Cooper's problem, one is to expect that if all particles interact with each other, the formation of correlated pairs of particles with momenta $(\hbar\boldsymbol{k}, -\hbar\boldsymbol{k})$ is still profitable. This expectation led Bardeen *et al.* (1957) to postulate a correlated wave function $|\psi\rangle$ for electrons in superconductors that is a properly antisymmetrized product of pair wave functions.

3.2.1 Generalized BCS wave function

Phrased in the language of second quantization, this ansatz, appropriately generalized to anisotropic and spin-dependent pairing, takes the form

$$|\psi\rangle = \prod_{\boldsymbol{k}}{}' \prod_{\alpha} \left(u_{\boldsymbol{k}\alpha\alpha} + \sum_{\beta} v_{\boldsymbol{k}\alpha\beta} a^{+}_{\boldsymbol{k}\alpha} a^{+}_{-\boldsymbol{k}\beta} \right) |0\rangle, \tag{3.11}$$

where the product over $\boldsymbol{k}$ is restricted to a half-space, say $\boldsymbol{k}\cdot\hat{\boldsymbol{x}} > 0$, to avoid double-counting, and $\alpha = \uparrow, \downarrow$. The wave function $|\psi\rangle$ reduces to that for the normal state if we put $u_{\boldsymbol{k}\alpha\alpha} = \theta(k - k_{\mathrm{F}})$ and $v_{\boldsymbol{k}\alpha\alpha} = \theta(k_{\mathrm{F}} - k)$, $v_{\boldsymbol{k}\alpha-\alpha} = 0$. The anticommutation relations of the field operators imply $u_{-\boldsymbol{k}\alpha\alpha} = u_{\boldsymbol{k}\alpha\alpha}$ and $v_{-\boldsymbol{k}\alpha\beta} = -v_{\boldsymbol{k}\beta\alpha}$. The operator $a^{+}_{\boldsymbol{k}\alpha} a^{+}_{-\boldsymbol{k}\beta}$ acting on the vacuum state $|0\rangle$ (the state without particles) creates a pair of (quasi)particles in single-particle states $(\boldsymbol{k}\alpha, -\boldsymbol{k}\beta)$. The probability amplitudes for the pair state to be occupied or empty, $v_{\boldsymbol{k}\alpha\beta}$ and $u_{\boldsymbol{k}\alpha\alpha}$ respectively, are considered as variational parameters. They may be determined by minimizing the energy expectation value formed with $|\psi\rangle$ or by diagonalizing the Hamiltonian directly, taking advantage of the particularly simple type of correlation admitted by (3.11). We shall follow the latter procedure.

Before doing so, let us remark on an unusual property of the wave function (3.11). By expanding the product, it is seen that $|\psi\rangle$ consists of a superposition of wave functions describing 0-, 2-, 4-, . . . , N-particle systems. In other words, the state $|\psi\rangle$ is *not* an eigenstate of the particle number operator N. On the other hand, in an isolated system the total number of particles is fixed. In this case the system has a particular symmetry, called "gauge invariance". This means that the Hamiltonian of the system is invariant under the global gauge transformation $a^{+}_{\boldsymbol{k}\alpha} \rightarrow \mathrm{e}^{\mathrm{i}\theta} a^{+}_{\boldsymbol{k}\alpha}$, where θ is a fixed, but *arbitrary*, phase variable. Clearly, the state $|\psi\rangle$, which may be characterized by the set of coefficients $\{u_{\boldsymbol{k}\alpha\alpha}, v_{\boldsymbol{k}\alpha\beta}\}$, is in general not gauge-invariant since it transforms into a state $|\psi(2\theta)\rangle$ specified by $\{u_{\boldsymbol{k}\alpha\alpha}, \mathrm{e}^{2\mathrm{i}\theta} v_{\boldsymbol{k}\alpha\beta}\}$. In the thermodynamic limit the states $|\psi(\phi)\rangle$ with different ϕ are mutually orthogonal, i.e. they represent *inequivalent* quantum-mechanical states. (By contrast, the gauge-transformed wave frunctions for the normal state differ only by an overall phase factor and are therefore unitarily equivalent, i.e. the ground state is unique.) Despite the inequivalence of the states $|\psi(\phi)\rangle$, they are still energetically degenerate since the energy does not depend on ϕ at all. This implies that the ground state is not unique but has a continuous degeneracy. Some of the most intriguing properties of pair-condensed states are a consequence of this fact, as will be shown later. Any one of the pair-correlated states $|\psi(\phi)\rangle$ is a state with *broken* gauge symmetry, because it no longer possesses the gauge symmetry of the Hamiltonian. Since an invariance under a phase transformation is mathematically equivalent to a U(1) symmetry, the broken gauge symmetry corresponds to a broken U(1) symmetry (see Section 6.2). The same concept holds in the case of a Bose–Einstein condensate, i.e. superfluidity in a Bose system; see e.g. the discussion by Huang (1987).

Because of the long-range order associated with the broken gauge symmetry, the phase ϕ of the pair-correlated state $|\psi(\phi)\rangle$ is defined throughout the system. It will become clear later (see Chapter 7) that the most important property of this macroscopic phase variable is its "stiffness" or "rigidity", i.e. the fact that it costs energy to change it in space or time, whereas its absolute value does not lead to observable consequences (as usual in quantum mechanics). An elementary but lucid discussion of phase rigidity in a Bose-condensed state (where again the same notion applies) and of the concept of rigidity as such, has been given by Anderson (1984).

The fact that the absolute value of the phase is not an observable may be used to construct a pair-condensed state with definite particle number by forming a phase-coherent linear superposition as

$$|\psi_{N_0}\rangle = \int_0^{2\pi} \mathrm{d}\phi \, \mathrm{e}^{\mathrm{i}\phi N_0/2} \, |\psi(\phi)\rangle, \tag{3.12}$$

where N_0 is the particle number (Anderson 1958a,b, 1966). This relation is analogous to a Fourier transformation relating position and momentum operators respectively. Since the latter operators are canonically conjugate, there exists an uncertainty relation between them. The same is true for the phase and the particle number, which obey the commutation relation $[\frac{1}{2}N, \phi] = \mathrm{i}$, with $N = 2\mathrm{i}\, \partial/\partial\phi$ in the ϕ representation. It is important to bear in mind that the integration over the phase ϕ in (3.12) only removes the *overall* phase but leaves the phase rigidity intact! It is the *rigidity,* not the existence of an overall phase, that is responsible for the phenomenon of superfluidity, as will be discussed in detail in Sections 7.1 and 7.2.

The above discussion is of course not limited to systems with broken gauge symmetry, but applies to all systems with a broken symmetry. Let us therefore analyse the situation in the case of a crystal, where things are more easily visualized. In the crystalline phase, with its strict periodicity, translational invariance is spontaneously broken. This implies that the conjugate variable to a displacement, the momentum, is no longer conserved. A state describing the system may be written as $|\psi(\mathbf{R}_i)\rangle$, where $\boldsymbol{R}_i$ are the lattice vectors. In the thermodynamic limit the states $|\psi(\boldsymbol{R}_i + \boldsymbol{x})\rangle$ with different $\boldsymbol{x}$, where $\boldsymbol{x}$ is some displacement, are inequivalent, i.e. mutually orthogonal. Superposition of such states by integrating $\mathrm{e}^{\mathrm{i}\boldsymbol{P}\cdot\boldsymbol{x}}\,|\psi(\boldsymbol{R}_i + \boldsymbol{x})\rangle$ over $\boldsymbol{x}$ leads to a state with definite momentum $\boldsymbol{P}$. This means that the overall position of the crystal (given by, say, the centre-of-mass coordinates) is now completely uncertain. But, at the same time, the *periodicity* of the crystal with its rigidity against displacements of the atoms from their positions $\boldsymbol{R}_i$, i.e. its long-range order, is of course still intact.

The properties of the BCS wave function (3.11) discussed above seem to require one to abandon the concept of a fixed phase variable. However, in a macroscopic system the fluctuations ΔN about the average particle number are proportional to $N^{1/2}$, so that fluctuations in the phase are proportional to $N^{-1/2}$ and hence are negligibly small. Therefore in the thermodynamic limit the phase *is* fixed. It is then simply more convenient to work with (3.11) rather than (3.12).

3.2.2 Diagonalization of the mean-field Hamiltonian

Let us now embark on the calculation of the parameters $u_{\boldsymbol{k}\alpha\alpha}$ and $v_{\boldsymbol{k}\alpha\beta}$. Quite generally, any expectation value of a product of creation (destruction) operators for fermion quasiparticles $a_{\boldsymbol{k}}^{+}$ ($a_{\boldsymbol{k}}$), taken in terms of the wave function (3.11), may be factorized into products of expectation values of *bilinear* combinations of field operators. These expectation values are found as

$$\langle \Psi | \, a_{\boldsymbol{k}\alpha}^{+} a_{\boldsymbol{k}'\beta} \, | \Psi \rangle = \delta_{\boldsymbol{k}\boldsymbol{k}'} \delta_{\alpha\beta} n_{\boldsymbol{k}\alpha} , \tag{3.13a}$$

$$\langle \Psi | \, a_{-\boldsymbol{k}\beta} a_{\boldsymbol{k}'\alpha} \, | \Psi \rangle = \delta_{\boldsymbol{k}\boldsymbol{k}'} F_{\boldsymbol{k}\alpha\beta} , \tag{3.13b}$$

where

$$n_{\boldsymbol{k}\alpha} = \sum_{\beta} v_{\boldsymbol{k}\alpha\beta} v_{\boldsymbol{k}\alpha\beta}^{*} , \tag{3.14a}$$

$$F_{\boldsymbol{k}\alpha\beta} = v_{\boldsymbol{k}\alpha\beta} u_{\boldsymbol{k}\beta\beta}^{*} = -F_{-\boldsymbol{k}\beta\alpha} . \tag{3.14b}$$

The correlation function $n_{\boldsymbol{k}\alpha}$ describes the distribution of particles in momentum space, while $F_{\boldsymbol{k}\alpha\beta}$ is referred to as the "pair amplitude".

The Hamiltonian of the interacting quasiparticle system is given by

$$H - \mu N = \sum_{\boldsymbol{k}\alpha} \xi_{\boldsymbol{k}\alpha} a_{\boldsymbol{k}\alpha}^{+} a_{\boldsymbol{k}\alpha} + \tfrac{1}{2} \sum_{\boldsymbol{k}\boldsymbol{k}'\boldsymbol{q}} \sum_{\alpha\beta\alpha'\beta'} \langle -\boldsymbol{k}\alpha, \boldsymbol{k} + \boldsymbol{q}\beta | \, V \, | \boldsymbol{k}'\alpha', -\boldsymbol{k}' + \boldsymbol{q}\beta' \rangle a_{\boldsymbol{k}'\alpha'}^{+} a_{-\boldsymbol{k}'+\boldsymbol{q}\beta'}^{+} a_{-\boldsymbol{k}\alpha} a_{\boldsymbol{k}+\boldsymbol{q}\beta} , \tag{3.15}$$

where $\xi_{\boldsymbol{k}\alpha} = \xi_{\boldsymbol{k}} - \alpha\mu_0 H$ for a finite magnetic field H and $\alpha = +1$ or -1 for $\uparrow$ or $\downarrow$ respectively. In the subspace of Hilbert space spanned by the pair wave functions (3.11) only pairs of particles with equal and opposite momenta are correlated. The fact that pair correlations are of dominant importance can be exploited to simplify the product of field operators in the interaction term of (3.15). This is achieved by replacing one pair of operators, $a_{\boldsymbol{k}}^{+} a_{\boldsymbol{k}'}^{+}$ or $a_{\boldsymbol{k}} a_{\boldsymbol{k}'}$, by its expectation value. Thus, for a given pair, the other pairs act effectively as a kind of mean field. The resulting mean-field Hamiltonian may be written as

$$H_{\mathrm{MF}} - \mu N = \sum_{\boldsymbol{k}\alpha} \xi_{\boldsymbol{k}\alpha} a_{\boldsymbol{k}\alpha}^{+} a_{\boldsymbol{k}\alpha} + \tfrac{1}{2} \sum_{\boldsymbol{k}\alpha\beta} (\Delta_{\boldsymbol{k}\alpha\beta}^{*} a_{-\boldsymbol{k}\beta} a_{\boldsymbol{k}\alpha} + a_{\boldsymbol{k}\alpha}^{+} a_{-\boldsymbol{k}\beta}^{+} \Delta_{\boldsymbol{k}\beta\alpha}) - \tfrac{1}{2} \sum_{\boldsymbol{k}\alpha\beta} \Delta_{\boldsymbol{k}\alpha\beta}^{*} F_{\boldsymbol{k}\alpha\beta} . \tag{3.16}$$

Here we have defined the "off-diagonal energy" or "off-diagonal mean field" $\Delta_{k\alpha\beta}$ by

$$\Delta_{k\alpha\beta} = \sum_{k'\alpha'\beta'} \langle -k\alpha, k\beta| V |k'\alpha', -k'\beta'\rangle F_{k'\alpha'\beta'} = -\Delta_{-k\beta\alpha}, \tag{3.17}$$

with the pair amplitude $F_{k\alpha\beta}$ given by (3.13b).

The mean-field Hamiltonian (3.16) is a bilinear form in the quasiparticle field operators and may be diagonalized by a canonical transformation (Bogoliubov 1958, Valatin 1958), i.e. the off-diagonal terms $a_{-k\beta}a_{k\alpha}$ and $a^+_{k\alpha}a^+_{-k\beta}$ may be eliminated. The new field operators $b_{k\alpha}$ and $b^+_{k\alpha}$ are defined by

$$b_{k\alpha} = \sum_{\beta} (u_{k\alpha\beta}a_{k\beta} - v_{k\alpha\beta}a^+_{-k\beta}), \tag{3.18a}$$

$$b^+_{k\alpha} = \sum_{\beta} (u^*_{k\alpha\beta}a^+_{k\beta} - v^*_{k\alpha\beta}a_{-k\beta}), \tag{3.18b}$$

where $u_{k\alpha\beta}$ and $v_{k\alpha\beta}$ are again variational parameters (in fact the same as those in (3.11)).

By means of the convention

$$u_{-k\alpha\beta} = u_{k\beta\alpha}, \quad v_{-k\alpha\beta} = -v_{k\beta\alpha}, \tag{3.19}$$

it is easily shown that the anticommutation relations of the new operators are indeed canonical, i.e.

$$\{b_{k\alpha}, b_{k'\beta}\} = \{b^+_{k\alpha}, b^+_{k'\beta}\} = 0, \tag{3.20a}$$

$$\{b_{k\alpha}, b^+_{k'\beta} = \delta_{kk'}\delta_{\alpha\beta}, \tag{3.20b}$$

if the 2×2 matrices $\mathbf{u}_k$ and $\mathbf{v}_k$ (defined by their elements $u_{k\alpha\beta}$ and $v_{k\alpha\beta}$ respectively) satisfy the conditions

$$\left.\begin{aligned} \mathbf{u}_k\mathbf{v}_k - \mathbf{v}_k\mathbf{u}_k &= \mathbf{0}, \\ \mathbf{u}_k\mathbf{u}^+_k + \mathbf{v}_k\mathbf{v}^+_k &= \mathbf{1}. \end{aligned}\right\} \tag{3.21}$$

With the help of these relations, the inverse transformation is found as

$$a_{k\alpha} = \sum_{\beta} (\mathbf{v}_k\mathbf{u}^+_k\mathbf{v}^{-1}_k)_{\alpha\beta}b_{k\beta} + \sum_{\beta} v_{k\alpha\beta}b^+_{-k\beta}. \tag{3.22}$$

Equation (3.22) may be simplified for Hermitian matrices $\mathbf{u}_k$, using (3.21), as

$$a_{k\alpha} = \sum_{\beta} (u_{k\alpha\beta}b_{k\beta} + v_{k\alpha\beta}b^+_{-k\beta}). \tag{3.23}$$

As will be shown later, the pair correlations in superfluid ^{3}He are such that (3.23) applies.

Substituting the field operators a_k and a^+_k in the Hamiltonian (3.16) with the aid of (3.23), one is led to a bilinear expression in the new fermionic field operators b_k and b^+_k. For (3.20a) to be fulfilled the coefficient of the operator product $b^+_k b^+_{-k}$ (and therefore, by Hermiticity, of $b_k b_{-k}$) must vanish by suitable choice of $\mathbf{u}_k$ and $\mathbf{v}_k$, yielding the condition

$$2\xi_k\mathbf{u}_k\mathbf{v}_k - \mathbf{v}_k\mathbf{\Delta}^+_k\mathbf{v}_k + \mathbf{u}_k\mathbf{\Delta}_k\mathbf{u}_k = 0, \tag{3.24}$$

where $\mathbf{\Delta}_k$ (etc.) is the 2×2 matrix defined by the elements $\Delta_{k\alpha\beta}$ (etc.). Equation (3.24), together with the normalization condition (3.21), may be easily solved for two cases that will later be found to be of special interest: (i) unitary $\mathbf{v}_k$ and spin-independent $\mathbf{u}_k$; and (ii) diagonal $\mathbf{v}_k$ and $\mathbf{u}_k$.

Writing $\mathbf{u}_k$ and $\mathbf{v}_k$ as

$$\left.\begin{aligned} u_{k\alpha\beta} &= \delta_{\alpha\beta} x_{k\alpha}/D_{k\alpha}, \\ v_{k\alpha\beta} &= -\Delta_{k\alpha\beta}/D_{k\alpha}, \end{aligned}\right\} \tag{3.25}$$

with $x_{k\alpha}$ and $D_{k\alpha}$ to be determined, (3.24) reduces to a quadratic equation for $x_{k\alpha}$, with the (positive) solution

$$x_{k\alpha} = \xi_{k\alpha} + E_{k\alpha}. \tag{3.26}$$

Here we have defined

$$E_{k\alpha} = [\xi_{k\alpha}^2 + (\mathbf{\Delta}_k \mathbf{\Delta}_k^+)_{\alpha\alpha}]^{1/2}. \tag{3.27}$$

By assumption, $\mathbf{\Delta}_k \mathbf{\Delta}_k^+$ is a diagonal matrix. From the normalization condition (3.21), one finds immediately

$$D_{k\alpha} = [2E_{k\alpha}(\xi_{k\alpha} + E_{k\alpha})]^{1/2}. \tag{3.28}$$

After substitution of $\mathbf{u}_k$ and $\mathbf{v}_k$ into the Hamiltonian, rearranging operator products to normal order and collecting terms, one finally obtains

$$H_{\mathrm{MF}} - \mu N = \sum_{k\alpha} \left[\tfrac{1}{2} (\mathbf{\Delta}_k \mathbf{\Delta}_k^+)_{\alpha\alpha} \left(\frac{1}{2E_{k\alpha}} - \frac{1}{\xi_{k\alpha} + E_{k\alpha}} \right) + E_{k\alpha} b_{k\alpha}^+ b_{k\alpha} \right]. \tag{3.29}$$

As the operator $b^+ b$ is positive, the ground state $|\Psi\rangle$ is characterized by the condition

$$b_{k\alpha} |\Psi\rangle = 0. \tag{3.30}$$

It is easily checked that the wave function given by (3.11) has this property.

3.2.3 Single-particle excitations

Excited states are obtained by operating with $b_{k\alpha}^+$ on $|\Psi\rangle$. The state defined by

$$|\Psi; k\alpha\rangle = b_{k\alpha}^+ |\Psi\rangle \tag{3.31}$$

describes a single elementary excitation, a so-called Bogoliubov quasiparticle. It is a momentum eigenstate with eigenvalue $\hbar \boldsymbol{k}$. The single-particle excitation energy is found from (3.29) to be

$$E_{k\alpha} = \left[\left(\frac{\hbar^2 k^2}{2m^*} - \mu \right)^2 + \sum_\beta \Delta_{k\alpha\beta} \Delta_{k\alpha\beta}^* \right]^{1/2} \tag{3.32}$$

At the Fermi surface, $\hbar^2 k^2/2m^* = \mu$, the excitation energy does not tend to zero as in the normal state, but rather remains finite: the energy spectrum

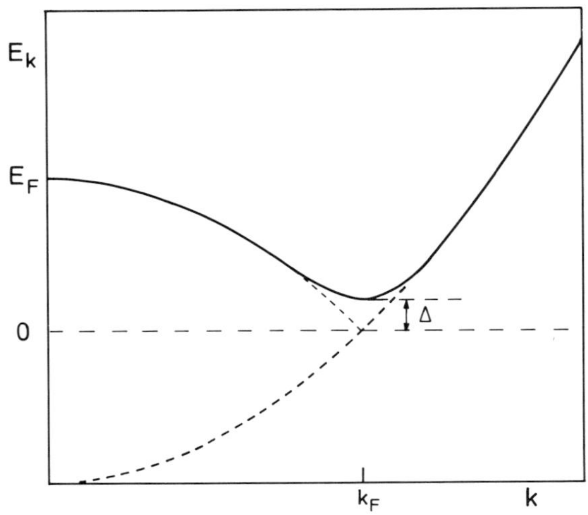

Figure 3.2 Energy spectrum of Bogoliubov quasiparticles.

has a gap depending in general on the direction $\hat{\boldsymbol{k}}$ in momentum space (see Fig. 3.2). The off-diagonal mean field $\Delta_{\boldsymbol{k}\alpha\beta}$ is therefore referred to as the "gap parameter".

3.2.4 Singlet versus triplet pairing

In usual superconducting materials the pairing of conduction electrons takes place in the spin-singlet state, i.e. the pair amplitude $F_{\boldsymbol{k}\alpha\beta}$ and the gap parameter $\Delta_{\boldsymbol{k}\alpha\beta}$ are both antisymmetric 2×2 matrices, leaving only one independent element. One may express this fact as

$$\boldsymbol{\Delta}_{\boldsymbol{k}} = \begin{pmatrix} 0 & \Delta \\ -\Delta & 0 \end{pmatrix} = \Delta \mathrm{i} \boldsymbol{\sigma}_2, \tag{3.33}$$

where $\boldsymbol{\sigma}_2$ is the second Pauli matrix.

In contrast, experimental evidence for superfluid ^{3}He unambiguously implies spin-triplet pairing. In this case $F_{\boldsymbol{k}\alpha\beta}$ and $\Delta_{\boldsymbol{k}\alpha\beta}$ are symmetric matrices, determined by three independent elements. This multiplicity, of course, corresponds to the three possible projections of $S_z = 0, \pm 1$ of a spin-$S = 1$ system. Rather than taking the amplitudes of $\Delta_{\boldsymbol{k}}$ and $F_{\boldsymbol{k}}$ corresponding to the three S_z projections, it is more convenient to combine the three spin components into a vector $\boldsymbol{d}(\boldsymbol{k})$ in spin space. This is achieved by expanding $\boldsymbol{\Delta}_{\boldsymbol{k}}$ in the three symmetric 2×2 matrices $\mathrm{i}\boldsymbol{\sigma}_\mu \boldsymbol{\sigma}_2$, $\mu = 1, 2, 3$, where $\boldsymbol{\sigma}_\mu$ are the Pauli matrices, as

$$\Delta_{\boldsymbol{k}\alpha\beta} = \sum_\mu d_\mu(\boldsymbol{k})(\boldsymbol{\sigma}_\mu \mathrm{i}\boldsymbol{\sigma}_2)_{\alpha\beta} = \begin{pmatrix} -d_1 + \mathrm{i}d_2 & d_3 \\ d_3 & d_1 + \mathrm{i}d_2 \end{pmatrix}. \tag{3.34a}$$

Conversely, this yields

$$d_\mu(\boldsymbol{k}) = -\tfrac{1}{2}\,\mathrm{tr}_\sigma\,(\boldsymbol{\Delta}_{\boldsymbol{k}}\boldsymbol{\sigma}_2\mathrm{i}\boldsymbol{\sigma}_\mu), \tag{3.34b}$$

$$|\boldsymbol{d}(\boldsymbol{k})|^2 = \tfrac{1}{2}\,\mathrm{tr}_\sigma\,(\boldsymbol{\Delta}_{\boldsymbol{k}}^+\boldsymbol{\Delta}_{\boldsymbol{k}}) \tag{3.34c}$$

where $\mathrm{tr}_\sigma\,(\ldots)$ is the trace over the 2×2 spin matrices.

It is easily verified from (3.34) that $\boldsymbol{d}(\boldsymbol{k})$ transforms like a vector under rotations in spin space, since

$$\mathbf{U}(\boldsymbol{R})\boldsymbol{\Delta}_{\boldsymbol{k}}\mathbf{U}^{\mathrm{T}}(\boldsymbol{R}) = \sum_\mu d'_\mu(\boldsymbol{k})(\boldsymbol{\sigma}_\mu\mathrm{i}\boldsymbol{\sigma}_2). \tag{3.35}$$

Here $\mathbf{U}(\boldsymbol{R}) = \exp\,(\frac{1}{2}\mathrm{i}\theta\sum_\mu \hat{n}_\mu\boldsymbol{\sigma}_\mu)$ is the unitary operator rotating two-component spinors about the axis $\hat{\mathbf{n}}$ through angle θ (the rotation operation $\boldsymbol{R}$), and $\boldsymbol{d}' = \boldsymbol{R}\boldsymbol{d}$ is the rotated vector ($\boldsymbol{R}$ is a rotation matrix). As the operation on $\boldsymbol{\Delta}$ from the right is performed using the transpose $\mathbf{U}^{\mathrm{T}}$ rather than the Hermitian conjugate, $\boldsymbol{\Delta}$ transforms like a two-particle wave function rather than an operator.

The physical interpretation of the "order-parameter vector" $\boldsymbol{d}(\boldsymbol{k})$ in spin space (defined at any point on the Fermi surface) is that it points along the direction of zero spin projection, i.e.

$$\sum_{\alpha'\beta'} \boldsymbol{d}(\boldsymbol{k})\cdot\boldsymbol{S}_{\alpha\beta\alpha'\beta'}\,\Delta_{\boldsymbol{k}\alpha'\beta'} = 0, \tag{3.36}$$

where $(S_{\alpha\beta\alpha'\beta'})_\mu = (\sigma_\mu)_{\alpha\alpha'}\delta_{\beta\beta'} + \delta_{\alpha\alpha'}(\sigma_\mu)_{\beta\beta'}$ is the total spin operator of the Cooper pairs and $(\sigma_\mu)_{\alpha\beta}$ is the α, β component of the Pauli matrix $\boldsymbol{\sigma}_\mu$.

In terms of $\boldsymbol{d}(\boldsymbol{k})$, the square of the energy gap is given by

$$(\boldsymbol{\Delta}_{\boldsymbol{k}}\boldsymbol{\Delta}_{\boldsymbol{k}}^+)_{\alpha\beta} = \boldsymbol{d}(\boldsymbol{k})\cdot\boldsymbol{d}^*(\boldsymbol{k})\delta_{\alpha\beta} + \mathrm{i}\sum_\mu\,[\boldsymbol{d}(\boldsymbol{k})\times\boldsymbol{d}^*(\boldsymbol{k})]_\mu(\sigma_\mu)_{\alpha\beta}. \tag{3.37}$$

States with the property $\boldsymbol{\Delta}\boldsymbol{\Delta}^+\propto\mathbf{1}$ are called "unitary". It follows that for unitary states $\boldsymbol{d}\times\boldsymbol{d}^* = 0$. The quantity $\boldsymbol{d}\times\boldsymbol{d}^*$ represents a kind of spin angular momentum of the Cooper pairs, at least locally on the Fermi surface (the average over the Fermi surface may still be zero). Note that the eigenvalues of $\boldsymbol{\Delta}\boldsymbol{\Delta}^+$ in the general case are given by (Takagi 1973)

$$|\Delta_{\boldsymbol{k}}|^2_{1,2} = \boldsymbol{d}(\boldsymbol{k})\cdot\boldsymbol{d}^*(\boldsymbol{k}) \pm \{[\boldsymbol{d}(\boldsymbol{k})\cdot\boldsymbol{d}^*(\boldsymbol{k})]^2 - \boldsymbol{d}^2(\boldsymbol{k})\boldsymbol{d}^{*2}(\boldsymbol{k})\}^{1/2}. \tag{3.38}$$

The self-consistent theory of a pair-correlated Fermi system described above may now be used to calculate physical quantities at zero temperature. However, it is more useful to extend the theory to finite temperatures first. The zero-temperature properties are then easily derived by taking the limit $T\to 0$.

3.3 PAIRING THEORY AT FINITE TEMPERATURE

The preceding treatment of the many-body problem is readily generalized to finite temperature T. For $T>0$ a certain fraction of pairs will be broken into single-particle excitations, such that the probability for pair formation will be reduced. Above a critical temperature T_c Cooper pairs cannot exist in equilibrium.

3.3.1 Effective Hamiltonian

The temperature-dependent probability amplitude for the existence of Cooper pairs is given by the generalization of (3.13b):

$$\langle a_{-\boldsymbol{k}\beta} a_{\boldsymbol{k}'\alpha} \rangle = \delta_{\boldsymbol{k}\boldsymbol{k}'} F_{\boldsymbol{k}\alpha\beta}. \tag{3.39}$$

Here the angular brackets denote the thermal average, defined by

$$\langle A \rangle = \mathrm{tr}\left\{ A \left[\exp\left(-\frac{H_{\mathrm{MF}} - \mu N}{k_{\mathrm{B}} T} \right) \right] \right\} \Big/ \mathrm{tr}\left[\exp\left(-\frac{H_{\mathrm{MF}} - \mu N}{k_{\mathrm{B}} T} \right) \right]. \tag{3.40}$$

The full Hamiltonian (3.15) is approximated by a temperature-dependent mean-field Hamiltonian of the form (3.16), with $\Delta_{\boldsymbol{k}\alpha\beta}$ given by (3.17) in terms of the temperature-dependent $F_{\boldsymbol{k}\alpha\beta}$ defined above. Passing through steps analogous to (3.18)–(3.29), one arrives at the diagonalized effective Hamiltonian

$$\begin{aligned} H_{\mathrm{MF}} - \mu N = \sum_{\boldsymbol{k}\alpha} \Big[& -\tfrac{1}{2} (\xi_{\boldsymbol{k}\alpha} + E_{\boldsymbol{k}\alpha})^{-1} (\boldsymbol{\Delta}_{\boldsymbol{k}} \boldsymbol{\Delta}_{\boldsymbol{k}}^{+})_{\alpha\alpha} \\ & -\tfrac{1}{2} \sum_{\beta} \Delta_{\boldsymbol{k}\alpha\beta}^{*} F_{\boldsymbol{k}\alpha\beta} + E_{\boldsymbol{k}\alpha} b_{\boldsymbol{k}\alpha}^{+} b_{\boldsymbol{k}\alpha} \Big]. \end{aligned} \tag{3.41}$$

The single-particle excitation energy $E_{k\alpha}$ and the Bogoliubov quasiparticle operators $b_{\boldsymbol{k}\alpha}$, $b_{\boldsymbol{k}\alpha}^{+}$ are given by (3.32) and (3.18), with, however, the temperature-dependent gap parameter $\Delta_{\boldsymbol{k}\alpha\beta}$. For a detailed derivation of (3.41) in the case of s-wave pairing see Eilenberger (1965).

It follows from the form of the effective Hamiltonian (3.41) that the Bogoliubov quasiparticles are independent fermions. At finite temperature these states are occupied according to Fermi statistics, which implies

$$\langle b_{\boldsymbol{k}\alpha}^{+} b_{\boldsymbol{k}'\beta} \rangle = \delta_{\boldsymbol{k}\boldsymbol{k}'} \delta_{\alpha\beta} f_{\boldsymbol{k}\alpha}, \tag{3.42a}$$

where

$$f_{\boldsymbol{k}\alpha} \equiv f(E_{\boldsymbol{k}\alpha}) = \left[\exp\left(\frac{E_{\boldsymbol{k}\alpha}}{k_{\mathrm{B}} T} \right) + 1 \right]^{-1}, \tag{3.42b}$$

$$\langle b_{\boldsymbol{k}\alpha} b_{\boldsymbol{k}'\beta} \rangle = \langle b_{\boldsymbol{k}\alpha}^{+} b_{\boldsymbol{k}'\beta}^{+} \rangle = 0. \tag{3.42c}$$

Note that in (3.42b) a chemical potential does not appear, since the number of quasiparticles is not conserved but is determined by T.

Making use of the canonical transformation (3.23), one can express thermal averages of the $a_{\boldsymbol{k}\alpha}$ operators in terms of the Bogoliubov quasiparticle distribution function $f_{\boldsymbol{k}\alpha}$ as

$$n_{\boldsymbol{k}\alpha} = \langle a_{\boldsymbol{k}\alpha}^{+} a_{\boldsymbol{k}\alpha} \rangle = \frac{1}{2}\left(1 + \frac{\xi_{\boldsymbol{k}\alpha}}{E_{\boldsymbol{k}\alpha}} \right) f_{\boldsymbol{k}\alpha} + \frac{1}{2}\left(1 - \frac{\xi_{\boldsymbol{k}\alpha}}{E_{\boldsymbol{k}\alpha}} \right) (1 - f_{\boldsymbol{k}\alpha}), \tag{3.43a}$$

$$F_{\boldsymbol{k}\alpha\beta} = \langle a_{-\boldsymbol{k}\beta} a_{\boldsymbol{k}\alpha} \rangle = -\frac{\Delta_{\boldsymbol{k}\alpha\beta}}{2E_{\boldsymbol{k}\alpha}}(1 - 2f_{\boldsymbol{k}\alpha}). \tag{3.43b}$$

We mention in passing that (3.43a, b) are not the most general expressions for $n_{\boldsymbol{k}\alpha}$ and $F_{\boldsymbol{k}\alpha\beta}$. Rather, they are valid only subject to the conditions of $\boldsymbol{\Delta}$ being either proportional to a unitary or to a spin-diagonal matrix.

3.3.2 Gap equation

Let us now turn to the determination of the gap parameter. Substituting $F_{\boldsymbol{k}\alpha\beta}$ from (3.43b) into (3.17), one finds a self-consistent equation for $\Delta_{\boldsymbol{k}\alpha\beta}$, the so-called gap equation,

$$\Delta_{\boldsymbol{k}\alpha\beta} = -\sum_{\boldsymbol{k}'} V_{\boldsymbol{k}-\boldsymbol{k}'} \frac{\Delta_{\boldsymbol{k}'\alpha\beta}}{2E_{\boldsymbol{k}'\alpha}} \tanh\left(\frac{E_{\boldsymbol{k}'\alpha}}{2k_{\mathrm{B}}T}\right), \tag{3.44}$$

Here the interaction function $V_{\boldsymbol{k}-\boldsymbol{k}'}$ is a simplified version of the matrix element used in (3.16), i.e.

$$\langle -\boldsymbol{k}\alpha, \boldsymbol{k}\beta | V | \boldsymbol{k}'\alpha', -\boldsymbol{k}'\beta' \rangle = V_{\boldsymbol{k}-\boldsymbol{k}'}\, \delta_{\alpha\alpha'}\, \delta_{\beta\beta'}. \tag{3.45}$$

For this to be justified, we have to limit ourselves to either pure spin-singlet or pure spin-triplet pairing, i.e. the interaction is then spin-independent, as expressed in (3.45).

The models for the effective pair interaction $V_{\boldsymbol{k}-\boldsymbol{k}'}$ considered in Section 2.5 suggest that the variation of $V_{\boldsymbol{k}-\boldsymbol{k}'}$ with $|\boldsymbol{k}|$ or $|\boldsymbol{k}'|$ occurs on a scale comparable to the Fermi momentum. The latter is large compared with the momenta of order Δ/v_{F} involved in the pair correlations. This is a consequence of the smallness of the ratio $T_{\mathrm{c}}/T_{\mathrm{F}}$, since Δ is of order $k_{\mathrm{B}}T_{\mathrm{c}}$ (see below). One may therefore neglect the dependence of $\Delta_{\boldsymbol{k}\alpha\beta}$ on $|\boldsymbol{k}|$ (Anderson and Morel 1961). The error induced by this approximation is of order $T_{\mathrm{c}}/T_{\mathrm{F}}$ small and appears to be tolerable. Another consequence of $T_{\mathrm{c}}/T_{\mathrm{F}} \ll 1$ is that angular-momentum components of $\Delta_{\boldsymbol{k}\alpha\beta}$ other than the one associated with the most attractive interaction parameter V_L are suppressed, provided that the corresponding V_l are not exceedingly close to V_L. However, there may arise small $l \neq L$ angular-momentum admixtures to the gap parameter even at temperatures above the corresponding critical temperature $T_{\mathrm{c}l}$, (Wojtanowski and Wölfle 1986b). This only occurs if the spin and orbital symmetries of the different l components match. We shall therefore assume that $V_{\boldsymbol{k}-\boldsymbol{k}'}$ may be approximated by the model form

$$V_{\boldsymbol{k}-\boldsymbol{k}'} = \begin{cases} (2L+1)V_L P_L(\hat{\boldsymbol{k}} \cdot \hat{\boldsymbol{k}}') & (|\xi_{\boldsymbol{k}}|, |\xi_{\boldsymbol{k}'}| < \epsilon_{\mathrm{c}}), \\ 0 & \text{otherwise.} \end{cases} \tag{3.46a}$$

In this case the Cooper pairs form in a state of definite (relative) orbital angular momentum L. Since the Cooper-pair wave function has to be antisymmetric, it follows that the spin state of the pairs is the singlet state for even L and the triplet state for odd L. Consequently, the gap matrix $\Delta_{\boldsymbol{k}\alpha\beta}$ is antisymmetric (symmetric) for even (odd) L (see (3.17)).

While in conventional superconductors the pair interaction is mediated by phonons and can in many cases be calculated in a controlled way (see e.g. Rainer 1986), the corresponding problem in liquid ^{3}He is much more complicated. Even though the strength of the pair interaction near the Fermi surface is related to the quasiparticle scattering amplitude (see Section 2.5.6), its shape and range as a function of energy is largely unknown. Thus the cut-off ϵ_c in (3.46a) also remains undetermined. It is, however, plausible that $T_Q \lesssim \epsilon_c \lesssim T_F$, where $T_Q \approx 40$ mK is the temperature above which the quasiparticle width $\hbar/\tau_N^0 > k_B T$, so that quasiparticles are ill-defined, and $T_F \approx 1$ K is the Fermi temperature. For a discussion in the context of heavy-electron superconductors, where a similar problem occurs, see Millis *et al.* (1988).

In the remainder of this section we shall restrict ourselves to the case of zero external magnetic field such that $\xi_{\boldsymbol{k}\alpha} \equiv \xi_{\boldsymbol{k}}$ (for finite fields, see Section 5.4). The gap equation (3.44), together with (3.46a), then reduces to

$$\Delta_{\boldsymbol{k}\alpha\beta} = -(2L+1)V_L N(0) \left\langle P_L(\hat{\boldsymbol{k}} \cdot \hat{\boldsymbol{k}}')\Delta_{\boldsymbol{k}'\alpha\beta} \int_0^{\epsilon_c} \mathrm{d}\xi_{\boldsymbol{k}'} \frac{\tanh(E_{\boldsymbol{k}'\alpha}/2k_B T)}{E_{\boldsymbol{k}'\alpha}} \right\rangle_{\hat{\boldsymbol{k}}'}, \tag{3.46b}$$

where $\langle\ \rangle_{\hat{\boldsymbol{k}}}$ is the angular average over $\hat{\boldsymbol{k}}$ and we have used the usual approximation $\sum_{|\boldsymbol{k}|} \approx N(0)\int_{-\epsilon_c}^{\epsilon_c} \mathrm{d}\xi$ valid for integrals peaked at the Fermi surface. For $\epsilon_c \gg \Delta$, the ξ integral on the right-hand side of (3.46b) is dominated by the ξ^{-1} dependence of the integrand, leading to a logarithmic dependence on ϵ_c.

The gap equation (3.46b) may also be written in an integrated form by multiplying both sides by $\boldsymbol{\Delta}_{\boldsymbol{k}}^{+}$, performing the trace and averaging over $\hat{\boldsymbol{k}}$. If, furthermore, the gap parameter is given by an arbitrary linear combination fo spherical harmonics $Y_{lm}(\Omega_{\hat{\boldsymbol{k}}})$ of a definite order l, i.e.

$$\Delta_{\boldsymbol{k}\alpha\beta}^{(l)} = \sum_{m=-l}^{l} a_{m,\alpha\beta} Y_{lm}(\Omega_{\hat{\boldsymbol{k}}}),$$

then, using the property

$$(2L+1)\langle P_L(\hat{\boldsymbol{k}} \cdot \hat{\boldsymbol{k}}')\Delta_{\boldsymbol{k}'\alpha\beta}^{(l)}\rangle_{\hat{\boldsymbol{k}}'} = \delta_{lL}\Delta_{\boldsymbol{k}\alpha\beta}^{(l)}, \tag{3.46c}$$

the integrated gap equation has the form

$$\sum_{\alpha} \left\langle (\boldsymbol{\Delta}_{\boldsymbol{k}}^{+}\boldsymbol{\Delta}_{\boldsymbol{k}})_{\alpha\alpha} \left[1 - |V_L| N(0) \int_0^{\epsilon_c} \mathrm{d}\xi \frac{\tanh(E_{\boldsymbol{k}\alpha}/2k_B T)}{E_{\boldsymbol{k}\alpha}}\right] \right\rangle_{\hat{\boldsymbol{k}}} = 0. \tag{3.46d}$$

Employing (3.46d), the average value of the gap parameter in the limits $T \to T_c$ and $T \to 0$ may easily be calculated (see below).

In conventional superconductors, the pair interaction is strongest in a relative s-wave state. The gap parameter $\Delta_{k\alpha\beta}$ is then completely specified by a single $\boldsymbol{k}$-independent function Δ. In this case the gap equation (3.46b) reduces to

$$1 = -V_0 N(0) \int_0^{\epsilon_c} \mathrm{d}\xi \frac{\tanh(E/2k_B T)}{E}, \tag{3.46e}$$

where $E = (\xi^2 + \Delta^2)^{1/2}$ and ϵ_c is given by the Debye frequency, $\epsilon_c = \hbar\omega_D$. This is a nonlinear algebraic equation for Δ, with a unique solution, which is tabulated (see e.g. Mühlschlegel 1959).

The situation is different for $L \neq 0$ pairing. In general, a solution of the gap equation exists for any angular dependence within the manifold of angular-momentum eigenfunctions considered. However, different sets of solutions may be discriminated by considering their free energy. The equilibrium state is given by any one of a degenerate set of solutions corresponding to the lowest free energy.

3.3.3 Critical temperature

For given L, all solutions $\Delta_{k\alpha\beta} \neq 0$ appear at temperatures below a common critical temperature T_c. The label L characterizes a given irreducible representation of the symmetry group of the system, in this case the group SO(3) of rotations in orbital space. In a more general situation, like for example in the case of electrons in a superconducting metal (where the symmetry group would be the point group of the crystal), all solutions of the gap equation belonging to a given representation of the group possess the same critical temperature T_c. Since the gap parameter goes to zero continuously as one approaches T_c from below, the gap equation may be linearized at T_c:

$$\Delta_{k\alpha\beta} = -\sum_{k'} V_{k-k'} \frac{\tanh(\xi_{k'}/2k_B T)}{2\xi_k} \Delta_{k'\alpha\beta}. \tag{3.47a}$$

In particular, (3.46b) yields

$$\Delta_{k\alpha\beta} = -(2L+1) V_L N(0) \langle P_L(\hat{\boldsymbol{k}} \cdot \hat{\boldsymbol{k}}') \Delta_{k'\alpha\beta} \rangle_{\hat{k}'} \int_0^{\epsilon_c} \mathrm{d}\xi \frac{\tanh(\xi/2k_B T_c)}{\xi}. \tag{3.47b}$$

Using (3.46c), this equation is seen to be identical with

$$1 = |V_L| N(0) \int_0^{\epsilon_c} \mathrm{d}\xi \frac{\tanh(\xi/2k_B T_c)}{\xi}. \tag{3.47c}$$

This clearly shows that states with an order parameter of definite L have the same T_c.

For $\tilde{\epsilon}_c \equiv \epsilon_c/2k_B T_c \gg 1$, the ξ integral in (3.47c) is calculated as

$$\int_0^{\tilde{\epsilon}_c} dz \frac{\tanh z}{z} = \int_1^{\tilde{\epsilon}_c} \frac{dz}{z} + c + O\left(\frac{e^{-2\tilde{\epsilon}_c}}{\tilde{\epsilon}_c}\right)$$
$$\approx \ln\left(C_0 \frac{\epsilon_c}{k_B T_c}\right),$$

where

$$c = \int_0^1 dz \frac{\tanh z}{z} - \int_1^\infty dz \frac{1 - \tanh z}{z}$$
$$= \ln \frac{4}{\pi} + \gamma$$

and $C_0 = \frac{1}{2}e^c = (2/\pi)e^\gamma$. Here $\gamma \approx 0.5772$ is Euler's constant, yielding $C_0 \approx 1.134$.

Equation (3.47b) yields a relation between the critical temperature T_c, the cutoff energy ϵ_c and the pair interaction constant V_L:

$$T_c = C_0 \frac{\epsilon_c}{k_B} \exp\left[-\frac{1}{N(0)\,|V_L|}\right]. \tag{3.48}$$

For $T_c \approx 2.6\,\text{mK}$, as appropriate at melting pressure, and a cutoff energy of $(0.04\text{–}1.0)k_B T_F$ (see below), we find from (3.48) that the coupling constant $N(0)\,|V_1|$ varies between 0.3 and 0.2. This is within the range of values for λ_1 deduced from the normal-state quasiparticle scattering amplitude (see Section 2.5).

3.3.4 Free energy

To a very good approximation, the entropy of a pair-correlated Fermi system is carried only by the single-particle excitations. The collective excitations (see Section 9.4) contribute only to higher order in T_c/T_F owing to phase-space restrictions. Hence the entropy is that of a system of independent fermions—the Bogoliubov quasiparticles—and is given by the familiar form

$$S = -k_B \sum_{\boldsymbol{k}\alpha} [f_{\boldsymbol{k}\alpha} \ln f_{\boldsymbol{k}\alpha} + (1 - f_{\boldsymbol{k}\alpha}) \ln(1 - f_{\boldsymbol{k}\alpha})]. \tag{3.49}$$

Substituting $E_{\boldsymbol{k}\alpha}$ from (3.32) and performing partial integrations with respect to $\xi_{\boldsymbol{k}}$, this may be written as

$$S = -\frac{1}{T} \sum_{\boldsymbol{k}\alpha} \xi_{\boldsymbol{k}}^2 \frac{\partial f_{\boldsymbol{k}\alpha}}{\partial E_{\boldsymbol{k}\alpha}}. \tag{3.50}$$

The internal energy is found from (3.41) by substituting the pair amplitude from (3.43b) and performing the thermal average as

$$\langle H-\mu N\rangle = \frac{1}{2}\sum_{k\alpha}(\boldsymbol{\Delta}_k\boldsymbol{\Delta}_k^+)_{\alpha\alpha}\left[\frac{1-2f_{k\alpha}}{2E_{k\alpha}}-\frac{1}{\xi_k+E_{k\alpha}}\right]+\sum_{k\alpha}E_{k\alpha}f_{k\alpha}. \quad (3.51)$$

The free energy is obtained by combining (3.50) and (3.51). Partially integrating the terms involving $f_{k\alpha}$ yields

$$F \equiv \langle H-\mu N\rangle - TS$$
$$= \sum_{\xi<0}\xi_k - \tfrac{1}{4}N(0)\sum_{\alpha}\langle(\boldsymbol{\Delta}_k\boldsymbol{\Delta}_k^+)_{\alpha\alpha}\rangle_{\hat{k}} + \tfrac{1}{2}\sum_{k\alpha}\xi_k^2\frac{\partial f_{k\alpha}}{\partial E_{k\alpha}}, \quad (3.52)$$

where terms of order Δ/ϵ_c have been neglected. Here the first term is the ground-state energy of the normal state. The second term may be identified with the pair-condensation energy at $T=0$, since the last term vanishes for $T\to 0$.

It should be reiterated that (3.52) has been derived within BCS weak-coupling theory. Since we shall later extend BCS theory to include corrections to this result (the so-called "strong-coupling corrections"), it is useful to construct a free-energy functional of the order parameter that attains its minimum for the equilibrium weak-coupling gap parameter. In Chapter 5 this functional will then be compared with the general free-energy functional. To construct the weak-coupling free-energy functional, we employ the integrated gap equation (3.46d). Eliminating the coupling constant V_L by subtracting (3.47c) in the form

$$1 = |V_L|\,N(0)\left[\int_0^{\epsilon_c}\mathrm{d}\xi\,\frac{\tanh(\xi/2k_BT)}{\xi}+\ln\frac{T}{T_c}\right], \quad (3.53a)$$

(3.46d) may be written as

$$\sum_{\alpha}\left\langle(\boldsymbol{\Delta}_k^+\boldsymbol{\Delta}_k)_{\alpha\alpha}\left\{\ln\frac{T}{T_c}-\int_0^{\infty}\mathrm{d}\xi\left[\frac{\tanh(E_{k\alpha}/2k_BT)}{E_{k\alpha}}-\frac{\tanh(\xi/2k_BT)}{\xi}\right]\right\}\right\rangle_{\hat{k}}=0. \quad (3.53b)$$

Note that in the weak-coupling limit the cutoff ϵ_c is no longer necessary and has been extended to infinity.

If one now adds $\frac{1}{2}N(0)$ times (3.53b) to the free-energy expression (3.52), the resulting functional

$$F\{\boldsymbol{\Delta}_k,\boldsymbol{\Delta}_k^+\} = F_N + \tfrac{1}{2}N(0)\sum_{\alpha}\langle\phi_1([\boldsymbol{\Delta}_k^+\boldsymbol{\Delta}_k]_{\alpha\alpha})-\phi_2([\boldsymbol{\Delta}_k^+\boldsymbol{\Delta}_k]_{\alpha\alpha})\rangle_{\hat{k}}, \quad (3.54a)$$

with

$$\phi_1([\boldsymbol{\Delta}_k^+\boldsymbol{\Delta}_k]_{\alpha\alpha}) = (\boldsymbol{\Delta}_k^+\boldsymbol{\Delta}_k)_{\alpha\alpha}\int\mathrm{d}\xi_k\left[\frac{\tanh(\xi_k/2k_BT)}{2\xi_k}-\frac{\tanh(E_{k\alpha}/2k_BT)}{2E_{k\alpha}}\right], \quad (3.54b)$$

$$\phi_2([\mathbf{\Delta}_k^+\mathbf{\Delta}_k]_{\alpha\alpha}) = \left(\tfrac{1}{2} - \ln\frac{T}{T_c}\right)(\mathbf{\Delta}_k^+\mathbf{\Delta}_k)_{\alpha\alpha} + \int \mathrm{d}\xi_k \xi_k^2\left[\frac{\partial f(\xi_k)}{\partial \xi_k} - \frac{\partial f(E_{k\alpha})}{\partial E_{k\alpha}}\right], \tag{3.54c}$$

is stationary for functions $\mathbf{\Delta}_k$ satisfying the gap equation. Here F_N is the free-energy density in the normal state. In writing (3.54a–c), we have assumed $\mathbf{\Delta}_k$ to lie within the L manifold of spherical harmonics. The functions $\phi_1(x)$ and $\phi_2(x)$ are both monotonically increasing and concave ($\phi''_{1,2}>0$). For small values of the argument, $\phi_1(x) \sim x^2$ and $\phi_2(x) \sim \ln(T_c/T)\,x$, whereas for large x, $\phi_1(x) \sim x \ln x$ and $\phi_2(x) \sim \ln(T_c/T)\,x$. Thus for $T < T_c$, the difference $\phi_1(x) - \phi_2(x)$ starts out linearly to negative values, goes through a minimum and becomes positive for large x. Clearly it is energetically advantagous for the quantity $\mathbf{\Delta}_k^+\mathbf{\Delta}_k$ to be as close as possible to the position $x_{\min}$ of the minimum of $\phi_1(x) - \phi_2(x)$. Any fluctuations of $\mathbf{\Delta}_k^+\mathbf{\Delta}_k$ as a function of $\boldsymbol{k}$ about the optimal value $x_{\min}$ cost additional condensation energy. This is shown schematically in Fig. 3.3. It follows that the minimum of the free energy is attained by the most uniform $\mathbf{\Delta}_k^+\mathbf{\Delta}_k$ as a function of $\boldsymbol{k}$.

Near the second-order transition, where the order parameter Δ is small, we may expand this free-energy functional in powers of Δ to derive the so-called Ginzburg–Landau functional (Landau and Lifschitz 1959a) in the weak-coupling limit

$$F = F_N + \tfrac{1}{2}N(0)\ln\frac{T}{T_c}\,\mathrm{tr}_\sigma\,\langle \mathbf{\Delta}_k^+\mathbf{\Delta}_k\rangle_{\hat{k}} + \tfrac{1}{4}\,\beta_0\,\mathrm{tr}_\sigma\,\langle(\mathbf{\Delta}_k^+\mathbf{\Delta}_k)^2\rangle_{\hat{k}}, \tag{3.55}$$

where

$$\beta_0 = -N(0)\int_{-\infty}^{\infty} \mathrm{d}\xi\,\frac{\mathrm{d}}{\mathrm{d}\xi^2}\left[\frac{\tanh(\xi/2k_B T_c)}{2\xi}\right]$$
$$= \frac{7}{8\pi^2}\,\zeta(3)N(0)(k_B T_c)^{-2}. \tag{3.56a}$$

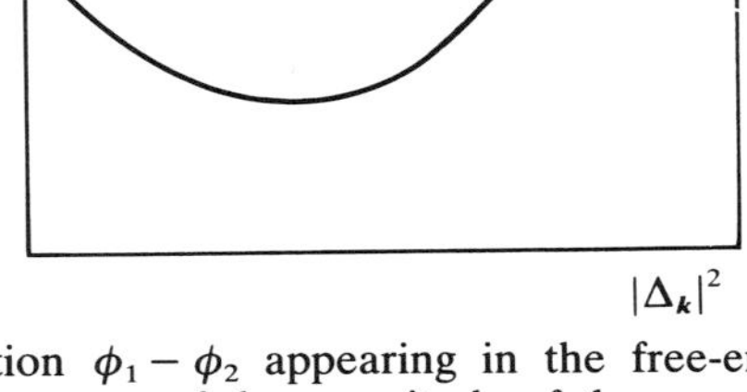

Figure 3.3 The function $\phi_1 - \phi_2$ appearing in the free-energy functional (3.54) versus the square of the magnitude of the gap along any direction $\boldsymbol{k}$ in momentum space.

In deriving (3.55), we have made use of the identity

$$\int \mathrm{d}\xi \frac{\mathrm{d}}{\mathrm{d}\xi^2}\left[\frac{\mathrm{d}f(\xi)}{\mathrm{d}\xi} + \frac{\tanh(\xi/2k_\mathrm{B}T)}{\xi}\right] = 0. \tag{3.56b}$$

The minimum of the Ginzburg–Landau free energy with respect to variations of the average magnitude of the gap parameter is easily found by splitting off the angular dependence as

$$\tfrac{1}{2}\,\mathrm{tr}_\sigma(\mathbf{\Delta}_{\boldsymbol{k}}^{+}\mathbf{\Delta}_{\boldsymbol{k}}) = \Delta^2(T)g(\hat{\boldsymbol{k}}), \tag{3.57}$$

where $\langle g(\hat{\boldsymbol{k}})\rangle_{\hat{\boldsymbol{k}}} = 1$, so that $\langle |\boldsymbol{d}(\hat{\boldsymbol{k}})|^2\rangle_{\hat{\boldsymbol{k}}} = \Delta^2(T)$ (see (3.34c)). Here $\Delta(T)$ is the root-mean-square value (*not* the maximum value) of the gap. Minimizing with respect to $\Delta(T)$, one obtains

$$\Delta^2(T) = \beta_0^{-1} N(0)\langle g^2(\hat{\boldsymbol{k}})\rangle_{\hat{\boldsymbol{k}}}^{-1}\left(1 - \frac{T}{T_\mathrm{c}}\right). \tag{3.58}$$

The free energy is found as

$$F = F_\mathrm{N} - \tfrac{1}{2}N(0)\left(1 - \frac{T}{T_\mathrm{c}}\right)\Delta^2(T) \tag{3.59}$$

and hence is a quadratic function of $T - T_\mathrm{c}$ near T_c, as appropriate for a second-order phase transition.

At zero temperature the ground-state energy may be calculated from (3.54) by dropping the terms involving $\mathrm{d}f(\xi)/\mathrm{d}\xi$ and $\mathrm{d}f(E)/\mathrm{d}E$ and putting $f(E) = 0$. One finds

$$E_0 = E_\mathrm{N} - \tfrac{1}{2}N(0)\Delta^2(0)\left\{1 - 2\ln\left[\frac{C_0\Delta(0)}{2k_\mathrm{B}T_\mathrm{c}}\right] - \langle g(\hat{\boldsymbol{k}})\ln g(\hat{\boldsymbol{k}})\rangle_{\hat{\boldsymbol{k}}}\right\}, \tag{3.60}$$

where C_0 has been defined after (3.47e). Using the integrated gap equation (3.46d) in the limit $T \to 0$,

$$2\ln\left[\frac{C_0\Delta(0)}{2k_\mathrm{B}T_\mathrm{c}}\right] + \langle g(\hat{\boldsymbol{k}})\ln g(\hat{\boldsymbol{k}})\rangle_{\hat{\boldsymbol{k}}} = 0, \tag{3.61}$$

where the coupling constant has again been eliminated by subtracting (3.47c), (3.60) may be simplified as

$$E_0 = E_\mathrm{N} - \tfrac{1}{2}\,N(0)\Delta^2(0). \tag{3.62}$$

The condensation energy is seen to be given by the angular average of the square of the gap. For given angular dependence of $\Delta_{\boldsymbol{k}}$, such as (3.57), the average magnitude of the gap is calculated from the gap equation (3.61) as

$$\Delta(0) = 1.76\,k_\mathrm{B}T_\mathrm{c}\exp\left[-\tfrac{1}{2}\langle g(\hat{\boldsymbol{k}})\ln g(\hat{\boldsymbol{k}})\rangle_{\hat{\boldsymbol{k}}}\right], \tag{3.63}$$

where we have used $2/C_0 \approx 1.76$.

3.4 THERMODYNAMIC PROPERTIES OF MODEL STATES

We are now in a position to calculate the most important static response functions of pair-correlated Fermi systems: (1) the specific heat; (2) the normal-fluid density (a response function characteristic of a superfluid system); and (3) the spin susceptibility. After deriving general expressions, we shall explicitly evaluate these quantities for the two model states that are thought to describe the two phases of superfluid ^{3}He observed in zero magnetic field. Considering that the observed magnetic properties of the superfluid ^{3}He are indicative of spin-triplet pairing and that one should expect the pair interaction components V_l (see (3.46a)) to fall off for increasing angular-momentum values, one is led to investigate p-wave pairing states.

3.4.1 p-wave pairing; the BW and ABM states

In the case that the Cooper pairs form a spin-triplet $L=1$ relative-angular-momentum state, the vector $\boldsymbol{d}(\boldsymbol{k})$ introduced in (3.34) is a linear combination of $L=1$ angular-momentum eigenstates $Y_{1m}(\hat{\boldsymbol{k}})$ $(m=0, \pm 1)$. Equivalently, it can be represented by a linear combination of the components $\hat{k}_j$ of the unit vector $\hat{\boldsymbol{k}}$ in momentum space:

$$d_\mu(\boldsymbol{k}) = \sum_{j=1}^{3} d_{\mu j}\hat{k}_j \quad \text{or} \quad \boldsymbol{d}(\boldsymbol{k}) = \boldsymbol{d}\hat{\boldsymbol{k}}. \tag{3.64}$$

The tensor quantity $\boldsymbol{d}$, with components $d_{\mu j}(T, P, H)$, is the proper order parameter describing pair correlations with p-wave orbital symmetry in a thermodynamic state characterized by temperature T, pressure P and magnetic field H. Note that the form of $d_{\mu j}$ is not fixed by the gap equation. Rather, for any given form of $d_{\mu j}$, the gap equation only determines the corresponding (T, P, H) dependence. To find out which of the possible states will be realized in nature, one therefore has to search for the states lowest in free energy. According to the preceding remarks, the free-energy minimum is obtained with the most uniform state in $\boldsymbol{k}$ space. In the absence of magnetic fields it turns out that this is the spherically symmetric state with total angular momentum $\boldsymbol{J}=\boldsymbol{L}+\boldsymbol{S}=0$, described by

$$d_{\mu j} = \mathrm{e}^{\mathrm{i}\phi}\Delta\delta_{\mu j}, \tag{3.65a}$$

such that the order-parameter *vector* (3.34b) is given by

$$\boldsymbol{d}(\boldsymbol{k}) = \mathrm{e}^{i\phi}\Delta\hat{\boldsymbol{k}}. \tag{3.65b}$$

This state was first discussed by Vdovin (1963) and Balian and Werthamer (1963) and is frequently referred to as the "BW state". It has the unique feature among $L \neq 0$ pairing states of having a uniform energy gap, since, according to (3.37),

$$(\mathbf{\Delta}_{\boldsymbol{k}}^{+}\mathbf{\Delta}_{\boldsymbol{k}})_{\alpha\beta} = \Delta^2\delta_{\alpha\beta} \tag{3.66}$$

is independent of $\hat{\boldsymbol{k}}$. A schematic drawing of the energy gap is shown in Fig. 3.4(a). The state is degenerate with the class of states formed by multiplying $d_{\mu j}$ by a constant phase factor and by rotating spin relative to position space (neglecting spin–orbit interaction for the present), since the free energy is invariant under these symmetry operations. Hence the general form of the BW state is

$$d_{\mu j} = \Delta e^{i\phi} R_{\mu j}, \tag{3.67}$$

where $R_{\mu j}$ is a rotation matrix, implying $\boldsymbol{d}(\boldsymbol{k}) = e^{i\phi}\Delta \boldsymbol{R}\hat{\boldsymbol{k}}$.

Note that this is no longer a state of definite total angular momentum J, but rather a superposition of $J = 0, 1, 2$ states. Owing to the isotropy of the single-particle spectrum, the BW state is in many respects similar to the s-wave pairing state.

The experimental observation of two superfluid phases in zero external magnetic field is in contradiction to the finding of weak-coupling p-wave-pairing theory that only one state, the BW state, is stable. As will be discussed in Chapter 5, at high pressure corrections to weak-coupling theory lower the free energy of the axially anisotropic state given by

$$d_{\mu j} = \Delta_0 \hat{d}_\mu (\hat{m}_j + i\hat{n}_j) \tag{3.68a}$$

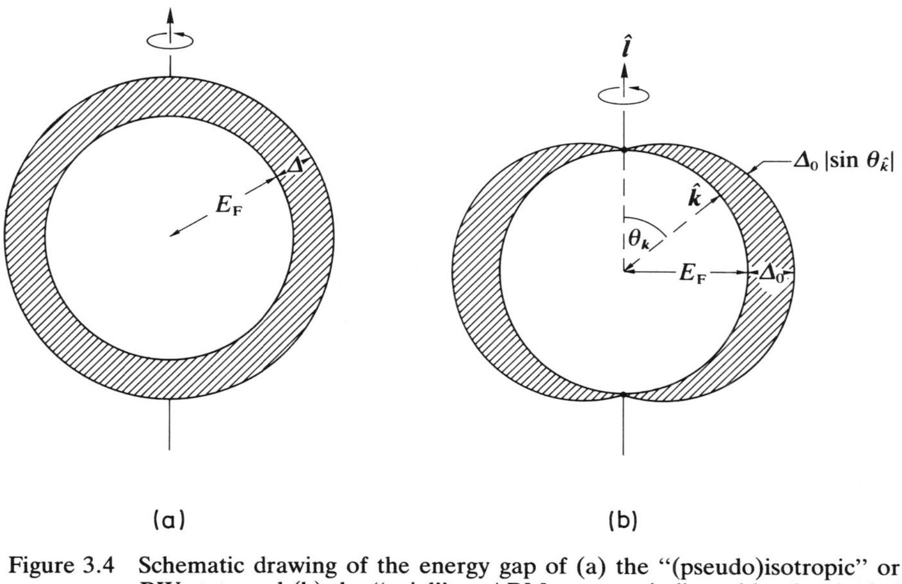

Figure 3.4 Schematic drawing of the energy gap of (a) the "(pseudo)isotropic" or BW state and (b) the "axial" or ABM state, as indicated by the shaded areas. In the ABM state the gap vanishes at two points along $\hat{\boldsymbol{l}}$. Note that E_F and Δ (or Δ_0) are not drawn to scale.

below that of the BW state. The order-parameter vector of this state has the form

$$\boldsymbol{d}(\boldsymbol{k}) = \Delta_0 \hat{\boldsymbol{d}}(\hat{\boldsymbol{k}} \cdot \hat{\boldsymbol{m}} + \mathrm{i}\hat{\boldsymbol{k}} \cdot \hat{\boldsymbol{n}}). \tag{3.68b}$$

This state, first discussed by Anderson and Morel (1961) and later shown to be realized in ^{3}He-A by Anderson and Brinkman (1973), is referred to as the "ABM state". The tensor $d_{\mu j}$ in this case is of separable form in spin and orbital variables, with a single preferred direction $\hat{\boldsymbol{d}}$ in spin space and $\hat{\boldsymbol{m}}$, $\hat{\boldsymbol{n}}$ in orbital space. Note that, according to (3.36), $\hat{\boldsymbol{d}}$ is perpendicular to the Cooper-pair spin **S**. Assuming $\boldsymbol{S}$ to point along the z direction, this implies $d_3 = 0$ in (3.34). Therefore the ABM state is a so-called "equal" spin pairing state, where only Cooper pairs with $S_z = \pm 1$ exist, while the $S_z = 0$ component is absent. The square of the gap parameter is

$$(\boldsymbol{\Delta}_{\boldsymbol{k}}^{+} \boldsymbol{\Delta}_{\boldsymbol{k}})_{\alpha\beta} = \Delta_0^2 [1 - (\hat{\boldsymbol{k}} \cdot \hat{\boldsymbol{l}})^2] \delta_{\alpha\beta}, \tag{3.69}$$

where

$$\hat{\boldsymbol{l}} = \hat{\boldsymbol{m}} \times \hat{\boldsymbol{n}} \tag{3.70}$$

is a third preferred direction in orbital space. The vectors $\hat{\boldsymbol{m}}$, $\hat{\boldsymbol{n}}$, $\hat{\boldsymbol{l}}$ form an orthogonal triad in orbital space. Note that the energy gap is axially symmetric about $\hat{\boldsymbol{l}}$, reaches a maximum value $\Delta_0(T)$ perpendicular to $\hat{\boldsymbol{l}}$ and vanishes along $\hat{\boldsymbol{l}}$ (see Fig. 3.4b).

The ABM state is therefore also called the "axial" state. Its gap parameter is intrinsically complex. In fact, as can be seen from (3.68), it is proportional to the angular-momentum eigenfunction $Y_{1,+1}$, which implies the existence of a finite orbital-angular-momentum projection, directed along $\hat{\boldsymbol{l}}$.

3.4.2 Gap parameter

As an essential ingredient for the calculations of thermodynamic properties, the temperature-dependent gap $\Delta(T)$ is needed. For the BW state, the gap equation reduces to that of s-wave pairing, which has been solved numerically (see e.g. Mühlschlegel 1959). The limiting behaviour near T_c and at low temperatures is given by (see (3.58b) and (3.63))

$$\Delta(T) \to a_B k_B T_c \left(1 - \frac{T}{T_c}\right)^{1/2} \quad \text{as } T \to T_c, \tag{3.71}$$

where

$$a_B = \left[\frac{8\pi^2}{7\zeta(3)}\right]^{1/2} \approx 3.06,$$

and

$$\Delta(0) = \frac{2}{C_0} k_B T_c \approx 1.76 k_B T_c \quad \text{as } T \to 0. \tag{3.72}$$

On the other hand, the gap parameter in the ABM state is strongly anisotropic:

$$|\Delta_{\boldsymbol{k}}| \equiv [\tfrac{1}{2}\,\mathrm{tr}_\sigma\,(\boldsymbol{\Delta}_{\boldsymbol{k}}^+\boldsymbol{\Delta}_{\boldsymbol{k}})]^{1/2} = \Delta_0 \sin\theta_{\boldsymbol{k}}, \tag{3.73}$$

where $\theta_{\boldsymbol{k}}$ is defined by $\cos\theta_{\boldsymbol{k}} = \hat{\boldsymbol{k}}\cdot\hat{\boldsymbol{l}}$ (see Fig. 3.4b).

Inserting (3.73) into (3.58) yields for the maximum of the gap near T_c

$$\Delta_0 = a_A k_B T_c\left(1-\frac{T}{T_c}\right)^{1/2}, \tag{3.74}$$

where

$$a_A = \left[\frac{10\pi^2}{7\zeta(3)}\right]^{1/2} \approx 3.42.$$

The limiting value of Δ_0 at low temperatures is given by (3.63) as

$$\Delta_0(0) \approx \tfrac{1}{2}\pi e^{5/6-\gamma} k_B T_c = 2.02 k_B T_c. \tag{3.75}$$

The average value Δ of the gap as defined in (3.57) is obtained from (3.74) and (3.75) by multiplying by $(\frac{2}{3})^{1/2}$, the square root of the average of $\sin^2\theta$.

3.4.3 Specific heat

The specific heat is most easily calculated from the entropy expression (3.49), yielding

$$C_V = -\frac{1}{T}\sum_{\boldsymbol{k}\alpha}\frac{\partial f_{\boldsymbol{k}\alpha}}{\partial E_{\boldsymbol{k}\alpha}}\left[E_{\boldsymbol{k}\alpha}^2 - \tfrac{1}{2}T\frac{\partial}{\partial T}(\boldsymbol{\Delta}_{\boldsymbol{k}}^+\boldsymbol{\Delta}_{\boldsymbol{k}})_{\alpha\alpha}\right]. \tag{3.76}$$

This expression may be evaluated analytically near T_c and for $T\to 0$. In particular, the term involving $\partial\Delta^2/\partial T$ gives rise to a discontinuity in C_V at T_c. Using the expression for Δ^2 near T_c, (3.58), and the property $\Sigma(-\partial f/\partial E) = 2N(0)$, one finds

$$C_V = C_N + \Delta C_V + O(\Delta^2) \quad (T \lesssim T_c), \tag{3.77}$$

where C_N is the specific heat in the normal state, given by (2.14), and

$$\Delta C_V = -\tfrac{1}{2}N(0)\frac{\partial}{\partial T}\mathrm{tr}_\sigma\,\langle\boldsymbol{\Delta}_{\boldsymbol{k}}^+\boldsymbol{\Delta}_{\boldsymbol{k}}\rangle_{\hat{\boldsymbol{k}}}. \tag{3.78}$$

For the BW state, substituting (3.71), the relative magnitude of the discontinuity is given by

$$\frac{\Delta C_V}{C_N} = \frac{12}{7\zeta(3)} \approx 1.43. \tag{3.79}$$

It is seen from (3.76) that C_V is a monotonically increasing concave function of T. At low temperatures the term involving $E_{\boldsymbol{k}}^2/T \sim \Delta_{\boldsymbol{k}}^2/T$ dominates. The

Fermi-distribution factor may be approximated by

$$-\frac{\partial f}{\partial E_{\boldsymbol{k}}} \approx \frac{1}{k_{\mathrm{B}} T} \exp\left(-\frac{\Delta_{\boldsymbol{k}} + \xi^2/2\Delta_{\boldsymbol{k}}}{k_{\mathrm{B}} T}\right). \tag{3.80}$$

Apparently the quasiparticle states are occupied according to a Boltzmann distribution of particles with momenta proportional to ξ and effective mass proportional to $\Delta_{\boldsymbol{k}}$. Hence there is a close analogy with the roton gas in ^{4}He. Performing the $\boldsymbol{k}$ sum in (3.80) yields for the specific heat of the BW state at low T

$$C_V = 2(2\pi)^{1/2} k_{\mathrm{B}} N(0) \Delta \left(\frac{\Delta}{k_{\mathrm{B}} T}\right)^{3/2} \exp\left(-\frac{\Delta}{k_{\mathrm{B}} T}\right). \tag{3.81}$$

The specific heat tends to zero exponentially for $T \to 0$ owing to the gap Δ in the excitation spectrum.

The specific-heat discontinuity for the ABM state is obtained by inserting (3.74) into (3.78), as

$$\frac{\Delta C_V}{C_{\mathrm{N}}} = \frac{10}{7\zeta(3)} \approx 1.19. \tag{3.82}$$

The value of $\Delta C_V / C_{\mathrm{N}}$ for the ABM state is smaller than for the BW state since in weak-coupling theory the ABM state is higher in free energy.

At low temperatures the behaviour of the ABM state is very different from the BW state since the ABM gap function is zero along the preferred direction $\hat{\boldsymbol{l}}$. At temperatures $T \ll \Delta_0 / k_{\mathrm{B}}$ quasiparticles are thermally excited mainly in small areas on the Fermi surface near the "poles" (i.e. $\boldsymbol{k} \approx \pm\hat{\boldsymbol{l}}$), of extension $\Delta k \sim (T/\Delta_0) k_{\mathrm{F}}$. The leading temperature power law at low temperatures including the precise prefactor may be calculated from (3.76) by transforming the integrals over energy $\xi_{\boldsymbol{k}}$ and angle θ into variables $t = \xi_{\boldsymbol{k}} / \Delta_{\boldsymbol{k}}$ and $v = E_{\boldsymbol{k}} / 2k_{\mathrm{B}} T$. The specific heat is thus found to vary as T^3,

$$C_V(T) = \frac{7\pi^2}{5} \left(\frac{T}{\Delta_0}\right)^2 C_{\mathrm{N}}(T), \tag{3.83}$$

rather than exponentially.

3.4.4 Normal-fluid density

In the pair-correlated state the one-to-one correspondence of quasiparticles and real particles discussed in Chapter 2 for the normal Fermi liquid no longer holds. The pair-correlated ground state cannot be reached from the noninteracting ground state by adiabatically switching on the interaction. As will be discussed in Section 7.1 and Chapter 9, the pair-correlated system may be described as consisting of two interpenetrating fluids: the fluid of thermal excitations or "normal component", and the pair condensate or "superfluid component" (see (7.15a,b)). Suppose that the thermal excitations are flowing with uniform velocity $\boldsymbol{v}_n$, while the condensate is at rest. The distribution function of the moving quasiparticle system is a shifted Fermi function:

$$f_{\boldsymbol{k}}^{\boldsymbol{v}_n} = f(E_{\boldsymbol{k}} - \hbar\boldsymbol{k}\cdot\boldsymbol{v}_n). \tag{3.84}$$

This is a consequence of the fact that the Bogoliubov quasiparticle states are momentum eigenstates, and hence their energy changes as $E_{\boldsymbol{k}} \to E_{\boldsymbol{k}} - \hbar\boldsymbol{k}\cdot\boldsymbol{v}_n$ under a Galilean transformation. The momentum density carried by the thermal excitations is accordingly

$$\boldsymbol{g}_n = \sum \hbar\boldsymbol{k} f_{\boldsymbol{k}}^{\boldsymbol{v}_n} = \hbar^2 \sum_{\boldsymbol{k}\alpha} \boldsymbol{k}\left(-\frac{\partial f_{\boldsymbol{k}}}{\partial E_{\boldsymbol{k}}}\right)\boldsymbol{k}\cdot\boldsymbol{v}_n + O(v_n^3). \tag{3.85}$$

One defines the normal-fluid density $\boldsymbol{\rho}_n$ by

$$\boldsymbol{g}_n = \boldsymbol{\rho}_n \boldsymbol{v}_n. \tag{3.86}$$

As seen by comparison with (3.85), $\boldsymbol{\rho}_n$ is a tensor quantity for anisotropic quasiparticle energy $E_{\boldsymbol{k}}$, with components given by $(f'_{\boldsymbol{k}} \equiv \partial f/\partial E_{\boldsymbol{k}})$

$$\hbar^2 \sum_{\boldsymbol{k}\alpha} k_i k_j (-f'_{\boldsymbol{k}}) \equiv \frac{m^*}{m} \rho^0_{nij}, \tag{3.87a}$$

where

$$\rho^0_{nij} = 3\rho \langle \hat{k}_i \hat{k}_j Y_0(\hat{\boldsymbol{k}}; T)\rangle_{\hat{\boldsymbol{k}}}. \tag{3.87b}$$

Here we have introduced a generalization of the so-called Yosida function Y_0 by

$$Y_0(\hat{\boldsymbol{k}}; T) = \int_{-\infty}^{\infty} \mathrm{d}\xi_{\boldsymbol{k}}\, (-f'_{\boldsymbol{k}}). \tag{3.88}$$

Y_0 is a monotonically increasing concave function of T, and has the limiting values

$$Y_0(\hat{\boldsymbol{k}}; T) = \begin{cases} 1 - \dfrac{7\zeta(3)}{4\pi^2} x^2 & (x \ll 1), \\ (2\pi x)^{1/2}\, \mathrm{e}^{-x} & (x \gg 1), \end{cases} \tag{3.89}$$

with $x = |\Delta_{\boldsymbol{k}}|/k_B T$, where we have used (3.56b) and (3.80) respectively. The function $Y_0(\hat{\boldsymbol{k}}; T)$ is a measure of the density of thermal excitations at the point $\hat{\boldsymbol{k}}$ on the Fermi surface. Note that (3.87a) does not reduce to $\rho\delta_{ij}$ for $T \to T_c$ if $m^*/m \neq 1$. This is because Fermi-liquid corrections have not yet been taken into account. They are necessary, however, to ensure Galilean invariance. As discussed in Section 2.2, the bare velocity field $\boldsymbol{v}_n$ is screened by backflow effects involving the Landau parameter F_1^s, and must be replaced by the effective field

$$\boldsymbol{v}_{\mathrm{n,eff}} = \boldsymbol{v}_{\mathrm{n}} - \frac{1}{3\rho}\frac{m}{m^*}F_1^{\mathrm{s}}\boldsymbol{g}_{\mathrm{n}}. \tag{3.90}$$

In writing (3.90), we have assumed that the Landau parameters are not affected by the pair correlations to order T_c/T_F. This may be understood by observing that the Fermi-liquid interaction involves polarization contributions from intermediate states with momenta all over the Fermi sphere. However, only a small portion of these states near the Fermi energy are modified by the pair correlations. Substituting $\boldsymbol{v}_{\mathrm{n}}$ by $\boldsymbol{v}_{\mathrm{n,eff}}$ in (3.86), the Fermi-liquid-corrected normal-fluid density is obtained as (Leggett 1965a,b, Combescot 1975e)

$$\boldsymbol{\rho}_{\mathrm{n}} = \frac{m^*}{m}\left(\mathbf{1} + \tfrac{1}{3}F_1^{\mathrm{s}}\frac{\boldsymbol{\rho}_{\mathrm{n}}^0}{\rho}\right)^{-1}\boldsymbol{\rho}_{\mathrm{n}}^0, \tag{3.91}$$

where in general $\boldsymbol{\rho}_{\mathrm{n}}^0$ and $\boldsymbol{\rho}_{\mathrm{n}}$ are tensors and $(\ldots)^{-1}$ is the matrix inverse for $\boldsymbol{\rho}_{\mathrm{n}}^0$ considered as a matrix. The superfluid density $\boldsymbol{\rho}_{\mathrm{s}}$ is then given by $\boldsymbol{\rho}_{\mathrm{s}} = \rho\mathbf{1} - \boldsymbol{\rho}_{\mathrm{n}}$ (see (7.15b)).

In the BW state the Yosida function $Y_0(\hat{\boldsymbol{k}}; T) \equiv Y_0(T)$, see (3.101c), is independent of $\hat{\boldsymbol{k}}$ and hence $\boldsymbol{\rho}_{\mathrm{n}}$ is isotropic:

$$\rho_{\mathrm{n}ij} = \delta_{ij}\rho\frac{(1 + \frac{1}{3}F_1^{\mathrm{s}})Y_0}{1 + \frac{1}{3}F_1^{\mathrm{s}}Y_0}. \tag{3.92}$$

$\boldsymbol{\rho}_{\mathrm{n}}$ decreases linearly for $T < T_c$ and vanishes exponentially fast for $T \to 0$. The Yosida function $Y_0(T)$ is shown in Fig. 3.5.

In the ABM state $\boldsymbol{\rho}_{\mathrm{n}}$ is an axially symmetric tensor, oriented along $\hat{\boldsymbol{l}}$. The components of $\boldsymbol{\rho}_{\mathrm{n}}$ near T_c are obtained by substituting the expansion (3.89)

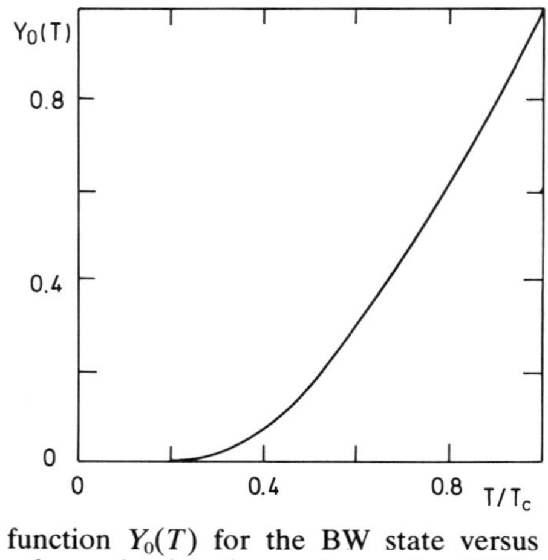

Figure 3.5 Yosida function $Y_0(T)$ for the BW state versus reduced temperature T/T_c. In the evaluation the weak-coupling BCS gap function $\Delta(T)$ has been used.

into (3.87), with the result

$$\rho_{n\perp} - \rho = 2(\rho_{n\parallel} - \rho) = -\tfrac{7}{5}\,\zeta(3)\frac{m}{m^*}\rho\left(\frac{\Delta_0}{\pi k_B T_c}\right)^2. \tag{3.93}$$

The anisotropy of $\boldsymbol{\rho}_n$ in the ABM state is even stronger at lower temperatures. As argued before, for $\Delta_0/k_B T \gg 1$, quasiparticles are excited mainly near the "pole caps" of the Fermi sphere. The weight functions $\hat{k}_i^2$ appearing in the angular average in (3.87b) are of order 1 for components parallel to $\hat{\boldsymbol{l}}$, but of order $(k_B T/\Delta_0)^2$ for the perpendicular components. Hence for $T \to 0$ the two independent components of $\boldsymbol{\rho}_n$ tend to zero as

$$\left.\begin{aligned} \rho_{n\parallel} &= \pi^2 \rho \frac{m^*}{m}\left(\frac{k_B T}{\Delta_0}\right)^2, \\ \rho_{n\perp} &= \frac{7\pi^4}{15}\rho\frac{m^*}{m}\left(\frac{k_B T}{\Delta_0}\right)^4, \end{aligned}\right\} \tag{3.94}$$

rather than vanishing exponentially. Accordingly, the anisotropy of the superfluid density $\boldsymbol{\rho}_s = \rho\mathbf{1} - \boldsymbol{\rho}_n$ vanishes at $T = 0$. For a discussion of $\boldsymbol{\rho}_n$ in other superfluid model states see Takagi (1974c).

3.4.5 Spin susceptibility

One expects the spin susceptibility of a general triplet state to be reduced from the Pauli value owing to the presence of Cooper pairs with magnetic quantum number $S_z = 0$, which do not couple to a magnetic field. Since the orientation of the Cooper-pair spins is described by the preferred direction in spin space, $\boldsymbol{d}$, the susceptibility $\boldsymbol{\chi}$ is a tensor in spin space, with axes determined by $\boldsymbol{d}$. The situation is slightly complicated by the fact that $\boldsymbol{d}$ is in general a function of position on the Fermi sphere. Let us therefore consider first the simpler case of the ABM state, where the preferred direction $\boldsymbol{d}$ in spin space is independent of $\boldsymbol{k}$ (see (3.68)).

When a weak uniform static magnetic field $\boldsymbol{H}$ acts on the system the magnetization $\boldsymbol{M}$ is generally determined by

$$\boldsymbol{M} = \boldsymbol{\chi}\boldsymbol{H}, \tag{3.95a}$$

where in the following $\boldsymbol{\chi}$ is a diagonal tensor with components $\chi_\perp$ and $\chi_\parallel$. The component of $\boldsymbol{M}$ along $\boldsymbol{H}$ is given by the number of up-spin particles minus the number of down-spin particles, multiplied by the magnetic moment per particle, μ_0:

$$M = \mu_0 \sum_{\boldsymbol{k}} (\langle a_{\boldsymbol{k}\uparrow}^+ a_{\boldsymbol{k}\uparrow}\rangle - \langle a_{\boldsymbol{k}\downarrow}^+ a_{\boldsymbol{k}\downarrow}\rangle). \tag{3.95b}$$

Let us first consider the case $\boldsymbol{d} \perp \boldsymbol{H}$. As can be seen from (3.34), $\Delta_{\boldsymbol{k}\alpha\beta}$ is diagonal and we may use the results derived in Section 3.3, with $\xi_{\boldsymbol{k}}$ replaced by $\xi_{\boldsymbol{k}\alpha} = \xi_{\boldsymbol{k}} - \alpha H\mu_0$. In particular, the number densities of up- and down-spin particles are given by (3.43a), and hence the magnetization is

$$M = -\mu_0 \sum_{k\alpha} \alpha \frac{\xi_{k\alpha}}{2E_{k\alpha}} \tanh\left(\frac{E_{k\alpha}}{2k_B T}\right). \tag{3.96}$$

Neglecting Fermi-liquid corrections, the susceptibility is obtained from (3.96) by differentiating with respect to H and putting $H=0$:

$$\chi_{\perp}^0 = 2\mu_0^2 N(0) \int d\xi \frac{d}{d\xi}\left(\frac{\xi}{2E} \tanh \frac{E}{2k_B T}\right) \equiv \chi_N^0. \tag{3.97}$$

Since in this case the occupation numbers $\langle a_{k\alpha}^+ a_{k\alpha}\rangle$ depend on H only via $\xi_{k\alpha}$, the result is the same as in the normal state. Physically speaking, there are no Cooper pairs with $S_z = 0$ if $\boldsymbol{d} \perp \boldsymbol{H}$, i.e. all pairs couple to the magnetic field. In (3.97) the superscript zero indicates the absence of Fermi-liquid corrections, which will be included later. If particle–hole-asymmetric terms (generated, for example, by the energy dependence of the density of states) are taken into account, one finds a small enhancement of the spin susceptibility of order T_c/T_F over its value in the normal state (Takagi 1974a).

The result is different for $\boldsymbol{d} \parallel \boldsymbol{H}$, since then $d_1 = d_2 = 0$ in (3.34a), i.e. $\Delta_{k\alpha\beta}$ is purely off-diagonal. This case is not covered by the discussion in Section 3.3. Solving (3.21) and (3.24), one finds the coefficients of the Bogoliubov–Valatin transformation as

$$u_{k\alpha\alpha} = \frac{\xi_k + E_k}{D_k}, \quad v_{k\alpha\beta} = \frac{-\Delta_{k\alpha\beta}}{D_k}, \tag{3.98}$$

with

$$D_k = [2E_k(E_k + \xi_k)]^{1/2}, \quad E_k = [\xi_k^2 + |\Delta_k|^2]^{1/2}.$$

Thus for $\mu_0 H \ll \Delta$ there is no explicit dependence on H! As before, the single-particle excitation energy can be read off from the transformed Hamiltonian as

$$E_{k\alpha} = E_k - \alpha\mu_0 H. \tag{3.99}$$

The occupation numbers are readily found by substituting the above $u_{k\alpha\alpha}$ and $v_{k\alpha\alpha}$ factors into (3.14a):

$$\langle a_{k\alpha}^+ a_{k\alpha}\rangle = \frac{1}{2}\left(1 + \frac{\xi_k}{E_k}\right) f_{k\alpha} + \frac{1}{2}\left(1 - \frac{\xi_k}{E_k}\right)(1 - f_{k-\alpha}). \tag{3.100}$$

The magnetization and the susceptibility without Fermi-liquid corrections are obtained as

$$M = \mu_0 \sum_k (f_{k\uparrow} - f_{k\downarrow}), \tag{3.101a}$$

$$\chi_{\parallel}^0 = 2\mu_0^2 \sum_k \left(-\frac{\partial f_k}{\partial E_k}\right) = \chi_N^0 Y_0, \tag{3.101b}$$

where the Yosida function $Y_0(T)$ is given by

$$Y_0(T) = \langle Y_0(\hat{\boldsymbol{k}}; T) \rangle_{\hat{\boldsymbol{k}}}, \tag{3.101c}$$

with $Y_0(\hat{\boldsymbol{k}}; T)$ defined by (3.88).

Note that for spin-singlet pairing $\Delta_{\alpha\beta}$ is purely off-diagonal (see (3.32)) and the preceding discussion applies, i.e. $\chi^0(T) = \chi_N^0 Y_0(T)$, a result due to Yosida (1958) for s-wave pairing.

We see that the spin susceptibility of the ABM state depends strongly on the relative orientation of $\boldsymbol{d}$ and $\boldsymbol{H}$. In its general form the susceptibility tensor reads

$$\chi_{\mu\nu}^0 = \chi_N^0 \{\delta_{\mu\nu} - \hat{d}_\mu \hat{d}_\nu [1 - Y_0(T)]\}. \tag{3.102}$$

We are now ready to tackle the conceptually more complicated case of the BW state. The clue is to first consider the contribution to the susceptibility tensor from a given point $\hat{\boldsymbol{k}}$ on the Fermi sphere. This may be calculated in the local frame of reference parallel to $\boldsymbol{d}(\boldsymbol{k})$, the local preferred direction in spin space. For the BW state, $\hat{d}_\mu(\hat{\boldsymbol{k}}) = \Sigma_j R_{\mu j} \hat{k}_j$, where $R_{\mu j}$ is the rotation matrix characterizing the BW order parameter. The contribution to $\chi_{\mu\nu}^0$ from each point $\hat{\boldsymbol{k}}$ has the form of (3.102), with, however, $\hat{\boldsymbol{d}}$ replaced by $\hat{\boldsymbol{d}}(\boldsymbol{k})$. The total spin susceptibility $\chi_{\mu\nu}^0$ is arrived at by summing over $\boldsymbol{k}$:

$$\chi_{\mu\nu}^0 = \chi_N^0 \langle \delta_{\mu\nu} - \hat{d}_\mu(\boldsymbol{k}) \hat{d}_\nu(\boldsymbol{k}) [1 - Y_0(\hat{\boldsymbol{k}}; T)] \rangle_{\hat{\boldsymbol{k}}}. \tag{3.103}$$

Here $Y_0(\hat{\boldsymbol{k}}; T)$ is defined by (3.88). For nonunitary states, $\boldsymbol{d}(\boldsymbol{k})$ may be a complex vector, in which case $\hat{d}_\mu \hat{d}_\nu$ should be replaced by $\mathrm{Re}\{\hat{d}_\mu \hat{d}_\nu^*\}$ in (3.103).

It is now a simple task to evaluate (3.103) for the BW state. Observing that $Y_0(\hat{\boldsymbol{k}}; T)$ is isotropic, one finds

$$\chi_{\mu\nu}^0 = \delta_{\mu\nu} \chi_N^0 [1 - \tfrac{1}{3}(1 - Y_0)], \tag{3.104}$$

a result that is easily interpreted as follows: the BW state is an equal mixture of $S_z = 0, \pm 1$ pairs, of which the $S_z = 0$ species does not contribute to the magnetization. Hence the susceptibility is given by its normal-state value minus one-third of the superfluid contribution $1 - Y_0$.

The above results are again subject to Fermi-liquid corrections. As discussed in Chapter 2, an applied magnetic field is screened by a polarization of the medium described by the Landau parameter F_0^a, resulting in the effective magnetic field

$$\boldsymbol{H}_{\mathrm{eff}} = \boldsymbol{H} - \frac{F_0^a \boldsymbol{M}}{\chi_N^0}. \tag{3.105}$$

In the presence of Fermi-liquid interactions the magnetization $\boldsymbol{M}$ induced by a magnetic field is equal to that of the noninteracting system in a magnetic field $\boldsymbol{H}_{\mathrm{eff}}$, or $\boldsymbol{M} = \boldsymbol{\chi}^0 \boldsymbol{H}_{\mathrm{eff}}$. Solving for $\boldsymbol{M}$ yields the renormalized spin susceptibility (Leggett 1965a,b)

$$\boldsymbol{\chi} = \left(\mathbf{1} + \frac{F_0^{\mathrm{a}}\boldsymbol{\chi}^0}{\chi_{\mathrm{N}}^0}\right)^{-1} \boldsymbol{\chi}^0, \tag{3.106}$$

where $\boldsymbol{\chi}$ and $\boldsymbol{\chi}^0$ are tensors in spin space.

For the BW state, the final result is

$$\boldsymbol{\chi}(T) = \chi_{\mathrm{N}}^0 \frac{2 + Y_0(T)}{3 + F_0^{\mathrm{a}}[2 + Y_0(T)]} \mathbf{1}. \tag{3.107}$$

For $T \to 0$ and $F_0^{\mathrm{a}} \approx -\frac{3}{4}$, (3.107) tends to about $\frac{1}{3}\chi_{\mathrm{N}}$, i.e. the reduction is twice the value of the unrenormalized result.

In general, there are also corrections to $\boldsymbol{\chi}$ from higher Landau parameters. For the BW state, owing to its high symmetry, only F_2^{a} enters (Czerwonko 1967) (see (10.56)).

3.5 EXPECTATION VALUE OF TWO-PARTICLE QUANTITIES

Following Leggett (1975a), we finally want to determine how the two-particle correlations involved in Cooper pairing influence one- and two-particle quantities in the superfluid. It turns out that in most cases the expectation value of a one-particle quantity acquires only a small correction of order $(T_{\mathrm{c}}/T_{\mathrm{F}})^2 \approx 10^{-6}$ caused by the slight energy variation of the one-particle density of states (namely if this quantity is an even function of the energy $\xi_{\boldsymbol{k}}$). Examples are the total number of particles and the total momentum. An important exception is the kinetic-energy part of the Hamiltonian (3.15), $\sum_{\boldsymbol{k}} \xi_{\boldsymbol{k}} n_{\boldsymbol{k}}$. The expectation value of this one-particle operator, $\sum_{\boldsymbol{k}} \xi_{\boldsymbol{k}} \langle n_{\boldsymbol{k}} \rangle$, is indeed substantially modified by the formation of the condensate owing to the linear dependence on $\xi_{\boldsymbol{k}}$.

In the case of two-particle quantities (those involving an interaction between two particles) we should, on the other hand, already intuitively expect that they are particularly sensitive to the pair correlation. Let us write such a quantity as

$$Q^{(2)} = \tfrac{1}{2} \sum_{n \neq m} Q(\boldsymbol{r}_n, \boldsymbol{r}_m; \boldsymbol{p}_n, \boldsymbol{p}_m; \sigma_\mu^{(n)}, \sigma_\nu^{(m)}), \ \mu, \nu = 1, 2, 3), \tag{3.108}$$

where we have specified that it generally depends on the position, momentum and spin of any two particles in the system. We denote its expectation value by $\langle Q^{(2)} \rangle$.

The most obvious example is the interaction part V of the Hamiltonian (3.15) itself. By means of the pair amplitude $F_{\boldsymbol{k}\alpha\beta}$, one finds from (3.15)

$$\langle V \rangle = \tfrac{1}{2} \sum_{\boldsymbol{k}\alpha\beta} \Delta_{\boldsymbol{k}\alpha\beta}^* F_{\boldsymbol{k}\alpha\beta}. \tag{3.109}$$

Using the gap equation (3.17) and the simplified spin-independent interaction (3.45), this gives

$$\langle V \rangle = \sum_{\boldsymbol{k}\alpha\boldsymbol{k}'\beta} V_{\boldsymbol{k}-\boldsymbol{k}'} F_{\boldsymbol{k}\alpha\beta} F_{\boldsymbol{k}'\alpha\beta}^*. \tag{3.110}$$

This result is not surprising—it merely expresses the fact that $\langle V \rangle$ is the sum over all momenta of the interaction weighted by the probability amplitudes $F_{\boldsymbol{k}\alpha\beta}$, $F^*_{\boldsymbol{k}'\alpha\beta}$ for the scattering from a pair state $\boldsymbol{k}$ to one with $\boldsymbol{k}'$.

More generally, we may assume Q to be a spin-dependent operator of the form

$$Q = \sum_{nm}\sum_{\mu\nu} Q_{\mu\nu}(\boldsymbol{r}_n - \boldsymbol{r}_m)\sigma_\mu^{(n)}\sigma_\nu^{(m)}, \tag{3.111}$$

involving particles n, m; the indices μ, ν specify the components of the spin, and we assume $Q_{\mu\nu}(\boldsymbol{r}_n - \boldsymbol{r}_m) = Q_{\mu\nu}(\boldsymbol{r}_m - \boldsymbol{r}_n) = Q_{\nu\mu}(\boldsymbol{r}_n - \boldsymbol{r}_m)$. This is not the most general form that Q can have, but it is sufficient for our purposes. We now insert (3.111) into (3.108), multiply by $\int d^3\boldsymbol{r} \int d^3\boldsymbol{r}'\, \delta(\boldsymbol{r} - \boldsymbol{r}_n)\, \delta(\boldsymbol{r}' - \boldsymbol{r}_m)$ and write the sum over n, m as $\sum_{n\neq m} = \sum_{n,m} - \sum_n$. Next, the spin density $\sigma_\mu(\boldsymbol{r})$ is introduced as

$$\sigma_\mu(\boldsymbol{r}) = \sum_n \sigma_\mu^{(n)}\, \delta(\boldsymbol{r} - \boldsymbol{r}_n) \tag{3.112}$$

$$= \tfrac{1}{2}\sum_{\alpha\beta} (\sigma_\mu)_{\alpha\beta}\psi^+_\alpha(\boldsymbol{r})\psi_\beta(\boldsymbol{r}), \tag{3.113}$$

where we have written $\sigma_\mu(\boldsymbol{r})$ in second-quantization form, with

$$\psi_\alpha(\boldsymbol{r}) = \sum_{\boldsymbol{k}} a_{\boldsymbol{k}\alpha}\, e^{i\boldsymbol{k}\cdot\boldsymbol{r}}. \tag{3.114}$$

This yields

$$Q^{(2)} = \tfrac{1}{8}\sum_{\mu\nu}\sum_{\alpha\beta\gamma\delta}\int d^3r \int d^3r'\, Q_{\mu\nu}(\boldsymbol{r} - \boldsymbol{r}')(\sigma_\mu)_{\alpha\beta}(\sigma_\nu)_{\gamma\delta}$$
$$\times\, \psi^+_\alpha(\boldsymbol{r})\psi^+_\gamma(\boldsymbol{r}')\psi_\delta(\boldsymbol{r}')\psi_\beta(\boldsymbol{r}). \tag{3.115}$$

The effect of the pair correlation on the expectation value of $Q^{(2)}$ involves only the *anomalous* averages $\langle\psi\psi\rangle$ etc., and we therefore find

$$\langle Q^{(2)}\rangle = \tfrac{1}{8}\sum_{\alpha\beta\gamma\delta}\sum_{\mu\nu}\int d^3r\, Q_{\mu\nu}(\boldsymbol{r})(\sigma_\mu)_{\alpha\beta}(\sigma_\nu)_{\gamma\delta}F^*_{\gamma\alpha}(\boldsymbol{r})F_{\delta\beta}(\boldsymbol{r}), \tag{3.116}$$

where

$$F_{\alpha\beta}(\boldsymbol{r}) = \sum_{\boldsymbol{k}} F_{\boldsymbol{k}\alpha\beta}\, e^{i\boldsymbol{k}\cdot\boldsymbol{r}}. \tag{3.117}$$

To simplify the notation in the present case of triplet pairing, we proceed as in (3.34) and introduce a *vector* $\boldsymbol{F}_{\boldsymbol{k}}$ in spin space via

$$F_{\boldsymbol{k}\alpha\beta} = \sum_\mu (\boldsymbol{\sigma}_\mu i\boldsymbol{\sigma}_2)_{\alpha\beta}F_{\boldsymbol{k}\mu} \tag{3.118a}$$

such that

$$F_{\boldsymbol{k}\mu} = -\tfrac{1}{2}\sum_{\alpha\beta} (\boldsymbol{\sigma}_2 i\boldsymbol{\sigma}_\mu)_{\alpha\beta}F_{\boldsymbol{k}\alpha\beta}, \tag{3.118b}$$

in exact analogy with $\Delta_{\boldsymbol{k}\alpha\beta}$ and $\boldsymbol{d}(\boldsymbol{k})$; $\boldsymbol{F}_{\boldsymbol{k}}$ is the Fourier transform of $\boldsymbol{F}(\boldsymbol{r})$ in position space. Inserting (3.117) and (3.118a) into (3.116) and performing the spin summation (with $\frac{1}{2}\operatorname{tr}_\sigma(\boldsymbol{\sigma}_\mu\boldsymbol{\sigma}_\nu\boldsymbol{\sigma}_\lambda\boldsymbol{\sigma}_\rho) = \delta_{\mu\nu}\delta_{\lambda\rho} - \delta_{\mu\lambda}\delta_{\nu\rho} + \delta_{\mu\rho}\delta_{\nu\lambda}$), we find

$$\langle Q^{(2)}\rangle = \tfrac{1}{4}\sum_{\mu\nu}\int d^3r\, Q_{\mu\nu}(\boldsymbol{r})\{|\boldsymbol{F}(\boldsymbol{r})|^2\,\delta_{\mu\nu} - 2\,\mathrm{Re}\,[F_\mu(\boldsymbol{r})F^*_\nu(\boldsymbol{r})]\}, \tag{3.119a}$$

which is identical with

$$\langle Q^{(2)}\rangle = \tfrac{1}{4}\sum_{kk'}\sum_{\mu\nu} \tilde{Q}_{\mu\nu}(\boldsymbol{k}-\boldsymbol{k}')\{\boldsymbol{F}_{\boldsymbol{k}}\cdot\boldsymbol{F}^*_{\boldsymbol{k}'}\,\delta_{\mu\nu} - 2\,\mathrm{Re}\,[F_{\boldsymbol{k}\mu}F^*_{\boldsymbol{k}'\nu}]\}, \quad (3.119\text{b})$$

where $\tilde{Q}_{\mu\nu}(\boldsymbol{k})$ is the Fourier transform of $Q_{\mu\nu}(\boldsymbol{r})$. Note that the simpler spin-singlet case, which is spherically symmetric, can be obtained from the above expressions by making the replacement $\frac{1}{4}\,Q_{\mu\nu}\rightarrow Q\delta_{\mu\nu}$ and $F_{\boldsymbol{k},\mu}\rightarrow F_{\boldsymbol{k}}$, where $F_{\boldsymbol{k}} = \langle a_{-\boldsymbol{k}\downarrow}a_{\boldsymbol{k}\uparrow}\rangle$. In terms of $F(\boldsymbol{r}) = \sum_{\boldsymbol{k}} F_{\boldsymbol{k}}\,\mathrm{e}^{\mathrm{i}\boldsymbol{k}\cdot\boldsymbol{r}}$, one finds

$$\langle Q^{(2)}\rangle_{\text{singl}} = \int \mathrm{d}^3r\, Q(\boldsymbol{r})\,|F(\boldsymbol{r})|^2. \quad (3.120)$$

The basic structure of (3.119a) and (3.120) is essentially the same, the additional complications in (3.119a) are only due to the spin-triplet pairing. Now $|F(\boldsymbol{r})|^2$ is just the probability density for a Cooper pair with separation $\boldsymbol{r}$, giving a very plausible interpretation of the result in (3.120) for the expectation value $\langle Q\rangle$ of the operator Q in a pair-correlated state. In this respect there is a certain similarity between a Cooper pair and a very large "diatomic molecule" (Leggett 1980a,b), where the expectation value of an operator is given by an expression identical with (3.120) but with $F(\boldsymbol{r})$ replaced by the pair wave function of the molecule. We shall later discuss this molecule picture in more detail (Chapter 12). If used with proper care, it proves to be helpful in calculating macroscopic properties of superfluid ^{3}He that involve the internal structure of the ^{3}He Cooper pair.

FURTHER READING

A selection of books on BCS theory

Abrikosov A A, Gorkov L P and Dzyaloshinski I E 1963 *Methods of Quantum Field Theory in Statistical Physics* (Prentice-Hall, Englewood Cliffs, New Jersey; reprinted 1975 Dover, New York)

Fetter A L and Walecka J D 1971 *Quantum Theory of Many-Particle Systems* (McGraw-Hill, New York)

de Gennes P G 1966 *Superconductivity of Metals and Alloys* (Benjamin, New York)

Lifshitz E M and Pitaevskii L P 1980 *Statistical Physics,* Part 2 (Vol. 9 of the Landau–Lifshitz *Course of Theoretical Physics*) (Pergamon, Oxford)

Rickayzen G 1965 *Theory of Superconductivity* (Wiley, New York)

Schrieffer J R 1964 *Theory of Superconductivity* (Benjamin, New York)

Tinkham M 1975 *Introduction to Superconductivity* (McGraw-Hill, New York)

Review articles on weak-coupling pairing theory of liquid ^{3}He

Anderson P W and Brinkman W F 1978 in *The Physics of Liquid and Solid Helium,* Part II, ed. K H Bennemann and J B Ketterson (Wiley, New York), p. 177

Lee D M and Richardson R C 1978 in *The Physics of Liquid and Solid Helium,* Part II, ed. K H Benneman and J B Ketterson (Wiley, New York), p. 287

Leggett A J 1975 *Rev. Mod. Phys.* **47** 331

Wheatley J C 1975 *Rev. Mod. Phys.* **47** 415

4

Basic Experimental Properties

4.1 EXPERIMENTAL TECHNIQUES FOR ATTAINING ULTRALOW TEMPERATURES

The experimental investigation of liquid and solid ^{3}He in the low-temperature, quantum regime could not have taken place without several important improvements in cooling techniques that took place in the 1950s and 1960s. First the ^{3}He–^{4}He dilution refrigerator, based on a cooling cycle involving the heat of solution of ^{3}He in liquid ^{4}He as proposed by London (1951), had been developed (London *et al.* 1962, Das *et al.* 1965, Neganov *et al.* 1966, Hall *et al.* 1966, Wheatley *et al.* 1971). It provides a means for removing relatively large amounts of heat down to temperatures of several millikelvin.

In order to go below these temperatures, one has to work with systems that order at even lower temperatures. These are spin systems of various types whose ordering temperature is below the temperature one wants to reach and that at the same time possess a large heat capacity, good thermal conductivity and low Kapitza boundary resistance to ^{3}He. A single-cycle cooling process is possible owing to the fact that, while the *total* entropy is conserved in an adiabatic process, the entropy of the spin system (i.e. the coolant) placed in a magnetic field increases upon lowering the magnetic field, such that the system absorbs heat (principle of adiabatic demagnetization). A method for cooling liquid ^{3}He widely used in the 1960s consisted of placing powdered cerium magnesium nitrate (CMN) directly in the ^{3}He sample chamber in order to circumvent problems of heat transfer (CMN is a paramagnetic salt in which electrons order magnetically at a temperature around 1 mK). After precooling the CMN–^{3}He mixture in a strong magnetic field to a starting temperature of typically 20 mK, the sample is thermally decoupled from the precooling stage. By adiabatic demagnetization, it is then possible to reach a temperature of about 2 mK, just sufficient to study the properties of superfluid ^{3}He near the transition temperature at higher pressures (Wheatley 1966, 1970, 1973, 1975).

Much lower temperatures may in principle be attained—and have actually been reached—by adiabatic demagnetization of a *nuclear* spin system. Here a suitable material is copper metal, owing to its good electronic thermal conductivity and the large spin–lattice relaxation. These properties provide the necessary thermal contact of the ^{3}He liquid sample with the copper conduction-electron system in the bulk (usually via the large area of a metallic "sponge" or sinter) and from there to the nuclear spin system. The first successful nuclear demagnetization experiment was performed as early as 1956 (Kurti *et al.* 1956), although the lattice and electron systems did not cool substantially. However, since then adiabatic nuclear demagnetization has been developed into a powerful cooling method, which is now exclusively used to attain temperatures in the millikelvin regime and below (Osgood and Goodkind 1967, Ahonen *et al.* 1974b,c). The ordering temperature of copper nuclei has been determined to be around 60 nK, i.e. 6×10^{-8} K (Lounasmaa 1984), which opens up a whole new temperature regime in the *micro*kelvin range. Since the nuclear moment of Cu is so much smaller than the electronic magnetic moment, the precooling temperature has to be rather low ($\leqslant$10 mK) and the magnetic field should be high ($\geqslant$80 kG) in order to remove a substantial fraction of the spin entropy. These extreme conditions may be somewhat relaxed if a material with a hyperfine enhanced nuclear magnetic moment, such as $PrNi_5$ or similar rare-earth compounds, is used (Andres and Bucher 1968, 1972).

Quite a different idea is to use the ^{3}He nuclear spin system itself as a refrigerant (Pomeranchuk 1950). Unfortunately, the cooling of liquid ^{3}He by nuclear demagnetization is impractical because the nuclear magnetization of the Fermi liquid ^{3}He is reduced by a factor T/T_F from its classical value. Hence the relative magnetization of liquid ^{3}He attainable in a laboratory is only of the order of 10^{-3} and is thus much too low to provide any significant cooling power.

In contrast, *solid* ^{3}He can be magnetized substantially at a precooling temperature of several millikelvin, at which it still behaves as a classical spin system. On the other hand, to reach thermal equilibration in an insulator like solid ^{3}He takes an increasingly longer time the lower the temperature. Therefore a metal like copper is much preferred as a refrigerant. Nevertheless, ^{3}He may be employed as a refrigerant in yet another, very specific, way. This makes use of the large difference in the entropies of solid and liquid ^{3}He. (In the solid the spins behave classically, the entropy being given by $S_{solid} = R \ln 2$; in contrast, the entropy of the liquid is $S_{liq} \propto T/T_F$ owing to the Pauli principle, hence $S_{liq} < S_{solid}$ at low enough temperatures.) This anomaly may be exploited for cooling near the melting curve, even without applying an external magnetic field. By simply compressing the liquid adiabatically, solid ^{3}He with high entropy is formed, thus decreasing the temperature of the remaining liquid. This is the so-called "Pomeranchuk effect", which may be employed to cool ^{3}He along the melting curve (Anufriyev 1965) until the spins in the solid order (at around 1 mK) and the entropy of the solid is reduced to a value below that of the liquid. This makes the method ideally suited for the investigation of the magnetic phase transitions in the solid, as well as of properties of the liquid at melting pressure. In fact, as already mentioned, the superfluid phases of ^{3}He were discovered in a Pomeranchuk experiment designed to investigate the magnetic phase transition in the solid (Osheroff, Richardson and Lee 1972a). The magnetic transition in the solid was only found later (Halperin *et al.* 1974).

One of the major experimental problems at these lowest temperatures is to establish an absolute temperature scale. To this day the interpretation of experimental results is hampered by uncertainties in temperature measurement. Various experimental temperature scales have been proposed. The majority are based on the magnetic properties of an electronic or nuclear spin system, for which one postulates the temperature dependence of the ideal model system and a value of T_c at the melting curve. Thus there exists a "La Jolla temperature scale" (Wheatley 1975) and a "Helsinki temperature scale" (Ahonen *et al.* 1976c). A second possibility is to derive a thermodynamic temperature scale from measurements on ^{3}He alone. This has been done along the melting curve, yielding the "Cornell temperature scale" (Halperin *et al.* 1975) and at lower pressures assuming ideal Fermi-liquid behaviour above T_c (Greywall 1986). The extremes, given by the La Jolla and Helsinki scales, differ by about 10% in the T_c value and up to 40% in the specific-heat results. A critical evaluation of the different temperature scales, including a detailed discussion of the possible origins of these discrepancies has been given by Greywall (1986). In this book we shall use Greywall's temperature scale.

4.2 THERMODYNAMIC PROPERTIES

4.2.1 Phase diagram

The first indication of the existence of two phase transitions in condensed ^{3}He at temperatures of about 2.6 mK and 1.8 mK respectively was a barely visible kink and a discontinuity in the pressure versus time curve (see Fig. 4.1) of a Pomeranchuk cell operated at a constant rate of volume change (Osheroff *et al.* 1972a). The two signatures, called A and B, clearly indicated a second-order transition (A) and a first-order transition (B) (Halperin *et al.* 1973). Using nuclear magnetic resonance (NMR) measurements, these were later unambiguously identified as coming from the liquid rather than the solid (note that liquid and solid ^{3}He coexist in a Pomeranchuk cell). In contrast with the A transition, the B transition did not always occur at the same temperature, but rather showed substantial supercooling. Upon warming and cooling several times, the B transition that occurred during a *cooling* cycle took place in a wide temperature range, depending on the history, whereas it happened at approximately the same

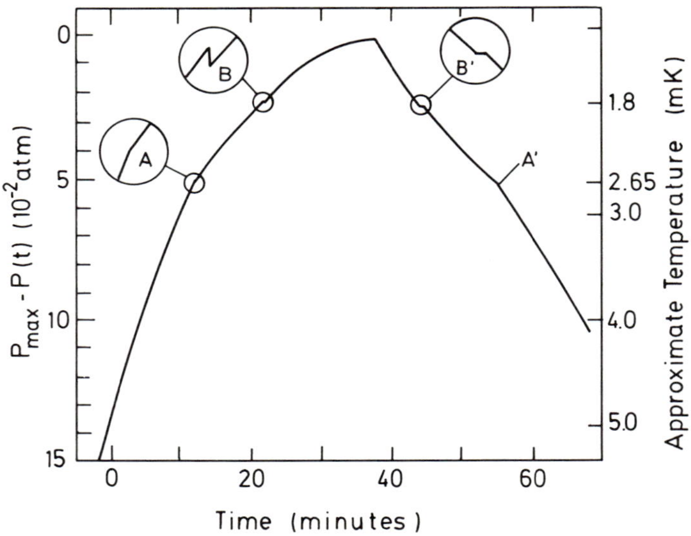

Figure 4.1 Pressure versus time trace when the volume in a Pomeranchuk cell was changed at a constant rate. The signatures A and B, shown magnified in the inserts, indicate the occurrence of two phase transitions in ^{3}He. (After Osheroff *et al.* (1972a).)

temperature during *warming*. It was also found that an applied magnetic field led to a very small splitting of the A transition into two transitions A_1 and A_2, whereas the B-transition temperature was strongly reduced (Gully *et al.* 1973). In a series of experiments using CMN as a refrigerant, the exploration of the phase diagram was extended to pressures below the melting pressure, down to somewhat below 10 bar (Greytak *et al.* 1973, Paulson *et al.* 1973a,b). Interestingly enough, it was found that the temperature region in which the A phase exists shrinks as the pressure is lowered, and disappears completely below a so-called "polycritical" point (PCP) at approximately 21 bar pressure. These measurements were then extended down to the saturated vapour pressure (which will sometimes be casually referred to as zero pressure) using the nuclear demagnetization technique (Ahonen *et al.* 1974b). Below 10 bar the transition temperature from the usual phase into the superfluid B phase decreases quite rapidly with decreasing pressure to a value of about 1 mK at s.v.p. A list of values of the transition temperature as a function of pressure in zero magnetic field as obtained by Greywall (1986) is presented in Table 4.1.

Application of a magnetic field has a drastic effect on the phase diagram, in particular below the PCP: (i) a distinctly new phase appears (referred to as the A_1 phase); (ii) the ordinary A phase (in this case referred to as the A_2 phase), is stabilized down to zero pressure. Hence the A_1 and A_2 phases now intervene between the normal Fermi liquid and the B phase (Paulson *et al.* 1974a,b). In analogy to the above nomenclature, the latter phase, i.e.

Table 4.1 List of smoothed values of the transition temperatures between the normal liquid and the superfluid state (T_c) and between the A and B phases (T_{AB}) as functions of pressure in zero magnetic field (Greywall 1986). The "polycritical point" lies at $P_{PCP} = 21.22$ bar and $T_c(P_{PCP}) = 2.273$ mK.

P (bar)	T_c (mK)	T_{AB} (mK)
0	0.929	—
1	1.061	—
2	1.181	—
3	1.290	—
4	1.388	—
5	1.478	—
6	1.560	—
7	1.636	—
8	1.705	—
9	1.769	—
10	1.828	—
11	1.883	—
12	1.934	—
13	1.981	—
14	2.026	—
15	2.067	—
16	2.106	—
17	2.143	—
18	2.177	—
19	2.209	—
20	2.239	—
21	2.267	—
21.22	2.273	2.273
22	2.293	2.262
23	2.317	2.242
24	2.339	2.217
25	2.360	2.191
26	2.378	2.164
27	2.395	2.137
28	2.411	2.111
29	2.425	2.083
30	2.438	2.056
31	2.451	2.027
32	2.463	1.998
33	2.474	1.969
34	2.486	1.941
34.338	2.491	1.933
34.358	—	1.932

the B phase in a magnetic field, is referred to as the B_2 phase. In principle, there is also a "B_1 phase" (namely the so-called planar phase; see the discussion in Chapters 5 and 6), but this phase is unstable with respect to the A_1/A_2 phases. Therefore in a magnetic field there exist three stable bulk phases: the A_1, A_2 and B_2 phases. As found in the earlier experiments along the melting curve, the transition temperature into the A phase hardly changes in a magnetic field, apart from a tiny splitting. Below the PCP it coincides with the transition temperature into the B phase in zero field. On the other hand, the first-order A–B transition temperature is shifted to lower temperatures, and eventually (i.e. in fields of 5–10 kG) is even below the lowest temperature attainable so far, depending on pressure.

Investigations of the A–B transition in a magnetic field have been reported by Feder *et al.* (1981) and Hoyt *et al.* (1981), as well as by Eisenstein *et al.* (1980b), who investigated the effect of a magnetic field on the flowing superfluid. Most recently, Kyynäräinen *et al.* (1989) have observed a splitting of the A–B transition at low pressures and magnetic fields of about 2 kG. This has been interpreted as evidence for the appearance of a yet unknown intermediate phase in a very narrow region (width of order 1 μK at 3.4 kG and $P = 2.3$ bar) between the A_2 and B_2 phases.

The magnetic-field dependence of the A_1 and A_2 transitions as a function of pressure has been measured by Israelsson *et al.* (1984) and by Sagan *et al.* (1984). The complete P–T–H phase diagram as known today is shown in Fig. 4.2.

4.2.2 Specific heat

A great deal of information about the nature of the superfluid states may be obtained by thermal measurements. In fact, as discussed above, it was the thermal anomalies associated with the phase transitions that actually led to the initial discovery of these phases. Subsequent specific-heat experiments (using CMN salt as a refrigerant) below the melting curve showed a specific-heat anomaly at the normal–superfluid transition reminiscent of the superconducting transitions in metals (Webb *et al.* 1973, Wheatley 1973, 1975). Later measurements of the specific heat along the melting curve in a Pomeranchuk cell yielded a similar result (Halperin *et al.* 1976). Precise measurements of the specific heat in the full pressure range have been performed by Alvesalo *et al.* (1980, 1981), Roach *et al.* (1982) and Greywall (1986). The experimental curve at melting pressure is shown in Fig. 4.3. There exists a sharp discontinuity at T_c of relative size $\Delta C/C_N$ as large as 2. This is larger than the BCS value of 1.43, which was expected for the BW state, and much larger than the value 1.19 appropriate for the ABM state. The discontinuity becomes smaller for decreasing pressure, and approaches the weak-coupling value 1.43 for pressures below 10 bar. A corresponding discontinuity is observed in the thermal expansion coefficient (Swift and Packard 1981, Roach *et al.* 1983). The character of the experimental specific-heat curve (Fig. 4.3) does not appear to change drastically at the

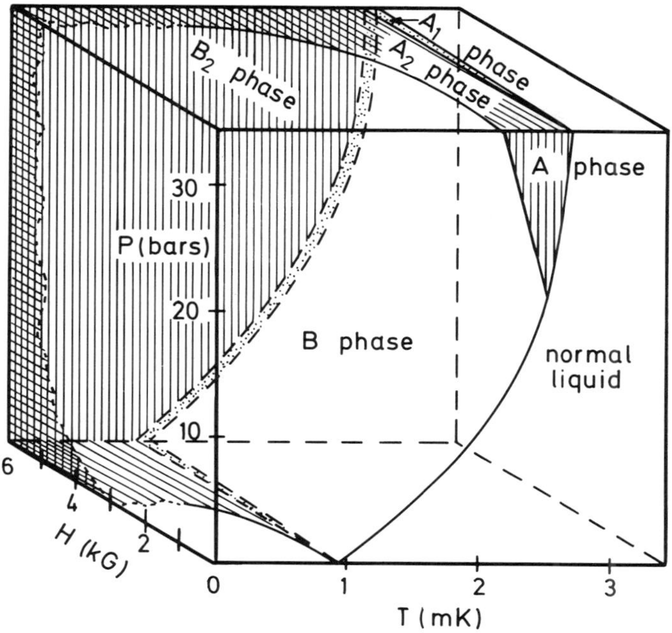

Figure 4.2 The P–T–H phase diagram of ^{3}He, showing the normal Fermi liquid as well as the superfluid A, A_1 and B phases. The superfluid A_2 and B_2 phases correspond to the A and B phases in the presence of a magnetic field. The values of the transition temperature shown in this figure are a composite of the results of several groups and should only be taken as semiquantitative (see the discussion in Section 4.1). A complete list of T_c values obtained by Greywall (1986) is shown in Table 4.1.

A–B transition, indicated by an arrow in the figure. In the temperature range shown the specific-heat data approximately follow the law $C(T) = C(T_C)(T/T_C)^3$. The exponential behaviour $C(T) \propto e^{-\Delta/k_B T}$ for the B phase at low temperatures, if it were identified with the BW state (see (3.81)), would only be expected to show up at somewhat lower temperatures. On the other hand, the low-temperature behaviour of the ABM-state specific heat, $C^{ABM} \propto T^3$, given by (3.83), is consistent with experiment, although the data do not extend to very low temperatures. There is a very small jump in the specific heat at the A–B transition given by $(C_B - C_A)/C_A = 0.091$. Also, the latent heat of the A–B transition is rather small, $l_{AB} = 15.4$ erg/mol for $T_{AB} = 2.1$ mK at the melting curve. It follows from the Clausius–Clapeyron equation that the associated volume change at the A–B transition is only one part in 10^8, the density of the A phase being larger. Substantial supercooling occurs at the A–B transition, whereas only modest superheating effects have been observed (Wheatley 1975, Kleinberg *et al.* 1974). The phase boundary between the A and B phases in the coexistence region is well defined, as observed in NMR experiments using a uniform magnetic-field gradient (Osheroff *et al.* 1972b).

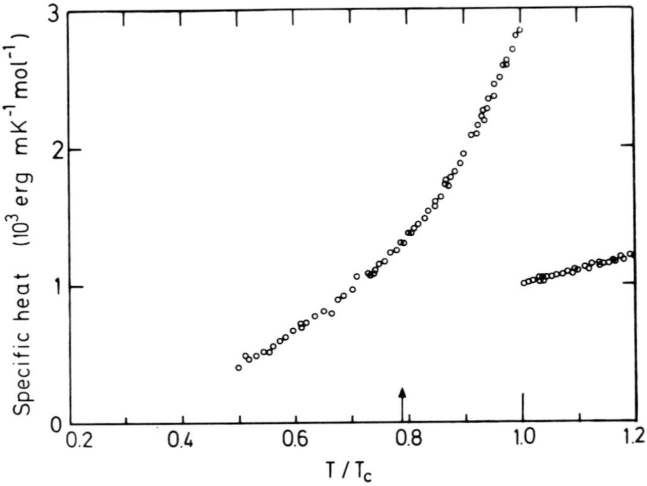

Figure 4.3 Specific heat in the superfluid phases measured in zero magnetic field at melting pressure. The B transition is indicated by an arrow. (After Halperin *et al.* (1976).)

As discussed above, the A transition splits into two transitions in an external magnetic field, whereas the B transition is moved to lower temperatures. The splitting is clearly observed in the specific-heat measurements. A discontinuity is found at each of the transitions, with the magnitude of the A_1 jump $\Delta C^{(A_1)}/C_N = 0.74$ at melting pressure being somewhat smaller than the A_2 jump. The discontinuities add up, however, to the jump observed in zero magnetic field.

4.3 MAGNETIC PROPERTIES

The first actual experiments performed on the superfluid phases of ^{3}He investigated the nuclear magnetic resonance of the ^{3}He nuclear spin system in a Pomeranchuk cell (Osheroff *et al.* 1972a,b). By applying a magnetic-field gradient to the sample and thereby shifting the local resonance frequency, it was possible to distinguish between contributions to the resonance signal from liquid and solid parts of the sample. Quite unexpectedly, it was found that below the transition into the A phase the resonance signal from the liquid started to shift from the Larmor frequency. The amoung of this shift was found to correspond to a local field of maximally 30 G following the law

$$\omega^2 = \omega_L^2 + \Omega^2(T) \tag{4.1}$$

(see also Fig. 4.4). Here ω_L is the Larmor frequency and $\Omega(T)$ is a

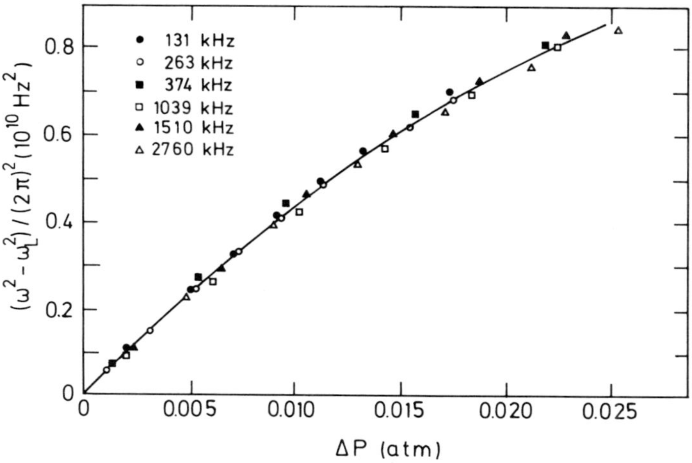

Figure 4.4 Transverse NMR shift as observed in ^{3}He-A in a Pomeranchuk experiment. (After Osheroff *et al.* (1972b).) The pressure difference ΔP may be translated into temperature using Table 4.1.

field-independent but temperature-dependent frequency. At the A–B transition the signal from the liquid appeared to shift back to the original Larmor frequency, but dropped by a factor of approximately two in magnitude. At still lower temperatures, in the B phase, the shift did not reappear, and the signal strength decreased further (but only by a small amount) down to the lowest accessible temperature. It was later found that the magnetic susceptibility in the A phase is not quite given by the temperature-independent Pauli susceptibility of the normal state, but is increased by a very small amount (Paulson *et al.* 1974b, Webb *et al.* 1974a, Webb 1977a,

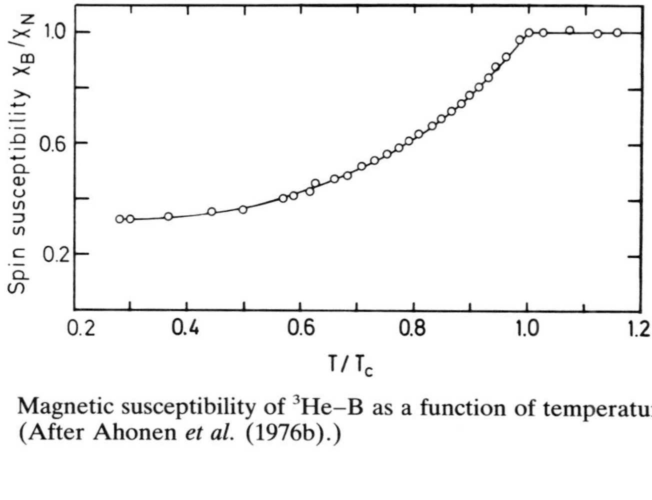

Figure 4.5 Magnetic susceptibility of ^{3}He–B as a function of temperature at 20 bar. (After Ahonen *et al.* (1976b).)

Sager *et al.* 1978a). The nuclear magnetic susceptibility of the B phase as a function of temperature is shown in Fig. 4.5 for two different pressures. The data seem to follow approximately a universal law as a function of T/T_c.

As shown by Leggett (1973a,b), an NMR shift of the observed magnitude may be explained by the coherent superposition of the spin–orbit forces caused by the nuclear dipole interactions within each Cooper pair. Given this theoretical explanation, a *longitudinal* resonance was predicted to be observable as well (Leggett 1973a,b). The expected oscillation of the longitudinal magnetization was first found by Osheroff and Brinkman (1974) and Bozler *et al.* (1974) in CW resonance experiments; for a review of early experiments see Richardson (1975). It was later discovered that the longitudinal oscillation could be excited by a sudden change in the applied magnetic field, a phenomenon called "ringing of magnetization" (Webb *et al.* 1974b). The ringing frequencies in the A and B phases are found to scale with the ratio of the magnetic susceptibilities, i.e. as

$$\frac{\Omega_{\mathrm{B}}^2(T)}{\Omega_{\mathrm{A}}^2(T)}=\frac{5}{2}\frac{\chi_{\mathrm{B}}(T)}{\chi_{\mathrm{A}}}, \tag{4.2}$$

except perhaps in the vicinity of the polycritical point (Osheroff *et al.* 1975, Ahonen *et al.* 1975c, Webb *et al.* 1975a). It also turns out that the longitudinal resonance frequency Ω_{A} in the A phase is equal to the characteristic frequency Ω in the expression (4.1) for the shifted transverse frequency. In Fig. 4.6 the resonance frequency data from longitudinal resonance experiments in the A and B phases and transverse resonance experiments in the A phase are plotted in scaled form, using (4.1) and (4.2).

A rich variety of nonlinear behaviour in the spin dynamics of superfluid ^{3}He has since been found. The longitudinal ringing frequency in both the A and B phases shows significant structure as a function of the change in magnetic field ΔH, which initiates the ringing (Webb *et al.* 1974a, 1975b, 1977, Webb 1977b). In the A (B) phase the frequency is substantially smaller than the $\Delta H \rightarrow 0$ limiting value at one (two) particular value(s) of the turn-off field ΔH and increases linearly with ΔH well above that point.

The nonlinear behaviour after application of a large spin-tipping pulse is found to be very complex. One observes tipping-angle-dependent frequency shifts, satellite resonance lines and unusual relaxation effects (Osheroff 1974, 1975, Osheroff and Corruccini 1975a,b, Corruccini and Osheroff 1975a,b, 1978, Osheroff and Brinkman 1974, Giannetta *et al.* 1981, Eska *et al.* 1982). These phenomena will be discussed in more detail in Chapter 8.

4.4 SOUND PROPAGATION

As discussed in Chapter 2, a Fermi liquid allows a well-defined density-oscillation mode, called zero sound. In normal liquid ^{3}He zero sound has been clearly observed for sufficiently high frequencies such that

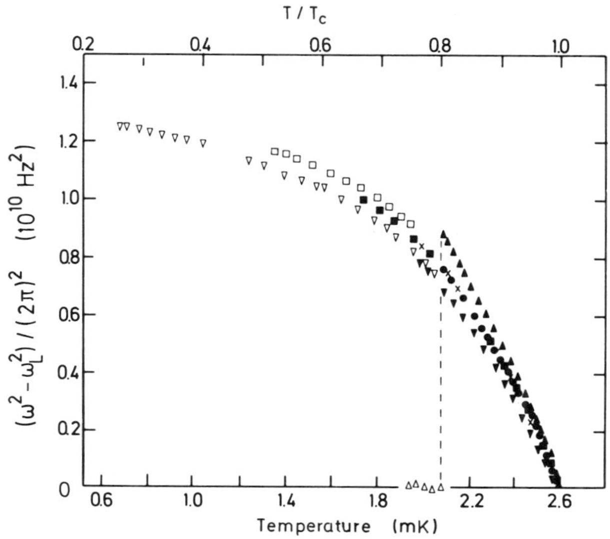

Figure 4.6 Longitudinal NMR frequencies as measured in the A and B phases and transverse NMR frequencies as measured in the A phase plotted in scaled form, using (4.1) and (4.2), as functions of temperature: △, ▲, transverse shift at melting pressure (Bozler *et al.* 1974); ×, transverse shift at melting pressure (Osheroff *et al.* 1975); □, ■, longitudinal shift at melting pressure (Osheroff *et al.* 1975); ●, longitudinal ringing at 33 bar (Webb *et al.* 1974b); ▽, ▼, transverse shift at 32 bar (Ahonen *et al.* 1976b) (open symbols, B phase (scaled); filled symbols, A phase). (After Wölfle (1979).)

the condition for its existence, $\omega \gg 1/\tau$, is satisfied (see Fig. 2.2). Here τ is a quasiparticle collision time. Since zero sound owes its existence to the stiffness of the Landau molecular field, which is the result of interaction processes involving energies on the scale of the Fermi energy, one should expect this mode to persist in the superfluid phases. Indeed, one of the first experiments done on superfluid ^{3}He in a Pomeranchuk cell showed that ultrasound with a frequency of 10 MHz propagates in the superfluid with essentially the same efficiency as in the normal fluid (Lawson *et al.* 1973, 1974). However, the sound attenuation, though small, develops a sharp peak as a function of temperature closely below T_c (see Fig. 4.7). Systematic measurements of the attenuation and the velocity of longitudinal sound in superfluid ^{3}He over a wide pressure range extending from about 9 bar up to the melting pressure were made later in an apparatus using CMN as a refrigerant (Paulson *et al.* 1973a,b, 1976a,c,d). Attenuation peaks were studied near the A–normal and B–normal transitions. The observation of these peaks is indicative of the existence of collective modes in the superfluid. For the three sound frequencies used (5, 15 and 25 MHz) the

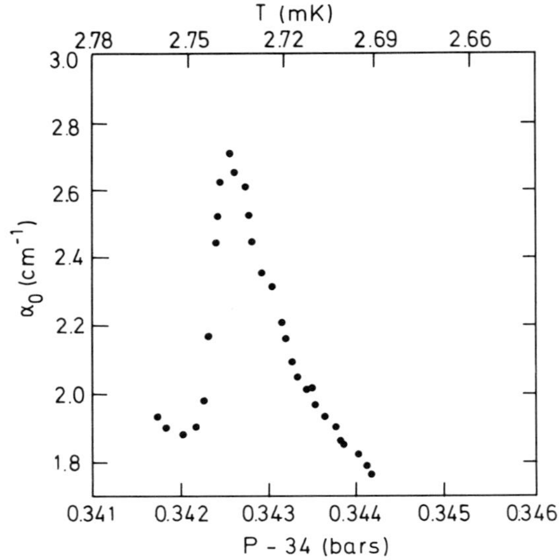

Figure 4.7 Sound attenuation in ^{3}He-B just below T_c as a function of melting pressure in a Pomeranchuk cell. (After Lawson *et al.* (1974).) The pressure translates into temperature as indicated on the top horizontal scale.

attenuation was found to increase strongly with frequency. Similar experiments conducted in the B phase showed interesting structure in the sound velocity near T_c (Roach *et al.* 1975a). In the B phase the sound propagation properties are isotropic. By contrast, in the A phase sound attenuation is found to depend sensitively on the direction of applied external fields, for instance magnetic fields, thermal currents, and also on the shape of the sample cell. This behaviour is interpreted as being due to the orienting effect of the aforementioned external fields on the anisotropic order parameter in the A phase. A systematic variation of the sound attenuation with the angle between a small magnetic field (approximately 20 G) and the sound propagation direction has been observed by Roach *et al.* (1975b) and Ketterson *et al.* (1975). These orientation effects are also evident from the fluctuations in the sound attenuation, which reflect fluctuations in the local anisotropy axis (Lawson *et al.* 1975a,b). The latter authors further investigated the sound attenuation in magnetic fields up to 7.5 kG, where the attenuation peaks were observed to split into two in accordance with the splitting of the A transition seen in thermodynamic experiments. The anisotropy of the ultrasound attenuation in the A phase has been used to investigate the dynamics of the local anisotropy axis (Paulson *et al.* 1977, Wheatley 1978). In these experiments it was found that the anisotropy axis generally relaxes to the new equilibrium position within a time that is simply related to the spin-relaxation time found in NMR experiments.

Later experiments showed the existence of a second small but very sharp attenuation peak at lower temperatures in the B phase (Giannetta *et al.* 1980, Mast *et al.* 1980). The group velocity of a sound pulse in the vicinity of this sharp attenuation feature was found to vary dramatically, the pulse being slowed down to a value as low as one-third of the phase velocity. In subsequent experiments the sharp resonance was shown to split into five peaks in an applied magnetic field of typically 100 G (Avenel *et al.* 1980).

Transverse zero sound has also been investigated. Measurements of the acoustic shear impedance show broad features, which may be interpreted as evidence for transverse zero sound (Roach and Ketterson 1976a,b).

Hydrodynamic, or low-frequency, sound propagation has been studied in order to obtain information on the viscous properties of the liquid (Eska *et al.* 1980). One finds that the attenuation and velocity change of 50–200 kHz sound is dominated by the friction of the liquid with the walls of the resonator cell. The temperature dependence of the viscosity coefficient extracted from these measurements agrees reasonably well with that obtained from shear-flow measurements.

It is well known from superfluid ^{4}He that the propagation of heat changes from diffusive to wave-like as one moves from the normal into the superfluid state. These entropy waves in the superfluid, called second sound, may be expected to be much less well defined in ^{3}He, where the entropy at low temperatures is reduced by a factor T/T_F from the classical value. This has in fact obstructed the observation of second sound. The first successful experimental observation of second sound took place in the A_1 phase, i.e. the phase that appears between the normal phase and the A_2 phase in a magnetic field (Corruccini and Osheroff 1980). There it was found to be strongly coupled to the magnetic properties, measured simultaneously by NMR. Later it also became possible to observe second sound at very low frequencies in zero field (Lu and Kojima 1985). A related phenomenon is the so-called "fountain pressure", i.e. the formation of a pressure head induced by a temperature gradient driving a counterflow of the normal and superfluid components (Shields and Goodkind 1977).

A well-defined hydrodynamic sound-like mode is observed in superfluid ^{3}He flowing in a confining geometry, for example inside the pores of a fine powder filling the container (Kojima *et al.* 1974, Yanof and Reppy 1974). Under these conditions, when the motion of the normal-fluid component is clamped by friction with the walls of the narrow channels, the superfluid component is able to sustain a sound-like excitation mode, so-called fourth sound. The velocity and damping of fourth sound provide information about the superfluid density component and the superfluid bulk viscosity. The existence of low-frequency wave-like motion of the superfluid in a fourth-sound experiment constitutes proof for the superfluid behaviour of the liquid on timescales of the order of the inverse sound-frequency.

4.5 HYDRODYNAMIC PROPERTIES AND SUPERFLUIDITY

In the early stages of experimental research on superfluid ^{3}He investigations of hydrodynamic flow properties were at the centre of interest, since they were known to elucidate whether or not an anisotropic-superfluid pairing model was qualitatively correct. A number of early experiments concentrated on the measurement of the normal-fluid density and viscosity.

4.5.1 Normal-fluid density and viscosity

In the first experiments designed to measure the viscosity a vibrating wire immersed in the liquid was used as a probe. The shift in resonant frequency and the damping of the vibrations as a function of temperature may be converted into a viscosity coefficient and the density fraction of the normal-fluid component (Alvesalo *et al.* 1973, 1974, 1975, Guernsey *et al.* 1976; see also Johnson *et al.* 1975). Another technique employs a torsional oscillator where either an object of cylindrical shape is suspended in the liquid (Parpia *et al.* 1976) or else a ring-shaped ^{3}He cell is rotated as a whole (Kiewiet *et al.* 1975, Main *et al.* 1976, Archie *et al.* 1979, 1981).

The common features of all these experiments are (i) the observation of a T^{-2} behaviour of the shear-viscosity coefficient in the normal phase (see Chapter 2), (ii) a rapid drop of the viscosity below T_c roughly following a $(T_c - T)^{1/2}$ law, and (iii) a smooth decrease of the normal-fluid density to very small values at temperatures below $0.5T_c$. The low-temperature limiting behaviour of the viscosity seems to indicate that the effective viscosity vanishes. Experiments on other transport coefficients will be discussed in Chapter 10.

A more precise measurement of the normal-fluid fraction is possible if the motion of the normal fluid is completely clamped by filling the torsional oscillator cell with a tightly packed fine-grained powder. The amount of liquid free to move, i.e. the superfluid component, reduces the moment of inertia of the torsional oscillator and may be measured through a shift of the resonant frequency (Main *et al.* 1976, Berthold *et al.* 1976). Precise values of the superfluid density fraction obtained in this way are shown in Fig. 4.8.

4.5.2 Persistent currents

Perhaps the most remarkable property of the superfluid phases of ^{3}He is, as indicated by their very name, the capability of flowing without friction. This property has been established beyond doubt for superconductors and superfluid ^{4}He, for which persistent (i.e. dissipation-free) currents flowing in a ring geometry have been observed to flow for up to several years. To measure the same property in the case of superfluid ^{3}He has turned out to

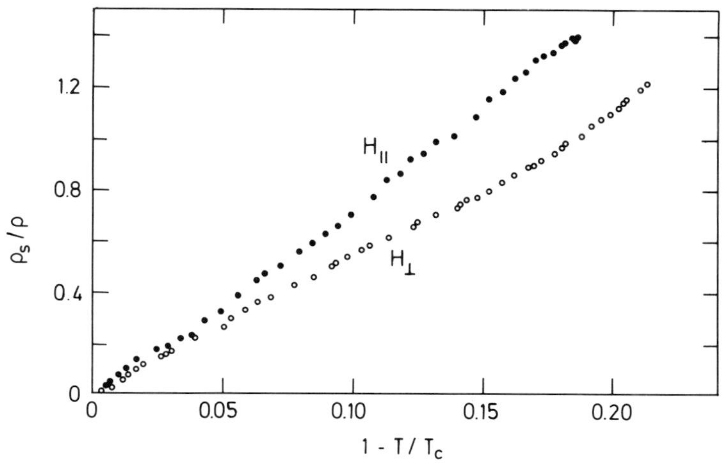

Figure 4.8 Superfluid density fraction ρ_s/ρ in ^{3}He-A as a function of reduced temperature at 27 bar. The measurements were taken at two different orientations of an external magnetic field relative to the flow direction. (After Berthold *et al.* (1976).)

be very difficult. The basic device used in such an experiment is a gyroscope, by which the angular momentum associated with the superflow is measured. In these experiments a torus filled with ^{3}He, packed with powder in order to clamp the motion of the normal fluid, is mounted on top of an axially flexible rod. The cryostat is rotated about the axis of the torus (z axis) at an angular velocity exceeding a certain critical value, such that a supercurrent is set up. After stopping the rotation, the supercurrent continues to flow and thereby gives rise to a macroscopic angular momentum. In principle, the persistent angular momentum is detected by exerting a periodic torque in the x direction and observing oscillations of the gyroscope along the y direction, at the eigenfrequency of these oscillations.

In this way the existence of persistent currents has been established for the B phase (Gammel *et al.* 1984, Pekola *et al.* 1984b, Pekola and Simola 1985). In fact, the persistent current was shown to last for at least 48 h without detectable dissipation. A similar experiment in the A phase failed to detect a persistent current. However, Gammel *et al.* (1985) succeeded in enhancing the sensitivity of the measurement by exploiting a subtle effect of superflow on the dissipation of torsional oscillations of the ^{3}He cell found previously by Gay *et al.* (1983). The Q factor of the torsional oscillations at resonance was indeed found to depend on the preparation of the fluid, i.e. the application of a rotation of the sample prior to the measurement. For angular velocities of rotation exceeding a certain critical value Ω_{min}, the minimum of the dissipation is shifted from $\Omega = 0$ to a finite value Ω_s. This is interpreted as the rotation speed where the superfluid is at rest with respect to the container wall, i.e. the average superfluid velocity $\bar{v}_s = R\Omega_s$, where R

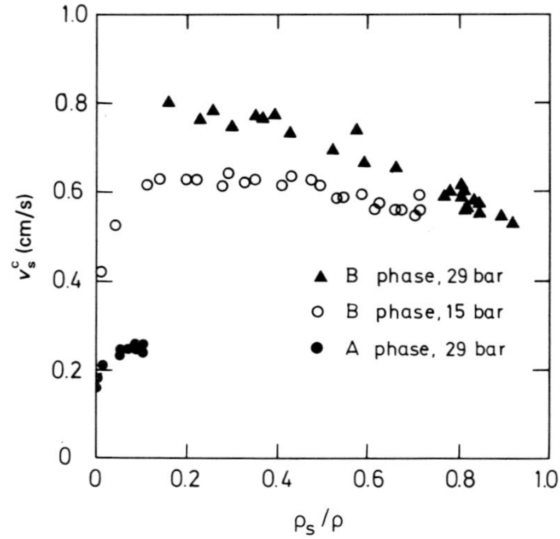

Figure 4.9 Temperature dependence of the critical superfluid velocities in the A and B phases plotted as v_s^c versus the superfluid density fraction ρ_s/ρ. (For the T dependence of ρ_s, see Fig. 4.8.) (After Gammel *et al.* (1985).)

is the radius of the ring channel. The behaviour as a function of Ω is reversible below a critical value Ω_c. This is interpreted as the onset of processes destroying superfluidity. Hence the so-called critical velocity can be deduced as $v_s^c = R\Omega_c$. Similar behaviour was found in the B phase in the earlier experiments. One finds values for the critical velocity of $v_s^c \approx 0.2$ cm/s in the A phase and $v_s^c \approx 0.6$–0.8 cm/s in the B phase, with a tendency to increase slightly with pressure (see Fig. 4.9). The values for the A phase are in agreement with earlier measurements for oscillatory flow through a single 18 μm diameter orifice (Parpia and Reppy 1979).

Perhaps the most direct manifestation of macroscopic quantum coherence in the superfluid is the Josephson effect, i.e. the tunnelling of condensate particles in a Bose liquid or of Cooper pairs through a so-called "weak link" (Josephson 1962, 1965). In a superfluid such a weak link is realized by an orifice connecting two reservoirs, whose extension is comparable to the coherence length. The experimental detection of the Josephson effect has proved to be difficult—even in superfluid ^{4}He. An unambiguous observation was only reported almost 25 years after the prediction of the effect (Avenel and Varoquaux 1985, 1986). Subsequently these authors were also able to observe a near-ideal Josephson effect in ^{3}He-B at zero pressure (Avenel and Varoquaux 1987).

4.6 ROTATING SUPERFLUID ^{3}He

Since the pioneering experiments on liquid ^{4}He by Andronikashvili beginning in the 1940s (Andronikashvili and Mamaladze 1966) it has been known that many fascinating and new features arise in a rotating superfluid. The principal property of the superfluid that comes into play here is that the superfluid flow field is irrotational, i.e. the superfluid velocity field is curl-free ($\nabla \times \boldsymbol{v}_s = 0$). Hence a superfluid is not capable of performing solid-body rotation like a normal fluid. Rather, by rotating a superfluid and thus imposing a certain angular-momentum density on it, one forces the liquid into an inhomogeneous state consisting of a regular lattice of parallel vortices, each carrying one quantum of circulation (for an explicit experimental verification, see Williams and Packard 1980, Yarmchuk and Packard 1982). The circulation is concentrated in the vortex cores, if the flow of the superfluid is indeed derivable from a potential ($\boldsymbol{v}_s \propto \nabla \Phi$), as in the case for superfluid ^{4}He. Similar behaviour is observed in a type-II superconductor, where the role of the mechanical rotation of the container is played by a magnetic field and where the vortices are flux lines. In this case the flux lattice may even be observed under a microscope (see Essmann and Träuble 1967).

Experiments on rotating superfluid ^{3}He indicate a rich variety of vortex structures and transitions among them (Hakonen *et al.* 1982a,b, 1983a, Ikkala *et al.* 1982). The experiments were performed on a rotating cryostat (Hakonen *et al.* 1983b) using rotation speeds of up to 3 rad/s and magnetic fields of 100–500 G. The experimental ^{3}He cell had the shape of a cylinder of 5 mm diameter and 30 mm length. The principal probe used to detect vortices is the shift in frequency and the shape of the NMR signal and, in particular, the appearance of satellite peaks as a function of rotation speed Ω. In the A phase the sharp NMR line measured in the stationary fluid is broadened upon rotation of the container, as shown in Fig. 4.10, and a satellite peak is seen to appear at the low-frequency side. In the B phase on the other hand, a large number of satellite peaks are already present in the NMR signal at $\Omega = 0$ (see Fig. 4.11). These peaks are shifted and broadened as a function of Ω (see Fig. 8.7). The frequency shifts vary significantly with temperature as shown in Fig. 8.8. At a temperature $T \approx 0.6T_c$ (at 29.3 bar pressure) an abrupt change in the frequency shifts takes place—indicative of a phase transition or structural change in the vortex system. Even more surprising, the frequency shifts depend on the *relative* orientation of the magnetic field (parallel or antiparallel) with respect to the rotation axis. Apparently a magnetic moment is induced by the rotation of the fluid ("gyromagnetism"). The location of the vortex transition as a function of pressure and temperature is shown in the phase diagram, Fig. 8.9.

In broad terms, the presence of vortices influences the NMR properties

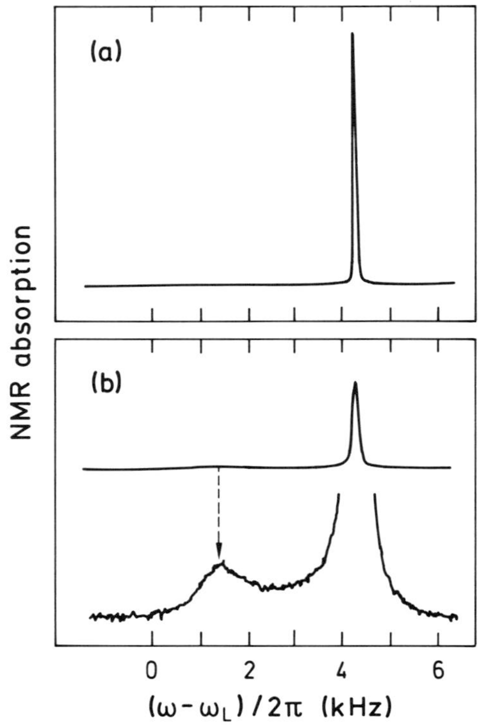

Figure 4.10 Effect of rotation on the NMR line in ^{3}He-A: (a) $\Omega = 0$; (b) $\Omega = 1.21$ rad/s. The arrow indicates the appearance of a satellite peak associated with the vortex formation. The absorption scale of the lower curve in (b) is magnified by a factor of 50. (After Hakonen *et al.* (1983a).)

by changing the textures in such a way that the torques exerted on the local spins are modified, causing frequency shifts. It may even happen that potential wells are created in which standing spin waves may exist, causing satellite NMR peaks to appear. The various possible vortex configurations and their NMR signatures will be discussed in detail in Sections 7.6 and 8.4 respectively.

Other experimental methods used to probe the vortex lattice include ion-mobility experiments (Simola *et al.* 1986, 1987) and hydrodynamic experiments such as measurements of the critical current (Pekola *et al.* 1984c). More details will be given in Section 7.6.

This concludes our discussion of basic experimental observation. Detailed comparison of experimental data and theory is made in the following chapters.

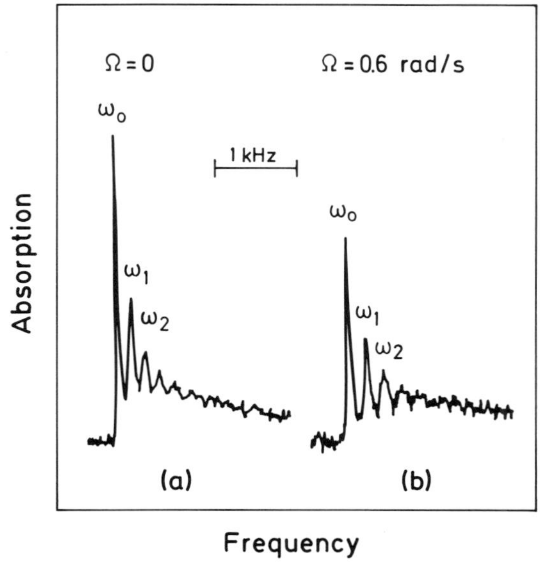

Figure 4.11 Effect of rotation on NMR spectra in ^{3}He-B at 29.3 bar: (a) $\Omega = 0$; (b) $\Omega = 0.6$ rad/s. (After Hakonen *et al.* (1983a).)

FURTHER READING

Hall H E and Hook J R 1986 in *Progress in Low Temperature Physics,* Vol. IX, ed. D F Brewer (North-Holland, Amsterdam), p. 143

Lee D M and Richardson R C 1978 in *The Physics of Liquid and Solid ^{3}Helium,* Part II, ed. K H Bennemann and J B Ketterson (Wiley, New York), p. 287

Osheroff D D 1978 *J. Physique* **39** Colloq. C-6, Vol. III, p. 1270 (Proceedings of the 15th International Conference on Low Temperature Physics, LT-15)

Wheatley J C 1975 *Rev. Mod. Phys.* **47** 415

Wheatley J C 1978 in *Progress in Low Temperature Physics,* Vol. VIIa, ed. D F Brewer (North-Holland, Amsterdam), p.1

5

Pair Correlations beyond Weak Coupling

The experimental observation of two phases, both p-wave pairing states as suggested by the details of the phase diagram, contradicts the result of weak-coupling theory that the BW state is lowest in free energy at all temperatures (see Section 3.4). On the other hand, weak-coupling theory is exact in the limit $T_c/T_F \rightarrow 0$. Since the various p-wave pairing states are close in free energy, it is not surprising that pressure-dependent corrections to weak coupling, which do exist, may reverse the order of states. At this stage, before indulging in elaborate microscopic theories, it is useful to try to get an overview of the possible states. Since we are dealing with a continuous phase transition, characterized by a continuously vanishing order parameter at the transition point, this can be achieved within the framework of the Ginzburg–Landau expansion of the free energy in powers of the order parameter (Landau and Lifshitz 1959a). In the temperature regime near the transition the expansion may be terminated at fourth-order terms. Quite generally, such an expansion is constructed by identifying those combinations of the order parameter that—for a given power of the order parameter—are invariant under all the symmetries of the system. The number of these invariants will be larger the more complex the order-parameter structure is. The numerical coefficients multiplying these invariants can only be determined from microscopic theory. This will be discussed in Section 5.2.

5.1 GENERAL GINZBURG–LANDAU EXPANSION OF THE FREE ENERGY

As discussed in Chapter 3, one should expect pairing to take place in states of pure relative orbital angular momentum, unless the pair interaction is such that two (or more) interaction parameters are close in value. Systematic discussions of the general Ginzburg–Landau free energy have been given for p-, d- and f-wave pairing. We restrict attention to p-wave pairing here, following the arguments of Mermin and Stare (1973) and Brinkman and Anderson (1973).

For general l pairing the order parameter is a complex tensor $d_{\mu j_1 \ldots j_l}$ of rank $l+1$, satisfying certain constraints for $l>1$ connected with the permutation symmetry of the j_i. The free energy is a real scalar quantity. Gauge invariance requires the expansion to be in powers of $\boldsymbol{dd}^*$ ($\boldsymbol{d}^*$ is the complex conjugate of $\boldsymbol{d}$). Rotational invariance in spin space and momentum space is guaranteed by contracting tensor indices in spin space and momentum space pairwise. Thus there is only one second-order invariant term, but there exist several fourth-order invariants; their actual number depends on l (Barton and Moore 1974a).

The p-wave pairing order parameter is a complex 3×3 matrix $d_{\mu j}$ (see (3.64). It is convenient to introduce the normalized order-parameter matrix $\boldsymbol{A}$ by

$$d_{\mu j} = 3^{1/2}\Delta(T)A_{\mu j}, \tag{5.1a}$$

such that

$$\operatorname{tr}(\boldsymbol{AA}^+) = 1, \tag{5.1b}$$

where Δ is the root-mean-square gap (see (3.57)),

$$\Delta(T) = \left(\tfrac{1}{2}\sum_{\alpha,\beta} \langle \Delta_{k\alpha\beta}\Delta^*_{k\alpha\beta}\rangle_{\hat{k}}\right)^{1/2}, \tag{5.2}$$

and tr (. . .) denotes the trace operation on 3×3 matrices. There are then five fourth-order invariants, which can be constructed graphically by representing $\boldsymbol{A}$ ($\boldsymbol{A}^*$) as an open (shaded) circle and sums over momentum space (spin-space) indices by a solid (broken) line (see Fig. 5.1). They are

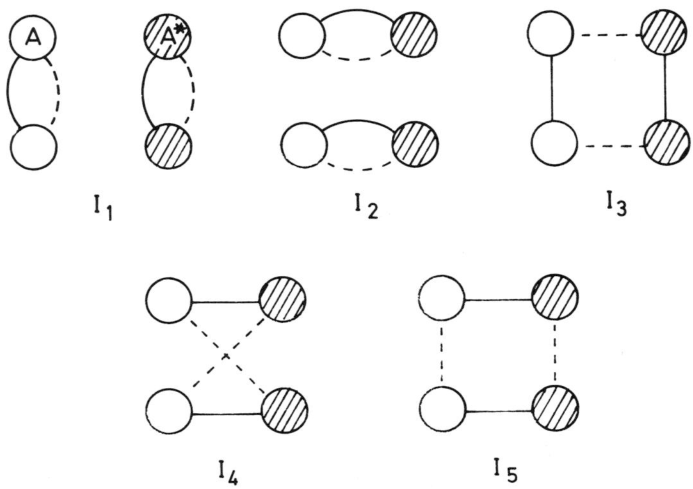

Figure 5.1 Graphical representation of the fourth-order invariants for p-wave pairing.

given by

$$\left.\begin{aligned} I_1 &= |\text{tr}\,(\boldsymbol{A}\boldsymbol{A}^{\text{T}})|^2, \\ I_2 &= [\text{tr}\,(\boldsymbol{A}\boldsymbol{A}^{+})]^2 \equiv 1, \\ I_3 &= \text{tr}\,[(\boldsymbol{A}\boldsymbol{A}^{\text{T}})(\boldsymbol{A}\boldsymbol{A}^{\text{T}})^*], \\ I_4 &= \text{tr}\,[(\boldsymbol{A}\boldsymbol{A}^{+})^2], \\ I_5 &= \text{tr}\,[(\boldsymbol{A}\boldsymbol{A}^{+})(\boldsymbol{A}\boldsymbol{A}^{+})^*]. \end{aligned}\right\} \tag{5.3}$$

Here $\boldsymbol{A}^{\text{T}}$ and $\boldsymbol{A}^{+}$ denote the transpose and Hermitian conjugate of $\boldsymbol{A}$ respectively. Note that various alternative conventions are used for labelling the quartic invariants (5.3).

The general free-energy functional per volume for p-wave pairing states up to fourth-order terms is thus obtained by adding up the invariants as

$$f \equiv \frac{F}{V} = f_{\text{N}} + \alpha\Delta^2 + \tfrac{1}{2}\Delta^4 \sum_{i=1}^{5} \beta_i I_i, \tag{5.4}$$

where V is the volume. Since in the case of homogeneous bulk phases the quantities F and f are related by a trivial factor, for the rest of this chapter the symbol F will denote the bulk free energy per unit volume. The coefficients α, β_i have to be calculated from microscopic theory. Note that we have already calculated these coefficients in weak-coupling theory (see (3.55) and below).

For an investigation of nonunitary p-wave states with equal spin pairing, such as the ABM state in a magnetic field to be discussed in Section 5.4, it is more convenient to express the Ginzburg–Landau free-energy functional in terms of the "spin-matrix gap parameter" $\boldsymbol{\Delta}_j$ (rather than the order parameter $A_{\mu j}$), defined by

$$\boldsymbol{\Delta}_j = 3^{1/2}\Delta \sum_{\mu} A_{\mu j} \boldsymbol{\sigma}_\mu \mathbf{i}\boldsymbol{\sigma}_2 \tag{5.5a}$$

such that (see (3.34a), (3.64))

$$\boldsymbol{\Delta}_{\boldsymbol{k}} = \sum_j \boldsymbol{\Delta}_j \hat{k}_j. \tag{5.5b}$$

The only second-order invariant is

$$I_0 = \tfrac{1}{2}\,\text{tr}_\sigma\,(\boldsymbol{\Delta}_j \boldsymbol{\Delta}_j^{+}) = 3\Delta^2, \tag{5.5c}$$

where summation over repeated indices is implied. Rotational invariance in spin space implies that only products of the type $(\boldsymbol{\Delta}\boldsymbol{\Delta}^{+})^n$ are allowed in any invariant expression. The five fourth-order invariants $\tilde{I}_i$ are obtained by taking all possible contractions of momentum-space indices:

$$\left.\begin{aligned}
\tilde{I}_1 &= \tfrac{1}{2}\sum_{jl} \operatorname{tr}_\sigma (\mathbf{\Delta}_j\mathbf{\Delta}_l^+\mathbf{\Delta}_j\mathbf{\Delta}_l^+) \equiv 9\Delta^4(2I_3 - I_1),\\
\tilde{I}_2 &= \left[\tfrac{1}{2}\sum_{j} \operatorname{tr}_\sigma (\mathbf{\Delta}_j\mathbf{\Delta}_j^+)\right]^2 \equiv 9\Delta^4 I_2,\\
\tilde{I}_3 &= \tfrac{1}{4}\sum_{jl} [\operatorname{tr}_\sigma (\mathbf{\Delta}_j\mathbf{\Delta}_l^+)\operatorname{tr}_\sigma (\mathbf{\Delta}_j\mathbf{\Delta}_l^+)] \equiv 9\Delta^4 I_3,\\
\tilde{I}_4 &= \tfrac{1}{4}\sum_{jl} [\operatorname{tr}_\sigma (\mathbf{\Delta}_j\mathbf{\Delta}_l^+)\operatorname{tr}_\sigma (\mathbf{\Delta}_l\mathbf{\Delta}_j^+)] \equiv 9\Delta^4 I_4,\\
\tilde{I}_5 &= \tfrac{1}{2}\sum_{jl} \operatorname{tr}_\sigma (\mathbf{\Delta}_j\mathbf{\Delta}_j^+\mathbf{\Delta}_l\mathbf{\Delta}_l^+) = 9\Delta^4(I_2 + I_4 - I_5).
\end{aligned}\right\} \tag{5.6}$$

In (5.6) we have also indicated the relation to the invariants I_i.

The free-energy functional thus takes the alternative form

$$F = F_{\mathrm{N}} + \tfrac{1}{3}\alpha I_0 + \tfrac{1}{18}\sum_{i=1}^{5} \tilde{\beta}_i \tilde{I}_i. \tag{5.7a}$$

Comparison of (5.7) with (5.4), using the definition of the invariants $\tilde{I}_i$ and I_i, yields the following relations among the coefficients $\tilde{\beta}_i$ and β_i:

$$\left.\begin{aligned}
\tilde{\beta}_1 = -\beta_1, \quad \tilde{\beta}_2 = \beta_2 + \beta_5, \quad \tilde{\beta}_3 = \beta_3 + 2\beta_1,\\
\tilde{\beta}_4 = \beta_4 + \beta_5, \quad \tilde{\beta}_5 = -\beta_5.
\end{aligned}\right\} \tag{5.7b}$$

In particular, the weak-coupling values of α and β_i may be determined by comparing (5.6) with (3.55) derived within the BCS weak-coupling approach; one finds

$$\left.\begin{aligned}
\alpha(T) = -N(0)(1 - T/T_{\mathrm{c}}),\\
\beta_2 = \beta_3 = \beta_4 = -\beta_5 = -2\beta_1 = \tfrac{6}{5}\beta_0,
\end{aligned}\right\} \tag{5.8}$$

with β_0 defined in (3.56). Here the fourth-order term $\langle \operatorname{tr}_\sigma (\mathbf{\Delta}_{\boldsymbol{k}}\mathbf{\Delta}_{\boldsymbol{k}}^+)^2\rangle_{\hat{\boldsymbol{k}}}$ in (3.55) is evaluated using

$$\operatorname{tr}_\sigma (\mathbf{\Delta}_{\boldsymbol{k}}\mathbf{\Delta}_{\boldsymbol{k}}^+)^2 = \sum_{ijlm} \operatorname{tr}_\sigma (\mathbf{\Delta}_i^+\mathbf{\Delta}_j\mathbf{\Delta}_l^+\mathbf{\Delta}_m)\, \hat{k}_i\hat{k}_j\hat{k}_l k_m \tag{5.9a}$$

and

$$\langle \hat{k}_i\hat{k}_j\hat{k}_l\hat{k}_m\rangle_{\hat{\boldsymbol{k}}} = \frac{1}{1\cdot 3\cdot 5}(\delta_{ij}\delta_{lm} + \delta_{il}\delta_{jm} + \delta_{im}\delta_{jl}). \tag{5.9b}$$

The resulting expression is in terms of $\tilde{I}_i$, (5.6); comparison with (5.7b) then yields (5.8).

The free-energy expression (5.4) may first be minimized with respect to Δ. The resulting maximum magnitude of Δ (for given matrix structure $\boldsymbol{A}$) is

$$\Delta_0(T) = a_0\kappa^{-1/2}k_{\mathrm{B}}T_{\mathrm{c}}\left(1 - \frac{T}{T_{\mathrm{c}}}\right)^{1/2}, \tag{5.10}$$

where $a_0 = (8\pi^2/7\zeta(3))^{1/2} \approx 3.06$. The strong-coupling parameter κ is defined as

$$\kappa = \frac{N(0)}{\alpha'} \sum_{i=1}^{5} \frac{\beta_i}{\beta_0} I_i, \tag{5.11}$$

where $\alpha' = T_c \, d\alpha/dT \,|_{T_c}$. For s-wave BCS pairing, $\kappa = 1$. However, in the present case one finds $\kappa \geqslant 1$ for the weak-coupling β_i values (Barton and Moore 1974b,c). The corresponding value of the free energy is given by

$$F - F_N = \tfrac{1}{2}\alpha\Delta^2 = -\tfrac{1}{2}a_0^2\kappa^{-1}N(0)(k_B T_c)^2\left(1 - \frac{T}{T_c}\right)^2. \tag{5.12}$$

The specific-heat discontinuity at T_c is seen to be determined by κ as

$$\frac{\Delta C}{C_N} = \frac{12}{7\zeta(3)}\kappa^{-1}, \tag{5.13}$$

where C_N is given by (2.14). Conversely, κ may be expressed in terms of $\Delta C/C_N$ in (5.10), yielding the useful result

$$\Delta(T) = \pi\left(\frac{2}{3}\frac{\Delta C}{C_N}\right)^{1/2} k_B T_c\left(1 - \frac{T}{T_c}\right)^{1/2} \tag{5.14}$$

There remains the task of finding the minimum of F with respect to variations of the actual (matrix) structure of $A_{\mu j}$, or, equivalently, of finding the maximum of $1/\kappa$.

The search for stable minima of (5.4), i.e. for possible superfluid phases, can proceed along quite different lines. For example, one may seek the solution of the idealized weak-coupling problem as discussed in Chapter 3 (Balian and Werthamer 1963, Ambegaokar and Mermin 1973); alternatively, one can seek the solution under certain constraints on the order parameter, such as unitarity (Mermin and Stare 1973, Varma and Werthamer 1974), or with the requirement that the resulting phases have certain magnetic properties (Mermin and Stare 1974) known from experiment. A different line of approach is due to Barton and Moore (1974b, 1975a), who concentrate on those phases whose order parameters do not depend on the parameters β_i. These phases are called "inert"—their order parameters do not change their structure upon a change of external parameters, only the overall magnitude of the gap will change.

Using methods of differential geometry, which will not be reproduced here, one can in fact derive a general equation for the extrema of the free energy (5.12) in terms of the matrix $\boldsymbol{A}$ (Jones 1977):

$$\beta_1[\boldsymbol{A}^* \operatorname{tr}(\boldsymbol{A}\boldsymbol{A}^T) - I_1\boldsymbol{A}] + \beta_3[\boldsymbol{A}\boldsymbol{A}^T\boldsymbol{A}^* - I_3\boldsymbol{A}] + \beta_4[\boldsymbol{A}\boldsymbol{A}^+\boldsymbol{A} - I_4\boldsymbol{A}] + \beta_5[\boldsymbol{A}^*\boldsymbol{A}^T\boldsymbol{A} - I_5\boldsymbol{A}] = 0, \tag{5.15}$$

with the constraint $\operatorname{tr}(\boldsymbol{A}\boldsymbol{A}^+) = 1$. In principle, this is a set of 18 real equations for the components $A_{\mu i}$. However, only 10 of these parameters are actually independent. This is due to the invariance of the free energy (5.4) under separate three-dimensional rotations in spin and orbital space and under an overall phase transformation (see Chapter 6): it implies that seven parameters (three Euler angles in spin space, three Euler angles in orbital space and one phase variable) can be chosen arbitrarily (see Eilenberger and Schopohl 1988). In addition, the constraint $\mathrm{tr}\,(\boldsymbol{AA}^+) = 1$ eliminates one more parameter, such that one is left with only 10 independent parameters. Equation (5.15) is therefore greatly overdetermined. The solution of this nonlinear equation leads to be a surprisingly complicated problem; in fact, for arbitrary β_i the problem has not yet been solved. For unitary p-wave states, however, a full analysis is possible, and will be given below.

The spectrum of the mean-field Hamiltonian (3.16) and, in particular, the resulting unitary states have been investigated by Solomon (1981) within a dynamical group model. A classification of f-wave pairing states has been discussed by Barton and Moore (1975a) and Mermin (1975, 1976). Such states have been found to be incompatible with experiment.

5.2 PHENOMENOLOGICAL CLASSIFICATION OF MODEL STATES

5.2.1 Unitary states

The signature of a unitary state is the condition

$$\boldsymbol{d}(\boldsymbol{k}) \times \boldsymbol{d}^*(\boldsymbol{k}) = 0$$

for the order parameter (see (3.37)), implying either (i) a real $\boldsymbol{d}(\boldsymbol{k})$, or (ii) complex $\boldsymbol{d} = c(\boldsymbol{k})\hat{\boldsymbol{d}}$, where $\hat{\boldsymbol{d}}$ is a $\boldsymbol{k}$-independent real unit vector.

In the case (i) we have $\boldsymbol{A}^+ = \boldsymbol{A}^{\mathrm{T}}$, and consequently $\boldsymbol{AA}^+$ is a real symmetric matrix, with real eigenvalues λ_n, $n = 1, 2, 3$. The free-energy density may be written in terms of these eigenvalues as

$$F = F_{\mathrm{N}} + \alpha\Delta^2 \sum_n \lambda_n + \tfrac{1}{2}\Delta^4\left[\beta_{12}\Big(\sum_n \lambda_n\Big)^2 + \beta_{345}\sum_n \lambda_n^2\right], \tag{5.16}$$

where we use the abbreviations $\beta_{12} = \beta_1 + \beta_2$ etc. The different states are characterized by the number of zero eigenvalues λ_n.

(1) Suppose that only one of the eigenvalues were finite: $\lambda_1 = \lambda$, $\lambda_2 = \lambda_3 = 0$. This is the so-called "polar" state because it is characterized by a single preferred direction both in spin space, $\boldsymbol{d}$, and orbital space, $\hat{\boldsymbol{w}}$:

$$\boldsymbol{d}(\boldsymbol{k}) = \Delta_0\hat{\boldsymbol{d}}(\hat{\boldsymbol{w}} \cdot \hat{\boldsymbol{k}}). \tag{5.17}$$

Remember that, in the absence of spin–orbit interaction, spin space and orbital space may be rotated with respect to each other without

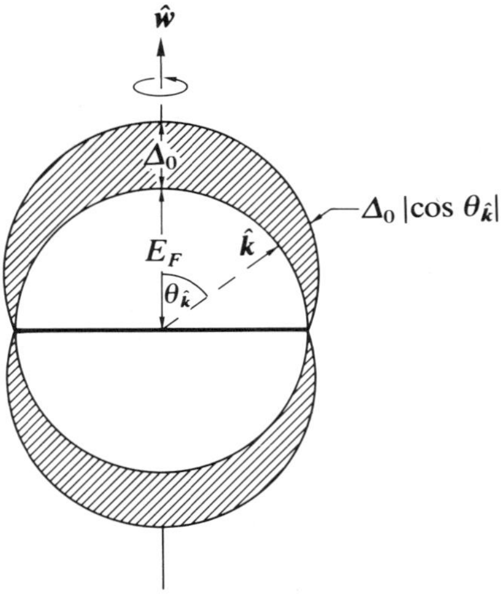

Figure 5.2 Schematic drawing of the energy gap of the polar state (5.17). The gap vanishes along the equatorial line. Note that E_F and Δ_0 are not drawn to scale.

changing the energy, so $\hat{\boldsymbol{d}}$ and $\hat{\boldsymbol{w}}$ need not be in any fixed relative configuration. The polar state has an energy gap $|\Delta_{\boldsymbol{k}}| = \Delta_0\,|\cos\theta_{\boldsymbol{k}}|$, where $\cos\theta_{\boldsymbol{k}} = \hat{\boldsymbol{w}}\cdot\hat{\boldsymbol{k}}$ (see Fig. 5.2). Its free energy is

$$F^{\text{polar}} = F_{\text{N}} - \frac{\alpha^2}{2\sum_i \beta_i}, \tag{5.18}$$

and $\kappa = \sum_i \beta_i/\beta_0$.

(2) The "planar" state is realized if two of the eigenvalues are nonzero. It follows from the symmetry of F in λ_1 and λ_2 that, for example, $\lambda_1 = \lambda_2 \equiv \lambda$, $\lambda_3 = 0$. Minimization of F yields

$$F^{\text{planar}} = F_{\text{N}} - \frac{\alpha^2}{2\beta_{12} + \beta_{345}}, \tag{5.19}$$

and $\kappa = (2\beta_{12} + \beta_{345})/2\beta_0$. The corresponding order parameter has no component in the third dimension, say along $\hat{\boldsymbol{l}}$; i.e. it is given by

$$\boldsymbol{d}(\boldsymbol{k}) = \Delta_0 \boldsymbol{R}[\hat{\boldsymbol{k}} - \hat{\boldsymbol{l}}(\hat{\boldsymbol{l}}\cdot\hat{\boldsymbol{k}})]. \tag{5.20}$$

Here $\boldsymbol{R}$ is an orthogonal 3×3 matrix, rotating spin space relative to orbital space. The energy gap has a torus-like shape, $|\Delta_{\boldsymbol{k}}| = \Delta_0\,|\sin\theta_{\boldsymbol{k}}|$, where $\cos\theta_{\boldsymbol{k}} = \hat{\boldsymbol{l}}\cdot\hat{\boldsymbol{k}}$.

(3) Finally, all three eigenvalues may be nonzero, and by symmetry $\lambda_1 = \lambda_2 = \lambda_3 \equiv \lambda$. The corresponding state is the BW state, already

discussed in Section 3.3. The free-energy minimum is given by

$$F^{\mathrm{BW}} = F_{\mathrm{N}} - \frac{\alpha^2}{2(\beta_{12} + \frac{1}{3}\beta_{345})}, \tag{5.21}$$

with $\kappa = (3\beta_{12} + \beta_{345})/3\beta_0$.

These three states exhaust the possibilities for case (i), i.e. for real $\boldsymbol{d}(\boldsymbol{k})$. By comparing the free-energy expressions (5.18), (5.19) and (5.21), one sees that F^{planar} lies between F^{polar} and F^{BW} for any values of the β_i, such that the planar state can never be the equilibrium state.

The order parameter for case (ii) takes the form

$$\boldsymbol{d}(\boldsymbol{k}) = (\boldsymbol{c} \cdot \boldsymbol{k})\hat{\boldsymbol{d}}, \tag{5.22}$$

with $\hat{\boldsymbol{d}}$ a real unit vector and $\boldsymbol{c}$ some vector with complex-valued components. The free-energy expression in terms of $\boldsymbol{c}$ is obtained from (5.4) as

$$F = F_{\mathrm{N}} + \tfrac{1}{3}\alpha(\boldsymbol{c} \cdot \boldsymbol{c}^*) + \tfrac{1}{18}[\beta_{13}\,|\boldsymbol{c} \cdot \boldsymbol{c}|^2 + \beta_{245}(\boldsymbol{c} \cdot \boldsymbol{c}^*)^2]. \tag{5.23}$$

In order to guarantee stability, we must assume that $\beta_{245} > 0$. Then, for $\beta_{13} > 0$, the minimum is reached if $|\boldsymbol{c} \cdot \boldsymbol{c}| = 0$, and hence $\boldsymbol{c} = \Delta_0(\hat{\boldsymbol{m}} + \mathrm{i}\hat{\boldsymbol{n}})$, where $\hat{\boldsymbol{m}}$ and $\hat{\boldsymbol{n}}$ are two real orthogonal unit vectors, $\hat{\boldsymbol{m}} \cdot \hat{\boldsymbol{n}} = 0$. This is the ABM state (or the axial state) discussed in Section 3.5. Its free energy is

$$F^{\mathrm{ABM}} = F_{\mathrm{N}} - \frac{\alpha^2}{2\beta_{245}}, \tag{5.24}$$

and $\kappa = \beta_{245}/\beta_0$.

For $\beta_{13} < 0$, the minimum of (5.23) occurs for real $\boldsymbol{c}$, which takes us back to the polar state. Note that the polar and the axial states are similar in that the energy gap is axially symmetric (see Figs. 3.4(b) and 5.1). This implies the existence of a single preferred direction in position space. On the other hand, their physical properties are quite different because only in the ABM state do the Cooper pairs have a finite angular-momentum projection.

So far we have only discussed unitary states. However, there is no a priori reason why nonunitary states should be unimportant. Such states may possess a spontaneous magnetization when particle–hole asymmetry is taken into account; this has not yet been observed. A detailed discussion of nonunitary states is beyond the scope of this book, but can be found in the literature (Barton and Moore 1974b). For p-wave pairing, there appear to be at least five stable states representing (local) minima of the free energy: three of these are unitary (the BW, ABM and polar states) and two are nonunitary (the β and A_1 states). The order-parameter structures of the latter two are given below.

In the β state

$$d_{\mu j} = \Delta_{\uparrow\uparrow}(\hat{d}_\mu + \mathrm{i}\hat{e}_\mu)\hat{l}_j, \tag{5.25}$$

where $\hat{\boldsymbol{d}}$ and $\hat{\boldsymbol{e}}$ are orthogonal unit vectors in spin space and $\Delta_{\uparrow\uparrow}$, as well as

$\Delta_{\uparrow\downarrow} = \Delta_{\downarrow\uparrow}$ and $\Delta_{\downarrow\downarrow}$ appearing below, are the matrix elements of the gap matrix $\boldsymbol{\Delta}$ (see also (6.15), (6.54)). This state may be obtained from the ABM state by interchanging spin and orbital spaces. Only spin-up pairs are formed, and the energy gap is found to be proportional to $|\cos\theta_{\boldsymbol{k}}| = \hat{\boldsymbol{k}}\cdot\hat{\boldsymbol{l}}$ as in the polar state.

In the A_1 state

$$d_{\mu j} = \tfrac{1}{2}\Delta_{\uparrow\uparrow}(\hat{d}_\mu + \mathrm{i}\hat{e}_\mu)(\hat{m}_j + \mathrm{i}\hat{n}_j) \tag{5.26}$$

(see also (6.15), (6.56)). Again only spin-up pairs are formed. The spin and orbital parts of $\boldsymbol{d}(\boldsymbol{k})$ separate as in the ABM state. In fact, as will be discussed in detail in Section 5.4, this state describes the so-called A_1 phase, which appears between the normal phase and the A phase in a strong magnetic field.

The five abovementioned states have a remarkable property: their structure, as characterized by the order parameter, is independent of the parameters β_i within their region of stability. States possessing this property have been called "inert" (Barton and Moore 1974a,b), in contrast with the "noninert" ones, whose structure changes continuously as a function of the β_i. Other inert states corresponding to stationary points of the free energy have been found to be saddle points; hence they are unstable (Jones 1977).

A systematic classification of possible inert and noninert states based on the symmetry properties of the free energy of superfluid ^{3}He has been derived by Bruder and Vollhardt (1986). It will be discussed in Chapter 6 within the context of broken symmetries. This analysis yields several more states, two of which are the A_2 and B_2 phases—both realized in superfluid ^{3}He in a magnetic field (see Section 5.4). These states are noninert. Their order parameters are given as follows (see also (6.48), (6.40)).

In the A_2 state

$$d_{\mu j} = (A\hat{d}_\mu + B\hat{e}_\mu + C\hat{f}_\mu)(\hat{m}_j + \mathrm{i}\hat{n}_j), \tag{5.27}$$

where

$$A = \tfrac{1}{2}(\Delta_{\uparrow\uparrow} + \Delta_{\downarrow\downarrow}), \quad B = \tfrac{1}{2}\mathrm{i}(\Delta_{\uparrow\uparrow} - \Delta_{\downarrow\downarrow}), \quad C = 0.$$

Here $\hat{\boldsymbol{f}} = \hat{\boldsymbol{d}}\times\hat{\boldsymbol{e}}$ is a third axis in spin space, perpendicular to both $\hat{\boldsymbol{d}}$ and $\hat{\boldsymbol{e}}$, defined in analogy with the unit vector $\hat{\boldsymbol{l}} = \hat{\boldsymbol{m}}\times\hat{\boldsymbol{n}}$ in orbital space. Note that $\hat{\boldsymbol{f}}$ indicates the direction of spin polarization $\boldsymbol{S}$ of the Cooper pairs. In equilibrium we expect $\boldsymbol{S}$ and hence $\hat{\boldsymbol{f}}$ to point along the direction of the magnetic field (cf. the remarks made on the choice of the spin quantization axis in Chapter 2, (2.17)). The sets of unit vectors $(\hat{\boldsymbol{d}}, \hat{\boldsymbol{e}}, \hat{\boldsymbol{f}})$ and $(\hat{\boldsymbol{m}}, \hat{\boldsymbol{n}}, \hat{\boldsymbol{l}})$ are orthogonal triads in spin and orbital spaces with arbitrary relative orientation. In Chapter 6 it will be shown that these triads are coupled by the dipole forces between the ^{3}He atoms such that in equilibrium $\hat{\boldsymbol{d}}$ and $\hat{\boldsymbol{l}}$ are oriented in parallel.

In the B_2 state

$$d_{\mu j} = \mathrm{e}^{\mathrm{i}\phi}[A(\hat{d}_\mu\hat{m}_j + \hat{e}_\mu\hat{n}_j) - B(\hat{d}_\mu\hat{n}_j - \hat{e}_\mu\hat{m}_j) + C\hat{f}_\mu\hat{l}_j], \tag{5.28}$$

where

$$A = \tfrac{1}{2}(\Delta_{\uparrow\uparrow} + \Delta_{\downarrow\downarrow}), \quad B = \tfrac{1}{2}\mathrm{i}(\Delta_{\uparrow\uparrow} - \Delta_{\downarrow\downarrow}), \quad C = \Delta_{\uparrow\downarrow}.$$

5.2.2 Relative stability of model states

In order to discuss the relative stability of model states, it is useful to invoke the concept of a parameter space of the β_i. The regions of stability of the model states form subspaces that completely cover this parameter space. In this space the actual physical system is represented by a single point $\boldsymbol{B} = (\beta_1, \beta_2, \beta_3, \beta_4, \beta_5)$. In the weak-coupling model it has the fixed coordinates $\boldsymbol{B}_{\rm WC} = \frac{6}{5}\beta_0(-\frac{1}{2}, 1, 1, 1, -1)$. Corrections to the weak coupling model will lead to a dependence of the position of $\boldsymbol{B}$ on external parameters (e.g. pressure). As these "strong-coupling" corrections can be expected to be small (owing to the smallness of T_c/T_F), the point $\boldsymbol{B}$ will stay relatively close to $\boldsymbol{B}_{\rm WC}$. In Section 5.3 we shall estimate the corrections $\Delta\beta_i$ as $0 < |\Delta\beta_i/\beta_i| \lesssim 0.5$. We can therefore confine our discussion to model states whose regions of stability are in the vicinity of this point. A detailed investigation of the boundaries of the stability region shows that $\boldsymbol{B}_{\rm WC}$ lies within the region of the BW state but close to that of the ABM state. To reach other stability regions (the closest one being that of the polar state, as indicated by the value of the total fourth-order contribution K; Table 5.1) requires the β_i to deviate from $\beta_i^{\rm WC}$ by amounts less than unity. Therefore the BW and ABM states are the most likely candidates for stable states in superfluid ^{3}He. To exemplify such a stability analysis, we now discuss the stability regions of the four unitary states.

Among the real states, the stability is governed by the relative size of β_{12} and β_{345}. If both $\beta_{12} > 0$ and $\beta_{345} > 0$, or else if $\beta_{12} < 0$ and $\beta_{345} > 3\,|\beta_{12}|$, the BW state is lowest in free energy. For $\beta_{345} < 0$ but $\beta_{12} > 0$, such that $-\beta_{12} < \beta_{345} < 0$, the polar state wins. All other regions are excluded by stability considerations. There is no regime where the planar state is stable. The ABM state is lower in free energy than the BW state if $\beta_{13} > 0$ and $0 < \beta_{245} < \beta_{12} + \frac{1}{3}\beta_{345}$. From these inequalities, using the weak-coupling values of the relevant parameter combinations, $\beta_{345}^{\rm WC} = 2\beta_{12}^{\rm WC} = \beta_{245}^{\rm WC} =$

Table 5.1 Invariants (5.3) of the free energy for several phases and the total fourth-order contribution $K = \sum_i \beta_i I_i / |\beta_1^{\rm WC}|$ using different sets of the parameters β_i as explained in the text.

	BW	ABM	Polar	β	A_1
I_1	1	0	1	0	0
I_2	1	1	1	1	1
I_3	$\frac{1}{3}$	0	1	1	1
I_4	$\frac{1}{3}$	1	1	1	1
I_5	$\frac{1}{3}$	1	1	0	0
$K_{\rm WC}$	$\frac{5}{3}$	2	3	6	4
$K_{\rm SF}$	$\frac{5}{3} + \frac{2}{3}\delta$	$2 - 2\delta$	$3 - 2\delta$	6	4
$K_{\rm SC}$, $P = 12$ bar	1.382	1.526	2.375	5.604	3.721
$K_{\rm SC}$, $P = 34.4$ bar	1.193	1.130	1.933	5.476	3.599

$2\beta_{13}^{WC} = \frac{6}{5}\beta_0$ and $\beta_{12}^{WC} + \frac{1}{3}\beta_{345}^{WC} = \beta_0$, one may conclude that it takes only a small change of parameters to enter into the region of stability of the ABM state. What then could be the possible nature of such changes?

It appears that for the main part they are due to a conceptually simple effect, first discussed by Anderson and Brinkman (1973): the fact that the pair interaction itself is modified by the pair correlations and thus acquires spin dependence and momentum dependence favouring certain gap structures and suppressing others. In Section 5.3 we discuss a simple model illustrating the basic mechanism. Unfortunately, it turns out that a theory that accounts fully for this effect is outside the realm of the quasiparticle picture and hence is outside the capability of present theories.

A different approach to strong-coupling effects was developed by Rainer and Serene (1976, 1980); see also Serene and Rainer (1977, 1978, 1979, 1983). They performed a systematic expansion in the small parameter T_c/T_F and showed that the first correction term may be entirely expressed in terms of the quasiparticle scattering amplitude for the normal state. The latter is to some extent experimentally accessible (see Chapter 2); this allows determination of the contribution of different physical processes such as spin fluctuations and transverse current fluctuations at small and large q. Unfortunately the next higher correction can no longer be expressed solely in terms of quasiparticle properties and hence cannot be calculated. An adequate account of this theory would carry us beyond the scope of this book. Suffice it to say that (i) the coefficient of the T_c/T_F correction term is found to be strongly enhanced (weakening the convergence of the T_c/T_F expansion), and (ii) a quantitative evaluation of the strong-coupling correction to the free energy (based on improved model forms of the quasiparticle scattering amplitude) yields good agreement with experiment (Sauls and Serene 1981a,b; see also Kuroda and Nagi 1976b,c, 1978, Bedell 1982).

It is nevertheless instructive to have a look at a simple spin-fluctuation model due to Leggett (1975a), which is in the spirit of the original derivation of the spin-fluctuation feedback effect by Anderson and Brinkman (1978) and which can qualitatively account for the stability of the ABM state. An entirely different explanation for the existence of another anisotropic phase besides the Balian–Werthammer states was proposed by Berezinskii (1974). It involves an order-parameter function $\Delta(\omega, \boldsymbol{p})$ that is an odd function of frequency ω and an even function of $\boldsymbol{p}$, and makes explicit use of the large spin susceptibility of the normal liquid.

5.3 SPIN FLUCTUATIONS AND THE STABILITY OF THE ABM STATE

It has been stressed in Chapter 2 that the effective quasiparticle interaction in ^3He differs greatly from the bare interaction potential owing to polarization effects of the medium. One may in fact (somewhat arbitrarily) break up the interaction into a direct part and a polarization-induced part:

$$V = V_{\text{direct}} + V_{\text{pol}}. \tag{5.29}$$

One of the more important polarization effects in liquid ^{3}He is due to spin fluctuations at small $q = |\boldsymbol{q}|$. However, this is by no means the only polarization effect, as the systematic treatment of Rainer and Serene (1976) demonstrates. Rather, a considerable contribution to the strong-coupling corrections and hence the induced interaction is due to nonmagnetic fluctuations (e.g. of the transverse current) at higher q values. This is also suggested by the results obtained within the Gutzwiller–Hubbard lattice-gas model (see Section 2.5.4), where the effect of localization (i.e. large $m^*/m = 1 + \frac{1}{3}F_1^s$) is found to be even more important for the static spin susceptibility than spin-fluctuation effects. Nevertheless, we restrict the present discussion to the effect of spin fluctuations, since this is easiest to visualize. For this, we recall that the molecular field enhances the spin polarization considerably because the Landau parameter $F_0^a \approx -\frac{3}{4} < 0$ such that $(1 + F_0^a)^{-1}$ is large. A detailed discussion of the effective interaction in liquid ^{3}He, including, for example, the role of density fluctuations, has been given by Leggett (1975a).

Suppose we add a quasiparticle with spin S to the system in equilibrium. The spin S polarizes the surrounding medium via the quasiparticle interaction (2.7) as if there were a magnetic field H' of magnitude

$$H' = -(\tfrac{1}{2}\mu_0\hbar)^{-1} f_0^a S. \tag{5.30}$$

The spin of a second quasiparticle nearby will be polarized by this effective field. It leads to a polarization S' as given by the magnetic susceptibility of the liquid, viz

$$S' = \gamma^{-1}\chi H' \tag{5.31}$$

The change in polarization of the second quasiparticle consequently acts back on the first one, giving rise to a shift in its energy of magnitude

$$\delta\epsilon = (\tfrac{1}{2}\hbar)^{-1} f_0^a S' = -\mu_0^{-2}(f_0^a)^2\chi. \tag{5.32}$$

This energy shift may be interpreted as an interaction potential between the quasiparticles. Specifying the initial and final spin projections of the first and second quasiparticles by (α, δ) and (β, γ) respectively, the pair-interaction matrix element derived from (5.32) takes the form

$$\langle \boldsymbol{k}\alpha, -\boldsymbol{k}\beta\, |V_{\text{pol}}|\, \boldsymbol{k}'\delta, -\boldsymbol{k}'\gamma \rangle = -\mu_0^{-2}[f_0^a(\boldsymbol{k} - \boldsymbol{k}')]^2 \sum_{\mu\nu} (\sigma_\mu)_{\alpha\delta}\chi_{\mu\nu}(\boldsymbol{k} - \boldsymbol{k}')(\sigma_\nu)_{\beta\gamma}. \tag{5.33}$$

The momentum dependence of the susceptibility $\boldsymbol{\chi}(\boldsymbol{q})$ and the interaction $f_0^a(\boldsymbol{q})$ comes in because in the interaction process the transferred momentum

$\boldsymbol{k}-\boldsymbol{k}'$ is carried away by the exchange of a spin fluctuation, which is described by $\boldsymbol{\chi}(\boldsymbol{k}-\boldsymbol{k}')$. Retardation effects (e.g. a frequency dependence of $\boldsymbol{\chi}$) are known to give rise to important strong-coupling corrections in superconductors. However, they can be expected to affect different model states in the same way. Therefore they may be neglected in the present context of a relative stability analysis.

For the purpose of calculating the strong-coupling corrections to the Ginzburg–Landau free energy, we need only the lowest-order term in the effective interaction. This is given by expanding the spin susceptibility (3.106) as

$$\chi_{\mu\nu}(\boldsymbol{q})=\delta_{\mu\nu}\chi_{\mathrm{N}}(\boldsymbol{q})+\left[1+\frac{F_0^{\mathrm{a}}(\boldsymbol{q})\chi_{\mathrm{N}}^0(\boldsymbol{q})}{\chi_{\mathrm{N}}^0}\right]^{-2}\delta\chi_{\mu\nu}^0(\boldsymbol{q}), \tag{5.34}$$

where

$$\chi_{\mathrm{N}}^0(\boldsymbol{q})\approx\frac{\chi_{\mathrm{N}}^0}{1+q^2/12k_{\mathrm{F}}^2} \tag{5.35}$$

is the $\boldsymbol{q}$-dependent Fermi-gas susceptibility.

It can be shown that the $\boldsymbol{q}$ dependence of the spin-susceptibility tensor is approximately equal for all spin components, and may be split off; hence

$$\delta\chi_{\mu\nu}^0(\boldsymbol{q})\approx g(\boldsymbol{q})\,\delta\chi_{\mu\nu}^0(\boldsymbol{q}=0). \tag{5.36}$$

The form factor $g(\boldsymbol{q})$ describes the spatial decay of pair correlations, which occurs on the scale of the inverse coherence length, such that

$$g(\boldsymbol{q})\approx[1+bq\xi_0+O(q^2/k_{\mathrm{F}}^2)]^{-1}, \tag{5.37}$$

where b is a coefficient of order unity. This result follows from a calculation of $\chi_{\mu\nu}^0\,(\boldsymbol{q})$ along the lines of the derivation of $\chi_{\mu\nu}(\boldsymbol{q}=0)$ discussed in Section 3.4 (Anderson and Brinkman 1975, 1978).

The change in the spin-susceptibility tensor for $\boldsymbol{q}=0$ can be obtained from (3.103) by expanding to order Δ^2:

$$\delta\chi_{\mu\nu}^0(\boldsymbol{q})\approx g(\boldsymbol{q})\chi_{\mathrm{N}}^0[\tfrac{7}{4}\zeta(3)](\pi k_{\mathrm{B}}T_{\mathrm{c}})^{-2}\langle\mathrm{Re}\,[d_\mu(\boldsymbol{k})d_\nu^*(\boldsymbol{k})]\rangle_{\hat{\boldsymbol{k}}}. \tag{5.38}$$

With the aid of (5.33)–(5.38), the change in the lth angular-momentum component of the pair interaction induced by pair correlations may be expressed as

$$\begin{aligned}\langle\alpha\beta\,|\delta V_l^{\mathrm{pol}}|\,\gamma\delta\rangle&\equiv\int_0^1\mathrm{d}(\cos\theta)\,P_l(\cos\theta)\langle\boldsymbol{k}\alpha,\,-\boldsymbol{k}\beta|\,\delta V^{\mathrm{pol}}\,|\boldsymbol{k}'\delta,\,-\boldsymbol{k}'\gamma\rangle\\&=\Lambda_l(\chi_{\mathrm{N}}^0)^{-1}\sum_{\mu\nu}(\sigma_\mu)_{\alpha\delta}\,\delta\chi_{\mu\nu}(\sigma_\nu)_{\beta\gamma},\end{aligned} \tag{5.39}$$

where $\cos\theta=\hat{\boldsymbol{k}}\cdot\hat{\boldsymbol{k}}'$, and Λ_l is defined as

$$\Lambda_l=-[2N(0)]^{-1}\int_0^{2k_{\mathrm{F}}^2}\frac{\mathrm{d}q^2}{2k_{\mathrm{F}}^2}g(q)\left[\frac{F_0^{\mathrm{a}}(q)}{1+F_0^{\mathrm{a}}(q)\chi_{\mathrm{N}}^0(q)/\chi_{\mathrm{N}}^0}\right]^2P_l\left(1-\frac{q^2}{2k_{\mathrm{F}}^2}\right). \tag{5.40}$$

Here we have made use of the fact that for particles on the Fermi surface the magnitude of the transferred momentum $\boldsymbol{q}$ is related to the angle θ by $q^2 = 2k_F^2(1 - \cos\theta)$.

The $\boldsymbol{q}$ dependence of the factor multiplying the Legendre function $P_l(x)$ under the $\boldsymbol{q}$ integral in (5.40) is smooth and presumably weak, if we assume the $\boldsymbol{q}$-dependent Landau parameter $F_0^a(\boldsymbol{q})$ to be a negative, slowly increasing function of $\boldsymbol{q}$. This would indicate that Λ_l is rapidly decreasing with l for $l > 1$, and that a sizeable spin fluctuation feedback effect should only be expected for $l = 1$ pairing.

A rough estimate of the magnitude of Λ_l can be obtained by approximating $F_0^a(\boldsymbol{q})$ and $\chi_N^0(\boldsymbol{q})$ by their $\boldsymbol{q} = 0$ values, and approximating the form factor by $g(\boldsymbol{q}) \approx (bq\xi_0)^{-1}$ for $q > \xi_0^{-1}$. In the case $l = 1$ the result is $|\Lambda_1| \approx 2[N(0)b\xi_0 k_F]^{-1}$.

The change in the free-energy functional induced by a change in the interaction is best obtained from the gap equation, where a correction term

$$\delta\Delta_{\boldsymbol{k}\alpha\beta} = -\sum_{\boldsymbol{k}'\gamma\delta} \langle\alpha\beta|\,\delta V_l^{\text{pol}}\,|\gamma\delta\rangle(2l+1)P_l(\hat{\boldsymbol{k}}\cdot\boldsymbol{k}')\frac{\tanh(\xi'/2k_B T_c)}{2\xi'}\Delta_{\boldsymbol{k}'\delta\gamma} \tag{5.41}$$

must be added to the right-hand side of the weak-coupling result (3.44). Performing the integral over $\xi_{\boldsymbol{k}'}$ with the help of (3.47), and taking spin and $\boldsymbol{k}$ components as (μ, j), the change in the order parameter $A_{\mu j}$ is found as

$$\delta A_{\mu j} = 2\Delta^2\beta_0\left|\frac{\Lambda_1}{N(0)V_1}\right|[\text{tr}\,(\boldsymbol{AA}^+)A_{\mu j} - (\boldsymbol{AA}^+\boldsymbol{A})_{\mu j} - (\boldsymbol{A}^*\boldsymbol{A}^{\text{T}}\boldsymbol{A})_{\mu j}], \tag{5.42}$$

where β_0 is defined in (3.56) and V_1 is the $l = 1$ pair interaction constant. If we compare this expression with the correction to the gap equation induced by small changes in the parameters β_i of the Ginzburg–Landau functional,

$$\begin{aligned}\delta A_{\mu j} = \Delta^2\,|V_1|\,\{&\delta\beta_1\,\text{tr}\,(\boldsymbol{AA}^{\text{T}})\,A_{\mu j}^* + \delta\beta_2\,\text{tr}\,(\boldsymbol{AA}^+)A_{\mu j}\\ &+ \delta\beta_3\,[(\boldsymbol{AA}^{\text{T}})A^*]_{\mu j} + \delta\beta_4\,(\boldsymbol{AA}^+\boldsymbol{A})_{\mu j} + \delta\beta_5\,(\boldsymbol{A}^*\boldsymbol{A}^{\text{T}}\boldsymbol{A})_{\mu j}\},\end{aligned} \tag{5.43}$$

we obtain the following corrections to the β_i:

$$\left.\begin{aligned}&\delta\beta_2 = -\delta\beta_4 = -\delta\beta_5 = \tfrac{6}{5}\beta_0\delta,\\ &\delta\beta_1 = \delta\beta_3 = 0,\end{aligned}\right\} \tag{5.44}$$

where the small positive dimensionless parameter δ is defined as

$$\delta = \frac{5}{3}\frac{|\Lambda_1|}{|V_1|^2\,N(0)} = \frac{6.3}{b}\frac{1}{[V_1 N(0)]^2}\frac{T_c}{T_F}. \tag{5.45}$$

Assuming a value of $|V_1 N(0)| \approx 0.2$, a rough estimate at melting pressure yields $\delta \approx 0.4b^{-1}$, with b defined in (5.37). The pressure dependence of δ is governed by the factor $(k_F\xi_0)^{-1} \approx T_c/T_F$, showing that δ increases with increasing pressure. Note that the coefficient of T_c/T_F in (5.45) is quite large.

Calculations of the parameters β_i within more refined spin-fluctuation models have been presented by Brinkman *et al.* (1974a), Tewordt *et al.* (1975a, 1978), Kuroda (1974, 1975), Kuroda and Nagi (1975, 1976a), Layzer and Fay (1974), Fay (1975) and Levin and Valls (1979a,b).

In Table 5.1 the values of the invariants I_i in the five inert p-wave states are tabulated. Also listed is the quantity $K = \sum_i \beta_i I_i / |\beta_1^{\mathrm{WC}}|$ for the following cases:

(i) the weak-coupling model (K_{WC}) with β values given in (5.8);
(ii) the simple spin-fluctuation model (K_{SF}) with β_i given by (5.44);
(iii) the strong-coupling model (K_{SC}) according to Sauls and Serene (1981), with $\{\beta_i/|\beta_1^{\mathrm{WC}}|\} = (-1.034, 1.920, 1.883, 1.801, -2.195)$ at $P = 12$ bar and $(-1.074, 1.897, 1.877, 1.702, -2.469)$ at $P = 34.4$ bar.

One finds that for values of δ greater than $\frac{1}{8}$ the ABM state is lower in free energy than the BW state. From the estimate of δ presented above, it seems likely that the ABM state is barely stable at high pressure, but is almost certainly unstable at low pressure. This is also borne out by the more detailed strong-coupling results shown in the table.

At lower temperatures the fourth-order Ginzburg–Landau expansion is no longer adequate. A rough idea of which of the two states in question is stable in the limit of low temperatures may be obtained by comparing the relative free-energy difference $\delta F = (F^{\mathrm{ABM}} - F^{\mathrm{BW}})/(F_{\mathrm{N}} - F^{\mathrm{BW}})$ as calculated in weak-coupling theory near T_{c} and in the low- temperature limit. In the Ginzburg–Landau regime δF can be calculated using (5.21) and (5.24): $\delta F = \frac{1}{6} \approx 0.167$. In the limit $T \to 0$ the weak-coupling free energy is given by (3.62). Using the $T = 0$ values of Δ, given by (3.72) and (3.75), one finds $\delta F \approx 1 - \frac{2}{3}(2.02/1.76)^2 = 0.113$. Hence in weak-coupling theory the BW state is seen to become less stable at lower temperatures in comparison with the ABM state. Since at high pressure the ABM state is only barely stabilized by strong-coupling effects near T_{c}, as argued above, it is surprising that the experiment seems to indicate that at low temperatures the BW state is always the stable one. This clearly indicates that strong-coupling corrections have an important dependence on temperature. Only above a critical pressure P_{c} (experimentally found to be about 21 bar) and at not too low temperatures is the ABM state stable. Upon cooling, a first-order transition to the BW state takes place at the thermodynamic transition temperature T_{AB}. To calculate T_{AB} (or rather the ratio $T_{\mathrm{AB}}/T_{\mathrm{c}}$, where T_{c} is the transition temperature for p-wave pairing), it is necessary to include sixth-order terms in Δ in the Ginzburg–Landau expression for the free energy. Such a calculation has been performed within the spin-fluctuation model by Brinkman *et al.* (1974a); see also Kuroda and Nagi (1975, 1976a).

The arguments given above strongly suggest that the ABM and BW states may be identified with the A and B phases of superfluid ^{3}He respectively. Historically, the most convincing argument for this identification came from the experimental observation of the predicted nuclear magnetic resonance shifts (see Chapter 8). Furthermore, the smoothness of the T_c curve at the polycritical point strongly suggests that the A and B phases belong to the same l manifold (i.e. $l = 1$). Indeed, T_c curves of phases with different l would be expected to intersect at this point with a finite angle (Leggett 1974b). So far essentially all experimental properties of the A and B phases are consistent with the calculated properties of the ABM and BW states. In the following we shall therefore use the terms "ABM state" and "A phase" as well as "BW state" and "B phase" synonymously.

The theory presented above is based on the mean-field approximation for the pair amplitude. According to a criterion first discussed by Ginzburg (1960), this is an excellent approximation except for a temperature region of order $(k_F\xi_0)^{-4} \approx (T_c/T_F)^4$ below T_c. Clearly, in the present case this region is extremely small (about $10^{-13}T_c$) owing to the large spatial extent of the Cooper pair. An estimate of the fluctuation contribution to the spin susceptibility has been given by Patton (1974), and fluctuation effects in small volumes of ^{3}He have been studied by Efetov and Salomaa (1981). The general effect of fluctuations on the mean-field solution has been considered by Jones *et al.* (1976), Bailin *et al.* (1977) and Sokolov (1979, 1980, 1983) using renormalization-group techniques. In an extremely narrow temperature regime at T_c, transitions that are second-order in mean-field theory were found to change to being weakly first-order.

As already mentioned, there is substantial supercooling and some superheating at the A–B transition. While the formation of small regions of A phase in the B phase upon warming is possibly related to little pockets of A phase in cavities or corners of the walls of the containing vessel, the formation of B phase upon cooling must occur spontaneously, triggered either by thermal fluctuations or by quantum-mechanical tunnelling. The free-energy barrier that the system must surmount in order to form a droplet of B phase in a region of A phase is determined by the competition between the free energies of the volume and of the corresponding surface of a droplet of B phase. The surface energy σ_{AB} has been measured by Osheroff and Cross (1979) and has been calculated variationally (Cross 1977b, Kaul and Kleinert 1980) and by numerical integration (Schopohl 1987, Salomaa 1988); this yields $\sigma_{AB} \approx 0.7\xi(T)f_B$, where f_B is the energy density of the bulk B phase. This is in very good agreement with the experimental result. The free-energy barrier is thus estimated to be of order $10^6\,k_BT$. On the basis of this estimate, the spontaneous formation of bubbles of B phase is completely ruled out: the system would never be able to make a first-order phase transition from the A phase to the B phase, no matter how low the temperature. Estimates of the quantum tunneling probability through the barrier likewise yield times that are far too long (Bailin and Love 1980). This has led to speculations that the phase transition may be triggered by an external source of energy, as could for instance be provided by cosmic-ray muons decaying within the sample region. The extremely localized amount of energy liberated in such a decay has been estimated to be sufficient to lift the system over the barrier (Leggett 1984).

A completely different mechanism for the nucleation of B phase has been proposed by Hong (1988), who postulates the spontaneous creation of topologically stable "Q balls" (Coleman 1985) at, or below, the thermodynamic A–B transition. These Q balls are time-dependent nonlinear solutions of the Ginzburg–Landau equations, which grow and transform into B phase upon cooling. The stability of these objects at finite temperature has yet to be clarified.

However, experimentally, the nucleation of the B phase upon cooling through the A phase seems to occur above a well-defined threshold temperature (Hakonen *et al.* 1983a). This behaviour suggests that the supercooled A phase is absolutely unstable below this temperature. Also, the distribution of nucleation temperatures observed by Hakonen *et al.* (1985a) is much narrower than that predicted from the original cosmic-ray hypothesis (Leggett 1984, 1985). The latter theory has since been refined and now gives a distribution consistent with experiment (Leggett and Yip 1989). However, a direct experimental verification of the cosmic-ray hypothesis is still lacking.

5.4 EFFECTS OF A MAGNETIC FIELD

In the presence of a static uniform magnetic field $\boldsymbol{H}$ there are two more invariants that have to be added to the free-energy functional (5.4) (Mermin and Stare 1973). There is a term quadratic in $\boldsymbol{H}$,

$$F^{(2)} = 3\tilde{\alpha}\Delta^2(T) \sum_{\mu\nu} \sum_{j} H_\mu A_{\mu j} H_\nu A^*_{\nu j}, \tag{5.46}$$

whose physical origin may be traced back to the dependence of the magnetic susceptibility on the order parameter. Quite generally, the free energy is lowered in a magnetic field (to order H^2) by

$$\Delta F = -\tfrac{1}{2}\boldsymbol{H}\boldsymbol{\chi}\boldsymbol{H}, \tag{5.47}$$

where $\boldsymbol{\chi}$ is the magnetic-susceptibility tensor. Substituting the weak-coupling form (3.102) for $\boldsymbol{\chi}$, expanded to second order in Δ, and subtracting the normal-state contribution $\Delta F_N = -\frac{1}{2}\chi_N H^2$, one is left with a contribution of precisely the form (5.46), with a coefficient

$$\tilde{\alpha} = \tfrac{2}{3}\beta_0\left(\frac{\mu_0}{1+F_0^a}\right)^2, \tag{5.48}$$

with β_0 given by (3.56).

It is interesting to note that there is yet another invariant, which is linear in $\boldsymbol{H}$:

$$F^{(1)} = \mathrm{i}\eta \sum_{\mu\nu\lambda} \sum_{j} \epsilon_{\mu\nu\lambda} H_\mu A_{\nu j} A^*_{\lambda j}. \tag{5.49}$$

The factor i makes the expression real for real η. As this term did not appear in the calculation of the susceptibility within the weak-coupling model, it must be of higher order in T_c/T_F. The origin of such a contribution can be traced back to the gap equation, if one takes into account the tiny shift in the Fermi levels of up- and down-spin quasiparticles induced by a magnetic field. The pair interaction and the density of states at the Fermi level will then be slightly different for opposite spins—an effect called "particle–hole asymmetry". This effect is only important in cases where the spin susceptibility is otherwise unaffected in lowest order in T_c/T_F, as is the case with the A phase, which assumes equal-spin-pairing form in a magnetic field (see Section 3.5).

Let us therefore discuss the A phase first. Assuming that the only effect of the magnetic field is to change the population of ($\uparrow\uparrow$) and ($\downarrow\downarrow$) Cooper pairs, while leaving the orbital structure unchanged, we make the following ansatz for the spin-matrix gap parameter (5.5a)

$$\boldsymbol{\Delta}_j = \mathrm{i}\begin{pmatrix} \Delta_{\uparrow\uparrow} & 0 \\ 0 & \Delta_{\downarrow\downarrow} \end{pmatrix} (\hat{m}_j + \mathrm{i}\hat{n}_j). \tag{5.50}$$

The corresponding order parameter $d_{\mu j}$ is of course still a separable form in orbital and spin indices. The essential difference is that the vector $\boldsymbol{d}$ now becomes complex,

$$d_{\mu j} = [\tfrac{1}{2}(\Delta_{\uparrow\uparrow} + \Delta_{\downarrow\downarrow})\hat{d}_\mu + \tfrac{1}{2}\mathrm{i}(\Delta_{\uparrow\uparrow} - \Delta_{\downarrow\downarrow})\hat{e}_\mu](\hat{m}_j + \mathrm{i}\hat{n}_j), \tag{5.51}$$

reflecting the finite spin polarization of Cooper pairs in a magnetic field. This state is referred to as the A_2 phase (see also (5.27)). Note that in the special case of zero magnetic field one has $\Delta_{\uparrow\uparrow} = \Delta_{\downarrow\downarrow} = \Delta_0$, and therefore (5.51) reduces to (3.68), i.e. the order parameter of the A phase. For $\Delta_{\downarrow\downarrow} = 0$, (5.51) yields the order parameter of the A_1 phase (see below). Since all spin components in the A_2 phase have the same orbital structure, the gap parameters of the A, A_1 and A_2 phases all have the same $\hat{\boldsymbol{k}}$ dependence, i.e. the axial structure given by (3.69).

Within weak-coupling theory, the up- and down-spin systems are completely independent, at least as far as the equilibrium properties are concerned. In particular, each subsystem has its own transition temperature T_c^0, given by the standard expression

$$T^0_{c\uparrow,\downarrow} = 1.13\frac{\epsilon_c}{k_B}\exp\left(-\frac{1}{\lambda_1^{\uparrow,\downarrow}}\right), \tag{5.52}$$

where the coupling constant $\lambda_1^{\uparrow,\downarrow}$ incorporates the changes induced by the shift of the Fermi levels:

$$\lambda_1^{\uparrow,\downarrow} = \lambda_1 \pm \mu_0 H \frac{\partial \lambda_1}{\partial E_F}, \quad \lambda_1 = N(0)\,|V_1|. \tag{5.53}$$

A rough estimate of the coefficient of $\mu_0 H$ in (5.53) is obtained from the variation of the free-fermion density of states $N(0) \propto E_F^{1/2}$ with energy, i.e. $\partial\lambda_1/\partial E_F = O(\lambda_1/E_F)$ (see also a calculation based on the paramagnon model by Levin (1975) and Levin and Valls (1981)). A more refined calculation of the linear T_c splitting within the so-called "induced interaction model" (see Chapter 2) has been presented by Bedell and Quader (1983, 1984) (see also Quader and Bedell 1985, Bedell and Sanchez-Castro 1986). A splitting of the A transition in a magnetic field of the order of magnitude predicted by (5.52) is observed experimentally, as discussed in Chapter 4. Pair formation takes place first for spin-up quasiparticles, until at a temperature $T_{c\downarrow}$ spin-down Cooper pairs are also formed (Ambegaokar and Mermin 1973). Note that we choose the spin-quantization axis to point in the opposite direction to the magnetic field (see the remarks preceding (2.17)).

In order to investigate the thermodynamic properties of the A phase in a magnetic field, i.e. of the A_2 phase, including strong-coupling effects, it is advantageous to make use of the free-energy functional in terms of the order parameter in spin-matrix form $\mathbf{\Delta}_j$, as defined in (5.5). The magnetic-field-dependent terms (5.46) and (5.49) can be cast into the form

$$\left.\begin{aligned} F^{(2)} &= -\tfrac{1}{4}\tilde{\alpha} \sum_{\alpha\beta} H_\alpha H_\beta \,\mathrm{tr}_\sigma\, (\boldsymbol{\sigma}_\alpha \mathbf{\Delta}_j [\boldsymbol{\sigma}_\beta, \mathbf{\Delta}_j^+]) \\ F^{(1)} &= \tfrac{1}{2}\eta \sum_{\alpha} H_\alpha \,\mathrm{tr}_\sigma\, (\boldsymbol{\sigma}_\alpha \mathbf{\Delta}_j \mathbf{\Delta}_j^+) \end{aligned}\right\} \tag{5.54}$$

where summation over j is implied. For the A_2-phase gap parameter (5.50), $F^{(2)}$ vanishes in accordance with the fact that to lowest order in T_c/T_F the magnetic susceptibility is equal to that of the normal state (ESP state).

Substituting the gap form (5.50) into (5.7) and (5.54), one arrives at a free-energy functional in terms of $\Delta_{\uparrow\uparrow}$ and $\Delta_{\downarrow\downarrow}$, of the form

$$F(H) = F_N(H) + \tfrac{1}{3}\alpha_\uparrow \Delta_{\uparrow\uparrow}^2 + \tfrac{1}{3}\alpha_\downarrow \Delta_{\downarrow\downarrow}^2 + \tfrac{1}{18}[\beta_{24}(\Delta_{\uparrow\uparrow}^4 + \Delta_{\downarrow\downarrow}^4) + 2(\beta_{24} + 2\beta_5)\Delta_{\uparrow\uparrow}^2 \Delta_{\downarrow\downarrow}^2], \tag{5.55}$$

where $\alpha_{\uparrow,\downarrow} = \alpha \pm \eta H = \alpha'(T - T_{c,\uparrow,\downarrow}^0)$.

The coefficients of the fourth-order terms in (5.55) may be determined within the spin-fluctuation model of Section 5.3 as $\beta_{24} = \frac{12}{5}\beta_0$ $(2.16\beta_0)$ and $\beta_{24} + 2\beta_5 = -\frac{12}{5}\beta_0\delta$ $(-0.80\beta_0)$, where the figures in parentheses are the results of the strong-coupling calculations quoted in Section 5.3. As anticipated, this implies a vanishing coefficient for the $\Delta_{\uparrow\uparrow}^2 \Delta_{\downarrow\downarrow}^2$ term in the weak-coupling model.

It follows from the free-energy expression (5.55) that below the critical temperature $T_{c\uparrow}^0$ the free energy is lowered by formation of spin-up pairs.

The order parameters are obtained as

$$\left.\begin{aligned}\Delta^2_{\uparrow\uparrow}(T) &= \frac{3N(0)}{\beta_{24}}\left(1-\frac{T}{T_{c\uparrow}}\right),\\ \Delta^2_{\downarrow\downarrow}(T) &= 0,\end{aligned}\right\} \tag{5.56}$$

for $T_{c\downarrow} < T < T_{c\uparrow}$ (note that we have defined $T_{c\uparrow} \equiv T^0_{c\uparrow}$). In this temperature regime the free energy is accordingly given by

$$F^{A_1} = F_N - \tfrac{1}{2}[N(0)]^2\beta_{24}^{-1}\left(1-\frac{T}{T_{c\uparrow}}\right)^2, \tag{5.57}$$

and the specific-heat discontinuity at $T_{c\uparrow}$ has the magnitude

$$\frac{\Delta C^{A_1}}{C_N} = \frac{5}{7\zeta(3)}\frac{\beta_{24}^{WC}}{\beta_{24}}. \tag{5.58}$$

This state may be identified with the A_1 phase. It has the peculiar property that only spin-up particles are paired, and hence it is an inherently magnetic superfluid. From (5.51), one infers that its order parameter is given by (see also (5.26))

$$d_{\mu j} = \tfrac{1}{2}\Delta_{\uparrow\uparrow}(\hat{d}_\mu + \mathrm{i}\hat{e}_\mu)(\hat{m}_j + \mathrm{i}\hat{n}_j). \tag{5.59}$$

The spin and orbital structures of this state are seen to be identical—both have the signature of an $l = 1$, $m = 1$ angular-momentum eigenfunction. As discussed below (5.51), the gap parameter has the orbital structure given by (3.69).

Formation of down-spin pairs becomes favourable below a temperature $T_{c\downarrow}$, which may in general deviate from $T^0_{c\downarrow}$ owing to the cross-term $\Delta^2_\uparrow\Delta^2_\downarrow$ in the free energy. Here $T_{c\downarrow}$ is obtained as

$$T_{c\downarrow} = \tfrac{1}{2}(1+p)T^0_{c\downarrow} + \tfrac{1}{2}(1-p)T^0_{c\uparrow}, \tag{5.60}$$

where $p = \beta_{245}/(-\beta_5)$. Within the simple spin-fluctuation model, p is found to be $p = (1-\delta)/(1+\delta) < 1$; the strong-coupling values of the parameters β close to melting pressure given in Section 5.3 yield $p \approx 0.5$. Hence it follows that $T_{c\downarrow}$ is somewhat higher than $T^0_{c\downarrow}$. The strong-coupling effects introduce some asymmetry into the T_c splitting.

For $T < T_{c\downarrow}$, the gap parameters for both spin-up and spin-down pairs are finite and are given by

$$\left.\begin{aligned}\Delta^2_{\uparrow\uparrow} &= 3N(0)\left[\frac{1}{2\beta_{245}}\left(1-\frac{T}{T_{c\downarrow}}\right) + \frac{1}{\beta_{24}}\left(\frac{T}{T_{c\downarrow}} - \frac{T}{T_{c\uparrow}}\right)\right],\\ \Delta^2_{\downarrow\downarrow} &= 3N(0)\frac{1}{2\beta_{245}}\left(1-\frac{T}{T_{c\downarrow}}\right).\end{aligned}\right\} \tag{5.61}$$

This is the A_2 phase. Its order parameter, given in (5.27) and (5.51), is the sum of terms corresponding to Cooper pairs with spin configuration $\uparrow\uparrow$ and $\downarrow\downarrow$.

Inserting the gap structure (5.50) into the general expression (5.7), the free energy of the A_2 phase below T_c is found as

$$F^{A_2} = F_N - \tfrac{1}{2}[N(0)]^2\left[\beta_{24}^{-1}\left(1-\frac{T}{T_{c\uparrow}}\right)^2 + (\beta_{245}^{-1}-\beta_{24}^{-1})\left(1-\frac{T}{T_{c\downarrow}}\right)^2\right]. \quad (5.62)$$

The second term in the square brackets in (5.62) gives rise to a specific-heat discontinuity at $T_{c\downarrow}$ of relative magnitude

$$\frac{\Delta C^{A_2}}{C_N} = \frac{5}{7\zeta(3)}\beta_{24}^{WC}(\beta_{245}^{-1}-\beta_{24}^{-1}). \quad (5.63)$$

Note that the specific-heat discontinuities ΔC^{A_1} and ΔC^{A_2} add up to ΔC, the discontinuity in zero field: $\Delta C/C_N = [5/7\zeta(3)]\beta_{24}^{WC}/\beta_{245}$, as given by (5.13).

Substituting values for the parameters β_i quoted in Section 5.3, it is seen that the specific-heat discontinuity at $T_{c\downarrow}$ is enhanced by a factor of more than two, while the discontinuity ΔC^{A_1} at $T_{c\uparrow}$ has approximately the weak-coupling value. These results are in satisfactory agreement with experiment (see Chapter 4).

The free-energy terms (5.49) originating from the T_c splitting, though seemingly linear in H, are of course of order H^2 in the limit of small field, which we only consider here. The magnetic susceptibility of the A_2 phase is hence not exactly equal to the normal-state value as predicted by weak-coupling theory assuming particle–hole symmetry, but has discontinuities at $T_{c\uparrow}$ and $T_{c\downarrow}$ of magnitude

$$\Delta\chi_{\uparrow,\downarrow} = \left(\frac{\partial T_{c\uparrow,\downarrow}}{\partial H}\right)^2 \frac{\Delta C_{\uparrow,\downarrow}}{T_{c\uparrow}}. \quad (5.64)$$

The result (5.64) follows from (5.62) by differenting twice with respect to the magnetic field. More generally, (5.64) should be recognized as exact thermodynamic relations. The continuity of the magnetization as well as the entropy across the second-order phase-transition lines $T_{c\uparrow,\downarrow}(H)$ implies that the discontinuities in the magnetic susceptibility $\Delta\chi$ and the specific heat ΔC are related by virtue of the identities (Leggett 1975a)

$$\left.\begin{aligned} \Delta\chi &= -\frac{\partial T_c}{\partial H}\Delta\left(\frac{\partial M}{\partial T}\right), \\ \Delta C &= -T\frac{\partial H}{\partial T_c}\Delta\left(\frac{\partial M}{\partial T}\right), \end{aligned}\right\} \quad (5.65)$$

which, when combined, yield (5.64). The susceptibility enhancement (5.64) has been observed experimentally.

Turning to the B phase, we again have to make a reasonable assumption about the distortion of the gap parameter in a magnetic field.

In zero field the order parameter of the B phase may be written as a symmetric superposition of three components describing the three spin substates $m_s = \pm 1$, 0. In each component the orbital wave function is chosen such that the spin and orbital angular-momentum projections (or rather the helicities) add up to zero. A magnetic field removes the spherical symmetry, since the energy of a Cooper pair in a magnetic field depends on its spin projection. Consequently, we are led to the following generalization of the B-phase order parameter (Hall and Hook 1986):

$$d_{\mu j} = e^{i\phi}[\tfrac{1}{2}\Delta_{\uparrow\uparrow}(\hat{d}_\mu + i\hat{e}_\mu)(\hat{m}_j - i\hat{n}_j) + \tfrac{1}{2}\Delta_{\downarrow\downarrow}(\hat{d}_\mu - i\hat{e}_\mu)(\hat{m}_j + i\hat{n}_j) + \Delta_{\uparrow\downarrow}\hat{f}_\mu \hat{l}_j] \quad (5.66a)$$

This state is referred to as the B_2 phase (see also (5.28)). Note the important difference from the A_2 phase, (5.27), (5.51), namely the different coupling of spin and orbital wave functions. In contrast with the B_2 phase, all spin components in the A_2 phase have the same orbital structure. In zero field one has $\Delta_{\uparrow\uparrow} = \Delta_{\downarrow\downarrow} = \Delta_{\uparrow\downarrow} = \Delta$, and in this case (5.66) acquires the usual BW structure.

From the preceding discussion of the A_2 phase, we may infer that the splitting of $\Delta_{\uparrow\uparrow}$ and $\Delta_{\downarrow\downarrow}$ is an effect of particle–hole asymmetry and hence is very small compared with the change in $\Delta_{\uparrow\downarrow}$ in a magnetic field. We shall therefore put $\Delta_{\uparrow\uparrow} = \Delta_{\downarrow\downarrow} = \Delta_\perp$ as well as $\Delta_{\uparrow\downarrow} = \Delta_\parallel$ in the following. The order parameter can then be expressed in the simpler form

$$d_{\mu j} = e^{i\phi}(\Delta_\perp R_{\mu j} + (\Delta_\parallel - \Delta_\perp)(\boldsymbol{R}\hat{\boldsymbol{H}})_\mu \hat{H}_j], \quad (5.66b)$$

with $R_{\mu j}$ the usual spin–orbit rotation matrix and $\hat{\boldsymbol{H}}$ a unit vector along the magnetic field.

Substituting the form (5.66) for $d_{\mu j}$ with the abovementioned simplification into the free-energy functional (5.4) yields

$$\begin{aligned} F = F_N &+ \tfrac{1}{3}\alpha(2\Delta_\perp^2 + \Delta_\parallel^2) + \tilde{\alpha}H^2\Delta_\parallel^2 \\ &+ \tfrac{1}{18}[\beta_{12}(2\Delta_\perp^2 + \Delta_\parallel^2)^2 + \beta_{345}(2\Delta_\perp^4 + \Delta_\parallel^4)], \end{aligned} \quad (5.67)$$

where we take the third axis along the magnetic-field direction. Since $\tilde{\alpha} > 0$ (see (5.48)), the magnetic field is seen to suppress the component $\Delta_\parallel$, expressing the fact that Cooper pairs with magnetic quantum number 0 are unfavoured in a magnetic field.

Therefore Cooper pairs are first formed below T_c (note that $\alpha(T_c) = 0$) in a state with $\Delta_\parallel = 0$. This is the planar state discussed above (see (5.19)). However, the planar state is higher in free energy than the A phase provided that $\beta_{245} < \beta_{12} + \frac{1}{2}\beta_{345}$. This can actually be shown to be the case, using the β_i values from Section 5.3. Thus in a magnetic field, which may even be arbitrarily weak, a small wedge of A phase (or rather A_1 and A_2 phase) appears between the normal phase and the B phase (see Fig. 4.2)

Below a temperature $T_c^\parallel < T_c$ given by

$$T_c^\parallel = T_c\left[1 - \left(1 + \frac{2\beta_{12}}{\beta_{345}}\right)h^2\right], \quad (5.68)$$

where

$$h^2 = \frac{7\zeta(3)}{4\pi^2}(1+F_0^a)^{-2}\left(\frac{\mu_0 H}{k_B T_c}\right)^2,$$

the coefficient α is sufficiently negative to overcome the positive magnetic-energy term in (5.67), and the formation of $m = 0$ pairs will occur too. We shall refer to this state as the B_2 phase. The gap parameters for $T < T_c^{\parallel}$ are found to be given by a similar expression as for the B phase:

$$\Delta_j^2 = \frac{N(0)}{\beta_{12} + \frac{1}{3}\beta_{345}}\left(1 - \frac{T}{T_c^j}\right) \quad (T < T_c^{\parallel}), \tag{5.69}$$

for $j = \perp, \parallel$. The T_c value of the perpendicular component is shifted upwards:

$$T_c^{\perp} = T_c\left(1 + \frac{\beta_{12}}{\beta_{345}}h^2\right). \tag{5.70}$$

This yields

$$\Delta_{\perp}^2 - \Delta_{\parallel}^2 \approx (\Delta_{\perp} - \Delta_{\parallel})\Delta_{\perp} = \tfrac{5}{12}\left(\frac{\gamma\hbar H}{1+F_0^a}\right)^2 \frac{\beta_{345}^{WC}}{\beta_{345}}\frac{T}{T_c}. \tag{5.71}$$

The B_2 phase is the stable low-temperature phase of superfluid ^{3}He for magnetic fields below a critical field $H_c^{B_2}(t)$. In this low-field regime a first-order transition takes place from the A_2 phase (stable at higher temperatures, as shown above) to the B_2 phase. The corresponding thermodynamic transition temperature is close to $T_c^{\parallel}$ for pressures below the polycritical point. For higher pressures a reliable prediction is difficult, since already the A–B transition temperature in zero magnetic field cannot be described within a fourth-order Ginzburg–Landau expansion. Nevertheless, the critical field $H_c^{B_2}(T)$ may be estimated from (5.68) as the field at given temperature, by replacing $T_c^{\parallel}$ with T, i.e.

$$H_c^{B_2}(T) = \left(\frac{k_B T_c}{\mu_0}\right)\frac{2\pi}{[7\zeta(3)]^{1/2}}(1+F_0^a)\left[\frac{\beta_{345}}{\beta_{345}+2\beta_{12}}\right]^{1/2}\left(1 - \frac{T}{T_c}\right)^{1/2} \tag{5.72}$$

Substituting the strong-coupling values for the β_i given in Section 5.3 and T_c values as quoted in Section 5.3, the critical field $H_c^{B_2}$ at $T = 0$ is estimated from (5.72) as 9 kG (5 kG) at melting pressure (s.v.p.). In magnetic fields stronger than $H_c^{B_2}$ the A_2 phase is predicted to be stable down to absolute zero.

The phase diagram of superfluid ^{3}He in a magnetic field has been discussed by Schopohl (1982) and Ashida and Nagai (1985), following earlier work by Carton (1975), Fetter (1975a), Kuroda and Nagi (1975, 1976c), Levin and Valls (1977), McInerney (1980, 1981), Jacak (1981), Nijhoff and Capel (1981, 1982), Nijhoff *et al.* (1985) and Capel *et al.* (1986).

FURTHER READING

Anderson P W and Brinkman W F 1978 in *The Physics of Liquid and Solid Helium,* Part II, ed. K H Bennemann and J B Ketterson (Wiley, New York), p. 177
Lee D M and Richardson R C 1978 in *The Physics of Liquid and Solid Helium,* Part II, ed. K H Bennemann and J B Ketterson (Wiley, New York), p. 287
Leggett A J 1975 *Rev. Mod. Phys.* **47** 331
Levin K and Valls O T 1983 *Phys. Rep.* **98** 1
Serene J W and Rainer D 1983 *Phys. Rep.* **101** 221
Wheatley J C 1975 *Rev. Mod. Phys.* **47** 415

6

Broken Symmetries and Macroscopic Order

In the previous chapter several phases of superfluid ^{3}He and their corresponding order parameters have been discussed on the basis of energy and stability criteria. We now want to investigate the actual structure of these order parameters in greater detail. In particular, we shall focus our attention on the symmetry properties connected with their structure. Knowledge of those properties is of particular importance if one wants to understand the dynamic and collective behaviour of the phases on a macroscopic level.

The superfluid phases of ^{3}He differ from the normal-liquid phase in the existence of Cooper pairs. Obviously there is a new, additional state of order that is not present in the normal phase. This fact may be quantitatively expressed by introducing an "order parameter" that has a finite value below T_c and goes to zero as the temperature is raised to T_c (the phase transition is of second order); above T_c it stays zero. A more common example of such a transition is the one from a paramagnetic to a ferromagnetic state of a metal when cooled below the Curie temperature. In the paramagnetic state, for $T > T_c$, the spins of the particles point randomly in all possible directions (Fig. 6.1a). Below the transition ($T < T_c$), i.e. in the ferromagnetic state, the spins are more or less perfectly aligned as shown in Fig. 6.1(b). To distinguish the two situations and to express this in more quantitative terms, one can proceed in different ways (see e.g. Mattuck and Johansson 1968).

First of all, one may introduce a spin–spin correlation function

$$C^{ss}(\boldsymbol{r} - \boldsymbol{r}') = \langle \boldsymbol{S}(\boldsymbol{r}) \cdot \boldsymbol{S}(\boldsymbol{r}') \rangle \tag{6.1a}$$

between the spin densities at positions $\boldsymbol{r}$ and $\boldsymbol{r}'$. It is a measure of the correlation between two spins, i.e. of the mutual dependence on each other's direction. If $\boldsymbol{S}(\boldsymbol{r})$ is pointing in a particular direction, the correlation function informs us about how the orientation of another spin at $\boldsymbol{r}'$ is

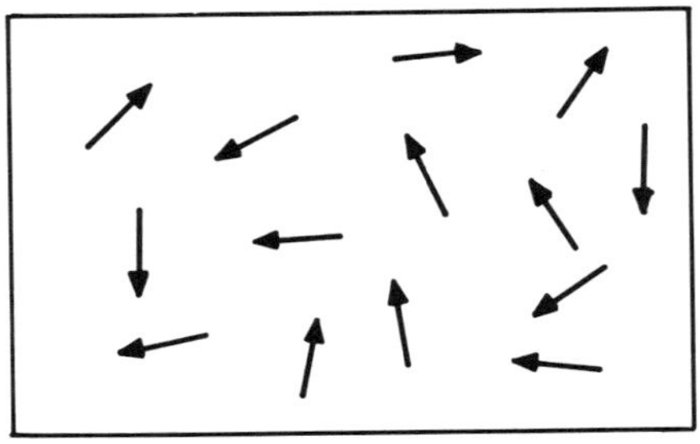

(a)

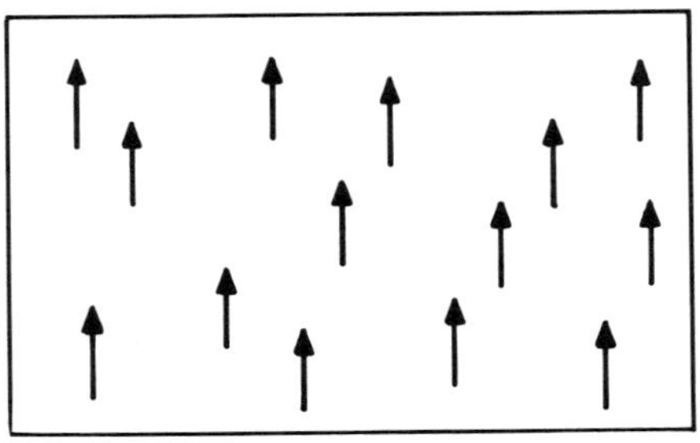

(b)

Figure 6.1 Orientation of spins: (a) in the paramagnetic state ($T > T_c$); (b) in the ferromagnetic state ($T \ll T_c$).

affected. We now want to know its behaviour as we increase the separation $|\boldsymbol{r}-\boldsymbol{r}'|$. In the paramagnetic case there is only very weak correlation between the spins. Although there is some mutual ordering of nearby spins, the correlation will go exponentially to zero above a characteristic correlation length:

$$\lim_{|\boldsymbol{r}-\boldsymbol{r}'|\to\infty} C^{\text{ss}}_{\text{para}}(\boldsymbol{r}-\boldsymbol{r}') = 0. \tag{6.1b}$$

There is no long-range order of the spins in the paramagnetic state. In contrast, below T_c, in the ferromagnetic regime, there is obviously a strong correlation between the spins because they are essentially aligned. In this case the correlation function does not go to zero for large separation of two spins, but rather approaches a *finite* value

$$\lim_{|\boldsymbol{r}-\boldsymbol{r}'|\to\infty} C^{\text{ss}}_{\text{ferro}}(\boldsymbol{r}-\boldsymbol{r}') > 0. \tag{6.1c}$$

This expresses the fact that there exists *long-range* order between the spins in the ferromagnetic state. Therefore a correlation function is a measure of the degree of order in the system.

A different way of quantifying the degree of order is to look at the average spin $\langle \boldsymbol{S} \rangle$ itself. This approach is closely connected to the investigation of the symmetries inherent in the system. In the paramagnetic phase the ground state of the system is left unchanged if we rotate all spins through the same angle around some rotation axis: it is invariant under spin rotations. In other words, there exists a symmetry under arbitrary spin rotations. Below T_c, in the ferromagnetic state, this is no longer true. Here a rotation of all spins leads to a perceptible change in the ground state. So the ground state does not show the same symmetry under spin rotations as the paramagnetic state: the symmetry appears to be "spontaneously broken", i.e. by the system itself, without external influence. On the other hand, the system is in principle not able to pick out any particular preferred direction for the spin. So, although there exists a preferred direction, it is still undetermined. Therefore the average $\langle \boldsymbol{S} \rangle$, when taken over many ensembles, is still zero as in the paramagnetic state. The ground state of the system is then degenerate. In particular, it costs no energy to turn all the spins into some direction. This allows for a collective excitation of the spins in the ferromagnetic state with energy $\hbar\omega = 0$ and wave vector $\boldsymbol{q} = 0$, i.e. a "spin wave", which simply rotates all the spins. Such a mode is usually referred to as a "Goldstone mode" because in any system with a broken global continuous symmetry there exists, in the case of short-range interactions, a gapless spin-zero excitation, a so-called "Goldstone boson". Hence in the present case of a ferromagnet the Goldstone mode is a spin wave connected with the spontaneously broken spin-rotation symmetry.

Here it becomes evident that the initial symmetry (i.e. that of the underlying Hamiltonian) has not simply "disappeared". Even though the ground state may have a reduced ("broken") symmetry, the corresponding Goldstone modes supplement it so as to restore the initial symmetry. The concept of broken symmetries and the associated Goldstone particles (or modes) are of great importance in many fields of physics, including condensed-matter physics and high-energy physics (see Section 6.2).

Returning to the ferromagnetic state, it is clear that the degeneracy of the ground state with respect to the actual direction of the magnetization will make the system susceptible to the influence of even infinitesimally small effects that lack symmetry under spin rotation. They may be due to interactions so far neglected (because they were unimportant for the actual formation of the ferromagnetic state) or to some external magnetic field. Although these effects can be arbitrarily weak, they are none the less able to select a particular preferred direction for the spins. In this case the Goldstone mode will no longer be a $\omega = 0$ mode but will acquire an energy gap proportional to the coupling strength of this interaction. Once a preferred direction has been determined, the average $\langle \boldsymbol{S} \rangle$ of the spins is no longer zero. The system then has a finite magnetization $\boldsymbol{M} \propto \langle \boldsymbol{S} \rangle$ and we can conveniently choose $\boldsymbol{M}$ itself as the order parameter of the system (rather than using the correlation function $C(\boldsymbol{r} - \boldsymbol{r}')$). Above T_c we have $|\boldsymbol{M}| = 0$, while below T_c, $|\boldsymbol{M}| > 0$ (see Fig. 6.1b).

The above discussion of symmetry breaking in a ferromagnet applies to an isotropic Heisenberg ferromagnet with $\boldsymbol{S} = (S_x, S_y, S_z)$ as well as to a planar ferromagnet with $\boldsymbol{S} = (S_x, S_y)$. The latter system (often referred to as the *XY* model) is similar in many respects to an isotropic superfluid in that the complex order parameter can be represented by a two-component vector in the complex plane.

The connection between a broken symmetry and long-range order has been illustrated in detail for a ferromagnet since this case is particularly simple. Nevertheless, the concept applies equally to a pair-correlated state, i.e. to superconductors and superfluid ^{3}He. In this case the new order is associated with the formation of Cooper pairs below the critical temperature T_c. In analogy to the ferromagnet (see (6.1a,c)), this order may be expressed mathematically by introducing a correlation function, which in the present case involves two Cooper pairs (CP) at positions $\boldsymbol{r}$ and $\boldsymbol{r}'$ respectively:

$$C^{\mathrm{CP,CP}}(\boldsymbol{r} - \boldsymbol{r}') = \langle \psi_\uparrow^+(\boldsymbol{r})\psi_\downarrow^+(\boldsymbol{r})\psi_\downarrow(\boldsymbol{r}')\psi_\uparrow(\boldsymbol{r}')\rangle \tag{6.2}$$

(for simplicity we restrict our discussion to conventional superconductors with $S = l = 0$ pairing). Here the field operators $\psi_\sigma^+(\boldsymbol{r})$ and $\psi_\sigma(\boldsymbol{r})$ respectively create and annihilate an electron with spin σ at position $\boldsymbol{r}$, and $\psi_\sigma^+(\boldsymbol{r}) = \sum_{\boldsymbol{k}} a_{\boldsymbol{k}\sigma}^+ \mathrm{e}^{-\mathrm{i}\boldsymbol{k}\cdot\boldsymbol{r}}$ etc., where $a_{\boldsymbol{k}\sigma}^+$ is the creation operator for a particle with momentum $\hbar\boldsymbol{k}$ and spin σ (see (3.11)). In the superconducting state the correlation function (6.2) approaches a finite value even in the limit of infinite separation of the pairs:

$$\lim_{|\boldsymbol{r}-\boldsymbol{r}'|\to\infty} C^{\mathrm{CP,CP}}(\boldsymbol{r} - \boldsymbol{r}') \neq 0. \tag{6.3}$$

This is in contrast with the normal state, where this expectation value is identically zero. The behaviour expressed in (6.3) is referred to as "off-diagonal long-range order" (Yang 1962). It expresses the fact that the off-diagonal elements of the two-particle density matrix (6.2) in the position-space representation (i.e. those for $\boldsymbol{r} \neq \boldsymbol{r}'$) show long-range order. The same concept applies to a Bose–Einstein condensate, which is the simplest example of off-diagonal long-range order (Penrose and Onsager 1956). The particular long-range order implied by (6.3) is a consequence of the fact that in the pair-correlated state the particle-nonconserving operator $a_{\boldsymbol{k}\uparrow}^+ a_{-\boldsymbol{k}\downarrow}^+$ has a finite expectation value—this is precisely the pair amplitude $F_{-\boldsymbol{k}\downarrow\uparrow}$ introduced in (3.13b). Hence, rather than using the correlation function (6.2), the degree of order in the pair condensate may equally well be described by the pair amplitude itself, which therefore takes the role of the so-called "order parameter". As discussed in detail in Section 3.2, the existence of such an order parameter implies a spontaneously broken gauge (i.e. U(1)) symmetry.

6.1 BROKEN SYMMETRIES AND GROUP THEORY

To gain a better understanding of the concept of "broken symmetry" and to formalize the connection between order-parameter structures and their symmetry, it is most useful to employ the language of group theory and topology (Michel 1980, Mineev 1980). It allows an elegant but straightforward treatment of the problems in question. In particular, we shall see that these methods provide the basis for an understanding of underlying phenomena common to seemingly different areas of physics.

From a group-theoretical point of view, "broken symmetry" means that below the transition the lowest state of the system is no longer invariant under the full group of symmetry transformations G, as it was above the transition. Certain symmetries (but, of course, not necessarily all) are "broken", such that this state is now only invariant under a subgroup H of the full group G of symmetry transformations. While the group H describes the remaining symmetries of the order parameter, the coset (or factor) space R defined by

$$R = G/H \tag{6.4}$$

characterizes the broken symmetries, i.e. those transformations that change the ground-state perceptibly. (Although they do change the state, they leave the energy invariant, because the corresponding states are degenerate with respect to the energy.)

Note that R will only be a group itself if H is a normal (or invariant) subgroup of G. (A normal subgroup H of G is defined by $gHg^{-1} = H$ for all $g \in G$.) None the less, even if this is not the case, $R = G/H$ may be called a "factorization" of G with respect to H and one may write $G = R \times H$. So the product $R \times H$ (meant as a product of sets) does not necessarily involve elements of two groups, but, in general, multiplies representatives of the set of transformations contained in R by elements of the group H.

The coset space R is called the "manifold of internal states" or "degeneracy space".

In most systems of condensed-matter physics that exhibit a broken symmetry this symmetry corresponds to rotations about an axis or to three-dimensional rotations. Rotations about an axis may be described by the Abelian group U(1), the *u*nitary group of rotations about a single axis. Equivalently, they are represented by the group SO(2), the *s*pecial *o*rthogonal group of rotations in a plane with determinant +1, having real elements; hence U(1) $\triangleq$ SO(2). Three-dimensional rotations are described by the non-Abelian group SO(3) (details will be discussed below).

Two examples from condensed-matter physics illustrate this point.

Example 1: BCS superconductor or superfluid ^{4}He. Here the order parameter is given by a single complex parameter $\psi = \psi_0 e^{i\phi}$, with amplitude ψ_0 and phase ("gauge") ϕ. Above T_c the system is invariant under an arbitrary

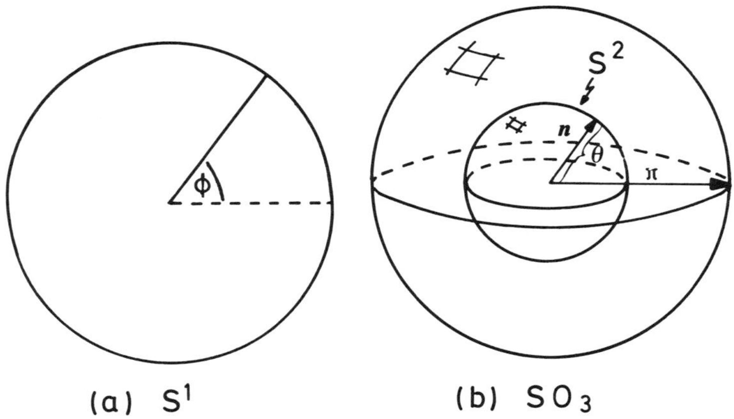

Figure 6.2 Schematic drawing of topological spaces: (a) the representation S^1 of the group U(1); (b) the representation SO_3 of the group SO(3), and of the coset space S^2.

change of the phase $\phi \to \phi'$, i.e. under a gauge transformation (see the discussion in Section 3.2.1). This invariance is equivalent to a U(1) symmetry i.e. $G = U(1)$. Below T_c a particular value of ϕ is spontaneously preferred; hence the U(1) symmetry is completely broken.

The group of transformations corresponding to rotations about a single axis, U(1), is uniquely characterized by a rotation angle that takes values between 0 and 2π. Therefore this group may be represented by mapping it onto the circumference of a (unit) circle: the topological space S^1 (see Fig. 6.2a). Note that we distinguish the abstract group U(1) from its topological representation S^1.

Example 2: Ferromagnet. The order parameter is given by the magnetization $\boldsymbol{M}$, a real three-component vector. In the paramagnetic phase, i.e. above the transition, the spin system is invariant under arbitrary global rotations in the three-dimensional spin space, i.e. $G = SO(3)$. Below the Curie temperature T_c, in the ferromagnetic regime, the spins are aligned along the preferred direction of the magnetization $\boldsymbol{M}$, i.e. along $\hat{\boldsymbol{M}} = \boldsymbol{M}/|\boldsymbol{M}|$. Now the lowest state of the system is only invariant under rotations *about* the single axis $\hat{\boldsymbol{M}}$. This remaining symmetry corresponds to the group of transformations U(1), i.e. $H = U(1)$. We see that in this case the symmetry $G = SO(3)$ is not completely broken but is reduced to $H = U(1)$. The broken symmetry is therefore given by $R = SO(3)/U(1)$.

Any three-dimensional rotation may be described by a rotation axis $\hat{\boldsymbol{n}}$ and an angle of rotation θ about this axis. The group of solid-body rotations SO(3) can therefore be represented as a topological space in the following way: any rotation about an axis $\hat{\boldsymbol{n}}$ by an angle θ is associated with a point in a solid sphere of radius π; this point is the endpoint of the vector $\boldsymbol{n} = \theta\hat{\boldsymbol{n}}$

originating at the centre of the sphere, pointing in the direction of $\hat{\boldsymbol{n}}$ and of length θ (see Fig. 6.2b). Diametrically opposite points on the surface of the sphere represent rotations about π and $-\pi$ respectively, and are therefore equivalent—hence these points have to be identified. This defines a doubly connected topological space referred to as SO_3.

In particular, three-dimensional rotations involving a *fixed* rotation angle θ can be represented by the points on the closed spherical surface of radius θ within the solid sphere SO_3 (see Fig. 6.2b). This surface is itself referred to as S^2, the "two-sphere".

Consequently, SO_3 factorizes as

$$SO(3)/U(1) = S^2. \tag{6.5}$$

We observe that the set of transformations contained in S^2 is not a group. For example, two successive rotations about the x and y axes by the same angle θ are in general not equivalent to a single rotation by θ.

The above factorization applies directly to the case of the para/ferromagnet. While $H = U(1)$ describes the remaining symmetry in the ferromagnetic state, the broken symmetry is characterized by the degeneracy space $R = S^2$ corresponding to the manifold of orientations of the magnetization $\boldsymbol{M}$.

6.1.1 The symmetry group describing superfluid ^{3}He

Owing to the fact that superfluid ^{3}He forms a spin-triplet p-wave condensate, the order parameter $A_{\mu j}$ has a considerably more complicated structure than conventional order parameters. If we only consider the interactions responsible for the formation of the condensed state, i.e. we neglect small residual interactions like the dipole interaction that are unimportant on this level, then the free energy of the system has to be invariant under separate three-dimensional rotations in spin space, in orbital space and under a gauge transformation (this requirement has in fact been used explicitly to derive the free-energy functional (5.4)). Hence the symmetry group that allows for these symmetries is given by a product of three independent symmetries

$$G = SO(3)_{\boldsymbol{L}} \times SO(3)_{\boldsymbol{S}} \times U(1)_{\phi}. \tag{6.6}$$

Here the indices $\boldsymbol{L}$, $\boldsymbol{S}$ and ϕ indicate orbital space, spin space and gauge respectively.

Clearly, this symmetry group is much richer in structure than those considered before. In fact, it incorporates the order-parameter symmetries of liquid crystals, magnets and isotropic superfluids all at the same time. This implies that the *broken* symmetries resulting from (6.6) will also be much more intricate than, say, in the case of the ferromagnet, where only a single degree of freedom, i.e. a single group, is relevant.

6.1.2 A two-dimensional model of superfluid ^{3}He

To understand the fundamental difference between the broken symmetries that may occur in the case of a single group like SO(3) and those in (6.6) consisting of several different symmetries, we first consider a somewhat simpler example. Let us have a look at a two-dimensional ($d = 2$) liquid with *two* different degrees of freedom given by, say, an angular momentum and a spin (Liu 1982). Both degrees of freedom are assumed to vary only in the plane; hence they may be described by a U(1) symmetry each. Consequently, if no symmetry were broken then the underlying symmetry would be given by

$$G = \mathrm{U}(1)_L \times \mathrm{U}(1)_S, \tag{6.7}$$

where L and S refer to the angular momentum and spin respectively.

We now think of the particles in the liquid as being characterized by a direction both with respect to their spin (↑) and their orbital angular momentum (⚲). Because of the U(1) symmetries, both directions may point in any direction in the plane.

In Fig. 6.3 we see all possibilities of ordering, i.e. symmetry breaking, of these degrees of freedom. Figure 6.3(a) shows the disordered state: the system is isotropic with respect to the orientation of both degrees of freedom. The system stays invariant under separate rotations in spin and orbital spaces; there is no long-range order and hence $G = H = \mathrm{U}(1)_L \times \mathrm{U}(1)_S$, $R = 1$. This case represents an example of an isotropic paramagnetic liquid. On the other hand, in Figs. 6.3(b–e) we see all possible different symmetry breakings, i.e. states with long-range order. In Fig. 6.3(b) only the spin degrees of freedom are ordered ("breaking of rotational symmetry in spin space"), while the orbital component is still disordered. So, while the $\mathrm{U}(1)_S$ is broken, the symmetry with respect to orbital rotations is still intact: $H = \mathrm{U}(1)_L$, $R = \mathrm{U}(1)_S$. If we identify ↑ with the spin of the particle, this system therefore represents an isotropic ferromagnetic liquid. Conversely in Fig. 6.3(c) only the orbital degrees of freedom are ordered while the spin part is disordered ("breaking of rotational symmetry in orbital space"). Here $\mathrm{U}(1)_L$ is broken while $\mathrm{U}(1)_S$ is preserved: $H = \mathrm{U}(1)_S$, $R = \mathrm{U}(1)_L$. One might take this as a model for some kind of liquid crystal, where the molecules are oriented along a preferred direction.

In Fig. 6.3(d) *both* degrees of freedom are ordered, i.e. both $\mathrm{U}(1)_L$ and $\mathrm{U}(1)_S$ are independently broken, so that no symmetry remains: $H = 1$, $R = \mathrm{U}(1)_L \times \mathrm{U}(1)_S$. In fact this is a two-dimensional model of superfluid ^{3}He-A, if we associate $\hat{\boldsymbol{d}}$ with ↑ and $\hat{\boldsymbol{l}}$ with ⚲ respectively. (The additional gauge symmetry also broken in the superfluid has not been included here.) So, $\hat{\boldsymbol{d}}$ and $\hat{\boldsymbol{l}}$ are both macroscopically ordered ("breaking of rotational symmetry in spin and orbital spaces separately").

One might think that these three states exhaust all possibilities of broken symmetries, but this is not the case. In Fig. 6.3(e) we finally show a

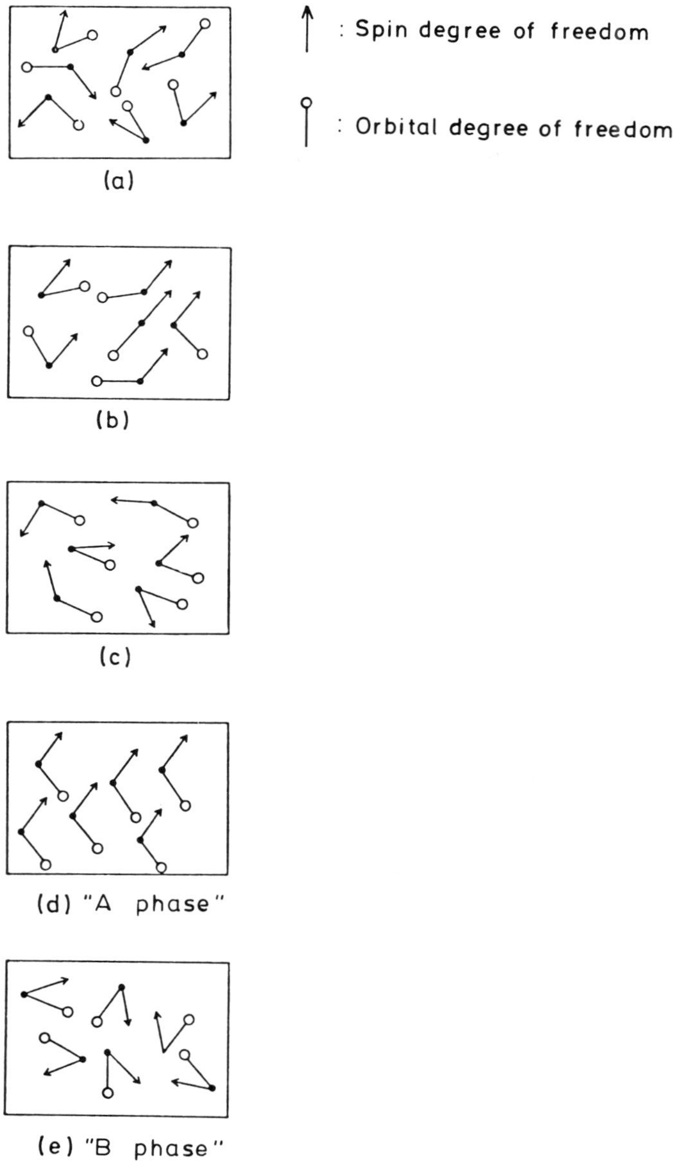

Figure 6.3 The possible states of order in a two-dimensional model liquid with two different degrees of freedom. (After Liu (1982).)

situation where neither of the two degrees of freedom is ordered—only their *relative* orientation is. It is the same for all particles, as indicated by the constant angle between ↑ and ⫯. Of course, the symmetries with respect to rotations in spin and orbital spaces are both broken (changing the direction of either ↑ or ⫯ changes the state perceptibly), but this time they are not independently broken—in fact, only a combined symmetry is broken. Equivalently, one may say that the state in Fig. 6.3(e) still *contains* a symmetry. This is easy to verify. Let us first turn the direction of ↑ through an angle ϕ and then turn ⫯ through the same angle. The final state is macroscopically indistinguishable from the initial state; hence there is a symmetry! Although it involves both spin and orbital spaces, the transformation described above corresponds to only a single U(1) symmetry, i.e. it is a *linear combination* of the two initial U(1) symmetries. We may express this fact by writing $H = \mathrm{U}(1)_{L+S}$ (this point will become clearer when formulated in the language of group theory; see Section 6.2). Consequently, the broken symmetry $R = G/H = \mathrm{U}(1)_{S,L}$ also involves both degrees of freedom.

This last case is a two-dimensional model of superfluid ^{3}He-B where the order parameter is given by a rotation matrix describing relative rotations between spin and orbital spaces. The rotation matrix is characterized by a rotation axis $\hat{\boldsymbol{n}}$ and a rotation angle θ. In our two-dimensional model the axis $\hat{\boldsymbol{n}}$ is perpendicular to the plane of the paper, while θ corresponds to the angle between ↑ and ⫯. (The broken gauge has again not been included.)

The broken symmetry leading to the ordered state shown in Figs. 6.3(d, e) is called "spontaneously broken spin–orbit symmetry" (SBSOS). This term was introduced by Leggett (1972), who was the first to realize the existence of this particular symmetry breaking. With this concept, he was able to explain the at-first inexplicably large frequency shifts observed in NMR experiments in the superfluid phases of ^{3}He (see Chapters 4 and 8). The original meaning of SBSOS expressed the fact that the relative orientation of the two directions characterizing the spin and the orbital parts of a Cooper pair shows long-range order. In this sense, both ^{3}He-A (Fig. 6.3d) and ^{3}He-B (Fig. 6.3e) possess this property. However, there is a subtle difference. In the A phase the symmetries with respect to rotations in spin and orbital spaces are broken independently, so that there exist two independent, preferred directions (long-range order of both $\hat{\boldsymbol{d}}$ and $\hat{\boldsymbol{l}}$). Consequently, their relative orientation will necessarily also be long-range ordered. This is not so in the B phase. There the new contents of SBSOS lies in the breaking of a *relative* symmetry (Liu and Cross 1978): again the symmetries with respect to rotations in spin and orbital spaces are both broken, but not independently—only a linear combination of these symmetries is broken. In this case the individual spin and orbital degrees of freedom are *not* long-range ordered, only their relative orientation is. It is essential to realize that we encounter here a case where a broken symmetry with respect to some degree of freedom does not lead to long-range order in that variable itself. In general, this is not an automatic consequence if we have more than one symmetry in the problem.

As discussed above, any change (i.e. rotation) of one degree of freedom may be compensated by a corresponding change of the other degree of freedom. There is then no way to distinguish the states "before" and "after". In the case of ^{3}He-B this fact implies that the application of a magnetic field, which will turn the spins, is equivalent to a corresponding mechanical rotation acting on the orbital component. Therefore a magnetic field can induce a response typical of a liquid crystal or, conversely, mechanical oscillations may excite collective spin motion (Liu and Cross 1978); see also Section 9.4. A system with a broken relative symmetry consequently behaves like one in which the symmetries of the constituent degrees of freedom are individually broken but that cannot distinguish between them—it confuses them.

6.1.3 Remaining symmetries of the ^{3}He-A, -A$_1$ and -B order parameters

For the sake of clarity, we have so far only discussed *two* different degrees of freedom and their symmetry breaking. Of course, in superfluid ^{3}He there exists a third broken symmetry, which is specifically connected with the actual superfluidity of ^{3}He: the broken symmetry of the overall phase of the order parameter. This leads to long-range order of this phase (see Section 3.2). So there are altogether three different continuous symmetries in superfluid ^{3}He, which can either be broken independently or in a relative sense. Fortunately, it is possible to formulate a systematic approach to the investigation of all possible symmetry breakings. This involves group-theoretical considerations that will be addressed in the next section. Using these methods, the findings discussed above can be derived rigorously. First, however, we want to do without this and find the remaining symmetries of the order parameters describing the superfluid A, A$_1$ and B phases, i.e. the symmetry operations under which the respective order parameters are still invariant.

(i) ^{3}He-B. The order parameter is given by (3.67), i.e. (see (5.1a))

$$A_{\mu j} = 3^{-1/2} e^{i\phi} R_{\mu j}(\hat{\boldsymbol{n}}, \theta), \tag{6.8}$$

where ϕ is the overall phase variable and $R_{\mu j}$ describes a relative rotation of spin and orbital spaces. Clearly, the effect of a rotation $R^{(\boldsymbol{L})}_{jk} \equiv O_{jk}$ of the orbital component of $A_{\nu k}$ can be undone by a rotation $R^{(\boldsymbol{S})}_{\mu\nu} = (\boldsymbol{ROR}^{-1})_{\mu\nu}$, i.e.

$$\sum_{\nu k} R^{(\boldsymbol{L})}_{jk} R^{(\boldsymbol{S})}_{\mu\nu} A_{\nu k} = A_{\mu j}. \tag{6.9}$$

Hence the order parameter is seen to be invariant under a simultaneous three-dimensional rotation in spin and orbital spaces. Consequently, the B-phase order parameter still shows an SO(3) invariance.

(ii) ^{3}He-A. The order parameter is given by (3.68), i.e.

$$A_{\mu j} = 2^{-1/2}\hat{d}_\mu(\hat{m}_j + \mathrm{i}\hat{n}_j), \quad \hat{\boldsymbol{l}} = \hat{\boldsymbol{m}} \times \hat{\boldsymbol{n}}. \tag{6.10}$$

Note that no explicit phase factor appears: the phase is implicitly contained in the orientation of the triad $(\hat{\boldsymbol{m}}, \hat{\boldsymbol{n}}, \hat{\boldsymbol{l}})$ along $\hat{\boldsymbol{l}}$. By performing a gauge transformation $A_{\mu j} \to \mathrm{e}^{\mathrm{i}\phi}A_{\mu j}$, an additional phase factor is introduced. On the other hand, a rotation of the orbital component of the order parameter about $\hat{\boldsymbol{l}}$ by an angle ϕ' leads to $\hat{\boldsymbol{m}} + \mathrm{i}\hat{\boldsymbol{n}} \to \mathrm{e}^{\mathrm{i}\phi'}(\hat{\boldsymbol{m}} + \mathrm{i}\hat{\boldsymbol{n}})$. Hence transformations of the gauge variable and of the orbital part around the z axis with $\phi' = -\phi$ just compensate each other. This particular linear combination of transformations thus leaves the order parameter invariant. It corresponds to a remaining U(1) symmetry of the order parameter involving both the gauge and the orbital variable, i.e. a relative symmetry. In addition, the order parameter remains unchanged under a rotation of the spin degrees of freedom about the direction of the real unit vector $\hat{\boldsymbol{d}}$. This is again equivalent to a U(1) symmetry, but now in spin space. In fact, the order parameter is also invariant under the replacement $(\hat{\boldsymbol{d}}, \hat{\boldsymbol{m}}, \hat{\boldsymbol{n}}) \to (-\hat{\boldsymbol{d}}, -\hat{\boldsymbol{m}}, -\hat{\boldsymbol{n}})$. This is a discrete symmetry, described by the group $\mathbb{Z}_2$ (the group of integers modulo 2, having two elements). Altogether, we find that the A-phase order parameter is still invariant under a $\mathrm{U}(1) \times \mathrm{U}(1) \times \mathbb{Z}_2$ symmetry, where one of the U(1) groups describes a relative symmetry.

(iii) ^{3}He-A$_1$. The order parameter is given by (5.59):

$$A_{\mu j} = \tfrac{1}{2}(\hat{d}_\mu + \mathrm{i}\hat{e}_\mu)(\hat{m}_j + \mathrm{i}\hat{n}_j). \tag{6.11}$$

A gauge transformation leads to $A_{\mu j} \to \mathrm{e}^{\mathrm{i}\phi}A_{\mu j}$ and a rotation of the orbital component about $\hat{\boldsymbol{l}} = \hat{\boldsymbol{m}} \times \hat{\boldsymbol{n}}$ to $\hat{\boldsymbol{m}} + \mathrm{i}\hat{\boldsymbol{n}} \to \mathrm{e}^{\mathrm{i}\phi'}(\hat{\boldsymbol{m}} + \mathrm{i}\hat{\boldsymbol{n}})$. Hence, just as in ^{3}He-A, a joint transformation of the gauge and the orbital part (the linear combination with $\phi' = -\phi$) leaves the order parameter invariant. This corresponds to a U(1) symmetry involving gauge and orbital degrees of freedom. On the other hand, since the spin part of the order parameter (6.11) is identical in structure with the orbital part, a gauge transformation can equally be compensated by a rotation of the *spin* component around $\hat{\boldsymbol{f}} = \hat{\boldsymbol{d}} \times \hat{\boldsymbol{e}}$ $(\hat{\boldsymbol{d}} + \mathrm{i}\hat{\boldsymbol{e}} \to \mathrm{e}^{\mathrm{i}\phi''}(\hat{\boldsymbol{d}} + \mathrm{i}\hat{\boldsymbol{e}}))$ if $\phi'' = -\phi$. This is yet another, different, U(1) symmetry, this time involving a linear combination of the gauge and the spin degree of freedom. (Note that this U(1) symmetry may equally be interpreted as a linear combination involving spin and orbital degrees of freedom, because the two can also compensate each other.) Hence the A$_1$-phase order parameter is seen to be invariant under two separate relative U(1) symmetries, i.e. under a $\mathrm{U}(1) \times \mathrm{U}(1)$ symmetry.

6.2 SYMMETRY AND ORDER-PARAMETER STRUCTURE†

We have now convinced ourselves that the order parameters of the A, A_1 and B phases still contain a remaining symmetry. Obviously the symmetry group *G*, (6.6), relevant to the overall structure of the order parameter $A_{\beta j}$ of superfluid ^{3}He is large enough to allow phases where the symmetry breaking does not destroy the full symmetry. This we found to be true even in the case of the para/ferromagnet, where the symmetry group $G = SO(3)$ is much smaller but where the ferromagnetic phase nevertheless has a remaining symmetry $H = U(1)$. On the other hand, the symmetry group relevant to the transition from normal liquid to superfluid ^{4}He, i.e. $G = U(1)$, is too small to allow continuous subgroups, and consequently the symmetry is completely broken below the transition. In the case of superfluid ^{3}He the symmetry group *G* is exceptionally large, leading to phases with a particularly complex structure. The observation that the A, A_1 and B phases still contain a symmetry $H \subset G$ may lead one to expect that physically relevant phases with a broken symmetry *R*, whose order parameter has a rich underlying symmetry *G*, will possess a remaining symmetry (i.e. $H \neq 1$).

Concerning superfluid ^{3}He, this expectation is supported by the following observation: it is easy to see that not only the order parameters of the A, A_1 and B phases but, in fact, those of *all* phases so far identified as stationary points of the Ginzburg–Landau free energy (5.4) (see Chapter 5) possess a remaining symmetry $H \subset G$. Hence in these phases the symmetry *G* is only partly broken. Consequently, it appears that an order parameter with a remaining symmetry corresponds to a phase with lower energy than one where all symmetries are broken. On the other hand, the question of which symmetry will actually be broken clearly depends on the details of the corresponding free energy. Therefore one should not expect the above statement to be generally true. Nevertheless, using somewhat simpler order parameters and free-energy expressions than those relevant for superfluid ^{3}He, one may give explicit examples where this is rigorously true (see e.g. Section 6.2.5).

Since an exact minimization of the Ginzburg–Landau functional (5.4) for general parameters β_i is not yet possible, it is clear that the symmetry concept of phases is extremely valuable: starting from the underlying symmetry group *G*, it allows for a systematic discussion of all phases with broken symmetry whose remaining symmetry $H \neq 1$ is a factor group of *G*. Once these phases and their corresponding order parameters have been obtained, one may compare their energy to find which is the stable one. Symmetry classifications of this kind have been used before. For

† This section contains a formal discussion of the symmetry classification of equilibrium order-parameter states. It is not a prerequisite for an understanding of the material in Section 6.3.

example, this approach has been applied to describe a ferroelectric phase transition in perovskites (Jarić 1982). Similarly, Kim (1985) illustrated the method by applying it to a Ginzburg–Landau potential describing $BaTiO_3$ in order to explain the three successive phase transitions known to occur in this system. Furthermore, such classifications have been shown to be very useful for studying possible phases of heavy-fermion superconductors (Anderson 1985, Volovik and Gorkov 1984, 1985, Ueda and Rice 1985, Blount 1985) and vortex states in superfluid ^{3}He and ^{4}He (Salomaa and Volovik 1985a,b, 1987). The latter will be discussed in Chapter 7.

In the case of superfluid ^{3}He we already know that the energetically stable phases (A, A_1, B) still have remaining symmetries, so we can be sure that they will appear in a symmetry classification as described above.

We now want to present a symmetry classification of the subgroups of the symmetry group relevant for superfluid ^{3}He, i.e. of $G = SO(3) \times SO(3) \times U(1)$. Such a classification was first considered by Golo and Monastyrsky (1977, 1978a–d) and Mineev (1980), yielding several states with a remaining symmetry H (see also Volovik and Gorkov 1984, 1985). In particular, Mineev (1980) identified all inert states (see Section 5.1) that are obtained by a factorization using continuous groups. This leads to five inert states (note that the "planar" state (5.20) cannot be constructed in this way since it contains a discrete symmetry; see below).

A symmetry classification of G, (6.6), including both continuous and discrete subgroups H, may be performed systematically (Bruder and Vollhardt 1986). It allows for explicit calculation of all bulk superfluid states with a remaining symmetry, i.e. states whose order parameter $A_{\mu j}$ is still invariant under a subgroup $H \subset G$. This classification will be presented next; continuous and discrete subgroups are discussed separately.

6.2.1 Continuous symmetries

Every group element $\mathsf{h} \in H$ corresponds to a transformation leaving $\boldsymbol{A} \equiv A_{\mu j}$ invariant (Mineev 1980)

$$\mathsf{h}\boldsymbol{A} = \boldsymbol{A}. \tag{6.12}$$

Introducing $\mathbf{T} = (\mathsf{T}_x, \mathsf{T}_y, \mathsf{T}_z)$, the generators of infinitesimal transformations of the group H, h may be written as

$$\mathsf{h} = \exp\left(-\frac{i}{\hbar}\boldsymbol{\alpha}\cdot\mathbf{T}\right) \tag{6.13a}$$

$$\approx 1 - \frac{i}{\hbar}\boldsymbol{\alpha}\cdot\mathbf{T}, \tag{6.13b}$$

where $\boldsymbol{\alpha}$ is an arbitrary vector of infinitesimal length. Equation (6.12) then reduces to

$$(\boldsymbol{\alpha} \cdot \mathsf{T})\boldsymbol{A} = 0, \tag{6.14a}$$

or

$$\mathsf{T}_i \boldsymbol{A} = 0, \quad i = x, y, z. \tag{6.14b}$$

The order parameter $A_{\mu j}$ has the transformation properties of a "bivector": it transforms like a vector with respect to its spin part (first index) and its orbital part (second index). Hence $A_{\mu j}$ may always be expressed as a linear superposition of tensor products between unit vectors $\hat{\boldsymbol{s}}^{(\nu)}$ and $\hat{\boldsymbol{o}}^{(i)}$ in spin and orbital spaces:

$$\boldsymbol{A} = \sum_{\nu i} \lambda_{\nu i} \hat{\boldsymbol{s}}^{(\nu)} \otimes \hat{\boldsymbol{o}}^{(i)}. \tag{6.15}$$

In the present text the sets of unit vectors $\hat{\boldsymbol{s}}^{(\nu)}$ and $\hat{\boldsymbol{o}}^{(i)}$ are denoted by $(\hat{\boldsymbol{d}}, \hat{\boldsymbol{e}}, \hat{\boldsymbol{f}})$ and $(\hat{\boldsymbol{l}}, \hat{\boldsymbol{m}}, \hat{\boldsymbol{n}})$ respectively. Equation (6.15), combined with the invariance properties of the free energy under the group of transformations (6.6) concerning rotations in spin and orbital spaces and a phase transformation, implies that T can itself be written as a linear combination of these transformations; for example,

$$\left.\begin{aligned} \mathsf{T}_x &= a_x \mathsf{L}^x + b_x \mathsf{S}^x, \\ \mathsf{T}_y &= a_y \mathsf{L}^y + b_y \mathsf{S}^y, \\ \mathsf{T}_z &= a_z \mathsf{L}^z + b_z \mathsf{S}^z + c\Phi. \end{aligned}\right\} \tag{6.16}$$

Here $\mathbf{L} = (\mathsf{L}^x, \mathsf{L}^y, \mathsf{L}^z)$ and $\mathbf{S} = (\mathsf{S}^x, \mathsf{S}^y, \mathsf{S}^z)$ are respectively the three-component spin and orbital-angular-momentum operators in the three-dimensional representation (i.e. 3×3 matrices). They are given by

$$(\mathsf{L}^i)_{jk} = -\mathrm{i}\hbar\epsilon_{ijk}, \tag{6.17a}$$

$$(\mathsf{S}^\mu)_{\nu\lambda} = -\mathrm{i}\hbar\epsilon_{\mu\nu\lambda}. \tag{6.17b}$$

Furthermore, $\Phi = -\mathrm{i}\hbar\, \partial/\partial\phi$ is the operator for a gauge transformation, with

$$\Phi \boldsymbol{A} = \boldsymbol{A}, \quad \Phi \boldsymbol{A}^* = -\boldsymbol{A}. \tag{6.18}$$

While $\mathbf{S}$ acts only on the first index of $A_{\mu j}$, $\mathbf{L}$ acts only on the second one and Φ acts only on the overall phase.

There are altogether seven generators $(\mathbf{L}, \mathbf{S}, \Phi)$—a fact that is reflected in the existence of seven arbitrary real parameters $\boldsymbol{a} = (a_x, a_y, a_z)$, $\boldsymbol{b} = (b_x, b_y, b_z)$ and c.

The relations (6.17) define Lie algebras among the L^i and S^μ components:

$$[\mathsf{L}^i, \mathsf{L}^j] = \mathrm{i}\hbar \sum_k \epsilon_{ijk} \mathsf{L}^k, \tag{6.19a}$$

$$[\mathsf{S}^\mu, \mathsf{S}^\nu] = \mathrm{i}\hbar \sum_\lambda \epsilon_{\mu\nu\lambda} \mathsf{S}^\lambda. \tag{6.19b}$$

We note that the Lie algebra (6.19) is common to both the rotation group SO(3) and the special unitary group SU(2), although they are different groups. The connection between the two groups is given by $SO(3) = SU(2)/\mathbb{Z}_2$, where $\mathbb{Z}_2$ describes a reflection, i.e. a discrete symmetry. This brings out the fact that SU(2) is a simply connected group while SO(3) is doubly connected (in Fig. 6.2(b) opposite points on the sphere are identified). Hence *locally* the two groups can be identified (and this is precisely the reason why they share the same Lie algebra under infinitesimal transformations). *Globally,* however, the difference is of great importance. Nevertheless, for an investigation of broken continuous symmetries in bulk superfluid states, they can be used alternatively.

Equation (6.14b) together with (6.16) and (6.17) yields

$$a_i \sum_l \epsilon_{ijl} A_{\mu l} + b_i \sum_\nu \epsilon_{i\mu\nu} A_{\nu j} + \mathrm{i}\delta_{iz} c A_{\mu j} = 0 \quad (i = x, y, z), \tag{6.20}$$

i.e. three sets of nine homogeneous equations for the nine complex components of the order parameter $A_{\mu j}$ in terms of the independent parameters a_j, b_μ and c. Clearly, this system of equations is overcomplete, which in general implies a solution $A_{\mu j} = 0$. Only by constructing relations, i.e. linear combinations, between the a_i, b_μ and c can we enforce $A_{\mu j} \neq 0$. This means that we break the underlying symmetry (at least parts of it). By constructing all possible linear combinations of the a_i, b_μ and c, we then find all continuous subgroups $H \neq 1$ of G, i.e. we construct all those order-parameter structures that still possess some of the original symmetry (6.6). In other words, in such states the symmetry G is not completely broken.

The order parameter $A_{\mu j}$ in (6.20) is calculated in a Cartesian coordinate system ($\mu, j = x, y, z$), i.e. the elements of $A_{\mu j}$ are given by A_{xx}, A_{xy} etc. This is merely done for reasons of convention. Clearly, any other coordinate system may equivalently be used; in fact, in the present case, a spherical basis is particularly suited. The different representations are connected via unitary transformations. In the case of a vector $\boldsymbol{a}_{\mathrm{Cart}} = (a_x, a_y, a_z)$ this is easily illustrated. Using

$$\mathbf{U} = \frac{1}{2^{1/2}} \begin{pmatrix} 1 & -\mathrm{i} & 0 \\ 0 & 0 & 2^{1/2} \\ 1 & \mathrm{i} & 0 \end{pmatrix}, \tag{6.21}$$

$\boldsymbol{a}_{\mathrm{spher}} = \mathbf{U}\boldsymbol{a}_{\mathrm{Cart}}$ is calculated as

$$\boldsymbol{a}_{\mathrm{spher}} = \begin{pmatrix} \dfrac{a_x - \mathrm{i}a_y}{2^{1/2}} \\ a_z \\ \dfrac{a_x + \mathrm{i}a_y}{2^{1/2}} \end{pmatrix} \tag{6.22a}$$

$$\equiv \begin{pmatrix} a_- \\ a_0 \\ a_+ \end{pmatrix}, \tag{6.22b}$$

where the notation in (6.22b) is self-explanatory. In the case of a matrix order parameter $\boldsymbol{A}$, the same concept applies, i.e.

$$\boldsymbol{A}_{\text{spher}} = \mathbf{U}\boldsymbol{A}_{\text{Cart}}\mathbf{U}^{+}. \tag{6.23}$$

Using a notation equivalent to (6.22b), the matrix elements of $\boldsymbol{A}_{\text{spher}}$ are given by A_{++}, A_{+-} etc.

We shall now list all possible linear combinations of L^i, S^μ, ϕ in the infinitesimal generators, i.e. all possible continuous subgroups H of G. Equation (6.20) is then used to calculate the components of $A_{\mu j}$. It should be noted that the order parameters thus obtained are only determined up to a rotation in spin and orbital space and an overall phase factor, because there is no fixed coordinate system (see the discussion below (5.15)).

First we start with the case where the linear combination leads to a remaining symmetry under *all* components of $\mathbf{T}$ in (6.16). This corresponds to a remaining SO(3) symmetry. From (6.20), it follows that a solution $A_{\mu j} \neq 0$ is only allowed if $c = 0$. In this case the T_i, (6.16), have to transform like the components of an angular momentum themselves, i.e. they have to obey (6.19), which leads to $a_i = 1$, $b_\mu = 1$ for all i, μ (note that other solutions, e.g. $a_x = 1$, $a_y = a_z = -1$, $b_\mu = 1$, may be brought into this form via rotations by π about one of the x, y, z axes, and are thus equivalent). So the first linear combination of the a_i, b_μ and c is given by the following.

(1) $c = 0$, $a_i = b_\mu = 1$. This implies

$$\mathbf{T} = \mathbf{L} + \mathbf{S}, \tag{6.24}$$

corresponding to $H = \mathrm{SO}(3)_{\boldsymbol{L}+\boldsymbol{S}}$. Here the order parameter is still invariant under joint three-dimensional rotations of $\boldsymbol{L}$ and $\boldsymbol{S}$. In this sense it is the most symmetric of all possible order parameters. Solving (6.20), one finds (see (3.65))

$$A_{\mu j} = 3^{-1/2}\delta_{\mu j}, \tag{6.25a}$$

which is a special case of the more general form (see the remarks made above) given in (3.67)

$$A_{\mu j} = 3^{-1/2}\mathrm{e}^{\mathrm{i}\phi}R_{\mu j}(\hat{\boldsymbol{n}}, \theta), \tag{6.25b}$$

where $R_{\mu j}$ is a rotation matrix describing relative rotations of spin and orbital spaces, represented by a rotation axis $\hat{\boldsymbol{n}}$ and a rotation angle θ. Note that this more general state (6.25b) is invariant under simultaneous rotations described by a matrix O_{lj} in orbit space and $(\boldsymbol{ROR}^{-1})_{\nu\mu}$ in spin space. This is, of course, the BW state, describing the B phase. The explicit phase factor $\mathrm{e}^{\mathrm{i}\phi}$ results from the separately broken gauge invariance. To determine the degeneracy space $R = G/H$, i.e. the broken symmetries, we factorize G with respect to H:

$$R = \frac{\mathrm{SO}(3)_{\boldsymbol{L}} \times \mathrm{SO}(3)_{\boldsymbol{S}} \times \mathrm{U}(1)_{\phi}}{\mathrm{SO}(3)_{\boldsymbol{L}+\boldsymbol{S}}} \tag{6.26a}$$

$$= \mathrm{SO}(3)_{\boldsymbol{L},\boldsymbol{S}} \times \mathrm{U}(1)_{\phi}. \tag{6.26b}$$

We observe that the factorization of G with respect to $SO(3)_{L+S}$, the group of all *joint* rotations of $\boldsymbol{L}$ and $\boldsymbol{S}$, leaves us with $SO(3)_{L,S}$, the group of all *relative* rotations of $\boldsymbol{L}$ and $\boldsymbol{S}$, as well as $U(1)_\phi$. Hence in the BW state the relative spin–orbit symmetry (as well as the gauge symmetry) is broken. This is precisely the "spontaneously broken spin–orbit symmetry (SBSOS)" introduced by Leggett (1972) (cf. the detailed discussion in Section 6.1 concerning this particular kind of symmetry breaking in a two-dimensional model liquid). We see that the notion of a "broken relative symmetry" (Liu and Cross 1978) occurs quite naturally within a group-theoretical formulation. With the exception of the BW phase with $H = SO(3)$, the subgroups of G are composed of U(1) subgroups only, i.e. there are no other solutions $A_{\mu j} \neq 0$ to (6.20) that involve all three components of $\mathbf{T}$. We may then restrict any further discussion to the invariance properties of $A_{\mu j}$ under T_z alone. Equation (6.20) then reduces to

$$a_z(\delta_{j1}A_{\mu 2} - \delta_{j2}A_{\mu 1}) + b_z(\delta_{\mu 1}A_{2j} - \delta_{\mu 2}A_{1j}) + \mathrm{i}cA_{\mu j} = 0. \tag{6.27}$$

In fact, it separates into four independent subsystems belonging to different sets of $A_{\mu j}$ components. For $A_{\mu j}$ to be nonzero, the determinant D of the matrix (6.27) has to vanish:

$$D = D_1 D_2 D_3 D_4 = 0. \tag{6.28a}$$

Here the D_i are the determinants of the submatrices belonging to the $A_{\mu j}$ components as indicated:

$$D_1 = [(a_z - b_z)^2 - c^2][(a_z + b_z)^2 - c^2] \quad (A_{11}, A_{12}, A_{21}, A_{22}), \tag{6.28b}$$

$$D_2 = b_z^2 - c^2 \quad (A_{13}, A_{23}), \tag{6.28c}$$

$$D_3 = a_z^2 - c^2 \quad (A_{31}, A_{32}), \tag{6.28d}$$

$$D_4 = \mathrm{i}c \quad (A_{33}). \tag{6.28e}$$

This way of writing allows for a simple and systematic investigation of all linear combinations of a_z, b_z and c in (6.16). Depending on which of the D_i we choose to be zero (nonzero), we can immediately determine which of the $A_{\mu j}$ components are nonzero (zero). From (6.28), we see that only linear combinations of the type $a_z = \pm c$, $a_z - b_z = \pm c$ etc. (and not, say, $a_z = \pm 2c$) yield nonvanishing $A_{\mu j}$. This is so because a_z, b_z and c correspond to angles of rotation around the z axis. To accomplish mutual compensation of rotations in the three different spaces (implying a residual symmetry), only particular choices of the a_z, b_z and c are possible.

To construct a linear combination of infinitesimal generators means that we impose conditions on them ("break" the symmetry). As will be shown below, in the case of (6.27) a *single* condition (say, $c = 0$), i.e. the weakest nontrivial restriction on (6.27), implies a remaining symmetry $H = U(1) \times U(1)$, leading to an inert state, discussed in Chapter 5. Their order-parameter matrices only depend on a single overall factor, which is fixed by normalization. Hence the order-parameter structure is constant and, in particular, independent of the parameters β_i. On the other hand, imposing more than one condition (say, $c=0$ *and* $a_z=b_z$) implies an even more restricted symmetry, $H=\mathrm{U}(1)$, leading to *non*inert states. Their order-parameter matrices depend on more than one parameter, such that even a slight variation of the β_i (due, for example, to a pressure change) influences their whole structure.

We see that the subgroups with the largest remaining symmetries lead to inert states. In fact, in group-theoretical language they are called "maximal" subgroups $H_{\max}$, because there are no other subgroups $H \subset G$ containing $H_{\max}$. Hence maximal subgroups produce inert states.

In the spherical representation of $\boldsymbol{A}$, (6.23), the invariance property (6.12) determining $A_{\mu\nu}\,(\mu,\,\nu=0,\,+,\,-)$, and hence the set of equations (6.27), has a particularly simple structure:

$$(a_z\nu + b_z\mu + c)A_{\mu\nu} = 0. \tag{6.29}$$

This is of course a result of the fact that in this basis the spherical harmonics $Y_{1,m}$ are eigenstates of L_z and S_z with eigenvalues $m\hbar$ (see (7.81)–(7.83)). The different order-parameter structures allowed by symmetry are again obtained by choosing all possible linear combinations of the infinitesimal generators a_z, b_z and c (Salomaa and Volovik 1987, Schakel and Bais 1989).

We now complete the list of possible states.

(2) $c=0$, a_z, b_z *arbitrary*. This implies

$$\mathsf{T}_z = a_z\mathsf{L}^z + b_z\mathsf{S}^z, \tag{6.30}$$

and hence $H=\mathrm{U}(1)_{S_z}\times\mathrm{U}(1)_{L_z}$. Because of $D_1, D_2, D_3 \neq 0, D_4=0$, one immediately obtains the polar state

$$A_{\mu j} = \mathrm{e}^{\mathrm{i}\phi}\begin{pmatrix} 0 & 0 & 0 \\ 0 & 0 & 0 \\ 0 & 0 & 1 \end{pmatrix}. \tag{6.31}$$

Its order parameter is still invariant under separate rotations around the preferred directions in spin and orbital spaces, which are both real unit vectors. Accordingly, its broken symmetries are given by

$$R = \frac{\mathrm{SO}(3)_{\boldsymbol{L}}}{\mathrm{U}(1)_{L_z}} \times \frac{\mathrm{SO}(3)_{\boldsymbol{S}}}{\mathrm{U}(1)_{S_z}} \times \mathrm{U}(1)_\phi, \tag{6.32a}$$

$$= \mathrm{S}^2_{\boldsymbol{L}} \times \mathrm{S}^2_{\boldsymbol{S}} \times \mathrm{U}(1)_\phi, \tag{6.32b}$$

i.e. besides the phase invariance, rotational invariance with respect to the directions of the preferred direction in spin and orbital space is also broken.

The polar state is unique in that it is the only inert state without a broken relative symmetry.

(3) $c=0$, $b_z=0$, a_z *arbitrary*. This implies

$$\mathsf{T}_z = a_z \mathsf{L}^z, \tag{6.33}$$

and hence $H = \mathrm{U}(1)_{L_z}$. Because $D_1, D_3 \neq 0$ and $D_2 = D_4 = 0$, the order parameter has the structure

$$A_{\mu j} = p_1 e^{i\phi} \begin{pmatrix} 0 & 0 & A \\ 0 & 0 & B \\ 0 & 0 & C \end{pmatrix}, \tag{6.34}$$

with A, B and C complex and $p_1 = [|A|^2 + |B|^2 + |C|^2]^{-1/2}$. The condition $b_z = 0$ implies that in this noninert state the symmetry with respect to rotation in spin space has been *completely* broken:

$$R = \frac{\mathrm{SO}(3)_{\boldsymbol{L}}}{\mathrm{U}(1)_{\boldsymbol{L}_z}} \times \mathrm{SO}(3)_{\boldsymbol{S}} \times \mathrm{U}(1)_\phi \tag{6.35}$$

$$= \mathrm{S}^2_{\boldsymbol{L}} \times \mathrm{SO}(3)_{\boldsymbol{S}} \times \mathrm{U}(1)_\phi. \tag{6.36}$$

In this respect it is less symmetric (i.e. more restrictive) than the polar state (6.30), which is seen to be a special case of (6.34), with $A = B = 0$.

(4) $c=0$, $a_z=0$, b_z *arbitrary*. This implies

$$\mathsf{T}_z = b_z \mathsf{S}^z, \tag{6.37}$$

and hence $H = \mathrm{U}(1)_{S_z}$. It corresponds to the previous state with orbital and spin components interchanged, i.e.

$$A_{\mu j} = p_1 e^{i\phi} \begin{pmatrix} 0 & 0 & 0 \\ 0 & 0 & 0 \\ A & B & C \end{pmatrix} \tag{6.38}$$

and $R = \mathrm{SO}(3)_{\boldsymbol{L}} \times \mathrm{S}^2_{\boldsymbol{S}} \times \mathrm{U}(1)$. Here the rotational symmetry in orbital space is completely broken.

(5) $c=0$, $a_z \mp b_z = 0$. This implies

$$\mathsf{T}_z = a_z(\mathsf{L}^z \pm \mathsf{S}^z), \tag{6.39}$$

and hence $H = \mathrm{U}(1)_{L_z+S_z}$. Because $D_1 = D_4 = 0$ and $D_2, D_3 \neq 0$, this leads to

$$A_{\mu j} = p_2 e^{i\phi} \begin{pmatrix} A & B & 0 \\ \pm B & \mp A & 0 \\ 0 & 0 & C \end{pmatrix}, \tag{6.40}$$

where A, B and C are complex numbers and $p_2 = [2(|A|^2 + |B|^2) + |C|^2]^{-1/2}$.

This state is a generalized representation of the so-called "ζ phase" (Barton and Moore 1974b). However, the name "oblate state" appears to be more useful, because it actually describes a squashed spherical state. In fact, (6.40) represents precisely the order-parameter structure of the B_2 phase (5.66), i.e. of the B phase in an external magnetic field (with $A = \frac{1}{2}(\Delta_{\uparrow\uparrow} + \Delta_{\downarrow\downarrow})$, $B = -\frac{1}{2}\mathrm{i}(\Delta_{\uparrow\uparrow} - \Delta_{\downarrow\downarrow})$, $C = \Delta_{\uparrow\downarrow}$). Furthermore, by flattening the sphere completely, i.e. choosing $B = C = 0$, one obtains the usual planar state. As will be discussed below, B and C are naturally forced to zero by demanding some additional *discrete* symmetry for $A_{\mu j}$. Therefore the actual planar state contains both continuous and discrete symmetries. The broken symmetries of (6.40) are found to be

$$R = \frac{\mathrm{SO}(3)_{\boldsymbol{L}} \times \mathrm{SO}(3)_{\boldsymbol{S}}}{\mathrm{U}(1)_{L_z+S_z}} \times \mathrm{U}(1)_{\phi} \tag{6.41}$$

$$= \mathrm{S}^2_{\boldsymbol{L}} \times \mathrm{S}^2_{\boldsymbol{S}} \times \mathrm{U}(1)_{L_z,S_z} \times \mathrm{U}(1)_{\phi}, \tag{6.42}$$

which includes the broken relative symmetry $\mathrm{U}(1)_{L_z,S_z}$ with respect to rotations in spin and orbital spaces.

(6) $a_z + c = 0$, b_z *arbitrary*. This implies

$$\mathsf{T}_z = a_z(\mathsf{L}^z - \Phi) + b_z\mathsf{S}^z, \tag{6.43}$$

and hence $H = \mathrm{U}(1)_{L_z-\phi} \times \mathrm{U}(1)_{S_z}$. Because $D_3 = 0$ and $D_1, D_2, D_4 \neq 0$, one finds

$$A_{\mu j} = \frac{1}{2^{1/2}}\begin{pmatrix} 0 & 0 & 0 \\ 0 & 0 & 0 \\ 1 & \mathrm{i} & 0 \end{pmatrix}, \tag{6.44}$$

which is of course the ABM state, describing the A phase. It is still invariant under rotations about one direction (the z direction) in spin space and under a joint transformation of the phase and a rotation around L_z. The broken symmetry is obtained as

$$R = \frac{\mathrm{SO}(3)_{\boldsymbol{S}}}{\mathrm{U}(1)_{S_z}} \times \frac{\mathrm{SO}(3)_{\boldsymbol{L}} \times \mathrm{U}(1)_{\phi}}{\mathrm{U}(1)_{L_z-\phi}} \tag{6.45}$$

$$= \mathrm{S}^2_{\boldsymbol{S}} \times \mathrm{SO}(3)_{\boldsymbol{L},\phi}. \tag{6.46}$$

Here and in the following we interpret $\mathrm{SO}(3)_{\boldsymbol{L},\phi}$ as $\mathrm{S}^2_{\boldsymbol{L}} \times \mathrm{U}(1)_{L_z,\phi}$. Hence invariance with respect to rotations of directions in spin and orbital spaces, as well as the relative gauge–orbital symmetry $\mathrm{U}(1)_{L_z,\phi}$, is broken (Liu and Cross 1979) (cf. the related discussion in Section 6.1 concerning a two-dimensional model liquid).

We note that in the above classification of continuous symmetries the discrete symmetry of the A phase under the replacement $(\hat{\boldsymbol{d}}, \hat{\boldsymbol{m}}, \hat{\boldsymbol{n}}) \rightarrow (-\hat{\boldsymbol{d}}, -\hat{\boldsymbol{m}}, -\hat{\boldsymbol{n}})$ discussed below (6.10) cannot be obtained (see next Section 6.2.2). Therefore the broken symmetries are not yet fully described by (6.46) (see (6.68b)).

(7) $a_z + c = 0$, $b_z = 0$. This implies

$$\mathsf{T}_z = a_z(\mathsf{L}^z - \Phi), \tag{6.47}$$

and hence $H = \mathrm{U}(1)_{L_z - \phi}$. Because $D_1 = D_3 = 0$, $D_2, D_4 \neq 0$, one finds

$$A_{\mu j} = \frac{p_1}{2^{1/2}} \begin{pmatrix} A & \mathrm{i}A & 0 \\ B & \mathrm{i}B & 0 \\ C & \mathrm{i}C & 0 \end{pmatrix}. \tag{6.48}$$

In this state spin-rotational symmetry is completely broken ($\boldsymbol{b} = 0$):

$$R = \mathrm{SO}(3)_{\boldsymbol{S}} \times \frac{\mathrm{SO}(3)_{\boldsymbol{L}} \times \mathrm{U}(1)_{\phi}}{\mathrm{U}(1)_{L_z - \phi}} \tag{6.49}$$

$$= \mathrm{SO}(3)_{\boldsymbol{S}} \times \mathrm{SO}(3)_{\boldsymbol{L},\phi}. \tag{6.50}$$

Because of the more restrictive symmetry requirements, its order parameter is less symmetric than either that of the A phase ($A = B = 0$) or that of the A_1 phase ($A = \mathrm{i}B$, $C = 0$). In fact, it is a general representation of the A_2 phase (5.51), which is obtained by choosing $A = \frac{1}{2}(\Delta_\uparrow + \Delta_\downarrow)$, $B = \frac{1}{2}\mathrm{i}(\Delta_\uparrow - \Delta_\downarrow)$ and $C = 0$. Note that the orientation of the spin-coordinate system in the present case ($\hat{\boldsymbol{d}} \parallel \hat{\boldsymbol{x}}$, $\hat{\boldsymbol{e}} \parallel \hat{\boldsymbol{y}}$) has been chosen differently from that for the A phase ($\hat{\boldsymbol{d}} \parallel \hat{\boldsymbol{z}}$), which is, however, of no importance.

(8) $b_z + c = 0$, $a_z = 0$. This implies

$$\mathsf{T}_z = b_z(\mathsf{S}^z - \Phi), \tag{6.51}$$

and hence $H = \mathrm{U}(1)_{S_z - \phi}$. It can be obtained from the previous state by interchanging orbital and spin spaces, i.e.

$$A_{\mu j} = \frac{p_1}{2^{1/2}} \begin{pmatrix} A & B & C \\ \mathrm{i}A & \mathrm{i}B & \mathrm{i}C \\ 0 & 0 & 0 \end{pmatrix}. \tag{6.52}$$

Here rotational symmetry in orbital space is completely broken ($\boldsymbol{a} = 0$), i.e. $R = \mathrm{SO}(3)_{\boldsymbol{L}} \times \mathrm{SO}(3)_{\boldsymbol{S},\phi}$.

(9) $b_z + c = 0$, *a_z arbitrary*. This implies

$$\mathsf{T}_z = a_z \mathsf{L}^z + b_z(\mathsf{S}^z - \Phi), \tag{6.53}$$

and hence $H = \mathrm{U}(1)_{L_z} \times \mathrm{U}(1)_{S_z - \phi}$. Because $D_2 = 0$ and $D_1, D_3, D_4 \neq 0$, one

finds

$$A_{\mu j}=\frac{1}{2^{1/2}}\begin{pmatrix}0 & 0 & 1\\ 0 & 0 & \mathrm{i}\\ 0 & 0 & 0\end{pmatrix}, \tag{6.54}$$

which is called the "β state". Its symmetry properties etc. are obtained by interchanging the orbital and spin parts of the ABM state (6.44) ($\boldsymbol{A}\rightarrow\boldsymbol{A}^{\mathrm{T}}$). Accordingly, in this case a relative gauge–spin symmetry is broken.

(10) $c=a_z-b_z$. This implies

$$\mathsf{T}_z=a_z(\mathsf{L}^z+\phi)+b_z(\mathsf{S}^z-\phi), \tag{6.55}$$

and hence $H=\mathrm{U}(1)_{L_z+\phi}\times\mathrm{U}(1)_{S_z-\phi}$. Here one has $D_1=0$ and D_2, D_3, $D_4\neq 0$, so that

$$A_{\mu j}=\frac{1}{2}\begin{pmatrix}1 & \mathrm{i} & 0\\ -\mathrm{i} & 1 & 0\\ 0 & 0 & 0\end{pmatrix}, \tag{6.56}$$

which is the representation of the A_1 phase. The broken symmetry is given by

$$R=\mathrm{S}_{\boldsymbol{L}}^2\times\mathrm{S}_{\boldsymbol{S}}^2\times\frac{\mathrm{U}(1)_{L_z}\times\mathrm{U}(1)_{S_z}\times\mathrm{U}(1)_{\phi}}{\mathrm{U}(1)_{L_z+\phi}\times\mathrm{U}(1)_{S_z-\phi}} \tag{6.57}$$

$$=\mathrm{S}_{\boldsymbol{L}}^2\times\mathrm{S}_{\boldsymbol{S}}^2\times\mathrm{U}(1)_{L_z,S_z,\phi}. \tag{6.58}$$

In this case a U(1) symmetry involving a linear combination of all *three* symmetries, i.e. a gauge–spin–orbit symmetry, is broken (Liu 1979a).

(11) $a_z+c=0$, $b_z+c=0$. This implies

$$\mathsf{T}_z=a_z(\mathsf{L}^z+\mathsf{S}^z-\phi), \tag{6.59}$$

and hence $H=\mathrm{U}(1)_{L_z+S_z-\phi}$. Here one has D_1, $D_4\neq 0$ and $D_2=D_3=0$, so that

$$A_{\mu j}=\frac{p_3}{2^{1/2}}\begin{pmatrix}0 & 0 & A\\ 0 & 0 & \mathrm{i}A\\ B & \mathrm{i}B & 0\end{pmatrix}, \tag{6.60}$$

where A and B are complex numbers and $p_3=[|A|^2+|B|^2]^{-1/2}$. This state has been called the "ε state" (Barton and Moore 1974b); it may be viewed as a linear combination of the A and β states. This noninert state is only invariant under joint U(1) rotations of spin, gauge and orbital variables. In a way it is complementary to the A_1 phase. Its broken symmetries are

$$R=\mathrm{S}_{\boldsymbol{L}}^2\times\mathrm{S}_{\boldsymbol{S}}^2\times\frac{\mathrm{U}(1)_{L_z}\times\mathrm{U}(1)_{S_z}\times\mathrm{U}(1)_{\phi}}{\mathrm{U}(1)_{L_z-S_z-\phi}} \tag{6.61}$$

$$= S^2_{\boldsymbol{L}} \times S^2_{\boldsymbol{S}} \times U(1)_{L_z,\Phi} \times U(1)_{S_z,\Phi}, \tag{6.62}$$

i.e. a relative spin–gauge and a relative gauge–orbit symmetry are broken.

(12) $a_z + 2b_z = 0$, $c + b_z = 0$. This implies

$$\mathsf{T}_z = c(2\mathsf{L}^z - \mathsf{S}^z + \Phi), \tag{6.63}$$

and hence $H = U(1)_{2L_z - S_z + \Phi}$. We are here dealing with a different linear combination of infinitesimal generators than in case (11) and hence with a different symmetry H. Because $D_1 = D_2 = 0$ and D_3, $D_4 \neq 0$, one has

$$A_{\mu j} = \frac{p_4}{2^{1/2}} \begin{pmatrix} A & -\mathrm{i}A & B \\ -\mathrm{i}A & -A & \mathrm{i}B \\ 0 & 0 & 0 \end{pmatrix}, \tag{6.64}$$

where A and B are complex numbers and $p_4 = [2\,|A|^2 + |B|^2]^{-1/2}$. This noninert state is a superposition of the A_1 and β states. As in the case of the ε state, it has two separate broken relative U(1) symmetries.

(13) $2a_z + b_z = 0$, $c + a_z = 0$. This implies

$$\mathsf{T}_z = -c(\mathsf{L}^z - 2\mathsf{S}^z - \Phi), \tag{6.65}$$

and hence $H = U(1)_{L_z - 2S_z - \Phi}$. It is obtained from (12) by interchanging the spin and orbital parts of the order parameter ($\boldsymbol{A} \rightarrow \boldsymbol{A}^{\mathrm{T}}$).

The last two states are similar to case (11) because the symmetry group H involves particular linear combinations of all three symmetries, i.e. particular joint rotations. Consequently, their broken relative symmetries will all be very similar.

We have thus found 13 order-parameter structures that still contain the symmetry of a continuous subgroup H of G: one with a remaining SO(3) symmetry, four with a U(1) × U(1) symmetry (they all lead to inert states) and eight with a remaining U(1) symmetry, yielding noninert states.

In Fig. 6.4 the consecutive factorization of the symmetry group G with respect to continuous subgroups is shown schematically. An arrow running diagonally always indicates that the factorization involves more than one group, such that the resulting state has a broken relative symmetry.

The figure also allows for a straightforward identification of the maximal subgroups, i.e. those subgroups of G that are not contained in any other subgroups themselves; they lead to inert states. There is one SO(3) subgroup, generating the B phase, and four U(1) × U(1) subgroups corresponding to the polar, A, β and A_1 phases. All other subgroups derive from these five maximal subgroups and lead to noninert states. Note that the classification yields all phases known to exist experimentally.

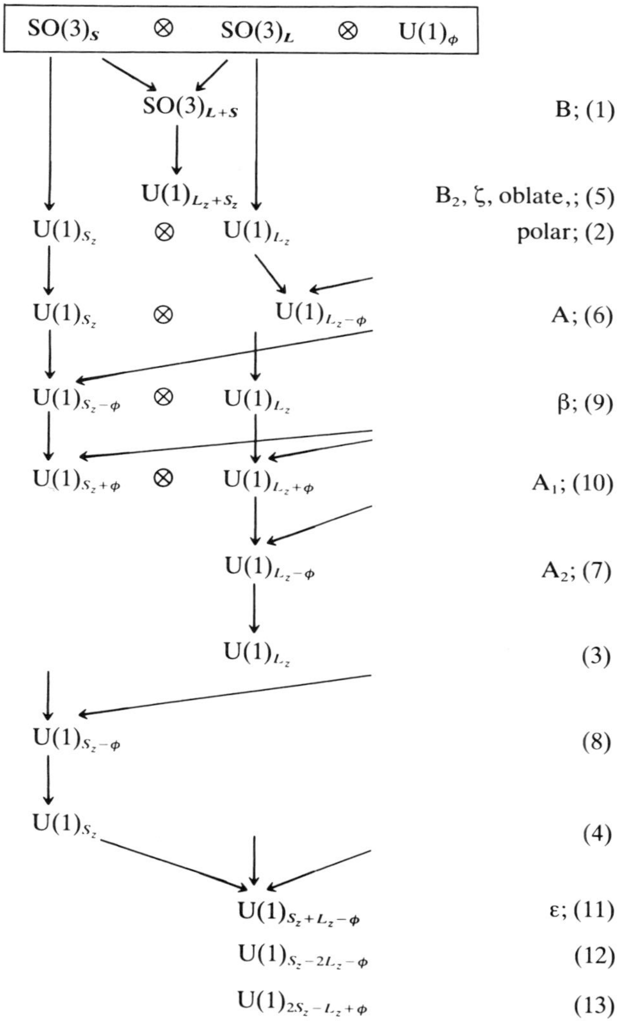

Figure 6.4 Consecutive factorization of the symmetry group $G = \mathrm{SO}(3)_L \times \mathrm{SO}(3)_S \times \mathrm{U}(1)_\phi$ with respect to continuous subgroups. Broken relative symmetries arise in factorizations involving more than one group, indicated by diagonal arrows. The numbers on the right refer to the states as they appear in the symmetry classification; conventional names of states are also included (Bruder and Vollhardt 1986).

6.2.2 Discrete symmetries

There is no a priori reason why superfluid phases in a spin-triplet p-wave state should only display broken continuous symmetries—they may also have broken discrete symmetries. As discussed above, the continuous subgroups of G may be specified by the infinitesimal generators of G: each element is written as an exponential of a linear combination of generators. There is an analogue in the theory of discrete groups: a discrete group is generated by a subset of elements, i.e. each element can be written as a product of powers of generators. The groups C_n for example, i.e. the rotations through $2\pi k/n$, $k = 1, \ldots, n-1$, are generated by one element only. However, because of the large number of different discrete subgroups of G, these generators turn out not to be so useful, i.e. it is more convenient to resort to a different approach.

Let H_d be a discrete subgroup of G, (6.6). Each element of G can be written as a triple $(r^o, r^s, e^{i\phi})$ describing a rotation r^o in orbital space, a rotation r^s in spin space and a gauge transformation $e^{i\phi}$. Clearly, r^o and r^s will have to be elements of discrete subgroups of SO(3), and $e^{i\phi}$ an element of a discrete subgroup of U(1).

The discrete subgroups of SO(3) and U(1) are well known (see e.g. Landau and Lifshitz 1958). For SO(3), they are given by

- (i) C_n — the rotations about the z axis through $2\pi k/n$, with $k = 1, \ldots, n-1$; the elements of C_n are denoted by C_n^k;
- (ii) D_n — generated by C_n and an additional rotation through π about an orthogonal axis;
- (iii) T, O, Y — the point groups of the tetrahedron, the cube and the icosahedron respectively.

For each discrete subgroup H_d of G, there exists a *minimal* direct-product subgroup $H_o \times H_s \times H_\phi$ of G that contains H_d. "Minimal" means that there is no smaller direct-product subgroup of $H_o \times H_s \times H_\phi$ that contains H_d. It is easy to construct H_o, H_s and H_ϕ by projection of H_d onto the factors $SO(3)_{\boldsymbol{L}}$, $SO(3)_{\boldsymbol{S}}$ and U(1) of G.

This construction may also be applied to continuous subgroups. If, for example, we take the symmetry group of the B phase, $SO(3)_{\boldsymbol{S}+\boldsymbol{L}}$, the trivial result is that $SO(3)_{\boldsymbol{L}} \times SO(3)_{\boldsymbol{S}}$ is the minimal direct product containing $SO(3)_{\boldsymbol{S}+\boldsymbol{L}}$.

It turns out that H_ϕ, the discrete group of gauge transformations, is essentially determined once H_o and H_s are given. It is therefore sufficient to elucidate the structure of the orbital and spin parts of H_d with respect to $H_o \times H_s$.

The discrete subgroups H_d of $H_o \times H_s$ can be thought of as follows: there exist normal subgroups H_o' of H_o and H_s' of H_s such that the factor groups H_o/H_o' and H_s/H_s' are isomorphic, i.e. in one-to-one correspondence. (Note that in the case of normal subgroups the coset space $R = G/H$ is itself a group.) The group H_d is then given by the set of all pairs (h_o, h_s) with $h_o \in H_o/H'_o$ and $h_s \in H_s/H'_s$, where h_o and h_s are isomorphic elements. Since H_o and H_s were chosen such that $H_o \times H_s$ was the minimal direct product containing H_d, we do not consider direct-product subgroups of $H_o \times H_s$ (for details, see Bruder and Vollhardt 1986).

Following this recipe, we can write down all discrete subgroups of $SO(3) \times SO(3)$: the invariant subgroups of C_n, D_n, T, O and Y are known, and it is easy to calculate H_o/H'_o and H_s/H'_s. Once all discrete subgroups of $SO(3) \times SO(3)$ have been found, we can proceed to determine the order-parameter states invariant under their action, together with the appropriate gauge transformations.

The states are obtained by solving systems of linear equations as in the case of continuous subgroups. If $A_{\mu j}$ is to be invariant under the transformations of the discrete group

$$H_d = \{(\mathrm{r}_k^o, \mathrm{r}_k^s, e^{i\phi_k})\}, \tag{6.66}$$

where k is an index denoting the elements of H_d, then one has to solve

$$\mathrm{r}_k^s \boldsymbol{A} (\mathrm{r}_k^o)^T e^{i\phi_k} = \boldsymbol{A} \quad \text{for all } k, \tag{6.67}$$

i.e. one has to determine $A_{\mu j}$ and ϕ_k for given $\mathrm{r}_k^o, \mathrm{r}_k^s \in SO(3)$. These equations can be solved for arbitrary r_k^o and r_k^s. Equation (6.67) is seen to break up into four systems of equations if the rotations are about the z axis. The orientation of the rotation axes is not important, the solution is essentially determined by the rotation angles α_k and β_k of r_k^o and r_k^s. (Different orientations of rotation axes lead to different, but equivalent, forms of the order parameter.) The solution of (6.67) is straightforward but tedious, and will not be reproduced here. The most important general result is that

$$|\phi_k| = \begin{cases} |\alpha_k \pm \beta_k| & \text{or} \\ |\alpha_k|, \ |\beta_k| & \text{or} \\ 0, \ \pi. \end{cases} \tag{6.68a}$$

This explains the earlier remark that the gauge group is determined by the spin and orbital parts of H_d. If, for example, $\alpha_k = \beta_k = \frac{1}{2}\pi$ then ϕ_k must take the values 0 or $\frac{1}{2}\pi$ or π unless $A_{\mu j} = 0$.

Having solved (6.67) for all possible discrete subgroups H_d of G, (6.6), one can now give a complete classification of all superfluid order-parameter states that have a residual symmetry. One of the results of the classification is that all the phases with a remaining continuous symmetry are also produced by discrete symmetry requirements. For U(1) symmetries this is to be expected since C_m approximates U(1) for large m. On the other hand, the non-Abelian group SO(3) cannot be approximated by a discrete group in such a way. It is therefore interesting to observe that small discrete groups can force the order parameter into a definite state. (An example is given by the polar phase: it is invariant under $D_\infty \times D_\infty$, but is already generated by $D_2 \times D_2$.) The reason for this lies in the $l=1$ structure of the ^{3}He order parameter: it is still quite unstructured (compared with, say, $l=5$), so that it can be approximated by rather small discrete groups.

It is also interesting to note that some of the phases that appear in the classification of continuous symmetries are found to be invariant under *additional discrete* symmetry transformations. For example, the A phase turns out to be invariant under $C_n \times D_m$ plus appropriate gauge transformations (n and m are arbitrary, which establishes an invariance under $C_\infty \times D_\infty$ plus gauge transformations). There appears here the additional $\mathbb{Z}_2$ symmetry under the replacement $(\hat{\boldsymbol{d}}, \hat{\boldsymbol{m}}, \hat{\boldsymbol{n}}) \rightarrow (-\hat{\boldsymbol{d}}, -\hat{\boldsymbol{m}}, -\hat{\boldsymbol{n}})$ that is discussed below (6.10) and (6.46). It emerges as an invariance under a combined *rotation* (e.g. by π around the z and x axes in orbital and spin space respectively). Hence the degeneracy space of the A phase may be written as

$$R_A = S_S^2 \times SO(3)_{L,\phi}/\mathbb{Z}_2. \tag{6.68b}$$

In addition to the order-parameter states already found through the symmetry classification of continuous subgroups, the classification using discrete symmetries yields a considerable number of new states. Some of those states only have a rather small remaining symmetry, corresponding to a small discrete subgroup H. More interesting, however, is the fact that the latter classification yields three more inert states, i.e. inert states that contain a discrete symmetry:

(1) the *planar* state given by

$$A_{\mu j} = \frac{1}{2^{1/2}} \begin{pmatrix} 1 & 0 & 0 \\ 0 & 1 & 0 \\ 0 & 0 & 0 \end{pmatrix}, \tag{6.69}$$

involving D_m symmetries ($m > 1$);

(2) the *bipolar* state

$$A_{\mu j} = \frac{1}{2^{1/2}} \begin{pmatrix} 1 & 0 & 0 \\ 0 & i & 0 \\ 0 & 0 & 0 \end{pmatrix}, \tag{6.70}$$

involving D_4 symmetries;

(3) the *α-state*

$$A_{\mu j} = \frac{1}{3^{1/2}} \begin{pmatrix} 1 & 0 & 0 \\ 0 & e^{\pi i/3} & 0 \\ 0 & 0 & e^{2\pi i/3} \end{pmatrix}, \tag{6.71}$$

involving tetrahedral symmetries.

These three inert states, as well as the five inert states with continuous symmetry, are already known from a different kind of investigation: Barton and Moore (1974b) found them to be stationary points of the Ginzburg–Landau free energy (5.4). It is interesting that these phases—previously determined by the zero points of systems of nonlinear equations—can be found simply by symmetry requirements. The symmetry classification is sometimes lengthy, but one does not encounter any substantial difficulties in carrying it through. In the next subsection we shall therefore have a closer look at this connection between algebra and analysis, i.e. between the symmetry and the free energy of the phases.

6.2.3 Symmetry and stationary points of a free energy

The symmetry classification described above is independent of any energy considerations. Hence it is not a priori clear whether the order-parameter structures found thereby are minima (or even stationary points) of any free-energy functional, for example of the Ginzburg–Landau free energy (5.4). On the other hand, the classification not only yielded the eight *inert* states but, in fact, *all* states known as minima or stationary points of (5.4) (Balian and Werthamer 1963, Ambegaokar and Mermin 1973, Mermin and Stare 1973, 1974, Barton and Moore 1974a, b, Jones 1977). All other order-parameter structures obtained in this way are potential candidates for new stationary points of the free energy. Therefore the question arises as to whether there is a connection between the symmetry of a state and the fact that its energy is stationary. In other words, one would like to know whether it is possible to predict the existence of stationary states once the underlying symmetry group of the free-energy functional is given. This is indeed the case, as will be discussed below.

The connection between the symmetry of an order parameter and its energy as given by some functional has already been studied within the context of high-energy physics and within a more mathematical context. For example, a method for determining the energy and residual symmetry of Higgs-potential minima has been described by Kim (1982). This was applied to a quartic Higgs potential with SU(5) symmetry to discuss the hierarchical symmetry breaking of SU(5) grand unification theory (Frautschi and Kim 1982). In a different context, Jarić (1980) showed that, given finite groups and a real-valued order parameter, the minima of a Ginzburg–Landau functional must always have a residual symmetry. Furthermore, Abud and Sartori (1983) proved that within a Ginzburg–Landau theory with algebraically independent fourth-order terms the minima correspond to states with a remaining symmetry.

A mathematical connection between the symmetry H of a state and its energy with respect to a given free-energy functional invariant under a group G may be constructed using methods borrowed from equivariant topology (Michel 1971, 1980). To this end, two theorems will be needed whose proof has been given by Michel (1980). In order to understand these theorems and their bearing on the Ginzburg–Landau theory of ^{3}He, a few mathematical terms have to be introduced (see also Abud and Sartori 1983).

Mathematically, the minimization of the Ginzburg–Landau functional involves the following ingredients:

(i) a manifold M of complex 3×3 matrices $A_{\mu j}$ with tr $\boldsymbol{AA}^+ = 1$;
(ii) a group G of transformations acting on the manifold;
(iii) a smooth real function F (the free energy) on the manifold that is invariant under the action of G, i.e. for $m \in M$ and arbitrary $g \in G$ one has $F(gm) = F(m)$.

We define the *orbit* $G(m)$ of a point m by $G(m) = \{gm \mid g \in G\}$, i.e. $G(m)$ is the set of all points obtained by letting all transformations in G act on m. The invariant function F is seen to be constant for all points of an orbit. If F is stationary at a point m_0, i.e. $\partial F/\partial A_{\mu j}|_{m_0} = 0$, then it will be stationary at all points of $G(m_0)$. Hence $G(m_0)$ is a *stationary orbit.*

Furthermore, the *isotropy group* (*little group*) G_m of a point m is defined as the subgroup of G leaving m invariant, $G_m = \{g \in G \mid gm = m\}$ (note that in our case the "points" of M are matrices). G_m corresponds to the subgroups H under which a given order parameter m is still invariant.

If m and n lie on the same orbit, i.e. if $n = gm$ for some $g \in G$, then it follows that $G_n = gG_mg^{-1}$. In this case the isotropy groups of n and m are *conjugate.*

All points on the orbit of m belong to isotropy groups conjugate to G_m. There may be points on other orbits having isotropy groups conjugate to G_m as well. This fact leads to the definition of the *stratum* $S(m)$ of m. It is the union of all orbits of points having isotropy groups that are conjugate to G_m, i.e. n is an element of $S(m)$ if there is an element $g \in G$ such that $G_n = gG_mg^{-1}$.

The stratum is a somewhat abstract notion. One may say that a stratum is the union of all orbits of the same symmetry type: all points of a stratum have conjugate and therefore isomorphic isotropy groups. This may be illustrated by a simple example: let the manifold be $M = \mathbb{R}^2$, i.e. a two-dimensional plane, and let the symmetry group acting on M be $G = \mathrm{U}(1)$, the rotations about the z axis. Clearly, the origin of the plane remains invariant under G: its orbit consists of one point only (see Fig. 6.5a) and its isotropy group is U(1). All other points of $M = \mathbb{R}^2$ are transported along orbits of constant distance from the origin, i.e. circles. These points are invariant under the identity element of G only; hence their invariance group is $G_m = E$. Hence the manifold $M = \mathbb{R}^2$ can be divided in two strata: one consisting of the origin ($x = y = 0$), the other one of the rest of $\mathbb{R}^2$.

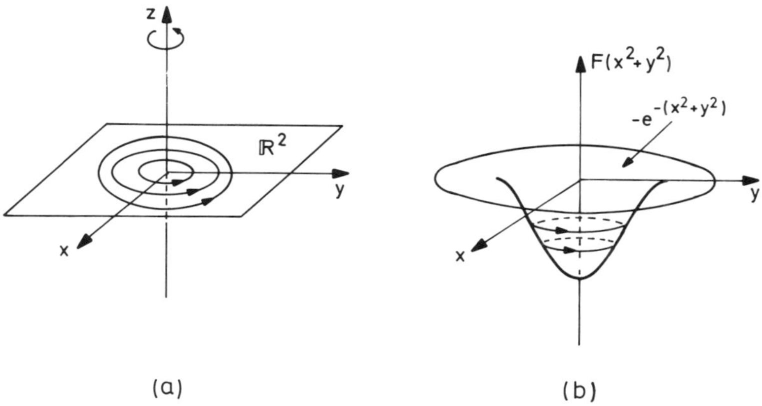

Figure 6.5 (a) Orbits of points in the (x, y) plane under rotations U(1) around the z axis. (b) Orbits of a function $F(x, y)$ invariant under U(1) (here $F = F(x^2 + y^2)$) (see text). In both cases the origin corresponds to a stationary orbit.

As shown by Michel (1971, 1980), the following theorem may be proved.

Theorem 1. If an orbit $G(m)$ lies isolated in its stratum $S(m)$, i.e. if there exists a neighbourhood U of $G(m)$ such that $U \cap S(m) = G(m)$, then $G(m)$ is a stationary orbit of *every smooth real function* on M that is invariant under G.

We note that if a stratum is composed of finitely many orbits then each of them is isolated in its stratum. In the case of the simple example mentioned above, this theorem implies that the orbit of the origin is stationary for every smooth real function F on $\mathbb{R}^2$ invariant under U(1). Since such a function is of the form $F = F(x^2 + y^2)$, we have proved that the origin $x = y = 0$ is a stationary point of *any* function $F(x^2 + y^2)$; see Fig. 6.5(b). While in the present case the result is trivial, the theorem is extremely valuable in the case of more complicated manifolds M and groups G.

In particular, it allows one to draw conclusions about possible superfluid states of ^{3}He (Bruder and Vollhardt 1986). We first consider the B phase. Its invariance group (isotropy group) is $H = SO(3)_{S+L}$, which determines the B-phase order parameter uniquely. It follows that the stratum of the B phase consists of one orbit only, i.e. the orbit is isolated in its stratum. Therefore Michel's theorem states that the B phase is a stationary point of every smooth real function that is invariant under G, (6.6)! This conclusion is independent of the form of the free-energy functional. It is not limited to a Ginzburg–Landau-type expression valid only close to T_c. At the same time, we know that the B-phase order parameter has the largest remaining symmetry, i.e. the least-broken symmetry of all states. In this respect it is interesting to note that Balian and Werthamer (1963) showed that within weak-coupling theory the B phase is indeed the stable state at all temperatures. Even if strong-coupling corrections are included, to make contact with a realistic situation, the B phase is known to occupy the largest part of the phase diagram (see Fig. 4.2) (this is true at least in the absence of a magnetic field, i.e. when G, (6.6), applies).

The fact that the stratum of the B phase consists of a single orbit does not only apply to this particular phase, but in fact is valid for *all* inert states: their isotropy groups H are all maximal. Hence the inert states are stationary points of *any* smooth free-energy function invariant under G, (6.6). This conclusion is again independent of the particular form of the free-energy function and is not limited to any kind of Ginzburg–Landau expansion (5.4). Michel's theorem thus makes remarkably general assertions about stationary points without the need to evaluate a single derivative or to solve a nonlinear equation.

If one searches for physically relevant states, it is not enough to know whether or not a particular state is a stationary point of a free energy—one has to find the states corresponding to stable minima of the energy F. This involves a comparison of the respective free energies and a calculation of (at least) the second derivative of F with respect to $A_{\mu j}$. Concerning the eight inert states, one finds that indeed three of them (the planar, bipolar and α states) are unstable saddle points in the space of parameters β_i, while the other five (the B, A, A_1, polar, β states) represent local minima (Jones 1977). In fact, even the latter conclusion is not rigorous, since the second derivative of F with respect to $A_{\mu j}$ yields relations among the β_i that only allow for necessary, but not for sufficient, conditions for a local minimum. At least, we may conclude that all the inert states discussed above that have a discrete symmetry are definitely unstable stationary points of the free energy (5.4).

Michel's theorem cannot be applied to the noninert states: since their structure depends on the coefficients in the free-energy functional, they elude a theorem making a statement about *all* G-invariant functions. None the less, there are special cases where it is possible to use a second theorem by Michel (1980). For our purposes, it can be formulated as follows.

Theorem 2. Let m be a point of the manifold M. If the isotropy group G_m is maximal in the set of isotropy groups then each G-invariant smooth real function F on M is stationary in at least two orbits of $S(m)$.

Now the stationary orbits depend on the function F. We note that there can be several maximal isotropy groups, because two isotropy groups may not be comparable at all (neither $G_n \subset G_m$ nor $G_m \subset G_n$). This theorem does not lead to new results for $G = SO(3) \times SO(3) \times U(1)$: the maximal isotropy groups like $SO(3)_{S+L}$ produce inert phases. Hence Michel's first theorem may be applied, which is much stronger than the second one. If, however, the symmetry group is restricted by an external magnetic field or by taking into account coupling of nuclear dipole moments then the second theorem can be useful.

6.2.4 Symmetry reduction due to a magnetic field or spin–orbit coupling

An external magnetic field defines a direction and hence breaks the symmetry with respect to rotations in spin space. The symmetry group of superfluid ^{3}He in a magnetic field, G_H, is therefore smaller than G, (6.6), i.e. $G_H \subset G$. The order parameter couples to the magnetic field via the terms $H_\mu(\boldsymbol{AA}^+)_{\mu\nu}H_\nu$ and $\mathrm{i}\epsilon_{\mu\nu\lambda}H_\mu(\boldsymbol{AA}^+)_{\nu\lambda}$; see (5.6), (5.49) (Mermin and Stare 1973). Therefore in spin space only rotations about the magnetic field leave the free energy invariant. The symmetry group G_H then takes the form

$$G_H = \mathrm{SO}(3)_{\boldsymbol{L}} \times \mathrm{U}(1)_{S_z} \times \mathrm{U}(1)_\Phi. \tag{6.72}$$

We may now use the symmetry classification as given before, taking into account those groups that only contain rotations about the z axis in spin space. In this case the B phase and the α, planar and bipolar states are no longer obtained. Nevertheless, concerning inert states, the polar, A, β and A_1 states are still produced explicitly by subgroups of G_H. According to Michel's first theorem, these states are stationary points of all free-energy functions invariant under G_H. As before, Michel's second theorem gives no result.

The dipole–dipole interaction between the nuclear spins of the ^{3}He atoms leads to a spin–orbit interaction (see below). It couples the orbital and spin degrees of freedom. Hence the symmetry group of ^{3}He subject to the dipole interaction is

$$G_\mathrm{D} = \mathrm{SO}(3)_{\boldsymbol{L}+\boldsymbol{S}} \times \mathrm{U}(1)_\Phi. \tag{6.73}$$

Since the pair interaction dominates the dipolar interaction by six orders of magnitude, one may still assume the former to determine the overall structure of the order parameter as $A_{\mu j}$. The dipolar interaction then acts as a small perturbation, choosing a particular state from a manifold of degenerate ones.

The symmetry classification of (6.73) is easily found from the results obtained before. We have to take those subgroups in which operations in spin space are compensated by identical operations in orbital space. One finds that the A_1 and B phases as well as the α and planar states are given explicitly by symmetry requirements. This implies that these states are stationary points of all free-energy functions invariant under G_D. Note that the A phase is no longer included in this list of stationary states. This is not surprising, since in the pure axial state up-spin and down-spin Cooper pairs are actually independent. Clearly, in reality this is not strictly true since the spin-nonconserving dipolar interaction couples the two populations. Nevertheless, the latter effect is very weak and may therefore be included in the axial state as a mere perturbation.

In order to find other states by Michel's second theorem, one has to know all maximal isotropy subgroups of G_{D}. We find that $\mathrm{U}(1)_{S_z+L_z-\phi}$ is a maximal isotropy group that does not lead to an inert state, but to the ε state, (6.60) (Barton and Moore 1974b):

$$A_{\mu j} \propto \begin{pmatrix} 0 & 0 & A \\ 0 & 0 & \mathrm{i}A \\ B & \mathrm{i}B & 0 \end{pmatrix}. \tag{6.74}$$

Michel's second theorem allows one to conclude that among all matrices of this form there are at least two stationary points for any free energy. If one inserts (6.74) in (5.4) and requires that $\partial F/\partial A_{\mu j}=0$ then one finds that the parameters A and B in (6.74) can be chosen such that this equation does indeed have two solutions.

Finally, a combination of the symmetry restrictions due to a magnetic field and the dipolar coupling leads to a symmetry group

$$G_{H,\mathrm{D}} = \mathrm{U}(1)_{L_z+S_z} \times \mathrm{U}(1)_{\phi}. \tag{6.75}$$

The subgroups are easy to find. One example is $\mathrm{U}(1)_{S_z+\phi} \times \mathrm{U}(1)_{L_z-\phi}$, leading to the A_1 phase. Hence the A_1 phase is a stationary state of the most general free-energy function that is invariant under $G_{H,\mathrm{D}}$.

There are two nontrivial maximal isotropy subgroups of $G_{H,\mathrm{D}}$: $\mathrm{U}(1)_{S_z+L_z}$, leading to the oblate state (6.40), and again $\mathrm{U}(1)_{S_z+L_z-\phi}$, i.e. the ε state. In this case Michel's second theorem can be applied and predicts at least two stationary points in the strata of the oblate and ε states. It is satisfactory to see the oblate state appearing at this point: the B_2 phase (6.40), i.e. the B phase in a magnetic field, lies in the stratum of the oblate state. If we consider the B phase in the presence of dipolar coupling and neglect the magnetic field for a moment, we are led to the oblate state as well: the dipolar interaction favours a state in which spin and orbital spaces are rotated against each other by a definite angle (see below). If we choose the z axis of spin space to be the rotation axis then the order parameter of the B phase is transformed to the oblate form. Hence the structure of the B_2 phase is determined by symmetry.

6.2.5 Symmetry classification of an exactly solvable case: d-wave pairing

As discussed above, the symmetry classification of G, (6.6), yields all order-parameter states found previously by minimization of the Ginzburg–Landau functional (5.4). This supports the observation that minima of the bulk free energy generally correspond to states with a remaining internal symmetry.

It is worthwhile verifying this point explicitly in the case of a different system also described by a tensor order parameter, whose free-energy functional is none the less simpler than (5.4) (Bruder and Vollhardt 1986). Of particular interest are the states thought to describe the bulk matter of neutron stars (see e.g. Pines and Alpar 1985). They are assumed to be due to Cooper pairing of neutrons in a $^3P_2$ state ($l=1$, $S=1$ and $J=2$ because of strong spin–orbit coupling). In order to describe a $J=2$ state, the order parameter has to be a traceless symmetric tensor $B_{\mu\nu}$. (This also holds for $l=2$ pairing in superfluid ^{3}He (Mermin and Stare 1973). Indeed, as shown by Sauls and Serene (1978), the Ginzburg–Landau functionals for the two problems are identical. In the case of such a pairing the free energy has to be invariant under joint three-dimensional rotations in spin and orbital spaces and a gauge transformation (Mermin 1974, 1978a), i.e.

$$G = SO(3)_{\boldsymbol{J}} \times U(1)_{\phi}. \tag{6.76}$$

In the Ginzburg–Landau regime the free energy compatible with this symmetry can be obtained from (5.4) by putting $\boldsymbol{A} \equiv \boldsymbol{B} = \tilde{\boldsymbol{B}}$ and using the identity

$$\operatorname{tr}(\boldsymbol{B}^*\boldsymbol{B})^2 = \tfrac{1}{2}\,|\operatorname{tr}\boldsymbol{B}^2|^2 + [\operatorname{tr}(\boldsymbol{B}^*\boldsymbol{B})]^2 - 2\operatorname{tr}[(\boldsymbol{B}^*)^2\boldsymbol{B}^2], \tag{6.77}$$

which is valid for traceless 3×3 matrices. This leads to a free-energy functional with three fourth-order terms:

$$F = \alpha\Delta^2 \operatorname{tr}(\boldsymbol{B}\boldsymbol{B}^*) + \tfrac{1}{2}\Delta^4\{\beta_1\,|\operatorname{tr}\boldsymbol{B}^2|^2 + \beta_2[\operatorname{tr}(\boldsymbol{B}\boldsymbol{B}^*)]^2 + \beta_3 \operatorname{tr}[\boldsymbol{B}^2(\boldsymbol{B}^*)^2]\}, \tag{6.78}$$

where we again assume $\boldsymbol{B}$ to be normalized ($\operatorname{tr}(\boldsymbol{B}\boldsymbol{B}^*) = 1$).

The minimization of (6.78) is nontrivial but can nevertheless be performed analytically (Mermin 1974). Depending on the β_i, there are three distinct absolute minima:

(i) for $\beta_3 > -\beta_1 + |\beta_1|$, the order parameter is given by

$$B_{\mu\nu} = \frac{1}{2}\begin{pmatrix} 1 & i & 0 \\ i & -1 & 0 \\ 0 & 0 & 0 \end{pmatrix}, \tag{6.79}$$

which has the form of the A_1 phase, (6.56), and whose energy is $F = -\alpha^2/2\beta_2$;

(ii) for $-6\beta_1 < \beta_3 < 0$, the matrix has the form

$$B_{\mu\nu} = \frac{1}{3^{1/2}}\begin{pmatrix} 1 & 0 & 0 \\ 0 & e^{-2\pi i/3} & 0 \\ 0 & 0 & e^{-4\pi i/3} \end{pmatrix}, \tag{6.80}$$

which is the α phase; its free energy is given by $F = -\alpha^2/[2(\beta_2 + \frac{1}{3}\beta_3)]$;

(iii) for $\beta_3 < -4\beta_1 - 2|\beta_1|$, the (highly degenerate) minimum is given by *any real* matrix $B_{\mu\nu}$ (up to a phase factor); its energy is found as $F = -\alpha^2/[2(\beta_1 + \beta_2 + \frac{1}{2}\beta_3)]$.

We now apply the symmetry classification to G, (6.76). This is again easily done by noting that the symmetry (6.76) corresponds to that of superfluid ^{3}He with a dipolar coupling. Its symmetry classification has been discussed before. There are three continuous subgroups: $U(1)_{J_z-\Phi}$, $U(1)_{2J_z-\Phi}$ and $U(1)_{J_z}$. The corresponding order parameters are found from (6.20) by setting $a_z = b_z = \frac{1}{2}c$, $a_z = b_z = c$ and $a_z = b_z$, $c = 0$ respectively. All three yield inert states. In the first case $\boldsymbol{B}$ has the form (6.79), which is the first of the global minima found by Mermin (1974). In the second case $\boldsymbol{B}$ is obtained as (6.60), the ε phase, with $A = B = 1$; although it is a stationary point, it does not correspond to a global minimum. In the third case $\boldsymbol{B}$ is given by a real traceless matrix with an overall phase factor, corresponding to the third stable phase. The discrete symmetries yield two more inert phases: the tetrahedral symmetries lead to the α phase, i.e. the second minimum of the free energy (6.80), and the D_m symmetries to the analogue of the planar phase, which, however, is always a saddle point of the free energy. There are some more discrete symmetries (more restrictive than the two mentioned above), which lead to noninert states.

We therefore find that all absolute minima of (6.78) are included in a symmetry classification, i.e. they can be obtained by symmetry alone. Here we have been able here to verify explicitly the importance of discrete symmetries.

6.2.6 Broken symmetries in high-energy physics

Symmetries and symmetry breaking are of fundamental importance in high-energy physics. This has to do with the close connection between symmetries and conserved quantities (Noether's theorem). If, experimentally, a conserved quantity is identified, the question arises as to what symmetry is responsible for it. For example, a one-parameter quantity (a charge, say) involves an underlying U(1) symmetry. One then has to reconstruct the initial underlying symmetry that by symmetry breaking leads to the experimentally observable lower symmetries. In this way high-energy physics may be understood as a sequence of symmetry breakings. We should note, however, that in high-energy physics the relevant symmetries are usually *local* (or gauge) symmetries rather than the global symmetries considered in condensed-matter physics. (This implies an Anderson–Higgs mechanism rather than Goldstone modes (Anderson 1963, Higgs 1964; see e.g. Huang 1982, Cheng and Li 1984).) On the other hand, even in high-energy physics, namely in the theory of strong interactions, there is an example of a globally broken symmetry that is very similar to those considered here.

Let us consider an isodoublet of quarks with only two flavours, namely an up and a down quark (u, d). Initially there is a chiral invariance between left handed (***L***) and right-handed (***R***) states (we neglect the weak interaction):

$$\begin{pmatrix} \mathrm{u} \\ \mathrm{d} \end{pmatrix}_{\boldsymbol{L}}, \quad \begin{pmatrix} \mathrm{u} \\ \mathrm{d} \end{pmatrix}_{\boldsymbol{R}}. \tag{6.81}$$

The corresponding symmetry, i.e. separate rotations with respect to left and right, is represented by†

$$G = \mathrm{SU}(2)_{\boldsymbol{L}} \times \mathrm{SU}(2)_{\boldsymbol{R}}. \tag{6.82}$$

Experimentally, however, we know that the invariance under the full symmetry group (6.82) is not observed—a fact attributed to a broken symmetry. Indeed, one imagines a condensation between quarks (u, d) and antiquarks (ū, d̄) to take place—very similar to Cooper pairing. In this way the symmetry (6.82) is broken down to a remaining $\mathrm{SU}(2)_{\boldsymbol{L}+\boldsymbol{R}}$, in analogy to the B phase. Here we again encounter a relative broken symmetry where the corresponding Goldstone bosons are the three pions.

A different example, now involving gauge symmetries in the theory of electroweak interactions, is a model due to Pati and Salam (1974), an extension of the Weinberg–Salam model. Again we consider left/right isodoublets of u, d quarks, and, in addition, a hypercharge. The left/right symmetries are both described by an SU(2), the latter one by a U(1) symmetry. Hence the underlying symmetry is

$$G = \mathrm{SU}(2)_{\boldsymbol{L}} \times \mathrm{SU}(2)_{\boldsymbol{R}} \times \mathrm{U}(1)_{Y/2} \tag{6.83}$$

(note that this is essentially identical with the symmetry group (6.6) describing superfluid ^{3}He). On an experimental level at low energies only the subgroup $H = \mathrm{U}(1)_{L_z+Y/2}$, ordinary QED, is realized. It involves the z components of the left-handed isospin (the $\mathrm{SU}(2)_{\boldsymbol{R}}$ is assumed to be completely broken) and the hypercharge, whose generator is the ordinary electric charge $Q = L_z + \frac{1}{2}Y$. This implies a broken relative symmetry involving two symmetries. It is the analogue of the phase (7) or (8) of superfluid ^{3}He discussed in the classification of continuous subgroups of G, (6.6).

† For simplicity, we have neglected two additional U(1) symmetries in (6.82): one related to baryon-number conservation and another connected with an axial symmetry. However, the latter is not realized in nature and is believed to be broken by strong interactions ("the U(1) problem"; Witten 1979).

6.3 ORIENTATION OF THE ORDER PARAMETER BY INTERNAL RESIDUAL INTERACTIONS AND EXTERNAL FIELDS

We now want to discuss some of the explicit consequences of macroscopic ordering in superfluid ^3He. For this, we have to understand how the long-range order in the system may actually be fixed—after all, the ordered equilibrium state is still highly degenerate (see the discussion at the beginning of the chapter). For example, although in the A phase we have a preferred direction $\hat{\boldsymbol{l}}$ in real space, the interactions that lead to the formation of this superfluid phase do not at the same time select its actual direction. On the other hand, the degeneracies may either be lifted by external fields or by some small interactions within the system, which were unimportant for the condensation mechanism itself, and which therefore have not been taken into account so far.

In the remainder of this chapter we shall discuss examples of both kinds of orientational effects. Of these, the spin–orbit interaction is the most important internal interaction. Several other, somewhat more "exotic", microscopic interactions, which may none the less be observable on a macroscopic scale, will be considered in Chapter 12.

The influence of a wall on the superfluid is yet another internal interaction leading to orientational effects. It is different from the other orienting forces in that the interaction is always strong and will, in fact, cause a complete change of the order-parameter structure near the wall.

With respect to the orientation of the order parameter by external fields, we shall consider the influence of magnetic and electric fields and of a superfluid velocity field.

6.3.1 Dipole interaction

The interaction responsible for the transition from the normal liquid to superfluid ^3He is invariant under separate rotations in orbital and spin spaces (hence the $SO(3) \times SO(3)$ symmetry in (6.6)). In reality both spaces are coupled by the spin–orbit interaction, which is provided by the dipole coupling of the nuclear spins (Leggett 1973a, b, 1974a). There is also a spin–orbit interaction due to the polarization of the electronic clouds of the ^3He atoms, but this is much smaller and may be neglected. The spin–orbit interaction will give rise to a relative orientation of the spin and orbital parts of the order parameter, but, because of its smallness, will only produce a minute change in the order-parameter structure itself.

Let us therefore study the dipole interaction, which, although perhaps the least obvious, has turned out to be the most important orienting force. In any case, it is different from the other orientational effects as it cannot be manipulated or switched off altogether. The nuclear-dipole interaction

Hamiltonian is given by the standard expression

$$H_D = \tfrac{1}{2}(\gamma\hbar)^2 \int d^3r \int d^3r' \left\{ \frac{\boldsymbol{\sigma}(\boldsymbol{r}) \cdot \boldsymbol{\sigma}(\boldsymbol{r})}{|\boldsymbol{r}-\boldsymbol{r}'|^3} - 3\frac{[(\boldsymbol{r}-\boldsymbol{r}')\cdot\boldsymbol{\sigma}(\boldsymbol{r})][(\boldsymbol{r}-\boldsymbol{r}')\cdot\boldsymbol{\sigma}(\boldsymbol{r}')]}{|\boldsymbol{r}-\boldsymbol{r}'|^5} \right\} \tag{6.84}$$

$$= \tfrac{1}{2}(\gamma\hbar)^2 \int d^3r \int d^3r' \sum_{\mu\nu} Q_{\mu\nu}(\boldsymbol{r}-\boldsymbol{r}')\sigma_\mu(\boldsymbol{r})\sigma_\nu(\boldsymbol{r}'), \tag{6.85}$$

with

$$Q_{\mu\nu}(\boldsymbol{r}) = \frac{1}{r^3}(\delta_{\mu\nu} - 3\hat{r}_\mu\hat{r}_\nu). \tag{6.86}$$

Here $\gamma = 2\mu_0/\hbar$ is the gyromagnetic ratio of the ^{3}He nucleus, with μ_0 the nuclear magnetic moment, $\boldsymbol{\sigma}(\boldsymbol{r})$ is the spin-density operator introduced in (3.112), (3.113). We now want to calculate the expectation value of the dipolar interaction energy density $\langle H_D \rangle / V \equiv f_D$. For this, we may use the results of Section 3.5, where the expectation value of two-particle quantities has been calculated (see (3.119a, b)). Inserting (6.86) into (3.119a) or using the Fourier transform of (6.86),

$$\tilde{Q}_{\mu\nu}(\boldsymbol{k}) = \int d^3r\, e^{i\boldsymbol{k}\cdot\boldsymbol{r}} Q_{\mu\nu}(\boldsymbol{r}) \tag{6.87a}$$

$$= -\tfrac{4}{3}\pi(\delta_{\mu\nu} - 3\hat{k}_\mu\hat{k}_\nu), \tag{6.87b}$$

together with (3.119b), we obtain (Leggett 1973a, b, 1974a, 1975a)

$$f_D = -\tfrac{1}{2}(\gamma\hbar)^2 \int d^3r \frac{1}{r^3}[|\boldsymbol{F}(\boldsymbol{r})|^2 - 3\,|\hat{\boldsymbol{r}}\cdot\boldsymbol{F}(\boldsymbol{r})|^2] \tag{6.88}$$

$$= -\tfrac{1}{2}(\gamma\hbar)^2 \sum_{\mu\nu} \sum_{\boldsymbol{k}\boldsymbol{k}'} \tilde{Q}_{\mu\nu}(\boldsymbol{k}-\boldsymbol{k}')F^*_{\boldsymbol{k}\mu}F_{\boldsymbol{k}'\nu}. \tag{6.89}$$

We note that $\tilde{Q}_{\mu\nu}(\boldsymbol{k})$, (6.87a, b), only depends on the direction of $\boldsymbol{k}$. Therefore it might appear as if the sums over the magnitudes of $\boldsymbol{k}$ and $\boldsymbol{k}'$ in (6.89) only involved the pair amplitudes and not $Q_{\mu\nu}$ itself. However, this is not quite true, since the argument of $Q_{\mu\nu}$ in (6.89) contains the difference between the two momenta, i.e. it involves $(\boldsymbol{k}-\boldsymbol{k}')/|\boldsymbol{k}-\boldsymbol{k}'|$, over which one has to sum. Nevertheless, to lowest order in the small quantity T_c/T_F, the dependence on the magnitudes of $\boldsymbol{k}$ and $\boldsymbol{k}'$ may be neglected, so that (6.89) takes the form

$$f_D = \frac{2\pi}{3}(\gamma\hbar)^2 \sum_{\mu\nu} \int \frac{d\Omega_{\hat{k}}}{4\pi} \int \frac{d\Omega_{\hat{k}'}}{4\pi} \tilde{Q}_{\mu\nu}(\boldsymbol{k}-\boldsymbol{k}')\Big(\sum_{|\boldsymbol{k}|} F^*_{\boldsymbol{k}\mu}\Big)\Big(\sum_{|\boldsymbol{k}'|} F_{\boldsymbol{k}\nu}\Big). \tag{6.90}$$

Inserting (3.43b) for $F_{\boldsymbol{k}\alpha\beta}$ into (3.118b) to obtain $F_{\boldsymbol{k}\mu}$, and making use of (3.34a) to replace $\Delta_{\boldsymbol{k}\mu\nu}$ by the vector $\boldsymbol{d}(\boldsymbol{k})$, the $|\boldsymbol{k}|$ integral over $F_{\boldsymbol{k}\mu}$ in (6.90) in the absence of a magnetic field is written as

$$\Lambda_\alpha(\hat{\boldsymbol{k}}) \equiv -\sum_{|\boldsymbol{k}|} F_{\boldsymbol{k}\alpha}$$
$$= \sum_\beta d_\beta(\hat{\boldsymbol{k}}) \sum_\mu [\tfrac{1}{2}(\boldsymbol{\sigma}_\beta \boldsymbol{\sigma}_\alpha)_{\mu\mu}]\left[N(0) \int_{-\epsilon_c}^{\epsilon_c} \mathrm{d}\xi \frac{\tanh(E_{\boldsymbol{k}\mu}/2k_B T)}{2E_{\boldsymbol{k}\mu}}\right]. \quad (6.91)$$

The last factor in (6.91), i.e. the ξ integral, has already appeared in the gap equation (3.46b). For unitary states this term is independent of μ, and in fact, even for nonunitary states, the μ dependence of $E_{\boldsymbol{k}\mu}$ is expected to be negligible (Leggett 1975a). In this case (6.91) reduces to

$$\Lambda_\alpha(\hat{\boldsymbol{k}}) = d_\alpha(\hat{\boldsymbol{k}}) I(|\Delta_{\boldsymbol{k}}(T)|, T), \quad (6.92)$$

where

$$I = N(0) \int_0^{\epsilon_c} \mathrm{d}\xi \frac{\tanh(E_{\boldsymbol{k}}/2k_B T)}{E_{\boldsymbol{k}}} \quad (6.93)$$

and $E_{\boldsymbol{k}} = [\xi^2 + |\Delta_{\boldsymbol{k}}(T)|^2]^{1/2}$.

As mentioned below (3.46b), for $\epsilon_c \gg |\Delta_{\boldsymbol{k}}(T)|$ the ξ integral is dominated by the ξ^{-1} dependence of the integrand at large ξ. This leads to a logarithmic dependence on ϵ_c, making the other energy scales, which are of order T_c, unimportant. Indeed, (6.93) hardly depends on $\Delta_{\boldsymbol{k}}$ at all, as can be verified by calculating the limiting values for $T \to T_c$ (see the evaluation below (3.47c)) and $T \to 0$ respectively:

$$I(0, T_c) = N(0) \ln\left(C_0 \frac{\epsilon_c}{k_B T_c}\right), \quad (6.94a)$$

$$I(|\Delta_{\boldsymbol{k}}(0)|, 0) = N(0) \ln\left[\frac{2\epsilon_c}{|\Delta_{\boldsymbol{k}}(0)|}\right]. \quad (6.94b)$$

Hence, while in the limit $T \to T_c$, (6.93) does not depend on the gap at all, it acquires only a weak logarithmic dependence for $T \to 0$. To a good approximation, we may therefore neglect its $\hat{\boldsymbol{k}}$ dependence altogether and use its value at T_c for all T. In this case the only $\hat{\boldsymbol{k}}$ dependence of (6.92) comes from $\boldsymbol{d}(\hat{\boldsymbol{k}})$, leaving us with

$$\Lambda_\alpha(\hat{\boldsymbol{k}}) \approx d_\alpha(\hat{\boldsymbol{k}}) I(0, T_c). \quad (6.95)$$

The dipole energy density (6.90) is then finally obtained as

$$f_D = \tfrac{2}{3}\pi(\gamma\hbar)^2 I^2(0, T_c) \int \frac{\mathrm{d}\Omega_{\hat{\boldsymbol{k}}}}{4\pi} \int \frac{\mathrm{d}\Omega_{\hat{\boldsymbol{k}}'}}{4\pi}$$
$$\times \{\boldsymbol{d}^*(\hat{\boldsymbol{k}}) \cdot \boldsymbol{d}(\hat{\boldsymbol{k}}') - 3[\hat{\boldsymbol{q}} \cdot \boldsymbol{d}^*(\hat{\boldsymbol{k}})][\hat{\boldsymbol{q}} \cdot \boldsymbol{d}(\hat{\boldsymbol{k}}')]\}, \quad (6.96)$$

where $\hat{\boldsymbol{q}} = (\boldsymbol{k} - \boldsymbol{k}')/|\boldsymbol{k} - \boldsymbol{k}'|$.

In the case that the order parameter is in a definite l state, such that $\boldsymbol{d}(\hat{\boldsymbol{k}})$ is a superposition of spherical harmonics of order l (see the discussion below

(3.46b)), one of the angular averages in (6.96) may be performed explicitly. Furthermore, using the T_c relation (3.48) to eliminate $I(0, T_c)$, (6.96) reduces to (Takagi 1974, Leggett 1975a)

$$f_D = \lambda_D N_F \langle 3\,|\hat{\boldsymbol{k}} \cdot \boldsymbol{d}(\hat{\boldsymbol{k}})|^2 - |\boldsymbol{d}(\hat{\boldsymbol{k}})|^2 \rangle_{\hat{\boldsymbol{k}}}. \tag{6.97}$$

Here we have introduced a dimensionless dipole coupling parameter

$$\lambda_D = \frac{3\pi}{8} \frac{1}{l(l+1)} \frac{\gamma^2\hbar^2/a^3}{E_F} \frac{1}{(N(0)\,|V_l|)^2}, \tag{6.98}$$

where V_l is the pair-interaction constant for given l and a is the average atomic spacing, $a = (3\pi^2)^{1/2} k_F^{-1}$, such that $\gamma^2\hbar^2/a^3$ is the dipole energy for two atoms at distance a. In particular, for p-wave pairing ($l = 1$) the dipole coupling parameter has the approximately pressure-independent value

$$\lambda_D \approx 5 \times 10^{-7} \tag{6.99}$$

if values of $N(0)\,|V_1| \approx 0.2$ and 0.3 are assumed at low pressure and at melting pressure respectively.

As discussed by Leggett (1974a, 1975a) and Takagi (1973), there may be renormalizations of (6.97) due to the quasiparticle nature of the ^{3}He atoms involved in the dipole interaction. This will lead to a prefactor, which is expected to be close (or even identical) to 1.

Equation (6.97) for the dipole energy density of a p-wave condensate can be expressed in terms of the order-parameter matrix $A_{\mu j}$, (5.1), as

$$f_D = \tfrac{3}{5} g_D(T) \sum_{ij} [A^*_{ii} A_{jj} + A^*_{ij} A_{ji} - \tfrac{2}{3} A^*_{ij} A_{ij}]. \tag{6.100}$$

The temperature-dependent dipole coupling constant g_D follows from the above as

$$g_D(T) \equiv \lambda_D N_F \Delta^2(T), \tag{6.101}$$

where $\Delta(T)$ is the average energy gap, (3.57a). It corresponds to the average dipole–dipole interaction energy of the two ^{3}He quasiparticles in the Cooper pairs at relative distance a (the average interparticle spacing), per unit volume. Indeed, using the definition of λ_D in (6.98), (6.101) may be rewritten as

$$g_D(T) \approx \frac{\gamma^2\hbar^2}{a^3} \left[\frac{\Delta(T)}{E_F}\right]^2 n, \tag{6.102}$$

where the first factor is the average dipole-coupling energy of two quasiparticles at distance a, $(\Delta/E_F)^2$ measures the probability for these two quasiparticles to form a Cooper pair, and n is the overall particle density.

In the normal state the dipole energy density f_D is zero since positive and negative contributions cancel on the average. As one enters the superfluid phase, $g_D(T)$ increases linearly with $T_c - T$, and at low temperatures approaches a temperature-independent value of approximately $\lambda_D N_F (k_B T_c)^2$.

It is a remarkable property of anisotropic pair-correlated states that the dipole interaction energy of atoms correlated into Cooper pairs adds up. This effect is due to the coherent orientation of the preferred axes in orbital and spin spaces of all the pairs in the superfluid state, i.e. it is a direct consequence of the spontaneously broken symmetries. Indeed, the broken symmetries set the stage for the amplification of the dipole interaction to a macroscopic size. All that the dipole interaction has to do is to select the energetically most favourable relative orientation of the already long-range-ordered preferred directions.

Let us now evaluate expression (6.100) for various model states. Note that in both (6.97) and (6.100) the last term in the angular average yields an orientation-independent contribution, since $\langle |\boldsymbol{d}(\hat{\boldsymbol{k}})|^2\rangle_{\hat{k}} = \Delta^2(T)$, $\mathrm{tr}\,(\boldsymbol{AA}^+) = 1$. Hence

$$f_{\mathrm{D}} = 3\lambda_{\mathrm{D}} N_{\mathrm{F}} \langle |\hat{\boldsymbol{k}} \cdot \boldsymbol{d}(\hat{\boldsymbol{k}})|^2\rangle_{\hat{k}} + \text{const} \tag{6.103a}$$

$$= \tfrac{3}{5} g_{\mathrm{D}}(T) \sum_{ij} [A_{ii}^* A_{jj} + A_{ij}^* A_{ji}] + \text{const.} \tag{6.103b}$$

$$= \tfrac{1}{5}\lambda_{\mathrm{D}} N_{\mathrm{F}} \sum_{ij} (d_{ii}^* d_{jj} + d_{ij}^* d_{ji}) + \text{const.} \tag{6.103c}$$

In the following only the orientation-dependent part of f_{D}, denoted by Δf_{D}, will be considered.

The A phase

First we substitute the A-phase parameter (6.10) into (6.103b). The dipole energy density in the A phase is found as

$$\Delta f_{\mathrm{D}} = -\tfrac{3}{5} g_{\mathrm{D}}(T)(\hat{\boldsymbol{d}} \cdot \hat{\boldsymbol{l}})^2 \tag{6.104a}$$

$$= -\tfrac{2}{5}\lambda_{\mathrm{D}} N_{\mathrm{F}} \Delta_0^2 (\hat{\boldsymbol{d}} \cdot \hat{\boldsymbol{l}})^2. \tag{6.104b}$$

So the energy minimum is given by $\hat{\boldsymbol{d}}$ oriented along $\pm\hat{\boldsymbol{l}}$. This is a plausible result, which may be readily understood, at least from a classical point of view (Leggett 1975a). Let us visualize the Cooper pair in the A phase as a pair of parallel magnetic dipoles rotating about the angular-momentum axis $\hat{\boldsymbol{l}}$. For $\hat{\boldsymbol{d}}$ parallel to $\hat{\boldsymbol{l}}$ (spins perpendicular to $\hat{\boldsymbol{l}}$), the dipoles spend at least some of their time in a relative "end-to-end" position, which in the classical case is the energetically favourable position. In contrast, for $\hat{\boldsymbol{d}}$ perpendicular to $\hat{\boldsymbol{l}}$ (i.e. $\boldsymbol{S}$ parallel to $\hat{\boldsymbol{l}}$), the dipoles are always in a "side-to-side" position, which is energetically less favourable.

In the A_1/A_2 phases, where $d_{\mu j}$ is given by (5.51), one finds

$$\Delta f_{\mathrm{D}}^{A_{1,2}} = -\tfrac{3}{5}\lambda_{\mathrm{D}} N_{\mathrm{F}} [\Delta_+^2 (\hat{\boldsymbol{d}} \cdot \hat{\boldsymbol{l}})^2 + \Delta_-^2 (\hat{\boldsymbol{e}} \cdot \hat{\boldsymbol{l}})^2], \tag{6.105}$$

where $\Delta_\pm = \tfrac{1}{2}(\Delta_{\uparrow\uparrow} \pm \Delta_{\downarrow\downarrow})$. Obviously, in the A_2 phase, where $\Delta_+ > \Delta_-$, the minimum of $f_{\mathrm{D}}^{A_{1,2}}$ is still attained for $\hat{\boldsymbol{d}}$ along the axis $\hat{\boldsymbol{l}}$. In the A_1 phase, however, when $\Delta_+ = \Delta_-$, the minimum of $f_{\mathrm{D}}^{A_1}$ is degenerate with respect to rotations of $\hat{\boldsymbol{d}}$ and $\hat{\boldsymbol{e}}$ (remember $\hat{\boldsymbol{d}} \cdot \hat{\boldsymbol{e}} = 0$) in the plane perpendicular to $\hat{\boldsymbol{l}}$.

The B phase

For the B phase we substitute (6.8), $A_{\mu j}=3^{-1/2}e^{i\phi}R_{\mu j}(\hat{\boldsymbol{n}}, \theta)$, into (6.103b). The rotation matrix $R_{\mu j}$, characterized by a rotation axis $\hat{\boldsymbol{n}}$ and a rotation angle θ, is given by

$$R_{\mu j}=(1-\cos\theta)\hat{n}_\mu\hat{n}_j+\cos\theta\,\delta_{\mu j}-\sin\theta\sum_k \epsilon_{\mu jk}\hat{n}_k\,. \tag{6.106}$$

This representation yields the relations

$$\sum_i R_{ii}(\hat{\boldsymbol{n}}, \theta)=1+2\cos\theta,$$

$$\sum_{ij} R_{ij}R_{ji}=\sum_i R_{ii}(\hat{\boldsymbol{n}}, 2\theta)=1+2\cos 2\theta.$$

Hence (6.103b) is written as

$$\Delta f_D^B=\tfrac{8}{5}g_D(T)(\cos\theta+\tfrac{1}{4})^2. \tag{6.107}$$

The dipole energy in the B phase is seen to be minimal for relative spin–orbit rotations through an angle $\theta_L=\cos^{-1}(-\frac{1}{4})\approx 104°$ *and* through $\theta_L'=2\pi-\theta_L$, where θ_L is the so-called "Leggett angle". The dipole potential as a function of θ is shown in Fig. 6.6. Note that f_D^B does not depend at all on the orientation of the rotation axis $\hat{\boldsymbol{n}}$. This is because in the B phase $\hat{\boldsymbol{n}}$ is the only vector quantity appearing in the calculation of the dipole energy, and hence the scalar property f_D^B cannot depend on it.

The situation is different if a magnetic field is applied, i.e. in the B_2 phase. Its order parameter is given by (5.66b) in terms of the matrix $\boldsymbol{R}(\hat{\boldsymbol{n}}, \theta)$, which rotates spin space relative to orbital space. From (6.103c)

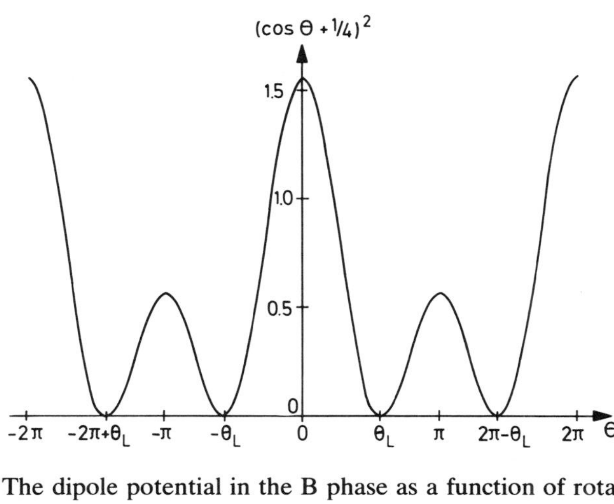

Figure 6.6 The dipole potential in the B phase as a function of rotation angle θ.

one finds

$$\Delta f_{\mathrm{D}}^{\mathrm{B}_2} = \tfrac{1}{5}\lambda_{\mathrm{D}} N_{\mathrm{F}} \{ \Delta_{\perp}^2 [(\operatorname{tr} \boldsymbol{R})^2 + \operatorname{tr}(\boldsymbol{R}^2)] + 2(\Delta_{\parallel} - \Delta_{\perp})\Delta_{\perp}[(\operatorname{tr}\boldsymbol{R})(\hat{\boldsymbol{H}}\boldsymbol{R}\hat{\boldsymbol{H}}) + \hat{\boldsymbol{H}}\boldsymbol{R}^2\hat{\boldsymbol{H}}] + 2(\Delta_{\parallel} - \Delta_{\perp})^2 (\hat{\boldsymbol{H}}\boldsymbol{R}\hat{\boldsymbol{H}})^2 \}, \tag{6.108}$$

where $\hat{\boldsymbol{H}}$ is a unit vector in the direction of the magnetic field. The traces and projections in (6.108) are readily evaluated by using (6.106). This yields

$$\Delta f_{\mathrm{D}}^{\mathrm{B}_2} = \tfrac{1}{5}\lambda_{\mathrm{D}} N_{\mathrm{F}} \{ f_0(\theta) + f_1(\theta)(\hat{\boldsymbol{n}}\cdot\hat{\boldsymbol{H}})^2 + f_2(\theta)(\hat{\boldsymbol{n}}\cdot\hat{\boldsymbol{H}})^4 \}, \tag{6.109}$$

where

$$\left.\begin{aligned} f_0(\theta) &= 8\Delta_{\perp}^2(\cos\theta + \tfrac{1}{4})^2 + 8(\Delta_{\parallel} - \Delta_{\perp})\Delta_{\perp}\cos\theta\,(\cos\theta + \tfrac{1}{4}) \\ &\quad + 2(\Delta_{\parallel} - \Delta_{\perp})^2\cos^2\theta, \\ f_1(\theta) &= 2(\Delta_{\parallel} - \Delta_{\perp})[\Delta_{\perp}(3 + \cos\theta - 4\cos^2\theta) \\ &\quad + 2(\Delta_{\parallel} - \Delta_{\perp})\cos\theta\,(1 - \cos\theta)], \\ f_2(\theta) &= 2(\Delta_{\parallel} - \Delta_{\perp})^2(1 - \cos\theta)^2. \end{aligned}\right\} \tag{6.110}$$

The dipole energy now depends on the relative orientation of $\hat{\boldsymbol{n}}$ and $\boldsymbol{H}$. The minimum of $f_{\mathrm{D}}^{\mathrm{B}_2}$ may be shown to be attained for $\hat{\boldsymbol{n}} = \pm\hat{\boldsymbol{H}}$. This is true for all values of $\Delta_{\perp} - \Delta_{\parallel}$ (ranging from zero in zero magnetic field to $\Delta_{\perp}$ as the B_2 transition is approached from below). Substituting $\hat{\boldsymbol{n}}\cdot\hat{\boldsymbol{H}} = 1$ into (6.109), one finds the dipole potential with respect to the rotation angle θ as

$$\Delta f_{\mathrm{D}}^{\mathrm{B}_2} = \tfrac{8}{5}\lambda_{\mathrm{D}} N_{\mathrm{F}}\left(\cos\theta + \frac{1}{4}\frac{\Delta_{\parallel}}{\Delta_{\perp}}\right)^2, \quad \hat{\boldsymbol{n}} = \pm\hat{\boldsymbol{H}}. \tag{6.111}$$

Clearly, this is minimized for a value of θ given by (Fetter 1975a, Tewordt and Schopohl 1979b, Schopohl 1982)

$$\theta_0(H) = \cos^{-1}\left(-\frac{1}{4}\frac{\Delta_{\parallel}}{\Delta_{\perp}}\right), \tag{6.112}$$

which means that the equilibrium spin–orbit rotation angle decreases with decreasing parallel gap component from $\theta_{\mathrm{L}} = 104°$ to $\theta = 90°$ when $\Delta_{\parallel} = 0$.

6.3.2 Orientation induced by a magnetic field

In general, the spin susceptibility is a tensor in spin space, with principal axes along the preferred directions. Therefore the magnetic energy density

$$f_H = -\tfrac{1}{2}\sum_{\mu\nu}\chi_{\mu\nu}H_\mu H_\nu \tag{6.113a}$$

depends on the orientation of the susceptibility tensor and hence on the order parameter itself. In equilibrium, and in the absence of other forces acting on the spin part of the order parameter, the tensor $\chi_{\mu\nu}$ will be oriented such that the axis associated with the largest eigenvalue will point along the magnetic field.

According to (3.102) and (3.106), the spin susceptibility of the A phase is given by

$$\chi_{\mu\nu} = \chi_N \delta_{\mu\nu} - \Delta\chi\, \hat{d}_\mu \hat{d}_\nu, \tag{6.113b}$$

where $\Delta\chi$ is the difference between the two eigenvalues, i.e.

$$\Delta\chi = \chi_N^0 \frac{1 - Y_0}{1 + F_0^a Y_0}. \tag{6.114}$$

Consequently, there is an energy that tends to orient $\hat{\boldsymbol{d}}$ perpendicular to the field:

$$\Delta f_H^A = \tfrac{1}{2}\Delta\chi\, (\hat{\boldsymbol{d}} \cdot \boldsymbol{H})^2. \tag{6.115}$$

Near T_c and at melting pressure the anisotropy coefficient has the approximate value $\frac{1}{2}\Delta\chi \approx 5.4 \times 10^{-7}(1 - T/T_c)\,\text{erg cm}^{-3}\,\text{G}^{-2}$. Comparison with the dipole orientation energy (6.104) yields the result that a magnetic field of about 30 G gives an orienting force on $\hat{\boldsymbol{d}}$ that is of the same order as the dipole energy (see also (7.28c)). This is approximately independent of temperature. Note that the temperature dependence of both energies is the same in the Ginzburg–Landau regime.

In the B phase there is no orientation of the order parameter by a magnetic field, because its magnetic susceptibility is isotropic with respect to directions in orbital space. Actually, this is even true for the B_2 phase, because the B_2 order parameter (5.66) is still degenerate with respect to arbitrary rotations in orbital space. There is, however, a small orientational effect due to the breaking of spin–orbit symmetry via the dipole interaction, as has been discussed in Section 6.3.1 (Engelsberg *et al.* 1974). The dipole energy in the B_2 phase depends on the relative direction of $\boldsymbol{H}$, as can be seen from (6.109). Hence there is an orientational energy density, given by the second term of (6.109), of order

$$\Delta f_H^{B_2} = \lambda_D N_F \frac{(\Delta_\parallel - \Delta_\perp)\Delta_\perp}{H^2} (\hat{\boldsymbol{n}} \cdot \boldsymbol{H})^2 \tag{6.116a}$$

$$= -\tfrac{5}{12}\lambda_D N_F \left(\frac{\gamma\hbar}{1 + F_0^a}\right)^2 (\hat{\boldsymbol{n}} \cdot \boldsymbol{H})^2, \tag{6.116b}$$

where we have used (5.71) in weak coupling. We see that this is a very tiny orientational effect, since the already small dipole energy is further reduced by yet another small factor. It should also be noted that this effect is temperature-independent to lowest order.

6.3.3 Orientation induced by an electric field

The order parameter may also be oriented by an electric field $\boldsymbol{E}$. The effect originates from the fact that the interaction between the induced electric dipole moments depends on the pair state in the superfluid (Delrieu 1974). The calculation parallels that for the nuclear magnetic dipole interaction, the magnetic moment $\frac{1}{2}\gamma\hbar\boldsymbol{\sigma}(\boldsymbol{r})$ being replaced by the induced electric dipole moment $\boldsymbol{p}=\alpha\boldsymbol{E}$, with α as the atomic polarizability. The result may be written in terms of the dimensionless dipole coupling parameter λ_D defined in (6.98) as (Delrieu 1974, Leggett 1975a, Brinkman and Cross 1978)

$$f_E=-\frac{2\alpha^2}{(\gamma\hbar)^2}R_E^2\lambda_D N_F\tfrac{6}{5}\sum_{\mu ij}A^*_{\mu j}E_jA_{\mu i}E_i, \tag{6.117}$$

where R_E^2 is an unknown renormalization factor.

For the A phase the orientation-dependent part reduces to

$$\Delta f_E^A=\frac{6}{5}\frac{\alpha^2}{(\gamma\hbar)^2}R_E^2\lambda_D N_F(\hat{\boldsymbol{l}}\cdot\boldsymbol{E})^2, \tag{6.118}$$

implying that $\hat{\boldsymbol{l}}$ is forced into the plane perpendicular to $\boldsymbol{E}$. Assuming $R_E=1$ for simplicity, one finds that a field as high as 20 kV/cm has to be used to produce an energy of the order of the dipole energy (see also Engelsberg 1977). Crossed $\boldsymbol{E}$ and $\boldsymbol{H}$ fields should then provide a convenient means for orienting $\hat{\boldsymbol{l}}$ and $\hat{\boldsymbol{d}}$ along $\boldsymbol{E}\times\boldsymbol{H}$ (Dahl 1977). Experimentally, the orienting effect appears to be much smaller (Gully *et al.* 1975, Paulson and Wheatley 1978d, Paalanen *et al.* 1979, Swift *et al.* 1980), suggesting a substantial reduction of R_E. There are suggestions that attribute this to the wave-vector dependence of the polarizability (Maki 1976a, b) and to the screening of density fluctuations by the Landau molecular field (Fomin *et al.* 1978). The latter would indeed introduce a reduction factor that is at least of order $(F_0^s)^{-1}\approx 10^{-2}$–$10^{-1}$. An improved molecular model of the polarization contribution has been discussed by Tankersley (1979).

In the B phase there is again no direct orientational effect by $\boldsymbol{E}$ owing to the isotropy of the order parameter. Only the distortion of the order-parameter structure induced by the electric field, and the ensuing dependence of the dipole energy on the relative orientation of $\hat{\boldsymbol{n}}$ and $\boldsymbol{E}$, gives rise to a small coupling (see below).

6.3.4 Effect of superflow

In the event that the phase under consideration is anisotropic in orbital space, like for example the A phase, superflow will exert an orienting force on the preferred direction in orbital space. Although we shall discuss the phenomenon of superflow at length in Chapter 7, we intend to anticipate a few basic notions here.

In a superfluid system the total mass current is the sum of contributions from the so-called normal component and the superfluid component $\boldsymbol{g}=\boldsymbol{\rho}_n\boldsymbol{v}_n+\boldsymbol{\rho}_s\boldsymbol{v}_s$ plus texture-induced terms. The two density components add up to the total density (see Chapter 9): $\boldsymbol{\rho}_n+\boldsymbol{\rho}_s=\rho\boldsymbol{1}$. We have already calculated the normal-fluid density in Section 3.4 and found that in the A phase it is a tensor quantity $\boldsymbol{\rho}_n$. Therefore the superfluid density must also be a tensor $\boldsymbol{\rho}_s$, since the total density is a scalar quantity. The kinetic-energy density of superflow

$$f_v^A = \tfrac{1}{2}\boldsymbol{v}_s\boldsymbol{\rho}_s\boldsymbol{v}_s \tag{6.119}$$

will therefore depend on the orientation of the vector $\hat{\boldsymbol{l}}$ with respect to the superflow. The orientation energy density is given by (de Gennes and Rainer 1974)

$$\Delta f_v^A = \tfrac{1}{2}(\rho_{s\parallel} - \rho_{s\perp})(\hat{\boldsymbol{l}}\cdot\boldsymbol{v}_s)^2, \tag{6.120}$$

where $\rho_{s\parallel,\perp}$ are the eigenvalues of $\boldsymbol{\rho}_s$ parallel and perpendicular to $\hat{\boldsymbol{l}}$. Employing (3.91), the coefficient in (6.120) is determined as

$$\tfrac{1}{2}(\rho_{s\parallel} - \rho_{s\perp}) = \tfrac{1}{2}(\rho_{n\perp} - \rho_{n\parallel}) < 0. \tag{6.121}$$

Superflow is seen to orient the vector $\hat{\boldsymbol{l}}$ parallel to the superfluid velocity $\boldsymbol{v}_s$. However, this is only true for rather small superfluid velocities and if the temperature is not too low (see Section 7.10).

In the B phase there is a much weaker coupling of superflow to the vector $\hat{\boldsymbol{n}}$, mediated by the dipole interaction, as will be discussed in Section 6.3.5.

6.3.5 Orientational effects in the B phase due to magnetic and electric fields and superflow

As discussed in Section 6.3.2, the pure B phase is unaffected by this external influence owing to its isotropy. Nevertheless, there is a very small orientational effect via the modification of the dipolar interaction energy induced by a distortion of the order-parameter structure (Engelsberg *et al.* 1974, Greaves 1976).

A similar effect occurs in an applied electric field (Gongadze *et al.* 1976, 1977). It follows from the electric field energy (6.117) that the orbital order-parameter components in the direction of the electric field are enhanced, while the perpendicular components are suppressed. The resulting change in the order parameter is of the form

$$\delta A_{\mu j} = \sum_i (a_E\delta_{ij} + b_E\hat{E}_i\hat{E}_j)A_{\mu i}, \tag{6.122}$$

with $b_E > 0$. The coefficients a_E and b_E may be determined in analogy with the calculation of the change of the order parameter in a magnetic field (see the last paragraphs of Section 5.4). For small values of the field $\boldsymbol{E}$, the orientational energy density is found to be

$$\Delta f_E^B = \tfrac{2}{5}b_E g_D[(\mathrm{tr}\,\boldsymbol{R})(\hat{\boldsymbol{E}}\boldsymbol{R}\hat{\boldsymbol{E}}) + \hat{\boldsymbol{E}}\boldsymbol{R}^2\hat{\boldsymbol{E}}]. \tag{6.123}$$

An analogous effect occurs in the presence of superflow. There the gap component in the direction of $\boldsymbol{v}_s$ is suppressed. From the energy contribution $\frac{1}{2}\rho_s v_s^2$, one obtains an order-parameter distortion

$$\delta A_{\mu j} = \sum_i [a_v \delta_{ij} - b_v \hat{v}_{s,i} \hat{v}_{s,j}] A_{\mu i}, \tag{6.124}$$

which gives rise to an orientational energy density of the form (6.123), with b_E and $\boldsymbol{E}$ replaced by $-b_v$ and $\boldsymbol{v}_s$ respectively.

In the presence of a magnetic field, cross-terms will appear connecting $\boldsymbol{v}_s$ and $\boldsymbol{H}$, as well as $\boldsymbol{E}$ and $\boldsymbol{H}$, i.e. orbital vectors and spin vectors (Fetter 1975a, 1976b, Dahl 1977). The corresponding orientational energy is obtained by substituting the gap distortion from superflow, (6.124), into the magnetic energy (6.116a), or vice versa.

All these orientational effects in the B phase may conveniently be collected into a single expression, provided that the fields are not too large, by scaling the fields $\boldsymbol{H}$, $\boldsymbol{E}$ and $\boldsymbol{v}_s$ appropriately (Brinkman and Cross 1978):

$$\Delta f^{\mathrm{B}}_{HvE} = -\epsilon_{\mathrm{B}}\left[\left(\frac{\hat{\boldsymbol{n}}\cdot\boldsymbol{H}}{H_c}\right)^2 + \frac{2}{5}\left(\frac{\boldsymbol{v}_s\boldsymbol{R}\boldsymbol{H}}{v_c \mathrm{H}_c}\right)^2 + \left(\frac{\hat{\boldsymbol{n}}\cdot\boldsymbol{v}_s}{v_c}\right)^2 - \frac{2}{5}\left(\frac{\boldsymbol{E}\boldsymbol{R}\boldsymbol{H}}{E_c H_c}\right)^2 - \left(\frac{\hat{\boldsymbol{n}}\cdot\boldsymbol{E}}{E_c}\right)^2\right]. \tag{6.125}$$

Here

$$\left.\begin{aligned} \epsilon_{\mathrm{B}} &= \frac{2}{5}\frac{g_{\mathrm{D}}^2}{\Delta\chi H^2}\frac{\Delta_\perp - \Delta_\parallel}{\Delta_\perp}, \\ \Delta\chi H_c^2 &= \tfrac{2}{5} g_{\mathrm{D}}, \qquad \rho_s v_c^2 = g_{\mathrm{D}}, \\ 4\alpha^2 E_c^2 &= \frac{(\gamma\hbar)^2}{R_E^2}, \qquad \Delta\chi = \chi_{\mathrm{N}} - \chi_{\mathrm{B}}. \end{aligned}\right\} \tag{6.126}$$

At intermediate pressures and near T_c, when ϵ_{B} tends to a finite value, the critical fields (see (6.99), (7.27), (7.28), (7.231b) are $H_c = 2[\lambda_{\mathrm{D}}(1 + F_0^a)]^{1/2} H_{\mathrm{A}} \approx 14\,\mathrm{G}$, $v_c = v_{s,\mathrm{D}}(m^*/6m)^{1/2} \approx 1.6\,\mathrm{mm/s}$, $E_c \approx 15\,\mathrm{kV/cm}$. For an experimental determination of ϵ_{B}, it is useful to relate ϵ_{B} to the transverse NMR shift $\Delta\omega_\perp$ discussed in (8.38), which yields

$$\epsilon_{\mathrm{B}} = \frac{5}{4}\frac{\Delta\omega_{\mathrm{L}}}{\omega_{\mathrm{L}}}\chi_{\mathrm{B}} H_c^2, \tag{6.127}$$

where ω_{L} is the Larmor frequency.

Let us now discuss the orientational energy density (6.125), but without the electric-field terms, whose effect has not yet clearly been observed. The remaining terms in (6.125) tend to orient $\hat{\boldsymbol{n}}$: (i) along $\boldsymbol{H}$ for $H \gg H_c$ and $v_s \ll v_c$; (ii) along $\boldsymbol{v}_s$ for $v_s \gg v_c$ and $H < H_c$; and (iii) for $v_s \gg v_c$, $H \gg H_c$, along a direction characterized by the property that a rotation about $\hat{\boldsymbol{n}}$ through the Leggett angle carries $\boldsymbol{v}_s$ to $\pm\boldsymbol{H}$. In the case that $H/H_c \gg v_s/v_c$, but arbitrary otherwise, only the terms involving $\boldsymbol{H}$ are important in (6.125). For parallel magnetic field and flow, the vector $\hat{\boldsymbol{n}}$ is oriented parallel to both. For perpendicular configuration of $\boldsymbol{H}$ and $\boldsymbol{v}_s$, say $\boldsymbol{H}$ along $\hat{\boldsymbol{z}}$ and $\boldsymbol{v}_s$ along $\hat{\boldsymbol{y}}$, the orientation depends on the ratio v_s/v_c, which determines whether the first or the second term in (6.125) dominates. Thus in the case that $v_s < v_c$, $\hat{\boldsymbol{n}}$ is oriented along $\boldsymbol{H}$, but for $v_s > v_c$, $\hat{\boldsymbol{n}}$ moves out of the common plane of $\boldsymbol{H}$ and $\hat{\boldsymbol{n}}$ into the direction $\hat{\boldsymbol{n}} = (n_x, -(\frac{5}{3})^{1/2} n_x n_z, n_z)$, where $n_x = \pm[(1 - n_z^2)/(1 + \frac{5}{3}n_z^2)]^{1/2}$ and $n_z = \pm[\frac{1}{5} + \frac{4}{5}(v_c/v_s)^2]^{1/2}$. At intermediate values of v_s the component of $\hat{\boldsymbol{n}}$ perpendicular to $\boldsymbol{H}$ is given by $n_\perp^2 = \frac{4}{5}(1 - v_c^2/v_s^2)$.

The four orienting forces acting on the order parameter in (6.125) generally compete against each other, i.e. each of them tries to force the vector $\hat{\boldsymbol{n}}$ locally into a given direction. However, rapid spatial variations of the order parameter are energetically unfavourable. So the actual energy cost is described by the "bending energy" of the order parameter, which will be discussed in the next chapter. It turns out that the bending energies involved in spatial variations of the order parameter on the scale of a few millimetres are of similar magnitude to the orientational energies in (6.125), i.e. of order $10^{-15}\,\mathrm{K}/k_B$ per atom, corresponding to $10^{-9}\,\mathrm{erg/cm^3}$. Therefore the order-parameter configuration will be smooth and long-range rather than rapidly varying.

6.3.6 Surface energies and boundary conditions

Finally, we want to discuss the influence of a wall on the order-parameter structure and on the macroscopically defined preferred directions. The interaction of the atoms in the liquid with a surface (for example with a solid wall, a free surface or an interface between different phases) is a complex subject, whose details are not yet well understood. Nevertheless, the overall orientational effects on the order parameter in the bulk due to surfaces are well established. At a surface the many-body correlations in the system are perturbed on a truly microscopic scale. After all, the surface potential and the binding energy per particle in the liquid are usually of the same order; furthermore, the surface potential changes over distances of the order of a few interparticle distances. As a consequence, any theoretical description of the influence of a surface necessarily involves the full range of energy and momentum transfers at the surface, making such a theory very complicated. This gives rise to considerable changes, for example in the gap equation at and near the surface. One may attempt to calculate these changes within a weak-coupling model calculation (Ambegaokar *et al.* 1974, Buchholtz and Rainer 1979, Buchholtz and Zwicknagl 1981, Buchholtz 1986). The important result of these calculations is that the longitudinal component of the order parameter $d_\mu^\perp = \sum_j \hat{s}_j d_{\mu j}$ (where $\hat{\boldsymbol{s}}$ is the surface normal) falls off more rapidly than the transverse components as the surface is approached.

This has been shown to occur for nonmagnetic surfaces in the case of specular scattering as well as for completely diffuse scattering. In fact, it may be expected to hold quite generally. The suppression of the order parameter only takes place within a layer of thickness $\xi(T)$ adjacent to the surface, where ξ is the temperature-dependent coherence length (7.18c).

However, even outside this layer, where the *magnitude* of the order parameter has recovered to its bulk value, its *orientation* is still affected by the surface. This is so because the orientation for $z \gtrsim \xi$ will minimize the longitudinal component of the order parameter to give a smooth match of the respective structures at $z \ll \xi$ and $z \gg \xi$, where z is the distance from the wall.

The A phase

The fact that the order parameter is anisotropic in this case makes it easy for it to avoid having a longitudinal component at the surface: it may simply orient itself in such a way that the vector $\hat{\boldsymbol{l}}$ is normal to the surface (Ambegaokar *et al.* 1974). The additional suppression of the order parameter due to surface roughness does not seem to change this qualitative effect much. The energy per surface area gained by orienting the order parameter in such a way that the longitudinal component vanishes (rather than assuming an arbitrary orientation where the *average* order parameter is substantially suppressed), is of order $f_s\xi$, where $f_s = f - f_N$ is the condensation energy density. On the other hand, the loss in bulk energy due to bending of the $\hat{\boldsymbol{l}}$ field is of order $f_s(\xi_0/L)^2L$, where L is the distance over which the bending occurs (size of the sample) and $f_s(\xi_0/L)^2$ is the typical order of the bending energy density. Since $\xi_0^2/\xi L \ll 1$, the latter energy loss is always smaller than the energy gain at the surface. At the surface $\hat{\boldsymbol{l}}$ is therefore rigidly oriented.

The B phase

In the pseudo-isotropic B phase, where all three orbital components of $d_{\mu j}$ are nonzero, a zero longitudinal component at the surface may not be generated by merely rotating the order parameter. Consequently, $d_\mu^\perp$ will be genuinely suppressed in a surface layer of thickness $\xi(T)$, and the orientation of the order parameter at the interface to the bulk region will depend on the way in which the order parameter changes as the surface is approached (Brinkman *et al.* 1974b, Fomin and Vuorio 1975, Smith *et al.* 1977). The position dependence of the order parameter may be determined approximately by requiring the order parameter $d_{\mu j}(\boldsymbol{r})$: (i) to have no longitudinal component right at the surface; and (ii) to minimize the Ginzburg–Landau energy (7.17) discussed in the next chapter. The minimization can then be performed by using a variational ansatz of the form (Smith *et al.* 1977)

$$d_{\mu j}(z) = \Delta \sum_i [\delta_{ij} - \hat{s}_i\hat{s}_j f(z)]R_{\mu j}(z), \tag{6.128}$$

where z is the distance along the surface normal $\hat{\boldsymbol{s}}$ (assuming a locally plane surface), $R_{\mu j}(z)$ is a position-dependent rotation matrix and $f(z)$ is a test function varying along the surface normal from unity at the surface to zero in the bulk. The variation of $R_{\mu j}(z)$ is governed by the change in the dipole energy (note that in the absence of spin–orbit coupling the spin part of the order parameter would be unaffected by a nonmagnetic surface). One may show that the rotation angle θ between spin and orbital spaces will stay at the Leggett angle θ_L (Brinkman *et al.* 1974b). Indeed, if $\theta \neq \theta_L$ then the maximum possible gain in dipole energy in the surface layer would be $g_D\xi$, while on the other hand the energy cost, i.e. the energy needed for θ to recover to the bulk rotation angle θ_L, would be $g_D\xi_D$; here $\xi_D = (N_F\Delta^2(T)/6g_D)^{1/2}\xi_0$ is the dipole healing length (see (7.27)). Since $\xi_D \gg \xi$, θ will stay at its bulk value.

The rotation axis $\hat{\boldsymbol{n}}$, on the other hand, *is* affected by the surface. In order to calculate the orienting effect of the surface on $\hat{\boldsymbol{n}}$, one has to substitute the model expression (6.128) into the general form (6.103) for the dipole energy density. This gives

$$\Delta f_{D,S} = -\tfrac{1}{4}g_D f(f+4)\left[(\hat{\boldsymbol{s}}\cdot\hat{\boldsymbol{n}})^2 - \frac{5f}{2(f+4)}(\hat{\boldsymbol{s}}\cdot\hat{\boldsymbol{n}})^4\right], \tag{6.129}$$

where $f(z)$ is still a position-dependent quantity. However, since the coefficient of the last term varies between 0 and $-\frac{1}{2}$, the minimum of $\Delta f_{D,S}$ is attained by parallel orientation of $\hat{\boldsymbol{n}}$ and $\hat{\boldsymbol{s}}$, irrespective of the precise value of $f(z)$. Hence the surface dipole energy tends to align $\hat{\boldsymbol{n}}$ with the surface normal $\hat{\boldsymbol{s}}$. The surface energy per unit area derives from (6.129) by integrating along the surface normal. Assuming the simple form $f(z) = e^{-z/\xi}$, the integral may be performed to yield the total dipole surface-energy contribution

$$F_{D,S} = -\tfrac{9}{8}g_D\xi \int_{\text{surface}} d^2r[(\hat{\boldsymbol{s}}\cdot\hat{\boldsymbol{n}})^2 - \tfrac{5}{18}(\hat{\boldsymbol{s}}\cdot\hat{\boldsymbol{n}})^4]. \tag{6.130}$$

The distortion of the order parameter near the surface also affects the orientational energy in a magnetic field. By substituting $d_{\mu j}(z)$, (6.128), into the energy contribution due to the magnetic field, (6.113), one finds the following result, which depends on the orientation of the order parameter with respect to the surface normal $\hat{\boldsymbol{s}}$ and the magnetic field $\boldsymbol{H}$ (Brinkman *et al.* 1974b):

$$F_{H,S} = -\epsilon_{H,S}\int d^2r(\hat{\boldsymbol{s}}\boldsymbol{R}\hat{\boldsymbol{H}})^2, \tag{6.131}$$

where $\epsilon_{H,S} \approx \xi\,\Delta\chi H^2$ and $\Delta\chi = \chi_N - \chi_B$ is the spin-susceptibility anisotropy. While the form of the surface-energy expression (6.129) and (6.131) follows directly from the symmetry of the phases (Fomin and Vuorio 1975), the exact values of the prefactors will depend on the details of the surface structure. In the case of nonplanar surfaces there are additional energy contributions depending on the spatial variation of $\hat{\boldsymbol{n}}$. The orientational energies (6.129) and (6.131) define a characteristic field $H_S \propto (g_D/\Delta\chi)^{1/2} \approx 25$ G; see (7.175). For $H \ll H_S$, the surface effect (6.129) dominates and $\hat{\boldsymbol{n}}$ is oriented perpendicular to the surface ($\hat{\boldsymbol{n}} \parallel \hat{\boldsymbol{s}}$). In the opposite case, $H \gg H_S$, $\hat{\boldsymbol{n}}$ will be oriented in such a way to make (6.131) minimal.

An explicit expression for the total magnetic field surface energy in terms of the vector $\hat{\boldsymbol{n}}$ is obtained by substituting the form of the rotation matrix $\boldsymbol{R}$

given by (6.106) with $\theta = \theta_L$ into (6.131), yielding

$$F_{H,S} = -(\tfrac{5}{4})^2 \epsilon_{H,S} \int d^2r[(\hat{s}\cdot\hat{n})(\hat{H}\cdot\hat{n}) + (\tfrac{3}{5})^{1/2}\hat{n}\cdot(\hat{s}\times\hat{H}) - \tfrac{1}{5}(\hat{s}\cdot\hat{H})]^2. \quad (6.132)$$

For later applications, we determine the minimum of (6.132) for general field orientations, perpendicular and parallel to the surface. For $\boldsymbol{H}$ along the axis of the surface normal, the dependence of $F_{H,S}$ on $\hat{n}$ is the same as that of the surface dipole energy (6.130), i.e. the surface tends to orient $\hat{n}$ perpendicular. On the other hand, for $\boldsymbol{H}$ in the plane of the surface, $\hat{n}$ is oriented such that a spin–orbit rotation by the Leggett angle θ_L about $\hat{n}$ takes $\hat{s}$ into $\hat{H}$. This is achieved when

$$\hat{n} = \pm(\tfrac{1}{5})^{1/2}(1, 3^{1/2}, \pm1), \quad (6.133)$$

where we have chosen $\hat{s} \parallel \hat{z}$, and $\hat{H} \parallel \hat{x}$. For $\boldsymbol{H}$ making an arbitrary angle ψ with the surface normal, minima of (6.132) are attained for the following directions of $\hat{n}$ (Ahonen *et al.* 1976c, Lin-Liu and Maki 1978):

$$\hat{n}_I = \pm5^{-1/2}(3^{1/2}\tan\tfrac{1}{2}\psi, (1+4\cos\psi)^{1/2}\tan\tfrac{1}{2}\psi, (1+4\cos\psi)^{1/2}), \quad (6.134a)$$

$$\hat{n}_{II} = \begin{cases} \pm5^{-1/2}(-3^{1/2}\cot\tfrac{1}{2}\psi, -(1-4\cos\psi)^{1/2}\cot\tfrac{1}{2}\psi, (1-4\cos\psi)^{1/2}) \\ \qquad \text{for } \cos^{-1}(\tfrac{1}{4}) \leqslant \psi \leqslant \tfrac{1}{2}\pi, \\ (\mp1, 0, 0) \quad \text{for } 0 \leqslant \psi < \cos^{-1}(\tfrac{1}{4}). \end{cases} \quad (6.134b)$$

Note that the solution $\hat{n}_{II}$ for $0 \leqslant \psi < \cos^{-1}(\tfrac{1}{4})$ is only a local—not a global—minimum.

FURTHER READING

Broken symmetry

Forster D 1975 *Hydrodynamic Fluctuation, Broken Symmetry and Correlation Functions* (Benjamin, Reading, Massachusetts)
Huang K 1982 *Quarks, Leptons and Gauge Fields* (World Scientific, Singapore)
Leggett A J 1973 *J. Phys.* **C6** 3187
Liu M 1982 *Physica* **109/110B + C** 1615 (Proceedings of the 16th International Conference on Low Temperature Physics, LT-16)
Mattuck R D and Johansson B 1968 *Adv. Phys.* **17** 509
Michel L 1980 *Rev. Mod. Phys.* **52** 617
Mineev V P 1980 in *Soviet Scientific Reviews,* Section A: *Physics Reviews,* Vol. 2, (Harwood Academic Publishers, Chur), p. 173

Orientational effects in superfluid ^{3}He

Brinkman W F and Cross M C 1978 in *Progress in Low Temperature Physics,* Vol. VIIA, ed. D F Brewer (North-Holland, Amsterdam), p. 105
Lee D M and Richardson R C 1978 in *The Physics of Liquid and Solid Helium,* Part II, ed. K H Bennemann and J B Ketterson (Wiley, New York), p. 287
Leggett A J 1975 *Rev. Mod. Phys.* **47** 331

7

Superflow and Textures

This chapter is devoted to the discussion of weakly inhomogeneous and—for the most part—time-independent states of superfluid ^{3}He. The great number of different phenomena known to exist in this context are a direct consequence of the rich structure of broken symmetries of the system discussed in the last chapter. Obviously the broken gauge invariance is of particular significance here, since it gives rise to the phenomenon of superfluidity—the very property after which the new phases of ^{3}He have been named. Nevertheless, the broken rotation symmetries in spin and orbital spaces also lead to very unusual properties, as has become clear in the discussion on orienting forces in the preceding chapter. In fact, in an actual sample of superfluid ^{3}He the spontaneously preferred directions of these degrees of freedom will never be uniform because the surrounding walls and any type of residual field and flow in the liquid tend to orient the order parameter locally in different, often competing, ways. On the other hand, the nonuniform orientation of the order parameter is opposed by the internal stiffness of the system. The latter is caused by the increase in energy due to "bending" of the order-parameter field. The competition between orientational and bending forces leads to a continuous configuration of the order-parameter field, called a "texture". Such textures are familiar from the physics of liquid crystals, where the preferred directions describe the orientation of molecules in the liquid (de Gennes 1974).

From the discussion in Chapter 6 of the broken symmetries and the resulting order-parameter structure of superfluid states in ^{3}He, we know that, with the exception of the B phase, the gauge symmetry and the rotational symmetries in spin and orbital spaces are not separately broken. They are connected via broken relative symmetries. This implies that, in general, superfluidity (which is due to broken gauge invariance) is intimately linked with the preferred directions in spin and orbital spaces. For example in the A phase, where a relative gauge–orbit symmetry is broken, superfluidity cannot be separated from the orientation of the orbital triad $(\hat{\boldsymbol{m}}, \hat{\boldsymbol{n}}, \hat{\boldsymbol{l}})$. In ^{3}He-A the broken relative symmetry thus connects superflow and textures. Hence the two notions are discussed together in one chapter.

In ^{3}He-B this coupling is only indirect (i.e. it is not caused by a broken symmetry) and is mediated by the weak dipole interaction.

In this chapter, which is by far the longest in the book, superflow will appear both as an *intrinsic* textural effect, i.e. as a ground-state property of textures, and as an *external* experimentally generated influence on existing textures. To this end, we shall first embark on a more or less separate discussion of the phenomena of superfluidity and textures, and only then shall we study the effect of external superflow.

By "superflow" one means a relative counterflow of the superfluid and the normal components of the density, or else the motion of the superfluid component by itself. Counterflow may be generated by applying thermal gradients across the sample, whereby a two-fluid convection as in superfluid ^{4}He is set up. A different technique, usually used to measure persistent currents, employs a toroidal flow channel, which is set into motion by rotation or torsional oscillations (see Hall and Hook 1986).

7.1 SUPERFLUIDITY

The property of superfluidity is itself perhaps the most impressive manifestation of the macroscopic quantum state of a pair-correlated Fermi system. Indeed, both ^{3}He-A and ^{3}He-B are known to flow through tiny pores or narrow slabs, apparently without friction (see Chapter 4). Nevertheless, the demonstration of a "permanent" (i.e. persistent) current, as is familiar from superconductors, was not possible for a long time. It took a considerable experimental effort to actually show that in the B phase such long-lived currents do exist (Gammel *et al.* 1984). In fact, persistent currents in the B phase were observed for more than 48 h (Pekola *et al.* 1984b, Pekola and Simola 1985). In the case of ^{3}He-A the situation is not as obvious. In fact, theoretical considerations and experimental observations appeared to indicate that supercurrents in ^{3}He-A might be unstable, although they were known to exist for timescales much longer than microscopic relaxation times. Nevertheless, even in ^{3}He-A clear evidence for the existence of persistent currents has been found (Gammel *et al.* 1985). In spite of the apparent experimental difficulty in showing the existence of dissipationless flow via the observation of permanent currents, the many other related properties of the new phases of liquid ^{3}He clearly justify the epithet "superfluid".

The observation of superflow strikingly contradicts our everyday experience of the behaviour of liquids. In terms of quantum-mechanical many-body theory, this phenomenon may be traced back to the macroscopic coherence property of the Cooper-pair wave function, which finds its mathematical expression in a complex-valued order parameter. In other words, there exist macroscopic quantum-mechanical phase variables. The gradients of these phases are closely related to the appearance of superflow.

In order to demonstrate this relationship, we consider a Galilean transformation from a rest frame into a frame of reference moving with velocity $\boldsymbol{u}$. Under this transformation, the momentum $\boldsymbol{p} = \hbar\boldsymbol{k}$ of a quasi-particle changes into $\hbar\boldsymbol{k}' = \hbar\boldsymbol{k} - m\boldsymbol{u}$, where m is the mass of the *bare* particles. This implies that the single-particle momentum eigenstate $\varphi_{\boldsymbol{p}}(\boldsymbol{r}) = \exp(\mathrm{i}\boldsymbol{k}\cdot\boldsymbol{r})$ transforms into $\varphi_{\boldsymbol{k}'} = \varphi_{\boldsymbol{k}}(\boldsymbol{r})\exp(-\mathrm{i}m\boldsymbol{u}\cdot\boldsymbol{r}/\hbar)$. The local value of the pair amplitude

$$F_{\boldsymbol{k}\alpha\beta}(\boldsymbol{r}) = \int \mathrm{d}^3r' \exp(-\mathrm{i}\boldsymbol{k}\cdot\boldsymbol{r}')\,\langle\psi_\alpha(\boldsymbol{r}+\tfrac{1}{2}\boldsymbol{r}')\psi_\beta(\boldsymbol{r}-\tfrac{1}{2}\boldsymbol{r}')\rangle \tag{7.1}$$

therefore changes into

$$F_{\boldsymbol{k}\alpha\beta}(\boldsymbol{r}) \to F_{\boldsymbol{k}\alpha\beta}(\boldsymbol{r})\exp(-2\mathrm{i}m\boldsymbol{u}\cdot\boldsymbol{r}/\hbar), \tag{7.2}$$

where a representation of the field operator $\psi_\alpha(\boldsymbol{r})$ in terms of plane-wave states $\psi_\alpha(\boldsymbol{r}) = \sum_{\boldsymbol{k}} a_{\boldsymbol{k}\alpha}\varphi_{\boldsymbol{k}}(\boldsymbol{r})$ has been used. The local value of the gap parameter $\Delta_{\boldsymbol{k}\alpha\beta}(\boldsymbol{r})$ may be expressed in terms of $F_{\boldsymbol{k}\alpha\beta}(\boldsymbol{r})$ via (3.17) and therefore transforms like $F_{\boldsymbol{k}\alpha\beta}(\boldsymbol{r})$.

In the event that the local value of the order parameter is given by a real quantity times a phase factor, as for the B phase,

$$d_{\mu j}(\boldsymbol{r}) = |\Delta(\boldsymbol{r})|\,R_{\mu j}(\boldsymbol{r})\exp[\mathrm{i}\phi(\boldsymbol{r})], \tag{7.3}$$

the phase $\phi(\boldsymbol{r})$ is seen to transform under a Galilean transformation as

$$\phi(\boldsymbol{r}) \to \phi(\boldsymbol{r}) - \frac{2m}{\hbar}\boldsymbol{u}\cdot\boldsymbol{r}. \tag{7.4}$$

From this relation, one concludes that the quantity

$$\boldsymbol{v}_{\mathrm{s}} = \frac{\hbar}{2m}\boldsymbol{\nabla}\phi(\boldsymbol{r}) \tag{7.5}$$

transforms as a velocity. It is referred to as the *superfluid velocity,* because it describes dissipationless flow. It is quite similar to the microscopic magnetization currents in atoms or molecules, but here acts on a *macroscopic level.* The stability and mechanisms for decay of superflow will be discussed in detail later in this chapter.

Superfluidity in states described by an intrinsically complex order parameter, such as in the A phase, is a different matter. As discussed in Section 6.1, the A-phase order parameter, given by

$$d_{\mu j}(\boldsymbol{r}) = |\Delta_0(\boldsymbol{r})|\,\hat{d}_\mu(\boldsymbol{r})[\hat{m}_j(\boldsymbol{r}) + \mathrm{i}\hat{n}_j(\boldsymbol{r})], \tag{7.6}$$

is invariant under a relative gauge–orbit transformation. Multiplication of $d_{\mu j}$ by a phase factor $\exp[\mathrm{i}\phi(\boldsymbol{r})]$ is equivalent to a rotation of $\hat{\boldsymbol{m}}$ and $\hat{\boldsymbol{n}}$ in their common plane by the angle $-\phi$ (see Section 6.1), i.e. $\hat{\boldsymbol{m}} + \mathrm{i}\hat{\boldsymbol{n}}$ transforms into $\hat{\boldsymbol{m}}' + \mathrm{i}\hat{\boldsymbol{n}}'$, where

$$\hat{\boldsymbol{m}}' + \mathrm{i}\hat{\boldsymbol{n}}' = \mathrm{e}^{-\mathrm{i}\phi}(\hat{\boldsymbol{m}} + \mathrm{i}\hat{\boldsymbol{n}}). \tag{7.7}$$

Applying the gradient operator to both sides and letting ϕ go to zero, we have

$$\boldsymbol{\nabla}(\hat{m}_j + \mathrm{i}\hat{n}_j) = -\mathrm{i}(\boldsymbol{\nabla}\phi)(\hat{m}_j + \mathrm{i}\hat{n}_j). \tag{7.8}$$

Upon taking the scalar product with $\hat{\boldsymbol{m}} - \mathrm{i}\hat{\boldsymbol{n}}$, this yields

$$\begin{aligned}\boldsymbol{\nabla}\phi &= -\tfrac{1}{2}\sum_j [\hat{m}_j(\boldsymbol{\nabla}\hat{n}_j) - \hat{n}_j(\boldsymbol{\nabla}\hat{m}_j)] \\ &= -\sum_j \hat{m}_j\boldsymbol{\nabla}\hat{n}_j. \end{aligned} \tag{7.9}$$

Hence, under a Galilean transformation, when the phase ϕ transforms according to (7.4), the quantity on the right-hand side of (7.9) transforms like $-(2m/\hbar)\boldsymbol{u}$, i.e. like a velocity. This allows for the definition of a superfluid velocity

$$v_{s,i} = \frac{\hbar}{2m}\hat{\boldsymbol{m}} \cdot \nabla_i\hat{\boldsymbol{n}}. \tag{7.10}$$

It is sometimes convenient to express the orientation of the orbital triad $(\hat{\boldsymbol{m}}, \hat{\boldsymbol{n}}, \hat{\boldsymbol{l}})$ of the A-phase order parameter relative to some fixed coordinate system $(\hat{x}, \hat{y}, \hat{z})$ in terms of Euler angles α, β, γ:

$$\hat{\boldsymbol{m}} = (\cos\alpha\cos\beta\cos\gamma - \sin\alpha\sin\gamma,\ \sin\alpha\cos\beta\cos\gamma + \cos\alpha\sin\gamma,\ -\sin\beta\cos\gamma), \tag{7.11a}$$

$$\hat{\boldsymbol{n}} = (-\cos\alpha\cos\beta\sin\gamma - \sin\alpha\cos\gamma,\ -\sin\alpha\cos\beta\sin\gamma + \cos\alpha\cos\gamma,\ \sin\beta\sin\gamma), \tag{7.11b}$$

$$\hat{\boldsymbol{l}} = (\cos\alpha\sin\beta,\ \sin\alpha\sin\beta,\ \cos\beta). \tag{7.11c}$$

While α and β are the usual azimuthal and polar angles respectively of spherical coordinates, the angle γ describes rotations *about* $\hat{\boldsymbol{l}}$ (see Fig. 7.1). Hence the orbital part of the A-phase order parameter may be written as

$$\hat{\boldsymbol{\Delta}} = 2^{-1/2}(\hat{\boldsymbol{m}} + \mathrm{i}\hat{\boldsymbol{n}}) \tag{7.12a}$$

$$= \mathrm{e}^{-\mathrm{i}\gamma}\hat{\boldsymbol{\Delta}}\,|_{\gamma=0}. \tag{7.12b}$$

In using the Euler angles, one has to beware of unphysical singularities in α when $\sin\beta = 0$ (Mermin 1978a). The superfluid velocity (7.10) then takes the form

$$\boldsymbol{v}_s = -\frac{\hbar}{2m}(\boldsymbol{\nabla}\gamma + \cos\beta\,\boldsymbol{\nabla}\alpha). \tag{7.13}$$

This shows that in the A-phase $\boldsymbol{v}_s$ is not simply given by the gradient of the overall phase of the order parameter as in superfluid ^{4}He or ^{3}He-B, but is more complicated owing to the *orientation* of the order parameter.

Clearly, the rotation angle γ plays the role of a phase of the order

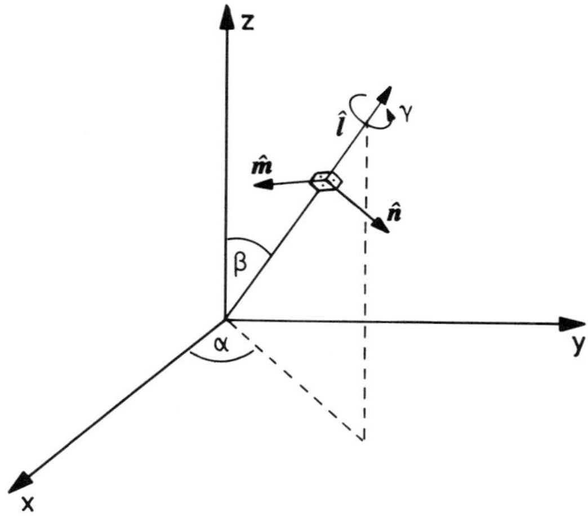

Figure 7.1 Euler angles α, β, γ characterizing the orbital triad ($\hat{\boldsymbol{m}}$, $\hat{\boldsymbol{n}}$, $\hat{\boldsymbol{l}}$) in ^{3}He-A.

parameter. Hence it may be identified as $\gamma = -\phi$, so that, for $\beta = \frac{1}{2}\pi$, (7.13) leads to the usual definition (7.5) of the superfluid velocity for isotropic superfluids.

The fact that a quantity $\boldsymbol{v}_s$ may be constructed from the order parameter at all is a strong indication, but no proof, of the existence of dissipationless flow of the condensate, i.e. superflow. The relevant quantity is in fact the *current density* associated with $\boldsymbol{v}_s$. It may be determined as follows. We perform a Galilean transformation of the condensate free-energy density f by an infinitesimal velocity $\boldsymbol{u}$, which yields

$$f \to f + \boldsymbol{g}_s \cdot \boldsymbol{u}. \tag{7.14a}$$

The coefficient of $\boldsymbol{u}$ in (7.14a) is the superfluid current density $\boldsymbol{g}_s$. The effect of the infinitesimal Galilean transformation is to multiply the order parameter by $1 - \mathrm{i}(2m/\hbar)\boldsymbol{u}\cdot\boldsymbol{r}$. This amounts to the replacement of $\boldsymbol{\nabla} d_{\mu j}$ by $[\boldsymbol{\nabla} - \mathrm{i}(2m/\hbar)\boldsymbol{u}]d_{\mu j}$. Hence $\boldsymbol{g}_s$ may be obtained from an expansion of the free-energy density in terms of gradients of the order parameter, the so-called "gradient free-energy density" f_{grad}, discussed in Section 7.2. To this end, we assume a fictitious gauge field $\boldsymbol{u}$ with charge m on the He atom. To obtain a gauge-invariant free energy, the gradient operator has to be replaced by the gauge-invariant operator $\boldsymbol{\nabla} - \mathrm{i}(2m/\hbar)\boldsymbol{u}$. Note that for a system of charged fermions the role of the velocity field $\boldsymbol{u}$ is played by the vector potential $\boldsymbol{A}$ (Ambegaokar *et al.* 1974). The mass supercurrent is then given by the term in the gradient free-energy density f_{grad} that is linear in $\boldsymbol{u}$:

$$\boldsymbol{g}_s = \frac{\delta f_{\text{grad}}}{\delta \boldsymbol{u}}. \tag{7.14b}$$

In order for Galilean invariance to be guaranteed, the total mass current $\boldsymbol{g}$, which is given by

$$\boldsymbol{g} = \boldsymbol{g}_s + \boldsymbol{g}_n, \tag{7.15a}$$

with $\boldsymbol{g}_n = \boldsymbol{\rho}_n \boldsymbol{v}_n$ obtained in (3.86), should change by the amount $\rho \boldsymbol{u}$ under transformation to a frame of reference moving with velocity $\boldsymbol{u}$. This implies that $\boldsymbol{g}_s$ is associated with a fraction of the total density ρ, called the "superfluid" density $\boldsymbol{\rho}_s$, such that the normal-fluid density $\boldsymbol{\rho}_n$ (given by (3.87b), (3.91)) and $\boldsymbol{\rho}_s$ add up to ρ:

$$\boldsymbol{\rho}_s = \rho \mathbf{1} - \boldsymbol{\rho}_n. \tag{7.15b}$$

In complete analogy to the above, one may introduce a spin superfluid velocity $\boldsymbol{v}_{sp}$ as the gradient of the relative phase of spin-up and spin-down Cooper pairs, multiplied by $\hbar/2m$. By introducing a gauge field for each spin component, one may show that the spin supercurrent is given by

$$\boldsymbol{j}^{sp}_{j\mu} = -\frac{\delta f_{grad}}{\delta(\nabla_j \theta^s_\mu)}, \tag{7.16}$$

where $\delta\theta^s_\mu$ is a rotation angle in spin space about the axis μ.

7.2 GRADIENT FREE ENERGY

The free energy of a weakly nonuniform state is well represented by the lowest-order term in an expansion of the free energy in terms of the spatial gradients of the order parameter. This is referred to as the "gradient free-energy density" f_{grad}. Since odd powers of the gradients are not compatible with the rotational and gauge invariance of the free energy, the expansion starts at second order. In the Ginzburg–Landau regime, close to the transition temperature, this expansion is particularly simple: only terms of second order in the order parameter need be kept. The general form of the gradient free-energy density is then obtained by combining the vector indices of the gradient operator ∇_j in all possible ways with the orbital tensor indices of the order parameter $d_{\mu j}$:

$$f_{grad} = \tfrac{1}{2}\sum_{jl\mu} [K_1(\nabla_j d_{\mu l})(\nabla_j d^*_{\mu l}) + K_2(\nabla_j d_{\mu l})(\nabla_l d^*_{\mu j}) + K_3(\nabla_j d_{\mu j})(\nabla_l d^*_{\mu l})]. \tag{7.17}$$

The simplifying feature of the Ginzburg–Landau (GL) expansion is that the coefficients are *normal-state* quantities, i.e. they do not depend on the order parameter, and hence are isotropic. In the general case, i.e. outside the GL regime, these coefficients will be anisotropic in orbital and spin spaces. The coefficients K_i have been calculated in weak-coupling theory, with the result (Wölfle 1974)

$$K_1 = K_2 = K_3 = \tfrac{1}{5} N_F \xi_0^2, \tag{7.18a}$$

where

$$\xi_0 = \left[\frac{7\zeta(3)}{48\pi^2}\right]^{1/2} \frac{\hbar v_F}{k_B T_c} \quad (\approx 120\text{ Å at } P_{\text{melt}}) \tag{7.18b}$$

is the zero-temperature coherence length characterizing the spatial extension of a Cooper pair. (Hence in superfluid ^{3}He ξ_0 is seen to be significantly shorter than in conventional clean superconductors, where $\xi_0 \approx 10^3$–10^4 Å.) For later use we also introduce the temperature-dependent coherence length

$$\xi(T) = \xi_0(1 - T/T_c)^{-1/2}. \tag{7.18c}$$

Actually, K_2 and K_3 are not exactly equal but differ by terms of order $(T_c/T_F)^2$ (Blount *et al.* 1974). The coefficients K_i need not be calculated directly. In fact, they follow from the expression for the superfluid current to be discussed below and the result for the normal-fluid density obtained in Section 3.4, by requiring that the sum of the superfluid and normal current contributions must obey Galilean invariance. Note that the difference of the second and third terms in (7.17) combines to a pure divergence and thus integrates to a surface term in the total free energy. None the less it may not be discarded when the supercurrent (a local quantity) is calculated from (7.17). In the weak-coupling regime the Ginzburg–Landau free energy may be derived to all orders in the gradients of $d_{\mu j}$ using the kinetic equation for quasiparticles presented in Chapter 10 (Combescot and Saslow 1977).

7.2.1 The A phase

In the A phase the spin and orbital parts of the order parameter factorize, i.e. $d_{\mu j} = d_\mu \Delta_j$, with $\boldsymbol{\Delta} = \Delta_0 \hat{\boldsymbol{\Delta}}$ given by (7.12a); note that this is a particularly simple form of (6.15). The gradient free-energy density in (7.17) may easily be expressed in terms of $\boldsymbol{\Delta}$ and $\hat{\boldsymbol{d}}$. In weak coupling ($K_1 = K_2 = K_3$) one obtains (Maki and Kumar 1977a,b)

$$\begin{aligned} f^{\mathrm{A}}_{\mathrm{grad}} = \tfrac{1}{2}K_1\{&3\,|\boldsymbol{\nabla}\cdot\boldsymbol{\Delta}|^2 + |\boldsymbol{\nabla}\times\boldsymbol{\Delta}|^2 + 2\,|(\boldsymbol{\Delta}\cdot\boldsymbol{\nabla})\hat{\boldsymbol{d}}|^2 \\ &+ |\boldsymbol{\Delta}|^2\,[|\boldsymbol{\nabla}\times\hat{\boldsymbol{d}}|^2 + (\boldsymbol{\nabla}\cdot\hat{\boldsymbol{d}})^2] \\ &+ 2\boldsymbol{\nabla}[(\boldsymbol{\Delta}\cdot\boldsymbol{\nabla})\boldsymbol{\Delta}^* - \boldsymbol{\Delta}(\boldsymbol{\nabla}\cdot\boldsymbol{\Delta}^*)]\}, \end{aligned} \tag{7.19}$$

where the last term is a pure divergence.

Let us now concentrate on the so-called *London limit,* i.e. the case where the order parameter attains its equilibrium structure everywhere in the bulk and where weak perturbations will not change its structure but will merely influence the preferred directions in orbital and spin spaces as well as the phase variable. In this case the general gradient free-energy density (7.17) may be rewritten in terms of gradients of the symmetry variables *only,* and is then referred to as the "bending energy" density f_{bend}. The corresponding variables are (i) the gradients of the orbital axis $\hat{\boldsymbol{l}}$, (ii) the gradients of the angle describing rotations about $\hat{\boldsymbol{l}}$, which may conveniently be replaced by the superfluid velocity $\boldsymbol{v}_s$ as defined by (7.10), and (iii) the gradients of the preferred direction in spin space $\hat{\boldsymbol{d}}$. The bending free-energy density is quite generally given by the sum of terms containing all possible combinations of $\boldsymbol{v}_s$, $\nabla\cdot\hat{\boldsymbol{l}}$, $\nabla\times\hat{\boldsymbol{l}}$ and ∇d_μ in the form (Blount *et al.* 1974, Cross 1975)

$$\begin{aligned} f_{\text{bend}} &= \tfrac{1}{2}\rho_{s\perp}(\hat{\boldsymbol{l}}\times\boldsymbol{v}_s)^2 + \tfrac{1}{2}\rho_{s\parallel}(\hat{\boldsymbol{l}}\cdot\boldsymbol{v}_s)^2 \\ &+ \frac{\hbar}{2m}[C_\perp\boldsymbol{v}_s\cdot(\nabla\times\hat{\boldsymbol{l}}) - (C_\perp - C_\parallel)(\boldsymbol{v}_s\cdot\hat{\boldsymbol{l}})\hat{\boldsymbol{l}}\cdot(\nabla\times\hat{\boldsymbol{l}})] \\ &+ \frac{1}{2}\left(\frac{\hbar}{2m}\right)^2\Big\{K_s(\nabla\cdot\hat{\boldsymbol{l}})^2 + K_t[\hat{\boldsymbol{l}}\cdot(\nabla\times\hat{\boldsymbol{l}})]^2 + K_b[\hat{\boldsymbol{l}}\times(\nabla\times\hat{\boldsymbol{l}})]^2 \\ &+ \rho_{sp\perp}\sum_\alpha(\hat{\boldsymbol{l}}\times\nabla\hat{d}_\alpha)^2 + \rho_{sp\parallel}\sum_\alpha[(\hat{\boldsymbol{l}}\cdot\nabla)\hat{d}_\alpha]^2\Big\}. \end{aligned} \tag{7.20}$$

Here a pure divergence term $\nabla[(\hat{\boldsymbol{l}}\cdot\nabla)\hat{\boldsymbol{l}} - \hat{\boldsymbol{l}}(\nabla\cdot\hat{\boldsymbol{l}})]$ has been omitted. The subscripts s, t and b on the coefficients K_s, K_t and K_b in (7.20) indicate that the respective energy contributions are due to a *s*play-, *t*wist- or *b*end-like spatial change of $\hat{\boldsymbol{l}}$, as illustrated in Fig. 7.2. This terminology is well established in liquid-crystal physics (de Gennes 1974). It should be noted that the appearance of $\hat{\boldsymbol{l}}$ in the last two terms of (7.20), describing the bending of $\hat{\boldsymbol{d}}$, does *not* imply any physical coupling (e.g. dipole coupling). The vector $\hat{\boldsymbol{l}}$ is only used as a reference direction to which the notation $\rho_{sp\perp}$, $\rho_{sp\parallel}$ refers. In the special case of dipole locking ($\hat{\boldsymbol{d}}\parallel\hat{\boldsymbol{l}}$), (7.20) becomes much simpler, since the terms involving $\hat{\boldsymbol{d}}$ may then be incorporated into those due to $\hat{\boldsymbol{l}}$ by the replacement

$$K_s\to K_s+\rho_{sp\perp},\quad K_t\to K_t+\rho_{sp\perp},\quad K_b\to K_b+\rho_{sp\parallel}. \tag{7.21a}$$

Hence the bending energy of $\hat{\boldsymbol{l}}$ is seen to increase. This reflects an increased stiffness of the $\hat{\boldsymbol{l}}$ texture due to the *spin* part of the order parameter!

It should be noted that the first two terms in (7.20) may also be written as

$$\tfrac{1}{2}\rho_{s\perp}(\hat{\boldsymbol{l}}\times\boldsymbol{v}_s)^2 + \tfrac{1}{2}\rho_{s\parallel}(\hat{\boldsymbol{l}}\cdot\boldsymbol{v}_s)^2 = \tfrac{1}{2}\rho_{s\perp}v_s^2 - \tfrac{1}{2}(\rho_{s\perp}-\rho_{s\parallel})(\hat{\boldsymbol{l}}\cdot\boldsymbol{v}_s)^2. \tag{7.21b}$$

There also exists a slightly different notation in the literature (e.g. Mermin and Ho 1976, Bhattacharyya *et al.* 1977, Hall and Hook 1986), which is related to that used above by

$$\left.\begin{aligned} \rho_s\equiv\rho_{s\perp},\ \ \rho_0\equiv\rho_{s\perp}-\rho_{s\parallel}, \\ C\equiv C_\perp,\ \ C_0\equiv C_\perp - C_\parallel. \end{aligned}\right\} \tag{7.21c}$$

In the Ginzburg–Landau (GL) regime the coefficients in (7.20) may be obtained by comparing with (7.17). In the weak-coupling limit one finds

$$\rho_{s\perp} = 2\rho_{s\parallel} = 2\left(\frac{2m}{\hbar}\right)^2 K_1\Delta_0^2 = \frac{2}{5}\left(\frac{2m}{\hbar}\right)^2 N_F\Delta_0^2\xi_0^2, \tag{7.22a}$$

$$C_\perp = -C_\parallel = \tfrac{1}{2}\rho_{s\parallel}, \tag{7.22b}$$

$$K_s = K_t = \tfrac{1}{3}K_b = \tfrac{1}{2}\rho_{s\parallel}, \tag{7.22c}$$

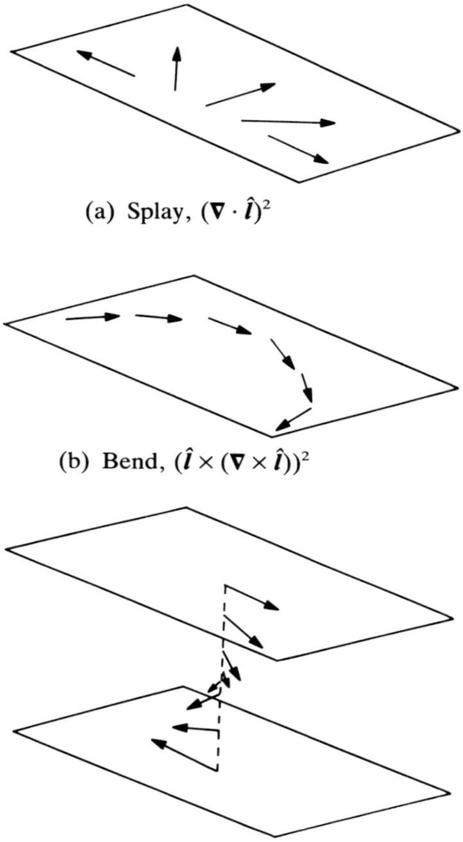

(a) Splay, $(\nabla \cdot \hat{\boldsymbol{l}})^2$

(b) Bend, $(\hat{\boldsymbol{l}} \times (\nabla \times \hat{\boldsymbol{l}}))^2$

(c) Twist, $(\hat{\boldsymbol{l}} \cdot (\nabla \times \hat{\boldsymbol{l}}))^2$

Figure 7.2 Three different kinds of deformation of a vector field $\hat{\boldsymbol{l}}$ and their corresponding contributions to the free energy (see (7.20)).

where Δ_0 is the maximum value of the gap. In this limit f_{bend} takes the form

$$f_{\text{bend}} = \rho_{s\|}\Big\{\boldsymbol{v}_s^2 - \tfrac{1}{2}(\hat{\boldsymbol{l}} \cdot \boldsymbol{v}_s)^2 + \frac{\hbar}{2m}[\tfrac{1}{2}\boldsymbol{v}_s \cdot (\nabla \times \hat{\boldsymbol{l}}) - (\boldsymbol{v}_s \cdot \hat{\boldsymbol{l}})\hat{\boldsymbol{l}} \cdot (\nabla \times \hat{\boldsymbol{l}})]$$
$$+\frac{1}{4}\left(\frac{\hbar}{2m}\right)^2[(\nabla \cdot \hat{\boldsymbol{l}})^2 + [\hat{\boldsymbol{l}} \cdot (\nabla \times \hat{\boldsymbol{l}})]^2 + 3[(\hat{\boldsymbol{l}} \cdot \nabla)\hat{\boldsymbol{l}}]^2$$
$$+4\sum_\alpha [(\hat{\boldsymbol{l}} \times \nabla)\hat{d}_\alpha]^2 + 2\sum_\alpha [(\hat{\boldsymbol{l}} \cdot \nabla)\hat{d}_\alpha]^2]\Big\}. \tag{7.22d}$$

Outside the GL regime the coefficients in (7.20) have been calculated by Cross (1975), Fetter (1979) and Fetter *et al.* (1983) in weak-coupling theory, including Fermi-liquid corrections. The coefficients $\rho_{s\perp,\|}$ may in fact be deduced from the result for the normal-fluid mass tensor in Section 3.4, as explained below. The remaining coefficients may be calculated along similar lines by considering the superfluid mass current and spin current as well as the orbital torque $\delta f_{\text{bend}}/\delta\hat{\boldsymbol{l}}$ discussed in Section 9.3. Keeping Landau parameters with $l \leqslant 1$ only, the result of this calculation is given by

$$\boldsymbol{\rho}_s = (\boldsymbol{D}_s)^{-1}\boldsymbol{\rho}_s^0, \tag{7.23a}$$

$$\boldsymbol{C} = \tfrac{1}{2}\rho_{s\parallel}^0(\boldsymbol{D}^s)^{-1}(\mathbf{1} - 2\hat{\boldsymbol{l}} \otimes \hat{\boldsymbol{l}}), \tag{7.23b}$$

$$\boldsymbol{\rho}_{sp} = \frac{\hbar}{2m}\frac{1 + \frac{1}{3}F_1^a}{1 + \frac{1}{3}F_1^s}(\boldsymbol{D}^a)^{-1}\boldsymbol{\rho}_s^0, \tag{7.23c}$$

$$K_s = \frac{1}{4}\frac{m}{m^*}\rho_{s\perp}^0, \tag{7.23d}$$

$$K_t = \frac{1}{12}\frac{m}{m^*}\left(\rho_{s\perp}^0 + 7\rho_{s\parallel}^0 - \frac{3\rho_{s\parallel}^0}{1 + \frac{1}{3}F_1^s\,\rho_{s\parallel}^0/\rho}\right), \tag{7.23e}$$

$$K_b = \frac{1}{4}\frac{m}{m^*}\left(3\rho_{s\parallel}^0 + \tfrac{8}{3}\rho\gamma - \frac{\rho_{s\parallel}^0}{1 + \frac{1}{3}F_1^s\rho_{n\parallel}^0/\rho}\right), \tag{7.23f}$$

where

$$\boldsymbol{D}^{s,a} = \mathbf{1} + \tfrac{1}{3}F_1^{s,a}\frac{\boldsymbol{\rho}_n^0}{\rho}, \tag{7.23g}$$

$$\gamma = 3\left\langle\frac{Y_0(\hat{\boldsymbol{k}}; T)(\hat{\boldsymbol{k}} \cdot \hat{\boldsymbol{l}})^4}{(\hat{\boldsymbol{k}} \times \hat{\boldsymbol{l}})^2}\right\rangle_{\hat{\boldsymbol{k}}}, \tag{7.23h}$$

$$\boldsymbol{\rho}_s^0 = \rho\mathbf{1} - \boldsymbol{\rho}_n^0. \tag{7.23i}$$

(Note that (7.23c) is only valid in the case that the spin polarization is perpendicular to $\hat{\boldsymbol{d}}$; for the general case see (10.60).) Here $\hat{\boldsymbol{l}} \otimes \hat{\boldsymbol{l}}$ is the tensor product, $\boldsymbol{\rho}_n^0$ is the normal-fluid mass density tensor in the absence of the Fermi-liquid interaction given by (3.87b), and $Y_0(\hat{\boldsymbol{k}}; T)$ has been defined in (3.88). Strong-coupling corrections to the above weak-coupling results have been calculated by Serene and Rainer (1977, 1983), Fetter (1979) and Fetter *et al.* (1983), and the effect of a magnetic field has been considered by Brand and Dörfle (1981).

The coefficient γ in (7.23h) is found to diverge as $\ln T$ in the limit $T \to 0$. This signals the breakdown of the gradient expansion at low temperatures. The physical reason for this divergence may be traced back to the existence of nodes in the energy gap of the ^{3}He-A order parameter. (Indeed the denominator in the angular brackets in (7.23h) has precisely the form of the square of the A-phase gap parameter and is thus the origin for the abovementioned $\ln T$ divergence.) As a consequence of the nodes, pair breaking, i.e. excitation of quasiparticles, is possible at arbitrarily low temperatures and, quite generally, forbids a well-defined analytic low-temperature expansion of any quantity in terms of T/T_c. We shall encounter this problem again in the discussion of the nonlinear contributions to the superfluid current.

7.2.2 The B phase

In spite of the apparent simplicity of the order parameter of ^{3}He-B, the corresponding bending free-energy density is considerably more complicated than in the A phase. This is of course due to the fact that in this case $A_{\mu j} = 3^{-1/2} e^{i\phi} R_{\mu j}(\hat{\boldsymbol{n}}, \theta)$ is proportional to a general rotation matrix characterized by an axis of rotation $\hat{\boldsymbol{n}}$ and a rotation angle θ (see (6.106)), where now both $\hat{\boldsymbol{n}}$ and θ vary with position. The full expression for $A_{\mu j}$ has to be inserted into (7.17) for the bending energy density. Using the relations

$$\hat{\boldsymbol{n}} \cdot \nabla_j \hat{\boldsymbol{n}}, = 0, \tag{7.23j}$$

$$\hat{\boldsymbol{n}} \cdot (\hat{\boldsymbol{n}} \cdot \boldsymbol{\nabla}) \hat{\boldsymbol{n}} = 0, \tag{7.23k}$$

$$\begin{aligned} \nabla_j R_{\mu l} = {} & (\nabla_j \theta)[\sin\theta\, (\hat{n}_\mu \hat{n}_l - \delta_{\mu l}) - \cos\theta \sum_k \epsilon_{\mu l k} \hat{n}_k] \\ & + (1 - \cos\theta)[(\nabla_j \hat{n}_\mu)\hat{n}_l + \hat{n}_\mu (\nabla_j \hat{n}_l)] - \sin\theta \sum_k \epsilon_{\mu l k} (\nabla_j \hat{n}_k), \end{aligned} \tag{7.23l}$$

and expressing $\frac{1}{2} K_i \Delta^2$ in the weak-coupling Ginzburg–Landau regime in terms of the isotropic superfluid density without Fermi-liquid corrections, ρ_s^0, (7.23i),

$$\tfrac{1}{2} K_i \Delta^2 = \frac{1}{10} \left(\frac{\hbar}{2m}\right)^2 \rho_s^0, \tag{7.23m}$$

where (see (3.92) with $F_1^s = 0$ and (3.89))

$$\frac{\rho_s^0}{\rho} = 1 - Y_0 \tag{7.23n}$$

$$= \frac{7\zeta(3)}{4\pi^2} \left(\frac{\Delta}{k_B T_c}\right)^2 \quad (T \lessdot T_c), \tag{7.23o}$$

we obtain the bending energy density in this limit as

$$\begin{aligned} f_{\text{bend}} = \frac{\hbar^2}{40m^2} \rho_s^0 \{ & 5(\boldsymbol{\nabla}\phi)^2 + 4(\boldsymbol{\nabla}\theta)^2 - 2(\hat{\boldsymbol{n}} \cdot \boldsymbol{\nabla}\theta)^2 \\ & + 2(1 - \cos\theta)\{4(\boldsymbol{\nabla} \times \hat{\boldsymbol{n}})^2 - (1 - \cos\theta)[\hat{\boldsymbol{n}} \cdot (\boldsymbol{\nabla} \times \hat{\boldsymbol{n}})]^2 \\ & + (3 - \cos\theta)(\boldsymbol{\nabla} \cdot \hat{\boldsymbol{n}})^2 - 2\sin\theta (\boldsymbol{\nabla} \cdot \hat{\boldsymbol{n}})\hat{\boldsymbol{n}} \cdot (\boldsymbol{\nabla} \times \hat{\boldsymbol{n}}) \\ & + (\boldsymbol{\nabla}\theta) \cdot (\boldsymbol{\nabla} \times \hat{\boldsymbol{n}}) - 2(\hat{\boldsymbol{n}} \cdot \boldsymbol{\nabla}\theta)\hat{\boldsymbol{n}} \cdot (\boldsymbol{\nabla} \times \hat{\boldsymbol{n}}) \\ & + 3\boldsymbol{\nabla}[(\hat{\boldsymbol{n}} \cdot \boldsymbol{\nabla})\hat{\boldsymbol{n}} - \hat{\boldsymbol{n}}(\boldsymbol{\nabla} \cdot \hat{\boldsymbol{n}})]\} \\ & - 2\sin\theta\, [(\hat{\boldsymbol{n}} \cdot \boldsymbol{\nabla}\theta)\boldsymbol{\nabla} \cdot \hat{\boldsymbol{n}} + (\hat{\boldsymbol{n}} \times \boldsymbol{\nabla}\theta) \cdot (\boldsymbol{\nabla} \times \hat{\boldsymbol{n}})]\}. \end{aligned} \tag{7.24a}$$

In particular, for constant ϕ and in the Leggett configuration ($\theta = \theta_L = \cos^{-1}(-\frac{1}{4})$), (7.24a) reduces to

$$\begin{aligned} f_{\text{bend}} \Big|_{\substack{\phi = \text{const}, \\ \theta = \theta_L}} = \frac{\hbar^2}{16m^2} \rho_s^0 \Big\{ & 4(\boldsymbol{\nabla} \times \hat{\boldsymbol{n}})^2 - \tfrac{5}{4}[\hat{\boldsymbol{n}} \cdot (\boldsymbol{\nabla} \times \hat{\boldsymbol{n}})]^2 + \tfrac{13}{4}(\boldsymbol{\nabla} \cdot \hat{\boldsymbol{n}})^2 \\ & - \frac{15^{1/2}}{2} (\boldsymbol{\nabla} \cdot \hat{\boldsymbol{n}})\hat{\boldsymbol{n}} \cdot (\boldsymbol{\nabla} \times \hat{\boldsymbol{n}}) + 3\boldsymbol{\nabla}[(\hat{\boldsymbol{n}} \cdot \boldsymbol{\nabla})\hat{\boldsymbol{n}} - \hat{\boldsymbol{n}}(\boldsymbol{\nabla} \cdot \hat{\boldsymbol{n}})] \Big\}. \end{aligned} \tag{7.24b}$$

In the case of constant angle θ, the general expression for the bending energy density containing Fermi-liquid corrections has been calculated by Cross (1975).

An alternative, useful form of the bending energy density is in terms of the *local* changes of the spin–orbit rotation, defined by the set of infinitesimal spin–orbit angles $\delta\theta^{\mathrm{so}}$ ($i = 1, 2, 3$) (Brinkman and Smith 1974, Brinkman and Cross 1978)

$$\delta\boldsymbol{\theta}^{\mathrm{so}} = 2\sin\tfrac{1}{2}\theta\,[\boldsymbol{R}(\hat{\boldsymbol{n}}, \theta)]^{1/2}\,\delta\hat{\boldsymbol{n}} + \hat{\boldsymbol{n}}\,\delta\theta. \tag{7.25a}$$

The general expression for the bending free energy density, valid also beyond the Ginzburg–Landau regime and including Fermi-liquid corrections, is then given by

$$f_{\mathrm{bend}} = \left(\frac{\hbar}{2m}\right)^2 \Big[\tfrac{1}{2}\rho_{\mathrm{s}}(\boldsymbol{\nabla}\phi)^2 + a\sum_{ij}(\nabla_i\theta_j^{\mathrm{so}})(\nabla_i\theta_j^{\mathrm{so}}) + b\sum_{ij}(\nabla_i\theta_j^{\mathrm{so}})(\nabla_j\theta_i^{\mathrm{so}}) + c(\boldsymbol{\nabla}\cdot\boldsymbol{\theta}^{\mathrm{so}})^2\Big], \tag{7.25b}$$

with coefficients (Cross 1975)

$$\left.\begin{aligned} a &= -(4+\delta)c,\\ b &= (1+\delta)c,\\ c &= -\frac{\rho_{\mathrm{s}}^0}{10}\left(\frac{1+\frac{1}{3}F_1^{\mathrm{a}}}{1+\frac{1}{3}F_1^{\mathrm{s}}}\right)\frac{1}{1+\frac{1}{3}F_1^{\mathrm{a}}(1-\frac{3}{5}\rho_{\mathrm{s}}^0/\rho)}\end{aligned}\right\} \tag{7.25c}$$

and

$$\delta = \frac{\frac{1}{3}F_1^{\mathrm{a}}\rho_{\mathrm{s}}^0/\rho}{1+\frac{1}{3}F_1^{\mathrm{s}}(1-\rho_{\mathrm{s}}^0/\rho)}. \tag{7.25d}$$

7.2.3 Healing lengths

As already mentioned, the bending energy density determines the scale for spatial variations of the order parameter. In the presence of competing orienting forces, the bending is minimized when the textures assume the smoothest possible configuration. This implies that every orienting energy that locally disturbs the uniform configuration of the order-parameter field is associated with a characteristic length ξ_{heal} over which the textures return to the undisturbed state ("healing length"). This length is simply obtained by equating the orientation energy density and the bending energy density f_{bend}. The latter corresponds to a spatial variation of the textures on a scale ξ_{L} and is of order

$$f_{\mathrm{bend}} \approx \tfrac{1}{2}K_1\Delta_0^2(T)\frac{1}{\xi_{\mathrm{L}}^2} \tag{7.26a}$$

$$= \tfrac{1}{10}N_{\mathrm{F}}\Delta_0^2(T)\left(\frac{\xi_0}{\xi_{\mathrm{L}}}\right)^2, \tag{7.26b}$$

where (7.26b) is valid in the Ginzburg–Landau regime and Δ_0 is the maximum value of the gap for any one of the phases considered. This energy density has to be set equal to the orientation energy density under investigation to that $\xi_L = \xi_{heal}$.

Dipole energy

Here we equate f_{bend} in (7.26) with $\frac{3}{5}g_D(T)$, the prefactor of the dipole energy density given by (6.100). This defines the dipole healing length ξ_D ("dipole coherence length"):

$$\xi_D = \left[\frac{5K_1\Delta_0^2(T)}{6g_D}\right]^{1/2} = \left(\frac{5K_1}{6\lambda_D N_F}\right)^{1/2}\frac{\Delta_0}{\Delta}. \tag{7.27a}$$

In particular, in the Ginzburg–Landau regime (see (7.26b) and (7.18b))

$$\left.\begin{aligned} \xi_D^B &= \left(\frac{1}{6\lambda_D}\right)^{1/2}\xi_0 \approx 7\ \mu\text{m}, \\ \xi_D^A &= \frac{1}{2}\left(\frac{1}{\lambda_D}\right)^{1/2}\xi_0 \approx 8\ \mu\text{m} \end{aligned}\right\} \tag{7.27b}$$

at melting pressure.

Magnetic energy

In the A phase the magnetic energy density is given by (6.115), $\Delta f_H = \frac{1}{2}\Delta\chi H^2 (\hat{\boldsymbol{H}}\cdot\hat{\boldsymbol{d}})^2$, where $\Delta\chi$ is the magnetic-susceptibility anisotropy. This defines a magnetic length

$$\xi_H^A = \left(\frac{K_1\Delta_0^2}{\Delta\chi H^2}\right)^{1/2} \tag{7.28a}$$

$$= \frac{H_A}{H}\xi_0 \quad (T \lessapprox T_c), \tag{7.28b}$$

where $H_A = 2\pi[6(1+F_0^a)/35\zeta(3)]^{1/2}k_B T_c/\gamma\hbar \approx 2\times 10^4$ G at melting pressure and $T = T_c$. For a typical magnetic field of 100 G, $\xi_H^A \approx 10^2\xi_0$. Note that the magnetic length ξ_H^A is equal to the dipole coherence length ξ_D at a field

$$H^* = 2\lambda_D^{1/2}H_A \approx 28\ \text{G}. \tag{7.28c}$$

Therefore this is the field that must be applied in order to overcome the alignment of the preferred directions $\hat{\boldsymbol{d}}$ and $\hat{\boldsymbol{l}}$ caused by the dipole interaction.

In the B phase (or rather the B_2 phase) the magnetic field orientation energy is smaller than that in the A phase by a factor λ_D (see (6.116)). Using

(6.116b), the corresponding healing length follows as

$$\xi_H^{\mathrm{B}} = \frac{c_{\mathrm{B}}}{\lambda_{\mathrm{D}}^{1/2}} \frac{\Delta}{k_{\mathrm{B}} T_{\mathrm{c}}} \frac{H_{\mathrm{A}}}{H} \xi_0 \tag{7.29a}$$

$$= 6^{1/2} c_{\mathrm{B}} \frac{\Delta}{k_{\mathrm{B}} T_{\mathrm{c}}} \frac{H_{\mathrm{A}}}{H} \xi_{\mathrm{D}}^{\mathrm{B}}, \tag{7.29b}$$

where

$$c_{\mathrm{B}} = [7\zeta(3)(1+F_0^{\mathrm{a}})/20\pi^2]^{1/2} \approx 0.10. \tag{7.30}$$

This length is therefore considerably larger than in the A phase except extremely close to T_{c}. Also, in contrast with the A phase, the magnetic length is much larger than ξ_{D} for all practical purposes.

Surfaces

While in the A phase there is a strict boundary condition $\hat{\boldsymbol{l}} \parallel \hat{\boldsymbol{s}}$, where $\hat{\boldsymbol{s}}$ is the surface normal, this is not so in the B phase (see (6.118)). To estimate the effect of a surface on a given texture, we have to relate the bending energy $F_{\mathrm{bend}}^{\mathrm{B}} = \int \mathrm{d}^3 r\, f_{\mathrm{bend}}^{\mathrm{B}}$, (7.17), to the surface energy $F_{\mathrm{D,S}}$ in (6.130). This leads to a surface healing length of order

$$\xi_{\mathrm{S}} = \frac{4}{9} \frac{K_1 \Delta^2}{g_{\mathrm{D}} \xi} \tag{7.31a}$$

$$= \frac{4}{45} \frac{\xi_0^2}{\xi} \frac{1}{\lambda_{\mathrm{D}}} \approx 0.2 \left(1 - \frac{T}{T_{\mathrm{c}}}\right)^{1/2} \mathrm{cm} \quad (T \lessapprox T_{\mathrm{c}}), \tag{7.31b}$$

which is quite large ($\xi_{\mathrm{S}} \approx 1$ mm at $T = 0.7T_{\mathrm{c}}$) except very close to T_{c}. This implies that textures in ^{3}He-B are not affected very much by a surface and can be very extended. In particular, in containers smaller than ξ_{S} the boundary is unimportant, i.e. it is less costly for a given texture to ignore the surface than to adjust its configuration in order to fit the boundary condition $\hat{\boldsymbol{n}} \parallel \hat{\boldsymbol{s}}$.

7.3 SUPERCURRENTS

We are now equipped to calculate the supercurrent, which, according to (7.14a), is obtained by differentiating the condensate free-energy density with respect to $\boldsymbol{v}_{\mathrm{s}}$. As indicated earlier, the superfluid mass-current density may equally well be obtained from the free-energy density by employing the covariance property of the latter under the operation $\boldsymbol{\nabla} \to \boldsymbol{\nabla} - \mathrm{i}(2m/\hbar)\boldsymbol{u}$, i.e. by making a Galilean transformation and retaining the linear terms in the corresponding velocity. The supercurrent density is then given by (7.14b), which yields

$$g_{\mathrm{s},i} = \left(\frac{\hbar}{2m}\right)^{-1} \mathrm{Im}\left\{\sum_{\mu j} [K_1 d_{\mu j}^* \nabla_i d_{\mu j} + K_2 d_{\mu j}^* \nabla_j d_{\mu i} + K_3 d_{\mu i}^* \nabla_j d_{\mu j}]\right\}. \tag{7.32}$$

In the case of superfluid ^{3}He-B the mass-supercurrent density is simply given by

$$\boldsymbol{g}_s = \rho_s \boldsymbol{v}_s, \tag{7.33}$$

where $\boldsymbol{\rho}_s = \rho_s \mathbf{1}$ (see (7.15b)), as one would expect for a superfluid with isotropic gap. The spin supercurrent is derived in Chapter 10.

7.3.1 The A phase

By contrast, the situation in the anisotropic A phase is much more interesting. Inserting the corresponding order parameter, $A_{\mu j} = 2^{-1/2}\hat{d}_\mu(\hat{m}_j + \mathrm{i}\hat{n}_j)$ into (7.31) and making use of $\hat{l}_i = \sum_k \epsilon_{ijk}\hat{m}_j\hat{n}_k$, the supercurrent density in the weak-coupling limit is cast into the form

$$g_{s,i} = \left(\frac{\hbar}{2m}\right)^{-1} \Delta_0^2 \sum_j [(2\delta_{ij} - \hat{l}_i\hat{l}_j)(\hat{\boldsymbol{m}} \cdot \nabla_j \hat{\boldsymbol{n}}) + (\tfrac{1}{2}\delta_{ij} - \hat{l}_i\hat{l}_j)(\nabla \times \hat{\boldsymbol{l}})_j]. \tag{7.34}$$

Note that $\boldsymbol{g}_s$ has been completely expressed in terms of the quantities $\hat{\boldsymbol{m}} \cdot \nabla_j \hat{\boldsymbol{n}}$ and $\hat{\boldsymbol{l}}$, and may be written as (Ambegaokar *et al.* 1974, Blount *et al.* 1974, Wölfle 1974)

$$\boldsymbol{g}_s = \boldsymbol{\rho}_s \boldsymbol{v}_s + \frac{\hbar}{2m} \mathbf{C}(\nabla \times \hat{\boldsymbol{l}}), \tag{7.35}$$

with

$$\rho_{s,ij} = \rho_{s\|}(2\delta_{ij} - l_i l_j), \tag{7.36a}$$

$$C_{ij} = -\rho_{s\|}(\tfrac{1}{2}\delta_{ij} - l_i l_j) \tag{7.36b}$$

and $\hat{\boldsymbol{l}}$ along the z axis. More generally, i.e. outside the Ginzburg–Landau regime, the superfluid mass-density tensor $\boldsymbol{\rho}_s$ and the tensor $\mathbf{C}$ are given by (Cross 1975)

$$\boldsymbol{\rho}_s = \begin{pmatrix} \rho_{s\perp} & 0 & 0 \\ 0 & \rho_{s\perp} & 0 \\ 0 & 0 & \rho_{s\|} \end{pmatrix}, \tag{7.37a}$$

$$\mathbf{C} = \begin{pmatrix} C_\perp & 0 & 0 \\ 0 & C_\perp & 0 \\ 0 & 0 & C_\| \end{pmatrix}. \tag{7.37b}$$

The first term in (7.34) is of the form of the usual phase-gradient term. For a uniform $\hat{\boldsymbol{l}}$ configuration, the A phase behaves as a conventional superfluid, but with an anisotropic superfluid density. In the general case, however, there is an additional contribution to the supercurrent proportional to $\nabla \times \hat{\boldsymbol{l}}$. Hence in general the superflow is not irrotational. In any nonuniform $\hat{\boldsymbol{l}}$

texture there will then flow a supercurrent that depends on the actual texture. (In fact, the $\hat{\boldsymbol{l}}$ texture in a vessel filled with superfluid ^{3}He-A will quite generally be nonuniform since $\hat{\boldsymbol{l}}$ is oriented perpendicular to the walls of the container.) Therefore, even in the ground state of ^{3}He-A, a supercurrent is expected to flow (see below).

We may use the result for $\boldsymbol{\rho}_n$ derived in Section 3.4 to determine $\boldsymbol{\rho}_s$ by making use of (7.15b). We thereby obtain the coefficients K_s etc. (Since (7.15b) is required by Galilean invariance, such a derivation is no longer possible if nonlinear terms are included; this will be discussed below.) One may easily convince oneself that the result is identical with (7.22a) and (7.22c).

Similarly, by differentiating f_{bend} with respect to the rotation angles of the spin axis $\hat{\boldsymbol{d}}$, the spin supercurrent is found as (Brinkman and Cross 1978)

$$j_\mu^\sigma = \boldsymbol{\rho}_{\text{sp}} \sum_\nu (\delta_{\mu\nu} - \hat{d}_\mu \hat{d}_\nu) \boldsymbol{v}_{\text{sp},\nu}, \tag{7.38}$$

where $\boldsymbol{v}_{\text{sp},\nu} = -(\hbar/2m)\boldsymbol{\nabla}\theta_\mu^{\text{s}}$ (for a microscopic derivation of this result, see (10.60) in Chapter 10). Magnetic-field-induced spin supercurrents in the A_1 phase have been discussed by Muzikar (1980, 1982).

We now want to discuss the expression for the superfluid mass-current density (7.35) in more detail. Equation (7.35) has a structure that is very similar to that for the total charge-current density $\boldsymbol{j}_{\text{tot}}$ of a system of atoms or molecules in classical electrodynamics. The latter is also composed of two parts: (i) a current $\boldsymbol{j}_c$ describing the actual charge transport; and (ii) an effective current $\boldsymbol{j}_{\text{eff}} = c\boldsymbol{\nabla} \times \boldsymbol{M}$, due to the magnetization $\boldsymbol{M}$ generated by the electronic currents within the atom or molecule, i.e. by the internal motion of the electrons. The total current density is then $\boldsymbol{j}_{\text{tot}} = \boldsymbol{j}_c + c\boldsymbol{\nabla} \times \boldsymbol{M}$. In ^{3}He-A the Cooper pair may be thought of as representing the atom and $\hat{\boldsymbol{l}}$ the magnetic orbital moment $\boldsymbol{M}$. Therefore the first term in (7.35) describes the usual flow of Cooper pairs, while the second one is an "orbital" supercurrent driven by the spatial variation of $\hat{\boldsymbol{l}}$. It must be pointed out, however, that the similarities concerning the second term are only superficial: in the case of the charge current $\boldsymbol{j}_{\text{tot}}$ there is no real change of charge density connected with $\boldsymbol{j}_{\text{eff}}$ since $\boldsymbol{\nabla} \cdot \boldsymbol{j}_{\text{eff}} = 0$, while the orbital supercurrent $\boldsymbol{g}_{\text{s,orb}} = \boldsymbol{C}(\boldsymbol{\nabla} \times \hat{\boldsymbol{l}})$ describes an actual mass transport ($\boldsymbol{\nabla} \cdot \boldsymbol{g}_{\text{s,orb}} \neq 0$) owing to the fact that $\boldsymbol{C}$ is a tensor quantity. In order to elaborate on this point a little further, we rewrite (7.34) and (7.35) as

$$\boldsymbol{g}_s = \boldsymbol{\rho}_s \boldsymbol{v}_s + \frac{\hbar}{2m} C_\perp (\boldsymbol{\nabla} \times \hat{\boldsymbol{l}}) - \frac{\hbar}{2m} C_0 \hat{\boldsymbol{l}} [\hat{\boldsymbol{l}} \cdot (\boldsymbol{\nabla} \times \hat{\boldsymbol{l}})], \tag{7.39}$$

with C_0 $(\equiv C_\perp - C_\parallel) = \rho_{s\parallel}$ in weak-coupling theory. As first noted by Volovik and Mineev (1981), the last term in (7.39) plays a special role: it gives rise to a source term in the conservation law for the momentum of the condensate. Hence at $T = 0$ and for $C_0 \neq 0$, $\boldsymbol{g}_s$ is *not* a conserved quantity.

Consequently, there must be a finite density of excitations carrying a nonzero momentum at $T=0$, such that the *total* momentum is conserved. In other words, even at $T=0$ there must be a nonvanishing normal density ρ_n; it will be calculated below. The C_0 term in (7.39), i.e. the fact that $C_0 \neq 0$, is due to the nodes in the energy gap of the A-phase order parameter.

As pointed out by Mermin and Muzikar (1980) and Mermin (1980), a Bose condensate of diatomic molecules whose size is smaller than their average separation L_a and that carry an intrinsic angular momentum $\boldsymbol{L} = \hbar\hat{\boldsymbol{l}}$ yields an expression for $\boldsymbol{g}_s$ as in (7.39) but with $C_0=0$. This would then correspond precisely to the charge current in classical electrodynamics discussed above. The difference originates from the fact that Cooper pairs—in contrast with ordinary diatomic molecules—have a very large extent $\xi_0 \gg L_a$ and therefore overlap substantially. So, while in a naive picture of nonoverlapping Bose-condensed diatomic molecules with angular momentum $\hbar\hat{\boldsymbol{l}}$ the total angular-momentum density is expected to be $\frac{1}{2}\hbar\rho\hat{\boldsymbol{l}}$ by simple superposition, the corresponding angular-momentum density for Cooper pairs is reduced by a factor of order $(L_a/\xi_0)^2 \approx 10^{-5}$ owing to almost complete cancellation among the mass-current contributions from overlapping Cooper pairs. This cancellation may also be understood as a compensation between the "bare" angular momentum $\frac{1}{2}\hbar\rho\hat{\boldsymbol{l}}$ of the condensate and the angular momentum of fermionic excitations at $T=0$, which lead to the finite normal component (Volovik 1986c, 1987). In fact, the angular-momentum density discussed above should be a *dynamical* quantity that makes its appearance as an inertial term in the equation of motion for $\hat{\boldsymbol{l}}$ (see Chapter 9). In a given sample there is also a *static* angular momentum since the boundary causes the $\hat{\boldsymbol{l}}$ texture to be nonuniform and thus creates an orbital superflow in the ground state (for a cylindrical geometry, see Miyake and Usui 1980a). Again part of the superflow is cancelled by the current of normal excitations (Combescot and Dombre 1986, Balatskii *et al.* 1986). The question of the so-called "intrinsic angular momentum" has been intensively discussed (Cross 1975, 1977a, Volovik 1975a,b, 1984b, 1986c, 1987, Ishikawa 1976, 1977a,b, 1980, Combescot 1978a,b, Fomin 1978d, McClure and Takagi 1979, Nagai 1979, Mermin and Muzikar 1980, Volovik and Mineev 1981, Volovik and Balatskii 1985, Stone *et al.* 1985, Balatskii and Mineev 1985, Combescot and Dombre 1986, Hall and Hook 1986, Balatskii *et al.* 1986, Schakel 1989). We shall return to this problem in Chapter 9, where we discuss the hydrodynamics of the A phase.

A similar question arises in the context of the mass supercurrent in the B phase in the presence of a magnetic field. In this case a contribution to the supercurrent has been found that may be interpreted as being caused by an intrinsic angular momentum $\sum_\mu M_\mu R_{\mu i}$ (where M_μ is the magnetization). The magnitude of this term is a subject of debate (Volovik and Mineev 1983, Mineev and Volovik 1984, Muzikar 1984, Dombre and Combescot 1985, Yip 1986).

7.3.2 Finite normal density at $T = 0$ in the A phase

The existence of nodes in the energy gap of the A-phase order parameter has many unusual and profound implications, for example for the very-low-temperature properties of this phase (remember that the A phase may be stabilized by a magnetic field down to $T = 0$). In particular, the bending free-energy density (7.20) is in general found to be a *nonanalytic* function of the gradients of the order parameter (Combescot and Dombre 1983a, 1985b, Nagai 1984, Muzikar and Rainer 1984, Choi and Muzikar 1986, Garg 1987). This has already been signalled by the appearance of the $\ln T$ divergence in (7.20) involving the quantity γ, (7.23h) (Cross 1975). As has been discussed, this divergence comes from the singular behaviour of the order-parameter phase along $\hat{\boldsymbol{l}}$. These nonanalyticities in the bending free-energy density show up in other quantities too, of which the superfluid current $\boldsymbol{g}_s$ is particularly interesting and has therefore been studied in great detail (Volovik and Mineev 1981, Combescot and Dombre 1983a,b, 1985, Muzikar and Rainer 1983). To include these nonanalytic terms in $\boldsymbol{g}_s$, one has to go beyond (7.35) since that contains the leading terms, which are analytic and of first order in the gradients. The nonanalytic terms are *nonlinear in* $\boldsymbol{v}_s$ and $\nabla \times \hat{\boldsymbol{l}}$, and are thus of second order in the gradients. Consequently, they are much smaller than the contribution from (7.35) and are mainly of theoretical interest. This is in contrast with the case of the bending free-energy density, where the leading nonanalytic terms are in fact larger than the leading analytic ones.

As first pointed out by Volovik and Mineev (1981), the nodes along $\hat{\boldsymbol{l}}$, i.e. the gaps in the excitation spectrum, imply a finite normal density $\boldsymbol{\rho}_n$ even at $T = 0$, which is due to the fact that right at the nodes pair breaking is possible even at $T = 0$. This is in striking contrast with the conventional belief that in a superfluid at $T = 0$ there are no normal excitations and therefore $\boldsymbol{\rho}_n = 0$.

We now want to calculate $\boldsymbol{\rho}_n$ following Volovik and Mineev (1981). Let us consider a nonuniform $\hat{\boldsymbol{l}}$ texture such that $\nabla \times \hat{\boldsymbol{l}} \neq 0$, i.e. the system carries a finite supercurrent. This supercurrent changes the quasiparticle energy spectrum because it causes a shift in the excitation energy from $E_{\boldsymbol{k}} = [\xi_{\boldsymbol{k}}^2 + |\Delta_{\boldsymbol{k}}|^2]^{1/2}$ to

$$E_{\boldsymbol{k}}(\hat{\boldsymbol{l}}) = E_{\boldsymbol{k}} + \boldsymbol{k} \cdot \boldsymbol{v}_s(\hat{\boldsymbol{k}}). \tag{7.40}$$

Equation (7.40) follows from (i) the fact that Bogoliubov quasiparticles are eigenstates of momentum $\boldsymbol{k}$, and (ii) from the transformation property of a system with given momentum under a Galilei transformation (for a microscopic derivation see Chapter 10). The quantity $\boldsymbol{v}_s(\hat{\boldsymbol{k}})$ is a kind of superfluid velocity defined on every point of the Fermi surface by the general relation (7.5)

$$\boldsymbol{v}_s(\hat{\boldsymbol{k}}) = \frac{\hbar}{2m} \nabla \phi(\hat{\boldsymbol{k}}). \tag{7.41}$$

Here $\phi(\hat{\boldsymbol{k}})$ is the $\hat{\boldsymbol{k}}$-dependent phase of the A-phase order parameter $d_\mu(\hat{\boldsymbol{k}})$, i.e.

$$\phi(\hat{\boldsymbol{k}}) = \tan^{-1}\frac{\hat{\boldsymbol{n}}\cdot\hat{\boldsymbol{k}}}{\hat{\boldsymbol{m}}\cdot\hat{\boldsymbol{k}}}, \tag{7.42}$$

which becomes singular for $\hat{\boldsymbol{k}} \parallel \hat{\boldsymbol{l}} = \hat{\boldsymbol{m}} \times \hat{\boldsymbol{n}}$. Substituting (7.42) into (7.41) yields

$$v_{s,i}(\hat{\boldsymbol{k}}) = v_{s,i} + \frac{1}{2}\frac{\hat{\boldsymbol{k}}\cdot\hat{\boldsymbol{l}}[(\hat{\boldsymbol{k}}\times\hat{\boldsymbol{l}})\cdot\nabla_i\hat{\boldsymbol{l}}]}{(\hat{\boldsymbol{k}}\times\hat{\boldsymbol{l}})^2}, \tag{7.43}$$

where $\boldsymbol{v}_s$ is given by (7.10). Note that the second term on the right-hand side of (7.43) is only nonzero for nonuniform $\hat{\boldsymbol{l}}$. We may now proceed to calculate $\boldsymbol{\rho}_n$ along the lines of the discussion presented in Section 3.4 by considering the momentum density of Bogoliubov quasiparticles in a frame of reference moving with uniform velocity $\boldsymbol{v}_n$:

$$\boldsymbol{P} = \sum_{\boldsymbol{k}} f[E_{\boldsymbol{k}}(\hat{\boldsymbol{l}}) - \boldsymbol{k}\cdot\boldsymbol{v}_n]\boldsymbol{k}, \tag{7.44}$$

where $f(E)$ is the Fermi function. At $T=0$, $f(E)$ reduces to a step function, i.e. the only contribution is from directions in $\boldsymbol{k}$ space where the argument $E_{\boldsymbol{k}} - \boldsymbol{k}\cdot\boldsymbol{v}_n$ is negative. We immediately see that in the B phase, where the gap is isotropic, this condition may only be fulfilled above a certain threshold of $|\boldsymbol{v}_s - \boldsymbol{v}_n|$, which is of the order of the so-called "pair-breaking critical velocity" (see Section 7.11). By contrast, in the A phase with a nonuniform $\hat{\boldsymbol{l}}$ texture, i.e. where the second term on the right-hand side of (7.43) is finite, the condition may be met for arbitrarily small $|\boldsymbol{v}_s - \boldsymbol{v}_n|$. This allows excitations with negative energy even at $T=0$. Expanding (7.44) to lowest order in $\boldsymbol{v}_n - \boldsymbol{v}_s$, a normal-fluid density $\boldsymbol{\rho}_n$ is defined by

$$\boldsymbol{P} = \boldsymbol{\rho}_n(\boldsymbol{v}_n - \boldsymbol{v}_s) \tag{7.45}$$

(note that $\boldsymbol{P}$ describes the flow of excitations; hence the prefactor of $\boldsymbol{v}_n - \boldsymbol{v}_s$ must be a normal-state quantity), with

$$\rho_{n,ij} = \sum_{\boldsymbol{k}} k_i k_j\, \delta[E_{\boldsymbol{k}}(\hat{\boldsymbol{l}})\,|_{\boldsymbol{v}_s=0}]. \tag{7.46}$$

Evaluation of (7.46) leads to a nonanalytic dependence on the gradient of $\hat{\boldsymbol{l}}$:

$$\rho_{n,ij} \approx \rho\xi_0\hat{l}_i\hat{l}_j\,|(\hat{\boldsymbol{l}}\cdot\boldsymbol{\nabla})\hat{\boldsymbol{l}}|, \tag{7.47}$$

which shows that for a texture with $|(\hat{\boldsymbol{l}}\cdot\boldsymbol{\nabla})\hat{\boldsymbol{l}}| \neq 0$ a finite normal density exists even at $T=0$. Mathematically speaking, the nonanalyticity stems from the inversion symmetry $\boldsymbol{v}_s(\hat{\boldsymbol{k}}) = \boldsymbol{v}_s(-\hat{\boldsymbol{k}})$. One may wonder why at $T=0$ there are order-parameter excitations present at all. Actually, these quasiparticle states can be interpreted as bound states in the potential well formed by the $\hat{\boldsymbol{l}}$ texture. The presence of these bound states should be expected to show up in many observable quantities. For example, they have been predicted to give rise to a distinct frequency structure in the ultrasonic attenuation (Maki and Combescot 1985) and to a linear specific heat for $T \to 0$. It should be clear, however, that these phenomena may only be observed at rather low temperature since the binding energies involved are very small.

The effect of the $\hat{\boldsymbol{l}}$ texture on the fermionic excitations has been discussed for the specific example of domain walls (see Section 7.8) in the $\hat{\boldsymbol{l}}$ texture (Ho *et al.* 1984; see also Combescot and Dombre 1985a). There it was found that the spectrum of the bound states does not show inversion symmetry. These findings have also been discussed in terms of supersymmetric quantum mechanics (Rozhkov 1986). A conceptually similar effect, namely the spin dependence of the quasiparticle spectrum of systems with complex order parameters, has been discussed by Balatskii (1988).

The qualitative discussion of the origin of ρ_n at $T=0$ given above has been put onto a firm microscopic basis (Muzikar and Rainer 1983, Combescot and Dombre 1983a, b, 1985, 1986, Dombre and Combescot 1984, Volovik 1985, Balatskii *et al.* 1986). There may be additional nonanalytic terms in the expression for the current that are of second and higher order in $\boldsymbol{\nabla} \cdot \hat{\boldsymbol{l}}$, as might be expected from expanding (7.44). It may be shown that the additional term in the total current caused by (7.45) is actually needed to restore Galilean covariance of the total current in the general nonlinear case (Volovik and Mineev 1981).

The fact that the finite ρ_n at $T=0$ is due to nonlinear effects in the gradients implies that this also has consequences for the formulation of nonlinear hydrodynamics (see Chapter 9).

The existence of normal excitations even at $T=0$ and of the anomalous C_0 term in the expression for the superfluid current (7.39), as well as of other effects, has been shown to be related to similar anomalies in quantum field theory (Volovik 1986a–d, 1987, Balatskii 1986, Volovik and Konyshev 1988, Schakel and Batenburg 1989, Schakel 1989).

7.3.3 Quantization of circulation

In both superfluid ^4He and ^3He-B the superfluid velocity is simply given by the gradient of the overall phase of the order parameter. In this case the rotation of $\boldsymbol{v}_s$ vanishes identically: $\boldsymbol{\nabla} \times \boldsymbol{v}_s = 0$. Such a relation is not particularly new—in classical hydrodynamics it is actually a necessary condition for frictionless flow. However, in this case the absence of internal friction is only a mathematical idealization, while the superfluids mentioned above represent *real* physical systems in which this relation actually holds. If there is no friction, there are no forces that transmit circulation to the particles. Hence there will be no creation of vortices—the flow is irrotational. The situation is different if the container of the fluid is rotating (see Section 7.6). For every closed curve $\mathscr{C}$ in the liquid, one may define the so-called "circulation" κ, defined by the line integral along $\mathscr{C}$, which can also be written as an area integral over $\nabla \times \boldsymbol{v}$:

$$\kappa = \oint_{\mathscr{C}} \boldsymbol{v} \cdot \mathrm{d}\boldsymbol{l} \tag{7.48}$$

$$= \int (\nabla \times \boldsymbol{v}) \cdot \mathrm{d}\boldsymbol{S}. \tag{7.49}$$

Since in a frictionless vortex-free liquid $\nabla \times \boldsymbol{v} = 0$, we find that for a simply connected container $\kappa = 0$. The situation is different if the container is multiply connected, as is the case for a torus or annulus (which are doubly connected). As soon as the liquid in the torus has a finite velocity, the line integral along a closed curve within the torus will necessarily be finite and may take *any* value, depending on the magnitude of $\boldsymbol{v}$. Let us now consider an isotropic superfluid in a torus with superfluid velocity $\boldsymbol{v}_s = (\hbar/2m)\nabla\phi$ (see (7.5)). Since the phase ϕ of the order parameter of the superfluid is a macroscopic quantity, it is well defined at every point on a contour $\mathscr{C}$ within the torus. On a closed contour it can then only change by multiples of 2π. In this situation the circulation

$$\kappa = \frac{\hbar}{2m} \oint_{\mathscr{C}} \nabla\phi \cdot \mathrm{d}\boldsymbol{l} \tag{7.50}$$

cannot assume arbitrary values, but in fact is *quantized.* The quantization condition is essentially the Bohr–Sommerfeld quantization rule for the electronic motion in an atom and is given by

$$\kappa = \oint_{\mathscr{C}} \boldsymbol{v}_s \cdot \mathrm{d}\boldsymbol{l} = n\kappa_0, \tag{7.51}$$

where $n = 0, \pm 1, \pm 2 \ldots,$ and $\kappa_0 = h/2m$ is called the "quantum of circulation". Thus a superfluid with isotropic order parameter in a ring-shaped container may carry *quantized currents.* Their quantization implies that they are stable and thus represent *persistent* currents. (Actually, they are only *meta*stable, meaning that such current-carrying states do not correspond to the absolute energy minimum of the system. Therefore, strictly speaking, the currents will decay, for example by spontaneous quantum-mechanical tunnelling. Nevertheless, their lifetime will be very long, usually of the order of cosmological times.)

In an anisotropic superfluid, like ^{3}He-A, the situation is very different. In this case the order parameter not only has a phase, which in this case is the rotation angle about $\hat{\boldsymbol{l}}$, but also an *orientation* given by the $\hat{\boldsymbol{l}}$ axis (we neglect for the moment $\hat{\boldsymbol{d}}$, the preferred direction in spin space). This has a far-reaching consequence: in general, the circulation will no longer be quantized. This may easily be verified by calculating $\nabla \times \boldsymbol{v}_s$ for ^{3}He-A.

Making use of (7.10), we find (Mermin and Ho 1976)

$$(\nabla \times \boldsymbol{v}_s)_i = \frac{\hbar}{2m} \sum_{jk} \epsilon_{ijk} (\nabla_j \hat{\boldsymbol{m}}) \cdot (\nabla_k \hat{\boldsymbol{n}})$$

$$= \frac{\hbar}{4m} \sum_{jk} \epsilon_{ijk} \hat{\boldsymbol{l}} \cdot (\nabla_j \hat{\boldsymbol{l}} \times \nabla_k \hat{\boldsymbol{l}}), \tag{7.52}$$

which may also be written as

$$\nabla \times \boldsymbol{v}_s = \frac{\hbar}{4m} \sum_{ijk} \epsilon_{ijk} \hat{l}_i (\nabla \hat{l}_j \times \nabla \hat{l}_k), \tag{7.53}$$

where we have made use of $\hat{\boldsymbol{l}} \doteq \hat{\boldsymbol{m}} \times \hat{\boldsymbol{n}}$ and the relation $(\nabla_j \hat{\boldsymbol{m}}) \cdot (\nabla_k \hat{\boldsymbol{n}}) = (\hat{\boldsymbol{l}} \cdot \nabla_j \hat{\boldsymbol{m}})(\hat{\boldsymbol{l}} \cdot \nabla_k \hat{\boldsymbol{n}})$. In terms of Euler angles, (7.11), this takes the form

$$\nabla \times \boldsymbol{v}_s = \sin \beta \, (\nabla \beta \times \nabla \alpha). \tag{7.54}$$

Equations (7.52)–(7.54) are equivalent expressions of the so-called "Mermin–Ho relation", which states that in general in the A phase $\nabla \times \boldsymbol{v}_s \neq 0$, depending on whether or not $\hat{\boldsymbol{l}}$ is uniform. (Mathematically speaking, this relation is a consequence of the fact that three-dimensional rotations do not commute (Liu and Cross 1978); for a discussion, see Section 9.2.) Correspondingly, the circulation κ is no longer quantized. Since the stability criterion for quantization of circulation fails, a bulk superfluid mass current in ^{3}He-A in a torus is globally unstable! Only *at* the wall, where $\hat{\boldsymbol{l}}$ is rigidly fixed, is quantization guaranteed. These points will be discussed further in Section 7.10.

An integral form of the relation (7.53) for the circulation around a closed contour $\mathscr{C}$ (on which the order parameter is assumed to be nonsingular) in a simply connected container may be written as (Ho 1978a)

$$\kappa = \oint_{\mathscr{C}} \boldsymbol{v}_s \cdot \mathrm{d}\boldsymbol{l} = \frac{\hbar}{2m} [S(\hat{\boldsymbol{l}}) + 4\pi n], \tag{7.55}$$

where n is an integer and $S(\hat{\boldsymbol{l}})$ is an area on the unit sphere, $0 \leq S \leq 4\pi$, which may visualized as follows (see Fig. 7.3). As we go along the contour $\mathscr{C}$, let the orientation of $\hat{\boldsymbol{l}}$ at any point on $\mathscr{C}$ be marked on the surface of a unit sphere (i.e. the position of the tip of $\hat{\boldsymbol{l}}$ on the unit sphere is recorded). As the contour is traversed, the $\hat{\boldsymbol{l}}$ vector traces a closed curve $L(\hat{\boldsymbol{l}})$ on the unit sphere with a particular orientation. Since $L(\hat{\boldsymbol{l}})$ is closed, it divides the surface area of the sphere into two parts: $S(\hat{\boldsymbol{l}})$ and $4\pi - S(\hat{\boldsymbol{l}})$, where $S(\hat{\boldsymbol{l}})$ is defined as the area to the left of the curve $L(\hat{\boldsymbol{l}})$. In this way, the area $S(\hat{\boldsymbol{l}})$ records the $\hat{\boldsymbol{l}}$ texture along $\mathscr{C}$. In principle, $S(\hat{\boldsymbol{l}})$ (and therefore κ) can have any value, and hence circulation is not quantized. The fact that this is due to the existence of an *orientation* of the order parameter (i.e. $\hat{\boldsymbol{l}}$), which may vary, is illustrated clearly by this construction of κ. Equation (7.55) was originally derived for a simply connected container. On the other hand,

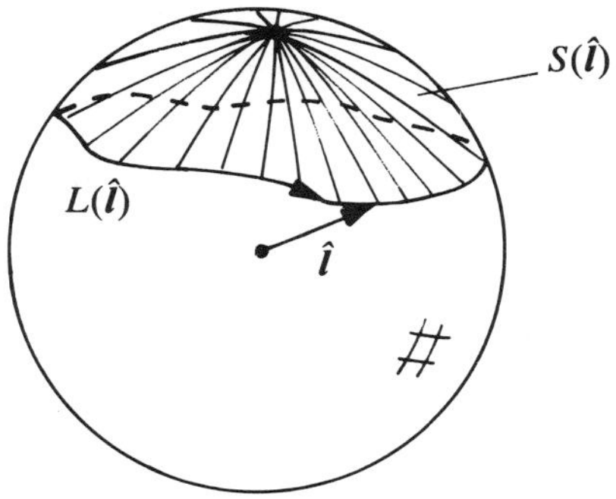

Figure 7.3 The map $L(\hat{l})$ of the orientation of the vector $\hat{l}$ on the surface of a unit sphere for a closed contour $\mathscr{C}$ in a sample of ^{3}He-A. The area $S(\hat{l})$ records the $\hat{l}$ texture on $\mathscr{C}$.

according to Hall and Hook (1986), it should even hold for multiply connected geometries like a torus.

In the absence of a quantization condition for the circulation, currents are quite generally expected to be unstable. This would exclude the existence of persistent currents. (We note, however, that the role of the surface of the container, where quantization *is* guaranteed, has so far not been considered.)

While general arguments based on the structure of the A-phase order parameter may imply the nonexistence of persistent superfluid currents, they do not yield information about how a current should actually decay. Before these questions can be addressed, we first have to investigate the topology of the order parameter in more detail. This will provide us with the necessary information about possible textures (singular or nonsingular), i.e. current-carrying states.

7.4 TOPOLOGICAL INVESTIGATION OF DEFECTS

In a finite sample of superfluid ^{3}He-A the vector fields $\hat{\boldsymbol{l}}(\boldsymbol{r})$ and $\hat{\boldsymbol{d}}(\boldsymbol{r})$ will necessarily be nonuniform owing to the boundary conditions at the surface confining the liquid. The same is true for $\hat{\boldsymbol{n}}(\boldsymbol{r})$ in the case of ^{3}He-B. The equilibrium configurations are determined in principle by the minima of the bending free energies (7.20) and (7.24), amended by the boundary conditions or surface energies. As a prerequisite for any calculation of the resulting textures, a certain knowledge of the overall character of the solution is desirable. For example, one would like to know what kind of texture may or may not exist. This will depend on the topology of the underlying order parameter. Of particular interest are singular textures, where the fraction of superfluid is forced to zero, i.e. the fluid becomes a normal liquid, inside a core region of extension $\xi(T)$. Well-known examples are the vortex lines in superfluid ^{4}He. As we shall see, such vortices may also exist in superfluid ^{3}He. But, in addition, in this superfluid there may also be defects *without* a normal core. An example of such a defect is a boundary separating regions of different uniform orientations of the order parameter, in analogy with domain walls in magnetically ordered systems.

Singular textures and other well-localized textural structures may arise spontaneously upon cooling through the superfluid transition or as a consequence of the application of some external field. After they have been created, they often decay quickly into smooth nonsingular configurations with a lower bending energy. In some cases, however, such an unfolding or unwinding of a singular configuration might only be possible at the expense of the system becoming normal in a macroscopic region, for example along a line or even a plane that extends throughout the system. Clearly, the corresponding potential barrier that the system must climb to eventually reach this state of lower total energy is much too high to be overcome by thermal fluctuations. In such a case a configuration is called "topologically stable". Even though it may not correspond to the absolute but only to a local minimum in energy, a topologically stable configuration will be (meta)stable over very long, perhaps cosmologically long, times.

The concept of topological stability is well illustrated by the example of a vortex line in superfluid ^{4}He or in s-wave superconductors. For these systems, the order parameter is represented by a complex, i.e. two-component, function $\psi = |\psi|\, e^{i\phi}$. If the free-energy density f_0 of the system only depends on the magnitude of ψ, i.e. $f_0 = f_0(|\psi|)$, then states with different values of the condensate phase ϕ, but with the same $|\psi|$, are degenerate. One may plot f_0 as a potential surface in the complex plane spanned by the variable ψ. The minimum of f_0 then corresponds to a full circle of radius $|\psi_0|$ along which the energy is degenerate with respect to ϕ. This case, i.e. the familiar "wine-bottle" or "mexican-hat" potential, is illustrated in Fig. 7.4. Hence the circle is called the "degeneracy space" R (or "order-parameter space", "order-parameter manifold", "manifold of internal states" etc.) as defined in (6.4). The degeneracy space is a subspace

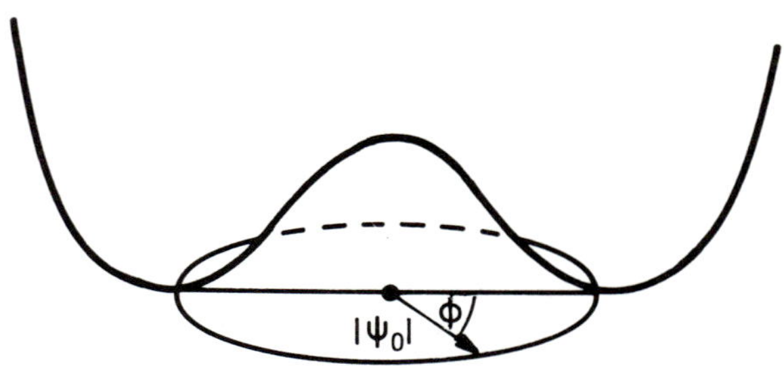

Figure 7.4 Wine-bottle potential of the free-energy density.

(i.e. coset space) of the overall symmetry group G under which the relevant free energy is invariant. In the present case this is the complex ψ plane; hence $G = U(1)$. In the presence of slight inhomogeneities the free energy will be increased by the gradient energy $f_{grad} \approx \xi_0^2 N_F |\nabla\psi|^2$. In particular, for a slowly varying order parameter, only the phase $\phi(\boldsymbol{r})$ varies in space, while the modulus $|\psi|$ is approximately constant. In this situation the order parameter of the system at any point $\boldsymbol{r}$ in position space is uniquely characterized by the phase $\phi(\boldsymbol{r})$, i.e. by a point corresponding to ϕ on the circle with radius $|\psi| = |\psi_0|$ in the complex ψ plane (since $e^{i\phi} = \cos\phi + i\sin\phi$, we naturally choose polar coordinates to represent the degeneracy space, i.e. ϕ corresponds to the polar angle). In this way, one achieves a mapping of a contour in position space into the abstract degeneracy space R of the system. In the example under consideration R is a circle (see Fig. 6.2a), i.e. the wine-bottle potential viewed from the top.

Using this example, we now want to envisage a particular order-parameter configuration where the phase variable $\phi(\boldsymbol{r})$ is *singular* along a line, i.e. where its value is ill defined. This singularity will force the system to become normal in a region of cross-section ξ_0^2 along the line, and thus increases the energy of the system by $\xi_0^2 L(\Delta f)$, where L is the length of the system and Δf is the difference in energy due to pair condensation. The system would like to get rid of the energetically unfavourable singularity, but can only attempt to do so by continuously deforming the phase-variable field $\phi(\boldsymbol{r})$. Whether or not it is possible to remove the singularity by purely local manipulations depends on what is called the "topological stability" of a singularity or defect.

As shown by Toulouse and Kléman (1976) and Volovik and Mineev (1976b, 1977a), defects may be classified and the stability of defects may be determined using the methods of homotopy theory, i.e. the theory of continuous mappings. This theory has been extremely useful in studying defects in condensed-matter physics, especially in systems with complicated underlying symmetries such as superfluid ^{3}He (Volovik and Mineev 1976b, 1977a, Mermin 1979, Volovik, 1979, Mineev 1980, Michel 1980, Anderson 1984). Considering maps of contours in real space into the respective order-parameter space, homotopy theory involves the study of properties of the contours thereby obtained in order-parameter space. In the case of superfluid ^{4}He or planar spins with a fixed length the underlying symmetry group $G = U(1)$ is completely broken ($H = 1$), yielding $R = U(1)/1 = U(1)$. The topological representation of the abstract group R is denoted by S^1, i.e. the order-parameter space is a circle. In the case of Heisenberg spins $\boldsymbol{S}$ in an isotropic ferromagnet ($G = SO(3)$, $H = U(1)$) the order-parameter space is $R = SO(3)/U(1) = S^2$, i.e. the surface of a sphere (see Section 6.1).

One may associate a "dimension" with every order-parameter space, which is given by the number n of coordinates of the order-parameter space itself. For example, in the case of superfluid ^{4}He or a superconductor (order parameter $\psi = \psi_0 e^{i\phi}$) or a planar ferromagnet (i.e. XY spins with order parameter $\boldsymbol{S} = (S_x, S_y)$), one has $n = 2$. Heisenberg spins have $n = 3$. Similarly, an Ising spin, which has only an up/down orientation, is described by $n = 1$ (note that in this particular case only a discrete symmetry is involved). Besides the dimension of the order-parameter space, there is also the dimension d of the real space in which the order parameter acts (e.g. on a line ($d = 1$) or a plane ($d = 2$)). So, in mapping a contour $\mathscr{C}$ in real space into a corresponding contour Γ in order-parameter space one maps a d-dimensional space into an n-dimensional space. To illustrate this point, we consider superfluid ^{4}He or a planar spin, both with $n = 2$, confined to a plane in real space ($d = 2$). We want to investigate the mapping of several explicit order-parameter configurations ("textures") into the order-parameter space, as shown in Fig. 7.5. (Note that the superfluid velocity field is orthogonal to the spin direction at every point.) To this end, we choose a closed contour $\mathscr{C}$ in real space surrounding the defect and, going along $\mathscr{C}$, record how it translates into a closed contour Γ in order parameter space. In particular, we measure the "circulation" in order-parameter space, by calculating the total change in the orientation of the order parameter on the contour. In the present case the order parameter is fully described by an angular variable ϕ (i.e. the phase variable in superfluid ^{4}He or the actual rotation angle of the planar spin $\hat{\boldsymbol{S}} = (\cos\phi, \sin\phi)$ in the plane). This yields a number

$$N = \frac{1}{2\pi} \oint_{\mathscr{C}} \boldsymbol{\nabla}\phi \cdot \mathrm{d}\boldsymbol{l}, \tag{7.56}$$

which is often called the "winding number" or "topological charge". Clearly, this is the same number as in the expression for the quantization of circulation, (7.51). It allows one to classify a given defect and thereby serves as an indicator for its stability.

In Fig. 7.5(a) the order-parameter field is supposed to have a singular defect at P, whose stability we want to investigate. We draw a closed contour $\mathscr{C}$ of arbitrary shape around P and, starting at an arbitrary point marked $\otimes$, traverse $\mathscr{C}$ in anticlockwise direction. On the way, we record the orientation of $\hat{\boldsymbol{S}}$ at every point on $\mathscr{C}$ and plot it on the circle in order-parameter space (with starting point P'). After having traversed $\mathscr{C}$ and thus coming back to P, we also return to P' in order-parameter space. Clearly, in the case of the defect in Fig. 7.5(a) the net change in the orientation of $\hat{\boldsymbol{S}}$ along $\mathscr{C}$ is zero; hence $N = 0$. Since the contour $\mathscr{C}$ may be arbitrarily close to P, this implies that the defect at P can be removed by continuously deforming (smoothing) the order-parameter field $\hat{\boldsymbol{S}}(\boldsymbol{r})$, i.e. by pure "local surgery" (Mermin 1979). Hence the defect at P is not stable. This example shows that $N = 0$ is only a *necessary*, but not a sufficient, condition for the absence of a singularity—after all, there *was* a defect at P before the texture was locally straightened out to get rid of it.

In Figs. 7.5(b–d) we see defects that do carry a winding number $N \neq 0$.

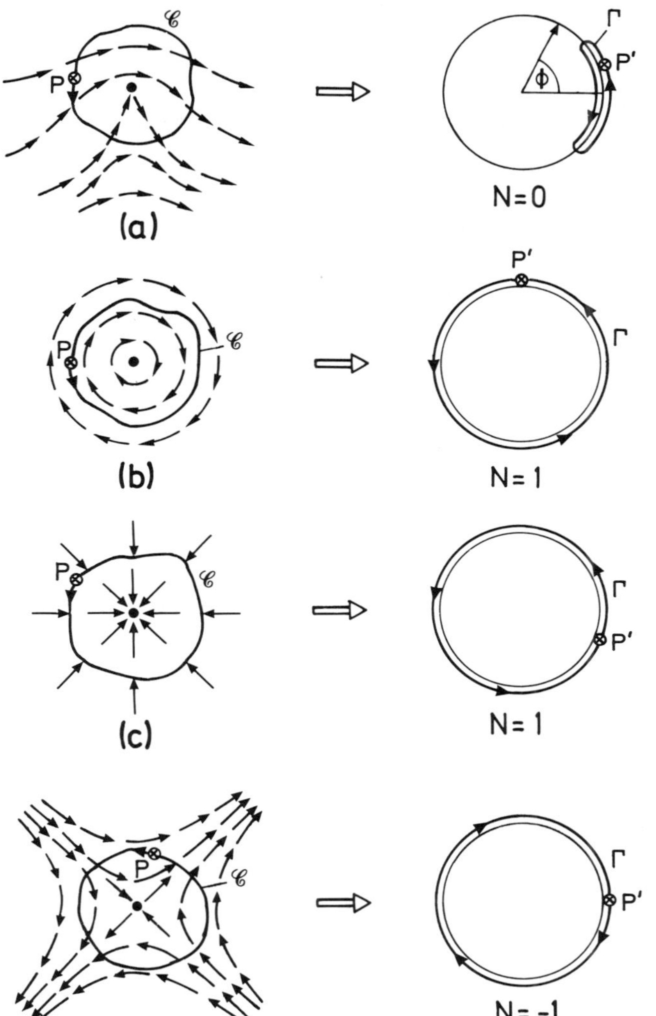

Figure 7.5 Contours 𝒞 around four defects in an order-parameter field involving planar spins, each starting at the point marked ⊗. The corresponding contours Γ in order-parameter space and the resulting winding numbers N are shown on the right.

This means that the respective contours Γ in order-parameter space wind at least once around the whole circle before they come back to the starting point. Therefore they cannot be contracted to zero. In other words, the singularities cannot be removed by local manipulations. To make them disappear would require "global surgery" in the system. Hence these singularities are called "topologically stable". In fact, a single topologically stable singularity may be detected at any distance from the defect.

It is interesting to note that the "antivortex" in Fig. 7.5(d) with $N = -1$ is *not* obtained by reversing the arrows of the vortex in Figs. 7.5(b,c) (such a configuration would also have $N = 1$!), but has a very different—namely hyperbolic—shape in real space.

Altogether, homotopy theory tells us that defects with $N = 0$ are topologically unstable, while those with $N \neq 0$ are topologically stable. It also states that defects with the same image contour, i.e. the same winding number N, are "topologically equivalent", and may be continuously transformed into each other. This may be verified in the case of the two vortices with $N = 1$ shown in Figs. 7.5(b,c). The necessary continuous (local) transformations are indicated in Fig. 7.6. On the other hand, if the winding numbers are different, the defects cannot be transformed into each other via continuous deformation.

Winding numbers are additive. Therefore two defects with N_1 and N_2 can, for example, be continuously transformed to yield a single defect with winding number $N = N_1 + N_2$. Likewise, different defects can be transformed into each other as long as the total winding number $\sum_i N_i$ is conserved. In particular, two defects with winding numbers N and $-N$ may add up to a nonsingular (uniform) configuration with $N = 0$. This is shown in Fig. 7.7 for the case of a vortex ($N = 1$) and an antivortex ($N = -1$). The two defects compensate each other, such that the resulting texture can be embedded into a uniform background.

Using topology, it is also possible to determine the dimensionality of defects. If, in the general case, the real space is d-dimensional while the order-parameter space is n-dimensional then it follows that defects of dimensionality $D = d - n$ are stable (Toulouse and Kléman 1976). In the example of two-dimensional superfluid ^{4}He discussed above ($d = n = 2$) this implies that only *pointlike* defects ($D = 0$) may exist. These are precisely the planar vortices shown in Figs. 7.5(b–d). Similarly, three-dimensional superfluid

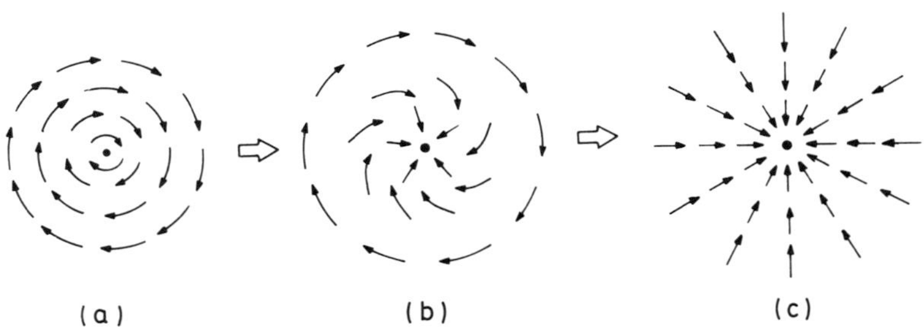

Figure 7.6 Local continuous transformation between defects with the same winding number $N = 1$, showing that the defects in (a) and (c) are topologically equivalent.

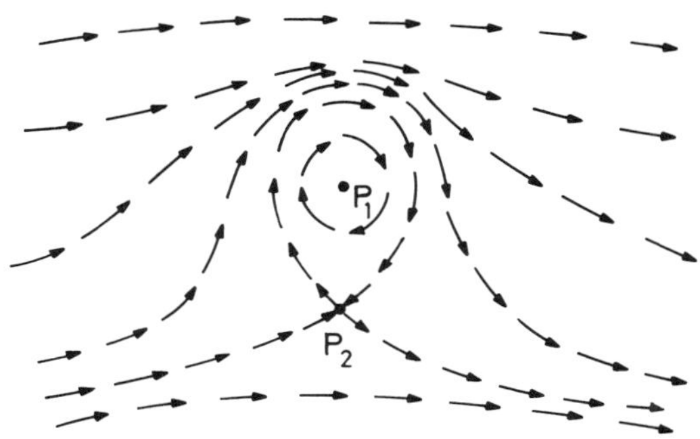

Figure 7.7 Annihilation of a defect at P_1 with $N = 1$ and one at P_2 with $N = -1$ (vortex–antivortex pair). The resulting texture may be embedded into a uniform background.

^{4}He ($d = 3$, $n = 2$, i.e. $D = 1$) allows *line* defects, which are the well-known vortex lines. In the case of Ising spins ($n = 1$) defects in $d = 1, 2, 3$ are pointlike, line-shaped and surface-like respectively (Fig. 7.8). In contrast, Heisenberg spins in a plane ($d = 2$, $n = 3$, i.e. $D < 0$) do not allow topological defects, since the spins may avoid a vortex configuration by escaping into the third dimension. This is also clear mathematically (see (7.58)): mapping a closed contour in two-dimensional real space into the two-dimensional order-parameter space leads to closed contours *on* the sphere that may be contracted to zero ("you can't lasso a balloon").

Even without much detailed analysis, it is clear that the structure of the order parameter of superfluid ^{3}He is in principle rich enough to allow all of the abovementioned defects, i.e. planar, linear and pointlike order-parameter configurations. What is not clear, however, is whether these defects will actually be stable. These questions may be addressed using similar topological methods to those discussed above.

To this end, one has to employ suitable contours around the singularities, which are then mapped into the respective degeneracy spaces. For singular points one needs a closed surface, for lines a closed loop and for a plane two points on either side of the plane (we see that the latter case is somewhat different from the first two in that it does not involve a continuous contour).

Following Volovik and Mineev (1976b, 1977a) and Mineev (1980), we now want to investigate the mappings of these contours into the respective degeneracy spaces for the A and B phases. In this way, all topologically distinct defects in superfluid ^{3}He are found; a similar discussion has been presented by Love and Zakrzewski (1979) for the planar state and by Schakel (1989) for the A_1/A_2 phases.

It should be borne in mind that the order parameter of a system and thus its topology generally depend on the relevant *lengthscale* of the object under investigation. In superfluid ^{3}He there are at least four such lengthscales: (i) the coherence length $\xi(T)$; (ii) the dipole coherence length ξ_D, (7.27a); (iii) the magnetic length ξ_H, (7.28a), (7.29a); (iv) the linear

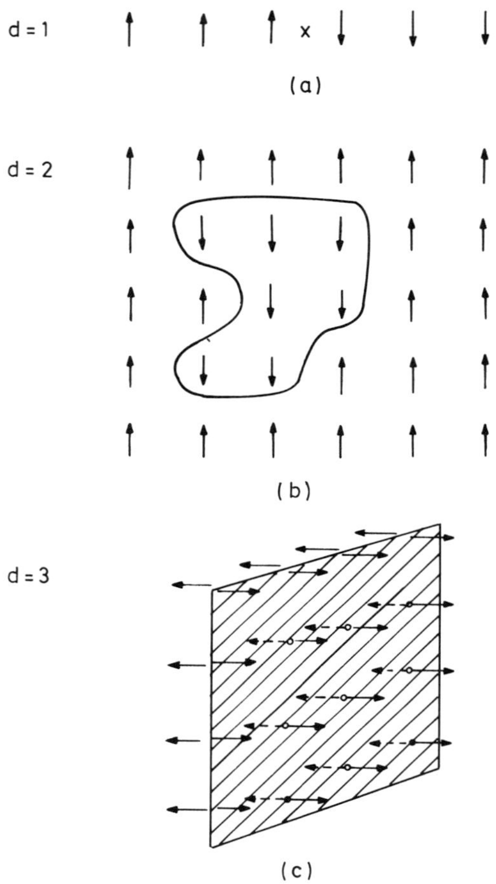

Figure 7.8 (a) Pointlike, (b) line-shaped and (c) surface-like defects for Ising spins in dimensions $d = 1$, 2 and 3 respectively.

dimension L of the container. Of these, the coherence length $\xi(T)$ is always considered the smallest. Even if external magnetic fields and the influence of walls on a texture are neglected, one is always left with one more lengthscale, namely ξ_D, which is intrinsic to the system. This implies that on lengthscales $r \ll \xi_D$ the dipole coupling of spin and orbital spaces can be neglected ("dipole-free" regime), but that for $r \gg \xi_D$ it is indeed important ("dipole-locked" regime). Correspondingly, the order-parameter space will be different (see Section 6.3), leading to different topological properties and thereby different singularities. We see here that the dipole-free regime—sometimes perhaps considered as "artificial" in view of the fact that the dipole interaction is always present—does have its physical importance, at least on lengthscales smaller than ξ_D.

Since we have already presented a detailed introductory discussion of the topological properties of vortices in superfluid ^{4}He, we shall now continue by investigating linear defects in superfluid ^{3}He-A and ^{3}He-B. After that, we shall consider the stability of point defects and planar defects.

7.5 LINEAR DEFECTS

7.5.1 Topological stability

The procedure of classifying linear defects in superfluid ^{3}He is the same as in superfluid ^{4}He or any other system. First one has to specify the degeneracy space R (see Chapter 6). Then one considers all possible closed contours in R, all of them being images of a contour $\mathscr{C}$ in real space encircling a linear defect (see Fig. 7.5). Each contour Γ belongs to a class of contours generated by continuous deformations of Γ. The set of classes of topologically inequivalent contours is in one-to-one correspondence with the topologically stable defects. Furthermore, the set of contour classes thus defined can be shown to form a group, the so-called "first homotopy group" or "fundamental group" π_1. The group product of two elements, i.e. of two classes represented by contours Γ_1 and Γ_2, is defined as the class of contours containing the joint contour $\Gamma_1 \circ \Gamma_2$, where $\Gamma_1 \circ \Gamma_2$ is obtained by first traversing contour Γ_1 and then Γ_2. In the following we shall only deal with Abelian fundamental groups. In this case the order of elements in the group multiplication is irrelevant. Consequently, multiplication of two group elements leads to precisely one new element. There is a one-to-one correspondence of the group elements of $\pi_1(R)$ and the topologically stable defects of the system. In the case of non-Abelian groups π_1, a given element Γ_1 and its conjugate elements $\Gamma_j \Gamma_1 \Gamma_j^{-1}$ (where Γ_j is any other element of the group) are equivalent, as they may be transformed into one another by shifting the starting point A of the contours (see Fig. 7.9). Consequently,

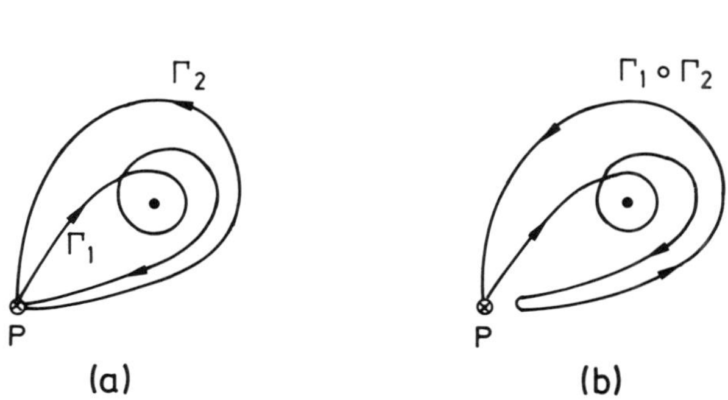

Figure 7.9 Group multiplication of two contours.

the set of topologically stable defects corresponds to the different classes of conjugate elements of π_1, rather than to the elements themselves.

The group multiplication defined above has a simple physical interpretation: it describes the coalescence of two line defects. Clearly, the order-parameter configuration that results when two defects coalesce can again be classified as a line defect. We note that this configuration may have *zero* topological charge, implying that the defects have annihilated each other. This property of defects, i.e. the existence of an inverse element, is a necessary condition for the application of a group-theoretical concept in this context.

Let us now turn to the application of the methods discussed above to superfluid ^{3}He (Volovik and Mineev 1976b, 1977a, Mermin *et al.* 1978, Mineev 1980). The first task is to find the degeneracy space R for a particular situation. In principle, the space R is given by the symmetry of the order parameter. In reality, however, there exist small residual interactions like the dipole interaction, which reduce the degeneracy space by fixing the relative orientation of the preferred directions of the order parameter. The relevance of this effect depends on the lengthscale of the configuration (see the discussion at the end of Section 7.4).

7.5.2 Topological properties of B-phase vortices

The order parameter is given by (6.8). Neglecting the dipole coupling, the residual symmetry of the order parameter is given by simultaneous rotations in orbital and spin spaces, represented by the group $SO(3)_{\boldsymbol{L}+\boldsymbol{S}}$ (see (6.24) and below). The degeneracy space is given by (6.26b), which contains all symmetry operations that change the order parameter, while leaving the free energy of a homogeneous order-parameter configuration invariant. This space is the direct product of the group of relative spin–orbit rotations and gauge transformations $R = U(1)_\phi \times SO(3)_{\boldsymbol{L},\boldsymbol{S}}$. As discussed in Section 6.1, the group of one-dimensional rotations U(1) may be represented by the topological space S^1 (circumference of a circle, see Fig. 6.2a) and the group of three-dimensional rotations SO(3) by the space SO_3 shown in Fig. 6.2(b). We are thus led to consider the fundamental group $\pi_1(S^1 \times SO_3)$. We note that in the B phase spin–orbit rotations and gauge transformations are not coupled. Hence line defects in the $\hat{\boldsymbol{n}}$ vector field and the rotation angle θ and line defects in the phase variable ϕ are unconnected. This implies that the fundamental group of a direct product of spaces is equal to the sum of the fundamental groups of the individual spaces:

$$\pi_1(S^1 \times SO_3) = \pi_1(S^1) + \pi_1(SO_3). \tag{7.57a}$$

We have in effect already derived the fundamental group of the circle S^1 (see Section 7.4): it is given by the group $\mathbb{Z}$ of all integers under by addition, corresponding to the topological charge $N = 0, \pm 1, \pm 2, \ldots$:

$$\pi_1(S^1) = \mathbb{Z}. \tag{7.57b}$$

This charge is simply the winding number of the phase, also called the "vorticity", as defined in (7.51) and (7.56). Thus, as expected, there exist vortices in ^{3}He-B just like in ^{4}He or in superconductors. In addition, one may also have line defects in the spin–orbit variables, classified by the elements of the fundamental group $\pi_1(SO_3)$. The topologically different classes of closed contours in the three-dimensional solid sphere SO_3 may easily be obtained by simple inspection.

(1) All contours that meet the surface of SO_3 an even number of times may be contracted into a point. They belong to the unit element of $\pi_1(SO_3)$ and describe the defect-free state.

(2) Any contour meeting the surface an odd number of times can be contracted into a contour starting at a point on the surface and ending on the opposite side of the surface (since they are equivalent), i.e. touching only once. An example of a contour initially meeting the surface three times and its contraction is shown in Fig. 7.10(a). The contracted contour is therefore attached to the surface and cannot be further contracted to a point—it corresponds to a line defect. All contours of this kind obviously belong to the same class, since points on the surface of the SO_3 sphere are freely movable (of course, their image points on the opposite side of the sphere have to follow along).

An example of each of the two types of contours is depicted in Fig. 7.10(b). This exhausts the possibilities for contours in SO_3. Consequently the fundamental group $\pi_1(SO_3)$ has only two elements. This result may also be obtained by formal mathematical methods, which give

$$\pi_1(SO_3) = \mathbb{Z}_2. \tag{7.57c}$$

That is, $\pi_1(SO_3)$ is isomorphic to $\mathbb{Z}_2$, the group of integers modulo 2. The topological charges are $N_2 = 0, 1$.

The stable vortex configuration in the spin–orbit variables with $N_2 = 1$ is obtained by minimizing the gradient free energy (7.17) under the constraint that the rotation angle θ changes from $+\pi$ to $-\pi$ as one encircles the defect. In order to make the bending energy as small as possible, it is therefore plausible to assume the $\hat{\boldsymbol{n}}$ vector field to be uniform (say, along the z axis), while the θ field increases uniformly with the azimuthal angle φ, independently of the distance to the axis of the defect. In this way, only a gradient along φ is created.

The order-parameter field of such a "disclination" is given by

$$A_{\mu j}(r, z, \varphi) = 3^{-1/2} e^{i\phi} R_{\mu j}(\hat{z}, \theta = \varphi - \pi). \tag{7.57d}$$

At larger distances from the defect core, $r \gg \xi_D$, the dipole energy forces θ to become equal to the Leggett angle θ_L. In this case the degeneracy

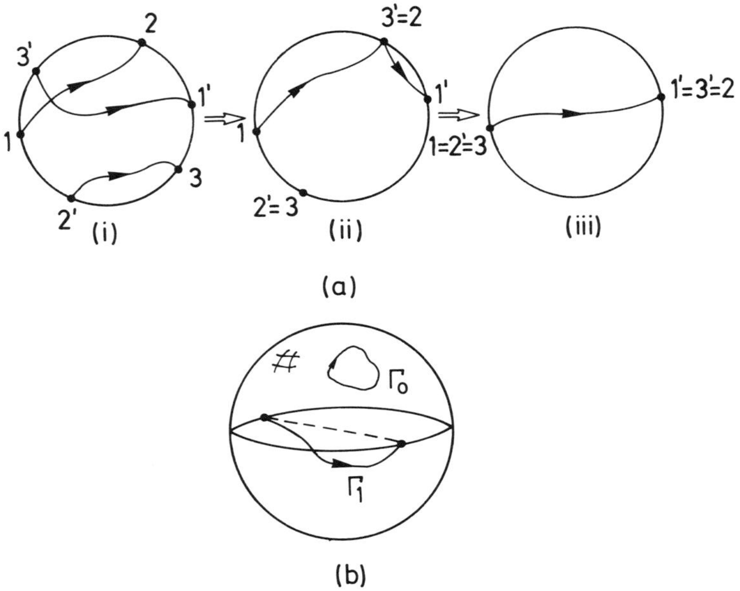

Figure 7.10 (a) Mapping of a closed contour into the three-dimensional solid sphere SO_3 (cross-section): (i) the initial contour touches the surface at three points $(1, 2, 3)$ (diametrically opposed points are indicated by $1'$, $2'$, $3'$); (ii) by moving 3 towards $2'$ (and hence $3'$ towards 2), the part $2'3$ of the contour may be shrunk to zero; (iii) finally the point $3' = 2$ is moved towards $1'$, leaving a contour that touches the surface only once and cannot be further contracted. (b) The two types of contours in SO_3 corresponding to the elements of the fundamental group $\pi_1(SO_3) = \mathbb{Z}_2$; Γ_0 may be contracted (defect-free state); Γ_1 cannot be contracted (existence of a defect).

space reduces to S^2, i.e. a spherical surface, of radius θ_L (see Fig. 6.2b). What is the fundamental group of S^2? As already mentioned, any closed contour on the surface of a sphere may be contracted into a point. Hence

$$\pi_1(S^2) = 0; \tag{7.58}$$

therefore the group consists only of the unit element—a stable linear defect involving only the $\hat{\boldsymbol{n}}$-vector system alone does not exist. The only stable defects on this lengthscale are the usual vortices associated with the phase ϕ.

One might think that there is a contradiction between the findings that there exists a disclination defect at $r \ll \xi_D$ but none at larger distances. In fact, what happens is that the disclination forms the edge of a domain wall in the θ field! The order-parameter configurations at $r \ll \xi_D$ and $r \gg \xi_D$ are sketched in Fig. 7.11, where the rotation matrix $R_{\mu j}$ at a given point in space is indicated by an arrow of direction $\hat{\boldsymbol{n}}$ and length θ. Whereas the order parameter field in Fig. 7.11(a) changes all along the contour and

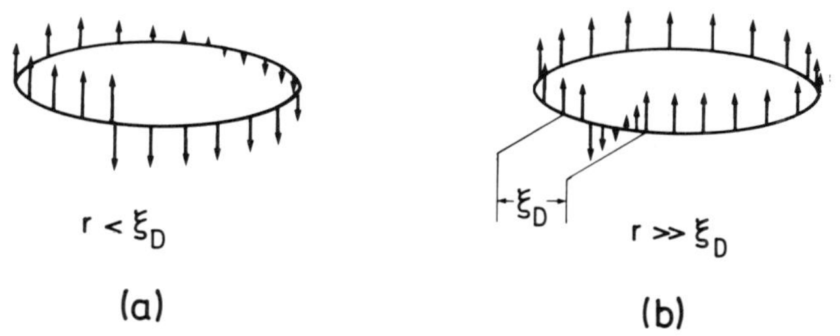

Figure 7.11 Contours of different diameter around a linear defect in the $(\hat{\boldsymbol{n}}, \theta)$ field in ^{3}He-B. Arrows indicate the direction of $\hat{\boldsymbol{n}}$, their length measures θ. (a) $r < \xi_D$; the existence of a defect is clearly indicated. (b) $r \gg \xi_D$; the defect is hidden.

thereby reflects the existence of the defect, the order-parameter field in Fig. 7.11(b) is uniform over most of the contour and thus hides the actual presence of the defect. The domain wall as well as the cross-section of the defect is of width ξ_D.

7.5.3 Topological properties of A-phase vortices at small distances

The order parameter in the A phase is given by (6.10). As discussed below (6.10) and (6.46), the residual symmetry of $A_{\mu j}$ for small distances, i.e. $r \ll \xi_D, \xi_H$, is described by the group $H = \mathrm{U}(1)_{S_z} \times \mathrm{U}(1)_{L_z - \phi} \times \mathbb{Z}_2$. Hence the degeneracy space is given by (6.68b). The effect of the discrete symmetry $\mathbb{Z}_2$ is to remove from the group SO(3) all rotations about the vector $\hat{\boldsymbol{l}}$ by $\pm\pi$. This implies that in the solid sphere SO_3 pairs of points on a common rotation axis (which may be taken to be the z axis) separated by π specify the same symmetry operation.

Mathematically, the fundamental group of R, $\pi_1(R)$, is found as the sum of the fundamental groups of S^2 and $\mathrm{SO}_3/\mathbb{Z}_2$. Because $\pi_1(\mathrm{S}^2) = 0$, (7.58), this implies

$$\pi_1((\mathrm{S}^2 \times \mathrm{SO}_3)/\mathbb{Z}_2) = \pi_1(\mathrm{SO}_3/\mathbb{Z}_2) \tag{7.59a}$$

$$= \mathbb{Z}_4, \tag{7.59b}$$

i.e. the group of integers modulo 4 (Volovik and Mineev 1977a, Cross and Brinkman 1977). This result may easily be understood: if the vector $\hat{\boldsymbol{d}}$ were fixed in space (or dipole-locked to $\hat{\boldsymbol{l}}$; see below), the degeneracy space would correspond to SO_3 only. In this case $\pi_1(\mathrm{SO}_3) = \mathbb{Z}_2$, i.e. there would be only two distinct textures: the uniform one ($N = 0$) and the vortex with charge $N = 1$. However, in the present case, where $\hat{\boldsymbol{d}}$ is free to move, the possibility that $\hat{\boldsymbol{d}}$ may change into $-\hat{\boldsymbol{d}}$ (leading to the extra $\mathbb{Z}_2$ in (7.59a)) implies two more possibilities in addition to the two mentioned above: $\hat{\boldsymbol{d}}$

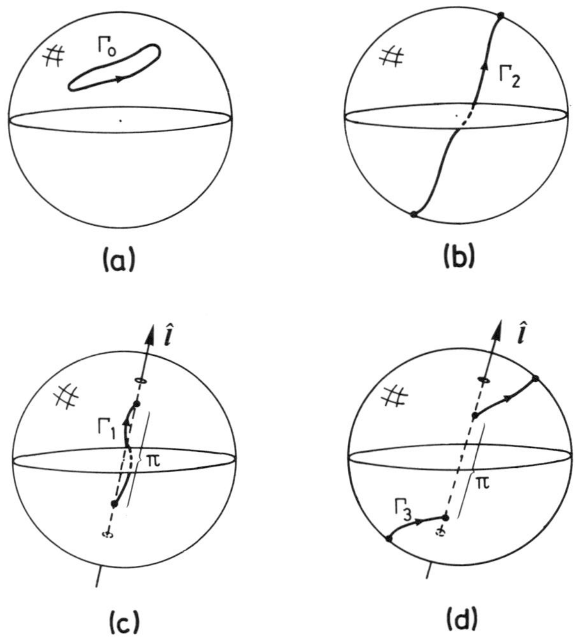

Figure 7.12 The four different contours in $SO_3/\mathbb{Z}_2$.

may continuously reverse into $-\hat{\boldsymbol{d}}$, while the orbital triad rotates about $\hat{\boldsymbol{l}}$ by π or $-\pi$ (Mermin 1978a). So there are altogether *four* different categories of contours in $SO_3/\mathbb{Z}_2$, which are illustrated in Fig. 7.12:

(i) closed contours Γ_0 in the interior, which may be contracted to a point (Fig. 7.12a). These correspond to the unit element of $\pi_1(SO_3/\mathbb{Z}_2)$ and hence to a nonsingular configuration;

(ii) contours Γ_2 starting and ending at points on opposite sides of the surface of the sphere SO_3 (Fig. 7.12b) (note, however, that the contour may touch the surface of the sphere at an arbitrary number of points, too);

(iii) contours Γ_1 starting and ending at points located on a common rotation axis (which may be taken to be the z axis) separated by π in the interior of the sphere (Fig. 7.12c);

(iv) contours Γ_3 connecting points on a common rotation axis (which may be taken to be the z axis) separated by π, touching the surface (Fig. 7.12d).

Just as in the familiar case of the phase vortex in the B phase, a topological quantum number N may be associated with the four cases listed above. The formal expression for N in terms of an integration over rotation angles along the closed contour is derived as follows. It is clear from the above that the angle of rotation θ plays the decisive role in this classification. Thus a given vortex belongs to class (i) or (ii) depending on whether it touches the surface of the SO_3 sphere an even or an odd number of times, i.e. whether the initial value θ_i and final value θ_f of the rotation angle differ by an even or odd multiple of 2π. Similarly, in cases (iii) and (iv), the "difference" between initial and final angles is $\theta_i - (-\theta_f) = \pi + 2\pi n$. The minus sign in front of θ_f arises because θ is positive by definition, i.e. counterclockwise rotations are parametrized by $(-\hat{\boldsymbol{n}}, +\theta)$ rather than $(+\hat{\boldsymbol{n}}, -\theta)$. In order to avoid the confusion that might arise from this inconvenient definition, and also for simple practical purposes, it is useful to describe the rotations in terms of Euler angles α, β, γ introduced in (7.11). The relation of θ to α, β, γ is easily obtained by comparing the traces of the respective rotation matrices, i.e. $R_{\mu j}$ as given by (6.106) and

$$R_{\mu j}(\alpha, \beta, \gamma) = \begin{pmatrix} \cos\alpha\cos\beta\cos\gamma - \sin\alpha\sin\gamma & \sin\alpha\cos\beta\cos\gamma + \cos\alpha\sin\gamma & -\sin\beta\cos\gamma \\ -\cos\alpha\cos\beta\sin\gamma - \sin\alpha\cos\gamma & -\sin\alpha\cos\beta\sin\gamma + \cos\alpha\cos\gamma & \sin\beta\sin\gamma \\ \cos\alpha\sin\beta & \sin\alpha\sin\beta & \cos\beta \end{pmatrix}. \tag{7.60}$$

One finds

$$\cos\tfrac{1}{2}\theta = \cos\tfrac{1}{2}(\alpha+\gamma)\cos\tfrac{1}{2}\beta. \tag{7.61}$$

The components of the unit vector $\hat{\boldsymbol{n}}$ (the rotation axis) are given in terms of the Euler angles by

$$\left.\begin{aligned} \begin{pmatrix} n_x \\ n_y \end{pmatrix} &= \frac{\sin\frac{1}{2}\beta}{\sin\frac{1}{2}\theta}\begin{pmatrix} \sin\frac{1}{2}(\alpha-\gamma) \\ \cos\frac{1}{2}(\alpha-\gamma) \end{pmatrix}, \\ n_x^2 + n_y^2 + n_z^2 &= 1, \end{aligned}\right\} \tag{7.62}$$

with θ given by (7.61). This will turn out to be a useful expression when we investigate the actual contour $\mathscr{C}$ in order-parameter space associated with a given texture.

Every time the contour Γ in SO_3 touches the surface, the angle θ is an odd multiple of π and $\cos\frac{1}{2}\theta = 0$. This implies that either $\alpha + \gamma$ or β has to be an odd multiple of π. In cases (i) and (ii) the initial and final values of α, β and γ may only differ by multiples of 2π, so that the integral of $d(\alpha + \beta + \gamma)$ along the closed contour must equal $2\pi N_0$, where N_0 is precisely the number of times the contour touches the surface of the sphere SO_3. In cases (iii) and (iv) the initial and final values of θ "differ" by an odd multiple of π. This implies that either β or $\alpha+\gamma$ must also change by an odd multiple of π upon traversing the closed contour (the other variable, $\alpha+\gamma$ or β respectively, changes by a multiple of 2π). We may therefore conclude that the contour integral over $\mathrm{d}(\alpha+\beta+\gamma)$ is connected with the topological charge N by

$$N=\frac{1}{2\pi}\oint \mathrm{d}(\alpha+\beta+\gamma) \quad (\mathrm{mod}\, 2). \tag{7.63}$$

The possible values are $N=0, \frac{1}{2}, 1, \frac{3}{2}$, where $\frac{3}{2}$ is equivalent to $-\frac{1}{2}$ etc. Two linear defects with charges N_1 and N_2 may combine to give a single line defect with charges $N=N_1+N_2$ (mod 2).

The superfluid velocity field associated with a vortex line is given in terms of the Euler angles (α, β, γ) by (7.13). Furthermore, the circulation κ about a vortex line is seen to depend on $\beta(\boldsymbol{r})$ and is found to be

$$\kappa=\frac{h}{2m}(N_\phi+\cos\beta\, N_\alpha), \tag{7.64}$$

where N_ϕ and N_α are winding numbers.

In order to illustrate the power of the results just derived, we shall now discuss several examples of vortex lines for each topological class. The true equilibrium configuration as well as the detailed dependence of the order parameter on position are determined by minimizing the free energy, including gradient terms (see Section 7.2). We shall use cylindrical coordinates (z, r, φ), with $\hat{\boldsymbol{z}}$, $\hat{\boldsymbol{r}}$ and $\hat{\boldsymbol{\varphi}}$ denoting the unit vectors along the vortex axis and in the radial and tangential direction respectively.

Topological class $N=0$

The most symmetric state in this class is the completely uniform texture given by $\hat{\boldsymbol{d}}=\hat{\boldsymbol{l}}=\hat{\boldsymbol{z}}$, $\hat{\boldsymbol{m}}+\mathrm{i}\hat{\boldsymbol{n}}=\mathrm{e}^{-\mathrm{i}\gamma}(\hat{\boldsymbol{x}}+\mathrm{i}\hat{\boldsymbol{y}})$, $\gamma=0$. Any texture for which the topological quantum number N defined by (7.63) is a multiple of 2 also belongs to this class. An example of this type is a pure phase vortex with an *even* number of circulation quanta $2k$, $k=0, \pm1, \pm2, \ldots$, characterized by

$$\left.\begin{aligned} &\hat{\boldsymbol{d}}=\hat{\boldsymbol{l}}=\hat{\boldsymbol{z}},\\ &\hat{\boldsymbol{m}}+\mathrm{i}\hat{\boldsymbol{n}}=\mathrm{e}^{-\mathrm{i}\gamma}(\hat{\boldsymbol{x}}+\mathrm{i}\hat{\boldsymbol{y}}), \quad \gamma=-2k\varphi.\end{aligned}\right\} \tag{7.65}$$

The Euler angles describing the rotation of the triad $(\hat{\boldsymbol{m}}, \hat{\boldsymbol{n}}, \hat{\boldsymbol{l}})$ along a contour encircling the vortex are given by $(\alpha, \beta, \gamma)=(0, 0, -2k\varphi)$. The winding number is therefore

$$N=\frac{1}{2\pi}\int_0^{2\pi}\mathrm{d}\varphi\frac{\mathrm{d}\gamma}{\mathrm{d}\varphi}=-2k, \tag{7.66}$$

which is equivalent to $N=0$.

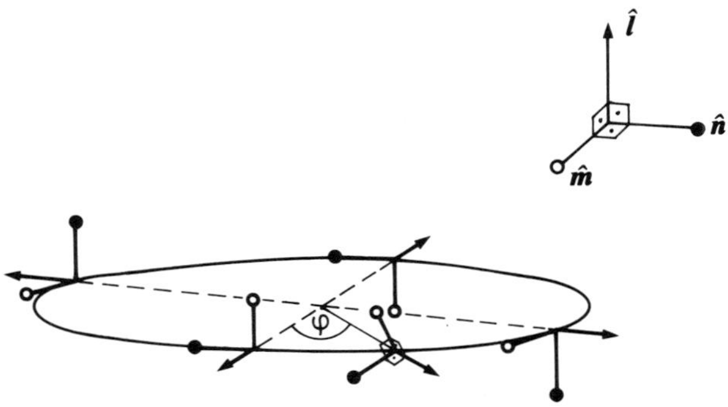

Figure 7.13 Example of a vortex structure with $N = 0$. It is an $\hat{\boldsymbol{l}}$ disgyration with phase vortex, (7.68), for $k = -1$. The topological charge is $N = 2$, which is equivalent to $N = 0$.

The superfluid velocity field is given by

$$\boldsymbol{v}_s = 2k \frac{\hbar}{mr} \hat{\boldsymbol{\varphi}}. \tag{7.67}$$

A second example is a vortex line that, in addition to a phase vortex, involves an "$\hat{\boldsymbol{l}}$ disgyration" (de Gennes 1973), where $\hat{\boldsymbol{l}}$ points radially outward (see Fig. 7.13). It is parametrized by

$$\left.\begin{aligned} \hat{\boldsymbol{l}} &= \hat{\boldsymbol{r}}, \\ \hat{\boldsymbol{m}} + \mathrm{i}\hat{\boldsymbol{n}} &= \mathrm{e}^{-\mathrm{i}\gamma}(\hat{\boldsymbol{z}} - \mathrm{i}\hat{\boldsymbol{\varphi}}), \quad \gamma = -(2k+1)\varphi, \end{aligned}\right\} \tag{7.68}$$

so that

$$\boldsymbol{v}_s = (2k+1) \frac{\hbar}{mr} \hat{\boldsymbol{\varphi}}. \tag{7.69}$$

Here the $\hat{\boldsymbol{d}}$ field may either be uniform or follow $\hat{\boldsymbol{l}}$ or assume some other configuration. In this case the Euler angles are found to vary as functions of the azimuthal angle φ as $(\alpha, \beta, \gamma) = (\varphi, -\frac{1}{2}\pi, -(2k+1)\varphi)$. The winding number is then

$$N = \frac{1}{2\pi} \int_0^{2\pi} \mathrm{d}\varphi \left(\frac{\mathrm{d}\alpha}{\mathrm{d}\varphi} + \frac{\mathrm{d}\gamma}{\mathrm{d}\varphi} \right) = -2k, \tag{7.70}$$

which is again equivalent to $N = 0$.

It follows that the two vortex structures (7.65) and (7.68), which possess a superfluid velocity field that *diverges* at the origin (and hence must have a normal-fluid core), may be continuously deformed into a *nonsingular* structure.

For example, the pure-phase vortex (7.65) with $k = -1$, i.e. with two quanta of circulation, may be continuously deformed into a nonsingular

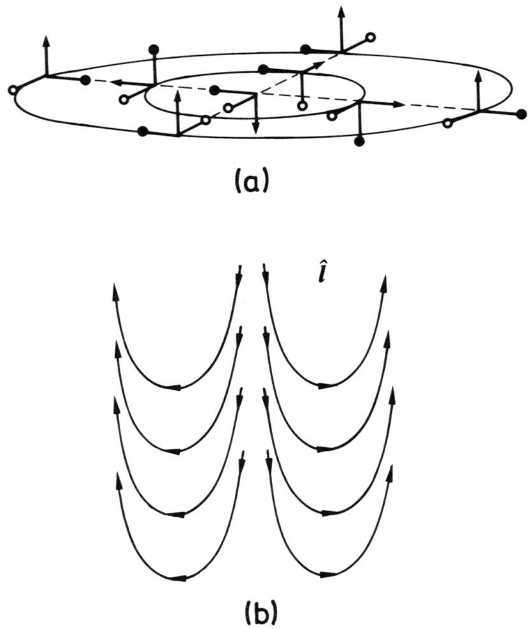

Figure 7.14 The Anderson–Toulouse vortex as an example of a nonsingular vortex (topological charge $N=0$). (a) Horizontal cross-section; on the outer circle the asymptotic $\hat{\boldsymbol{l}}$ configuration is assumed. The orientation of the orbital triad upon traversal of the vortex on two concentric circular contours is indicated. (b) Vertical cross-section; only the orientation of the vector $\hat{\boldsymbol{l}}$ is shown.

texture by rotating $\hat{\boldsymbol{l}}$ in the $(\hat{\boldsymbol{r}}, \hat{\boldsymbol{z}})$ plane as one moves radially towards the core, such that in the centre of the vortex $\hat{\boldsymbol{l}}$ is antiparallel to the asymptotic $\hat{\boldsymbol{l}}$ field, i.e. $\hat{\boldsymbol{l}}(r=0) = -\hat{\boldsymbol{z}}$. This reorientation is depicted in Fig. 7.14. In this way the superfluid velocity vanished continuously as $r \rightarrow 0$, and the energetically unfavourable normal core is avoided. This type of nonsingular structure is called an "Anderson–Toulouse (AT) vortex" (Anderson and Toulouse 1977; see also Chechetkin 1976a). A convenient parametrization is given by

$$\left.\begin{aligned} \hat{\boldsymbol{d}} = \hat{\boldsymbol{l}} &= \sin\eta(r)\,\hat{\boldsymbol{z}} + \cos\eta(r)\,\hat{\boldsymbol{r}}, \\ \hat{\boldsymbol{m}} + \mathrm{i}\hat{\boldsymbol{n}} &= \mathrm{e}^{-\mathrm{i}\gamma}[\cos\eta(r)\,\hat{\boldsymbol{z}} - \sin\eta(r)\,\hat{\boldsymbol{r}} - \mathrm{i}\hat{\boldsymbol{\varphi}}],\ \gamma = \varphi, \end{aligned}\right\} \tag{7.71}$$

where $\eta(r)$ is the rotation angle of $\hat{\boldsymbol{l}}$, with η increasing monotonically from $-\frac{1}{2}\pi$ at $r=0$ to $\frac{1}{2}\pi$ as $r \rightarrow \infty$, i.e. $(\eta + \frac{1}{2}\pi) \propto r$ as $r \rightarrow 0$. We have taken $\hat{\boldsymbol{d}}$ to be parallel to $\hat{\boldsymbol{l}}$, anticipating the case of dipole locking to be discussed below. The Euler angles associated with (7.71) are given by $(\varphi, \eta - \frac{1}{2}\pi, \varphi)$, leading to $N = 2 \equiv 0$.

The superfluid velocity field associated with (7.71) is found to be

$$\boldsymbol{v}_s = -\frac{\hbar}{2mr}[1 + \sin \eta(r)]\hat{\boldsymbol{\varphi}}. \tag{7.72}$$

It clearly has the desired property $\boldsymbol{v}_s \to 0$ as $r \to 0$. The vorticity is continuously distributed over the cross-section of the vortex. We shall return to the energetics of such textures in Section 7.5.7.

Another important type of nonsingular texture is obtained by removing the singularity of the phase-disgyration vortex (7.68) with $k = -1$ by the same procedure as described above. This results in the so-called "Mermin–Ho (MH) vortex" (Anderson and Brinkman 1975, Mermin and Ho 1976) shown in Fig. 7.15. It is characterized by (7.71) with, however, the boundary condition as $r \to \infty$ changed to $\eta = 0$. Hence it corresponds to the inner core of the AT vortex (see Fig. 7.15). As discussed in Section 7.9, a state very close to the MH texture is expected to occur as the ground-state configuration in a long cylindrical vessel, where $\hat{\boldsymbol{l}}$ is forced to point radially outward, owing to the boundary condition at the walls of the cylinder.

There is yet another important nonsingular texture where $\hat{\boldsymbol{l}}$ points along

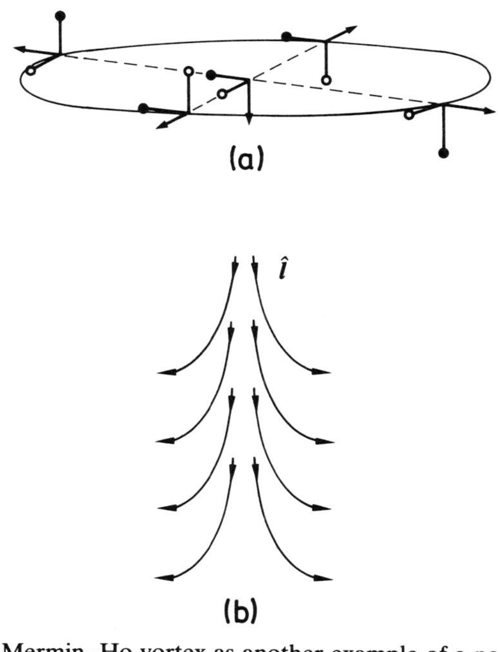

Figure 7.15 The Mermin–Ho vortex as another example of a nonsingular vortex. It corresponds to the inner part of the Anderson–Toulouse vortex shown in Fig. 7.14. (a) Horizontal cross-section. (b) Vertical cross-section only, showing the $\hat{\boldsymbol{l}}$ configuration.

the axis at the centre and, as $r \to \infty$, lies in the horizontal plane. This so-called "mixt–twist" (MT) (Ho 1978c) or "hyperbolic" texture (Maki 1983a, Zotos and Maki 1984, Maki and Zotos 1985a) is characterized by the Euler angles $(\alpha, \beta, \gamma) = (-\varphi + \alpha_0, \beta(r), -\varphi)$, with $\beta(0) = \pi$ and $\beta = \frac{1}{2}\pi$ as $r \to \infty$, which integrates to $N = -2 \equiv 0$. Its projection onto a plane perpendicular to $\hat{\boldsymbol{z}}$ looks like the hyperbolic vortex with $N = -1$, as shown in Fig. 7.5(d) or 7.7, and the corresponding superfluid velocity field $\boldsymbol{v}_s = (\hbar/2mr)(1 + \cos\beta)\hat{\boldsymbol{\varphi}}$ looks asymptotically like that of a singly quantized vortex.

All three vortices—the AT, MH and MT textures—may be continuously deformed into the uniform texture, which has the lowest energy. Their existence thus requires the presence of external fields or a surface. A further topological classification of defects *within* a given class, say $N = 0$, can be performed by taking into account the constraints on the order parameter due to the action of external fields (Mineev and Volovik 1978).

Topological class $N = 1$

From the above discussion of (unstable) singular $N = 0$ vortices, it should be clear that textures with $N = 1$ are obtained by replacing the phase winding number $2k$ in the expression for the phase vortex, (7.65), by any odd integer $2k + 1$, and the winding number $2k + 1$ in the expression for the radial disgyration, (7.68), by any even number $2k$. For a radial disgyration with $k = 0$, see Fig. 7.16. In addition to the radial disgyration in the $\hat{\boldsymbol{l}}$ field, (7.68), one may have tangential disgyrations, parametrized by

$$\left.\begin{aligned} \hat{\boldsymbol{l}} &= \hat{\boldsymbol{\varphi}}, \\ \hat{\boldsymbol{m}} + \mathrm{i}\hat{\boldsymbol{n}} &= \mathrm{e}^{-\mathrm{i}\gamma}(\hat{\boldsymbol{z}} + \mathrm{i}\hat{\boldsymbol{r}}), \quad \gamma = -2k\varphi, \end{aligned}\right\} \tag{7.73}$$

with a superfluid velocity field $\boldsymbol{v}_s = k(\hbar/mr)\hat{\boldsymbol{\varphi}}$. The Euler angles associated with this structure are given by $(\alpha, \beta, \gamma) = (\varphi + \frac{1}{2}\pi, -\frac{1}{2}\pi, -2k\varphi)$, which leads via (7.63) to a winding number $N = -2k + 1$, equivalent to $N = 1$.

All of these vortex structures have a singularity at the origin—in the superfluid velocity or in the $\hat{\boldsymbol{l}}$ texture—which drives the liquid to become normal within a core region of radius equal to the coherence length ξ. The vorticity $\boldsymbol{\nabla} \times \boldsymbol{v}_s$, is distributed in different ways: in the case of a pure phase

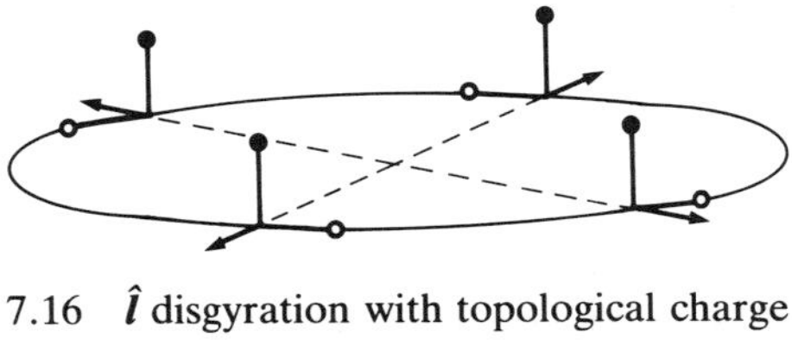

Figure 7.16 $\hat{\boldsymbol{l}}$ disgyration with topological charge $N = 1$.

vortex it is concentrated within the normal core region (since the flow outside the core is potential), whereas for a disgyration the vorticity is distributed continuously.

The energetics of vortices will be discussed below. One finds that among all $N = 1$ vortices the radial disgyration, given by (7.68) with $2k + 1 = 0$, has the lowest energy (Volovik and Mineev 1977a, Brinkman and Cross 1978).

Topological class $N = \pm\frac{1}{2}$

The most symmetric vortex structure in this class is the one where $\hat{\boldsymbol{l}}$ is uniform ($\hat{\boldsymbol{l}} \parallel \hat{\boldsymbol{z}}$, say), with the orbital triad rotating about $\hat{\boldsymbol{l}}$ along the contour (equivalent to a phase transformation) by an odd multiple of π, such that $(\hat{\boldsymbol{m}} + \mathrm{i}\hat{\boldsymbol{n}})|_{\varphi=2\pi} = -(\hat{\boldsymbol{m}} + \mathrm{i}\hat{\boldsymbol{n}})|_{\varphi=0}$. This change in sign is then compensated by a simultaneous rotation of $\hat{\boldsymbol{d}}$ into $-\hat{\boldsymbol{d}}$ (Volovik and Mineev 1976b, Cross and Brinkmann 1977):

$$\left.\begin{aligned} \hat{\boldsymbol{m}} + \mathrm{i}\hat{\boldsymbol{n}} &= \mathrm{e}^{-\mathrm{i}\gamma}(\hat{\boldsymbol{x}} + \mathrm{i}\hat{\boldsymbol{y}}), \quad \gamma = -(2k \pm \tfrac{1}{2}), \\ \hat{\boldsymbol{d}} &= \cos\left[(k' + \tfrac{1}{2})\varphi\right]\hat{\boldsymbol{x}} + \sin\left[(k' + \tfrac{1}{2})\varphi\right]\hat{\boldsymbol{y}}, \end{aligned}\right\} \tag{7.74}$$

where k and k' are integers. These vortices carry a half-integral number of circulation quanta. At the same time the $\hat{\boldsymbol{d}}$ field forms a disclination, a type of texture known to exist in nematic liquid crystals (de Gennes 1974). But, unlike the latter case, disclinations in ^{3}He-A can only exist in conjunction with half-quantum phase vortices.

7.5.4 Topological properties of A-phase vortices at large distances

Let us next consider the topology of vortices in ^{3}He-A at large distances. In zero magnetic field there is only one relevant lengthscale: ξ_D. In this case, i.e. for $r \gg \xi_D$, one is in the dipole-locked regime and the degeneracy space is reduced to a common SO_3 describing rotations of the orbital triad with $\hat{\boldsymbol{d}}$ locked to $\hat{\boldsymbol{l}}$. Clearly, this eliminates the possibility of vortices with $N = \pm\frac{1}{2}$, which involve *distinct* rotations of $\hat{\boldsymbol{d}}$ and the orbital triad. Hence the topological charge may only assume integer values ($N = 0, 1 \pmod 2$). It means that there are now only two different contours rather than four as in the dipole-free situation. This is also expressed by the fact that the fundamental group is given by (7.57c), i.e. $\pi_1(SO_3) = \mathbb{Z}_2$. Consequently, for $r \gg \xi_D$, defects with $N = \pm\frac{1}{2}$ are effectively hidden in the same way as in the B phase, i.e. the defect forms the edge of a domain wall.

Weak magnetic field

In the presence of a magnetic field one has to distinguish whether $H > H^*$ or $H < H^*$ (where $H^* \approx 28$ G is given by (7.28c)), i.e. whether the magnetic length ξ_H is smaller or larger than ξ_D. For weak magnetic fields such that

$\xi_H \gg \xi_D$, a new regime of distances enters, i.e. $r \gg \xi_H \gg \xi_D$, where the magnetic orientation energy confines the vector $\hat{\boldsymbol{d}}$ in the plane perpendicular to $\boldsymbol{H}$, and where at the same time $\hat{\boldsymbol{l}}$ and $\hat{\boldsymbol{d}}$ are dipole-locked. The degeneracy space then consists of two circles

$$R_{\mathrm{A}}^{\mathrm{D},H} = \mathrm{S}^1 \times \mathrm{S}^1, \tag{7.75a}$$

where the first S^1 corresponds to the overall phase (i.e. a rotation about $\hat{\boldsymbol{l}}$) and the second to a joint rotation of $\hat{\boldsymbol{d}} \parallel \hat{\boldsymbol{l}}$ in the plane. (Note that in this situation $\boldsymbol{\nabla} \times \boldsymbol{v}_{\mathrm{s}} = 0$, as for example in superfluid ^{4}He, because $\hat{\boldsymbol{l}}$ is confined to lie in a plane; see Section 7.10). Since the first homotopy group of S^1 is $\pi_1(\mathrm{S}^1) = \mathbb{Z}$ (see (7.57b)), we obtain two different kinds of topological charge, both characterized by $\mathbb{Z}$. These are the winding numbers N_1 and N_2 characterizing vortices due to the phase and the $\hat{\boldsymbol{l}} \parallel \hat{\boldsymbol{d}}$ field respectively. The latter vortex is characterized by

$$\left.\begin{aligned} \hat{\boldsymbol{l}} = \hat{\boldsymbol{d}} &= \sin(N_1\varphi)\hat{\boldsymbol{x}} + \cos(N_1\varphi)\,\hat{\boldsymbol{y}}, \\ \hat{\boldsymbol{m}} + \mathrm{i}\hat{\boldsymbol{n}} &= \mathrm{e}^{-\mathrm{i}\gamma}\{\hat{\boldsymbol{z}} + \mathrm{i}[\cos(N_1\varphi)\,\hat{\boldsymbol{x}} - \sin(N_1\varphi)\hat{\boldsymbol{y}}], \\ \gamma &= -N_2\varphi, \\ \boldsymbol{v}_{\mathrm{s}} &= N_2\frac{\hbar}{2m}\hat{\boldsymbol{\varphi}}. \end{aligned}\right\} \tag{7.75b}$$

An example for $N_1 = 2$ is shown in Fig. 7.17. This structure therefore carries N_2 quanta of vorticity but, at the same time, has a topological charge $N = N_1 + N_2$. Note that in general $N_2 \neq N$. Hence the vorticity and the topological charge do not coincide!

In the core region, where $\xi_H \gg \xi_D \gg r$, we already know that there are only two topological classes $N = 0,\ 1$. This implies that the vortex core structure will be continuous or discontinuous depending on whether $N_1 + N_2$ is even or odd. In the case where $N_1 + N_2$ is an even number the vortex thus has a "soft core" of extension ξ_H. Examples of possible vortex structures

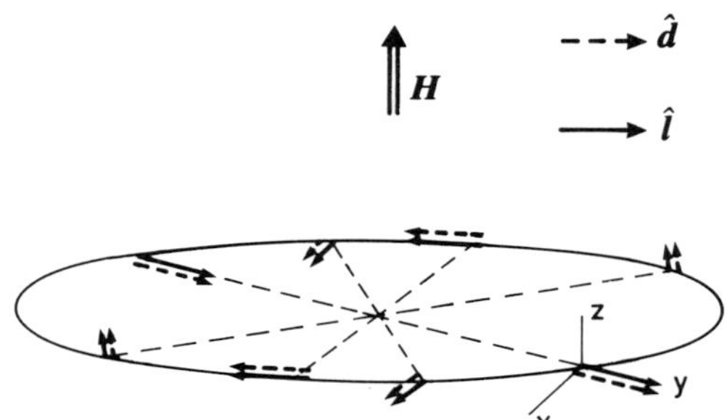

Figure 7.17 Vortex in a weak magnetic field $\boldsymbol{H}$ with $\hat{\boldsymbol{d}} \parallel \hat{\boldsymbol{l}} \perp \boldsymbol{H}$ and winding number $N_1 = 2$ in the $\hat{\boldsymbol{d}}$, $\hat{\boldsymbol{l}}$ field. The change in orientation of the orbital triad, which determines the winding number N_2 of the phase, is not shown.

are the AT texture and the MH texture. In general, the vorticity is distributed over the cross-sections of radius ξ_H. In the opposite case, where $N_1 + N_2$ is an odd number, the outer and inner vortex core regions $\xi_D < r < \xi_H$ and $\xi < r < \xi_D$ belong to the class $N = 1$, and there is a "hard core", i.e. a normal-fluid region for $r < \xi$. So such a vortex has *three* cores altogether (Volovik and Hakonen 1981).

Strong magnetic field

The vortex core structure changes if a strong magnetic field is applied, i.e. for $H \geqslant 30$ G. In the outermost shell of the vortex core, for $\xi_H < r < \xi_D$, the dipole orientation energy is unimportant, but $\hat{\boldsymbol{d}}$ is confined to lie in the plane perpendicular to the magnetic field. Therefore the degeneracy space is

$$R_{\mathrm{A}}^{H} = (\mathrm{S}^1 \times \mathrm{SO}_3)/\mathbb{Z}_2. \tag{7.75c}$$

According to (7.57b) and (7.59b), the first homotopy group of this space consists of the groups $\mathbb{Z}$ and $\mathbb{Z}_4$, i.e. $\pi_1(R_{\mathrm{A}}^{H}) = \mathbb{Z} \times \mathbb{Z}_4$, describing the defect classes of disclinations of winding number N in the $\hat{\boldsymbol{d}}$ field and the orbital triad. In the inner core region, $r < \xi_H$, the $\hat{\boldsymbol{d}}$ field in a disclination may avoid the singularity by escaping into the third dimension. The loss in magnetic orientation energy is then overcompensated by the gain in bending energy. It may be shown that the energetically most favourable core structure in the class of nonsingular $\hat{\boldsymbol{l}}$ textures is given by the AT texture, rather than the MH texture (Volovik and Hakonen 1981, Zotos and Maki 1985, Fetter 1985a).

The vortex containing *one* quantum of vorticity again has three core regions. In the outermost region, $r > \xi_D$, $\hat{\boldsymbol{l}}$ and $\hat{\boldsymbol{d}}$ are dipole-locked and confined to the plane perpendicular to the magnetic field. Inside the outer core, i.e. for $\xi_H < r < \xi_D$, $\hat{\boldsymbol{d}}$ is still locked in the plane, while $\hat{\boldsymbol{l}}$ transforms into a disgyration, which contains a normal-fluid core of size ξ. The $\hat{\boldsymbol{d}}$ field is found to deviate little from its uniform configuration, even inside the magnetic core region, i.e. for $r < \xi_H$ (Seppälä and Volovik 1983, Vulovic *et al.* 1984).

7.5.5 Symmetry classification of vortices

The topological classification of defects presented above allows one to group textures into distinct classes, which contain many qualitatively different order-parameter configurations. The actual configuration realized at given temperature, pressure and magnetic field and for given boundary conditions is of course the one that minimizes the appropriate free energy (which has to include the gradient free energy). The Euler–Lagrange equations of this problem generally take the form of a set of coupled nonlinear partial differential equations that cannot be solved in closed form. However, in some cases one may again invoke symmetry arguments to try to single out symmetric structures, which are likely to have lower energy than a completely unsymmetric structure (for the reasoning behind this, we refer back to Section 6.2). One such case is a single vortex in an infinitely extended medium. As shown by Salomaa and Volovik (1983b, 1985a, 1987) and Balinskii *et al.* (1984), a symmetry classification of vortex structures within one topological class can be performed as in the case of the classification of uniform equilibrium order-parameter states for the bulk phases. For this, one needs to investigate the structure of the symmetry group associated with a vortex and its possible subgroups.

Unfortunately, there is no *guarantee* that highly symmetric (or even the most symmetric) order-parameter structures are energetically favoured. This we already saw in the case of the symmetry classification of the bulk phases: the inert states, being represented by maximal subgroups, were always found to be extrema, but not necessarily minima, of the free energy (see the discussion on symmetry and the stationary points of the free energy in Section 6.2.3). The symmetry classification of vortex states is even more subtle, since we are no longer investigating homogeneous phases but have to include boundary conditions etc. In fact, we shall see later that in many cases the most symmetric vortex states given by the maximal subgroups of the relevant symmetry group are not the ones with lower energy but that somewhat less symmetric states are energetically favoured. Nevertheless, a group-theoretical classification of vortices is a convenient and useful approach, since it allows for a systematic investigation of vortices with respect to different symmetry classes.

Following Salomaa and Volovik (1987), the group of symmetry operations that leave the order parameter of an axially symmetric vortex in superfluid ^{3}He invariant is given by

$$G_v = \mathrm{D}_{\infty h} \times \mathrm{t}^z \times \mathscr{T} \times \mathrm{U}(1). \tag{7.76}$$

Here $\mathrm{D}_{\infty h}$ comprises the group of rotations about a given axis $\hat{z}$, rotations about any axis perpendicular to $\hat{z}$ through angle π, and space inversion. The group t^z describes translations along $\hat{z}$, and $\mathscr{T}$ is the time-inversion symmetry group. Their elements t^z and ℓ imply the following transformations of the order-parameter matrix $\boldsymbol{A}$:

$$\mathrm{t}^z \boldsymbol{A}(\boldsymbol{r}) = \boldsymbol{A}(\boldsymbol{r} - z), \tag{7.77a}$$

$$\ell \boldsymbol{A} = \boldsymbol{A}^*. \tag{7.77b}$$

(Note that $\mathscr{T}$ is broken in the case of magnetic ordering.) The group U(1) in (7.76) represents the usual gauge group whose generator ϕ acts on $\boldsymbol{A}$ according to (6.18). Superfluidity is connected with a broken U(1) symmetry.

For a symmetry classification of vortex states based on the group G_v, one may now proceed as in the case of the homogeneous bulk phases (see

Section 6.2). To this end, one has to find the subgroups $H \subset G_v$ by breaking the symmetry contained in G_v. Of particular interest are the maximal subgroups H of G_v, which describe those states with the highest remaining symmetry.

In the present case there are two continuous subgroups, whose generators are given by

$$\mathsf{T}_1 = \mathsf{J}^z - n\phi, \quad n = 0, \pm 1, \pm 2, \ldots, \tag{7.78a}$$

$$\mathsf{T}_2 = \mathsf{p}_z - q\phi. \tag{7.78b}$$

Here J^z and $\mathsf{p}_z = -\mathrm{i}\,\partial/\partial z$ are respectively the generators of a rotation about $\hat{z}$ and a translation along $\hat{z}$ and q is a real number. To find the order parameter that is invariant under these symmetries, the equations

$$\mathsf{T}_1 \boldsymbol{A} = 0, \tag{7.79a}$$

$$\mathsf{T}_2 \boldsymbol{A} = 0 \tag{7.79b}$$

have to be solved. The first symmetry, (7.78a), consists of a combination of a rotation and a phase transformation. The integer n may be viewed as the eigenvalue of the total angular momentum of the vortex. Equation (7.79a) implies that the rotation only leads to an overall phase factor, which is then compensated by the gauge transformation. In the second case, (7.78b), a translation *along* the symmetry axis produces a phase factor, which is then compensated by the gauge transformation, such that (7.79b) is valid. In this situation the order parameter varies periodically along $\hat{z}$.

Apart from the continuous symmetry operations described by (7.78a,b), the order parameter may also possess discrete symmetries, constructed from the discrete symmetry elements of G_v, i.e. time inversion ℓ, space parity P, and rotations $\mathsf{O}^{(J)}_{x,\pi}$, by π around any perpendicular axis $\hat{x}$. We note that ℓ and $\mathsf{O}^{(J)}_{x,\pi}$ cannot be symmetries of a vortex with finite circulation since they reverse the sense of circulation. They may, however, appear in combination. The following operators represent one possible choice (Salomaa and Volovik 1983b, 1985b):

$$\mathsf{P}_1 = \mathsf{P} \quad \text{or} \quad \mathsf{P}\mathsf{U}_\pi, \tag{7.80a}$$

$$\mathsf{P}_3 = \ell\mathsf{O}^{(J)}_{x,\pi} \quad \text{or} \quad \ell\mathsf{O}^{(J)}_{x,\pi}\mathsf{U}_\pi, \tag{7.80b}$$

$$\mathsf{P}_2 = \mathsf{P}_3\mathsf{P}_1, \tag{7.80c}$$

where U_π is a gauge transformation by π, i.e. multiplication by $e^{i\pi}$ (hence $\mathsf{U}_\pi^2 = 1$). Which one of the two possibilities given in (7.80a,b) actually applies is determined by the transformation properties of the order parameter in the asymptotic region far away from the vortex axis.

7.5.6 Axisymmetric vortices

Let us now consider vortex structures invariant under the operation T_1, (7.78a), which implies axial symmetry (we assume translational invariance along the vortex axis). The order parameter $A_{\mu j}$ must then obey (7.79a).

The general structure of the solution of the order parameter is obtained as a superposition of products of the eigenfunctions of the operators contributing to T_1. In particular, the rotation operator J^z is composed of three parts

$$\mathsf{J}^z = \mathsf{L}^z_{\mathrm{cm}} + \mathsf{L}^z + \mathsf{S}^z. \tag{7.81}$$

The first term on the right-hand side,

$$\mathsf{L}^z_{\mathrm{cm}} = \frac{\hbar}{\mathrm{i}}\frac{\partial}{\partial \varphi}, \tag{7.82}$$

is the generator of centre-of-mass rotations around the z axis, where φ is the azimuthal angle in a cylindrical coordinate system. The other two terms, L^z and S^z, are the z components of the (intrinsic) orbital-angular-momentum and spin operators respectively, given by (6.17a) and (6.17b). (Note that in ^{3}He-B J^z actually has the form $\mathsf{J}^z = \mathsf{L}^z_{\mathrm{cm}} + \mathsf{L}^z + \sum_\mu R_{\mu z}\mathsf{S}^\mu$; for (7.81) to be correct, the coordinate system in spin space has to be chosen appropriately, i.e. it must be rotated by $R_{\mu j}$ with respect to the frame of reference (Salomaa and Volovik 1985a).)

The eigenfunctions of the operators in (7.81) are defined by (see also the discussion below (6.21) and (6.29))

$$\mathsf{L}^z_{\mathrm{cm}}\,|X_n\rangle = \hbar n\,|X_n\rangle, \quad n \text{ integer}, \tag{7.83a}$$

$$\mathsf{L}^z\,|Y_{1,m}\rangle = \hbar m\,|Y_{1,m}\rangle, \quad m = 0, \pm 1, \tag{7.83b}$$

$$\mathsf{S}^z\,|\chi_{1,\mu}\rangle = \hbar \mu\,|\chi_{1,\mu}\rangle, \quad \mu = 0, \pm 1, \tag{7.83c}$$

and are given explicitly in the appropriate representations in position space, momentum space and spin space as

$$\langle \varphi \mid X_n\rangle = \mathrm{e}^{\mathrm{i}n\varphi}, \tag{7.84a}$$

$$\langle \hat{\boldsymbol{k}} \mid Y_{1,m}\rangle = \begin{cases} \hat{k}_z & (m=0), \\ 2^{-1/2}(\hat{k}_x \pm \mathrm{i}\hat{k}_y) & (m = \pm 1), \end{cases} \tag{7.84b}$$

$$\langle \alpha\beta \mid \chi_{1,\mu}\rangle = \begin{cases} \mathrm{i}(\sigma_2\sigma_3)_{\alpha\beta} & (\mu = 0), \\ 2^{-1/2}\mathrm{i}[\sigma_2(\sigma_1 \pm \mathrm{i}\sigma_2)]_{\alpha\beta} & (\mu = \pm 1), \end{cases} \tag{7.84c}$$

where σ_i are the Pauli matrices.

It is convenient to parametrize the order parameter in terms of these eigenfunctions as

$$\Delta_{\alpha\beta}(\hat{\boldsymbol{k}}; \boldsymbol{r}) = \sum_{\mu m} a_{\mu m}(\boldsymbol{r})\langle \alpha\beta \mid \chi_{1,\mu}\rangle\langle \hat{\boldsymbol{k}} \mid Y_{1,m}\rangle. \tag{7.85}$$

The matrix components $a_{\mu m}$ of a vortex carrying angular momentum $\mathsf{J}^z = \hbar n$ then take the form

$$a_{\mu m}(\boldsymbol{r}) = \Delta_0 C_{\mu m}(r)\mathrm{e}^{\mathrm{i}(n-\mu-m)\varphi}, \tag{7.86}$$

where (r, z, φ) are cylindrical coordinates, with r the distance from the axis and Δ_0 is the maximum value of the gap. Any component $C_{\mu m}$ for which

$n - \mu - m \neq 0$ has to vanish inside a normal core of extension ξ in order to avoid the singularity at the origin. This is the familiar behaviour of the order parameter inside a vortex, known for example from superfluid ^{4}He. Here, on the other hand, the components $C_{\mu m}$ for which $n - \mu - m = 0$ need *not* vanish, so that the core may actually be superfluid—albeit of a different type.

The restrictions imposed on $C_{\mu m}$ in the presence of one or several of the discrete symmetries P_1, P_2, P_3 (see (7.80)) are found as follows. Under the parity transformation P, the orbital wave function $Y_{1,m}$ changes sign, and the azimuthal angle transforms as $\varphi \to \varphi + \pi$. Consequently,

$$\mathsf{P}C_{\mu m} = (-1)^{\mu+m-n-1}C_{\mu m}, \qquad (7.87)$$

implying that, if the symmetry operation P_1 is given by $\mathsf{P}_1 = \mathsf{P}\mathsf{U}_\pi^{n+1}$, the vortex structure with $\mathsf{P}_1 C_{\mu m} = C_{\mu m}$ is characterized by the five components with $\mu + m$ even, while the remaining four components vanish. Similarly, in the case with $\mathsf{P}_1 = \mathsf{P}\mathsf{U}_\pi^n$, $C_{\mu m} = 0$ for $\mu + m$ even, and hence four components are nonzero.

The resulting order-parameter structures are nonunitary. Consequently, they possess a nonzero value of $\boldsymbol{d} \times \boldsymbol{d}^*$ (see (3.37)), which gives rise to a spontaneous magnetization density in the vortex core region (Salomaa and Volovik 1983b, Passvogel *et al.* 1984).

The situation is different if the P_1 symmetry conflicts with underlying topological constraints. For example, the $N = \frac{1}{2}$ vortices and the continuous vortices in the A phase do not allow for a P_1 symmetry and must therefore be described by a larger number of components.

The effect of the time-reversal symmetry $\mathcal{T}$ on the order parameter is expressed by

$$t\{C_{\mu m}\mathrm{e}^{\mathrm{i}(n-\mu-m)\varphi}\} = (-1)^{\mu+m}C^*_{-\mu,-m}\mathrm{e}^{-\mathrm{i}(n-\mu-m)\varphi}, \qquad (7.88)$$

where we have used $Y^*_{1,m} = (-1)^m\, Y_{1,-m}$ etc. The simultaneous rotation by π about the x axis in orbital and spin space affects $C_{\mu m}$ as

$$\mathsf{O}^{(J)}_{x,\pi}\{C_{\mu m}\mathrm{e}^{\mathrm{i}(n-\mu-m)\varphi}\} = C_{-\mu,-m}\mathrm{e}^{-\mathrm{i}(n+\mu+m)\varphi}. \qquad (7.89)$$

Combining (7.87)–(7.89), one therefore obtains

$$\mathsf{P}_2 C_{\mu m} = C^*_{\mu m}, \qquad (7.90)$$

$$\mathsf{P}_3 C_{\mu m} = (-1)^{\mu+m}C^*_{\mu m}. \qquad (7.91)$$

The most symmetric vortices are thus characterized by four or five real components $C_{\mu m}$, depending on the asymptotics.

For example in the B phase far from the vortex core, one has $C_{+-} = C_{00} = C_{-+}$, where the components share the common phase factor $\mathrm{e}^{in\varphi}$. The appropriate definition of P_1 compatible with the above structure is $\mathsf{P}_1 = \mathsf{P}\mathsf{U}_\pi{}^{n+1}$. Hence the most symmetric vortex in the B phase is characterized by *five* real components $C_{\mu m}$, $\mu + m$ odd (Ohmi *et al.* 1983). It may be shown that this vortex carries n quanta of circulation, which is equal to the angular-momentum quantum number. This is necessarily so because the B phase does not possess an intrinsic angular momentum.

In the A phase the situation is different, as we shall see in the next subsection. There, the number of circulation quanta n' is related to n, (7.83a), by (Fujita and Tsuneto 1975; see also Fujita 1975)

$$n = n' + \hat{z} \cdot \hat{l}(\infty). \tag{7.92}$$

Thus the intrinsic angular momentum $\hat{l}$ adds to the external one.

In summary, using the discrete symmetry operations defined above, one may identify four types of axisymmetric vortex:

(i) o-vortices; these have all the discrete symmetries, P_1, P_2 and P_3;
(ii) v-vortices; these are vortex structures with broken P_1 symmetry, but with the P_2 symmetry still intact;
(iii) w-vortices; the only residual symmetry is P_3;
(iv) uvw-vortices; these are axisymmetric vortices without additional discrete symmetries.

We shall now take a closer look at some specific vortex structures in both the A and B phases. For this, it is necessary to determine their respective energies.

7.5.7 Energetics of vortices

The energy of any given vortex is given by the sum of the bulk and the gradient energy, supplemented by orientational energies if necessary. Hence it may be calculated by substituting the corresponding order parameter into the expressions for the free energies

$$F_{\text{bulk}} = \int d^3r\, f, \tag{7.93a}$$

$$F_{\text{grad}} = \int d^3r\, f_{\text{grad}} \tag{7.93b}$$

given by the space integrals of the respective free-energy densities f, (5.4), and f_{grad}, (7.17), respectively, and subtracting the energy of the uniform state. In the case of vortices without a normal core it is possible to use the much simpler *bending* free energy

$$F_{\text{bend}} = \int d^3r\, f_{\text{bend}} \tag{7.93c}$$

instead of the *gradient* free energy, since here only the hydrodynamic components of the order parameter change in space.

An important special case is that of axially symmetric vortices. As discussed in the preceding subsections, the order parameter is then described by the components $C_{\mu m}$, which depend only on the distance r from the vortex axis. In this situation it is useful to employ cylindrical coordinates and to scale the distance from the vortex centre in units of a characteristic length $\tilde{\xi}(T)=5^{-1/2}\xi(T)$, see (7.18c), by introducing the dimensionless length variable $\tilde{r}=r/\tilde{\xi}$. Expressing (5.4) and (7.17) in terms of $C_{\mu m}$, the free energy per unit length of a vortex with quantum number n, (7.83a), in the Ginzburg–Landau regime, $\mathscr{F}$, then follows as (Fetter 1986, Salomaa and Volovik 1987)

$$\mathscr{F}=\frac{\pi}{5}\left(\frac{\hbar}{2m}\right)^2\rho_s\int_0^\infty d\tilde{r}\,\tilde{r}(\tilde{f}+\tilde{f}_{\text{grad}})+\mathscr{F}_{\text{orient}}, \tag{7.94}$$

where

$$\begin{aligned}\tilde{f}=\sum_{\mu m}\Big[&-C_{\mu m}C^*_{\mu m}+b_1\,|C^*_{\mu m}C^*_{-\mu,-m}|^2+b_2(C_{\mu m}C^*_{\mu m})^2\\ &+\sum_{\nu,n}(b_3C_{\mu m}C^*_{\mu n}C_{\nu,-m}C^*_{\nu,-n}\\ &+b_4C^*_{\mu m}C_{\nu m}C^*_{\nu n}C_{\mu n}+b_5C_{\mu m}C_{-\mu n}C^*_{\nu m}C^*_{-\nu n})\Big],\end{aligned} \tag{7.95}$$

$$\begin{aligned}\tilde{f}_{\text{grad}}=\sum_{\mu m k}&\left[\left(|m|\frac{\partial}{\partial\tilde{r}}C_{\mu m}-m\frac{n-\mu-m}{\tilde{r}}C_{\mu m}\right)\right.\\ &\left.\times\left(|k|\frac{\partial}{\partial\tilde{r}}C^*_{\mu k}-k\frac{n-\mu-k}{\tilde{r}}C^*_{\mu k}\right)\right]\\ &+\sum_{\mu m}\left[\left|\frac{\partial}{\partial\tilde{r}}C_{\mu m}\right|^2+\left(\frac{n-\mu-m}{\tilde{r}}\right)^2|C_{\mu m}|^2\right].\end{aligned} \tag{7.96}$$

Here $\mathscr{F}_{\text{orient}}$ is the integral over the relevant orientational energy densities (due to dipole interaction, magnetic field etc.) in the (φ, r) plane. Furthermore,

$$b_i\equiv\frac{1}{6\kappa}\frac{\beta_i}{\beta_0}, \tag{7.97}$$

where β_i are the β parameters of the free-energy density (5.4), and κ and β_0 are defined in (5.11) and (3.56) respectively. The order-parameter components $C_{\mu m}$ in (7.95) and (7.96), defined in (7.86), are dimensionless. The prefactor ρ_s in (7.94) is a superfluid density in the Ginzburg–Landau regime:

$$\rho_s=\frac{7\zeta(3)}{4\pi^2}\frac{m}{m^*}\left(\frac{\Delta_0}{k_B T_c}\right)^2\rho. \tag{7.98}$$

The minimization of $\mathscr{F}$ with respect to the components $C_{\mu m}$ generally leads to a system of nine coupled Euler–Lagrange equations, each a nonlinear second-order ordinary differential equation. These equations may only be solved numerically. An alternative approach, requiring less computational effort, is to employ a variational method. For this, one chooses a set of test functions for the $C_{\mu m}(\bar{r})$, which depend on a number of variational parameters λ_i, substitutes these into the functional and calculates the minimum of the resulting free energy $\mathscr{F}_{\text{var}}(\{\lambda_i\})$ with respect to variations of λ_i. Both methods have been employed to determine the $C_{\mu m}$ for those vortex structures considered in the symmetry classification scheme.

In the case of *nonaxisymmetric* vortices the energy is calculated from the general expressions for the free-energy densities in the A and B phases, i.e. the bulk free-energy density (5.4) and the gradient free-energy density (7.17).

7.5.8 Vortices in the A phase

Axisymmetric vortices

According to the above classification of axisymmetric vortices, there exist four different types, which may be distinguished by the remaining discrete symmetries.

(*i*) *o-vortex.* This is the most symmetric vortex, since it possesses all the discrete symmetries P_1, P_2 and P_3. The symmetry P_1 requires $\hat{\boldsymbol{l}}$ and $\hat{\boldsymbol{d}}$ to be uniform, oriented along the vortex axis far from the core region. In this case one has a pure phase vortex. If we consider a vortex whose orbital angular momentum in the asymptotic region (where only C_{0+} and C_{0-} are nonvanishing) is characterized by the quantum number n, (7.83a), (7.84a), then a parity operation implies

$$\mathsf{P}C_{0\pm} = \mathrm{e}^{in\pi}C_{0\pm}. \tag{7.99}$$

Therefore we shall require $C_{\mu m}$ to be invariant under the operation $\mathsf{P}_1 = \mathsf{P}(\mathsf{U}_\pi)^n$. This would leave us with four real components $C_{\mu m}$ for this type of vortex. However, in the region $r < \xi_{\mathrm{D}}$, where the dipole coupling may be neglected, the symmetries in spin and orbital spaces in (7.81) are also uncoupled, so that the most symmetric vortex has two *separate* continuous symmetries with the generators: $\mathsf{L}^z_{\text{cm}} + \mathsf{L}^z - n\phi$ and S^z. The latter implies $\hat{\boldsymbol{d}} \parallel \hat{z}$, which leaves us with only *two* components: C_{0+} and C_{0-}. The general expression for the order parameter of this vortex structure is given by (Salomaa and Volovik 1985b)

$$A_{\mu j} = 2^{-1/2}\hat{z}_\mu[C_{0+}(r)(\hat{x}_j + \mathrm{i}\hat{y}_j)\mathrm{e}^{\mathrm{i}(n-1)\varphi} + C_{0-}(r)(\hat{x}_j - \mathrm{i}\hat{y}_j)\mathrm{e}^{\mathrm{i}(n+1)\varphi}]. \tag{7.100}$$

The two possible boundary conditions at infinity are

(i) $C_{0+}(\infty) = 1$, $C_{0-}(\infty) = 0$ for the vortex with $n' = n - 1$ quanta of circulation and with asymptotic orientation $\hat{\boldsymbol{l}}(\infty) = \hat{z}$; and
(ii) the vortex with $n' = n + 1$ quanta and $\hat{\boldsymbol{l}}(\infty) = -\hat{z}$ and where C_{0+} and C_{0-} are interchanged.

These vortices have a normal core except for the cases $n' = -2$, $n = -1$ and $n' = 2$, $n = 1$.

Vortices of the above type belong to the classes with topological quantum numbers $N = 0, 1$. They are not allowed within the class of $N = \frac{1}{2}$ vortices, because these vortices necessarily involve a nonuniform $\hat{\boldsymbol{d}}$ field, which is incompatible with the P_1 symmetry. However, even within the $N = 0, 1$ classes, pure phase vortices as given by (7.100) are energetically unstable. In the case $N = 0$ this is evident, since in this case a continuous deformation into a nonsingular structure lowers the vortex energy considerably.

(*ii*) *v-vortex.* Here the P_1 symmetry is broken. A classification according to the residual (i.e. P_2) symmetry has been performed by Salomaa and Volovik (1983b) and Seppälä *et al.* (1984). An example of a nonsingular v-vortex of the AT type (see (7.71)) is given by

$$\hat{\boldsymbol{l}} = \sin\eta(r)\,\hat{\boldsymbol{z}} \pm \cos\eta(r)\,\hat{\boldsymbol{r}}, \tag{7.101a}$$

$$\boldsymbol{v}_{\mathrm{s}} = -\frac{\hbar}{2mr}[1 + \sin\eta(r)]\hat{\boldsymbol{\varphi}}, \tag{7.101b}$$

where $\eta(0) = -\frac{1}{2}\pi$ and $\eta(\infty) = \frac{1}{2}\pi$. The vortex carries two quanta of circulation. The fact that in the above state the P_1 symmetry is spontaneously broken leads to the existence of two degenerate states (indicated by the $\pm$ signs), which transform into each other by a P_1 operation.

(*iii*) *w-vortex.* A nonsingular vortex of this type, where P_3 is the remaining symmetry, is given by the twisted $\hat{\boldsymbol{l}}$ texture

$$\hat{\boldsymbol{l}} = \sin\eta(r)\,\hat{\boldsymbol{z}} \pm \cos\eta(r)\,\hat{\boldsymbol{\varphi}}, \tag{7.102a}$$

$$\boldsymbol{v}_{\mathrm{s}} = -\frac{\hbar}{2mr}[1 + \sin\eta(r)]\hat{\boldsymbol{\varphi}}, \tag{7.102b}$$

with $\eta(0) = -\frac{1}{2}\pi$, $\eta(\infty) = \frac{1}{2}\pi$.

(*iv*) *uvw-vortex.* An example of a nonsingular texture of this type where all discrete symmetries are broken is

$$\hat{\boldsymbol{l}} = \sin\eta(r)\,\hat{\boldsymbol{z}} \pm \cos\eta(r)[\cos\alpha(r)\,\hat{\boldsymbol{\varphi}} \pm \sin\alpha(r)\,\hat{\boldsymbol{r}}], \tag{7.103a}$$

$$\boldsymbol{v}_{\mathrm{s}} = -\frac{\hbar}{2mr}[1 + \sin\eta(r)]\hat{\boldsymbol{\varphi}}, \tag{7.103b}$$

with the same boundary conditions on $\eta(r)$ as in cases (ii) and (iii). The vortex state is fourfold-degenerate; the states transform into each other under the action of P_1, P_2 and P_3, which form the group $\mathbb{Z}_2 \times \mathbb{Z}_2$.

Nonaxisymmetric vortices

In spite of their high symmetry, the axisymmetric vortices in ^{3}He-A do *not* have the lowest energy and therefore are not stable. Hence the symmetry classification approach is not very helpful here. To calculate the energy of vortices in the A phase in the London limit (see above (7.19)), one has to consider the free-energy density composed of the bending free-energy density (7.20), and the energy densities (6.104) and (6.115), due to the dipole interaction and an external magnetic field respectively.

As first noticed by Volovik and Mineev (1977a), the pure phase vortex with $\hat{\boldsymbol{d}}=\hat{\boldsymbol{l}}=\hat{\boldsymbol{z}}$, for example, is unstable against a nonaxisymmetric deformation into a pure phase vortex with $\hat{\boldsymbol{l}}=\hat{\boldsymbol{d}}=\hat{\boldsymbol{x}}$ owing to the anisotropy of the superfluid density tensor. In fact, the energies per unit length, $\mathscr{F}_{\mathrm{pv}}(n, \hat{\boldsymbol{d}}, \hat{\boldsymbol{l}})$, of the pure *p*hase *v*ortices (suffix "pv") with n quanta of circulation (see e.g. (7.65) with $\gamma=-n\varphi$) in the dipole-locked region $\xi_{\mathrm{D}}<r<R$ (here R is the distance between vortices or the size of the container—whichever is smaller) are found from (7.20) as

$$\mathscr{F}_{\mathrm{pv}}(n, \hat{\boldsymbol{d}}=\hat{\boldsymbol{l}}=\hat{\boldsymbol{z}})=n^2\rho_{\mathrm{s},\perp}k_0, \tag{7.104a}$$

$$\mathscr{F}_{\mathrm{pv}}(n, \hat{\boldsymbol{d}}=\hat{\boldsymbol{l}}=\hat{\boldsymbol{x}})=\tfrac{1}{2}n^2(\rho_{\mathrm{s},\parallel}+\rho_{\mathrm{s},\perp})k_0, \tag{7.104b}$$

where $k_0=\frac{1}{2}(\hbar/2m)^2\ln(R/\xi_{\mathrm{D}})$. Since the components of the superfluid density tensor parallel and perpendicular to $\hat{\boldsymbol{l}}$ are related by $\rho_{\mathrm{s},\parallel}<\rho_{\mathrm{s},\perp}$ the energy of the nonaxisymmetric vortex is seen to be lower than the energy of the o-vortex (Volovik and Hakonen 1981). The reason is that superflow tends to align the vector $\hat{\boldsymbol{l}}$, and hence an $\hat{\boldsymbol{l}}$ configuration perpendicular to the axis (in the plane of the flow) is preferable.

In the soft-core region $r<\xi_{\mathrm{D}}$ the $\hat{\boldsymbol{d}}$-vector field is decoupled from the $\hat{\boldsymbol{l}}$ texture and may assume a uniform configuration even for highly nonuniform $\hat{\boldsymbol{l}}$ textures. On the other hand, the gradient energy of the orbital part of the order parameter may be lowered by allowing the vorticity to be carried over from the phase field φ to the $\hat{\boldsymbol{l}}$ texture in the form of a disgyration (see e.g. (7.73) with $\gamma=-n\varphi=0$). The energies per unit length, $\mathscr{F}_{\mathrm{dis}}(n=0, \hat{\boldsymbol{d}}, \hat{\boldsymbol{l}})$, of the disgyrations with $\hat{\boldsymbol{l}}=\hat{\boldsymbol{r}}$ and $\hat{\boldsymbol{l}}=\hat{\boldsymbol{\varphi}}$, i.e. for singly quantized vortices, are given by

$$\mathscr{F}_{\mathrm{dis}}(n=0, \hat{\boldsymbol{d}}=\mathrm{const}, \hat{\boldsymbol{l}}=\hat{\boldsymbol{r}})=K_{\mathrm{s}}k_2, \tag{7.105a}$$

$$\mathscr{F}_{\mathrm{dis}}(n=0, \hat{\boldsymbol{d}}=\mathrm{const}, \hat{\boldsymbol{l}}=\hat{\boldsymbol{\varphi}})=K_{\mathrm{b}}k_2, \tag{7.105b}$$

where K_{s} and K_{b} are the stiffness parameters in the expression for the bending free energy (7.20) and $k_2=\frac{1}{2}(\hbar/2m)^2\ln(\xi_{\mathrm{D}}/\xi)$. In the Ginzburg–Landau regime K_{s} and K_{b} are given by (7.22c); since $K_{\mathrm{s}}=\frac{1}{3}K_{\mathrm{b}}$, the radial disgyration costs only one-third of the energy of the nonaxisymmetric phase vortex. The broken axisymmetry of the singly quantized vortex is hence restored inside the soft core. As shown by Muzikar (1978, 1979), the radial disgyration changes continuously into a polar state in the core regime $r\approx\xi$, before the liquid goes normal inside the hard core.

The (singular) vortices discussed above belong to the topological class with quantum number $N=1$. Let us now turn to the class with $N=0$, containing vortices without a normal core. At first one might think that these vortex structures should be lower in energy, since they do not have a normal core. However, the texture required to avoid the normal core is rather complicated and thus implies an increase in bending energy. Even in the most favourable case (the nonaxisymmetric AT texture; $\hat{\boldsymbol{l}} = \hat{\boldsymbol{x}}$) this leads to an energy per unit circulation quantum that is *twice* as high as the bending energy of the nonaxisymmetric phase vortex outside the soft core. Were it not for the energy of the soft-core region, the singular vortex would always be lower in energy. The energies per unit of circulation outside the soft-core region ($r > \max(\xi_D, \xi_H)$) of (i) the MH texture with $\hat{\boldsymbol{l}} = \hat{\boldsymbol{r}}$, (ii) the MH texture with $\hat{\boldsymbol{l}} = \hat{\boldsymbol{\varphi}}$, and (iii) the nonaxisymmetric doubly quantized AT texture, which are all nonsingular textures, have a ratio of 4.5:3.5:3. (Salomaa and Volovik 1987).

Approximate numerical calculations show that the singular non-axisymmetric vortex has a slightly lower energy than the AT texture for vortex separations $r_\Omega > r_{\Omega_c}$, i.e. for low rotation speeds $\Omega \leqslant \Omega_c$ (see Section 7.6 on the rotating system) (Seppälä and Volovik 1983, Vulovic *et al.* 1984).

The simplest representation of *doubly* quantized nonsingular vortices in large magnetic fields ($\xi_H \leqslant \xi_D$), thought to be stable for $\Omega > \Omega_c$, is the w-vortex characterized by

$$\hat{\boldsymbol{d}} = \hat{\boldsymbol{y}}, \quad \hat{\boldsymbol{l}} = \sin\eta(r)\,\hat{\boldsymbol{y}} \pm \cos\eta(r)\,[\sin\varphi\,\hat{\boldsymbol{x}} - \cos\varphi\,\hat{\boldsymbol{z}}], \tag{7.106a}$$

$$\hat{\boldsymbol{m}} + \mathrm{i}\hat{\boldsymbol{n}} = \mathrm{e}^{-\mathrm{i}\gamma}[\sin\eta(r)\,(-\sin\varphi\,\hat{\boldsymbol{x}} + \cos\varphi\,\hat{\boldsymbol{z}}) \pm \cos\eta(r)\,\hat{\boldsymbol{y}} + \mathrm{i}(\sin\varphi\,\hat{\boldsymbol{z}} + \cos\varphi\,\hat{\boldsymbol{x}})], \tag{7.106b}$$

$$\boldsymbol{v}_s = \frac{\hbar}{2mr}[1 + \sin\eta(r)]\hat{\boldsymbol{\varphi}}, \tag{7.106c}$$

with $\eta(r \ll \xi_D) = -\frac{1}{2}\pi$; $\eta(r \gg \xi_D) = \frac{1}{2}\pi$ and $\hat{\boldsymbol{l}}(\infty) = \hat{\boldsymbol{y}}$. A v-vortex of similar structure is also possible. A calculation of the energy of these vortices give preference to the w-vortex (7.106) (Seppälä *et al.* 1984, Zotos and Maki 1984, Maki and Zotos 1985a).

Consequences of the broken discrete symmetries

The spontaneous breakdown of discrete symmetries in the soft core of the ^{3}He-A vortices is associated with the appearance of (i) a spontaneous electric polarization D_z along the vortex axis (in the case of broken P_1 and P_3 symmetries), (ii) an axial supercurrent (broken P_1 and P_2 symmetries) and (iii) an axial spin supercurrent (broken P_1 and P_3 symmetries) (Volovik 1984a). Thus an axial supercurrent is flowing in the w-vortex (7.106). On the other hand, a v-vortex may carry electric polarization via a polarization of the ^{3}He atomic shells. In particular, at the boundary where one of the two doubly degenerate solutions (see (7.101)) having opposite orientations of the electric dipole moment switches over into the other, there resides a charge e^*, which has been estimated as $e^* \approx 10^{-5}e$. The electric polarization as well as the axial spin current may lead to observable effects if the interaction between the vortices leads to an ordering of the usually randomly distributed orientations.

The energy of an array of w-vortices (7.106) may be lowered by forming a helical arrangement of the preferred direction in the (x, y) plane along the axis of the vortex (Salomaa and Volovik 1987), i.e. the $\hat{\boldsymbol{l}}$ texture outside the vortex core is described by

$$\hat{\boldsymbol{l}}_{\text{bulk}} = \sin(qz)\,\hat{\boldsymbol{x}} + \cos(qz)\,\hat{\boldsymbol{y}}. \tag{7.107}$$

Again the interaction between the vortices forms a "ferromagnetic" arrangement, i.e. the vortices are all of the same type (+ or − sign in (7.106)). The existence of this structure would also be reflected in the dispersion relation for "Kelvin waves", i.e. the wave-like excitation of a vortex along its axis (Sonin and Fomin 1985).

Half-quantum vortices

As discussed in Section 7.5.3, vortices in the topological class $N = \frac{1}{2}$ carry half a quantum of circulation. Their existence is made possible because the $\hat{\boldsymbol{d}}$-vector field compensates for the missing half quantum. A different, but somewhat simpler, interpretation of the half-quantum vortex can be given in the weak-coupling limit, where the subsystems of up and down spins are uncoupled: it corresponds to a vortex with a single unit of vorticity for one of the spin components and zero vorticity for the other component (Leggett 1989). The corresponding textures require $\hat{\boldsymbol{d}}$ and $\hat{\boldsymbol{l}}$ to be nonparallel over a region of macroscopic size; as a consequence of the dipole interaction, this region is confined to a domain wall bounded by the half-quantum vortex.

The half-quantum vortices are very costly in dipole energy. The domain wall may be avoided altogether in a restricted geometry, for example a narrow slab with spacing $d < \xi_D$ between the parallel plates (Salomaa and Volovik 1985c, Volovik and Salomaa 1985b; see also Maki 1986b).

The interaction with the walls orients $\hat{\boldsymbol{l}}$ parallel to the surface normal, whereas application of a magnetic field of strength $H^* \approx 30$ G parallel to the surface normal forces the $\hat{\boldsymbol{d}}$ texture into the (x, y) plane. In this way, the dipole energy is fixed at its maximum value. With these boundary conditions, a half-quantum vortex (7.74) is possible. Estimates of the energies show that the half-quantum vortex can compete with the single-quantum phase vortex, but the energies are very close.

Another configuration in which half-quantum vortices may be stabilized is that of pairs of vortices with opposite values of the period of the $\hat{\boldsymbol{d}}$ field (i.e. with $k' = 0$ and $k' = 1$ in (7.74)). Since these are connected by a domain wall, their energy is proportional to their linear separation R, just as in high-energy physics, where quarks are bound by gluons. On the other hand, the bending energy prevents the vortices from coming too close together; therefore there exists an equilibrium distance R_0, which has been calculated by Salomaa and Volovik (1985c).

Yet another consequence of the existence of half-quantum vortices is an increased attentuation of certain order-parameter collective modes, to be discussed in detail in Chapter 11. One of these modes, the so-called "clapping" mode, represents an oscillation of the $\hat{\boldsymbol{m}}$ and $\hat{\boldsymbol{n}}$ vectors. In the presence of a vortex of circulation n' the amplitude of the clapping mode obeys the equation (Khazan 1985)

$$\frac{\partial^2 \psi}{\partial t^2} = c^2 \left(\frac{1}{\mathrm{i}} \boldsymbol{\nabla} - \boldsymbol{A} \right)^2 \psi - (\omega_0^2 + U)\psi, \tag{7.108}$$

where ω_0 and c are the energy gap and the velocity of this mode and U and $\boldsymbol{A}$ are external potentials generated by the vortex. The "vector potential" $\boldsymbol{A}$ is given by

$$\boldsymbol{A} = \frac{n'}{r} \hat{\boldsymbol{\varphi}}. \tag{7.109}$$

Hence the motion of the collective boson is similar to that of a charged particle in a magnetic field $\boldsymbol{H} = \boldsymbol{\nabla} \times \boldsymbol{A} = 2\pi n' \hat{\boldsymbol{z}}\, \delta^{(2)}(\boldsymbol{r})$, confined to the vortex core where $\delta^{(2)}(\boldsymbol{r})$ is the two-dimensional delta function. The familiar Aharonov–Bohm interference effects (Aharonov and Bohm 1959) lead to an additional scattering of the collective bosons in the case of half-integer quantum number n', which should be observable in the width of the corresponding collective mode peak of the ultrasound absorption (see Section 11.4).

7.5.9 Vortices in the B phase

Axisymmetric vortices

The most symmetric vortex consistent with the asymptotic boundary condition $a_{+-}(\infty) = a_{00}(\infty) = a_{-+}(\infty) = \Delta(T)\mathrm{e}^{\mathrm{i}n\varphi}$ necessary to generate the bulk B phase far away from the core is the o-vortex. It is characterized by five real components C_{+-}, C_{00}, C_{-+}, C_{++} and C_{--}.

The vortex with $n = 1$ circulation quantum has a normal core because, as discussed below (7.86), the phase factors multiplying all the five components $C_{\mu m}$ have singularities at the vortex centre. The radial dependence of this vortex has been calculated in full generality by Ohmi *et al.* (1983) and by Salomaa and Volovik (1983b) and for the simpler version with $C_{++} = C_{--} = 0$ by Passvogel *et al.* (1983). It is found that C_{+-}, C_{00} and C_{-+} decrease monotonically as one moves towards the centre. Until $r \gtrsim 2.5\xi$ they stay at their bulk value ($r \to \infty$), while for $r \lesssim 2.5\xi$ they decrease linearly with r. C_{++} and C_{--} are rather small by comparison (see Fig. 7.18). The bulk free-energy-density contribution behaves roughly as $|C_{\mu m}|^2$, i.e. it tends to zero for $r \to 0$, whereas the gradient-free-energy contribution increases for decreasing r, reaching a maximum at $r = 0$. These properties are similar to those of the phase vortices in superfluid ^{4}He.

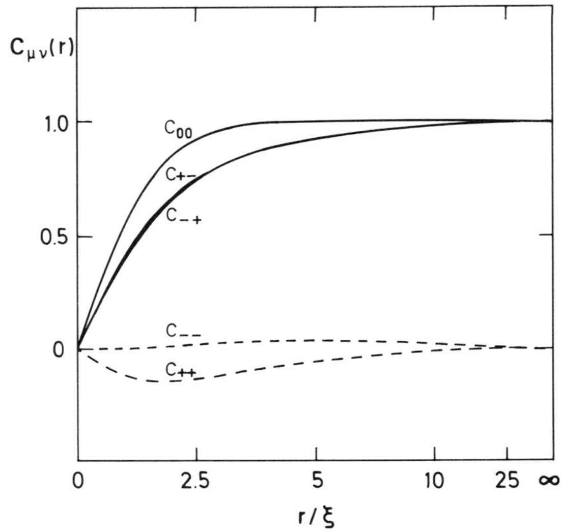

Figure 7.18 Radial dependence of the real order-parameter components characterizing the o-vortex, i.e. the most symmetric singly quantized vortex, in the B phase (weak-coupling limit). For $r/\xi \leqslant 5$ the scale is linear but goes as $(r/\xi)^{-1}$ for $r/\xi \geqslant 5$. (After Salomaa and Volovik (1983b).)

In spite of its high residual symmetry, the o-vortex is, however, probably not the vortex with the lowest energy in the region of free-energy parameters β_i where the B phase is stable. In fact, it only corresponds to a saddle point of the Ginzburg–Landau free energy and is found to be unstable with respect to a spontaneous breaking of the P_1 symmetry (with P_2 still intact), i.e. it transforms into a v-vortex (Salomaa and Volovik 1983b, 1985a, Passvogel *et al.* 1984). It should be noted, however, that the above results depend crucially on the free-energy parameters used in the calculation. In fact, using a more accurate set of parameters β_i but a variational technique to determine the vortex structure, Fetter and Theodorakis (1984) found the o-vortex to be stable at high pressures. In this context it is interesting to keep in mind that a transition from the o-structure to the v-structure would be of second order (Salomaa and Volovik 1983b).

The v-vortex is characterized by nine real components $C_{\mu m}$. As discussed above, the components with $n - \mu - m = 0$ are not multiplied by a phase factor, and hence need not vanish at $r = 0$. This implies that in the case of the $n = 1$ vortex the components C_{0+} and C_{+0} are nonzero inside the vortex core. Comparing with the results of Chapter 6, one finds that C_{0+} and C_{+0} correspond to the A phase and the β phase respectively. While the A phase is unitary, the β phase is not, i.e. it possesses a spontaneous magnetization. This has actually been observed in rotating ^{3}He-B, as will be discussed below in Section 7.6. Furthermore, the broken parity gives rise to a small electric dipole moment of the v-, w- and uvw-vortices (Salomaa and Volovik 1987). If polarized by a strong electric field, these dipole moments may give rise to peculiar orientation and flow effects.

The fact that one of the superfluid components in the core of the v-vortex has the symmetry of the A phase is of particular interest in the neighbourhood of the A–B phase transition. As may be expected, one finds that the A-phase component C_{0+} grows at the expense of the component C_{+0}, and the radius of the A-phase core increases as the A–B transition is approached (at T_{AB}, one finds $C_{+0}=0$).

It has been suggested (Salomaa and Volovik 1987) that the v-vortex acts as a nucleation centre for the formation of the A phase upon warming up from the B phase (for a discussion of the A–B phase transition, see Section 5.3). There exists a point of absolute instability for the superheated B phase, beyond which the v-vortex radius grows very large, thus effecting the B→A phase transition. The corresponding catastrophe line in the phase diagram can be estimated by assuming the simple but qualitatively correct expression for the energy per unit length of a v-vortex

$$E = \pi r_{\text{core}}^2(f_A - f_B) + 2\pi r_{\text{core}}\sigma_{AB} + \pi\rho_s\left(\frac{\hbar}{2m}\right)^2 \ln\frac{R}{r_{\text{core}}}. \tag{7.110}$$

Here the first term accounts for the energy of the A-phase core region relative to the bulk phase, the second term is the energy of the interface between the A and B regions (σ_{AB} is the A–B surface energy per unit area) and the last term is the familiar hydrodynamic energy expression for a vortex of radius R (see (7.104) and (7.143)). Minimization of (7.110) with respect to r_{core} leads to a quadratic equation for r_{core}, with real solutions provided that the condition

$$f_B - f_A \leqslant \left(\frac{2m}{\hbar}\right)^2 \frac{\sigma_{AB}^2}{2\rho_s} \tag{7.111}$$

is satisfied. As the condition is approached, the one solution for r_{core} corresponding to the minimum of E grows large, indicating that the superheated B phase surrounding the vortex becomes absolutely unstable as compared with the A phase in the core region. A somewhat similar investigation has been performed by Leggett and Yip (1989), who considered the opposite situation of nucleation of the B phase from a linear defect in the supercooled A phase. They find, however, that no instability will occur at any temperature T_{AB}.

It is interesting to note that the order parameter describing a v-vortex falls off asymptotically as $1/r$, as compared with the $1/r^2$ behaviour of a phase vortex in an isotropic superfluid (or the o-vortex) (Passvogel *et al.* 1984, Hasegawa 1985). This fact is related to the additional degeneracy of the order parameter in the present case, i.e. its covariance with respect to spin–orbit rotations described by the rotation matrix $R_{\mu j}$. One may indeed show that the minimization of the free energy with respect to $R_{\mu j}$ leads to a $1/r$ behaviour of the angle of rotation θ about the tangential direction $\hat{\boldsymbol{\varphi}}$ (Salomaa and Volovik 1987).

Nonaxisymmetric singly quantized vortices

At low pressures the coefficients in the Ginzburg–Landau theory are such that the axisymmetric v-vortex becomes unstable against breaking of the axial symmetry (Thuneberg 1986a, Salomaa and Volovik 1986a). In fact, a direct minimization of the Ginzburg–Landau functional for singly quantized vortices yields a nonaxisymmetric vortex structure that may be characterized as a v-vortex with C_2 symmetry rather than C_∞ symmetry about the vortex axis (Thuneberg 1986a, 1987a). This new vortex structure has been termed a "double-core vortex". In fact, it may be interpreted as a bound state of *two* half-quantum vortices, whose detailed structure is discussed on p. 250. It is found to be stable at pressures below about 25 bar in the Ginzburg–Landau regime. At higher pressures the v-vortex structure has lower energy. The free energies of the singular vortex (i.e. normal core), v-vortex (i.e. A-phase core) and double-core vortex as functions of pressure are shown in Fig. 7.19.

Earlier investigations of nonaxisymmetric vortices of the form

$$A_{\mu j} = \frac{\Delta(r)}{3^{1/2}\Delta(\infty)} e^{i\phi} \sum_k R_{\mu k}(\delta_{jk} - f(r)\hat{s}_j\hat{s}_k) \tag{7.112}$$

by Theodorakis and Fetter (1983) yielded a free-energy minimum for $\hat{\boldsymbol{s}} = \hat{\boldsymbol{x}}$, perpendicular to the vortex axis. The stability of axisymmetric vortices with respect to nonaxisymmetric perturbations has been investigated by Volovik and Salomaa (1985a). These authors considered nonaxisymmetric contributions $\delta A_{\mu j}^{(T_1)}$ to the solution $A_{\mu j}$ of the axisymmetry condition (7.79a), which are specified by the equations

$$\mathsf{T}_1\, \delta A_{\mu j}^{(T_1)} = T_1\, \delta A_{\mu j}^{(T_1)}. \tag{7.113}$$

Here T_1 is an eigenvalue of the operator T_1, defined by (7.78a) with (7.81). (Note also the discussion below (7.82) on the relative orientation of the coordinate systems in spin and orbital spaces of (7.81) in the B phase.) They found that the axisymmetric o-vortex as well as the v-vortex are stable with respect to small nonaxial perturbations for any eigenvalue Q. However, it was later found that the v-vortex is unstable with respect to *large*-amplitude perturbations of quantum number $|Q| = 2$ (Salomaa and Volovik 1986a). The order-parameter structure of this vortex state is given approximately by

$$a_{\mu m}(r, \varphi) = e^{i(n-\mu-m)\varphi} \Delta \sum_Q C_{\mu m}^{(Q)}(r) e^{iQ\varphi}, \tag{7.114}$$

where the sum extends over $Q = 0$, ± 2. For the singly quantized vortex ($n = 1$), the presence of the $Q = -2$ component means that the

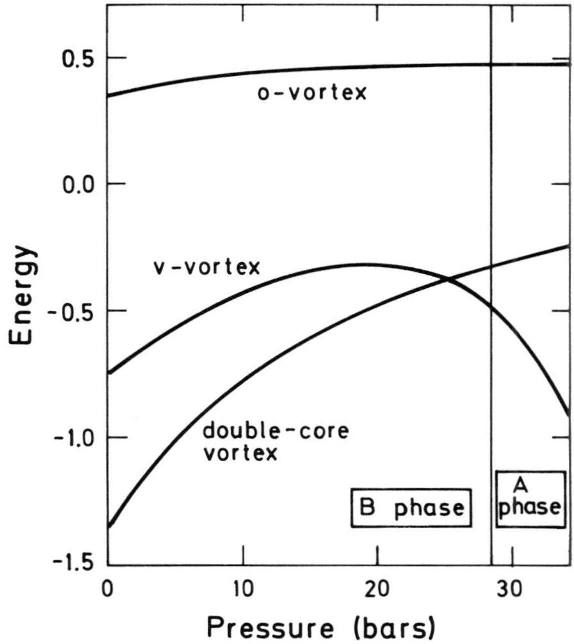

Figure 7.19 Energies of the axisymmetric o- and v-vortices and the double-core vortex, as functions of pressure. Pressures are given on a theoretical scale. (After Thuneberg (1986a).)

condition for the existence of a superfluid core, $n - \mu - m + Q = 0$, is now satisfied for $\mu + m = -1$. This implies finite components C_{0-} and C_{-0} (in addition to C_{0+} and C_{+0} in the case $Q = 0$) for $Q = -2$. Numerical evaluation shows that at the vortex core one has $C_{+0} \approx C_{-0}$ and $C_{0+} + C_{0-} \approx C_{+0} + C_{-0} \ll C_{0+} - C_{0-}$, i.e. a structure characteristic of the planar state, or else the axi-planar state. These states are closer to the B phase in structure, and the respective vortex core may therefore be expected to be energetically more favourable than the A-phase/β-phase core of the axisymmetric v-vortex.

It appears then that the vortex structure is driven to nonsymmetry by the gain in condensation energy in the vortex core. The system uses the available complexity in order-parameter structure to lower the vortex energy as much as possible. First, it manages to avoid the costly normal vortex core in favour of a superfluid core by breaking parity symmetry. At low pressure, where the particular type of superfluid core possible for the axisymmetric v-vortex is disadvantageous compared with the bulk B phase, a further transition to a nonaxisymmetric state may occur, allowing a still richer order-parameter structure in the vortex core. Other vortex states have been proposed, including nonaxisymmetric states of higher (e.g.

fourfold) symmetry and structures with broken translational symmetry along the vortex axis (Salomaa and Volovik 1987). It has been suggested that a vortex with a fourfold symmetry axis may occur as a metastable state at the vortex-core transition (Volovik and Salomaa 1987).

Vortex core structure and the topology of nodes on the Fermi surface

As pointed out by Volovik and Mineev (1982), the structure of the vortex core may be considered from quite a different point of view. Similarly to nonsingular textures in the A phase, the order parameter in the B phase can avoid having a singularity at the centre of a vortex by "flaring out" in some way, thereby absorbing the vorticity into a *nonsingular* configuration (cf. the AT nonsingular vortex in the A phase). In the B phase this is not possible in configuration space, but it is indeed possible in "phase space", i.e. in the six-dimensional position–momentum space. For the singularity at the vortex centre to be removed, it is not necessary to form a normal core. It is in fact sufficient if the order parameter (as a function of momentum) develops *nodes on the Fermi surface.* These nodes of the gap represent point singularities of the momentum-dependent phase $\phi(\hat{\boldsymbol{k}})$ (see (7.41) in the case of the A phase). Volovik and Mineev (1982) call them "boojums" in analogy with the point singularities in the $\hat{\boldsymbol{l}}$-vector field of the A phase discussed by Mermin (1977), which exist in *real space,* i.e. on the surface of a container wall (see Section 7.9.5). In fact, $\phi(\hat{\boldsymbol{k}})$ changes by 2π upon circling the node once, i.e. these nodes carry "vorticity in momentum space". One may say that vorticity is flowing from configuration space into momentum space as one goes from outside to the centre of the axisymmetric or the nonaxisymmetric v-vortex. The topology of the nodes of the gap provides a way of classifying the different structures of the vortex core. The position of the nodes (described by two unit vectors $\hat{\boldsymbol{l}}_\uparrow$ and $\hat{\boldsymbol{l}}_\downarrow$ for the two spin components of these, in general, nonunitary states) is a function of position in configuration space. The mechanism of phase slippage is then similar to the one known from the A phase (Anderson and Toulouse 1977). This allows the vortex core structure to be described as a texture in the $\hat{\boldsymbol{l}}_\uparrow$ and $\hat{\boldsymbol{l}}_\downarrow$ fields. In the case of the axisymmetric vortex one finds that a sequence of transitions occurs as the vortex axis is approached from outside (Salomaa and Volovik 1986b): first from the B phase to a planar-type phase with $\hat{\boldsymbol{l}}_\uparrow$ and $\hat{\boldsymbol{l}}_\downarrow$ antiparallel and oriented along $\hat{\boldsymbol{\varphi}}$; then to the axi-planar phase with $\hat{\boldsymbol{l}}_\uparrow$ and $\hat{\boldsymbol{l}}_\downarrow$ at a finite angle in the tangential plane; and finally, in the central region, to the A phase with $\hat{\boldsymbol{l}}_\uparrow$ and $\hat{\boldsymbol{l}}_\downarrow$ parallel, along the z axis.

The singly quantized nonaxisymmetric v-vortex turns out to be characterized by a completely different set of $\hat{\boldsymbol{l}}_{\uparrow,\downarrow}$ textures. It has the form of a composite object, with two centres, each carrying *half* a quantum of circulation. Inside the core of the two half-quantum vortices the vectors $\hat{\boldsymbol{l}}_{\uparrow,\downarrow}$ flare out from the plane perpendicular to the axis, whereas outside they are confined to this plane. A generalization of the Mermin–Ho theorem to this case (Volovik and Mineev 1982) therefore allows one to conclude that the vorticity is concentrated in the core regions. Note that the planar state, characterized by an antiparallel orientation of $\hat{\boldsymbol{l}}_\uparrow$ and $\hat{\boldsymbol{l}}_\downarrow$ in the plane, is invariant under interchange of $\hat{\boldsymbol{l}}_\uparrow$ and $\hat{\boldsymbol{l}}_\downarrow$, since it is a unitary state. Such an invariance allows for a half-quantum vortex. This is seen to be similar to the case of the A phase, where the invariance of the order parameter under a simultaneous rotation $\hat{\boldsymbol{d}} \to -\hat{\boldsymbol{d}}$, $\phi \to \phi + \pi$ gives rise to vortices with half a quantum of circulation. In contrast with the A phase, however, half-quantum vortices in the B phase may only occur in bound pairs, i.e. they are "confined", to use the language of particle physics. There appear to exist many interesting connections between the physics of vortices and other textures and problems of quantum field theory in the context of particle physics (Volovik 1986a–c, Balachandran 1986, Garg *et al.* 1986, Combescot and Dombre 1986).

The fact that according to the above discussion the axisymmetric and the nonaxisymmetric v-vortices carry different topological charges, implies that they may not be continuously transformed into one another. One should therefore expect a vortex-core transition from one type to the other to be of first order. This is borne out by numerical evaluation (Thuneberg 1986a, Salomaa and Volovik 1986a).

Other vortex structures

So far, we have only considered singly quantized vortices. Doubly quantized vortices have been considered by Salomaa and Volovik (1984, 1985a). It is found that the o-vortex is the stable structure in the weak-coupling approximation. A lattice of such doubly quantized o-vortices has a higher free energy per unit volume than one consisting of singly quantized vortices for vortex densities corresponding to angular velocities less than about 160 rad/s.

There are still other possible broken symmetries. For example, the translational symmetry along the vortex axis may be broken. Helical structures have been investigated by Salomaa and Volovik (1983a, 1985a), who even included different structures for the orbital and spin coordinates. However, the energies of these different structures have not been calculated in detail.

Finally, we should mention that so far we have only considered straight-line vortex configurations. At finite temperatures or, more generally, in excited states one expects vortex lines to bend and fluctuate. A configuration of particular interest is a vortex ring, forming a localized defect in the superfluid. Vortex rings have been studied extensively in superfluid ^{4}He (Tough 1982). The possibility of vortex rings formed out of nonsingular textures in the A phase has been considered by Fujita *et al.* (1978b,c).

7.5.10 Magnetic properties of vortices in the B phase

Axisymmetric vortices

The distortion of the order parameter in the core region of a vortex line gives rise to a change in the coupling of the condensate to a magnetic field $\boldsymbol{H}$: first, the spin susceptibility changes locally at the vortex core, and, secondly, a nonunitary component of the order parameter in this region leads to a magnetic moment.

In order to understand the consequences of the change in spin susceptibility, let us assume the vortex to be oriented along an axis $\hat{\boldsymbol{\Omega}}$, e.g. along the axis of rotation in a rotating cryostat. For axisymmetric vortices the preferred axis $\hat{\boldsymbol{\Omega}}$ in orbital space thus defined corresponds to a preferred axis $\boldsymbol{R}\hat{\boldsymbol{\Omega}}$ in spin space. Here $\boldsymbol{R}$ is the rotation matrix characterizing the bulk B phase surrounding the vortex. Consequently, there will be a change in the free energy per unit length of the vortex (Gongadze *et al.* 1981, Hakonen and Volovik 1982, Maki and Nakahara 1983, Salomaa and Volovik 1985a)

$$\mathscr{F}_H^{\text{vortex}} = \frac{1}{2}\int \mathrm{d}^2r\, \Delta\chi_{\mu\nu}(\boldsymbol{r})H_\mu H_\nu, \tag{7.115}$$

where $\Delta\chi_{\mu\nu}(\boldsymbol{r})$ is the local change in the spin-susceptibility tensor, given by

$$\Delta\chi_{\mu\nu}(\boldsymbol{r}) = \Big(\sum_i R_{\mu i}\Omega_i\Big)\Big(\sum_j R_{\nu j}\Omega_j\Big)\,\Delta\chi^{\text{vortex}}(\boldsymbol{r}). \tag{7.116}$$

In the Ginzburg–Landau regime $\Delta\chi^{\text{vortex}}$ is bilinear in the order-parameter components $C_{\mu m}$:

$$\Delta\chi^{\text{vortex}}(\boldsymbol{r}) = \tfrac{1}{2}(\chi_\mathrm{N} - \chi_\mathrm{B})\bar{\lambda}(r), \tag{7.117}$$

where

$$\bar{\lambda}(r) = \sum_m [|C_{0,m}(r)|^2 - \tfrac{1}{2}\,|C_{+,m}(r)|^2 - \tfrac{1}{2}\,|C_{-,m}(r)|^2]. \tag{7.118}$$

In the presence of an array of vortices of density n_v the contribution of vortices to the magnetic-anisotropy energy density may be written as

$$f_H^{\text{vortex}} = \tfrac{2}{5}\lambda\,\Delta f_{H,\parallel}\,(\hat{\boldsymbol{H}}\boldsymbol{R}\hat{\boldsymbol{\Omega}})^2, \tag{7.119}$$

with $\Delta f_{H,\parallel} = \lambda_\mathrm{D} N_\mathrm{F}(\Delta_\parallel - \Delta_\perp)\Delta_\perp$ (i.e. (6.116a) for $\hat{\boldsymbol{n}} \parallel \hat{\boldsymbol{H}}$) and

$$\lambda = \frac{5}{2}\frac{H^2}{\Delta f_{H,\parallel}}\,n_\mathrm{v}\int_{A_\mathrm{v}} \mathrm{d}^2r\,\Delta\chi^{\text{vortex}}(\boldsymbol{r}). \tag{7.120}$$

The integral in (7.120) extends over the area A_v occupied by a single vortex i.e. roughly a disc of radius r_Ω, with r_Ω the spacing between vortices. There are two qualitatively different contribution to this integral: (i) from the core region, where the order parameter changes in all its components; and (ii) from the region outside, where the order parameter changes with respect to rotations and gauge transformations. In the latter region the state of the system is rigorously described by hydrodynamics, and the corresponding contribution to f_H^{vortex} is proportional to $[\boldsymbol{HR}(\boldsymbol{v}_s - \boldsymbol{v}_n)]^2$, as given by (6.125) (Gongadze *et al.* 1981).

The second contribution to the free energy mentioned in the beginning of this subsection is due to the spontaneous magnetization density $\boldsymbol{M}(\boldsymbol{r})$ per unit area of the vortex:

$$\mathcal{F}_M = -\boldsymbol{H} \cdot \int \mathrm{d}^2 r \, \boldsymbol{M}(\boldsymbol{r}). \tag{7.121}$$

The magnitude of the magnetic-moment density in the Ginzburg–Landau regime can be calculated from the free-energy contribution (5.49) by identifying the quantity multiplying the magnetic field as the magnetic moment of the condensate. In the B phase the magnetic-moment density associated with a vortex core oriented along $\hat{\boldsymbol{\Omega}}$ is directed along the axis $\hat{\boldsymbol{m}} = \boldsymbol{R} \cdot \hat{\boldsymbol{\Omega}}$, as discussed above. It has the magnitude

$$M(\boldsymbol{r}) = g_{1,H} \Delta^2(T) \sum_m (|C_{+,m}(\boldsymbol{r})|^2 - |C_{-,m}(\boldsymbol{r})|^2), \tag{7.122}$$

where

$$g_{1,H} = \tfrac{1}{3} \gamma N(0) \frac{\eta}{T_c}. \tag{7.123}$$

Here η is a dimensionless parameter, which can be determined experimentally from the T_c splitting of the A transition in a magnetic field; this yields $\eta = 4 \times 10^{-3}$ at zero pressure to 2.2×10^{-2} at $P = 29$ bar (Israelsson *et al.* 1984).

The change in the free-energy density in the presence of an array of vortices of area density n_v is then given by

$$f_{\text{gm}}^{\text{vortex}} = \tfrac{4}{5} H_M \frac{\Delta f_{H,\parallel}}{H^2} \boldsymbol{HR\hat{\Omega}}, \tag{7.124}$$

where the subscript "gm" stands for "gyromagnetic". The magnetic field H_M can be expressed as

$$H_M = -\frac{5}{4} \frac{H^2}{\Delta f_{H,\parallel}} n_v \int_{A_v} \mathrm{d}^2 r \, M(\boldsymbol{r}), \tag{7.125}$$

with $M(\boldsymbol{r})$ given by (7.122). Estimates of H_M in the temperature region near T_c yield values of order 1 G at vortex densities corresponding to a rotation speed of 1 rad/s (Salomaa and Volovik 1987).

The existence of a magnetization inside the core of a quantized vortex was first discussed by Sauls *et al.* (1982) for superfluid neutron stars.

Calculations of the vortex core magnetization for superfluid ^{3}He were performed by Hakonen *et al.* (1983c), Ohmi *et al.* (1983) and Salomaa and Volovik (1983a, 1985a). A similar effect in the bulk superfluid is found to be much smaller (Mineev 1986).

The two magnetic energy contributions (7.119) and (7.124) provide an additional orienting force on the vector $\hat{\boldsymbol{n}}$, which adds to the usual term, given by (6.116). In the dipole-locked case, i.e. for the spin–orbit angle $\theta = \theta_L = \cos^{-1}(-\frac{1}{4})$ and for an axial magnetic field, $\boldsymbol{H} = \pm H\hat{\boldsymbol{\Omega}}$, the scalar product $\boldsymbol{H}\mathbf{R}\hat{\boldsymbol{\Omega}}$ is found to depend on the angle between the rotation axis $\hat{\boldsymbol{n}}$ and the axis $\hat{\boldsymbol{\Omega}}$ as $\boldsymbol{H}\mathbf{R}\hat{\boldsymbol{\Omega}} = \pm H(1 - \frac{5}{4}\sin^2\beta)$, where we used the explicit expression (6.106) for $\mathbf{R}$. The angular-dependent part of the sum of the energy-density contributions (6.116), (7.119) and (7.124) is then given by

$$\Delta f_H + f_H^{\text{vortex}} + f_{\text{gm}}^{\text{vortex}} = \Delta f_{H,\parallel} \sin^2\beta\left(1 - \lambda + \tfrac{5}{8}\lambda \sin^2\beta \mp \frac{H_M}{H}\right). \quad (7.126)$$

The change in orientation energy induced by the presence of vortices has observable consequences in NMR experiments, and in fact constitutes one of the major probes of vortices in rotating ^{3}He-B.

In the case that the $\hat{\boldsymbol{n}}$ texture is not disturbed too much from its uniform configuration along the magnetic field, i.e. for small angles β, vortices are seen to lead only to a renormalization of the anisotropy energy and consequently of the magnetic healing length defined in (7.29):

$$\xi_H \to \xi_H\left(1 - \lambda \mp \frac{H_M}{H}\right)^{-1/2}. \quad (7.127)$$

If, however, the density of vortices is large enough that $\lambda \gtrsim 1$, the minimum of the orientation energy (7.126) is attained for nonzero values of β given by (see Gongadze *et al.* (1981), except for the H_M term)

$$\sin^2\beta = \frac{4}{5}\frac{\lambda - 1 \mp H_M/H}{\lambda}. \quad (7.128)$$

Depending on the ratio of magnetic healing length ξ_H to container radius R, the interaction with the walls of the container may force the $\hat{\boldsymbol{n}}$ texture to be axially symmetric (if $R \ll \xi_H$). This costs additional bending energy, which is balanced by the wall orientation energy. Otherwise, the axisymmetry of the vortex texture outside the core is spontaneously broken; that is, $\hat{\boldsymbol{n}} = \hat{\boldsymbol{z}}\cos\beta + \hat{\boldsymbol{x}}\sin\beta$. The resulting state is degenerate.

In high magnetic fields the nonuniformity of the orientation force on the $\hat{\boldsymbol{n}}$ texture caused by a vortex leads to a nonuniform $\hat{\boldsymbol{n}}$ texture as a function of the distance from the vortex axis (Sonin 1983a, 1984a). Under conditions where the distance r_Ω between neighbouring vortices is much larger than the vortex core radius ξ, the susceptibility anisotropy due to the distortion of the order parameter inside the vortex core may be approximated by a two-dimensional delta-function potential $\delta^{(2)}(\boldsymbol{r})$. The free-energy-density contribution is then given by

$$f_H^{\text{vortex}}(\boldsymbol{r}) = \tfrac{2}{5}\lambda \frac{\Delta f_{H,\parallel}}{H^2}(\boldsymbol{H}\boldsymbol{R}\hat{\boldsymbol{\Omega}})^2 \frac{1}{n_\text{v}} \sum_j \delta^{(2)}(\boldsymbol{r} - \boldsymbol{r}_j). \tag{7.129}$$

The sum extends over all vortex cores, located at positions $\boldsymbol{r}_j$. A spatial average of (7.129) reduces to the expression (7.119) discussed before. From (7.129), it is clear that *at* the vortex core $\hat{\boldsymbol{n}}$ tends to be oriented such that the vector $\boldsymbol{R}\hat{\boldsymbol{\Omega}}$ is perpendicular to $\boldsymbol{H}$, while far away from the core the magnetic field wants to orient $\hat{\boldsymbol{n}}$ along $\pm\boldsymbol{H}$. These two forces will in general induce a nonuniform $\hat{\boldsymbol{n}}$ texture. The extent of spatial variation is limited, however, by the bending energy of the $\hat{\boldsymbol{n}}$ texture involved. Since in small magnetic fields the orientation energies are proportional to H^2, in such fields the bending energy dominates and the texture will be uniform as described above. At some sufficiently high field $H > H_\text{v}$, however, the orientational forces will generate a texture described by

$$\boldsymbol{R}\hat{\boldsymbol{\Omega}} = \hat{\boldsymbol{z}} \cos \alpha(\boldsymbol{r}) + \hat{\boldsymbol{x}} \sin \alpha(\boldsymbol{r}), \tag{7.130}$$

with α near $\frac{1}{2}\pi$ at the vortex core and $\alpha = 0$ outside.

The spatial dependence of α is found by minimizing the free-energy expression obtained as a sum of (7.12), (7.129) and (6.116):

$$f_\text{bend} + f_H^\text{vortex} + \Delta f_H$$

$$= \tfrac{2}{5}\Delta f_{H,\parallel}\left[\tfrac{3}{4}\xi_H^2(\boldsymbol{\nabla}\alpha)^2 + \frac{\lambda}{n_\text{v}}\cos^2\alpha \sum_j \delta^{(2)}(\boldsymbol{r} - \boldsymbol{r}_j) + 2(1 - \cos\alpha)\right]. \tag{7.131}$$

The transition of the uniform texture with $\alpha = 0$ at low fields to a nonuniform texture occurs when the linearized equation for the fluctuation amplitudes $\alpha(\boldsymbol{r})$ first acquires a negative eigenvalue. This equation is obtained by functionally differentiating the space integral of the free-energy density (7.131) with respect to $\alpha(\boldsymbol{r})$. The instability occurs when

$$\xi_H^2 = \frac{2\lambda}{3\pi n_\text{v}} \ln \frac{r_\Omega}{\xi}. \tag{7.132}$$

The corresponding magnetic-field value $H_\text{v} \approx 1$ kG is calculated from ξ_H via (7.29). It is interesting to note that H_v depends only logarithmically on the density of vortices n_v, since λ/n_v is *independent* of n_v.

As is evident from (7.130), the $\hat{\boldsymbol{n}}$ texture for $H > H_\text{v}$ will be non-axisymmetric. In this respect it is similar to the uniform nonaxisymmetric texture appearing at high vortex density (i.e. where $\lambda > 1$) discussed above. However, in the present context the symmetry breaking relates to a *single* vortex, in contrast with the former situation, where it was associated with the collective behaviour of the vortex array.

The transition from the uniform $\hat{\boldsymbol{n}}$ texture to a nonuniform one at $H = H_\text{v}$ may be detected through an effective weakening of the orientational force constant λ. At $H \gg H_v$ the area around the vortex cores where $\lambda(\boldsymbol{r})$ is nonzero is only a small fraction of the total area. This implies that the vortex cores are essentially decoupled from the bulk of the sample as far as the orientation of $\hat{\boldsymbol{n}}$ is concerned.

Nonaxisymmetric vortices

The breaking of axisymmetry in the core of the v-vortex discussed in the previous subsection introduces additional orientational effects owing to the existence of a further preferred direction $\hat{\boldsymbol{b}}$ in the plane orthogonal to the vortex axis.

The magnetic-anisotropy energy of a v-vortex (axis along $\hat{\boldsymbol{z}}$) with completely broken axisymmetry (i.e. with a vector symmetry along $\hat{\boldsymbol{x}}$, say) is determined by the spin-susceptibility tensor

$$\chi_{\mu\nu} = (\boldsymbol{R}^{-1}\boldsymbol{\chi}\boldsymbol{R})_{\mu\nu} = \begin{pmatrix} \chi_{xx} & 0 & \chi_{xz} \\ 0 & \chi_{yy} & 0 \\ \chi_{xz} & 0 & \chi_{zz} \end{pmatrix}. \tag{7.133}$$

For a v-vortex with C_2 symmetry (see Section 6.2), i.e. a "director symmetry", one has $\chi_{xz} = 0$, since in this case a contribution involving *both* preferred directions, $\hat{\boldsymbol{z}}$ and $\hat{\boldsymbol{x}}$, would be odd under the parity transformation. The magnetic-anisotropy energy for the vector case is given by (Theodorakis and Fetter 1983, Salomaa and Volovik 1983a)

$$\begin{aligned} f_H^{\text{vortex}} = \tfrac{2}{5}\Delta f_{H,\parallel}\, \{&\lambda_{11}(\hat{\boldsymbol{H}}\boldsymbol{R}\hat{\boldsymbol{\Omega}})^2 + \lambda_{12}(\hat{\boldsymbol{H}}\boldsymbol{R}\hat{\boldsymbol{\Omega}})(\hat{\boldsymbol{H}}\boldsymbol{R}\hat{\boldsymbol{b}}) \\ &+ \tfrac{1}{2}\lambda_{22}[(\hat{\boldsymbol{H}}\boldsymbol{R}\hat{\boldsymbol{b}})^2 - [\hat{\boldsymbol{H}}\boldsymbol{R}(\hat{\boldsymbol{\Omega}} \times \hat{\boldsymbol{b}})]^2]\}, \end{aligned} \tag{7.134}$$

where

$$\tfrac{2}{5}\Delta f_{H,\parallel}\, \lambda_{11} = (\chi_{zz} - \tfrac{1}{2}\chi_{yy} - \tfrac{1}{2}\chi_{xx})H^2, \tag{7.135a}$$

$$\tfrac{2}{5}\Delta f_{H,\parallel}\, \lambda_{12} = 2\chi_{xz}H^2, \tag{7.135b}$$

$$\tfrac{2}{5}\Delta f_{H,\parallel}\, \lambda_{22} = (\chi_{xx} - \chi_{yy})H^2. \tag{7.135c}$$

The expressions reduce to (7.119) in the case of axisymmetric vortices (or certain nonaxisymmetric vortices of higher symmetry), where $\chi_{xz} = 0$, $\chi_{xx} = \chi_{yy}$ and $\lambda_{11} = \lambda$.

The mixed term in (7.134), proportional to λ_{12}, causes the angle β between $\hat{\boldsymbol{n}}$ and $\hat{\boldsymbol{H}}$ to be nonzero in the bulk liquid. For $|\lambda_{12}| \ll 1$ and $\lambda_{22} \ll \lambda_{11} < 1$, one finds

$$\sin^2\beta = \frac{\lambda_{12}^2}{10(1-\lambda_{11})^2}. \tag{7.136}$$

As discussed in Section 8.4, this would be detectable in an NMR experiment. On the other hand, the term proportional to λ_{22} in (7.134) is sensitive to the relative orientation of $\hat{\boldsymbol{b}}$ in the vortex array. The effective value of λ for uniformly aligned orientation of $\hat{\boldsymbol{b}}$ is related to the value for completely random orientation by

$$\lambda_{\text{uniform}} = \lambda_{\text{random}} + \tfrac{1}{2}\,|\lambda_{22}|. \tag{7.137}$$

Hence the value of λ may be different according to the preparation of the sample.

7.6 ROTATING SUPERFLUID ^{3}He

The most effective, convenient and controlled way to generate vortices in superfluid ^{3}He is to rotate the container at a given constant angular velocity $\boldsymbol{\Omega}$. Upon rotation of the container, the normal-fluid fraction near the container walls will be carried along by friction with the moving wall, until the system reaches an equilibrium state characterized by solid-body rotation of the system of thermal excitations, i.e.

$$\boldsymbol{v}_{\text{n}} = \boldsymbol{\Omega} \times \boldsymbol{r}. \tag{7.138}$$

The normal-fluid flow has vorticity $\boldsymbol{\nabla} \times \boldsymbol{v}_{\text{n}} = 2\boldsymbol{\Omega}$. The energy of the superfluid component is increased by the normal-fluid flow, which acts like a vector potential on a charged superfluid. The superfluid tends to screen this external field (the analogue of the Meissner effect in a superconductor), but can do so only incompletely because the normal flow has vorticity while the superflow (in the lowest-energy configuration) is irrotational. As a consequence, vortices are formed, which carry the required vorticity. The mutual repulsive interaction of vortices leads to the formation of a regular vortex lattice.

7.6.1 Isotropic superfluid

Let us first consider the effect of rotation on an isotropic superfluid, such as superfluid ^{4}He. In the rotating state (see Feynman 1955, Vinen 1961, Andronikashvili and Mamaladze 1966, Langer and Reppy 1970, Tough 1982, Glaberson and Donnelly 1986) the free-energy density is given by

$$f(\boldsymbol{\Omega}) = f_0 + \tfrac{1}{2}(\rho_{\text{n}} v_{\text{n}}^2 + \rho_{\text{s}} v_{\text{s}}^2) - \boldsymbol{\Omega} \cdot \boldsymbol{L}, \tag{7.139}$$

where $\boldsymbol{L} = \boldsymbol{r} \times \boldsymbol{g}$ is the angular-momentum density and $\boldsymbol{g} = \rho_{\text{n}}\boldsymbol{v}_{\text{n}} + \rho_{\text{s}}\boldsymbol{v}_{\text{s}}$ is the mass-current density. The minimum of f with respect to variations of $\boldsymbol{v}_{\text{n}}$ is found by taking $\delta f/\delta \boldsymbol{v}_{\text{n}} = 0$ and is indeed given by $\boldsymbol{v}_{\text{n}} = \boldsymbol{\Omega} \times \boldsymbol{r}$. Using this result to eliminate $\boldsymbol{\Omega}$ from f, one finds

$$f(\boldsymbol{\Omega}) = f_0 + \tfrac{1}{2}\rho_{\text{s}}(\boldsymbol{v}_{\text{s}} - \boldsymbol{v}_{\text{n}})^2 - \tfrac{1}{2}\rho v_{\text{n}}^2. \tag{7.140}$$

The superfluid velocity $\boldsymbol{v}_{\text{s}}$ is in principle obtained by minimizing (7.140), i.e. the term $(\boldsymbol{v}_{\text{s}} - \boldsymbol{v}_{\text{n}})^2$. This means that the superfluid tries to follow the normal fluid as much as possible. However, because of the irrotational character of $\boldsymbol{v}_s$, the two velocities $\boldsymbol{v}_s$ and $\boldsymbol{v}_n$ cannot be equal.

The next best possibility is for the superfluid velocity to follow the normal flow *on the average*. Indeed, superflow can build up vorticity in a multiply connected geometry by flowing around core regions of normal liquid, i.e. by forming a system of rectilinear vortices parallel to the rotation axis. The density of vortices can be determined by requiring the average vortices of the superfluid and the normal fluid in a given area A to be equal. The integrated superfluid vorticity is given by the line integral along the contour $\mathscr{C}$ bounding the area A as (see (7.51))

$$\kappa = \oint_{\mathscr{C}} \boldsymbol{v}_s \cdot d\boldsymbol{l} = p\kappa_0 N, \tag{7.141}$$

where κ_0 is the quantum of circulation ($\kappa_0 = h/m_4$ in superfluid ^{4}He, while $\kappa_0 = h/2m_3$ in superfluid ^{3}He), p is the number of quanta per vortex, and N is the number of vortices enclosed by the contour. The integrated vorticity of the normal fluid is then $2\Omega A$ and the density of vortices is thus given by

$$n_v = \frac{N}{A} = \frac{2\Omega}{p\kappa_0}. \tag{7.142}$$

As will be discussed below, the energy of a vortex with p quanta of circulation varies as p^2, implying that for given vorticity, vortices with the lowest possible circulation are energetically favourable. For an isotropic superfluid, this implies that only simply quantized vortices appear.

For very slow rotation, such that the total normal-fluid vorticity is less than a single quantum of circulation, the superfluid flow is irrotational. Only above a critical angular velocity $\Omega_{c1} \approx \kappa_0/R^2$, where R is the radius of the container, does the system start to form vortices. The critical angular velocity is very small, of order 10^{-3} rad/s for R of order 1 cm. Upon increasing Ω, there first occurs a series of discrete transitions, whereby vortices are generated one by one. For $\Omega \gg \Omega_{c1}$, the system behaves similarly to an unbounded homogeneous fluid. Only at much higher angular velocities $\Omega > \Omega_{c2} \approx \kappa_0/\xi^2_{\text{core}}$, when the vortex core regions start to overlap substantially, does the system return to the normal state.

In the region $\Omega_{c1} \ll \Omega \ll \Omega_{c2}$ the vortices form a regular lattice that rotates rigidly with the container. In the rotating frame the relative velocity $\boldsymbol{v}_s - \boldsymbol{v}_n$ is a periodic function on the lattice, as is the free energy (Tkachenko 1965). It is then sufficient to consider a unit cell of the lattice containing only a single vortex, supplemented by the appropriate boundary condition. The free energy (7.140) associated with a p-fold quantized phase vortex may be calculated as follows. Let us assume for simplicity a circular boundary of the Wigner–Seitz cell, instead of the hexagonal or square boundary of a realistic lattice. The boundary condition at $r = r_0$, i.e. at the surface of the Wigner–Seitz cell, is given by $\boldsymbol{v}_s - \boldsymbol{v}_n = 0$, i.e. superfluid and normal fluid are in complete equilibrium. The radius of the Wigner–Seitz cell follows from (7.142) as $r_0 = (p\kappa_0/2\pi\Omega)^{1/2}$ (here $N=1$). The superfluid velocity around the vortex is given by $\boldsymbol{v}_s = (p\kappa_0/2\pi r)\hat{\boldsymbol{\varphi}}$ (see e.g. (7.67)), where r is the distance from the centre of the vortex, while the normal-fluid velocity is $\boldsymbol{v}_n = \boldsymbol{\Omega} \times \boldsymbol{r} = \Omega r\hat{\boldsymbol{\varphi}} = (p\kappa_0/2\pi r_0^2)r\hat{\boldsymbol{\varphi}}$. The free-energy change per Wigner–Seitz cell caused by the superfluid condensation is then given by

$$\Delta\mathscr{F}(\boldsymbol{\Omega}) = \int_{\xi}^{r_0} \mathrm{d}r\, 2\pi r \tfrac{1}{2}\rho_s(\boldsymbol{v}_s - \boldsymbol{v}_n)^2 \tag{7.143a}$$

$$= \rho_s \frac{(p\kappa_0)^2}{4\pi}\left[\ln\left(\frac{r_0}{\xi_0}\right) - \frac{3}{4}\right]. \tag{7.143b}$$

The normal core region of the vortex has been excluded from the integration. A more rigorous evaluation of $\Delta\mathscr{F}$ shows that the triangular vortex lattice is the stable configuration (Tkachenko 1965).

7.6.2 Superfluid ³He

The considerations of the previous subsection can be directly extended to the more complicated case of an anisotropic superfluid such as the superfluid phases of ³He (for reviews see Fetter 1986, Salomaa and Volovik 1987). In the rotating frame the free-energy density is given by the sum of the bulk term (5.4) and the gradient term (7.17), subject to a local Galilean transformation by $\boldsymbol{u} = \boldsymbol{\Omega} \times \boldsymbol{r}$. The latter amounts to replacing the gradient operators in (7.17) by $\mathbf{D} = \boldsymbol{\nabla} \pm \mathrm{i}(2m/\hbar)\boldsymbol{\Omega} \times \boldsymbol{r}$, depending on whether $\mathbf{D}$ acts on the order parameter $d_{\mu j}$ or its complex conjugate $d_{\mu j}^*$ (Fujita *et al.* 1978a, Williams and Fetter 1979). Outside the vortex cores, where the gradient energy may be replaced by the bending energy (7.20), (7.24), this is equivalent to replacing $\boldsymbol{v}_s$ by $\boldsymbol{v}_s - \boldsymbol{\Omega} \times \boldsymbol{r}$. The minimum of the free energy is then given as before by the order-parameter configuration where the superfluid performs a solid-body-like rotation, i.e. where the average local superfluid velocity satisfies the condition

$$\langle \boldsymbol{v}_s \rangle = \boldsymbol{\Omega} \times \boldsymbol{r}. \tag{7.144}$$

In contrast with an isotropic superfluid, where the superfluid component may carry vorticity only in the form of normal-core vortices, superfluid ³He also allows for the possibility of continuously distributed vorticity associated with nonsingular textures. As we have seen, the latter may extend over much larger regions than the diameter of a normal core, i.e. than the coherence length. For example, the region where the vorticity of nonsingular vortices in the A phase in low magnetic fields is concentrated is of the size of the magnetic healing length ξ_H, which may be much larger than the dipole length ξ_D. In this case the vorticity of neighbouring cores already starts to overlap at rather low angular velocities Ω. The appropriate boundary condition on the velocity field $\boldsymbol{v}_s(\boldsymbol{r})$ in this situation is obtained by requiring the circulation of $\boldsymbol{v}_s$ along the boundary of an elementary cell of the vortex lattice to be quantized (see (7.47), (7.48) and (7.51)):

$$\kappa = \int_{A_{\text{cell}}} d\boldsymbol{S} \cdot (\nabla \times \boldsymbol{v}_s) \tag{7.145a}$$

$$\equiv A_{\text{cell}} \langle (\nabla \times \boldsymbol{v}_s)_z \rangle \tag{7.145b}$$

$$= 2\Omega A_{\text{cell}} = n\kappa_0. \tag{7.145c}$$

Here L is the boundary and A_{cell} is the surface area of an elementary cell. As discussed above, nonsingular textures can only have an even number n of circulation quanta.

The problem of finding the type of vortex lattice realized at a given rotation speed Ω and for given thermodynamic conditions of pressure, temperature and magnetic field therefore amounts to finding the vortex, with circulation κ inside an elementary cell, that has the lowest energy per quantum of circulation. Just as in the case of the calculation of the stable equilibrium state discussed in Chapter 5, the determination of the equilibrium vortex structure is very difficult. Even if the problem of minimizing the free energy could be solved with arbitrary precision, the uncertainties in the free-energy parameters would preclude a definitive result in cases where two or several vortex structures are found to be very close in energy. Given this situation, it is necessary to rely on experimental observations in order to identify a given vortex structure. At present, the principal experimental tool is NMR, although other probes such as persistent currents, the motion of ions, hydrodynamic experiments or the propagation of sound through the vortex lattice have also been applied. The relevant theory will be discussed in Chapters 8, 10 and 11.

For the moment, let us summarize the results on the stability of vortices obtained in the preceding subsections and compare with experiment.

The B phase

In the B phase the calculation of vortex energies, as far as it has been carried out, has yielded two stable vortex structures: the axisymmetric v-vortex at high pressures ($\gtrsim$20 bar) and the double-core vortex discussed above (7.112) at low pressures. These results apply to the Ginzburg–Landau regime, i.e. not too far from the phase boundary, and to the low-density vortex system. The vortex lattice is presumably of hexagonal symmetry in both cases. It is conceivable that the phase boundary between the above two vortex states bends upward and meets the melting curve, but this has not been calculated yet. In this case, the two vortex states might be identified with the two regions observed in NMR experiments (see Fig. 8.9). A comparison of these data with theory is made in Section 8.5. Given the large number of possible vortex types, there may exist further states in regions of the phase diagram that have not yet been experimentally investigated. In particular, it is likely that a different vortex state is stabilized at high vortex density, or equivalently high rotation speed.

The A phase

Here the texture depends sensitively on the strength of an applied magnetic field. For low magnetic fields, continuous nonsingular vortex textures appear to be more favourable in energy than singular vortices. The boundary condition imposed by the periodicity of the $\hat{\boldsymbol{l}}$ field in the rotating frame requires $\hat{\boldsymbol{l}}$ to be along the axis of rotation for axisymmetric vortices in a lattice with one vortex per elementary cell. One possible texture of this kind is a triangular lattice of doubly quantized AT vortices (Volovik and Kopnin 1977). Another possibility is to combine an MH and an MT texture into a pair of interpenetrating square lattices with four vortices per unit cell, carrying four quanta of circulation (Tsuneto *et al.* 1978, Fujita *et al.* 1978a, Ho 1978a,c, Nakahara *et al.* 1979). In this way, the MH and MT textures, which do not satisfy the boundary condition of $\hat{\boldsymbol{l}}$ being parallel to the axis as required for a Bravais lattice of MH or MT textures, may form a continuous periodic $\hat{\boldsymbol{l}}$ texture where, at the cell boundary, $\hat{\boldsymbol{l}}$ is normal to the boundary and in the plane (see Fig. 7.20). Fujita *et al.* (1978a) have numerically evaluated the dipole-locked free energy of such a vortex lattice by explicitly

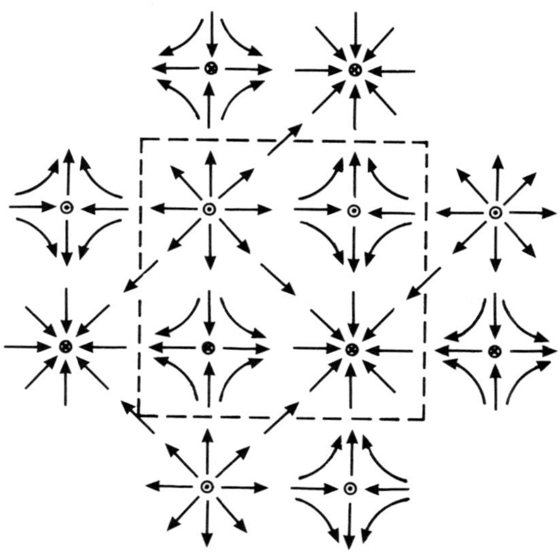

Figure 7.20 Lattice of continuous nonsingular vortices in ^{3}He-A made up of alternating MH and MT textures. The unit cell, indicated with dashed lines, contains four vortices and carries four quanta of circulation ($\odot$, $\hat{\boldsymbol{l}}$ up; $\otimes$, $\hat{\boldsymbol{l}}$ down).

incorporating the fourfold lattice symmetry. A simpler, approximate, evaluation, in which the MH and MT textures are treated separately, each confined to circular cells of area $\frac{1}{2}A_{\text{cell}}$, has been carried out by Fetter *et al.* (1983), and yields similar results. For currently available rotation speeds Ω ($\leqslant 1$ rad/s), the Wigner–Seitz-cell radius ($\approx 100\,\mu$m) far exceeds the dipole length ξ_D ($\approx 8\,\mu$m), so that deviations from dipole locking are small. In this regime, both of the above calculations conclude that the MH–MT lattice has lower free energy per quantum of circulation than other candidates such as the doubly quantized AT-vortex lattice or singular vortices. The symmetry of vortex lattices is discussed in group-theoretical terms by Sonin and Fomin (1985).

At higher rotation speeds, where the distance between vortices becomes comparable to the dipole length, a triangular lattice of singly quantized phase vortices becomes favoured. One example is the "z vortex" of Fetter *et al.* (1983), which has $\hat{\boldsymbol{l}}$ and $\hat{\boldsymbol{d}}$ dipole-locked along $-\hat{\boldsymbol{z}}$ for $r > \xi_D$. For $r < \xi_D$, however, it is energetically advantageous for $\hat{\boldsymbol{l}}$ to deviate from the uniform $\hat{\boldsymbol{d}}$ orientation and to tilt towards the centre into a radial disgyration pattern. As mentioned above, the singularity in the $\hat{\boldsymbol{l}}$ field is avoided by the order parameter changing its structure from the axial into the polar state for $r \leqslant \xi$. The angular velocity above which the z-vortex lattice should be stable is temperature-dependent and has been estimated as 30 rad/s at $T \approx 0.99T_c$, with increasing values at lower temperatures. The theory of vortices in the neighbourhood of the upper critical angular velocity has been considered by Schopohl (1980).

The situation is entirely different in a strong magnetic field $\boldsymbol{H}$, parallel to the rotation axis, such that the vector $\hat{\boldsymbol{d}}$ is forced into the plane perpendicular to $\boldsymbol{H}$. The bending energy of the $\hat{\boldsymbol{d}}$ field acts in such a way as to keep $\hat{\boldsymbol{d}}$ uniform (along $\hat{\boldsymbol{y}}$, say) except for a small region of radius ξ_H^A, (7.28a,b). Inside a soft core of radius ξ_D the $\hat{\boldsymbol{l}}$ field may deviate from the uniform orientation along $\hat{\boldsymbol{d}}$ ($=\hat{\boldsymbol{y}}$), in order to transfer some of the kinetic energy of flow stored in the phase variable field into the $\hat{\boldsymbol{l}}$ field. Vortex structures of this type have been discussed already in Section 7.5. There is a singly quantized singular nonaxisymmetric vortex and a doubly quantized nonsingular vortex (given by (7.106)). Detailed numerical calculations show the singular vortex to have slightly lower energy than the nonsingular one for angular velocities Ω less than about 3 rad/s (Seppälä and Volovik 1983, Vulovic *et al.* 1984; see also Passvogel *et al.* 1982). As will be discussed in the next chapter, NMR experiments indicate the presence of nonsingular vortices. More work is needed to clarify this question.

It is worth mentioning that a doubly quantized nonaxisymmetric vortex may be viewed as a bound pair of an MH and an MT texture (Zotos and Maki 1984, Maki and Zotos 1985a), somewhat similar to the interpretation of the nonaxisymmetric vortex in the B phase. The picture has been used to construct variational trial wave functions. It has also been used to calculate the energy of orientation of the vortex-pair axis with respect to the triangular lattice (Ohmi 1984). At present these effects are too small to be observed.

The effect of tilting the magnetic field away from the rotation axis has also been considered (Volovik and Hakonen 1981, Maki and Zotos 1985b). One finds that the degeneracy of the orientation of $\hat{\boldsymbol{d}}$ in the (x, y) plane is lifted and $\hat{\boldsymbol{d}}$ is oriented perpendicular to both $\boldsymbol{H}$ and $\boldsymbol{\Omega}$. Also, the separation of the two vortices in the bound pair decreases with increasing tipping angle. The corresponding change in the NMR response is consistent with the observations.

The transition from the low-field square lattice of MH–MT vortices to the high-field doubly quantized triangular lattice of bound pairs has also been investigated (Fetter 1985a, Zotos and Maki 1985). As the magnetic field is increased from low values, the vector $\hat{\boldsymbol{d}}$ is forced into the (x, y) plane and pulls the vector $\hat{\boldsymbol{l}}$ with it (in the region $r > \xi_D$). Recalling that the vector $\hat{\boldsymbol{l}}$ is in the (x, y) plane at the Wigner–Seitz zone boundary, it is seen that the magnetic field tends to flatten the $\hat{\boldsymbol{l}}$ texture. At some point, when the magnetic field is of order $H^* \approx 30$ G (see (7.28c)), the MH–MT structure might undergo a first-order transition into the high-field phase. However, other types of transition involving a distortion of the lattice are also possible. Ion-mobility experiments show the existence of a low-field regime at $H < 4$ G, with low mobility along the vortex cores, which is consistent with the assumption of a continuous vortex texture. They also show a high-field regime with high mobility, which is indicative of a singular vortex state (Simola *et al.* 1986, 1987). As discussed above (see Section 7.5.8), the singular vortex structure is expected to have a polar phase core. A calculation of the ion mobility in such a vortex core is in agreement with experiment (Salmelin and Salomaa 1987a,b). More work remains to be done before the stable vortex structures are fully understood in all parts of parameter space.

7.7 POINT DEFECTS

The structure of the order parameter of superfluid ^{3}He is in general rich enough to allow not only linear defects but also pointlike defects (Anderson and Brinkman 1975, Volovik and Mineev 1976b, 1977a,b, Cross and Brinkman 1977, Bailin and Love 1978d). In analogy to the investigation of line defects by mapping a contour $\mathscr{C}$ around a defect into a corresponding contour Γ in the degeneracy space R, homotopy theory may also be used to study the stability of point defects (Volovik and Mineev 1977a,b, Mermin 1979, Mineev 1980). In this case one has to consider a closed surface $\mathscr{S}$ enclosing the defect and its map Σ in the degeneracy space. (This will again be a closed surface.) Different point singularities will in general lead to different mappings Σ in R. One may then study their relationship by investigating their behaviour under continuous deformations. Those closed contours that may be continuously deformed into each other belong to the same homotopy class, implying that the enclosed defects are topologically equivalent. If, in particular, a surface Σ is contractable to a point then it corresponds to a trivial ("zero") topological class. On the other hand, if such a contraction is not possible (because, for example, the surface is hooked at a cavity in R not belonging to R itself) then this surface belongs to a nonzero topological class.

In analogy to the first homotopy group involving linear contours, one may again define a multiplication between mappings Σ_1 and Σ_2, where $\Sigma = \Sigma_1 \circ \Sigma_2$ is a surface combining both closed surfaces Σ_1 and Σ_2. The set of classes of homotopic mappings of the surface $\mathscr{S}$ into the degeneracy space R forms a group. It is called the "second homotopy group" of R and is denoted by $\pi_2(R)$. As we know, the degeneracy spaces of interest in the various phases of superfluid ^{3}He are composed of S^1, S^2 and SO_3. Of the relevant homotopy groups, $\pi_2(S^1)$, involving the gauge variable in ^{3}He-B, is trivial, i.e. $\pi_2(S^1) = 0$, since there can be no three-dimensional point singularity in real space in an order-parameter field with only two components. Furthermore $\pi_2(SO_3) = 0$, because a closed surface in a simply connected space may always be contracted to a point. So in effect one is left with $\pi_2(S^2)$, i.e. the set of classes of mappings of a closed surface $\mathscr{S}$ around a point defect into a sphere, i.e. another closed surface. The possibilities arising in this case may be visualized as follows (Brinkman and Cross 1978).

(i) The closed surface $\mathscr{S}$ may be mapped *onto* the surface of the sphere $R = S^2$ by having it lie flat on the sphere (like a completely deflated rubber balloon stuck to the outside of a ball). In this case the surface can be contracted to a point, and consequently the point defect in $\mathscr{S}$ may be removed by purely local deformations.

(ii) Alternatively, the mapped surface Σ may *enclose* the sphere S^2 (now the sphere is *in* the balloon). In this case Σ may actually surround the sphere more than once (for example by pulling it such that it wraps around S^2 more than once).

Mathematically speaking, this implies $\pi_2(S^2) = \mathbb{Z}$. So, in analogy to the topological charge of line defects given by the winding number, there is a similar number characterizing point singularities: the "degree of mapping". It measures the number of times a mapping encloses the sphere S^2. Surfaces with different degrees of mappings belong to different homotopy classes.

7.7.1 The B phase

The order parameter (6.8) is determined by the gauge and the rotation matrix $\boldsymbol{R}(\hat{\boldsymbol{n}}, \theta)$. For a closed surface surrounding a point singularity at distance $r \gg \xi_D$, the rotation angle θ corresponds to the Leggett angle θ_L.

Hence the solid sphere SO_3 is reduced to a sphere S^2 (see Fig. 6.2) such that we are only dealing with the degeneracy space $R = S^1 \times S^2$. The mapping of a closed surface around the point singularity is then given by

$$\begin{aligned} \pi_2(S^1 \times S^2) &= \pi_2(S^1) + \pi_2(S^2) \\ &= \pi_2(S^2) = \mathbb{Z}. \end{aligned} \tag{7.146}$$

This implies that in the dipole-dominated region the $\hat{\boldsymbol{n}}$ texture may contain a stable pointlike defect (sometimes referred to as a "monopole" or "hedgehog") (Anderson and Brinkman 1975, Volovik and Mineev 1976b). To investigate the structure of the centre of the monopole, we decrease the size of the sphere around the centre of the point defect until its radius $r \ll \xi_D$. In this case the dipole restriction $\theta = \theta_L$ is lifted and the $\hat{\boldsymbol{n}}$ field regains its degeneracy space SO_3. We now have $\pi_2(SO_3) = 0$. Hence in the absence of the dipole coupling the mapping is trivial: a point defect in the $\hat{\boldsymbol{n}}$ field is then found to be unstable. This is achieved by the additional degree of freedom in the variable θ, which will go to zero at the centre of the monopole. Consequently, a point defect in the $\hat{\boldsymbol{n}}$ field will be stable but will have a nonsingular core.

7.7.2 The A phase

In the dipole-locked region with $\hat{\boldsymbol{d}} \parallel \hat{\boldsymbol{l}}$, i.e. for closed surfaces further than ξ_D away from a point singularity ($r \gg \xi_D$), the degeneracy space is $R = SO_3$. Since $\pi_2(SO_3) = 0$, point singularities are unstable in the dipole-dominated A phase.

In the dipole-free situation, as is valid for surfaces surrounding a point defect at distances $r \ll \xi_D$, the degeneracy space is $R = (S^2 \times SO_3)/\mathbb{Z}_2$ (see (6.68b)). The mapping $\pi_2(R)$ is (Volovik and Mineev 1977a, Cross and Brinkman 1977)

$$\begin{aligned} \pi_2[(S^2 \times SO_3)/\mathbb{Z}_2] &= \pi_2(S^2 \times SO_3) \\ &= \pi_2(S^2) + \pi_2(SO_3) \\ &= \mathbb{Z}, \end{aligned} \tag{7.147}$$

since $\pi_2(SO_3) = 0$. Hence it is only nontrivial because of the topology of $\hat{\boldsymbol{d}}$, i.e. the spin part of the order parameter. So we find that a stable point defect may only exist in the $\hat{\boldsymbol{d}}$- but *not* the $\hat{\boldsymbol{l}}$-vector field, and only at lengthscales smaller than ξ_D. At larger lengthscales it becomes unstable. These findings do not contradict each other. Similarly to the situation in the B phase, where line defects in the $\hat{\boldsymbol{n}}$ field may establish themselves as edges of planar defects, the point singularity in the $\hat{\boldsymbol{d}}$ field in the present case must be thought of as the termination point of a nonsingular line defect in the $\hat{\boldsymbol{d}}$ field as shown in Fig. 7.21. A sphere enclosing P at a distance less than ξ_D will only see the singularity at P, while a large sphere, besides cutting the

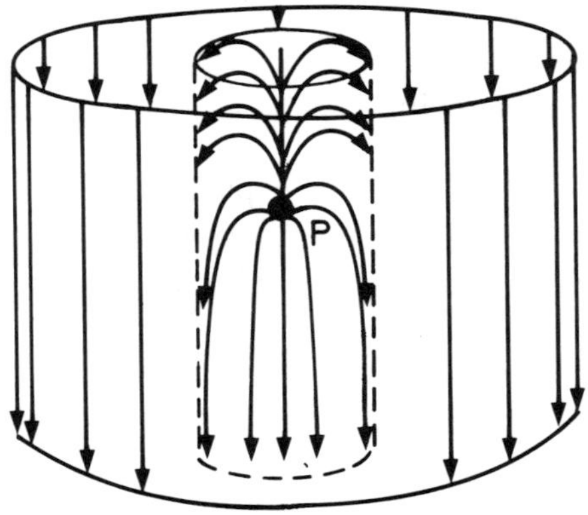

Figure 7.21 Point singularity in the $\hat{\boldsymbol{d}}$-vector field of ^{3}He-A. It is the termination point of a nonsingular line defect in the $\hat{\boldsymbol{d}}$ field.

line defect somewhere above *P*, essentially passes through a uniform $\hat{\boldsymbol{d}}$ texture.

As discussed above, the orbital part of the ^{3}He-A order parameter does not allow stable point singularities. Hence any (hypothetical) pointlike defect in the $\hat{\boldsymbol{l}}$ field ("$\hat{\boldsymbol{l}}$ monopole") is topologically unstable. In fact, any $\hat{\boldsymbol{l}}$ monopole would always be connected with a line singularity carrying two quanta of circulation (Volovik and Mineev 1976a); see Section 7.9.5. (Equally well, one may say that a singular vortex line with two quanta of circulation (i.e. a 4π vortex) may terminate at a singular point to escape into the third dimension.) Hence an $\hat{\boldsymbol{l}}$ monopole does not appear in a topological classification of point defects since it cannot actually be surrounded by a closed contour without cutting through the normal-fluid vortex core, where the order parameter is zero (Brinkman and Cross 1978). Such a (hypothetical) object is unstable since it transforms into a nonsingular 4π vortex, which may end in a nonsingular point defect and which can be removed by continuous transformations (Anderson and Toulouse 1977). Finally, the 4π vortex connected with any $\hat{\boldsymbol{l}}$ monopole is energetically unfavourable. It is expected that the system gets rid of it by contracting the vortex line to zero, for example by pulling the $\hat{\boldsymbol{l}}$ monopole to the wall of the container, where it becomes a *surface* defect now generally referred to as a "boojum" (Mermin 1977b, 1978a,b). This will be discussed further in Sections 7.9.5 and 7.12.3.

As discussed by Volovik and Mineev (1977b), the order parameter of the A phase also allows stable non-singular point-like defects, which may be embedded in a *uniform* $\hat{\boldsymbol{l}}$ field ("particle-like solitons"; see the discussion in Section 7.8.1). They are of finite size and thereby carry finite momentum, energy and angular momentum. This is in analogy to (singular) vortex rings in superfluid ^{4}He. Hence there is a Magnus force acting on the soliton. The same is true for a solid body of size $R > \xi_0$ immersed in ^{3}He-A (see also Section 10.4.7).

7.8 PLANAR DEFECTS

7.8.1 Topological classification

Although planar defects have not been explicitly addressed so far, they have already made their appearance in Section 7.5. There we saw that certain line defects may be viewed as the edges, i.e. termination lines, of domain walls. In this situation the domain wall is characterized by the same topological invariant as the line singularity (Volovik and Mineev 1977a, Mineev and Volovik 1978).

Without having to go into any detailed topological study of planar defects in superfluid ^{3}He, we are able to conclude that in the absence of external fields and dipole coupling, planar defects may not exist. To have stable defects of dimension $D = 2$ (see Section 7.4) in a three-dimensional space, order parameters with only a *single* degree of freedom, i.e. with $n = 1$ as for an Ising spin, are necessary. Since in superfluid ^{3}He the order-parameter space is at least two-dimensional, singular surfaces may not be formed.

On the other hand, external fields (e.g. a magnetic field) and the dipole interaction will introduce further symmetry breakings that may, in fact, make order-parameter degrees of freedom Ising-like.

The A phase

For example, in the A phase the dipole energy aligns $\hat{\boldsymbol{d}}$ and $\hat{\boldsymbol{l}}$ but does not distinguish between $\hat{\boldsymbol{d}} \parallel \hat{\boldsymbol{l}}$ and $\hat{\boldsymbol{d}} \parallel -\hat{\boldsymbol{l}}$. This *discrete* symmetry breaks the invariance with respect to arbitrary relative $\hat{\boldsymbol{d}}$, $\hat{\boldsymbol{l}}$ rotations but still allows a discrete degeneracy of the ground states. So we see that the $\hat{\boldsymbol{d}}$-vector field may act as a backbone on which the $\hat{\boldsymbol{l}}$-vector field becomes Ising-like, and vice versa. In this situation stable planar defects may exist by virtue of the finite dipole coupling. Indeed, it is now possible for the system to assume different ground states at different positions in real space. In this case there will be a boundary region where the asymptotic configurations meet (see Fig. 7.22), in analogy to a domain wall in a ferromagnet. These defects are nonsingular and have an asymptotically uniform texture. Hence they are localized objects, which may be added together without creating a substantial interaction between them. Domain walls in superfluid ^{3}He have been introduced and extensively discussed by Maki and coworkers (see Maki 1978b, 1986a), who also coined the term "(planar) soliton" to describe this kind of nonsingular planar defect. The terminology originates from the close similarities in the properties of such a domain wall and those of solitons, as known from the theory of nonlinear dynamics (e.g. the sine–Gordon equation; see below and Section 8.4.4). In both cases the corresponding differential equations have localized solutions, which resemble classical particles and whose stability is guaranteed by the conservation of a topological charge. Such solitons may be formed in both the A and the B phases of superfluid ^{3}He. They will be discussed in detail in the end of this section. In fact, topologically stable defects are sometimes grouped into *solitons* and *singularities,* irrespective of their dimensionality, in order to differentiate between nonsingular asymptotically uniform textures and defects causing a long-range distortion of the respective vector field. Therefore the line defect in Fig. 7.21 is also referred to as a "linear soliton" and the point defect in ^{3}He-A discussed at the end of Section 7.7.2 as a "pointlike" or "particle-like soliton" (Volovik and Mineev 1977b, Mineev and Volovik 1978).

As in the case of line and point defects, planar defects may also be classified in terms of their topological properties (Mineev and Volovik 1978). To this end, we consider two points P_1 and P_2 on either side of the planar defect and map the line between P_1 and P_2 into the corresponding order-parameter space (see Fig. 7.22a). This leads to the definition of a topological charge N that measures the reorientation between $\hat{\boldsymbol{d}}$ and $\hat{\boldsymbol{l}}$ on the line P_1–P_2. Assuming that a magnetic field $\boldsymbol{H}$ along $\hat{\boldsymbol{z}}$ confines $\hat{\boldsymbol{d}}$ to the (x, y) plane, this topological charge (or "winding number") is defined by

$$N = \frac{1}{\pi} \sum_i \int_{P_1}^{P_2} \mathrm{d}\xi_i \, \hat{\boldsymbol{H}} \cdot [\hat{\boldsymbol{l}} \times (\nabla_i \hat{\boldsymbol{l}}) - \hat{\boldsymbol{d}} \times (\nabla_i \hat{\boldsymbol{d}})], \tag{7.148}$$

with $\xi_i = x, \, y$ and where the integral is along any path joining P_1 and P_2.

Owing to the existence of both an $\hat{\boldsymbol{l}}$- *and* a $\hat{\boldsymbol{d}}$-vector field, several kinds of domain walls appear to be possible, depending on which vector field changes its orientation across the wall (see Figs. 7.22b–g).

Figures 7.22(b, c) correspond to $\hat{\boldsymbol{l}}$ solitons, i.e. $\hat{\boldsymbol{l}}$ does the winding while $\hat{\boldsymbol{d}}$ stays (more or less) constant. According to (7.148), the two domain walls carry opposite topological charge. Note, however, that by turning both $\hat{\boldsymbol{l}}$ and $\hat{\boldsymbol{d}}$ by π at every point in the entire sample, such that their relative angle is conserved, one is led from Fig. 7.22(b) to Fig. 7.22(c). Such a transformation does not cost energy as long as there are no walls to fix the the direction of $\hat{\boldsymbol{l}}$ somewhere in the sample. This implies that a planar defect with $N = 1$ is topologically equivalent to one with $N = -1$. (This can also been seen by simply turning Fig. 7.22(b) upside down, thus changing $N = 1$ into $N = -1$.) Mathematically speaking, the mapping of contours P_1 and P_2 through a planar defect in ^{3}He-A into the appropriate degeneracy space corresponds to $\mathbb{Z}_2$: there are only two classes of domain walls in the A phase, corresponding to $N \in \mathbb{Z}_2$, i.e. $N = 0, 1$ (Mineev and Volovik 1978). In particular, this implies that in an infinite system $\hat{\boldsymbol{d}}$ and $\hat{\boldsymbol{l}}$ solitons are *topologically equivalent,* as can be verified for example by observing that the $\hat{\boldsymbol{l}}$ soliton in Fig. 7.22(c) may be transformed into the $\hat{\boldsymbol{d}}$ soliton in Fig. 7.22(e) by aligning $\hat{\boldsymbol{l}}$ but keeping the angle between $\hat{\boldsymbol{d}}$ and $\hat{\boldsymbol{l}}$ fixed. Furthermore, the structures in Figs 7.22(f, g) do not carry a topological change because they can be continuously transformed into the homogeneous texture.

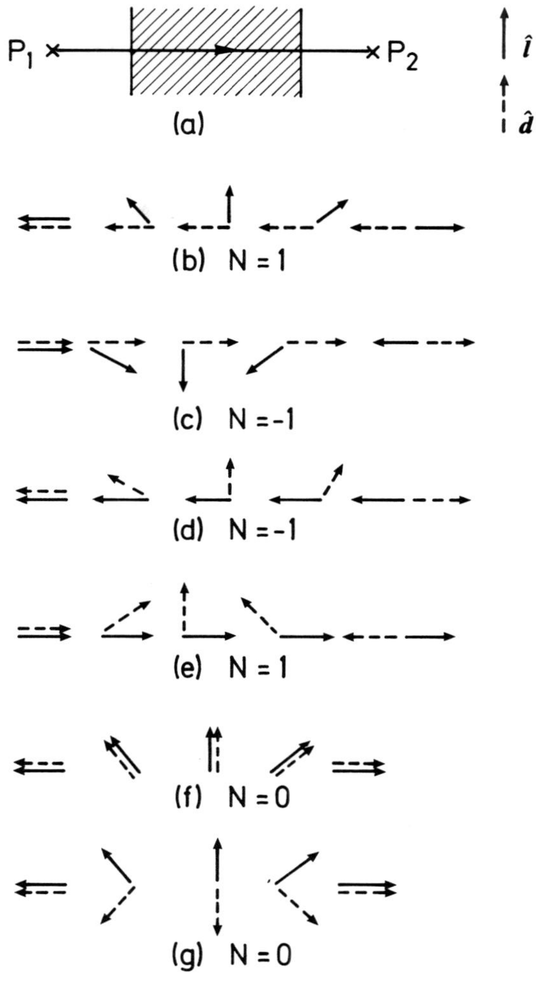

Figure 7.22 Planar domain wall in ^{3}He-A. (a) Contour across the domain wall, joining the asymptotic points P_1 and P_2. (b)–(g) Different reorientations of the $\hat{\boldsymbol{d}}$- and $\hat{\boldsymbol{l}}$-vector fields across the domain wall carrying a topological charge N as indicated.

The fact that planar domain walls in ^{3}He-A may only carry charges $N = 0, 1$ implies that two solitons can annihilate each other. In particular, a periodic structure of solitons, obtained for example by a consecutive winding of $\hat{\boldsymbol{l}}$ or $\hat{\boldsymbol{d}}$ as shown in Fig. 7.23, is topologically unstable and is equivalent to at most a single domain wall. As already mentioned, these conclusions are not valid if $\hat{\boldsymbol{l}}$ is pinned by boundary conditions, in which case a homogeneous rotation of $\hat{\boldsymbol{l}}$, as required for the annihilation, will be prohibited by an energy barrier.

Figure 7.23 Periodic winding of the $\hat{\boldsymbol{d}}$- or $\hat{\boldsymbol{l}}$-vector field in ^{3}He-A, corresponding to a periodic "lattice" of domain walls.

The B phase

In the B phase the dipole energy by itself does not lead to discrete degeneracies that would allow planar defects. Indeed, the dipole interaction couples spin and orbital spaces only by fixing the angle of relative rotations at the Leggett angle θ_L, while leaving the rotation axis $\hat{\boldsymbol{n}}$ unaffected. On the other hand, an external magnetic field will align $\hat{\boldsymbol{n}}$ parallel *or* antiparallel to itself, leading again to a discrete degeneracy of the ground state. In this situation planar solitons are possible. For the investigation of planar defects in the B phase, it is therefore essential to include both the dipole interaction *and* a magnetic field (Mineev and Volovik 1978).

To study the mapping of a path leading from P_1 to P_2 (say, along the $\hat{\boldsymbol{z}}$ axis) through a planar defect centred at $z = 0$ (see Fig. 7.22a), we have to distinguish between different lengthscales relative to the intrinsic lengths ξ_H and ξ_D. Here we assume $\xi_H \gg \xi_D$.

(i) Let P_1 and P_2 correspond to the asymptotic points $z = +\infty$ and $z = -\infty$ respectively. At P_1 and P_2 we are in the dipole-locked regime *and* $\hat{\boldsymbol{n}}$ is oriented by $\boldsymbol{H}$, where both $\hat{\boldsymbol{n}} \parallel \boldsymbol{H}$ and $-\hat{\boldsymbol{n}} \parallel \boldsymbol{H}$ are allowed. In this case the degeneracy space is $R_B^{D,H} = S^1 \times \mathbb{Z}_2$, corresponding to two points $\boldsymbol{n} = \pm\hat{H}\theta_L$ on the sphere S^2 with $\theta = \theta_L$ in the solid sphere SO_3 (see Fig. 7.24). (Here we included the S^1 of the overall phase for completeness, although it is irrelevant for the present discussion.)

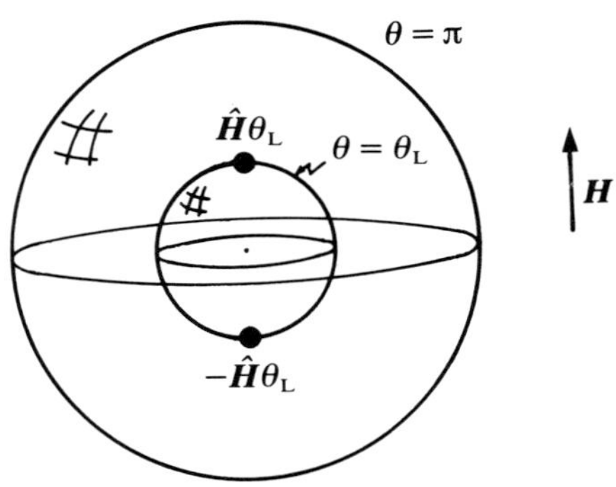

Figure 7.24 Degeneracy space of ^{3}He-B corresponding to different lengthscales relative to the intrinsic healing lengths ξ_D and ξ_H.

(ii) As we approach the planar defect, we come into a region $\xi_H \gg |z| \gg \xi_D$ where the influence of the magnetic field disappears, so that the degeneracy space $R_B^D = S^1 \times S^2$ is regained (Figs. 6.2(b) and 7.24).

(iii) In the centre of the domain wall, $\xi_H \gg \xi_D \gg |z|$, the dipole-free magnetic-field-free case is realized, so that $R_B = S^1 \times SO_3$.

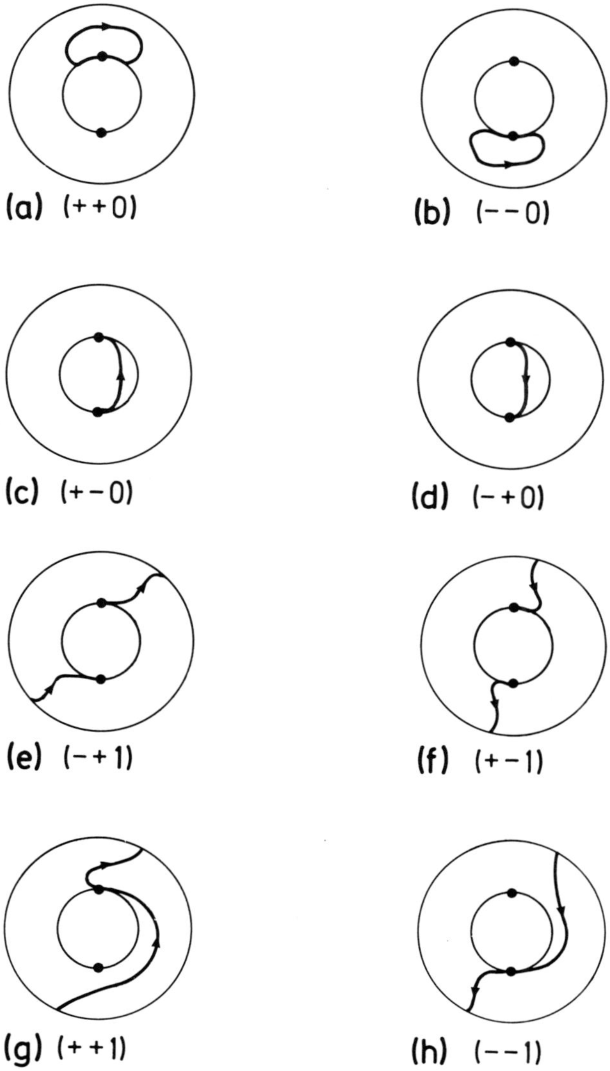

Figure 7.25 Different mappings of the path P_1P_2 in Fig. 7.22(a) into the degeneracy space of ^{3}He-B, corresponding to different domain walls.

The topologically stable classes of planar defects in the B phase are then obtained by starting at one of the points $\boldsymbol{n} = \pm\hat{\boldsymbol{H}}\theta_L$ and also ending at one of them. The possible paths are shown in Fig. 7.25. There are altogether eight different classes of paths, i.e. eight different domain walls. The classification is represented by a triplet, $(+\,-\,0)$ etc., where the first two indices indicate the direction of $\hat{\boldsymbol{n}}$ relative to $\boldsymbol{H}$ at the starting point and end point respectively, and the third index, $N=1$ or $N=0$, shows whether or not the path went through $\theta=\pi$. The paths in Figs. 7.25(a, b) may be contracted to a point and thus do not form a stable domain wall. Figures 7.25(c, d) correspond to a transition from a configuration with $\hat{\boldsymbol{n}} \parallel \boldsymbol{H}$ to one with $-\hat{\boldsymbol{n}} \parallel \boldsymbol{H}$ ("$\hat{\boldsymbol{n}}$ soliton"), i.e. a domain wall due to a reorientation of $\hat{\boldsymbol{n}}$ (this structure is topologically equivalent to one where θ varies between θ_L and $2\pi-\theta_L$ by going through $\theta=0$; however, they differ in energy and extension). The contours in Figs. 7.25(e, f) imply a planar defect in which the angle θ goes from θ_L on one side of the wall to $2\pi-\theta_L$ on the other side by passing through $\theta=\pi$ in the middle of the wall. The structures corresponding to Figs. 7.25(c–f) reflect the degeneracy of the dipole potential in ^{3}He-B (see Fig. 6.6). Finally, in the domain walls corresponding to Figs. 7.25(g, h) θ also goes through π but only passes through one of the points θ_L or $2\pi-\theta_L$. They may be constructed from the domain walls in Figs. 7.25(f, d) and 7.25(e, c) respectively.

The index $N=0,\,1$ characterizing the domain wall is the same as the index classifying the line defects in ^{3}He-B, where the homotopy group $\pi_1(SO_3)=\mathbb{Z}_2$ is relevant; hence $N\in\mathbb{Z}_2$ Therefore two domain walls with $N=1$ can annihilate each other by virtue of $N_1+N_2=2\equiv 0 \pmod 2$.

The A_1 phase

The order parameter of ^{3}He-A$_1$, (5.5a), does not allow the kind of solitons discussed above since the dipole energy (6.105) is minimized whenever $\hat{\boldsymbol{l}}$ lies in the plane defined by the complex spin vector $\hat{\boldsymbol{d}}+\mathrm{i}\hat{\boldsymbol{e}}$. This implies a *continuous* degeneracy of the ground state rather than a discrete degeneracy as encountered in the A phase. Consequently, domain walls in the A$_1$ phase lack topological stability.

7.8.2 Planar solitons in the A phase

The term "soliton" was first introduced in the present context by Maki and coworkers in their investigation of magnetic excitations in ^{3}He-A and ^{3}He-B (Maki and Tsuneto 1975, Maki and Ebisawa 1975, Maki and Kumar 1976a). Indeed, the spin dynamics in the A phase (i.e. the dynamics of the vector $\hat{\boldsymbol{d}}$) allows localized magnetic excitations, i.e. sine–Gordon solitons (see Section 8.4.4). These are propagating domain walls in the $\hat{\boldsymbol{d}}$ field where the orientation of $\hat{\boldsymbol{d}}$ changes from $\hat{\boldsymbol{d}} \parallel \hat{\boldsymbol{l}}$ to $-\hat{\boldsymbol{d}} \parallel \hat{\boldsymbol{l}}$ as shown in Fig. 7.21. They have the following properties: (i) they are localized objects; (ii) for small perturbations, they behave as classical particles; (iii) they carry a topological charge whose conservation is guaranteed by topological conservation.

The soliton solution for the $\hat{\boldsymbol{d}}$ domain wall only holds if the underlying $\hat{\boldsymbol{l}}$ texture is *uniform* throughout the system. Clearly, the dipole coupling of the $\hat{\boldsymbol{d}}$ soliton to $\hat{\boldsymbol{l}}$ will distort any given uniform $\hat{\boldsymbol{l}}$ texture, and thereby eliminates the necessary condition for the existence of the $\hat{\boldsymbol{d}}$ soliton itself. Hence the $\hat{\boldsymbol{d}}$ soliton is unstable.

As discussed in Section 7.8.1, a domain wall in the $\hat{\boldsymbol{l}}$ field is topologically equivalent to a $\hat{\boldsymbol{d}}$ domain wall. Which one of the domain walls will actually be formed then depends on the details of the energies involved (this is so at least in an open system, where boundary conditions may be neglected). In turns out that in an unbounded system a domain wall in the $\hat{\boldsymbol{l}}$ field with a slight adjustment of $\hat{\boldsymbol{d}}$ ("composite soliton") has the lowest energy (Maki and Kumar 1977a,b, 1978). Although such a domain wall is not a soliton solution in the strict mathematical sense, it is nevertheless referred to as a "soliton" because it is also localized and topologically stable.

The composite soliton is thought to be the stable end product of a decaying $\hat{\boldsymbol{d}}$ soliton (Maki 1977b, 1986, Kumar 1977), where the latter may be created by pulsed NMR (see Section 8.4.5). For such a decay to occur, the $\hat{\boldsymbol{d}}$ soliton has to transfer its winding number to the $\hat{\boldsymbol{l}}$ texture. Details of the creation of composite solitons and of the scenario of the transfer of the winding from $\hat{\boldsymbol{d}}$ to $\hat{\boldsymbol{l}}$ are, however, not yet known. In fact, $\hat{\boldsymbol{d}}$ solitons themselves have so far not been observed at all. A more detailed study also has to take into account the effect of boundaries. Since $\hat{\boldsymbol{l}}$ is fixed on the surface of a container, the creation of a domain wall in $\hat{\boldsymbol{l}}$ with given winding number, starting from a more or less uniform $\hat{\boldsymbol{l}}$ texture with zero winding, can only take place if the texture carrying the *opposite* winding is somehow spread over the rest of the sample.

Energetics of solitons

To determine the spatial structure of a soliton, we follow Maki and Kumar (1977a,b) and start from the bending free energy $F_{\text{bend}} = \int d^3r\, f_{\text{bend}}$ in weak coupling, (7.19), amended by the dipole interaction energy $\Delta F_{\text{D}} = \int d^3r\, \Delta f_{\text{D}}$, (6.104). The more general case with F_{bend}, (7.20), valid at all temperatures and including strong-coupling and Fermi-liquid corrections, has been discussed by Maki (1986a). The orbital triad $(\hat{\boldsymbol{m}}, \hat{\boldsymbol{n}}, \hat{\boldsymbol{l}})$ and $\hat{\boldsymbol{d}}$ are parametrized by Euler angles, with $\hat{\boldsymbol{m}}$, $\hat{\boldsymbol{n}}$ and $\hat{\boldsymbol{l}}$ given by (7.11) and with $\gamma = -\phi$ (see the discussion below (7.13)); the vector $\hat{\boldsymbol{d}}$ is written as

$$\hat{\boldsymbol{d}} = (\sin\theta\cos\psi, \sin\theta\sin\psi, \cos\theta). \tag{7.149}$$

We now assume a strong static magnetic field $\boldsymbol{H} \parallel \hat{\boldsymbol{z}}$ to confine $\hat{\boldsymbol{d}}$ and (by dipolar coupling) $\hat{\boldsymbol{l}}$ to the (x, y) plane, i.e.

$$\hat{\boldsymbol{l}} = (\cos\alpha, \sin\alpha, 0), \tag{7.150a}$$

$$\hat{\boldsymbol{\Delta}} = 2^{-1/2} e^{-i\gamma}(-i\sin\alpha, i\cos\alpha, -1), \tag{7.150b}$$

$$\hat{\boldsymbol{d}} = (\cos\psi, \sin\psi, 0). \tag{7.151}$$

The total free energy is then given by

$$\begin{aligned} F = \tfrac{1}{2}A\int d^3r\, [&|\nabla\alpha|^2 + 2(\hat{\boldsymbol{l}}\cdot\nabla\alpha)^2 + 4\,|\nabla\gamma|^2 \\ &- 2(\hat{\boldsymbol{l}}\cdot\nabla\gamma)^2 + 4\,|\nabla\psi|^2 - 2(\hat{\boldsymbol{l}}\cdot\nabla\psi)^2 \\ &+ 2\alpha_z(\hat{\boldsymbol{l}}\cdot\nabla\gamma) - 6\gamma_z(\hat{\boldsymbol{l}}\cdot\nabla\alpha) \\ &- 2\xi_D^{-2}\cos^2(\alpha-\psi)], \end{aligned} \tag{7.152}$$

where $\alpha_z = \partial\alpha/\partial z$ etc. and $A = \frac{1}{2}K_1\Delta_0^2$ and where the pure divergence in (7.19) has been neglected. We are now looking for planar solutions of (7.152). To this end, we note that α, γ and ψ are functions of only a single variable $s = \hat{\boldsymbol{k}}\cdot\boldsymbol{r}$, with $\hat{\boldsymbol{k}} = (\hat{k}_x, \hat{k}_y, \hat{k}_z)$. Then (7.152) reduces to

$$\begin{aligned} F = \tfrac{1}{2}A\sigma(\hat{\boldsymbol{k}})\int ds\, [&(1+2a^2)\alpha_s^2 + 2(2-a^2)(\gamma_s^2+\psi_s^2) \\ &- 4k_z a\alpha_s\gamma_s - 2\xi_D^{-2}\cos^2(\alpha-\psi)], \end{aligned} \tag{7.153}$$

with $a = \hat{\boldsymbol{l}}\cdot\hat{\boldsymbol{k}} = \hat{k}_x\cos\alpha + \hat{k}_y\sin\alpha$, and $\sigma(\hat{\boldsymbol{k}})$ is the surface area of the domain wall, with normal vector $\hat{\boldsymbol{k}}$. The free-energy functional (7.153) has to be minimized with respect to the functions $\alpha(s)$, $\gamma(s)$ and $\psi(s)$. This may be done by solving the Euler–Lagrange equations

$$\frac{\partial}{\partial s}\frac{\partial f(s)}{\partial(dy_i/ds)} - \frac{\partial f(s)}{\partial y_i} = 0, \tag{7.154}$$

where $F = \int ds\, f(s)$ and $y_i = \alpha, \gamma, \psi$. Clearly, the overall phase γ is a cyclic variable (since f only depends on γ_s) and may be eliminated, yielding

$$\frac{F}{\sigma(\hat{\boldsymbol{k}})} = \tfrac{1}{2}A\int ds\left[\left(1+2a^2-\frac{2k_z^2a^2}{2-a^2}\right)\alpha_s^2 + 2(2-a^2)\psi_s^2 - 2\xi_D^{-2}\cos^2(\alpha-\psi)\right]. \tag{7.155}$$

If the $\hat{\boldsymbol{l}}$ texture is uniform, with $\hat{\boldsymbol{l}} \parallel \hat{\boldsymbol{x}}$ (owing to $\hat{\boldsymbol{l}}$ being clamped in a narrow slab, or imposed as an initial condition in a time-dependent situation), ψ is determined from (7.155) through

$$\frac{d^2\psi}{ds^2} - \frac{1}{2\xi_D'^2}\sin 2\psi = 0. \tag{7.156}$$

This is a time-independent sine-Gordon equation (see (8.139)) with the

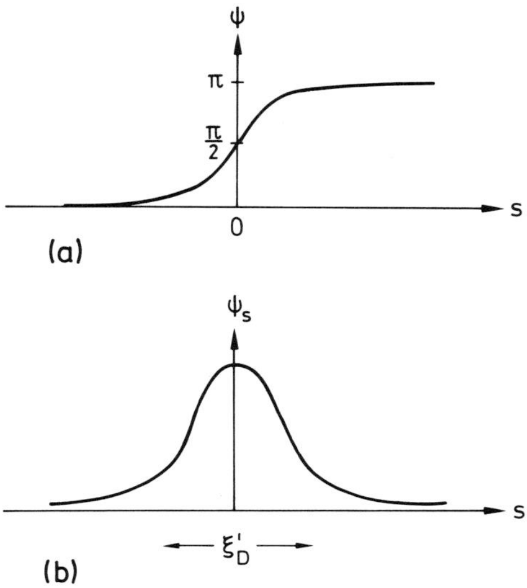

Figure 7.26 (a) The angular variable ψ describing the orientation of $\hat{\boldsymbol{d}}$ in a $\hat{\boldsymbol{d}}$ soliton. (b) The change in ψ, i.e. ψ_s, across the soliton clearly shows the localized nature of the soliton.

solution

$$\psi = 2\tan^{-1}\left[\exp\left(\frac{s}{\xi'_{\mathrm{D}}}\right)\right] \tag{7.157}$$

i.e. $\sin\psi = \operatorname{sech}(s/\xi'_{\mathrm{D}})$ and $\xi'_{\mathrm{D}} = [2(1-\frac{1}{2}\hat{k}_x^2)]^{1/2}\xi_{\mathrm{D}}$. The solution (7.157) describes a $\hat{\boldsymbol{d}}$ soliton at rest, i.e. a kink in the $\hat{\boldsymbol{d}}$ field, where the angle ψ varies from 0 at $s = -\infty$ to π at $s = +\infty$. The change in the orientation,

$$\psi_s = \frac{1}{\xi'_{\mathrm{D}}}\operatorname{sech}\frac{s}{\xi'_{\mathrm{D}}}, \tag{7.158}$$

occurs in a narrow localized region of width ξ'_{D} (see Fig. 7.26). The energy of the $\hat{\boldsymbol{d}}$ soliton comes out as

$$\frac{F_{\hat{d}}}{\sigma(\hat{\boldsymbol{k}})} = 2^{5/2}A(1-\tfrac{1}{2}\hat{k}_x^2)^{1/2}\xi_{\mathrm{D}}^{-1}. \tag{7.159}$$

Clearly, the $\hat{\boldsymbol{d}}$ soliton assumes the lowest energy for $\hat{\boldsymbol{k}} \parallel \hat{\boldsymbol{l}}$, i.e. $\hat{k}_x = 1$. The $\hat{\boldsymbol{d}}$ soliton carries a spin current with polarization along $\hat{\boldsymbol{z}}$ (Maki 1977b), i.e. a magnetization pulse.

Composite solitons

We shall now consider planar textures where both $\hat{\boldsymbol{l}}$ and $\hat{\boldsymbol{d}}$ cooperate in the reorientation, which have been named "composite" solitons (Maki and Kumar 1977a,b). For a planar texture with its surface normal parallel to $\boldsymbol{H}$

("twist" structure), i.e. $\hat{\boldsymbol{k}} = (0, 0, 1)$ and hence $a = \hat{\boldsymbol{l}} \cdot \hat{\boldsymbol{k}} = 0$, (7.155) reduces to

$$\frac{F}{\sigma(k_z)} = \tfrac{1}{2}A \int \mathrm{d}z \, [\tfrac{1}{5}u_z^2 + \tfrac{4}{5}v_z^2 - 2\xi_{\mathrm{D}}^{-2}\cos^2 v], \tag{7.160}$$

where we have introduced new variables

$$\left.\begin{aligned} u &= \alpha + 4\psi, \\ v &= \alpha - \psi. \end{aligned}\right\} \tag{7.161}$$

The Euler–Lagrange equations for u and v,

$$\left.\begin{aligned} \tfrac{2}{5}v_{zz} - \tfrac{1}{2}\xi_{\mathrm{D}}^{-2}\sin 2v &= 0, \\ u_{zz} &= 0, \end{aligned}\right\} \tag{7.162a}$$

are seen to decouple. They have the solution

$$\left.\begin{aligned} &v = 2\tan^{-1}\left\{\exp\left[\pm\left(\frac{5}{2}\right)^{1/2}\frac{z}{\xi_{\mathrm{D}}}\right]\right\}, \\ &u = \text{constant}. \end{aligned}\right\} \tag{7.162b}$$

For the general solution, valid also beyond the Ginzburg–Landau regime, see (8.146a,b). Hence the *relative* coordinate v changes from 0 to π, while u, the centre-of-mass coordinate, is constant. This means that $\hat{\boldsymbol{l}}$ only has to do $\tfrac{4}{5}$ of the reorientation, with $\hat{\boldsymbol{d}}$ doing the remaining $\tfrac{1}{5}$ as one goes through the domain wall. This reorientation is shown in Fig. 7.27. The correspond-

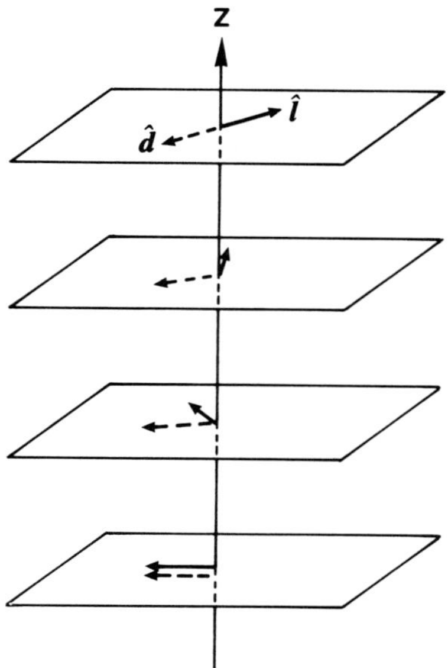

Figure 7.27 Twist composite soliton in ^{3}He-A; the change in orientation occurs along $\hat{z}$, with $\hat{\boldsymbol{l}}$ doing $\tfrac{4}{5}$ of the reorientation.

ing free energy is found to be

$$F_{\text{comp}}^{\text{twist}} = 5^{-1/2} F_{\boldsymbol{d}}^{\text{twist}}, \tag{7.163}$$

and is hence considerably lower than that of the pure $\hat{\boldsymbol{d}}$-twist soliton given by (7.159) with $\hat{k}_x = 0$. This is due to the fact that the width of the composite domain wall, $(\frac{2}{5})^{1/2}\xi_D$, is itself a factor $5^{-1/2}$ smaller than that of the pure $\hat{\boldsymbol{d}}$ domain wall, which is also reflected in the energy, given by $4A\xi_D^{-2}$ times the domain-wall width. Twisting the $\hat{\boldsymbol{d}}$ texture is obviously more costly in energy than twisting $\hat{\boldsymbol{l}}$. This is clear from the free-energy expression (7.20), where, according to (7.21), the coefficients of the splay and twist term of the $\hat{\boldsymbol{d}}$ texture are about a factor of $\rho_{\text{sp}\perp}/K_{\text{s,t}} \approx 4$ larger than the corresponding coefficients for the $\hat{\boldsymbol{l}}$ terms. The equilibrium configuration of a twist composite soliton in the immediate vicinity of the transition into the A_1 phase has been investigated by Gongadze *et al.* (1978b).

In general the vector $\hat{\boldsymbol{k}}$, i.e. the orientation of the domain wall, is expected to be determined by the shape of the container in conjunction with the orientation of the magnetic field (Maki and Kumar 1978). For example, in a cylinder the domain-wall surface is smallest for $\hat{\boldsymbol{k}}$ along the cylinder axis. If the magnetic field is also along the axis ($\hat{\boldsymbol{k}} \parallel \boldsymbol{H}$) then the twist composite soliton will be favoured, while for $\boldsymbol{H}$ perpendicular to the axis a composite soliton with $\hat{\boldsymbol{k}} \parallel \hat{\boldsymbol{x}} \perp \boldsymbol{H}$ and a "splay" configuration in $\hat{\boldsymbol{l}}$ (see Fig. 7.36a) has a lower energy. Since the free energy (7.155) can no longer be minimized analytically in this case, these configurations have been investigated variationally. Using the parametrization (7.161), the splay composite soliton is well represented by

$$\left.\begin{aligned} \sin v &= \operatorname{sech} \eta z, \\ \alpha &= tv + \tfrac{1}{2}\pi(1-t), \end{aligned}\right\} \tag{7.164}$$

with $t = 0.681$ and $\eta = 0.70\xi_D$. Of all domain walls, the twist composite soliton has the lowest energy.

The effect of boundaries on domain walls in a restricted geometry of width of order ξ_D has been investigated by Lin-Liu and Maki (1980b, 1981) and Maki and Bruinsma (1980) for parallel-plate and circular-cylinder geometries respectively.

A soliton in ^{3}He-A represents a localized reorientation of the preferred directions of the order parameter and thus carries a local deficit of nuclear dipole energy. Consequently, such a structure acts as a potential well for spin oscillation. This leads to well-defined satellite peaks in the NMR spectrum, which will be discussed in Section 8.4.5.

Soliton lattice

We have so far considered domain walls as single entities in an otherwise homogeneous texture. Clearly, there may be more than one domain wall in a sample. In particular, if composite solitons do indeed originate from $\hat{\boldsymbol{d}}$ solitons then many such solitons will be created after a magnetic pulse. Since solitons experience a strong mutual repulsion, they are expected to form an array of equally spaced domain walls (Bruinsma and Maki 1979c). Indeed, the Euler–Lagrange equations (7.154) for the free energy (7.160) have a stable periodic solution

$$\left.\begin{aligned} \cos v &= \mathrm{sn}\left[\left(\frac{5}{2}\right)^{1/2}\frac{1}{k}\frac{z}{\xi_{\mathrm{D}}}\,\middle|\, k\right],\\ u &= \text{constant},\end{aligned}\right\} \tag{7.165}$$

where sn is a Jacobian elliptic function (Abramowitz and Stegun 1965) of parameter $m = k^2$. This solution describes an evenly spaced soliton *lattice* with lattice constant

$$L_0 = (\tfrac{8}{5})^{1/2}\xi_{\mathrm{D}} k K(k), \tag{7.166}$$

where $K(k)$ is a complete elliptic integral of the first kind. In the limit $k \to 1$ it reduces to

$$\cos v = \tanh\left[\left(\frac{5}{2}\right)^{1/2}\frac{z}{\xi_{\mathrm{D}}}\right], \tag{7.167}$$

i.e. the equation for a single soliton, which is identical with (7.162b). For $k \to 0$, a harmonic solution

$$\cos v = \sin\left[\left(\frac{5}{2}\right)^{1/2}\frac{z}{\xi_{\mathrm{D}}}\right] \tag{7.168}$$

is obtained, where $\hat{\boldsymbol{l}}$ and $\hat{\boldsymbol{d}}$ rotate oppositely at constant rate.

For other orientations of the magnetic field relative to the domain wall, soliton lattices with different $\hat{\boldsymbol{l}}$, $\hat{\boldsymbol{d}}$ configurations may be determined variationally as in the case of a single soliton.

7.8.3 Planar solitons in the B phase

As discussed earlier in this section, domain walls in bulk ^{3}He-B are only stable if the vector $\hat{\boldsymbol{n}}$ is asymptotically aligned, say by an external magnetic field. The symmetry of ^{3}He-B then allows qualitatively different domain walls, as shown in Fig. 7.25 (Mineev and Volovik 1978); they are nonsingular planar defects ("planar solitons"). As in the case of ^{3}He-A, planar solitons in the B phase were introduced and extensively discussed by Maki and coworkers (Maki and Tsuneto 1975, Maki and Ebisawa 1975, Maki and Kumar 1976a,b, 1977c, Maki and Lin-Liu 1978, Maki 1977a,b,c, 1986a).

Of all orientational energies in ^{3}He-B, the dipole energy is by far the largest. Therefore it is energetically favourable to keep the rotation angle θ at its equilibrium value $\theta = \theta_L$. The remaining degree of freedom is then the rotation axis $\hat{\boldsymbol{n}}$ itself, which may change its orientation between the asymptotic configurations $\hat{\boldsymbol{n}} \parallel \boldsymbol{H}$ and $-\hat{\boldsymbol{n}} \parallel \boldsymbol{H}$. This leads to a "planar $\hat{\boldsymbol{n}}$ soliton", whose structure is determined as follows (Maki and Kumar 1976a, 1977c). We parametrize $\hat{\boldsymbol{n}}$ by

$$\hat{\boldsymbol{n}} = (\cos\varphi \sin\eta, \sin\varphi \sin\eta, \cos\eta) \tag{7.169}$$

and assume that φ and η depend only on a single variable $s = \hat{\boldsymbol{k}} \cdot \boldsymbol{r}$, with $\hat{\boldsymbol{k}} = (\hat{k}_x, \hat{k}_y, \hat{k}_z)$ the unit vector perpendicular to the domain wall. Then the energy $F_{\hat{n}} = \int d^3r\,(f_{\text{bend}} + \Delta f_H)$ due to bending, (7.24b), and magnetic orientation with $\boldsymbol{H} \parallel \hat{\boldsymbol{z}}$, (6.116b), takes the form

$$\begin{aligned}\frac{F_{\hat{n}}}{\sigma(\hat{\boldsymbol{k}})} = \frac{\hbar^2}{10m^2}\rho_s^0 \int ds\,\{\tfrac{5}{2}\{\eta_s^2 + \varphi_s^2 \sin^2\eta \\ &- \tfrac{1}{16}[(3^{1/2}a + 5^{1/2}b)\eta_s + (5^{1/2}a - 3^{1/2}b)\varphi_s \sin\eta]^2\} \\ &- \tfrac{1}{4}\xi_H^{-2}\cos^2\eta\},\end{aligned} \tag{7.170}$$

where

$$\left.\begin{aligned} a &= \cos\eta\,(\hat{k}_x \cos\varphi + \hat{k}_y \sin\varphi) - \hat{k}_z \sin\eta, \\ b &= \hat{k}_x \sin\varphi - \hat{k}_y \cos\varphi, \end{aligned}\right\} \tag{7.171}$$

$\sigma(\hat{\boldsymbol{k}})$ is the surface area of the domain wall, and ξ_H is the magnetic healing length for ^{3}He-B, as defined in (7.29).

In the case of $\hat{\boldsymbol{k}} \parallel \hat{\boldsymbol{z}}$, implying a splay-like structure of the domain wall, i.e. for $\hat{\boldsymbol{k}}$ parallel to $\boldsymbol{H}$, (7.170) reduces to

$$\begin{aligned}\frac{F_{\hat{n}}^{\text{splay}}}{\sigma(\hat{\boldsymbol{k}})} = \frac{\hbar^2}{10m^2}\rho_s^0 \int dz\,\{\tfrac{5}{2}[\eta_z^2 + \varphi_z^2 \sin^2\eta \\ &- \tfrac{1}{16}\sin^2\eta\,(3^{1/2}\eta_z + 5^{1/2}\varphi_z \sin\eta)^2] \\ &- \tfrac{1}{4}\xi_H^{-1}\cos^2\eta\}.\end{aligned} \tag{7.172}$$

The Euler–Lagrange equations (7.154) then yield

$$\eta_z = \pm(\tfrac{5}{8})^{1/2}\xi_H^{-1}\left[\frac{1 - \frac{5}{16}\sin^2\eta}{1 - \frac{1}{2}\sin^2\eta}\right]^{1/2} \sin\eta, \tag{7.173a}$$

$$\varphi_z = 15^{1/2}\frac{\sin\eta}{16 - 5\sin^2\eta}\eta_z, \tag{7.173b}$$

which give the explicit profile of the $\hat{\boldsymbol{n}}$ soliton: $\hat{\boldsymbol{n}}$ changes from antiparallel to $\boldsymbol{H}$ at $z = -\infty$ to parallel at $z = +\infty$ and is perpendicular to $\boldsymbol{H}$ at $z = 0$; this splay-like behaviour, described by η, is quite similar to the reorientation shown in Fig. 7.2(b) and occurs in a confined region of width ξ_H. However, superimposed on this profile is a reorientation of the xy component of $\hat{\boldsymbol{n}}$, described by φ, which changes from $\varphi \approx -0.33$ at $z = -\infty$ to $\varphi \approx 0.33$ at $z = +\infty$, with $\varphi = 0$ at $z = 0$. This gives an additional wave-like component to the reorientation of $\hat{\boldsymbol{n}}$. Hence we see that, unlike the solitons in ^{3}He-A, the $\hat{\boldsymbol{n}}$ soliton has an internal *three*-dimensional structure. We also note that the width of the $\hat{\boldsymbol{n}}$ soliton is determined by ξ_H, which is much larger than the corresponding lengths in ^{3}He-A. Hence the $\hat{\boldsymbol{n}}$ soliton is a very extended object.

Other configurations of $\hat{\boldsymbol{n}}$ solitons, as well as the influence of surfaces on $\hat{\boldsymbol{n}}$ solitons in a parallel-plate geometry, when the surface orientation energies (6.130) and (6.131) dominate, have been discussed by Maki and Kumar (1977c). The contribution of solitons in the B phase to the thermodynamic properties has been considered by Kumar (1978).

7.9 SURFACE-INDUCED TEXTURES

We now want to examine the role of surfaces and boundaries in determining the textures in a container of given shape. After all, any experimental sample has a finite size, so that surfaces will always have some effect. In particular, in the absence of external fields, boundaries of a sample will be essentially responsible for the equilibrium texture.

As in the bulk situation, textures at a surface may contain defects. We shall therefore first consider the types of defects and singularities related to the existence of a boundary. This involves three different aspects (Anderson and Palmer 1977): first, one has to classify the topologically stable defects that are permitted by the order-parameter structure; secondly, while in the infinitely extended bulk defects always correspond to excited states of the system, the presence of a surface may *force* the existence of a defect even in the ground state. One therefore has to ask whether in a given container a defect-free texture is possible at all. Thirdly, one has to know how a surface defect connects with the interior of the sample.

The first aspect may be answered by using topological methods (i.e. homotopy theory) similar to those employed in the bulk situation (Volovik and Mineev 1976b, Volovik 1979, Bailin and Love 1978a,c,e). This requires a mapping of a contour on the surface, surrounding a defect, into the degeneracy space $\tilde{R}$ of the surface. In general, the latter is a subspace of the degeneracy space R of the bulk, depending on the restrictions on the order parameter imposed by the surface.

For the second point, the topological properties of the container itself have to be investigated. For this, it is necessary to know the "connectedness" n of the container, i.e. it is simply connected ($n = 0$, e.g. a sphere), doubly connected ($n = 1$, e.g. a torus), triply connected ($n = 2$, e.g. a pretzel) etc. More precisely, one needs to know the so-called "Euler characteristic" of the surface, given by

$$E = 2(1 - n). \qquad (7.174)$$

It can be proved that the sum of the topological quantum numbers of all point singularities of a vector field on the surface (for example the $\hat{\boldsymbol{l}}$ field in the A phase) is equal to E (Mermin 1977b). For example, for a sphere ($E = 2$), the total topological quantum number is 2. Hence in a sphere the equilibrium texture of ^{3}He-A must have two quanta of surface circulation (see below). On the other hand, textures in a torus ($E = 0$) may be defect-free. In fact, a torus is the *only* container with this property!

Concerning the third aspect, a surface point (line) defect may, but need not, be the termination point (line) of a vortex line (domain wall) in the bulk. This in fact depends on the details of the energetics involved; although a bulk defect will cost additional energy, it may help to reduce the bending energy in the neighbourhood of the surface defect, as is the case in tiny spheres (see below).

Even though most experimental cells have a simply connected geometry and hence necessarily contain surface defects, the importance of these defects for the overall configuration of the bulk texture depends in an essential way on the shape of the cell. For example, in a flat disc-shaped cell the volume occupied by surface defects may essentially be neglected, in contrast with the case of a sphere. This situation will be discussed in Section 7.9.5.

From the discussion of the various healing lengths in superfluid ^{3}He (see Section 7.2.3), we know that in the B phase the lengths are relatively large, i.e. often comparable to the usual container sizes. (We note that the much shorter dipolar length ξ_D is almost always unimportant in ^{3}He-B because one is in a dipole-locked situation ($\theta = \theta_L$)). In addition, the relevant surface energies are rather small, so that boundary conditions on $\hat{\boldsymbol{n}}$ are not strict. Consequently, an $\hat{\boldsymbol{n}}$ texture is usually smoothly varying over the entire sample.

Textures in the A phase are generally more intricate than in the B phase because the order parameter is characterized by *two* vector fields $\hat{\boldsymbol{l}}$ and $\hat{\boldsymbol{d}}$. Furthermore, the characteristic lengths involved, i.e. the dipolar length ξ_D and the magnetic healing length ξ_H^A (with $\xi_H^A < \xi_D$ except for rather small fields), are much shorter than the relevant lengths in the B phase, i.e. ξ_H^B or the surface healing length ξ_S. This implies that in containers whose dimensions are large compared with ξ_D the liquid will be in equilibrium with respect to both the dipole energy and the susceptibility anisotropy. Yet another complication is that spin and mass superflow are not separated as in the B phase. Therefore superfluid or normal flow will not only influence the vector $\hat{\boldsymbol{l}}$ but, by dipole coupling, will also affect the magnetic properties. This implies that, in general, heat flow will induce NMR shifts.

7.9.1 ^{3}He-B in a slab

The $\hat{\boldsymbol{n}}$ textures in the B phase are determined by the bulk orientation effect provided by a magnetic field, which tends to align $\hat{\boldsymbol{n}}$ with the axis of the magnetic field (see (6.116)), as well as by surface orientation energies as discussed below (6.128) and, of course, by the bulk bending energy density (7.24a). There are two contributions to the surface energy, both due to the orienting effect of the surface on the slightly distorted, i.e. anisotropic, order parameter. In low magnetic fields the distortion of the order parameter is mainly due to the dipole interaction. The corresponding surface energy (6.130) tends to align $\hat{\boldsymbol{n}}$ along the axis of the surface normal. At higher magnetic fields, exceeding a critical value

$$H_S = \left(\frac{9}{8}\frac{g_D}{\Delta\chi}\right)^{1/2} = \frac{5^{1/2}3}{4}(1+F_0^a)^{1/2}H^* \tag{7.175}$$

of order 25 G (see (7.28c)), which is temperature-independent in the Ginzburg–Landau regime, the order-parameter distortion due to the magnetic-anisotropy energy dominates. The corresponding surface energy (6.132) is minimized (i) for $\hat{\boldsymbol{n}}$ parallel to the axis of the surface normal if the magnetic field is also perpendicular to the surface, and (ii) for $\hat{\boldsymbol{n}} = \pm 5^{-1/2}(\hat{\boldsymbol{H}} + 3^{1/2}\hat{\boldsymbol{s}} \times \hat{\boldsymbol{H}} \pm \hat{\boldsymbol{s}})$ when the magnetic field lies in the plane of the surface (see (6.133)).

Following Smith *et al.* (1977), we shall now consider the case of a rather strong magnetic field $H \gg H_S$ parallel to the surface (with surface normal $\hat{\boldsymbol{s}} \parallel \hat{\boldsymbol{z}}$) and small plate separation $L \ll \xi_H$, where ξ_H is the magnetic healing length discussed in (7.29). Under these circumstances, the bending of the $\hat{\boldsymbol{n}}$ texture will be small. As a first step, we therefore determine the minimum of the total energy in the limit of infinite stiffness of the texture, when only the magnetic orientation energy (6.116) and the surface energy (6.130) contribute. The result, using (6.133), is

$$(\hat{\boldsymbol{n}}\cdot\hat{\boldsymbol{H}})^2\big|_{\text{inf}} = \frac{1}{5}\left[1 + \frac{1}{3}\frac{\lambda_D}{(1+F_0^a)^2}\left(\frac{L}{\xi}\right)\frac{(\gamma\hbar)^2 N_F}{\chi_N - \chi_B}\right], \tag{7.176a}$$

where the small correction term due to the bulk orientation energy may be neglected.

The bending of the $\hat{\boldsymbol{n}}$ texture caused by the different preferred directions in the bulk and at the surface may be calculated in lowest order by expanding about the limit of infinite stiffness. One finds that the projection of $\hat{\boldsymbol{n}}$ onto the surface normal does not change in this approximation, but the projection onto the magnetic field does, and is given by

$$(\hat{\boldsymbol{n}}\cdot\hat{\boldsymbol{H}})^2 = (\hat{\boldsymbol{n}}\cdot\hat{\boldsymbol{H}})^2\big|_{\text{inf}} + \frac{1}{200}\left(\frac{L}{\xi_H}\right)^2\left[1-\left(\frac{2z}{L}\right)^2\right] + O\left[\left(\frac{L}{\xi_H}\right)^4\right], \tag{7.176b}$$

where z is the length coordinate between the plates, $-\frac{1}{2}L \leqslant z \leqslant \frac{1}{2}L$. As will be shown in Section 8.4.6, the parabolic shape of $(\hat{\boldsymbol{n}}\cdot\hat{\boldsymbol{H}})^2$ acts as a confining potential for an oscillating magnetization wave. Under appropriate conditions, the potential will have bound-state solutions or "standing spin waves", which have been observed as satellite peaks in the NMR signal (Osheroff *et al.* 1977).

The qualitative behaviour of the $\hat{\boldsymbol{n}}$ textures in high field at larger plate separation ($L \gg \xi_H$) may be described as follows: in the bulk of the sample the $\hat{\boldsymbol{n}}$ texture is uniform, and is oriented parallel to the axis of the field; this uniform region is connected to surface layers of width ξ_H, where $\hat{\boldsymbol{n}}$ changes over into the surface-dominated orientation.

The behaviour at low magnetic fields $H \ll H_S$ (parallel to the surface) is more complicated. Here the surface energy is independent of the magnetic field, whereas the bulk field orientation energy is proportional to H^2L. It follows that for small L the surface energy dominates and $\hat{\boldsymbol{n}}$ is oriented uniformly parallel to the surface normal. For increasing L, a second-order phase transition occurs at a critical value L_c, beyond which the orientation of $\hat{\boldsymbol{n}}$ starts to deviate from the directions of the surface normal (Fomin and Vuorio 1975). Similarly, for small plate separation ($L < \xi_H$) and increasing magnetic field parallel to the surface, there is a second-order transition from the surface-dipole-energy-dominated regime, where $\hat{\boldsymbol{n}}$ is perpendicular to the surface, to the magnetic-surface-energy-dominated regime, where $\hat{\boldsymbol{n}}$ is tilted from the surface normal. This transition has been observed by Webb *et al.* (1977). The complete phase diagram in the plate-separation–magnetic-field plane has been determined by Fomin and Vuorio (1975).

The $\hat{\boldsymbol{n}}$ textures in a tilted magnetic field for both a single planar wall and a slab have been considered by Lin-Liu and Maki (1978), using a variational method.

7.9.2 ^{3}He-B in a cylinder

In contrast with a slab geometry, which at least in principle allows a uniform $\hat{\boldsymbol{n}}$ texture, $\hat{\boldsymbol{n}}$ configurations in a cylinder are necessarily nonuniform. An explicit computation of the equilibrium texture is therefore considerably more complex. An external magnetic field, even it is is along the cylinder axis, introduces further complications.

In zero magnetic field the $\hat{\boldsymbol{n}}$ texture is determined by the surface dipole energy, which seeks to orient $\hat{\boldsymbol{n}}$ along the surface normal (i.e. radially outward), and the bulk bending energy. For small cylinder radii $R \ll \xi_S$, where ξ_S is the surface healing length (7.31a) (which is of the order of a millimetre), the surface causes only a slight deviation from a uniform $\hat{\boldsymbol{n}}$ texture. Hence there are two possibilities: $\hat{\boldsymbol{n}}$ perpendicular or parallel to the cylinder axis. The former case is easily seen to be more favourable since the boundary condition $\hat{\boldsymbol{n}} \parallel \hat{\boldsymbol{s}}$ is at least partially satisfied, in contrast with the latter case. This might be expected to change in a finite magnetic field along the cylinder axis, which favours $\hat{\boldsymbol{n}}$ lying in this direction too. By comparing the surface dipole energy (6.130) with the bulk magnetic orientation energy (6.116b), one finds that for $H < H_S$ the "in-plane" texture (Fig. 7.28a) is

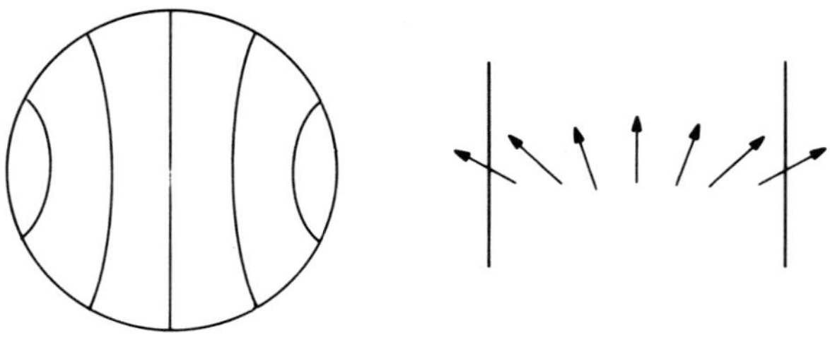

Figure 7.28 $\hat{\boldsymbol{n}}$ texture in a cylinder: (a) "in-plane" texture viewed along the cylinder axis; (b) "flare-out" texture viewed perpendicular to the axis.

energetically favoured. The explicit configuration of this texture has been obtained variationally (Smith *et al.* 1977). To this end, $\hat{\boldsymbol{n}}$ is assumed to lie in the (x, y) plane, with $\boldsymbol{H}$ along the cylinder axis,

$$\hat{\boldsymbol{n}} = (\cos\varphi, \sin\varphi, 0), \tag{7.177}$$

and φ is given by the ansatz

$$\varphi = \tan^{-1}\frac{2xy}{R_0^2 + x^2 - y^2}, \tag{7.178}$$

which defines a family of curves as shown in Fig. 7.28(a). Here $R_0 > R$ is a variational parameter. The bending free energy $F_{\text{bend}} = \int d^3r\, f_{\text{bend}}$, (7.24b), and the surface dipole energy F_{DS}, (6.130), are then found to be minimized by

$$R_0 \approx 6(R\xi_{\text{S}})^{1/2}, \quad R \ll \xi_{\text{S}}. \tag{7.179}$$

This specifies the in-plane $\hat{\boldsymbol{n}}$ texture for a given radius R.

For larger cylinders ($R > \xi_{\text{S}}$), the surface energy is more important than the bending free energy. Hence the boundary condition $\hat{\boldsymbol{n}} \parallel \hat{\boldsymbol{s}}$ is favoured. Since this is difficult in the in-plane texture, the vector $\hat{\boldsymbol{n}}$ flares out into the third dimension along the cylinder axis ("flare-out" texture) as shown in Fig. 7.28(b). To determine its configuration, $\hat{\boldsymbol{n}}$ is espressed in terms of angles φ and η as in (7.169) and η is written as $\eta = \beta r/R$, with $\beta \leqslant \frac{1}{2}\pi$ as a variational parameter; φ and r are taken to be cylindrical coordinates. Minimization of the free energy for large cylinder radii, $R \gg \xi_{\text{S}}$, yields

$$\beta_{\min} = \tfrac{1}{2}\pi, \tag{7.180}$$

i.e. $\hat{\boldsymbol{n}} \parallel \hat{\boldsymbol{s}}$ at the cylinder surface. Comparison of this free energy with that for the in-plane texture shows that the flare-out texture is lower in energy for radii $R \geqslant R^* = 81\xi_{\text{S}}$ (Smith *et al.* 1977).

Clearly, a magnetic field along the cylinder axis, which favours the flare-out texture anyway, will reduce this crossover value of R. For small fields ($H \ll H_S$), perturbation theory shows that the reduction is proportional to H^2:

$$R^* \approx 81\xi_S\left[1 - \frac{25}{1+F_0^a}\left(\frac{H}{H_S}\right)^2\right]. \tag{7.181}$$

At higher magnetic fields ($H > H_S$), where the magnetic surface energy dominates, explicit calculations become quite cumbersome. The qualitative behaviour may none the less be guessed: while $\hat{\boldsymbol{n}}$ is parallel to $\hat{\boldsymbol{H}}$ over most of the sample, there is a surface layer of thickness ξ_H where $\hat{\boldsymbol{n}}$ changes to the surface-controlled configuration in a magnetic field, exhibiting a spiral-like configuration. This texture may be parametrized by

$$\hat{\boldsymbol{n}} = \sin\beta\cos\alpha\,\hat{\boldsymbol{r}} + \sin\beta\sin\alpha\,\hat{\boldsymbol{\varphi}} + \cos\beta\,\hat{\boldsymbol{z}} \tag{7.182}$$

in cylindrical coordinates, where the angles α and β are functions of r only. At the surface $\alpha = \frac{1}{2}\pi$ and $\beta = \cos^{-1} 5^{-1/2} \approx \frac{1}{3}\pi$. The detailed dependence of α and β on r has been calculated both analytically (Hakonen and Volovik 1982, Maki and Nakahara 1983) and numerically (Jacobsen and Smith 1983). It is found that β decreases rapidly from its value at the boundary to zero at the centre. For small r, it varies quadratically, with a prefactor proportional to e^{-R/ξ_H}. On the other hand, α depends only weakly on r (see Fig. 7.29), changing from $\alpha(R) = \frac{1}{3}\pi$ to $\alpha(0) \approx \pi$ if $R/\xi_H \gg 1$. Behaviour of this type had been found earlier by Spencer and Ihas (1982) in a transverse NMR experiment by deconvoluting the NMR spectra. To this end, they first determined the angle $\beta(r)$ from the NMR shift and shape and then

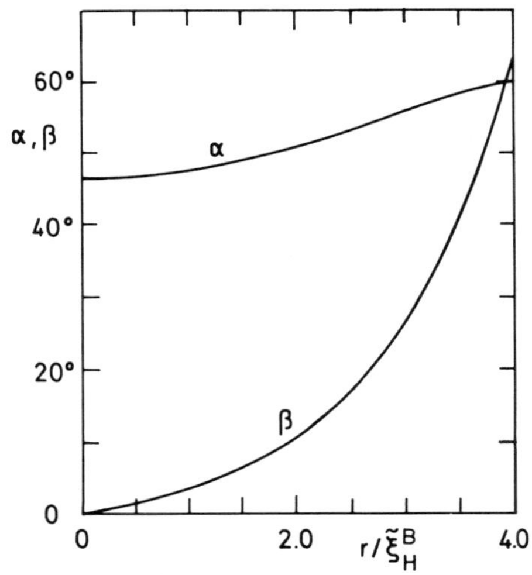

Figure 7.29 Radial dependence of the angles α and β parametrizing the $\hat{\boldsymbol{n}}$ texture in (7.182) for ^{3}He-B in a cylinder of radius $R = 4\tilde{\xi}_H^B$; here $\tilde{\xi}_H^B = (\frac{65}{8})^{1/2}\xi_H^B$, with ξ_H^B defined in (7.29). (After Jacobsen and Smith (1983).)

calculated $\alpha(r)$ by minimizing the free energy with respect to α. These authors even determined the temperature dependence of $\alpha(r)$ and $\beta(r)$ and of the magnetic healing length ξ_H.

7.9.3 ^{3}He-A in a slab

In the absence of external orientational effects (e.g. magnetic field, superflow) or textural defects, the $\hat{\boldsymbol{l}}$ field in a slab will assume a uniform configuration with $\hat{\boldsymbol{l}} \parallel \hat{\boldsymbol{d}}$ aligned perpendicular to the slab. The uniformity will be disturbed by an external magnetic field (unless $\boldsymbol{H}$ lies parallel to the slab), which orients the vector $\hat{\boldsymbol{d}}$ and thereby pulls on $\hat{\boldsymbol{l}}$ via the dipole coupling. The situation is indeed quite similar to the one discussed for the B phase involving the vector $\hat{\boldsymbol{n}}$. The maximum effect occurs if $\boldsymbol{H}$ is perpendicular to the slab. Depending on the width of the slab, there will be a critical field at which the $\hat{\boldsymbol{l}}$ texture undergoes an orientational transition from a uniform to a nonuniform configuration (de Gennes and Rainer 1974, Takagi 1975a, Fetter 1976b, 1977, 1987) in analogy with the Freedericksz transition known from liquid crystals (de Gennes 1974).

We assume that the magnetic field $\boldsymbol{H}$ is perpendicular to the slab, $\boldsymbol{H} \parallel \hat{\boldsymbol{z}}$. We expect the resulting texture to be planar, for example $\hat{\boldsymbol{l}}$ and $\hat{\boldsymbol{d}}$ to lie in the (x, z) plane. The relevant free energy is then given by the sum of the integrals over the energy densities due to bending, (7.19), dipole coupling, (6.104), and magnetic orientation, (6.115). Using the parametrization of $\hat{\boldsymbol{\Delta}}$, (7.12a), with $\alpha = \gamma = 0$, and of $\hat{\boldsymbol{d}}$, (7.149), with $\theta = \frac{1}{2}\pi$, and assuming spatial variations only along $\hat{\boldsymbol{z}}$ $(-\frac{1}{2}L \leq z \leq \frac{1}{2}L)$, we obtain in weak coupling

$$F = A\sigma(\hat{k}_z)\int_{-L/2}^{L/2} \mathrm{d}z\, [\tfrac{1}{2}(1 + 2\cos^2\beta)\beta_z^2 + (1 + \sin^2\beta)\theta_z^2 - \xi_D^{-2}\cos^2(\beta - \theta) + \xi_H^{-2}\cos^2\theta], \tag{7.183}$$

where $\sigma(\hat{k}_z)$ is the surface area with normal along $\hat{\boldsymbol{z}}$. For not too narrow slabs (plate separation $L \gg \xi_D$), $\hat{\boldsymbol{d}}$ and $\hat{\boldsymbol{l}}$ are dipole-locked $(\beta = \theta)$, i.e.

$$F = \tfrac{5}{2}A\sigma(\hat{k}_z)\int \mathrm{d}z\, (\beta_z^2 + \tfrac{2}{5}\xi_H^{-2}\cos^2\beta). \tag{7.184}$$

We note that while the orientation energy is of order $(\beta/\xi_H)^2$ for small β and thus goes to zero proportionally to H^2 as $H \to 0$, the bending energy is at least of order $(\beta/L)^2$. Consequently, for small H such that $\xi_H > L$, bending is too costly and the texture remains uniform. The Euler–Lagrange equation for β then yields

$$\beta_{tt} + \sin 2\beta = 0, \tag{7.185}$$

where $t = z/5^{1/2}\xi_H$. Considering t as a dimensionless time, (7.185) describes a mathematical pendulum subject to the boundary condition $\beta = 0$ at

$t_{\pm}=\pm\frac{1}{2}L/5^{1/2}\xi_H$. Clearly, a solution $\beta\neq 0$ is only possible for "oscillation periods" $T=2\pi/2^{1/2}>2(t_+-t_-)=2L/5^{1/2}\xi_H$. This means that for magnetic fields $H<H_{\mathrm{F}}$, where

$$H_{\mathrm{F}}=(\tfrac{5}{2})^{1/2}\pi\frac{\xi_0}{L}H_{\mathrm{A}}, \tag{7.186}$$

with H_{A} defined in (7.28b), the $\hat{\boldsymbol{l}}$ texture is uniformly oriented perpendicular to the surface, while for $H>H_{\mathrm{F}}$ bending sets in continuously (Fetter 1976b). Note that the critical field decreases with increasing slab width as $H_{\mathrm{F}}\propto 1/L$. The implicit solution for β as a function of z is given by an elliptic integral

$$\int_0^{\beta(z)}\frac{\mathrm{d}\beta'}{[\sin^2\beta(0)-\sin^2\beta']^{1/2}}=\pi\left(\frac{z}{L}+\frac{1}{2}\right)\frac{H}{H_{\mathrm{F}}}. \tag{7.187}$$

For H just above H_{F}, one finds $\beta(0)=2(H/H_{\mathrm{F}}-1)^{1/2}$, i.e. β has infinite slope and, for $H\rightarrow\infty$, tends to $\frac{1}{2}\pi$. The transition has been observed by Hook *et al.* (1986, 1987) in two different experiments: by use of a torsional oscillator and by zero-sound propagation.

In general, and in particular for slab widths $L\lesssim\xi_{\mathrm{D}}$, $\hat{\boldsymbol{d}}$ and $\hat{\boldsymbol{l}}$ will not be dipole-locked. In this case both $\hat{\boldsymbol{d}}$ and $\hat{\boldsymbol{l}}$ have to be treated as separate variables. The Euler–Lagrange equations for $\hat{\boldsymbol{l}}(\beta)$ and $\hat{\boldsymbol{d}}(\theta)$ follow from (7.183) as

$$\beta_{zz}(2+\cos 2\beta)+(\beta_z^2+\theta_z^2)\sin 2\beta-\xi_{\mathrm{D}}^{-2}\sin 2(\theta-\beta)=0, \tag{7.188a}$$

$$\theta_{zz}(3-2\cos 2\beta)+2\theta_z\beta_z\sin 2\beta+\xi_{\mathrm{D}}^{-2}\sin 2(\beta-\theta)+\xi_H^{-2}\sin 2\theta=0, \tag{7.188b}$$

with the boundary conditions $\beta(\pm\frac{1}{2}L)=0$. In contrast with the dipole-locked case, a general analytic solution of these coupled nonlinear differential equations is not available. On the other hand, certain limiting cases may easily be discussed. In the limit of plate separation $L<\xi_{\mathrm{D}}$, $\hat{\boldsymbol{l}}$ and $\hat{\boldsymbol{d}}$ must be uniform with $\hat{\boldsymbol{l}}$ perpendicular to the plates ($\beta=0$). In this case the free energy in (7.183) reduces to

$$F\propto\int\mathrm{d}z\,(\xi_{\mathrm{D}}^{-2}-\xi_H^{-2})\sin^2\theta. \tag{7.189}$$

For small magnetic fields, such that $\xi_H>\xi_{\mathrm{D}}$, the minimum is given by $\theta=0$, i.e. $\hat{\boldsymbol{d}}\parallel\hat{\boldsymbol{l}}$, while for fields $H>H^*$ (i.e. $\xi_H<\xi_{\mathrm{D}}$) the configuration $\hat{\boldsymbol{d}}\perp\hat{\boldsymbol{l}}$ is favoured. The corresponding critical field is of course given by $H^*\approx 28\,\mathrm{G}$, (7.28c). For general magnetic fields and plate separations, the phase boundaries of the two configurations $\hat{\boldsymbol{d}}\parallel\hat{\boldsymbol{l}}$ and $\hat{\boldsymbol{d}}\perp\hat{\boldsymbol{l}}$ may be determined from a linear stability analysis of (7.188) (Ambegaokar and Rainer 1974, Fetter 1976b), as shown in Fig. 7.30. Variational calculations of the nonuniform $\hat{\boldsymbol{d}}$, $\hat{\boldsymbol{l}}$ configuration in this phase diagram have been performed by Fetter (1977) (see also Bruinsma and Maki 1979). It is found that for strong magnetic fields $H\gg H^*$ and $L<L^*$, where $L^*=3^{1/2}\pi\xi_{\mathrm{D}}$,

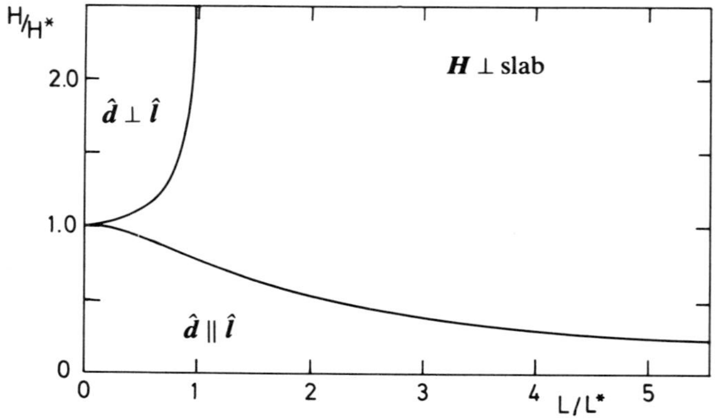

Figure 7.30 Phase boundaries between different $\hat{d}$, $\hat{l}$ configurations of ^{3}He-A in a slab of width L (in units of $L^* = 3^{1/2}\pi\xi_D$) with the magnetic field $\boldsymbol{H}$ oriented perpendicular to the slab. (After Fetter (1976b).)

a uniform configuration with $\hat{d} \perp \hat{l} \parallel \boldsymbol{H}$ is stable (see Fig. 7.30). For somewhat larger plate separations, a configuration with $\hat{d} \parallel \hat{l} \perp \boldsymbol{H}$ (except for a narrow surface layer) is formed. On the other hand, for small magnetic fields $H < H^*$, and $L \lesssim (H^*/H)\xi_D$, the uniform configuration $\hat{d} \parallel \hat{l} \parallel \boldsymbol{H}$ is realized. For intermediate values of H and L, the vector $\hat{l}$ is described by a more complicated deformed texture. A quantitative analysis for narrow slabs, taking into account strong-coupling corrections and temperature dependences, has been presented by Fetter (1987).

The more general case of a tilted magnetic field has been addressed by Fetter (1977). In this situation sharp transitions no longer occur.

Abrupt textural changes have been predicted by Gongadze *et al.* (1978a) to occur in narrow gaps of width $L \gtrsim \xi_D$ with a strong magnetic field applied perpendicular to the slab, when the $A \rightarrow A_1$ transition is approached by changing the temperature.

The $\hat{l}$ and $\hat{d}$ textures in a slab induce shifts in NMR that allow one to make contact with experiment. These will be discussed in Section 8.4.3. The $\hat{l}$ texture in such a geometry may be inferred by measuring the frequency shift and the damping of a torsional oscillator. Comparing the measured values with the results of a model calculation for specific $\hat{l}$ textures allows one to distinguish between various qualitatively different configurations, such as textures with and without domain walls (Hook 1989, Hook *et al.* 1989).

7.9.4 ^{3}He-A in a cylinder

In a cylindrical geometry with radius $R \gg \xi(T)$ the boundary conditions force $\hat{l}$ to point radially outwards at the wall, i.e. along a *curved* surface.

The wall configuration being fixed, the question is now how $\hat{\boldsymbol{l}}$ is oriented in the interior of the cylinder and how the vector $\hat{\boldsymbol{d}}$ behaves in this situation. For cylinder radii $\xi(T) \ll R \ll \xi_D$, i.e. smaller than the dipole healing length, $\hat{\boldsymbol{l}}$ and $\hat{\boldsymbol{d}}$ are uncoupled and $\hat{\boldsymbol{d}}$ may be assumed uniform throughout the cylinder. On the other hand, for $R \gg \xi_D$, $\hat{\boldsymbol{l}}$ and $\hat{\boldsymbol{d}}$ will be dipole-locked in the sample. Furthermore, in the presence of a magnetic field (e.g. along the cylinder axis) $\hat{\boldsymbol{d}}$ will be oriented to lie in the plane perpendicular to both the magnetic field and the cylinder axis.

For zero magnetic field, the situation is quite similar to the one for ^{3}He-B in a cylinder, where $\hat{\boldsymbol{n}}$ may either assume a planar configuration or else flare out into the third dimension to reduce the bending energy. The same applies to the $\hat{\boldsymbol{l}}$ texture in a cylinder. Candidates for planar textures are (see Fig. 7.31) (a) the $\hat{\boldsymbol{l}}$ disgyration, where $\hat{\boldsymbol{l}}$ points radially outward, forming a singular line in the centre, with $\hat{\boldsymbol{d}}$ uniform (de Gennes 1973, Ambegaokar *et al.* 1974), and (b) the "double-half" or "circular" disgyration (sometimes also referred to as the "Pan-Am texture"), having two singular lines on opposite points of the cylinder's surface (Maki 1978).

The three-dimensional ("flare-out") texture corresponds to the Mermin–Ho vortex discussed in Section 7.5 (Anderson and Brinkman 1975, Mermin and Ho 1976). This texture has been studied in detail by Buchholtz and Fetter (1976, 1977a,b). Using a trial state parametrized in cylindrical coordinates as

$$\hat{\boldsymbol{l}} = \hat{\boldsymbol{d}} = \cos\chi\,\hat{\boldsymbol{z}} + \sin\chi\,\hat{\boldsymbol{r}}, \qquad (7.190)$$

$$\hat{\boldsymbol{\Delta}} = 2^{-1/2} e^{i\phi}(\cos\chi\,\hat{\boldsymbol{r}} - \sin\chi\,\hat{\boldsymbol{z}} + i\hat{\boldsymbol{\varphi}}), \qquad (7.191)$$

the bending free energy is obtained from (7.20) as

$$F_{MH} = \pi A \int_0^R \mathrm{d}r\, r \left[5\left(\frac{\mathrm{d}\chi}{\mathrm{d}r}\right)^2 + \frac{1}{r^2}(9 - 8\cos\chi - \cos^2\chi) \right], \qquad (7.192)$$

where $A = \frac{1}{2} K_1 \Delta_0^2$ and we have limited ourselves to the weak-coupling

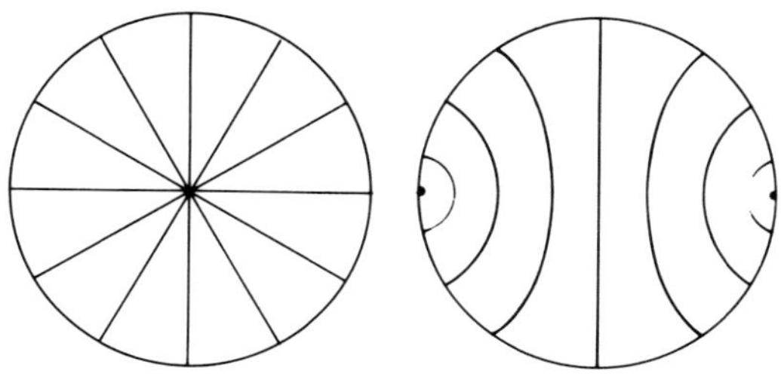

Figure 7.31 Planar $\hat{\boldsymbol{l}}$ textures in a cylinder: (a) radial disgyration; (b) "double-half" or circular disgyration.

regime. The Euler–Lagrange equation for χ may be solved exactly to yield

$$\cot \tfrac{1}{2}\chi = \frac{1}{5^{1/2}}\left(\frac{1}{\rho^{1/2}} - \rho^{1/2}\right), \tag{7.193}$$

where $\rho = \alpha r/R$ and $\alpha = [2/(3-5^{1/2})]^2$. Hence $\chi(r)$ is seen to increase from $\chi(0) = 0$ to $\chi(R) = \frac{1}{2}\pi$ almost linearly. The bending free energy per unit length is thus integrated to

$$\begin{aligned} F_{\mathrm{MH}} &= 2\pi 5^{1/2} A\{5^{1/2}2 - 3 + 4\ln[\tfrac{1}{4}(3+5^{1/2})]\} \\ &\approx 11.4\pi A. \end{aligned} \tag{7.194}$$

We note that this energy is *independent* of the cylinder radius R.

As noted in Section 7.5.3, there is a finite azimuthal superfluid mass current associated with this texture; it is given by

$$g_{\hat{\varphi}} = \frac{8A}{\hbar}\left[\frac{4(1-\cos\chi)}{r} - \frac{\mathrm{d}}{\mathrm{d}r}\cos\chi\right] \tag{7.195}$$

and vanishes smoothly at $r = 0$. This implies a finite angular momentum per unit length and particle (total number N), given by the surface integral

$$\begin{aligned} L_z &= \frac{1}{N}\int \mathrm{d}^2 r\, m r g_{\hat{\varphi}} \\ &= 0.782\hbar\frac{\rho_{\mathrm{s}\parallel}}{\rho}, \end{aligned} \tag{7.196}$$

giving a definite value to this quantity, which had been discussed earlier by Mermin and Ho (1976).

Concerning planar textures, the $\hat{\boldsymbol{l}}$ disgyration was first discussed in detail by Fishman and Privorotskii (1976), who found it to be stable against infinitesimal perturbations. Its parametrization

$$\left.\begin{aligned} \hat{\boldsymbol{l}} &= \hat{\boldsymbol{d}} = \hat{\boldsymbol{r}}, \\ \hat{\boldsymbol{\Delta}} &= 2^{-1/2}(-\hat{\boldsymbol{z}} + \mathrm{i}\hat{\boldsymbol{\varphi}}) \end{aligned}\right\} \tag{7.197}$$

yields a bending free energy per unit length

$$F_{\mathrm{dis}} = \pi A\int_{\xi}^{R} \mathrm{d}r\, r\left(\frac{5}{r^2}\right) \tag{7.198a}$$

$$= 5\pi A \ln\frac{R}{\xi}, \tag{7.198b}$$

where $\xi(T)$ was chosen as a lower cutoff (Buchholtz and Fetter 1976, 1977a). It depends logarithmically on R owing to the singular line along the axis where depairing occurs. This texture does not involve superfluid mass or spin currents. Also, it does not have a surface circulation as has the MH texture. Equating the results for the free energies of the MH texture and the $\hat{\boldsymbol{l}}$ disgyrations shows that for $R \lesssim 10\xi(T)$, i.e. in very narrow pores, the disgyration is energetically favoured.

Investigation of the "double-half" disgyration with

$$\left.\begin{aligned} \hat{\boldsymbol{l}} = \hat{\boldsymbol{d}} &= (\cos\alpha, \sin\alpha, 0), \\ \hat{\boldsymbol{\Delta}} &= 2^{-1/2}(-\cos\alpha, \sin\alpha, \mathrm{i}) \end{aligned}\right\} \tag{7.199}$$

in Cartesian coordinates yields

$$\tan\alpha = \frac{2xy}{R^2 - x^2 + y^2} \tag{7.200}$$

and leads to the same bending free energy (7.198b) as for the $\hat{\boldsymbol{l}}$ disgyration (Maki 1978). Since possible planar configurations may be classified in terms of the number of disgyrations in the cylinder, these two examples belong to the same class with the lowest energy.

Textures induced by two coaxial cylinders have been considered by Arai and Soda (1983). They found that the ratio of the radii of the outer and inner cylinder determines whether a planar or nonplanar texture is assumed.

Planar textures in containers with the same topology as that of a cylinder (e.g. rectangular tubes), as well as the interaction of disgyrations with each other, have been investigated by Maki (1978). Their explicit calculation is simplified by the observation that for two-dimensional textures with $\hat{\boldsymbol{d}} \parallel \hat{\boldsymbol{l}}$, which depend only on one angle α, the local minimum of the free energy is given by $\nabla^2\alpha = 0$, i.e. a Poisson equation. As noted by Golo *et al.* (1979), the Ginzburg–Landau equations for planar textures of superfluid ^{3}He are equivalent to a completely integrable Hamiltonian system for which general solutions may be obtained.

In the presence of an external magnetic field along the axis of the cylinder ($R \gg \xi_D$) the problem becomes considerably more complicated since the additional variable prohibits a simple scaling as in (7.193). As H increases, $\hat{\boldsymbol{d}}$ is forced into the plane perpendicular to the cylinder axis, and dipole locking of $\hat{\boldsymbol{l}}$ and $\hat{\boldsymbol{d}}$ will no longer be energetically favourable. For $H \ll H^*$, with H^* the field where magnetic and dipole coupling are of the same order, the MH texture will only be slightly distorted, with the central region retaining its smooth configuration (Buchholtz and Fetter 1977a,b). In this case the free energy and the angular momentum increase proportionally to H^2. For $H \gtrsim H^*$, a discrete transition to a planar structure is expected, i.e. to an $\hat{\boldsymbol{l}}$ disgyration with $\hat{\boldsymbol{d}} \parallel \hat{\boldsymbol{l}}$, to minimize the dipole energy. However, these two structures are topologically distinct (cf. the difference in surface circulation) and cannot be continuously deformed into one another. As discussed by Nakahara and Ohmi (1979), there is a barrier between the two structures, making the (deformed) MH structure metastable even for magnetic fields where the energy of the disgyration is smaller. The response of the MH texture to a time-dependent rotational perturbation has been studied by Ohmi *et al.* (1981).

As usual, the nonuniformity of the $\hat{\boldsymbol{l}}$ and $\hat{\boldsymbol{d}}$ configurations induced by the cylinder will show up in an NMR experiment via shifts in the signals. This will be calculated in Section 8.4.3.

7.9.5 ^{3}He-A in a sphere

In a spherical container the surface (with normal $\hat{\boldsymbol{s}}$) has an even stronger effect on the orientation of the $\hat{\boldsymbol{l}}$-vector field than in the geometries considered so far (plates and cylinders), since the boundary condition $\hat{\boldsymbol{l}} \parallel \hat{\boldsymbol{s}}$ must now be satisfied simultaneously in *every* direction. As discussed at the beginning of this section, topology alone tells us that in a sphere (Euler characteristic $E = 2$) the equilibrium texture necessarily has to contain surface singularities with a total of two quanta (i.e. 4π) of surface circulation (Anderson and Palmer 1977, Mermin 1977a,b, Volovik 1978b). This is easily understood from the surface geometry of a sphere. Consider a three-dimensional sphere with the orbital triad ($\hat{\boldsymbol{m}}, \hat{\boldsymbol{n}}, \hat{\boldsymbol{l}}$) positioned at the north pole and pointing upwards. We now move the triad to the south pole on lines of constant longitude (i.e. azimuthal angle). When we arrive at the south pole, $\hat{\boldsymbol{l}}$ points downwards of course, but at the same time $\hat{\boldsymbol{m}}$ and $\hat{\boldsymbol{n}}$ have an orientation that depends on which path on the sphere was chosen, i.e. on the azimuthal angle. The total difference in orientation of the triad is equivalent to a rotation of the triad *about* $\hat{\boldsymbol{l}}$ by 4π, which is precisely the surface circulation quoted above. Since a rotation about $\hat{\boldsymbol{l}}$ corresponds to a phase change, any $\hat{\boldsymbol{l}}$ texture in a sphere (or in any closed simply connected container, for that matter) has to have a singularity somewhere. If this singularity were located in the bulk liquid then there would have to be one or two vortex lines with a total of 4π circulation connecting the singularity with the wall of the container, as shown in Figs. 7.32(a, b) (Volovik and Mineev 1976a, 1977a,b, Blaha 1976, Anderson and Palmer 1977, Mermin 1977a,b, Hu *et al.* 1977, Maki 1977a).

What then does the $\hat{\boldsymbol{l}}$ texture in a spherical container look like? Intuitively, one might expect that—owing to the spherical symmetry of the problem—the equilibrium configuration would have to be the one where $\hat{\boldsymbol{l}}$ points radially inwards (or outwards), forming a "monopole" at the centre of the sphere (Fig. 7.32a). However, this is not correct—in fact, even the assumption of spherical symmetry is incorrect, because the order parameter in orbital space is *not* just given by $\hat{\boldsymbol{l}}$ but by a triad, which is the reason for the 4π circulation connected with the $\hat{\boldsymbol{l}}$ texture. The necessary phase vortex line(s) discussed above in fact break the spherical symmetry. Furthermore, these vortex line(s) have a normal core and are thus associated with an energy that increases as $R \ln(2R/\xi)$, where R is the radius of the container. So they are energetically disadvantageous. To reduce their length, the monopole has to move to the wall (Fig. 7.32b), thus forming a surface

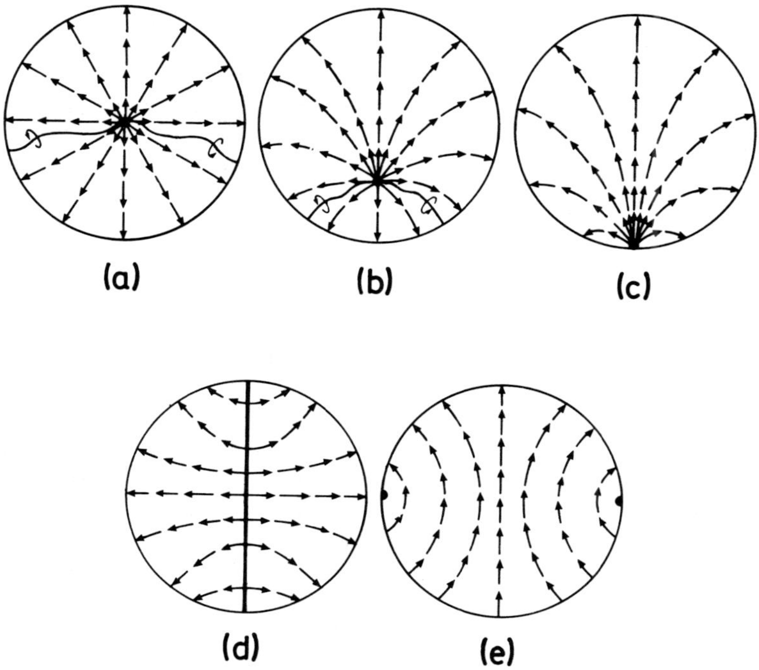

Figure 7.32 $\hat{\boldsymbol{l}}$ configurations in a sphere (cross-section through centre of sphere): (a) $\hat{\boldsymbol{l}}$ monopole with two singular phase vortices carrying a total of 4π circulation; (b) $\hat{\boldsymbol{l}}$ monopole is pulled to the wall to reduce energy of vortices; (c) final $\hat{\boldsymbol{l}}$ configuration ("boojum"); (d) relaxed monopole with a disgyration across the north and south poles; (e) half-disgyration along the equatorial line.

singularity as shown in Fig. 7.32(c) (Maki 1977a; Mermin 1977b, Anderson and Palmer 1977); see also the discussion at the end of Section 7.7.2. This surface singularity has been called a "bouquet", a "fountain" and a "flower", but is today generally referred to as a "boojum"†. However, there are several other configurations that are also possible candidates in a spherical container, for example the one where the monopole has relaxed into an $\hat{\boldsymbol{l}}$ disgyration across the sphere (Fig. 7.32d) (Hu *et al.* 1977), whereby the bending energy is reduced, or one in which a line singularity along the equator of the sphere is formed ("half-disgyration", see Fig. 7.32e) (Anderson and Palmer 1977, Mermin 1977a). Which of these textures is actually formed depends in the details of the associated bending energies of $\hat{\boldsymbol{l}}$ and $\hat{\boldsymbol{d}}$, and therefore depends critically on the *size* of the container, i.e. the

† This rather bizarre nomenclature was introduced by Mermin (1977b), who also gave a thorough review of the semantic background (Mermin 1981). The word "boojum" is borrowed from Lewis Carroll's tale *The Hunting of the Snark,* and is supposed to describe what is left of the unstable $\hat{\boldsymbol{l}}$ monopole after it has "softly and suddenly vanished away".

radius R. For $R \ll \xi_D$, the dipole coupling is unimportant and $\hat{\boldsymbol{d}}$ may be taken to be uniform, while for $R \gg \xi_D$, $\hat{\boldsymbol{l}}$ and $\hat{\boldsymbol{d}}$ are dipole-locked. A detailed calculation of the energy of the different ground-state configurations shown in Figs. 7.32(c–e) for container radii $R \gtrsim 3\xi$ has been performed by Hirschfeld and Stein (1983). For a large sphere ($R \gg \xi_D$) with $\hat{\boldsymbol{l}} \parallel \hat{\boldsymbol{d}}$, and assuming cylindrical symmetry, the vector $\hat{\boldsymbol{l}}$ may be parametrized in spherical coordinates as

$$\hat{\boldsymbol{l}}(r, \theta) = \cos \chi(r, \theta)\, \hat{\boldsymbol{r}} + \sin \chi(r, \theta)\, \hat{\boldsymbol{\theta}}, \tag{7.201}$$

where (r, θ, ϕ) are spherical coordinates with origin at the south pole of the sphere. In the case of a *plane* surface with a surface singularity of the form of a boojum (Fig. 7.32c) that flares out into the half-space, there is no intrinsic length, and hence χ should only be a function of θ. The bending free energy, determined by (7.20) with $\hat{\boldsymbol{d}} \parallel \hat{\boldsymbol{l}}$ and boundary conditions $\chi(0) = 0$ and $\chi(\frac{1}{2}\pi) = \frac{1}{2}\pi$, can then be minimized exactly to yield an explicit solution for $\chi(\theta)$ (Hu 1978; see also Hu *et al.* 1978). The resulting boojum texture may be considered as an MH texture in a cylinder, (7.191), where the lower end is shrunk to a point and where the walls are deformed into a plane surface. In the present case of a *spherical* container the boundary conditions change to $\chi(2R \cos \theta, \theta) = \theta$, which, however, does not change the solution much, and $\chi(r, \theta) = \theta$ is found to be a good trial solution.

For spheres with $R \ll \xi_D$, the dipole energy is negligible, but one must now go beyond the bending free energy and consider both the gradient free energy and the bulk free energy, determined by (7.19) and (5.4) respectively, in order to account for the vortex cores. Using trial order parameters for the energy minimization, Hirschfeld and Stein (1983) conclude that (i) for $R \gtrsim 46\xi$, i.e. for spheres that are not too small, the boojum configuration (Fig. 7.32c) is energetically favoured; (ii) for $3\xi \lesssim R \lesssim 46\xi$, the "relaxed monopole" (Fig. 7.32d) is stable; and (iii) for $R \lesssim 3\xi$, the "half-disgyration" (Fig. 7.32e) is lowest in energy. Therefore the pure $\hat{\boldsymbol{l}}$ monopole configuration in Fig. 7.32(a) is indeed expected to be unstable for any size of sphere.

The possibility of the formation of a monopole–antimonopole pair, which is necessarily connected by a vortex line of two circulation quanta, in a finite spherical container has been considered by Hu *et al.* (1977) and Soni (1978).

In an hour-glass-shaped container or an ellipsoid with circular cross-section an $\hat{\boldsymbol{l}}$ configuration with a half-disgyration around the equational line (where the curvature is largest) should be expected to be most favourable (Anderson and Palmer 1977).

7.10 STABILITY OF SUPERFLOW AND RELATED TEXTURAL TRANSITIONS IN ^{3}He-A

In ^{3}He-B the superfluid velocity $\boldsymbol{v}_s$, (7.5), has exactly the same form as in superfluid ^{4}He: it is simply given by the gradient of the overall phase. This is due to the fact that in the B phase the U(1) gauge symmetry is independently broken from the equally broken spin–orbit symmetry. Therefore there is no coupling of $\boldsymbol{v}_s$ to the spin–orbit degrees of freedom $\hat{\boldsymbol{n}}$ and θ. (This only comes in via the very weak dipole interaction.)

The situation in ^{3}He-A is very different. Here the gauge symmetry is of course also broken, but is at the same time connected to the orbital degrees of freedom by virtue of the broken relative gauge–orbit symmetry (see Section 6.1). Hence there exists both a superfluid velocity $\boldsymbol{v}_s$ *and* an orbital degree of freedom $\hat{\boldsymbol{l}}$, which are inseparably linked via the Mermin–Ho relation (7.53). As a consequence, the two terms in the expression for the superfluid current, (7.35), are not independent. This crucial fact had not been realized in the early days of superfluid ^{3}He, when it was essentially taken for granted that $\boldsymbol{v}_s$ and $\hat{\boldsymbol{l}}$ were unrelated. (In this case ^{3}He-A would have essentially been a superfluid nematic liquid crystal with $\hat{\boldsymbol{l}}$ as the orbital direction (Mermin 1978b).) However, $\hat{\boldsymbol{l}}$ is part of the orbital triad ($\hat{\boldsymbol{m}}$, $\hat{\boldsymbol{n}}$, $\hat{\boldsymbol{l}}$), and hence rotations around $\hat{\boldsymbol{l}}$ may be distinguished. As already discussed in Section 7.3.3, the linkage between $\boldsymbol{v}_s$ and $\hat{\boldsymbol{l}}$ prohibits the quantization of superflow, even though the order parameter is single-valued.

7.10.1 Stability of superflow in the bulk liquid

Topological (global) stability

The answer to the question of whether or not superflow in a superfluid phase is quantized, and consequently whether or not this superflow is stable for *topological* reasons, may be obtained directly from the given order-parameter structure. Following Mermin (1978a,b), this is achieved by a suitable adaption of the topological methods used to investigate the stability of line defects described in detail in Section 7.5. To this end, we consider superfluid flow in a (large) torus. In such a doubly connected geometry periodic boundary conditions are conveniently guaranteed.

Let us first *neglect* any surface effects. In this situation the order-parameter configuration may be assumed to be the same on any given radial cross-section of the torus (shaded area in Fig. 7.33). (If it were not then it could be deformed so that this was the case.) So, if at all, the configuration within the torus will depend only on the angle φ changing from 0 to 2π around the torus. Every point on a radial cross-section then carries the same information, and the cross-section may just as well be shrunk to a point ($r \to 0$ in Fig. 7.33). The torus is then reduced to a circular loop.

To determine whether two given states of flow in the torus, corresponding to two order-parameter configurations around the torus, may be continuously deformed into each other (i.e. whether they are stable with respect to each other), we have to proceed exactly as in the case of the stability investigation of line defects: we map the order-parameter configuration along this circular contour into the respective order-parameter space. These results have already been obtained in Section 7.5.

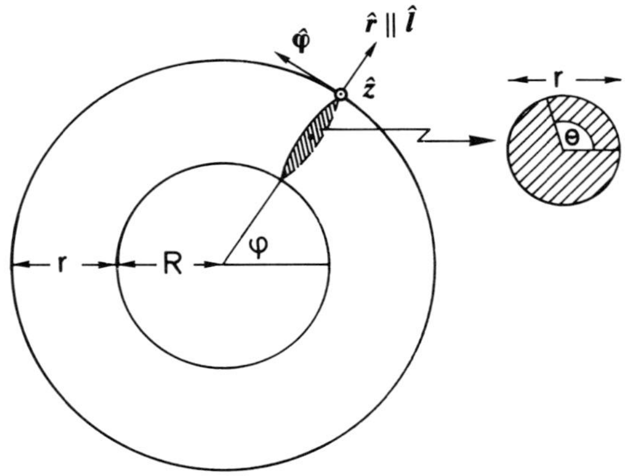

Figure 7.33 Torus with inner radius R (top view). A radial cross-section of the torus (shaded area) is also shown separately for clarity. Cylindrical coordinates (with unit vectors $\hat{\boldsymbol{r}}$, $\hat{\boldsymbol{\varphi}}$, $\hat{\boldsymbol{z}}$), with $\hat{\boldsymbol{z}}$ pointing out of the plane, are indicated at the outer circumference.

In ^{3}He-B this mapping led to (7.57a), i.e. essentially $\pi_1(S^1)$, the set of classes of mappings of the circular contour into a circle (the topological space corresponding to the U(1) symmetry). This is exactly as in superfluid ^{4}He and yields the group $\mathbb{Z}$ of integers, corresponding to the topological charges $N = 0, \pm 1, \pm 2, \ldots$. Hence circulation is quantized and, in principle, may take any integer value. Superflow is therefore stable for strong topological reasons—it is *globally* stable.

On the other hand, in ^{3}He-A this is generally very different. Assuming $\hat{\boldsymbol{d}}$ and $\hat{\boldsymbol{l}}$ to be dipole-locked (as will be the case in a torus with channel thickness $r \gg \xi_D$) and no external magnetic field, the degeneracy space is $R = SO_3$ and hence the homotopy group is $\pi_1(SO_3) = \mathbb{Z}_2$, (7.57c), as discussed in Section 7.5.4. The topological charges are $N_2 = 0, 1$, implying that a stable state of superflow may at most carry a *single quantum of circulation*! As already expressed by the Mermin–Ho relation (7.53), superflow in ^{3}He-A is not quantized. It may therefore be *continuously* decreased by an appropriate spatial variation of the three-dimensional orbital triad, until it is reduced to zero, or at most one quantum of circulation, depending on whether the original state carried an even or an odd number of circulation quanta.

This result is not changed much if there is no dipole locking, as for example in a very "thin" torus with channel thickness $r < \xi_D$. The degeneracy space is then a little larger, leading to the mapping (7.59a) with *four* topological charges: $N = 0, \pm\frac{1}{2}, 1$. This allows two more current-carrying states: both with *half* a quantum of circulation.

One therefore finds that, in the absence of an external magnetic field and if surface effects are neglected, superflow in ^{3}He-A is topologically, i.e.

globally, unstable—it dissipates. Possible dissipation mechanisms will be discussed in Section 7.12.

The situation is changed if an external magnetic field $\boldsymbol{H}$ is applied, which confines $\hat{\boldsymbol{d}}$ to the plane perpendicular to it. Because of the dipole coupling, $\hat{\boldsymbol{l}}$ will also lie in this plane. Now the orbital triad is no longer able to perform three-dimensional motion, which implies $\nabla \times \boldsymbol{v}_s = 0$, i.e. the current is quantized. Indeed, the degeneracy space is reduced to (7.75a), i.e. two circles. The topological charges given by $\pi_1(S^1)$, (7.57b), are characterized by $\mathbb{Z}$ as in superfluid ^{4}He or ^{3}He-B. So a magnetic field stabilizes superflow in ^{3}He-A via the dipole–dipole interaction of the ^{3}He-nuclei!

The surface and its orientational effects on $\hat{\boldsymbol{l}}$ have been neglected so far. Owing to the boundary condition that $\hat{\boldsymbol{l}}$ be normal to a surface, the orbital triad cannot move freely at a surface and hence a continuous decay of surface superflow cannot take place. Indeed, it is easily shown that right at the surface the circulation is quantized. To this end, we consider the inner or outer circumference of the torus in Fig. 7.33. It lies in the plane perpendicular to $\hat{\boldsymbol{z}}$, and the vector $\hat{\boldsymbol{l}}$ will always point radially outward. The orbital triad may then be parametrized in cylindrical coordinates as

$$\left.\begin{aligned} \hat{\boldsymbol{m}} &= \cos\varphi\,\hat{\boldsymbol{\varphi}} - \sin\varphi\,\hat{\boldsymbol{z}}, \\ \hat{\boldsymbol{n}} &= \sin\varphi\,\hat{\boldsymbol{\varphi}} + \cos\varphi\,\hat{\boldsymbol{z}}, \\ \hat{\boldsymbol{l}} &= \hat{\boldsymbol{r}}. \end{aligned}\right\} \tag{7.202}$$

The superfluid velocity $\boldsymbol{v}_s$, (7.10), along the circumference is given by

$$\boldsymbol{v}_s = \frac{\hbar}{2m}\frac{\hat{\boldsymbol{\varphi}}}{r}, \tag{7.203}$$

just as in superfluid ^{4}He, and consequently the circulation κ, (7.55), is quantized:

$$\kappa = N\frac{h}{2m}. \tag{7.204}$$

This result holds for any contour along the surface.

So, while topology guarantees the stability of superflow *at* the surface, it fails to do so in the bulk. This suggests that the current will decrease in magnitude as one goes from the surface into the interior of the liquid. *How* this is achieved is a very delicate matter and is still not fully understood (see Section 7.12). While it is clearly advantageous for the system to reduce the current to lower the flow energy, the contortions through which the orbital triad will have to go in order to achieve this will add to the gradient energy. Besides that, the topological arguments cannot tell whether or not the orbital triad can freely move to reduce the current. For example, the anisotropy of the superfluid density prefers $\hat{\boldsymbol{l}}$ pointing along $\boldsymbol{v}_s$, as expressed by (6.120). In this case it is not clear whether the system first has to climb a small barrier in energy space, i.e. whether it needs an activation energy, before the decay can eventually start. If so, this would imply "local" stability (in energy space) of superflow. Such behaviour is indeed found for certain parameter values of the prefactors in the gradient energy and will be discussed next.

Local stability of the uniform $\hat{\boldsymbol{l}}$ texture

Superflow in a superfluid with anisotropic superfluid density introduces an new preferred direction to the system. It will thereby influence the orientation of Cooper pairs (Glassgold and Sessler 1961). In particular, in the A phase a superfluid velocity field $\boldsymbol{v}_s$ may be expected to orient the vector $\hat{\boldsymbol{l}}$ parallel to itself (see (6.120), (7.21b)), since the component of the superfluid density tensor along $\hat{\boldsymbol{l}}$, $\rho_{s\parallel}$, is smallest (de Gennes and Rainer 1974, Leggett 1975a). Furthermore, this orientation may be taken to be uniform because then there are no contributions to the bending free energy.

Although these arguments, taken by themselves, are all correct, the bending free energy of ^{3}He-A, (7.20), is clearly complicated enough to defy an analysis of the stability of the $\hat{\boldsymbol{l}} \parallel \boldsymbol{v}_s$ configuration that simply discusses individual terms. The detailed investigation of this problem was initiated by Hook and Hall (1977, unpublished), who tested this hydrodynamic stability of the uniform $\hat{\boldsymbol{l}}$ texture, finding an instability of the $\hat{\boldsymbol{l}} \parallel \boldsymbol{v}_s$ texture in the Ginzburg–Landau regime. Even though their analysis was based on incorrect equations, it was of considerable significance since it provoked further studies of the problem, which finally led to its clarification (Bhattacharyya *et al.* 1977a, Cross and Liu 1978).

Following Bhattacharyya *et al.* (1977a), we study the stability of an equilibrium configuration in the presence of (arbitrarily small) uniform flow along $\hat{z}$. Periodicity along $\hat{z}$—as realized in a torus—is assumed. (In superfluid ^{4}He this doubly connected geometry leads to the topological stability of superflow as discussed above.) The flow is assumed to take place between plane-parallel walls of separation d whose surface normal is along $\hat{\boldsymbol{y}}$. At the walls $\hat{\boldsymbol{l}} \parallel \hat{\boldsymbol{y}}$. Far away from the surfaces we assume that the flow has aligned the vector $\hat{\boldsymbol{l}}$:

$$\hat{\boldsymbol{l}}_0 = \hat{\boldsymbol{z}}, \quad \boldsymbol{v}_s = \frac{\hbar}{2m} \bar{v}_s \hat{\boldsymbol{z}}. \tag{7.205}$$

We note that $\bar{v}_s^{-1}$ defines a characteristic length, with $\bar{v}_s^{-1} \approx 10\,\mu\text{m}$ being of the order of the dipole coherence length ξ_D for $v_s = 1$ mm/s. For $\bar{v}_s^{-1} \gtrsim d$, the surfaces will dominate the direction of $\hat{\boldsymbol{l}}$, such that $\hat{\boldsymbol{l}} \parallel \hat{\boldsymbol{y}}$. Only for $\bar{v}_s^{-1} \ll d$, i.e. $v_s \gg (\hbar/2m)d^{-1}$, which is assumed here, will alignment of $\hat{\boldsymbol{l}}$ into the flow direction ($\hat{\boldsymbol{l}} \parallel \hat{z}$) be possible. Although this condition limits the superfluid velocity from below, v_s can be made arbitrarily small by making d large enough. On the other hand, in the case that v_s becomes so small that

$\bar{v}_s^{-1} \gtrsim R$, where R is the diameter of the torus (see Fig. 7.33), the effects of curvature must be included (Bailin and Love 1978b, Harris 1980).

The stability test should only consider deviations from (7.205) that vanish at the surface. This is achieved by limiting the range of transverse wave vectors from below, which leads to a second-order stability test.

Introducing small fluctuations u, v and w,

$$\delta\hat{\boldsymbol{l}} \equiv \hat{\boldsymbol{l}} - \hat{\boldsymbol{l}}_0 = (u, v, w), \tag{7.206}$$

with $w \approx -\frac{1}{2}(u^2 + v^2)$, (7.206) may be written as

$$\delta\boldsymbol{l} = \boldsymbol{l}_1 - \tfrac{1}{2}l_1^2\hat{\boldsymbol{z}}, \tag{7.207a}$$

where

$$\boldsymbol{l}_1 = (u, v, 0). \tag{7.207b}$$

The deviation $\delta\hat{\boldsymbol{l}}$ leads to a contribution $\delta\boldsymbol{v}_s$ to the superfluid velocity via (7.53):

$$\nabla \times \delta\boldsymbol{v}_s \approx \frac{\hbar}{2m}(\nabla u \times \nabla v). \tag{7.208}$$

Using the identity $\nabla u \times \nabla v = \nabla \times (u\nabla v)$, $\delta\boldsymbol{v}_s$ is found to have the general form

$$\delta\boldsymbol{v}_s = \frac{\hbar}{2m}[\nabla\gamma + \tfrac{1}{2}(u\nabla v - v\nabla u)], \tag{7.209}$$

where γ is a rotation angle about $\hat{\boldsymbol{l}}$. Equations (7.207) and (7.209) are now substituted into the bending free energy $F_{\text{bend}} = \int d^3r\, f_{\text{bend}}$, given by (7.20), which is then expanded in terms of the fluctuations. Assuming that fluctuations are characterized by a wave vector $\boldsymbol{k} = (k_x, k_y, k_z)$, we express $\boldsymbol{l}_1$ and γ by their Fourier transforms, and thus—after minimization with respect to $\gamma(\boldsymbol{k})$—obtain the following quadratic contribution to F_{bend}:

$$\begin{aligned}\delta F_{\text{bend}} = \frac{1}{2}\left(\frac{\hbar}{2m}\right)^2 \sum_{\boldsymbol{k}} \Big\{ & K_s\,|\boldsymbol{q}\cdot\boldsymbol{l}_1|^2 + K_t\,|\boldsymbol{q}\times\boldsymbol{l}_1|^2 \\ & + (K_b k_z^2 + \rho_0\bar{v}_s^2)l_1^2 + (2C_0 + \rho_{s\|})\bar{v}_s\boldsymbol{k}\cdot\operatorname{Im}(\boldsymbol{l}_1^*\times\boldsymbol{l}_1) \\ & - \frac{1}{\rho_s q^2 + \rho_{s\|}k_z^2}[\rho_0^2\bar{v}_s^2\,|\boldsymbol{q}\cdot\boldsymbol{l}_1|^2 + C_0^2k_z^2\,|\boldsymbol{q}\times\boldsymbol{l}_1|^2 \\ & + \rho_0 C_0\bar{v}_s q^2\boldsymbol{k}\cdot\operatorname{Im}(\boldsymbol{l}_1^*\times\boldsymbol{l}_1)]\Big\}.\end{aligned} \tag{7.210}$$

Here $\boldsymbol{q} = (k_x, k_y, 0)$ is the transverse part of $\boldsymbol{k}$ with respect to $\boldsymbol{v}_s \parallel \hat{\boldsymbol{z}}$ and we have used the notation of (7.21c). It should be noted that (7.210) is correct in both the dipole-free and the *rigidly* dipole-locked situations. (In the latter case the coefficients K_s, K_t and K_b have to be interpreted as indicated in (7.21a).) If the $\hat{\boldsymbol{d}}$, $\hat{\boldsymbol{l}}$ texture is dipole-locked at $v_s = 0$ then this remains true even for (infinitely) small flow velocities v_s as are considered in the present investigation. For higher flow velocities, however, this is no longer true since then $\hat{\boldsymbol{d}}$ and $\hat{\boldsymbol{l}}$ will acquire different directions. This case will be treated in the next subsection.

Equation (7.210) now has to be minimized with respect to $\boldsymbol{l}_1$, i.e. the fluctuations $u(\boldsymbol{k})$ and $v(\boldsymbol{k})$. This is considerably simplified by taking into account that $|\boldsymbol{q}| \approx d^{-1}$, such that we may neglect q relative to k_z. The minimization of (7.210) then only involves the third and fourth terms in (7.210), and δF_{bend} is found to have a minimum at

$$k_z^0 = \frac{C_\perp - C_\parallel + \frac{1}{2}\rho_{s\parallel}}{K_b \bar{v}_s}. \tag{7.211}$$

For δF_{bend} to be positive at $\boldsymbol{k} = k_z^0 \hat{\boldsymbol{z}}$, this leads to the stability criterion for the $\hat{\boldsymbol{l}} \parallel \boldsymbol{v}_s$ texture (Bhattacharyya *et al.* 1977a, Cross and Liu 1978)

$$s \equiv \frac{(C_\perp - C_\parallel + \frac{1}{2}\rho_{s\parallel})^2}{(\rho_{s\perp} - \rho_{s\parallel})K_b} < 1. \tag{7.212}$$

By contrast, if $s > 1$, one has $\delta F_{\text{bend}} < 0$, indicating an instability of the $\hat{\boldsymbol{l}} \parallel \boldsymbol{v}_s$ texture at $|\boldsymbol{k}| = k_z$ characterized by a circularly polarized vector $\boldsymbol{l}_1$.

The denominator of the stability parameter s in (7.212) is determined by the prefactor of the two contributions to the bending free energy (7.20), (7.21b) that favour the $\hat{\boldsymbol{l}} \parallel \boldsymbol{v}_s$ configuration. Indeed, for strong anisotropy of the superfluid density $\boldsymbol{\rho}_s$ and large stiffness against bending of $\hat{\boldsymbol{l}}$, these two parameters make s small and thereby stabilize the uniform $\hat{\boldsymbol{l}} \parallel \boldsymbol{v}_s$ texture. On the other hand, the terms $\frac{1}{2}\rho_{s\parallel}(\hat{\boldsymbol{l}} \cdot \boldsymbol{v}_s)^2$ and $-(C_\perp - C_\parallel)(\hat{\boldsymbol{l}} \cdot \boldsymbol{v}_s)[\hat{\boldsymbol{l}} \cdot (\boldsymbol{\nabla} \times \hat{\boldsymbol{l}})]$ are found to disfavour this configuration, their prefactors appearing in the numerator of (7.212). The cause of this behaviour will be discussed in the next subsection.

For transverse dimensions of the system $d \gg \xi_D$ and for $v_s \gg (\hbar/2m)^{-1}$ (e.g. for $v_s \gg 10^{-2}$ mm/s at $d = 1$ mm) the vector $\hat{\boldsymbol{l}}$ is dipole-locked to $\hat{\boldsymbol{d}}$ (see the discussion below (7.201)). Using the weak-coupling results for the parameters appearing in (7.212) in the Ginzburg–Landau regime as given by (7.22) with (7.21a), one finds

$$s \Big|_{\substack{\text{weak coupling,} \\ \text{dipole-locked}}} = \tfrac{9}{10} < 1. \tag{7.213}$$

This indicates that in weak coupling and close to T_c the configuration $\hat{\boldsymbol{l}} \parallel \boldsymbol{v}_s$ is indeed stable, although not by much. Therefore strong-coupling corrections (Serene and Rainer 1977, 1983), although numerically small, might already lead to a violation of (7.212). On the other hand, for $T \to 0$ the quantities in the numerator of s, (7.212), all approach a constant, while $\rho_{s\perp} - \rho_{s\parallel}$ vanishes proportionally to T^2 (see (3.94) with (7.15b)), and K_b, (7.23f), diverges logarithmically as discussed in Section 7.2.1. Overall, $s(T)$ is seen to diverge for $T \to 0$, so that the stability condition (7.212) is violated below some instability temperature T_i. Estimates of T_i by Cross (1975), using the expressions (7.33), which assume weak coupling but include Fermi-liquid corrections, yielded a T_i that at melting pressure is well within the range of temperatures for which the A phase is stable (Cross and Liu 1978). A similar analysis by Fetter (1979) yielded $T_i = 0.82T_c$ and, neglecting Fermi-liquid corrections, $T_i = 0.88T_c$ (see also Kleinert 1978).

As pointed out by Bhattacharyya *et al.* (1977a), the weak-coupling value of s in (7.212) in the *absence* of dipole interaction is found from (7.22a–c) as

$$s\Big|_{\substack{\text{weak coupling,}\\ \text{dipole-free}}} = 1.5, \tag{7.214}$$

implying instability of the $\hat{\boldsymbol{l}} \parallel \boldsymbol{v}_s$ texture at all temperatures. This indicates that it is the additional stiffness given to the $\hat{\boldsymbol{l}}$ texture by the *spin part* of the ^{3}He-A order parameter via the dipole–dipole interaction of the ^{3}He nuclei that stabilizes the $\hat{\boldsymbol{l}} \parallel \boldsymbol{v}_s$ configuration and thus prohibits the decay of a weak superflow.

The above analysis shows that, in spite of the global instability of (arbitrarily weak) superflow in ^{3}He-A, the $\hat{\boldsymbol{d}} \parallel \hat{\boldsymbol{l}} \parallel \boldsymbol{v}_s$ configuration—and hence superflow—is stable close to T_c, since this configuration corresponds to a local minimum of the free energy. This stability is only due to a numerical coincidence of the coefficients in the free energy near T_c, and will fail below some instability temperature T_i, or (if the stabilizing support of the $\hat{\boldsymbol{d}}$ texture is removed) by the action of a magnetic field along $\boldsymbol{v}_s$. The latter possibility was investigated by Takagi (1975) and Saslow and Hu (1978), who derived a generalization of the stability criterion (7.212) in the presence of a magnetic field $\boldsymbol{H} \parallel \boldsymbol{v}_s$ as

$$s(T) < 1 - \left(\frac{H}{H_v}\right)^2, \tag{7.215}$$

where

$$H_v = \left(\frac{\rho_{s\perp} - \rho_{s\parallel}}{\Delta\chi}\right)^{1/2} v_s \tag{7.216}$$

is a characteristic magnetic field. This condition can be met at all temperatures by choosing H appropriately, and hence flow can be destabilized at any temperature. In particular, for $s(T) < 1$, the critical field *above* which the flow becomes unstable is

$$H = (1-s)^{1/2} H_v. \tag{7.217}$$

For $\boldsymbol{H} \perp \boldsymbol{v}_s$, flow is found to be stabilized in accordance with the topological arguments given already.

7.10.2 The helical instability of the uniform $\hat{\boldsymbol{l}}$ texture

The question is then whether below $T_{\rm i}$ (or, more generally, for $s > 1$) a current-carrying state is totally unstable, such that a catastrophic decay of superflow sets in, or whether there is a "displacive" transition to a new, nonuniform—but still stable—state of superflow. In the dipole-free case Bhattacharyya *et al.* (1977a) explicitly constructed a continuous family of nonsingular configurations that lead from the original state of flow to one with no flow, with the free energy monotonically decreasing. On the other hand, since $\hat{\boldsymbol{l}}$ and $\hat{\boldsymbol{d}}$ are dipole-locked, at least for small superfluid velocities, the additional stiffness of the $\hat{\boldsymbol{l}} \parallel \boldsymbol{v}_{\rm s}$ configuration due to the $\hat{\boldsymbol{d}}$ texture obstructs such a monotonic decay (a similar argument applies in the case of a finite magnetic field, which fixes the vector $\hat{\boldsymbol{d}}$; see Liu 1979b). Indeed, Fetter (1978a,b, 1979) and Kleinert *et al.* (1978, 1979) showed that, beyond the threshold given by (7.212), the texture undergoes a transition to a static *helical* configuration with $\hat{\boldsymbol{l}} \parallel \hat{\boldsymbol{d}}$ winding around the direction of superflow. This implies that superfluid currents may flow even beyond the stability region $s < 1$.

The ultimate reason for the formation of such a helical texture lies in the presence of the term $-(C_\perp - C_\parallel)(\boldsymbol{v}_{\rm s} \cdot \hat{\boldsymbol{l}})[\hat{\boldsymbol{l}} \cdot (\boldsymbol{\nabla} \times \hat{\mathbf{l}})]$ in the bending free-energy density (7.20). Since the factor $\hat{\boldsymbol{l}} \cdot (\boldsymbol{\nabla} \times \hat{\boldsymbol{l}})$ describes a torque on $\hat{\boldsymbol{l}}$ (which is why the contribution to $f_{\rm bend}$ containing this term is sometimes called the "torque term"), one has competition between a configuration with $\boldsymbol{v}_{\rm s} \parallel \hat{\boldsymbol{l}}$ constant in space, making $|\boldsymbol{v}_{\rm s} \cdot \hat{\boldsymbol{l}}|$ maximal but $\hat{\boldsymbol{l}} \cdot \boldsymbol{\nabla} \times \hat{\boldsymbol{l}} = 0$, and one where $\boldsymbol{v}_{\rm s}$ and $\hat{\boldsymbol{l}}$ are not aligned, with $\hat{\boldsymbol{l}}$ varying in space. In the latter case $|\hat{\boldsymbol{v}}_{\rm s} \cdot \hat{\boldsymbol{l}}| < 1$ but now $|\hat{\boldsymbol{l}} \cdot (\boldsymbol{\nabla} \times \hat{\boldsymbol{l}})| > 0$, so that the torque term leads to a finite negative contribution that decreases the overall free energy. A spatial variation of $\hat{\boldsymbol{l}}$ will therefore lead to a (perpendicular) torque on $\hat{\boldsymbol{l}}$, causing a helical conformation. We note that the torque term is the only term with a *linear* coupling between $\hat{\boldsymbol{l}}$ and $\boldsymbol{v}_{\rm s}$, which may therefore distinguish between $\hat{\boldsymbol{l}} \parallel \boldsymbol{v}_{\rm s}$ and $\hat{\boldsymbol{l}} \parallel -\boldsymbol{v}_{\rm s}$. While in the uniform state this is irrelevant ($\boldsymbol{\nabla} \times \hat{\boldsymbol{l}} = 0$), a spatial variation of $\hat{\boldsymbol{l}}$ as in the helical texture will lead to a breakdown of chiral symmetry.

Following Fetter (1978a, 1979), we initially assume a uniform $\hat{\boldsymbol{l}}$ texture and superfluid velocity $\boldsymbol{v}_{\rm s,0}$ as given by (7.205); $\hat{\boldsymbol{l}}$ and $\hat{\boldsymbol{d}}$ are supposed to be dipole-locked. Using the parametrization of $\hat{\boldsymbol{l}}$ in (7.11), the Euler angles α and γ are found to increase linearly with z,

$$\alpha = -k_\alpha z, \quad \gamma = -k_\gamma z, \tag{7.218}$$

so that (see (7.13))

$$\boldsymbol{v}_{\rm s,0} = \frac{\hbar}{2m} \bar{v}_{\rm s,0} \hat{\boldsymbol{z}}, \quad \bar{v}_{\rm s,0} = k_\alpha + k_\gamma. \tag{7.219}$$

For superflow in a torus, the periodic boundary conditions imply topological constraints on the order parameter to guarantee its single-valuedness. In particular, the "winding numbers" k_α and k_γ are in general quantized (Fetter 1979, Kleinert *et al.* 1979, Kleinert 1979a,b, Fetter and Williams 1979, 1981). Transitions between states of superflow in a toroidal geometry with different winding numbers are thus prohibited for topological reasons. While for $\beta \neq 0, \pi$ the values of k_α and k_γ are separately quantized, for $\beta = 0, \pi$ only the combination $k_\alpha \pm k_\gamma$ is quantized. In an "open" geometry, with a relative superflow generated for example by applying a thermal gradient, the situation is quite different: in this case there are no topological requirements on the order parameter and it is the current g_z itself that is a fixed quantity. This difference has been discussed in detail by Fetter and Williams (1979, 1981).

Below the instability temperature T_i discussed above a helical texture develops. With β as the opening angle of the helix at a given temperature $T < T_i$, the superfluid velocity (7.13) is given by

$$\left.\begin{aligned} \boldsymbol{v}_s &= \boldsymbol{v}_{s,0} - \Delta\boldsymbol{v}_s, \\ \Delta\boldsymbol{v}_s &= \frac{\hbar}{2m} k_\alpha (1 - \cos\beta)\hat{\boldsymbol{z}}, \end{aligned}\right\} \tag{7.220}$$

i.e. it is found to be reduced by $\Delta\boldsymbol{v}_s$; $\boldsymbol{v}_s$ is still along $\hat{\boldsymbol{z}}$ but is no longer quantized. On the other hand, the superfluid current (7.39) takes the form $\boldsymbol{g}_s = g_{s,z}\hat{\boldsymbol{z}} + g_{s\perp}\boldsymbol{l}_\perp$, where $\boldsymbol{l}_\perp = (\sin\beta\cos\alpha, \sin\beta\sin\alpha, 0)$ and

$$g_{s,z} = \frac{\hbar}{2m}[\bar{v}_{s,0}(\rho_{s\parallel} + \rho_0 \sin^2\beta) - C_0 k_\alpha \cos\beta \sin^2\beta], \tag{7.221a}$$

$$g_{s\perp} = \frac{\hbar}{2m}[-\bar{v}_{s,0}\rho_0\cos\beta + k_\alpha(C_\perp - C_0\sin^2\beta)], \tag{7.221b}$$

with $\rho_0 = \rho_{s\perp} - \rho_{s\parallel}$ and $C_0 = C_\perp - C_\parallel$. The current is seen to have a solenoidal direction.

The opening angle β is now determined by minimization of the bending free-energy density (7.20) with fixed $\bar{v}_s$ and k_α. Just below T_i, i.e. for temperatures T with $T_i - T \ll T_i$, one finds

$$\beta^2 \approx a\left[\frac{\rho_{0h} - \rho_0}{\rho_{0h}} - \left(\frac{k_\alpha - k_{\alpha h}}{k_{\alpha h}}\right)^2\right], \tag{7.222}$$

where

$$a = \left[\frac{2K_t}{K_b} - \frac{\rho_{s\parallel}}{2K_b} - \frac{3\rho_{s\parallel}}{2(C_0 + \frac{1}{2}\rho_{s\parallel})}\right]^{-1}$$

and

$$\rho_{0h} = \left.\frac{(C_0 + \frac{1}{2}\rho_{s\parallel})^2}{K_b}\right|_{T_i}, \tag{7.223a}$$

$$k_{\alpha\mathrm{h}} = \bar{v}_{\mathrm{s}} \frac{C_0 + \frac{1}{2}\rho_{\mathrm{s}\|}}{K_{\mathrm{b}}}\bigg|_{T_{\mathrm{i}}}. \tag{7.223b}$$

The parameter $k_{\alpha\mathrm{h}}$ is given by the value of k_α at T_{i} that minimizes the free energy at the point where β becomes finite. For the helix to be in static equilibrium, the right-hand side of (7.222) must be positive.

The equation

$$\rho_0(T_{\mathrm{i}}) = \rho_{0\mathrm{h}} \tag{7.224}$$

determines the temperature T_{i} at which the helix first appears. Within weak coupling, but including Fermi-liquid corrections, one finds $T_{\mathrm{i}} \approx 0.82 T_{\mathrm{c}}$ and

$$\left.\begin{aligned} \rho_{0\mathrm{h}} &\approx 1.15\rho_{\mathrm{s}\|}, \\ k_{\alpha\mathrm{h}} &\approx 0.72\bar{v}_{\mathrm{s}} \end{aligned}\right\} \tag{7.225}$$

at T_{i}. A detailed stability analysis of the helical texture in the regime of parameters where $s > 1$ has been performed by Fetter (1979) and Kleinert *et al.* (1979). Helical textures are found to be stable for all $T < T_{\mathrm{i}}$. The topological aspects of the stability of a supercurrent in a torus when a helical texture is present have been discussed by Kleinert (1979a,b).

The above results are only valid for very small superfluid currents and a correspondingly large "pitch" $\bar{v}_{\mathrm{s}}^{-1}$ (i.e. a long-wavelength helical distortion), such that dipole locking is maintained. If the flow is increased so that $\bar{v}_{\mathrm{s}}^{-1}$ becomes of the order of the dipole length ξ_{D}, a dipole-locked configuration of $\hat{\boldsymbol{l}}$ and $\hat{\boldsymbol{d}}$ can no longer be assumed and the stability analysis of the uniform $\hat{\boldsymbol{l}}$ configuration has to be extended to a more general $\hat{\boldsymbol{l}}$, $\hat{\boldsymbol{d}}$ configuration (Lin-Liu *et al.* 1978, 1979). Such an investigation shows that helical configurations of $\hat{\boldsymbol{l}}$ and $\hat{\boldsymbol{d}}$ with the same winding period, but different opening angles, may be stabilized at *all* temperatures, in particular near T_{c}. For this, the superflow either has to be strong enough such that $\bar{v}_{\mathrm{s}}^{-1} \approx \xi_{\mathrm{D}}$, or a magnetic field $\boldsymbol{H}$ has to be applied. As discussed above (7.217), a magnetic field $\boldsymbol{H} \parallel \boldsymbol{v}_{\mathrm{s}}$ destabilizes the uniform $\hat{\boldsymbol{d}} \parallel \hat{\boldsymbol{l}} \parallel \boldsymbol{v}_{\mathrm{s}}$ texture by moving $\hat{\boldsymbol{d}}$ away from $\hat{\boldsymbol{l}}$, and thereby triggers the transition to a helical state (see also Saslow and Hu 1978). The stability analysis shows that quite generally the longitudinal fluctuations are responsible for the instability of the uniform texture.

Following Lin-Liu *et al.* (1979), we shall now calculate the phase boundary between the uniform $\hat{\boldsymbol{l}} \parallel \boldsymbol{v}_{\mathrm{s}}$ configuration and the helical state in the Ginzburg–Landau regime and for weak coupling. An open geometry with a constant external superflow is assumed. (Note that the length $\xi_\perp$ used by these authors corresponds to $2^{1/2}\xi_{\mathrm{D}}$.) The relevant free energy is given by the integral over the energy densities due to bending, (7.20) with (7.22), dipole coupling, (6.104), and magnetic orientation, (6.115). Using the parametrization of $\hat{\boldsymbol{l}}$, $\boldsymbol{v}_{\mathrm{s}}$ and $\hat{\boldsymbol{d}}$ as given by (7.11c), (7.13) and (7.149) respectively, and assuming a spatial dependence of all variables along $\boldsymbol{v}_{\mathrm{s}} \parallel \hat{\boldsymbol{z}}$

only, the free energy is found as

$$F = \frac{1}{4}\left(\frac{\hbar}{2m}\right)^2 \rho_{s\|}\sigma(\hat{k}_z)\int \mathrm{d}z\, \{2(1+\sin^2\beta)\gamma_z^2 + (1+\cos^2\beta)\alpha_z^2 + 4\cos\beta\,\alpha_z\gamma_z$$

$$+ (1+2\cos^2\beta)\beta_z^2 + 2(1+\sin^2\beta)(\theta_z^2+\psi_z^2\sin^2\theta)$$

$$- 2\xi_D^{-2}[\cos\beta\cos\theta + \sin\beta\sin\theta\cos(\alpha-\psi)]^2$$

$$+ 2\xi_H^{-2}(\hat{\boldsymbol{H}}\cdot\hat{\boldsymbol{d}})^2\}, \qquad (7.226)$$

where the dipole and magnetic healing lengths are given by (7.27) and (7.28) respectively, and $\sigma(\hat{k}_z)$ is the cross-section of the system. The free-energy functional $F\{\alpha, \beta, \gamma; \theta, \psi\} = \int \mathrm{d}^3r f$ now has to be minimized with respect to $\alpha(z)$, $\beta(z)$ etc. This is done either by solving the Euler–Lagrange equations (7.154) or by making use of variational functions. The phase $\gamma(z)$ is seen to be a cyclic variable, for which the Euler–Lagrange equations reduce to $(\mathrm{d}/\mathrm{d}z)\,\partial f/\partial\gamma_z = 0$, i.e.

$$\frac{\mathrm{d}}{\mathrm{d}z}[\gamma_z(1+\sin^2\beta) + \alpha_z\cos\beta] = 0. \qquad (7.227)$$

Clearly, $-(\hbar/2m)^{-1}\,\partial f/\partial\gamma_z$ is precisely the definition (7.14b) of the supercurrent in the z direction, $g_{s,z}$, so that (7.227) becomes

$$\gamma_z = -\left[\left(\frac{\hbar}{2m}\rho_{s\|}\right)^{-1} g_{s,z} + \alpha_z\cos\beta\right]\frac{1}{1+\sin^2\beta}. \qquad (7.228)$$

Together with the definition of $\boldsymbol{v}_s$, (7.13), this yields

$$g_{s,z} = \rho_{s\|}\left[v_{s,z}(1+\sin^2\beta) + \frac{\hbar}{2m}\alpha_z\cos\beta\sin^2\beta\right], \qquad (7.229)$$

in agreement with (7.221a). Equation (7.228) is now used to replace γ_z in (7.226). To minimize F further after the elimination of γ_z, we have to bear in mind that we may no longer use f itself, but have to work with the appropriate Legendre transform (the "Routhian" of classical mechanics)

$$\mathcal{G} = f - \gamma_z\frac{\partial f}{\partial\gamma_z}, \qquad (7.230)$$

which we shall refer to as the "Gibbs potential"; the energy itself is then given by $G = \int \mathrm{d}^3r\,\mathcal{G}$.

Since the dipole coherence length ξ_D is the only intrinsic, relevant lengthscale in the present problem, we shall take an overall factor ξ_D^{-2} out of $\mathcal{G}$, thereby scaling all lengths by ξ_D. Furthermore, the supercurrent may then be written in terms of a dimensionless quantity

$$p = \frac{g_{s,z}}{\rho_{s\|}v_{s,D}}, \qquad (7.231a)$$

where

$$v_{s,D} \equiv \frac{\hbar}{2m} \xi_D^{-1} \tag{7.231b}$$

is a characteristic velocity ($\bar{v}_s = \xi_D^{-1}$ in (7.205)), determined by the dipole length ξ_D, (7.27),

$$v_{s,D} \approx 1.3 \text{ mm/s}. \tag{7.231c}$$

Altogether, the Gibbs potential $\mathscr{G}$, (7.230), takes the form

$$\begin{aligned}\mathscr{G} = A\xi_D^{-2}\biggl\{ & -\frac{p^2}{1+\sin^2\beta} + \frac{1}{2}\frac{\sin^2\beta\,(3-\sin^2\beta)}{1+\sin^2\beta}\,\alpha_s^2 \\ & -2p\frac{\cos\beta}{1+\sin^2\beta}\,\alpha_s + \tfrac{1}{2}(3-2\sin^2\beta)\beta_s^2 \\ & +(1+\sin^2\beta)(\theta_s^2+\psi_s^2\sin^2\theta) \\ & -[\cos\beta\cos\theta+\sin\beta\sin\theta\cos(\alpha-\psi)]^2 + h^2(\hat{\boldsymbol{H}}\cdot\hat{\boldsymbol{d}})^2\biggr\},\end{aligned} \tag{7.232}$$

where

$$A = \frac{1}{2}\left(\frac{\hbar}{2m}\right)^2 \rho_{s\|}, \quad s = \frac{z}{\xi_D}, \quad h = \frac{H}{H^*}, \tag{7.233}$$

with the characteristic magnetic field H^* defined in (7.28c). The third term on the right-hand side of (7.232) is the "torque term" discussed above, which is responsible for the winding of $\hat{\boldsymbol{l}}$ with nonzero inclination angle β. It is the only one that depends linearly on $\cos\beta$, p and α_s. Hence it depends on the sense of winding of $\hat{\boldsymbol{l}}$: if we take $p > 0$ (flow in the positive z direction) then either $\alpha_s < 0$ and $\hat{\boldsymbol{l}}$ is essentially in the $\hat{\boldsymbol{z}}$ direction (i.e. $\beta \gtrsim 0$), or $\alpha_s > 0$ and $\hat{\boldsymbol{l}}$ points essentially in the $-\hat{\boldsymbol{z}}$ direction ($\beta \lesssim \pi$). This reflects both the discrete degeneracy of the orientational energies and the broken chiral symmetry. Although this does not affect the energy, it will matter in NMR, which is sensitive to the helicity. Here we choose $\alpha_s < 0$, and hence $\hat{\boldsymbol{l}}$ essentially along $\hat{\boldsymbol{z}}$.

We shall now investigate the helical texture, both in a magnetic field and in superflow. In general, $\hat{\boldsymbol{l}}$ and $\hat{\boldsymbol{d}}$ will not be parallel and will form a "double helix" as shown in Fig. 7.34. We shall concentrate on the two cases $\boldsymbol{v}_s \parallel \boldsymbol{H}$ and $\boldsymbol{v}_s \perp \boldsymbol{H}$ and the Ginzburg–Landau regime.

The helical texture for $\boldsymbol{v}_s \parallel \boldsymbol{H}$

In this configuration the magnetic-field contribution to the energy (i.e. the last term in (7.232)) takes the form

$$h^2(\hat{\boldsymbol{H}}\cdot\hat{\boldsymbol{d}})^2 = h^2\cos^2\theta. \tag{7.234}$$

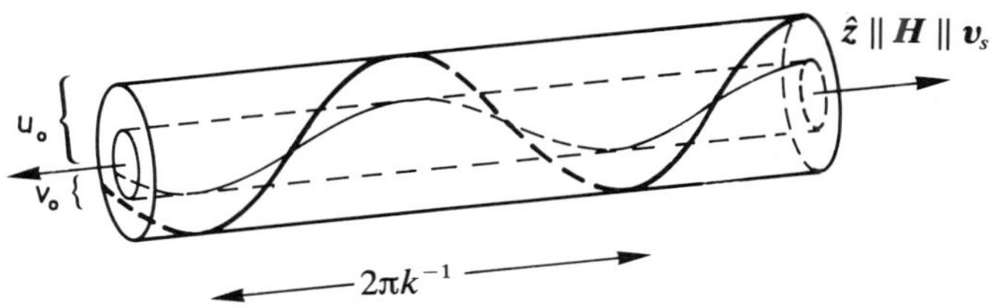

Figure 7.34 Helical texture involving the vectors $\hat{\boldsymbol{l}}$ and $\hat{\boldsymbol{d}}$ in the presence of superflow and magnetic field $\boldsymbol{H} \parallel \boldsymbol{v}_s$. The curves are meant to show the traces of the endpoints of $\hat{\boldsymbol{l}}$ (thick line) and $\hat{\boldsymbol{d}}$ (thin line) along $\hat{z}$, which are located on cylinders with radii u_0 and v_0 respectively. The wavelength ("pitch") of the helix, $2\pi k^{-1}$, is indicated.

The magnetic field and the superflow then compete for the orientation of $\hat{\boldsymbol{d}}$ and $\hat{\boldsymbol{l}}$: while the former favours $\boldsymbol{H} \perp \hat{\boldsymbol{d}}$ and the latter $\boldsymbol{v}_s \parallel \hat{\boldsymbol{l}}$, the dipole coupling favours $\hat{\boldsymbol{l}} \parallel \hat{\boldsymbol{d}}$.

To investigate the helical texture in the case of small opening angles β and θ, we may expand $\mathscr{G}$ in terms of β and θ. This is substantially simplified if we cast $\hat{\boldsymbol{l}}$ and $\hat{\boldsymbol{d}}$ into the forms

$$\hat{\boldsymbol{l}} = \begin{pmatrix} u_0 \cos kz \\ -u_0 \sin kz \\ 1 - u_0^2 \end{pmatrix}, \quad \hat{\boldsymbol{d}} = \begin{pmatrix} v_0 \cos kz \\ -v_0 \sin kz \\ 1 - v_0^2 \end{pmatrix}; \tag{7.235a}$$

i.e.

$$\alpha = \psi = kz, \tag{7.235b}$$

$$\sin\beta = u_0, \quad \sin\theta = v_0, \tag{7.235c}$$

where u_0 and v_0 are the projections of $\hat{\boldsymbol{l}}$ and $\hat{\boldsymbol{d}}$ in the (x, y) plane (see Fig. 7.34). This means that we have chosen spatially constant opening angles, as is natural for the particular configuration under investigation. Furthermore, $\hat{\boldsymbol{d}}$ and $\hat{\boldsymbol{l}}$ are chosen to have the same periodicity, with wavelength $\lambda = 2\pi k^{-1}$, to keep the dipolar energy minimal. Expanding (7.232) up to fourth order in u_0 and v_0, we may now proceed to minimize $\mathscr{G}$ with respect to u_0 and v_0. For v_0, we obtain

$$v_0 = \frac{u_0}{1 - h^2 + \bar{k}^2}, \tag{7.236}$$

where

$$\bar{k} = k\xi_D. \tag{7.237}$$

The inclination of $\hat{\boldsymbol{d}}$ is seen to be different from that of $\hat{\boldsymbol{l}}$ except for $h = \bar{k}$. Elimination of v_0 from $\mathscr{G}$ yields

$$\mathscr{G} = \mathscr{G}_0 + \mathscr{G}_1 u_0^2 + \mathscr{G}_2 u_0^4, \tag{7.238}$$

where

$$\mathscr{G}_0 = A\xi_D^{-2}(-p^2 + 2p\bar{k} + h^2) \tag{7.239a}$$

is the contribution of the uniform configuration ($\beta = \theta = 0$) and

$$\mathscr{G}_1 = A\xi_D^{-2}\left(p^2 + \tfrac{3}{2}\bar{k}^2 - 3p\bar{k} - \frac{h^2 - \bar{k}^2}{1 - h^2 + \bar{k}^2}\right), \tag{7.239b}$$

$$\mathscr{G}_2 = -2A\xi_D^{-2}\left[p^2 + 2\bar{k}^2 - \tfrac{11}{4}p\bar{k} - \frac{\bar{k}^2}{(1 - h^2 + \bar{k}^2)^2} - \frac{(h^2 - \bar{k}^2)^2}{(1 - h^2 - \bar{k}^2)^3}\right]. \tag{7.239c}$$

This implies the following.

(1) When $\mathscr{G}_1(k) > 0$ for all k, the uniform texture with $\hat{\boldsymbol{l}} \parallel \hat{\boldsymbol{d}} \parallel \boldsymbol{v}_s$ is locally stable.

(2) When $\mathscr{G}_1(k) \geqslant \mathscr{G}_1(k_m) = 0$ and the equality is satisfied by a single k_m, and furthermore if $\mathscr{G}_2(k_m) > 0$, the uniform texture becomes unstable against the appearence of the helical texture with wavelength $\lambda_m = 2\pi k_m^{-1}$. Therefore this condition gives the phase boundary between the uniform and helical textures.

(3) For $\mathscr{G}_1(k_m) < 0$ and $\mathscr{G}_2(k_m) > 0$, the helical texture with wave vector k_m and $u_0^2 = |\mathscr{G}_1|/\mathscr{G}_2$ is locally stable.

(4) For $\mathscr{G}_1(k_m) < 0$, $\mathscr{G}_2(k_m) < 0$, both the uniform and helical textures are unstable.

These criteria allow one to construct the stability region of the uniform texture $\boldsymbol{v}_s \parallel \hat{\boldsymbol{l}} \parallel \hat{\boldsymbol{d}}$ (region I in Fig. 7.35) as well as the phase boundary with the helical state. Region I is *completely* surrounded by region II, where the helical texture is stable. Near the origin ($p = h = 0$), where $\hat{\boldsymbol{l}}$ and $\hat{\boldsymbol{d}}$ are essentially dipole-locked (see (7.236)), the phase boundary is given by

$$p = 10^{1/2}h. \tag{7.240}$$

This agrees with the result of the stability analysis in the strictly dipole-locked case, (7.217), for which the stability parameter in weak coupling is $s = \frac{9}{10}$ (see (7.213)) and where, according to (7.28) and (7.232a), $H_v = H_A\xi_0(\hbar/2m)^{-1}v_s = H^*(v_s/v_{s,D})$, so that $H_v/H = p/h$.

Figure 7.35 also shows the phase boundary that limits the region where the uniform texture $\boldsymbol{v}_s \perp \hat{\boldsymbol{l}} \parallel \hat{\boldsymbol{d}}$ is stable, i.e. where the magnetic field wins against the alignment effect of the flow. This configuration is stable for (see e.g. Fetter 1981)

$$p < p_c \equiv \left(\frac{2}{1 + h^2}\right)^{1/2} h. \tag{7.241}$$

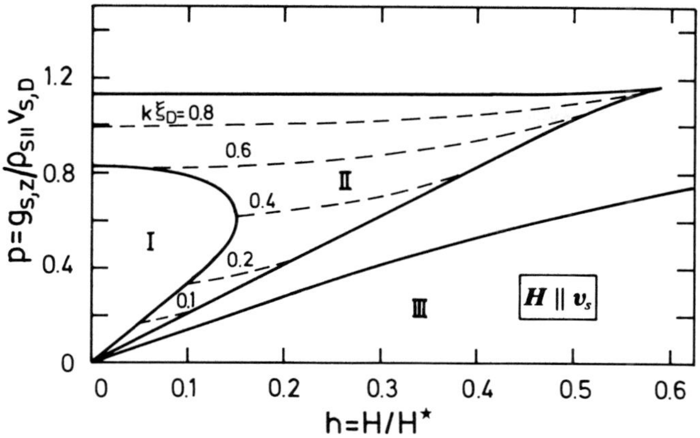

Figure 7.35 Stability regions of the uniform texture $\hat{\boldsymbol{d}} \parallel \hat{\boldsymbol{l}} \parallel \boldsymbol{v}_s$ (I) and of the helical texture (II) in the presence of superflow and magnetic field ($\boldsymbol{H} \parallel \boldsymbol{v}_s$) for $T \lesssim T_c$. The critical currents terminating regions I and II at $H=0$ are given by $p_{c,h_1}=0.826$ and $p_{c,h_2}=1.13$, respectively. Also shown are contours of constant k, the optimal wave vector of the helix with pitch $2\pi k^{-1}$, in units of ξ_D. (After Vollhardt *et al.* (1979).)

The stability region of the uniform texture is limited even in the absence of a magnetic field. For $p > p_{c,h_1} = 0.826$ (i.e. $v_{s,h_1} \approx 1.0\,\text{mm/s}$) a helical texture is stable, where $\hat{\boldsymbol{d}}$ and $\hat{\boldsymbol{l}}$ are no longer parallel to each other. In the vicinity of the phase boundary the inclination angles β and θ of $\hat{\boldsymbol{l}}$ and $\hat{\boldsymbol{d}}$ respectively are small. Minimizing $\mathcal{G}$ with respect to u_0 and expanding $\mathcal{G}_1$ in terms of a small deviation $h - h_i$ of the magnetic field h from its value at the boundary, $h_i(p)$, one obtains

$$\sin^2\beta = a'\frac{h-h_i}{h_i} \quad (h > h_i), \tag{7.242a}$$

where

$$a' = h_i\left(\frac{\partial \mathcal{G}_1}{\partial h}\right)_{h_i}\frac{1}{\mathcal{G}_2}. \tag{7.242b}$$

may be calculated along the boundary. This dependence is seen to be very similar to that given by (7.222). In particular, for very small p and h, i.e. close to the origin of Fig. 7.35, one finds

$$a' = \tfrac{10}{29}. \tag{7.243}$$

The wavenumber k (i.e. the inverse wavelength) grows linearly with p for small p. The general dependence has to be calculated numerically (Lin-Liu *et al.* 1979).

A small misalignment between the magnetic field and the flow decreases the slope of the phase boundary near the origin, i.e. it decreases the stability region of the helix in that part of the phase diagram, and lowers p_{c,h_1}, the critical current at $h = 0$.

Assuming that the helical instability is always associated with the longitudinal fluctuations, it is straightforward to obtain the full stability region of the helical texture in the phase diagram (Vollhardt *et al.* 1979). While this assumption is indeed correct in the vicinity of the entire phase boundary, the importance of transverse fluctuations away from the boundary cannot be excluded. If they are relevant, this will lead to a reduction of the stability region. To determine the region of stability of the helical texture, the parametrization of $\hat{\boldsymbol{l}}$ and $\hat{\boldsymbol{d}}$ as in (7.235) may still be used, but now u_0 and v_0 are no longer small in general. Furthermore, in the case of an open experimental geometry (e.g. a cylinder with open ends) the wavenumber k of α and ψ, (7.235b), must itself be assumed to be a variable, with respect to which the energy should be minimized. (This is in contrast with the toroidal geometry discussed below (7.219), where k is quantized and cannot simply adjust.) The possible helical solutions with constant β and θ, and the optimal k are then obtained from

$$\frac{\partial \mathcal{G}}{\partial \beta} = 0, \quad \frac{\partial \mathcal{G}}{\partial \theta} = 0, \quad \frac{\partial \mathcal{G}}{\partial k} = 0. \tag{7.244}$$

The stability criteria for such a texture are given in terms of the eigenvalues λ_i of the eigenequations

$$\boldsymbol{Mf} = \lambda \boldsymbol{f}, \tag{7.245}$$

where $f = (x_1, x_2, x_3)$ and $M_{ij} = (\partial^2 \mathcal{G}/\partial x_i\, \partial x_j)_0$, with $x_1 = \delta\beta$, $x_2 = \delta\theta$ and $x_3 = \delta k$, and the subscript zero implies that the value is taken at the optimal values of β, θ and k. The texture is stable if all eigenvalues are non-negative. In this way, one obtains the stability region II of the helical texture shown in Fig. 7.35. The outer phase boundaries at which the stability regime terminates have a roughly triangular shape with a sharp tip at $p = 1.16$, $h = 0.59$. For $h \lesssim 0.4$, and thus in particular for $H \to 0$, the lower boundary is given by $p = 2.10h$ and the upper one by $p_{c,h_2} = 1.13 \approx$ constant (i.e. $v_{s,h_2} \approx 1.5$ mm/s). In Fig. 7.35 the optimal wave vector k (inverse pitch) of the helix is also shown. At p_{c,h_2} the wave vector of the helix is $k \approx 0.9\xi_D^{-1}$. The contours of constant β and θ in the stability regime have also been calculated (Vollhardt *et al.* 1979). They show that the inclination angles of $\hat{\boldsymbol{l}}$ and $\hat{\boldsymbol{d}}$ take maximal values at $\beta_{max} \approx 39°$, $\theta_{max} \approx 33°$. In particular, at the upper boundary it is found that for small h the instability is purely due to $\hat{\boldsymbol{l}}$, while for larger h the vector $\hat{\boldsymbol{d}}$ is also involved. On the other hand, at the lower boundary the instability is caused by a joint $\hat{\boldsymbol{l}}$–$\hat{\boldsymbol{d}}$ motion for small h and mainly by $\hat{\boldsymbol{d}}$ near the tip.

The existence of a critical current p_{c,h_2} associated with $k^{-1} \approx \xi_D$ above which the helical texture is unstable is not surprising. The helix is only stabilized by the dipolar coupling of $\hat{\boldsymbol{l}}$ to $\hat{\boldsymbol{d}}$; if $\hat{\boldsymbol{d}}$ varies on a lengthscale of order ξ_D then the associated bending energy becomes too large and it is no longer favourable for $\hat{\boldsymbol{d}}$ to follow $\hat{\boldsymbol{l}}$. At that point $\hat{\boldsymbol{l}}$, and thus the helix itself, become unstable.

An investigation of the stability regions of the uniform and the helical texture extending to lower temperatures has been performed by Schopohl and Tewordt (1980). For decreasing temperatures, it is found that region I in Fig. 7.35 shrinks until it vanishes altogether at $T = T_i$, (7.224) (see also Fetter and Williams 1981). On the other hand, while the stability region of the helical-texture region II in Fig. 7.35 also decreases in size for $T \to T_i$, its shape is almost unchanged.

The stability of helical textures in a toroidal geometry, where the pitch of the helix may be assumed to be fixed by topological constraints, has been investigated by Fetter and Williams (1979, 1981). They find that, as the magnetic field H is increased above H_I where the helix is first formed, the opening angle β increases until at some critical field H_1, where $\beta \approx 35°$, the helical texture becomes unstable. When H is further increased, a new region of stability of the helical texture is found for $H > H_2$, where the opening angle β now exceeds $\frac{1}{2}\pi$. The helical texture in the first regime ($H_I < H < H_1$) is referred to as a "small-angle helix", and the one for $H > H_2$ as a "large-angle helix". The effects of boundaries on helical textures, i.e. their pitch, have also been considered (Fetter 1982). The effect of fluctuations on the wide-angle helical texture, causing a decay of the equilibrium texture to different values of the opening angle, has been discussed by Duru and Ünal (1982).

The phase diagram in Fig. 7.35 has been investigated experimentally by exploiting the dependence of ultrasound attenuation on the orientation of the vector $\hat{\boldsymbol{l}}$ (see (11.108)). A first indication of the phase boundary between regions I and II in Fig. 7.35, i.e. an abrupt change in the sound attenuation as a function of magnetic field for given superflow consistent with (7.242) and (7.243), was observed by Kleinberg (1979a,b). Extensive investigations of the phase boundary were later performed by Bates *et al.* (1983, 1984, 1986), who found a well-defined reproducible transition between the uniform $\hat{\boldsymbol{l}}$ texture and what is believed to be the helical texture. In particular, the initial slope of the phase boundary is again in agreement with theory. However, the bending back of the boundary (Fig. 7.35), predicted to occur for velocities of the order of 1 mm/s, seems to occur at considerably higher values.

The helical texture for $\boldsymbol{v}_s \perp \boldsymbol{H}$

In a configuration $\boldsymbol{H} \perp \boldsymbol{v}_s$ the magnetic field and the superflow do not compete with each other, but rather stabilize the uniform configuration in which $\hat{\boldsymbol{l}}$ and $\hat{\boldsymbol{d}}$ are aligned by the flow. An investigation of a helical instability in this new situation may be performed along the lines discussed above for $\boldsymbol{v}_s \parallel \boldsymbol{H}$ (Lin-Liu *et al.* 1979). The magnetic-field term in (7.232) is now given by $h^2 \sin^2 \theta \sin^2 \psi$, assuming $\boldsymbol{H}$ to be along $\hat{\boldsymbol{y}}$. Since $\boldsymbol{H}$ breaks the axial symmetry, it is easier for $\hat{\boldsymbol{d}}$ (and hence $\hat{\boldsymbol{l}}$) to move in the plane perpendicular to $\boldsymbol{H}$ than out of it. There is then an anisotropy, i.e. an "easy" and a "hard" direction for the motion of $\hat{\boldsymbol{d}}$ and $\hat{\boldsymbol{l}}$. Any texture where $\hat{\boldsymbol{l}}$ winds around the superflow will therefore be forced into an *elliptic* configuration. The parametrization of $\hat{\boldsymbol{l}}$ and $\hat{\boldsymbol{d}}$ in (7.235a) then has to be suitably generalized to

$$\hat{\boldsymbol{l}} = \begin{pmatrix} u_0 \cos kz \\ u_0' \sin kz \\ 1 - u_0^2 - u_0'^2 \end{pmatrix}, \quad \hat{\boldsymbol{d}} = \begin{pmatrix} v_0 \cos kz \\ v_0' \sin kz \\ 1 - v_0^2 - v_0'^2 \end{pmatrix}, \tag{7.246}$$

where the free energy now has to be minimized with respect to u_0, u_0', v_0 and v_0'. Eliminating v_0, v_0' and u_0' by minimizing the quadratic terms in $\mathscr{G}$, (7.232), an expression for $\mathscr{G}$ up to fourth order in u_0 of the form (7.238) may be obtained. This can then be further analysed, as described below (7.238), to determine the stability region of the uniform texture $\hat{\boldsymbol{l}} \parallel \hat{\boldsymbol{d}} \parallel \boldsymbol{v}_s$ as well as the phase boundary with a helical texture. It is found that a helical texture only develops *above* a critical strength of the supercurrent given by $p_{c,h_1} = 0.826$ at $H = 0$, i.e. the critical current found already in the case $\boldsymbol{v}_s \parallel \boldsymbol{H}$ for $H = 0$. For magnetic fields $H > 0$, this critical current increases and reaches $p_{c,\infty} = 1.26$ as $H \to \infty$. As expected, the stability region of the uniform texture is much more extended than in the case $\boldsymbol{v}_s \parallel \boldsymbol{H}$. The opening angles of the elliptically shaped helix just beyond the phase boundary, i.e. the projections of $\hat{\boldsymbol{l}}$ and $\hat{\boldsymbol{d}}$ in the (x, y) plane, and the pitch of the helix are determined by an expression similar to (7.242).

7.10.3 Textures and superflow in the presence of boundaries

As discussed in Section 7.9, boundaries orient the vector $\hat{\boldsymbol{l}}$ and may thus lead to a competition to the orientational effect due to superflow. One example is superflow in a slab geometry (with width L) parallel to the surfaces. In the absence of superflow the vector $\hat{\boldsymbol{l}}$ is uniformly oriented normal to the surface. This situation is not changed by a small supercurrent owing to the fact that $\hat{\boldsymbol{l}}$ is *rigidly* fixed at the surface such that any bending energy is at least of order $(\alpha/L)^2$, while the energy gain associated with the orientation due to the superflow is of order $(\alpha \bar{v}_s)^2$ small. Here α denotes a small deviation from the perpendicular configuration ($\alpha = 0$). At a critical velocity $v_s \approx (\hbar/2m)L^{-1}$ a transition into a nonuniform texture takes place. This situation is very similar to the effect of a magnetic field on the $\hat{\boldsymbol{l}}$ texture in a slab (see Section 7.9.3). The nature of this transition has been discussed by Fetter (1976b, 1978c), Hu and Ham (1978), Hu (1979a), Ham and Hu (1980) and Dow and Hook (1984). In particular, Hu (1979a) found that for $L \ll \xi_D$ the transition is first-order and nearly first-order for $L \gg \xi_D$. While in the Ginzburg–Landau regime the resulting texture is generally found to be nonplanar, varying only in the direction perpendicular to the slab, for low enough temperatures a two-dimensional periodic structure is obtained.

A simultaneously applied external magnetic field leads to a further competition, depending on its orientation with respect to both the superflow and the surface. The simplest situation is that with $\boldsymbol{H} \parallel \boldsymbol{v}_s$, in which case the magnetic field helps to stabilize the uniform $\hat{\boldsymbol{l}}$ texture. On the other hand, for $\boldsymbol{H}$ perpendicular to the slab, both the magnetic field and the flow destabilize the uniform $\hat{\boldsymbol{l}}$ texture. The case of a half-infinite space has been considered by Bruinsma and Maki (1979b). The nature of the nonuniform texture beyond the transition point in this situation has been investigated by Hu (1979b) using a linear stability analysis. The results suggest the formation of a periodic array of supercurrent rolls when a weak (strong) magnetic field is applied to a thick (thin) slab.

The effect of superflow on textures in a cylindrical geometry, with the flow parallel to the axis, has been investigated by Buchholtz and Fetter (1977a). They assumed a dipole-locked configuration $\hat{\boldsymbol{l}} \parallel \hat{\boldsymbol{d}}$, as is valid for weak superflow and cylinder radii $R \gtrsim \xi_D$, and calculated the change of the angle χ of $\hat{\boldsymbol{l}}$ with the axis, (7.190), in perturbation theory. The case of flow in narrow cylinders ($R \ll \xi_D$), where $\hat{\boldsymbol{d}}$ may be assumed to be uniform, has been studied in detail by Bruinsma and Maki (1979d). They found that superflow induces a slight twist in the Mermin–Ho texture, and observed that this is the case even in the absence of external flow. In the case of the $\hat{\boldsymbol{l}}$ disgyration (7.197), expected to be stable for very narrow cylinders, a second-order Freederickz-like transition to a nonplanar state is predicted, where $\hat{\boldsymbol{l}}$ acquires a finite component in the direction of superflow. There is a critical value of the superflow where the Mermin–Ho texture becomes energetically favourable. In contrast with the two textures considered above, the double-half disgyration (7.199) is found to undergo a helical transition in a large superflow.

7.10.4 The effect of superflow on domain walls

The coupling of a superfluid velocity field to the orbital component in the A phase implies that external superflow has an immediate orientational influence on already existing nonuniform $\hat{\boldsymbol{l}}$ textures. Even in the absence of surfaces, such nonuniform configurations may exist in the form of defects of the order parameter. We now want to discuss how domain walls ("solitons", see Section 7.8), i.e. nonsingular defects, are affected by superflow. This investigation will also shed light on textural instabilities in the A phase that have not been considered so far, and will be relevant in view of possible dissipative mechanisms for superflow involving $\hat{\boldsymbol{l}}$ (see Section 7.11).

Concerning the B phase, where orientational effects are much weaker (see Section 6.43), the effect of crossed magnetic field and superflow on the

$\hat{\boldsymbol{n}}$ soliton has been investigated by Lin-Liu and Maki (1980a). In particular, in the high-field limit (H of the order of a kilogauss) and for superfluid velocities $v_s \gtrsim 1$ mm/s, they identified three types of $\hat{\boldsymbol{n}}$ solitons, with distinctly different spatial conformations.

In the case of A-phase solitons let us consider $\hat{\boldsymbol{l}}$, $\hat{\boldsymbol{d}}$ configurations in the presence of a magnetic field $\boldsymbol{H}$ confining $\hat{\boldsymbol{d}}$ perpendicular to $\boldsymbol{H}$. For $\boldsymbol{H} \parallel \hat{\boldsymbol{z}}$, the free energy per unit area is then given by (7.155), which allows domain-wall solutions. The specific structure of the domain wall depends on the relative direction of $\hat{\boldsymbol{k}}$, the surface normal, with respect to $\boldsymbol{H}$. In particular, as discussed in Section 7.8.2, domain walls in a cylinder are expected to be oriented perpendicular to the axis in order to keep their size minimal. Depending on whether the magnetic field is perpendicular or parallel to the axis, a "splay" or "twist" composite soliton respectively is most likely to be formed. Following Vollhardt and Maki (1979b), we shall now consider the effect of a constant superflow with $\boldsymbol{v}_s \parallel \hat{\boldsymbol{k}}$ on these two domain walls, since this is a realistic experimental configuration for flow in a cylinder. (Note that these authors characterize the superflow by a quantity q related to (7.231a) by $p = q\xi_D$, since their lengthscale is given by $\xi_\perp \equiv 2^{1/2}\xi_D$.)

Splay composite soliton ($\boldsymbol{v}_s \perp \boldsymbol{H}$)

In the absence of superflow, this domain wall is given by (7.164) in parametrized form. In order to be able to use the general expression for the free-energy density in the presence of a superflow along $\hat{\boldsymbol{z}}$, (7.232), we assume the magnetic field $\boldsymbol{H}$ to lie along $\hat{\boldsymbol{y}}$. The splay composite soliton in this field configuration (see Fig. 7.36a) is then obtained from (7.161) and (7.164) by the replacement of the angles ($\alpha \to \beta$, $\psi \to \theta$) and coordinates ($x \to z$, $y \to x$, $z \to y$).

A superflow in the $\hat{\boldsymbol{z}}$ direction will have two main effects on the domain wall: (i) outside the domain wall the configuration will be aligned by the flow to minimize the $(\hat{\boldsymbol{l}} \cdot \boldsymbol{v}_s)^2$ terms in the free energy; (ii) the torque term wants to tilt $\hat{\boldsymbol{l}}$ out of the plane perpendicular to $\boldsymbol{H}$ at the expense of the dipole energy. These effects are shown in Fig. 7.36(b). The first effect will lead to a uniform texture $\hat{\boldsymbol{l}} \parallel \hat{\boldsymbol{d}} \parallel \boldsymbol{v}_s$ far away from the domain wall, as long as the superflow does not exceed the critical value $p_{c,h_1} = 1.26$ at which a transition to an elliptically shaped helix is expected to occur (see the discussion below (7.246)).

For superflow $p < p_{c,h_1}$, the two effects may be incorporated into the variational parametrization of the splay composite soliton (i.e. (7.164) with the abovementioned replacement of angles) by the generalization

$$\sin v = \operatorname{sech} \eta z, \tag{7.247a}$$

$$\beta = tv + (1-t)\sin^{-1}(\operatorname{sech} \eta\rho z), \tag{7.247b}$$

$$\alpha = 2(a - b \sin\beta), \tag{7.247c}$$

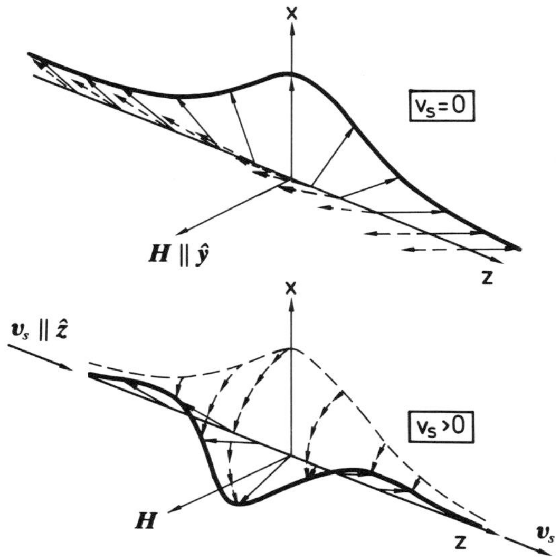

Figure 7.36 (a) Splay composite soliton in a strong magnetic field $\boldsymbol{H} \parallel \hat{\boldsymbol{y}}$, with surface normal $\hat{\boldsymbol{k}} \parallel \hat{\boldsymbol{z}}$, without external superflow. The vectors $\hat{\boldsymbol{l}}$ and $\hat{\boldsymbol{d}}$ are represented by solid and broken arrows respectively. (b) Deformation of the splay composite soliton under the influence of an external superflow $\boldsymbol{v}_s \parallel \hat{\boldsymbol{z}}$. (After Vollhardt and Maki (1979b).)

where $v = \beta - \theta$ and $\psi = 0$. Here t, η, ρ, a and b are variational parameters, with $\rho = a = b = 0$ for $v_s = 0$: (i) at $v_s = 0$, t determines the asymptotic orientation of $\hat{\boldsymbol{l}}$ and $\hat{\boldsymbol{d}}$ outside the domain wall; for $v_s \neq 0$, it describes the crossover to the uniform configuration $\hat{\boldsymbol{d}} \parallel \hat{\boldsymbol{l}} \parallel \hat{\boldsymbol{z}}$; (ii) η^{-1} is the length over which the angle between $\hat{\boldsymbol{l}}$ and $\hat{\boldsymbol{d}}$ changes from 0 to π; (iii) $(\rho\eta)^{-1}$ stands for the second lengthscale introduced by the superflow; (iv) $2a$ is the azimuthal angle of $\hat{\boldsymbol{l}}$ in the asymptotic regime; (v) $2b$ describes the degree of winding of $\hat{\boldsymbol{l}}$ out of the plane for a given inclination angle β.

The variational ansatz (7.247) may now be used to minimize the domain-wall energy relative to the uniform state

$$F = \int \mathrm{d}^3 r\, (\mathscr{G} - \mathscr{G}_{\mathrm{uni}}), \tag{7.248}$$

with $\mathscr{G}$ given by (7.232) and with $\mathscr{G}_{\mathrm{uni}}$ the free-energy density of the uniform state. For small superflow p, one indeed finds the qualitative effects on the domain wall described above. As p increases, $\hat{\boldsymbol{l}}$ begins to tilt more and more out of the plane perpendicular to $\boldsymbol{H}$ (see Fig. 7.36b) and F monotonically *decreases*; at a tilt angle $\alpha \approx \pi$ the domain wall has doubled its width to accommodate the strong winding. Then, at a critical supercurrent $p_{\parallel} = 0.54$ ($v_{s,c} \approx 0.7$ mm/s), the solution (7.247) becomes unstable against unlimited winding of $\hat{\boldsymbol{l}}$ around the superflow and the texture delocalizes. For $p > p_{\parallel}$ the energy F is negative.

It is interesting to note that the splay composite soliton is in fact *globally* unstable in the presence of superflow. This can be seen from the variational ansatz

$$\left.\begin{aligned} \sin\beta &= \operatorname{sech}\eta z, \\ \alpha &= -b^3 \operatorname{sech}(\eta z/b), \end{aligned}\right\} \tag{7.249}$$

with $\hat{\boldsymbol{d}}$ constant along $\hat{\boldsymbol{x}}$. For $b \gg 1$ and finite η, this ansatz describes a localized $\hat{\boldsymbol{l}}$ texture that is strongly wound up around the direction of superflow. The domain-wall energy F, (7.248), may easily be calculated in terms of (7.249). After minimization with respect to η, one finds that in the limit $b \rightarrow \infty$ the torque term (the one linear in p) makes the largest contribution, i.e.

$$F \propto b(\zeta - pb^2), \tag{7.250}$$

where ζ is a constant of order unity. Furthermore, the domain-wall width is given by

$$\eta^{-1} \propto b^2 \xi_D, \tag{7.251}$$

i.e. it becomes arbitrarily large. Hence, for any superflow $p > 0$, the domain wall energy is unbounded from below, corresponding to an arbitrarily strongly wound up, *delocalized* $\hat{\boldsymbol{l}}$ configuration. At this point the assumption of a *constant* external superflow cannot be maintained because dissipation occurs. It should be pointed out that, for weak superflow, the delocalized state ($b \rightarrow \infty$) appears to be separated by an energy barrier from the localized domain wall with $b \approx 1$, thus ensuring the local stability of the domain wall for small currents.

Twist composite soliton ($\boldsymbol{v}_s \parallel \boldsymbol{H}$)

In the case of zero external superflow the configuration of this domain wall is given by the exact analytic solution (7.162b) (see Fig. 7.27). Application of a superfluid mass current perpendicular to $\hat{\boldsymbol{l}}$ and the wall will (i) give rise to a finite component of $\hat{\boldsymbol{l}}$ in the direction of flow and (ii) introduce a new lengthscale associated with this nonplanarity. The first effect not only improves the energy via the $(\hat{\boldsymbol{l}} \cdot \boldsymbol{v}_s)^2$ term, but also allows the torque term to act and thereby reduce the energy. These new degrees of freedom are incorporated in the variational ansatz

$$\left.\begin{aligned} \sin(\alpha/\tau) &= \operatorname{sech}\eta z, \\ \cos\beta &= -\lambda \operatorname{sech}\rho\eta z, \\ \sin(\alpha - \psi) &= \operatorname{sech}\eta z, \\ \theta &= \tfrac{1}{2}\pi. \end{aligned}\right\} \tag{7.252}$$

In the absence of superflow one has $\lambda = 0$, $\rho = 1$, $\tau = \frac{4}{5}$ and $\eta = (\frac{5}{2})^{1/2}\xi_D^{-1}$;

for finite flow, λ and ρ parametrize the two effects described above. The form (7.252) is now used to minimize the domain-wall energy (7.248) in terms of the variational parameters τ, η, λ and ρ for given superflow. Note that the superfluid *velocity* $v_{s\perp}$ corresponding to a superflow p in the configuration where $\boldsymbol{v}_s \perp \hat{\boldsymbol{l}}$ is given by (see (7.231))

$$v_{s,\perp} = p\frac{\rho_{s\parallel}}{\rho_{s\perp}}v_{s,D} = \tfrac{1}{2}pv_{s,D}. \tag{7.253}$$

As p is increased from zero, the pure-twise $\hat{\boldsymbol{l}}$ texture is deformed in the centre of the wall, with $\hat{\boldsymbol{l}}$ acquiring a component along the direction of flow. At the same time, the domain-wall energy decreases slightly. There is a critical value of the current $p_{\perp 1} = 0.45$ ($v_{s,\perp 1} \approx 0.3$ mm/s) where the twist solution becomes unstable and a new stable texture with $\lambda = \tau = 1$, $\rho = 0$, $\eta = \infty$ (but $\rho\eta$ finite), having a much lower energy, is formed. This new domain wall only involves a *planar* $\hat{\boldsymbol{l}}$ texture ($\alpha = \psi = 0$), where $\hat{\boldsymbol{l}}$ changes from $\hat{\boldsymbol{l}} \perp \boldsymbol{v}_s$ outside the wall through $\hat{\boldsymbol{l}} \parallel \boldsymbol{v}_s$ at the centre of the wall in a pure bendlike fashion (Fig. 7.37a). This configuration is energetically favourable since the torque term can act by twisting half of the entire domain wall at the centre where $\hat{\boldsymbol{l}} \parallel \hat{\boldsymbol{z}}$, without creating much bending energy. The domain-wall energy of this texture is obtained from (7.232) and (7.248) as

$$\frac{F}{F_{\text{unit}}} = \frac{1}{2^{5/2}}\int_{-\infty}^{\infty} \mathrm{d}s\left[(3-2\sin^2\beta)\beta_s^2 - p^2\frac{\cos^2\beta}{1+\sin^2\beta} + 2\cos^2\beta\right] - \tfrac{1}{2}\pi p, \tag{7.254a}$$

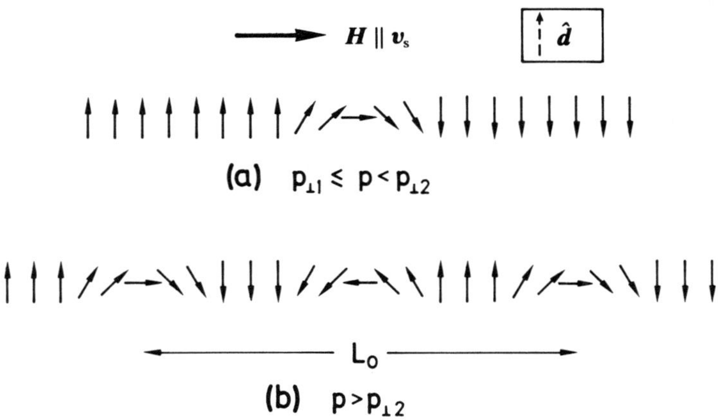

Figure 7.37 (a) The planar, purely bendlike domain wall in the presence of superflow $\boldsymbol{v}_s \parallel \boldsymbol{H}$, into which the twist composite soliton transforms at a superfluid current $p_{\perp 1}$. The vector $\hat{\boldsymbol{d}}$ is constant thoughout the system, with $\hat{\boldsymbol{d}} \parallel \hat{\boldsymbol{l}} \perp \boldsymbol{H}$ outside the domain wall. (b) One-dimensional lattice of planar domain walls in $\hat{\boldsymbol{l}}$ with period L_0 for superflow $p > p_{\perp 2}$. (After Vollhardt (1979).)

where

$$F_{\text{unit}} \equiv 8A\sigma(\hat{k}_z)\xi_{\text{D}}^{-1}. \tag{7.254b}$$

The Euler–Lagrange equation for β is easily integrated and yields β implicitly through

$$\frac{z}{\xi_{\text{D}}} = \frac{1}{2^{1/2}} \int_0^{\beta} \mathrm{d}\beta' \left[\frac{1+\cos^2\beta'}{P(\sin\beta')}\right]^{1/2}, \tag{7.255a}$$

with

$$P(x) \equiv (1-x^2)\left(1 - \frac{1}{2}\frac{p^2}{1+x^2}\right), \tag{7.255b}$$

while the domain-wall energy reduces to

$$\frac{F}{F_{\text{unit}}} = \frac{1}{2^{1/2}} \int_0^1 \mathrm{d}t \left[\frac{3-2t^2}{1-t^2} P(t)\right]^{1/2} - \tfrac{1}{2}\pi p. \tag{7.256}$$

This energy decreases almost linearly for increasing superflow p—mainly owing to the last term, the torque term, in (7.256)—and vanishes at $p = p_{\perp 2} = 0.625$ ($v_{\text{s},\perp 2} \approx 0.4$ mm/s). We note that $p_{\perp 2}$ is considerably smaller than the critical value $p_{\text{c}} = 2^{1/2}$, (7.241), above which the uniform texture $\boldsymbol{v}_{\text{s}} \perp \hat{\boldsymbol{l}} \parallel \hat{\boldsymbol{d}}$ in a strong magnetic field is unstable, such that the variational ansatz (7.252) is valid.

The flow-induced domain-wall lattice

For superflow $p > p_{\perp 2}$, the energy F of a single domain wall relative to the uniform texture becomes negative. This implies that the uniform texture becomes globally unstable against the spontaneous formation of domain walls (Vollhardt and Maki 1979a,b, Vollhardt 1979); for the topological constraints, see Section 7.8.1. Limiting our consideration to a pure $\hat{\boldsymbol{l}}$ texture ($\alpha = \psi = 0$) with localized twist at the centre of the domain wall, the free energy per unit length of a regular array ("lattice") of purely one-dimensional domain walls with period L_0 (see Fig. 7.37b) is given by (7.254) with the replacement

$$\int_{-\infty}^{\infty} \mathrm{d}s \rightarrow \frac{1}{L_0} \int_{-L_0/2}^{L_0/2} \mathrm{d}s. \tag{7.257}$$

Here we have put $\psi_z = \pi\delta(z - \frac{1}{2}L_0)$ and have assumed β to be a periodic function of z with period L_0. The Euler–Lagrange equation for β is then found as

$$\beta_s^2 = 2\frac{k^2 + P(\sin\beta)\cos^2\beta}{1 + 2\cos^2\beta}, \tag{7.258}$$

where k is an integration constant, with $k=0$ for $p \leqslant p_{\perp 2}$. The domain-wall energy per unit length is consequently given by

$$\frac{F/L_0}{F_{\text{unit}}}=\frac{1}{2^{3/2}L_0}\int_0^{\pi} \mathrm{d}\beta\,\{(3-2\sin^2\beta)[k^2+P(\sin\beta)]\}^{1/2}-\frac{\pi}{2}\frac{p}{L_0}-\frac{1}{4}\frac{k^2}{\xi_{\mathrm{D}}}, \tag{7.259}$$

while the period L_0 is determined by

$$\frac{L_0}{\xi_{\mathrm{D}}}=\frac{1}{2^{1/2}}\int_0^{\pi}\mathrm{d}\beta\left[\frac{3-2\sin^2\beta}{k^2+P(\sin\beta)}\right]^{1/2}. \tag{7.260}$$

Minimization of F with respect to k yields

$$\frac{F/L_0}{F_{\text{unit}}}=-\frac{1}{4\xi_{\mathrm{D}}}k^2(p) \quad (p_{\perp 2}\leqslant p<p_{\mathrm{c}}). \tag{7.261}$$

Here k is a nonanalytic function of the superflow for p approaching $p_{\perp 2}$ from above; it is given implicitly by

$$p=\frac{1}{2^{1/2}\pi}\int_0^{\pi}\mathrm{d}\beta\,\{(3-2\sin^2\beta)[k^2+P(\sin\beta)]\}^{1/2} \quad (p_{\perp 2}\leqslant p<p_{\mathrm{c}}). \tag{7.262}$$

Equation (7.261) clearly shows that the domain-wall energy is negative for $p \geqslant p_{\perp 2}$. The period L_0, being the inverse domain-wall density, becomes finite for $p>p_{\perp 2}$ and diverges at $p_{\perp 2}$. At $p=p_{\mathrm{c}}=2^{1/2}$, where the uniform texture in a strong magnetic field with $\boldsymbol{v}_{\mathrm{s}} \perp \hat{\boldsymbol{l}} \parallel \hat{\boldsymbol{d}}$ becomes unstable, the lattice has reached a period of $L_0 \approx 2\xi_{\mathrm{D}}$.

A rather similar, although opposite, behaviour has been observed in cholesteric crystals in a magnetic field (de Gennes 1974). In the absence of the field, a helix is formed. Applying a finite field along the winding axis increases the domains where the molecules are favourably aligned, and hence the period L_0 increases. There is then an upper critical field H_{c} above which the helix is suppressed ($L_0 \rightarrow \infty$) and the molecules are aligned by the field. The behaviour of L_0 for fields H *increasing* towards H_{c} is very similar to that found here for superflow p *decreasing* towards $p_{\perp 2}$.

The formation of a $\hat{\boldsymbol{d}}$-soliton lattice due to superflow in ^{3}He-A near the transition to the A_1 phase has been discussed by Poluéktov (1982).

7.10.5 Flow-induced dynamical textures

There exist relatively few investigations of the nature of the textural state beyond the stability regime III (Fig. 7.35) of the uniform textures and the helical texture (region II). Fetter and Williams (1979, 1981) have studied the case of a fixed uniform superflow in a parallel magnetic field (a change of state corresponding to a horizontal line in the phase diagram in Fig. 7.35). The helical texture is found to become unstable at a critical field beyond which the system exhibits runaway growth. In an applied heat flow the texture is found to become time-dependent, oscillating periodically but anharmonically in time. On the other hand, in a toroidal geometry, where the superflow is not fixed externally (see the discussion below (7.219)), the texture apparently evolves into a stable wide-angle helix.

The effect of a finite angle χ between the superflow and the magnetic field has been considered by Fetter (1981), for the case of a fixed nonzero magnetic field and a superflow increasing from zero. For sufficiently small values of the superflow, a uniform texture is formed, with $\hat{\boldsymbol{l}}$ oriented at some angle determined by the relative strength of the two orienting forces. At a critical value of the superflow $g_{z,c} \propto 1 - \lambda\chi^{2/3}$, which depends nonanalytically on χ, the uniform texture becomes unstable (for $\chi = 0$, this is the phase boundary of region III in Fig. 7.35). The wave vector at which the instability first appears increases as $|k_c| \propto \chi^{2/3}$. For small χ, one finds a catastrophic instability (inverted bifurcation).

In zero magnetic field and for superfluid currents $p > p_{c,h_2}$, where the helical texture is found to be unstable, transitions to flow-driven well-defined time-dependent textures are known to occur (Hook 1978, Hook and Hall 1979). These have been investigated in detail by Dow (1984) and Dow and Hook (1985), using numerical integration of the partial differential equations for $\hat{\boldsymbol{l}}$ and $\boldsymbol{v}_s$ in space and time (see also Hall and Hook 1986). As discussed in Chapter 9 (see (9.87) and below) and in particular in Chapter 10 (below (10.164)), the motion of the vector $\hat{\boldsymbol{l}}$ is damped by the normal-fluid component. This is due to the fact that, according to (3.32) and (3.69), $\hat{\boldsymbol{l}}$ defines the local anisotropy axis of the excitation energy of quasiparticles, so that any motion of $\hat{\boldsymbol{l}}$ takes the normal component out of equilibrium and thus leads to rapid relaxation. This effect is described by the so-called "orbital viscosity coefficient" μ_l, (10.166) (Cross and Anderson 1975). Written in terms of the Euler angles α, β, γ, (7.11), the equations of motion for $\hat{\boldsymbol{l}}$, (9.87), take the form

$$\mu_l \sin^2\beta \frac{\partial\alpha}{\partial t} = -\left[\frac{\delta F}{\delta\alpha} - \boldsymbol{\nabla}\cdot\frac{\delta F}{\delta(\boldsymbol{\nabla}\alpha)}\right], \tag{7.263a}$$

$$\mu_l \frac{\partial\beta}{\partial t} = -\left[\frac{\delta F}{\delta\beta} - \boldsymbol{\nabla}\cdot\frac{\delta F}{\delta(\boldsymbol{\nabla}\beta)}\right], \tag{7.263b}$$

$$0 = -\left[\frac{\delta F}{\delta\gamma} - \boldsymbol{\nabla}\cdot\frac{\delta F}{\delta(\boldsymbol{\nabla}\gamma)}\right], \tag{7.263c}$$

where F is the bending energy (7.226) with $\xi_H = \infty$. Assuming a purely one-dimensional spatial dependence for $\hat{\boldsymbol{l}}$ (e.g. along $\hat{\boldsymbol{z}}$) with $\hat{\boldsymbol{d}} \parallel \hat{\boldsymbol{z}}$ fixed, and imposing boundary conditions $\hat{\boldsymbol{l}} \parallel \hat{\boldsymbol{z}}$ at $z = 0$ and $z = z_1 \gg \xi_D$, these equations have been integrated for given superflow along $\hat{\boldsymbol{z}}$. In this way, one finds that in the region extending beyond the stability regime of the static helical texture a regular array of domain walls ("kinks") is formed. These domain walls separate regions where $\hat{\boldsymbol{l}}$ is oriented parallel and antiparallel respectively to the flow and they precess periodically in time about $\hat{\boldsymbol{z}}$. The motion of $\hat{\boldsymbol{l}}$ is associated with energy dissipation due to the orbital viscosity μ_l (see Section 7.12).

The existence of flow-induced precessing domain walls has also been established analytically (Fetter 1988). In this work the boundary condition on $\hat{\boldsymbol{l}}$ is replaced by a criterion of dynamical pattern selection, namely the principle of maximum entropy production. Under these more realistic conditions, stable solutions of the type described by Dow and Hook (1985) are indeed found to be possible, but only in a range of supercurrents given by $1.55 \leqslant p \leqslant 2.45$ in dimensionless units (see (7.231a)), i.e. at a certain distance from the phase boundary of the helical texture. The behaviour for $p > 2.45$ is not yet known, but it appears as if the texture continues to exist although the angle β is no longer single-valued (Fetter 1989).

One may wonder whether the precessing domain-wall solutions discussed above are also possible in a flow channel, where $\hat{\boldsymbol{l}}$ is locked perpendicular to the wall. In fact, as shown by Ho (1978b), there do exist configurations of $\hat{\boldsymbol{l}}$ near the surface that allow the system to absorb the continuous winding of $\hat{\boldsymbol{l}}$ in the centre portion of the channel (see also Hu 1979c).

The effect of thermal fluctuations on the stability of the helical texture near the phase boundary to the dynamically unstable region and the associated decay of superflow have been discussed by Kleinert (1979a,b).

7.11 PAIR-BREAKING CRITICAL CURRENTS

The first discussion of critical currents in a superfluid was given by Landau (1941) (see Khalatnikov 1965). Imagine a body of relatively large mass M moving at velocity $\boldsymbol{u}_{\mathrm{i}}$ through a volume of the superfluid at rest in the laboratory frame. Suppose an excitation with momentum $\hbar\boldsymbol{k}$ and energy $E_{\boldsymbol{k}}$ is created by the interaction of the body with the liquid. For this process to take place, energy and momentum must be conserved, i.e.

$$\tfrac{1}{2}Mu_{\mathrm{i}}^2 = \tfrac{1}{2}Mu_{\mathrm{f}}^2 + E_{\boldsymbol{k}}, \tag{7.264a}$$

$$M\boldsymbol{u}_{\mathrm{i}} = M\boldsymbol{u}_{\mathrm{f}} + \hbar\boldsymbol{k}, \tag{7.264b}$$

where $\boldsymbol{u}_{\mathrm{f}}$ is the velocity of the body after creating the excitation. Since $M \gg m$, where m is the mass of the superfluid particles, the change in the kinetic energy of the body can be neglected, and hence

$$E_{\boldsymbol{k}} = \hbar\boldsymbol{k}\cdot\boldsymbol{u}_{\mathrm{i}} \leqslant \hbar\,|\boldsymbol{k}|\,u_{\mathrm{i}}, \tag{7.265}$$

i.e. $u_i \geqslant E_{\boldsymbol{k}}/\hbar\,|\boldsymbol{k}|$. This leads to the so-called "Landau critical velocity"

$$v_{\mathrm{sc}} = \min\left(\frac{E_{\boldsymbol{k}}}{\hbar\,|\boldsymbol{k}|}\right), \tag{7.266}$$

which u_i must exceed in order to create any excitations at all in the system. Superfluid flow (i.e. a persistent current) in an isotropic superfluid is only possible if $v_{sc} > 0$. For free particles, where $E_{\boldsymbol{k}} = (\hbar \boldsymbol{k})^2/2m$, v_{sc} is strictly zero. On the other hand, in the case of superfluid ^{4}He the excitation spectrum consists of a phonon and a roton part, so that there are *two* critical velocities

$$v_{\mathrm{sc,roton}} \approx \frac{\Delta_{\mathrm{r}}}{p_0}, \tag{7.267a}$$

$$v_{\mathrm{sc,phonon}} = \lim_{|\boldsymbol{k}| \to 0} \frac{E_{\boldsymbol{k}}}{\hbar\,|\boldsymbol{k}|} = c_1, \tag{7.267b}$$

where Δ_{r} is the roton gap, p_0 is the roton momentum and c_1 is the sound velocity, with $c_1 \gg v_{\mathrm{c,roton}}$. In a Fermi superfluid such as a superconductor or superfluid ^{3}He the spectrum of fermionic excitations $E_{\boldsymbol{k}}$ (i.e. of Bogoliubov quasiparticles) is given by (3.32). For a (pseudo)isotropic superfluid, $E_{\boldsymbol{k}}$ is shown in Fig. 3.2. Clearly, for v_{sc} given by (7.266), one finds

$$v_{sc} \approx \frac{\Delta}{p_{\mathrm{F}}} \tag{7.268a}$$

$$= \tfrac{1}{2}\frac{\Delta}{E_{\mathrm{F}}}\, v_{\mathrm{F}} \ll v_{\mathrm{F}}, \tag{7.268b}$$

similarly to the roton critical velocity, where $p_{\mathrm{F}} = \hbar k_{\mathrm{F}}$ is the Fermi momentum. Note that the critical velocity for bosonic excitations (e.g. phonons) is much higher and is therefore irrelevant since $c_1 \approx v_{\mathrm{F}}$.

In the A phase the excitation spectrum has nodes at $\boldsymbol{k} \parallel \pm \hat{\boldsymbol{l}} k_{\mathrm{F}}$ such that $v_{sc}^{\mathrm{A}} = 0$, i.e. an arbitrarily small velocity is sufficient to excite quasiparticles near the nodes of the energy gap. This does not imply, however, that a superflow in the A phase is completely unstable: what happens is that, at $T = 0$ in the presence of a superfluid velocity field $\boldsymbol{v}_{\mathrm{s}}$, quasiparticles are only excited near the nodes of the gap within a Fermi-surface area of radius proportional to v_{s}, leading to a finite density of excitations proportional to v_{s}^2. Once the available states are filled, a new equilibrium state is reached and further decay is not possible (Volovik 1984b). The only effect of the excitation is to reduce the superfluid density component ρ_{s} by a fraction proportional to v_{s}^2. At finite temperatures the quasiparticles excited by the superflow may scatter into states *above* the Fermi energy, thus giving way for further excitations induced by the superflow. This process is clearly nonlinear in v_{s}, and thus superflow is practically stable for velocities $v_{\mathrm{s}} \ll \Delta_0/p_{\mathrm{F}}$. However, it should be noted that this does still not guarantee superflow in the A phase, since, as discussed in Sections 7.10 and 7.12, there are still other dissipation mechanisms involving textural effects (i.e. "textural critical velocities") at these low velocities.

In contrast with the situation in the A phase, the Landau criterion applies to the B phase without reservation. A quantitative calculation of the pair-breaking critical velocity in this phase requires proper treatment of the effect of a finite superfluid velocity on the energy gap itself. Indeed, a superfluid velocity defines a new symmetry axis and thus leads to an axial distortion of the previously isotropic gap. This feeds back into the gap equation, causing an overall reduction of the magnitude of the gap.

7.11.1 The B phase

The general structure of the order parameter $A_{\mu j}$ in the presence of superflow along the z axis is given by

$$A_{\mu j} = \frac{1}{3^{1/2}\Delta} e^{i\phi(\boldsymbol{r})} \left[\boldsymbol{R} \begin{pmatrix} \Delta_\perp & 0 & 0 \\ 0 & \Delta_\perp & 0 \\ 0 & 0 & \Delta_\parallel \end{pmatrix} \right]_{\mu j}, \tag{7.269}$$

where $\boldsymbol{R}$ is an arbitrary rotation matrix. The order-parameter components $\Delta_\perp$ and $\Delta_\parallel$ are obtained by minimizing the total free energy F_{tot} composed of the free-energy functional for p-wave pairing states, (5.4), and the gradient free energy (7.17),

$$F_{\text{tot}} = \int d^3 r \, (f + f_{\text{grad}}), \tag{7.270}$$

where

$$f_{\text{grad}} = \frac{3}{2} \left(\frac{2m}{\hbar} \right)^2 K \Delta_\parallel^2 v_s^2, \tag{7.271}$$

with $K \equiv \frac{1}{3}(K_1 + K_2 + K_3)$. In the Ginzburg–Landau regime one finds (Fetter 1975a)

$$\left(\frac{\Delta_\perp}{\Delta} \right)^2 = 1 + \frac{3}{2} \frac{2\beta_{12} - \beta_{345}}{\beta_{345}} \frac{K}{|\alpha|} \left(\frac{2m}{\hbar} \right)^2 v_s^2, \tag{7.272a}$$

$$\left(\frac{\Delta_\parallel}{\Delta} \right)^2 = 1 - \frac{3}{2} \frac{4\beta_{12} + 3\beta_{345}}{\beta_{345}} \frac{K}{|\alpha|} \left(\frac{2m}{\hbar} \right)^2 v_s^2, \tag{7.272b}$$

with $\Delta \equiv \Delta_\perp(v_s = 0) = \Delta_\parallel(v_s = 0)$ and α defined in (5.8). In the weak-coupling limit, where the β_i are given by (5.8), $\Delta_\perp$ is found to be unchanged, whereas $\Delta_\parallel$ is strongly reduced and vanishes at a critical velocity $v_{sc,\parallel} \propto (1 - T/T_c)^{1/2}$. At this point a transition into the planar state (5.20) takes place. For still higher values of v_s, even this planar state will become unstable and the system turns normal. The nature of this transition may be expected to depend sensitively on strong-coupling and Fermi-liquid interaction effects. The supercurrent (7.32) itself is obtained as

$$\boldsymbol{g}_s = \rho_s \boldsymbol{v}_s \left(1 - \frac{b^2}{|1 - T/T_c|} v_s^2 \right), \tag{7.273}$$

where

$$b^2 = 3 \frac{8\beta_{12} + 11\beta_{345}}{5\beta_{345}} \frac{K}{N_F} \left(\frac{2m}{\hbar} \right)^2. \tag{7.274}$$

For small superfluid velocities, one finds the usual linear dependence of g_s on v_s. For somewhat higher values, the nonlinearities in v_s, which are due to the reduction of $\Delta_\parallel$, cause g_s to go through a maximum before the transition to the planar state takes place. We define the "pair-breaking critical current" $g_{s,c}$ as the *maximum* of the supercurrent, given by $dg_s/dv_s = 0$, i.e.

$$g_{s,c} = \frac{2}{3^{3/2}}\frac{1}{b}\rho_s\left(1-\frac{T}{T_c}\right)^{1/2} \tag{7.275a}$$

$$\propto \left(1-\frac{T}{T_c}\right)^{3/2}. \tag{7.275b}$$

This maximum is attained at the "pair-breaking critical velocity"

$$v_{s,c} = \frac{1}{3^{1/2}b}\left(1-\frac{T}{T_c}\right)^{1/2}. \tag{7.276}$$

For general temperatures, a full solution of the gap equation in the presence of superflow is required. For this, we note that in the presence of superflow v_s the quasiparticle spectrum $E_{\boldsymbol{k}}$ is shifted to $E_{\boldsymbol{k}} + \boldsymbol{k}\cdot\boldsymbol{v}_s$ (see (7.40)). Restricting ourselves to weak coupling, the gap equation is then obtained by making the above replacement for $E_{\boldsymbol{k}}$ and substituting the form of the order parameter

$$|\Delta_{\boldsymbol{k}}(\boldsymbol{v}_s)|^2 = \Delta_\perp^2[1-(\hat{\boldsymbol{k}}\cdot\hat{\boldsymbol{v}}_s)^2] + \Delta_\parallel^2(\hat{\boldsymbol{k}}\cdot\hat{\boldsymbol{v}}_s)^2 \tag{7.277}$$

into the gap equation (3.46b). Eliminating the pair-coupling constant by means of the gap equation at $v_s = 0$, one finds the two equations (Vollhardt *et al.* 1980, Kleinert 1980b)

$$\int_0^\infty d\xi \int_0^1 d\mu\, f_i(\mu)\left\{\frac{1}{E}\left[\tanh\frac{E+E_s\mu}{2k_BT} + \tanh\frac{E-E_s\mu}{2k_BT}\right] - \frac{2}{E_0}\right\} = 0, \quad i = \parallel, \perp, \tag{7.278}$$

where $\mu = \hat{\boldsymbol{k}}\cdot\hat{\boldsymbol{v}}_s$ and

$$E_s = v_s p_F, \tag{7.279a}$$

$$f_\parallel(\mu) = 2\mu^2, \tag{7.279b}$$

$$f_\perp(\mu) = 1-\mu^2, \tag{7.279c}$$

$$E^2 \equiv E_{\boldsymbol{k}}^2(\boldsymbol{v}_s) = \xi^2 + \Delta_\perp^2(v_s, T)(1-\mu^2) + \Delta_\parallel^2(v_s, T)\mu^2, \tag{7.279d}$$

$$E_0^2 = \xi^2 + \Delta^2(v_s = 0, T = 0). \tag{7.279e}$$

A numerical solution of (7.278) reveals that the qualitative behaviour found in the Ginzburg–Landau regime continues to hold at least down to $T/T_c \approx 0.5$, while at even lower temperatures nonmonotonic behaviour is found.

The superfluid current g_s is in general given by

$$\boldsymbol{g}_s = \rho\boldsymbol{v}_s - g_n\hat{\boldsymbol{v}}_s, \tag{7.280}$$

where the second term on the right-hand side of (7.280) is the nonlinear normal current (see (3.85))

$$g_n = \sum_{\boldsymbol{k}} \hbar \boldsymbol{k} \cdot \hat{\boldsymbol{v}}_s f(E_{\boldsymbol{k}}(\boldsymbol{v}_s) - \boldsymbol{k} \cdot \boldsymbol{v}_s) \tag{7.281a}$$

$$= \hbar k_F N_F \int_0^\infty d\xi \int_0^1 d\mu \frac{\mu}{2} \left(\tanh \frac{E + E_s \mu}{2k_B T} - \tanh \frac{E - E_s \mu}{2k_B T} \right). \tag{7.281b}$$

In particular, at $T=0$ pair breaking is only possible for superfluid velocities

$$v_s > v_{c,0} \equiv \frac{\Delta(v_s = 0, T = 0)}{p_F}, \tag{7.282}$$

in accordance with the Landau criterion. Indeed, at $T=0$, g_n has the form (Vollhardt *et al.* 1980)

$$g_n \big|_{T=0} = \frac{\rho}{p_F} \theta(v_s - v_{c,0}) \frac{(E_s^2 - \Delta_\parallel^2)^{3/2}}{E_s(E_s^2 + \Delta_\perp^2 - \Delta_\parallel^2)}, \tag{7.283}$$

where $\theta(x)$ is the unit step function. This implies that at $T=0$ and for $v_s < v_{c,0}$ the superfluid current is given by $g_s = v_s \rho$ and thus increases linearly with v_s until it reaches the value $v_{s,0}\rho$. As v_s is increased above $v_{c,0}$, pair breaking sets in and begins to reduce the superfluid current. The maximum, i.e. critical, current is therefore reached only slightly *above* $v_{c,0}$ and is itself just a little larger than $v_{c,0}\rho$.

The full temperature dependence of the critical current $g_{s,c}$ has to be obtained numerically from (7.280) and is shown in Fig. 7.38. It is apparent that the $(1 - T/T_c)^{3/2}$ dependence of the Ginzburg–Landau regime extends to rather low temperatures. This temperature dependence of the critical

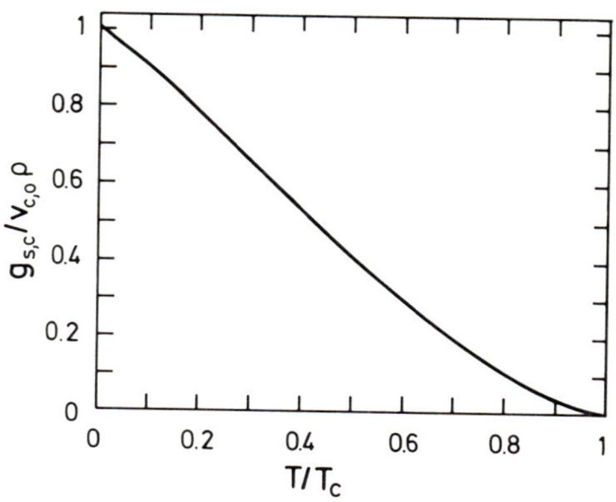

Figure 7.38 Pair-breaking critical current $g_{s,c}$ for the B phase as a function of reduced temperature in the weak-coupling limit. The current is measured in units of $\rho v_{c,0}$, with ρ the normal-state density and $v_{c,0} = \Delta(v_s = 0, T = 0)/p_F$. (After Vollhardt *et al.* (1980).)

processes in ^{3}He-B, except for thermally activated dissipation of superflow for temperatures near T_c and for superfluid velocities (or current densities) near the critical values, as adapted from superfluid ^{4}He (Pekola *et al.* 1988; see also Soda and Arai 1981). In the absence of a more detailed theory, it is therefore interesting to at least discuss the experimental situation.

In persistent-current experiments using a toroidal flow channel of 3 mm radius, packed with a powder of $d = 20\,\mu$m plastic grains, a temperature-independent critical velocity of the form (7.287):

$$v_{sc} = \alpha \frac{\kappa_0}{d} \ln \frac{d}{\beta \xi_0} \tag{7.294}$$

was observed for $T \lesssim 0.9T_c$, with $\alpha \approx 0.6$ and $\beta \approx 19$, where now $\kappa_0 = 2\pi\hbar/2m_3$ (Pekola *et al.* 1984a,b, Pekola and Simola 1985). The pressure dependence of v_{sc} seems to be well accounted for by the pressure dependence of ξ_0 in the above expression. For $T > 0.9T_c$, v_{sc} was found to decrease considerably. This is perhaps related to thermal activation processes as discussed above.

The influence of the size of the flow channel on v_{sc} has been studied in torsional oscillator experiments (Parpia and Reppy 1979, Crooker *et al.* 1981, Crooker 1983) where the flow channel contained a small aperture with a diameter varying from 2 to 100 μm. Again the critical velocity was found to be temperature-independent at low *T*, being given approximately by $v_{sc} \approx 4\kappa_0/\pi d$. A number of other experiments are consistent with these findings (Gammel *et al.* 1984, Gammel and Reppy 1984; see also Dahm *et al.* 1980, Hutchins *et al.* 1981a,b, Brewer 1983, Ling *et al.* 1984a).

A series of experiments investigating U-tube flow through circular channels of radii $R \approx 100$–$200\,\mu$m and lengths of about 5–10 mm showed an *R*-independent critical velocity much higher than given by (7.287), which might indicate a transition to a vortex state with higher dissipation levels (Eisenstein *et al.* 1980b, Eisenstein and Packard 1982). There was one experiment (Manninen and Pekola 1982, 1983) in which the channel diameter was only 0.8 μm, so that the critical velocity for vortex creation (7.287) became comparable to the depairing critical velocity (7.268), (7.276). Indeed, the only critical velocity observed had the characteristics of the depairing critical velocity as discussed in the last subsection.

Results of an experimental investigation of flow through submicrometre-diameter channels have been reported by Pekola *et al.* (1988). These clearly demonstrate the presence of thermally induced dissipation and are in good agreement with theory, as discussed in detail by these authors.

There has been little or no attempt to explain the dissipation occuring *above* the critical current discussed above. One might expect that well above v_{sc} a vortex tangle is formed. The resulting pressure gradient in a flow experiment would then be given by the mutual frictional force of vortices with the normal component, i.e. by (7.289). The calculation of the precise effect of this force depends on the experiment under consideration. It is interesting to note that (7.289), with a value of L given by (7.290), gives a pressure difference of the correct order of magnitude to explain the dissipation observed in those experiments in which a larger value of v_{sc} than given by (7.287) was observed (Eisenstein and Packard 1982, Brewer 1983).

Evidence for the presence of remanent vorticity in uniformly rotating ^{3}He-B at superfluid velocities considerably below the critical velocity has been obtained by Hall *et al.* (1984), who found a component of the dissipation increasing linearly with angular velocity Ω, as is expected from the vortex friction force (7.289).

There also exists the possibility of critical superfluid *spin* currents in the B phase (Fomin 1987a, 1988). On the basis of his earlier work on the spin dynamics of superfluid ^{3}He (Fomin 1980, 1981b) (see Section 8.3.3), Fomin derived expressions for the superfluid spin current and equations of motion for the corresponding phase variable or spin superfluid velocity. The current is found to become unstable beyond a critical value of the superfluid velocity, given approximately by

$$v_{sc}^{spin} \approx \frac{\hbar}{mc_{sp}} [\omega_L(\omega_p - \omega_L)]^{1/2}, \tag{7.295}$$

where c_{sp} is the spin-wave velocity, ω_L is the Larmor frequency and ω_p is the spin precession frequency. The theoretical predictions are in semiquantitative agreement with experiment (Borovik-Romanov *et al.* 1987a,b, 1989).

7.12.3 The A phase

As discussed in Section 7.10, bulk superflow in the A phase is topologically unstable in the absence of a magnetic field. It is only as a result of a numerical coincidence of the parameters in the free energy that above a critical temperature T_i superflow is stabilized against textural deformation by coupling of the vector $\hat{\boldsymbol{l}}$ to the spin part of the order parameter. On the other hand, on the *surface* superflow is quantized and hence is stable. The walls therefore have a stabilizing effect, so that superflow is expected to exist at least within a surface layer close to the wall. Even this is only true as long as there are no surface singularities. Since we know that all vessels, except for those with the topology of a torus, must necessarily contain such singularities (see Section 7.9), the stability of superflow again becomes questionable. As pointed out by Mermin (1977b, 1978a,b), surface singularities such as the "boojum", discussed above (7.201) in Section 7.9.5, can reduce the supercurrent in a channel by a periodic deformation of the texture (see also Ho 1978a,b, Hu 1979c). A boojum is a point singularity in the $\hat{\boldsymbol{l}}$ texture at the surface (Figs. 7.32c, 7.39a), which carries two quanta of circulation and is nonsingular as one goes away from the surface (Fig. 7.39b). If the boojum moves once around the flow channel (i.e. on the surface, perpendicular to the direction of superflow; see Fig. 7.39c) then the

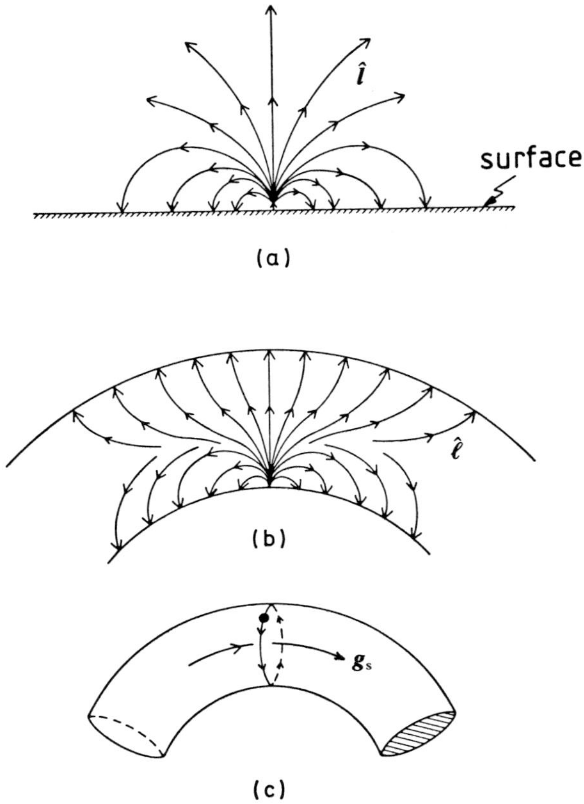

Figure 7.39 (a) Surface singularity in the $\hat{l}$ texture ("boojum"). (b) Cross-section through a boojum in a tube or torus. (c) Degradation of a supercurrent g_s in the tube or torus by motion of the boojum around the flow channel, perpendicular to the flow. By traversing the channel once, the superflow is reduced by two quanta of circulation. (After Mermin (1987b).)

supercurrent in the channel is reduced by exactly two quanta of circulation. Unless the boojum is pinned at the wall, one may expect it to move periodically around the channel, thereby causing a periodic variation of the $\hat{l}$ texture, and thus leading to a continuous reduction of the supercurrent. The existence of persistent currents in ^{3}He-A therefore depends sensitively on whether surface singularities exist and whether they are free to move.

Persistent currents inside a torus packed with powder (particle size approximately 25 μm) have indeed been detected experimentally. For this, a rotating cryostat was used, where the dissipation of small-amplitude oscillatory flow was measured as a function of the angular velocity Ω of the steady-state rotation (Gammel *et al.* 1985). While in the absence of a superfluid current the dissipation was found to be minimal at $\Omega = 0$, increasing as Ω^2, the presence of a supercurrent caused the minimum to shift to a value Ω_{min} corresponding to the rotation rate for which the superflow and the walls of the toroidal channel move in phase. Such irreversible persistent currents were generated by rotating the cryostat with an angular velocity exceeding a critical value Ω_c. The critical velocity was then determined as $v_{sc} = R\Omega_c$, with R the radius for the torus. The critical velocities found thereby were as large as 2.5 mm/s at lower T, and dropped towards T_c. Since the observed value of v_{sc} is one order of magnitude smaller than the estimate of v_{sc} for singular vortex creation, the relevant dissipation processes is assumed to be associated with the dissipative motion of the $\hat{\boldsymbol{l}}$ texture.

Orbital dissipation

A simple model of orbital dissipation in ^{3}He-A has been proposed by Hall and Hook (1986) (see also Volovik 1980, Gay *et al.* 1983), which will be discussed below. It involves a simplified treatment of the dynamics of $\hat{\boldsymbol{l}}$ textures to be discussed in detail in Chapter 9. Here it is necessary to anticipate some of the results derived there.

The starting point is the picture of superflow collapse by creation of coreless vortices. Consider superflow through a toroidal channel. The constant superfluid velocity at the surface, where superflow is topologically stable, is thought to be screened by a layer of coreless vortices from the region of collapsed superflow in the interior (Ho 1978a,b). This situation is reminiscent of the Meissner effect in ordinary superconductivity. The thickness λ of the vortex layer ("penetration depth") is given approximately by the relation (7.287), which specifies the typical size of vortices for a given velocity difference $\boldsymbol{v}_s - \boldsymbol{v}_n$ between the interior and the surface layer:

$$\lambda = \alpha \frac{\hbar}{2m} \frac{1}{|\boldsymbol{v}_s - \boldsymbol{v}_n|}, \tag{7.296}$$

where α is a constant of order unity. The spatial variation of the relative superflow $\boldsymbol{g}_s = \boldsymbol{g} - \rho \boldsymbol{v}_n$ is then obtained as the solution to a type of London equation, i.e.

$$\lambda^2 \nabla^2 \boldsymbol{g}_s = \boldsymbol{g}_s. \tag{7.297}$$

As a further simplification, one may assume $\boldsymbol{g}_s$ to be expressed in terms of a locally averaged superfluid velocity $\langle \boldsymbol{v}_s \rangle$ as

$$\boldsymbol{g}_s = \rho_{s\perp}(\langle \boldsymbol{v}_s \rangle - \boldsymbol{v}_n), \tag{7.298}$$

where the perpendicular component of the superfluid mass tensor $\boldsymbol{\rho}_s$ is substituted in view of the predominantly perpendicular orientation of $\hat{\boldsymbol{l}}$ to the flow direction.

The rate at which $\boldsymbol{g}_s$ can adjust itself to the steady state described by

(7.297) is controlled by an orbital relaxation time τ_{orb}. A phenomenological model equation combining (7.297) and these relaxational processes has the form

$$\lambda^2\nabla^2(\langle \boldsymbol{v}_s\rangle - \boldsymbol{v}_n) = \langle \boldsymbol{v}_s\rangle - \boldsymbol{v}_n + \tau_{\text{orb}}\frac{d\langle \boldsymbol{v}_s\rangle}{dt}. \tag{7.299}$$

In order to estimate τ_{orb}, we have to make use of the equation of motion for $\hat{\boldsymbol{l}}$ from Chapter 9, (9.87). The $\hat{\boldsymbol{l}}$ bending forces in this equation may be approximated by assuming the scale of variation of $\hat{\boldsymbol{l}}$ to be given by λ, which makes these forces of the same order as the flow terms. Hence one has

$$\left|\frac{d\hat{\boldsymbol{l}}}{dt}\right| \approx \frac{1}{\mu_l}\rho_{s\perp}(\langle \boldsymbol{v}_s\rangle - \boldsymbol{v}_n)^2, \tag{7.300}$$

where μ_l is the orbital viscosity coefficient calculated in Chapter 10, (10.170). The orbital relaxation rate is obtained from (7.300) as

$$\frac{1}{\tau_{\text{orb}}} = \frac{1}{\beta}\frac{\rho_{s\perp}}{\mu_l}(\langle \boldsymbol{v}_s\rangle - \boldsymbol{v}_n)^2, \tag{7.301}$$

where β is a coefficient of order unity. The rate is thus proportional to the square of the superfluid counterflow velocity. This is in agreement with the finding that the dissipation observed in the rotating-cryostat experiment (Gammel *et al.* 1985) is proportional to Ω^2 as discussed above.

Equation (7.299) may be viewed as an averaged version of the superfluid acceleration equation, as given in Chapter 9, (9.84). It is then natural to add force terms corresponding to the gradient of the chemical potential μ, i.e. the pressure gradient $\nabla P = \rho\nabla\mu$, to (7.299), which extends the range of applicability to the case of driven flow as well. Collecting all the terms, one finally obtains a nonlinear differential equation describing the space and time development of the locally averaged superfluid velocity field due to orbital motion (Hall and Hook 1986):

$$\left(\frac{\alpha\hbar}{2m}\right)^2\nabla^2(\langle \boldsymbol{v}_s\rangle - \boldsymbol{v}_n) = (\langle \boldsymbol{v}_s\rangle - \boldsymbol{v}_n)^3 + \beta\frac{\mu_l}{\rho_{s\perp}}\left(\frac{\partial\langle \boldsymbol{v}_s\rangle}{\partial t} - \frac{1}{\rho}\nabla P\right). \tag{7.302}$$

This equation must be supplemented by boundary conditions. In the simplest case the superfluid velocity at the surface may be assumed to be constant. In the presence of surface singularities such as the "boojum" discussed above, which may deplete the superflow by precessing around the channel, an appropriate boundary condition on $d\langle \boldsymbol{v}_s\rangle/dt$ must be specified.

There are now several limiting cases of interest. If the thickness of the vortex layer at the surface is small compared with the channel diameter, $\kappa_0/2\pi v_{\max} \ll d$ (where $v_{\max}$ is the maximum value of $|\langle \boldsymbol{v}_s\rangle - \boldsymbol{v}_n|$ attained at the centre of the channel), then the spatial variation of $\langle \boldsymbol{v}_s\rangle$ may be neglected, i.e. the left-hand side of (7.302) can be put equal to zero. The resulting equation for $\langle \boldsymbol{v}_s \rangle$ has also been derived by Volovik (1978a). The same approximate equation is obtained if the surface relaxation of superflow by precessing singularities is as effective as the bulk dissipation, so that superflow is dissipated uniformly over the cross-section of the channel.

A different situation arises for narrow channels and small velocities, such that $\lambda \gg d$. In this case the orbital relaxation rate is governed by the bending force induced by the variation of $\hat{\boldsymbol{l}}$ over the distance d,

$$\frac{1}{\tau_{\text{orb}}} \approx \frac{\hbar}{m} \frac{\rho_{s\perp}}{\mu_l d} |\langle \boldsymbol{v}_s \rangle - \boldsymbol{v}_n|, \tag{7.303}$$

which is larger than (7.301) by a factor λ/d. An estimate of the torque term in the acceleration equation for the superfluid, (9.84) (Chapter 9), then yields

$$\frac{\hbar}{2m} \hat{\boldsymbol{l}} \cdot \left(\frac{\partial \hat{\boldsymbol{l}}}{\partial t} \times \nabla_i \hat{\boldsymbol{l}} \right) \approx \left(\frac{\hbar}{m} \right)^2 \frac{\rho_{s\perp}}{\mu_l d^2} |\langle \boldsymbol{v}_s \rangle - \boldsymbol{v}_n|, \tag{7.304}$$

leading to the following equation of motion for $\langle \boldsymbol{v}_s \rangle$:

$$\frac{\partial \langle \boldsymbol{v}_s \rangle}{\partial t} = \frac{1}{\rho} \boldsymbol{\nabla} P - \beta' \left(\frac{\hbar}{m} \right)^2 \frac{\rho_{s\perp}}{\mu_l d^2} (\langle \boldsymbol{v}_s \rangle - \boldsymbol{v}_n), \tag{7.305}$$

where β' is a constant of order unity. A simple dimensional consideration shows that the gradient term on the left-hand side of (7.302) gives rise to a term with a similar structure as that of the last term in (7.305) (i.e. with ∇^2 replaced by $1/d^2$), so that (7.305) is qualitatively correct even in situations where $|\langle \boldsymbol{v}_s \rangle - \boldsymbol{v}_n|$ varies appreciably over the cross-section of the channel. The effect of orbital relaxation on the decay of vorticity in ^{3}He-A has also been considered by Kopnin (1978) and Hakonen and Mineev (1987).

Experimentally, the effects of orbital relaxation on superflow have been observed in many different cases. The damping of torsional oscillations of the superfluid in a *toroidal flow channel* has been shown to be partly due to orbital relaxation as well as normal-fluid friction (Gay *et al.* 1981, 1983). Earlier experiments, which employed wider flow channels, had only shown the effect of stationary bending of the $\hat{\boldsymbol{l}}$ texture, since the condition $\omega\tau_{\text{orb}} < 1$ necessary for orbital relaxation was not satisfied (Main *et al.* 1976, 1977, Bagley *et al.* 1978). One of the intriguing observations of the later experiments was the strong history dependence. This has been related to the existence of varying amounts of surface singularities. With this assumption and solving (7.302), one obtains a semiquantitative description of the data (Hall and Hook 1986). Within the model, it turns out that the surface singularities are formed for values of the superfluid velocity exceeding $v_{sc} \approx 2$ mm/s, with the density of defects growing linearly with $v_s - v_{sc}$. It has been suggested that the critical velocity is associated with the characteristic dipole-unlocking velocity $v_{s,D}$, (7.231b). Experiments on a flow channel with an 18 μm orifice yielded similar results (Parpia and Reppy 1979).

A second group of experiments investigated the properties of *driven* superflow. There is evidence for orbital dissipation in all of them, but several features are not yet understood. Furthermore, the effect of dissipation due to singular vortices may be estimated from (7.289) to be of comparable magnitude. The experiments of Dahm *et al.* (1980) measured the response of a volume of fluid driven by a diaphragm following a sudden change in the applied voltage, which leads to a pressure difference ΔP. The data seem to be explained well by the model discussed above (i.e. (7.302) with the left-hand side put equal to zero), when values of the parameters $\mu_l/\rho_{s\perp} \approx (1 - T/T_c)^{1/2}\,\text{cm}^2/\text{s}$ and $\beta = 1.8$ are used and with the assumption of a sufficient density of surface singularities. A critical velocity $v_{sc} \approx 0.5(1 - T/T_c)^{1/2}$ mm/s was observed. In later experiments by Ling *et al.* (1984a,b,c), in which the force on the diaphragm was ramped in time, much higher critical velocities, as high as 0.5–1 cm/s were observed. In similar experiments by Manninen and Pekola (1982, 1983) *two* critical currents were found at temperatures $T \gtrsim T_{AB}$. It is possible that the dissipation at lower temperatures (occurring in supercooled A phase) is different from that for $T \gtrsim T_{AB}$ owing to the presence of B phase outside the channel, if a dissipation mechanism involving the precession of singular vortices at the ends of the channel is assumed (Thuneberg and Kurkijärvi 1981).

A critical velocity v_{sc} of the order of 0.5 mm/s was also observed by Paalanen and Osheroff (1980) in a flow experiment with a strong magnetic field applied parallel to the flow. A filter paper with 8 µm pores, used to increase the flow impedance of the channel, was positioned across one end of the channel. The pressure difference ΔP measured over the length of the channel varied linearly with v_s at $v_s > v_{sc}$ and dropped near T_c as $(1 - T/T_c)^{1/2}$. If ΔP is calculated within the theory appropriate for wide channels, (7.302), discussed above, one finds a result proportional to v_s^3 and $(1 - T/T_c)^{-1/2}$, in contradiction to the experimental results. This clearly indicates that other dissipative mechanisms have to be taken into account to explain the experiment.

While it seems that the simple relaxational motion of the $\hat{\boldsymbol{l}}$ texture, when driven by superflow, can be accounted for—at least qualitatively—there are obviously other types of motion (e.g. periodic or quasiperiodic) that are not yet understood. (See for example the early experimental observation of periodic motions of $\hat{\boldsymbol{l}}$ by Paulson *et al.* (1976d), Krusius *et al.* (1978) and Wheatley (1978).) A step forward in understanding these phenomena is the theoretical discovery of a periodic motion of domain walls ("kinks") in the $\hat{\boldsymbol{l}}$ texture, discussed in Section 7.10.5.

Although a certain qualitative understanding of flow dissipation in the A phase has been achieved, more experimental and theoretical work remains to be done to clarify the detailed structure of the complex processes that may occur in this context.

FURTHER READING

Textures and defects

Anderson P W and Brinkman W F 1978 in *The Physics of Liquid and Solid Helium,* Part II, ed. K H Bennemann and J B Ketterson (Wiley, New York), p. 177
Brinkman W F and Cross M C 1978 in *Progress in Low Temperature Physics,* Vol. VIIA, ed. D F Brewer (North-Holland, Amsterdam), p. 105
de Gennes P G 1974 *The Physics of Liquid Crystals* (Clarendon Press, Oxford)
Maki K 1986 in *Solitons,* ed. S E Trullinger, V E Zakharov and V L Pokrovskii (North-Holland, Amsterdam), p.435
Mermin N D 1979 *Rev. Mod. Phys.* **51** 591
Mineev V P 1980 in *Soviet Scientific Reviews,* Section A: *Physics Reviews,* Vol. 2 (Harwood Academic Publishers, Chur), p. 173
Toulouse G and Kléman M 1976 *J. Physique Lett.* **37** L149
Volovik G E 1984 *Usp. Fiz. Nauk* **143** 73 [*Sov. Phys. Usp.* **27** 363 (1984)]
Volovik G E and Mineev V P 1977 *Zh. Eksp. Teor. Fiz.* **72** 2256 [*Sov. Phys. JETP* **45** 1186 (1977)]

Properties of superflow

Cross M C 1983 in *Quantum Fluids and Solids—1983* (AIP Conf. Proc. No. 103), ed. E D Adams and G G Ihas (AIP, New York), p. 325
Hall H E and Hook J R 1986 in *Progress in Low Temperature Physics,* Vol. IX, ed. D F Brewer (North-Holland, Amsterdam), p. 143
Mermin N D 1978a in *Quantum Liquids,* ed. J Ruvalds and T Regge (North-Holland, Amsterdam), p. 195
Mermin N D 1978b *J. Physique* **39** Colloq. C-6, Vol. III, p. 1283 (Proceedings of the 15th International Conference on Low Temperature Physics, LT-15)

Rotating superfluids

Fetter A L 1986 in *Progress in Low Temperature Physics,* Vol. X, ed. D F Brewer (North-Holland, Amsterdam), p. 1
Feynman R P 1972 *Statistical Mechanics* (Benjamin, Reading, Massachusetts)
Mineev V P, Salomaa MM and Lounasmaa O V 1986 *Nature* **324** 333
Salomaa M M and Volovik G E 1987 *Rev. Mod. Phys.* **59** 533
Sonin E B 1987 *Rev. Mod. Phys.* **59** 87

8

Spin Dynamics

Experiments probing the spin dynamics of the superfluid phases of ^{3}He provided the first specific information on the structure of the ordered state. At first the observed shift of the transverse NMR frequency seemed inexplicably large. After all, the usual spin exchange interaction is isotropic in spin space and therefore cannot give rise to any resonance shift. Spin–orbit interactions, on the other hand, which are by their nature spin-nonconserving and which thus can lead to such a shift, are very small. Of these, the nuclear dipole–dipole interaction is—relatively speaking—the largest. It was well known that in the normal phase the average force exerted on the total spin by dipole interaction is zero to first order. This is so because the contributions of all possible relative orientations of total spin and effective orbital angular momentum of pairs of dipoles average to zero in lowest order.

In the pair-correlated state, however, the relative orientation of the spin and orbital degrees of freedom of the Cooper pairs is fixed, as pointed out by Leggett (1972, 1973a), who coined the term "spontaneously broken spin–orbit symmetry" or SBSOS (see Chapter 6): owing to the broken relative symmetry, the spin–orbit forces acting between the partners of the Cooper pairs add up coherently (see also Anderson 1973, Anderson and Varma 1973). Thus the nuclear-dipole force and the pair correlations conspire to produce sizeable nuclear-spin resonance shifts. They can be measured conveniently and actually may be considered as a fingerprint of the particular pair-correlated state.

As the spin resonance frequencies are generally sensitive to the preferred directions in both spin space and orbital space, NMR experiments can serve as a probe of the textures of the order-parameter field. This technique has been successfully used for determining macroscopic textures and, to some extent, textural objects such as domain walls and other nonsingular or singular defects.

The preferred directions of the order parameter in spin space are of course related to the Cooper-pair spins. Large-amplitude motions of the total spin of the system will therefore induce changes in the preferred direction, which in turn act back on the spin itself. This coupling gives rise to a rich and unusual nonlinear spin dynamics, offering an ideal test ground for concepts in the rapidly growing field of nonlinear dynamics in general. In the simplest case, the uniform A phase, the longitudinal spin dynamics is identical with that of a Josephson oscillator, with the dipole interaction providing the weak link between the two superfluids of up- and down-spin Cooper pairs. Tunnelling of Cooper pairs between the two spin subsystems also provides the dominating mechanism for spin relaxation, as pointed out by Leggett and Takagi (1976). As a consequence, relaxation effects are orders of magnitude stronger in the superfluid phases than in the normal phase.

8.1 DERIVATION OF THE EQUATIONS OF MOTION

It follows from the preceding discussion that the spin motion can be affected by the dipole interaction. For this to happen, the corresponding change of the spin part of the order parameter has to change the expectation value of the dipole Hamiltonian. The orienting effect of the dipole forces has been discussed in Section 6.3. There it was shown that in the A phase the orientation-dependent dipole energy density Δf_D is proportional to $-(\hat{\boldsymbol{d}} \cdot \hat{\boldsymbol{l}})^2$, whereas in the B phase only the rotation angle, but not the rotation axis of a spin rotation, affects f_D.

Owing to the smallness of the dipole energy in comparison with the characteristic energies of the pair-correlated Fermi liquid (Fermi energy ϵ_F and condensation energy Δ), the timescale for dipole-interaction-induced effects is much longer than typical microscopic times such as the quasiparticle relaxation time τ. On the former timescale almost all degrees of freedom are in equilibrium, the only exceptions being the conserved densities. In the present case these are the spin density and the quasiconserved variables associated with the spontaneously broken symmetry, i.e. the preferred directions. The spin dynamics for frequencies well below the gap frequency $\Delta/\hbar$ and for frequencies well below the quasiparticle relaxation rate τ^{-1} is therefore described by a set of quasihydrodynamic equations for the total spin $\boldsymbol{S}$ and the order-parameter vector $\boldsymbol{d}(\boldsymbol{k})$. (Note that in the following $\boldsymbol{d}(\boldsymbol{k})$ will be taken to depend only on the direction of $\boldsymbol{k}$, i.e. $\boldsymbol{k} = k_F\hat{\boldsymbol{k}}$.)

There remains the question of the extent to which the dynamics of the orbital preferred directions has to be included into the theory. In the A phase, for example, one could imagine incorporating a variable like an orbital angular-momentum density of Cooper pairs. As a matter of fact, these variables are not conserved, but rather their motion is damped on the timescale of the quasiparticle relaxation rate, τ, as will be discussed in Section 10.3. For the purpose of the spin dynamics, one may assume the orbital variables to be in the equilibrium state defined by the time averaging over the respective motion.

We are now in a position to derive the equations of motion for the total spin $\boldsymbol{S}$ and the order-parameter vector $\boldsymbol{d}(\boldsymbol{k})$ (Leggett 1974a,b; see also Varma and Werthamer 1974). The dynamics of $\boldsymbol{S}$ and $\boldsymbol{d}(\boldsymbol{k})$ are governed by the effective Hamiltonian

$$H_S = \tfrac{1}{2}\gamma^2 \boldsymbol{S}\boldsymbol{\chi}^{-1}\boldsymbol{S} - \gamma \boldsymbol{S} \cdot \boldsymbol{H} + H_{\mathrm{D}}. \tag{8.1}$$

Here $\boldsymbol{\chi}$ is the susceptibility tensor. In all cases to be discussed below, $\boldsymbol{\chi}$ may be replaced by the one eigenvalue χ observed in a static experiment. The dipole Hamiltonian H_{D} is a bilinear functional of $\boldsymbol{d}$ and $\boldsymbol{d}^*$, given explicitly by (6.96) and (6.97).

We now consider $\boldsymbol{S}$ and $\boldsymbol{d}$, $\boldsymbol{d}^*$ as Heisenberg operators. The components of $\boldsymbol{S}$ satisfy the well-known commutation relations

$$[S_\mu, S_\nu] = \mathrm{i}\hbar \sum_\lambda \epsilon_{\mu\nu\lambda} S_\lambda. \tag{8.2}$$

Since $\boldsymbol{d}$ is a vector in spin space and $\boldsymbol{S}$ is the generator of rotations in spin space, an infinitesimal rotation in spin space by $\delta\boldsymbol{\theta}^{\mathrm{s}}$ leads to a rotation $\boldsymbol{d} \to \boldsymbol{d}'$, where $\boldsymbol{d}'$ is given either by

$$\boldsymbol{d}'(\boldsymbol{k}) = \boldsymbol{d}(\boldsymbol{k}) - \delta\boldsymbol{\theta}^{\mathrm{s}} \times \boldsymbol{d}(\boldsymbol{k}) \tag{8.3}$$

or

$$\boldsymbol{d}'(\boldsymbol{k}) = \exp\left(\frac{\mathrm{i}}{\hbar}\delta\boldsymbol{\theta}^{\mathrm{s}} \cdot \boldsymbol{S}\right) \boldsymbol{d}(\boldsymbol{k}) \exp\left(-\frac{\mathrm{i}}{\hbar}\delta\boldsymbol{\theta}^{\mathrm{s}} \cdot \boldsymbol{S}\right). \tag{8.4}$$

Expanding the latter expression in powers of $\delta\boldsymbol{\theta}^{\mathrm{s}}$ and comparing the linear term with (8.3) yields the commutation relations

$$[S_\mu, d_\nu(\boldsymbol{k})] = \mathrm{i}\hbar \sum_\lambda \epsilon_{\mu\nu\lambda} d_\lambda(\boldsymbol{k}). \tag{8.5}$$

The relations (8.5) also hold with $\boldsymbol{d}^*$. If we confine ourselves to lowest order in the dipole coupling, we do not need the commutation relations of the $\boldsymbol{d}$ and $\boldsymbol{d}^*$ among themselves in the following.

With the aid of the commutation relations, the equations of motion for $\boldsymbol{S}$ and $\boldsymbol{d}(\boldsymbol{k})$ may be expressed as

$$\dot{\boldsymbol{S}} = -\frac{\mathrm{i}}{\hbar}[\boldsymbol{S}, H_{\mathrm{S}}] = -\boldsymbol{S} \times \frac{\delta H_{\mathrm{S}}}{\delta \boldsymbol{S}} - \boldsymbol{R}_{\mathrm{D}}, \tag{8.6a}$$

$$\dot{\boldsymbol{d}}(\boldsymbol{k}) = -\frac{\mathrm{i}}{\hbar}[\boldsymbol{d}(\boldsymbol{k}), H_{\mathrm{S}}] = -\boldsymbol{d}(\boldsymbol{k}) \times \frac{\delta H_{\mathrm{S}}}{\delta \boldsymbol{S}} + O(H_{\mathrm{D}}), \tag{8.6b}$$

where $\boldsymbol{R}_{\mathrm{D}}$ is the so-called dipole torque, which is due to the coherent anisotropic spin–orbit interaction "pulling" on $\boldsymbol{S}$ once $\boldsymbol{S}$ moves. It is given by the change in the dipole energy density, $\delta f_{\mathrm{D}} = \delta\boldsymbol{\theta}^{\mathrm{s}} \cdot \boldsymbol{R}_{\mathrm{D}}$, when the spin system is rotated by $\delta\boldsymbol{\theta}^{\mathrm{s}}$. Hence

$$\boldsymbol{R}_{\mathrm{D}} = -\frac{1}{\hbar}\left\langle \boldsymbol{d}(\boldsymbol{k}) \times \frac{\delta H_{\mathrm{D}}}{\delta \boldsymbol{d}(\boldsymbol{k})}\right\rangle_{\hat{\boldsymbol{k}}} + \text{c.c.} \tag{8.7}$$

Note that a corresponding term in (8.6b) only leads to an additional dependence on $\boldsymbol{d}$ itself of order H_{D}, and can therefore be neglected except at ultralow frequencies (Leggett 1984a). The angular brackets denote an average over the directions $\hat{\boldsymbol{k}}$. Substituting the explicit forms for $\delta H_{\mathrm{S}}/\delta \boldsymbol{S}$ and $\delta H_{\mathrm{D}}/\delta \boldsymbol{d}(\boldsymbol{k})$, we finally obtain the celebrated Leggett equations (Leggett 1974a,b)

$$\dot{\boldsymbol{S}} = \gamma \boldsymbol{S} \times \boldsymbol{H} + \boldsymbol{R}_{\mathrm{D}}, \tag{8.8a}$$

$$\dot{\boldsymbol{d}}(\boldsymbol{k}) = \boldsymbol{d}(\boldsymbol{k}) \times \gamma\left(\boldsymbol{H} - \frac{\gamma}{\chi}\boldsymbol{S}'\right), \tag{8.8b}$$

with the explicit expression for the dipole torque in a general triplet pairing state obtained from (8.7) and (6.103a) as

$$\boldsymbol{R}_{\mathrm{D}} = -2\pi(\gamma\hbar)^2 I^2(0, T_{\mathrm{c}})\langle\langle[\hat{\boldsymbol{q}} \times \boldsymbol{d}(\boldsymbol{k})][\hat{\boldsymbol{q}} \cdot \boldsymbol{d}^*(\boldsymbol{k}')] + \text{c.c.}\rangle_{\hat{\boldsymbol{k}}}\rangle_{\hat{\boldsymbol{k}}'}, \tag{8.9a}$$

where $\hat{\boldsymbol{q}} = (\hat{\boldsymbol{k}} - \hat{\boldsymbol{k}}')/|\hat{\boldsymbol{k}} - \hat{\boldsymbol{k}}'|$ and $I(0, T_{\mathrm{c}})$ is defined in (6.94a). For an order parameter in a definite l state this reduces to

$$\boldsymbol{R}_{\mathrm{D}} = 6\lambda_{\mathrm{D}} N_{\mathrm{F}} \,\mathrm{Re}\, \langle[\hat{\boldsymbol{k}} \times \boldsymbol{d}(\boldsymbol{k})][\hat{\boldsymbol{k}} \cdot \boldsymbol{d}^*(\boldsymbol{k})]\rangle_{\hat{\boldsymbol{k}}}, \tag{8.9b}$$

which for unitary states ($\boldsymbol{d} \times \boldsymbol{d}^* = 0$; see below (3.37)) takes a particularly simple form; see (8.40).

8.2 NUCLEAR MAGNETIC RESONANCE UNDER LINEAR SPATIALLY HOMOGENEOUS CONDITIONS

The Leggett equations (8.8a,b) are nonlinear owing to the appearance of bilinear terms both in $\boldsymbol{d}$ and $\boldsymbol{d}^*$ in the dipole torque and in $\boldsymbol{d}$ and $\boldsymbol{S}$ in the $\boldsymbol{d}$-vector torque. For small deviations of $\boldsymbol{S}$ and $\boldsymbol{d}(\boldsymbol{k})$ from their equilibrium values $\boldsymbol{S}_0 = \chi\boldsymbol{H}/\gamma$ and $\boldsymbol{d}_0(\boldsymbol{k})$ in a static magnetic field $\boldsymbol{H}_0$, one can linearize (8.9) in the deviations $\delta\boldsymbol{S} = \boldsymbol{S} - \boldsymbol{S}_0$ and $\delta\boldsymbol{d}(\boldsymbol{k}) = \boldsymbol{d}(\boldsymbol{k}) - \boldsymbol{d}_0(\boldsymbol{k})$. Such deviations may be caused for example by a small time-dependent magnetic field $\delta\boldsymbol{H}$ or a small change in the static field $\boldsymbol{H}_0$. Taking the time derivative of (8.8a), $\delta\boldsymbol{d}(\boldsymbol{k})$ may be eliminated by virtue of the equation of motion for $\delta\boldsymbol{d}(\boldsymbol{k})$, (8.8b). This leads to a second order differential equation in time:

$$\delta\ddot{\boldsymbol{S}} - \gamma\,\delta\dot{\boldsymbol{S}} \times \boldsymbol{H}_0 + \boldsymbol{\Omega}^2\delta\boldsymbol{S} = \gamma\boldsymbol{S}_0 \times \delta\boldsymbol{H}. \tag{8.10a}$$

In particular, in zero magnetic field (8.10a) takes the form

$$\delta\ddot{\boldsymbol{S}} + \boldsymbol{\Omega}^2\delta\boldsymbol{S} = 0, \tag{8.10b}$$

which is equivalent to the equation of a three-dimensional anisotropic harmonic oscillator with eigenfrequencies given by the eigenvalues $\Omega_{\mu\nu}$ of the tensor $\mathbf{\Omega}$. For general triplet pairing, $\mathbf{\Omega}$ is given by (Leggett 1974a)

$$\Omega^2_{\mu\nu} = 8\lambda_D N_F \frac{\gamma^2}{\chi} \operatorname{Re} \langle\langle \delta_{\mu\nu}[\hat{\boldsymbol{q}} \cdot \boldsymbol{d}(\boldsymbol{k})][\hat{\boldsymbol{q}} \cdot \boldsymbol{d}(\boldsymbol{k}')] - d_\mu(\boldsymbol{k})\hat{q}_\nu \hat{\boldsymbol{q}} \cdot \boldsymbol{d}^*(\boldsymbol{k}') - [\hat{\boldsymbol{q}} \times \boldsymbol{d}(\boldsymbol{k})]_\mu [\hat{\boldsymbol{q}} \times \boldsymbol{d}^*(\boldsymbol{k}')]_\nu \rangle_{\hat{\boldsymbol{k}}} \rangle_{\hat{\boldsymbol{k}}'}. \tag{8.11}$$

The interpretation of $\Omega^2_{\mu\nu}$ as oscillator frequencies suggests a relation to the curvature at the minimum of the dipole energy potential. Indeed, comparing (8.10a) with the equation for $\ddot{\boldsymbol{S}}$ obtained from (8.6a) by taking the second (quantum-mechanical) time derivative, one finds that the third term on the left-hand side of (8.10a) may be expressed by the time derivative of $\boldsymbol{R}_D$, and hence that $\Omega^2_{\mu\nu}$ is determined by the double commutator (see Leggett 1975a)

$$\Omega^2_{\mu\nu} = \frac{\gamma^2}{\hbar^2 \chi} [[S_\mu, H_D], S_\nu]_{eq}. \tag{8.12}$$

Here the subscript "eq" indicates that, after performing the commutators, $\boldsymbol{d}^*$ and $\boldsymbol{d}$ have to be replaced by their equilibrium values. The double commutator in (8.12), however, is nothing else but the coefficient of the quadratic term in the expansion of the dipole energy about the minimum. This can be seen by considering the change in the dipole energy under infinitesimal rotations in spin space:

$$\begin{aligned} H'_D &= \exp\left(\frac{i}{\hbar} \delta\boldsymbol{\theta}^s \cdot \boldsymbol{S}\right) H_D \exp\left(-\frac{i}{\hbar} \delta\boldsymbol{\theta}^s \cdot \boldsymbol{S}\right) \\ &= H_D + \frac{1}{2\hbar^2} \sum_{\mu\nu} [[S_\mu, H_D], S_\nu]\, \delta\theta^s_\mu\, \delta\theta^s_\nu, \end{aligned} \tag{8.13}$$

where it is taken into account that the linear term is absent at the minimum. From (8.13), one finds, by comparison with (8.12), a useful relation between the (squared) eigenfrequencies and the dipole energy density $f_D \equiv \langle H_D \rangle / V$:

$$\Omega^2_{\mu\nu} = \frac{\gamma^2}{\chi} \frac{\partial^2 f_D}{\partial\theta^s_\mu \partial\theta^s_\nu}, \tag{8.14}$$

implying that the oscillator frequencies Ω are indeed determined by the curvature of the dipole energy at the minimum as a function of the rotation angles θ^s_μ.

Let us now discuss the eigenfrequencies of (8.10a). We take the z axis along the magnetic field $\boldsymbol{H}_0$ and assume one of the principal axes of $\Omega^2_{\mu\nu}$ to be along $\boldsymbol{H}_0$ (this is the case when the orientation of $\boldsymbol{d}$ is dominated by the magnetic field and the dipole force). Then the equations for δS_z and δS_x, δS_y are seen to decouple. Solving the one for δS_z, one obtains the remarkable result that there is a *longitudinal* spin resonance (Leggett 1974a), i.e. an oscillation of S_z about the equilibrium value $S_0 = \chi H_0$, of frequency

$$\omega_\parallel = \Omega_{zz}. \tag{8.15}$$

One should keep in mind that in the normal liquid (or in any other ordinary system, for that matter) a longitudinal resonance does not occur. There, only an exponential relaxation of the excess magnetization takes place. The longitudinal mode corresponds to harmonic oscillations of the vector $\boldsymbol{d}$ near the bottom of the potential well provided by the dipole energy. The longitudinal resonant frequency is seen to be independent of magnetic field. This resonance can be observed in a CW experiment in which an RF magnetic field $\delta\boldsymbol{H}_0$ is applied parallel to $\boldsymbol{H}_0$. The existence of this longitudinal resonance is entirely due to the broken relative spin–orbit symmetry (SBSOS) discussed above and in Chapter 6.

The transverse resonance usually observed in NMR experiments is of course also affected by the dipole torque. The transverse components of the spin satisfy the coupled equations

$$\left.\begin{aligned}(\omega^2 - \Omega_{xx}^2)\,\delta S_x - \mathrm{i}\omega\omega_\mathrm{L}\,\delta S_y &= 0,\\ (\omega^2 - \Omega_{yy}^2)\,\delta S_y + \mathrm{i}\omega\omega_\mathrm{L}\,\delta S_x &= -\mathrm{i}\omega\omega_\mathrm{L}\chi\,\delta H_x,\end{aligned}\right\} \tag{8.16}$$

where $\omega_\mathrm{L} = \gamma H_0$ is the Larmor frequency and we have taken the RF field to be along the x axis. The solution of (8.16) is

$$\delta S_x = \left(\frac{\omega_+^2}{\omega^2 - \omega_+^2} - \frac{\omega_-^2}{\omega^2 - \omega_-^2}\right)\frac{\omega_\mathrm{L}^2}{\omega_+^2 - \omega_-^2}\,\gamma\,\delta H_x, \tag{8.17a}$$

$$\delta S_y = -\mathrm{i}\frac{\omega_-^2\,\Omega_{xx}^2}{\omega\omega_\mathrm{L}}\,\delta S_x, \tag{8.17b}$$

Thus there are two elliptically polarized transverse oscillations at two resonant frequencies determined by

$$\omega_\pm^2 = \tfrac{1}{2}\{\omega_\mathrm{L}^2 + \Omega_{xx}^2 + \Omega_{yy}^2 \pm [(\omega_\mathrm{L}^2 + \Omega_{xx}^2 + \Omega_{yy}^2)^2 - 4\Omega_{xx}^2\Omega_{yy}^2]^{1/2}\}. \tag{8.18}$$

The prefactors of the pole terms in (8.17a,b) determine the intensity ("spectral weight") of the respective resonance lines (see also Maki and Ebisawa 1973, 1974b, 1975b, Takagi 1974b, Tewordt *et al.* 1975b).

8.2.1 Limit of zero magnetic field

Let us now apply these results to the A and B phases. In the A phase the preferred direction $\hat{\boldsymbol{d}}$ is independent of $\boldsymbol{k}$ and must be oriented perpendicular to the static magnetic field $\boldsymbol{H}_0$ in equilibrium. The dipole energy is minimal for the preferred orbital direction $\hat{\boldsymbol{l}}$ pointing along $\hat{\boldsymbol{d}}$, say in the x direction. From (6.104), the dipole energy density may be expressed in terms of the rotation angles θ_μ^s as

$$\Delta f_\mathrm{D}^\mathrm{A} = \tfrac{3}{5}\,g_\mathrm{D}(T)[(\theta_y^\mathrm{s})^2 + (\theta_z^\mathrm{s})^2], \quad \theta_\mu^\mathrm{s} \ll 1. \tag{8.19}$$

Substituting (8.19) into the general expression (8.14), one finds for the A phase

$$\Omega_{xx}^2 = 0, \tag{8.20a}$$

$$\Omega_{yy}^2 = \Omega_{zz}^2 = \frac{6}{5}\frac{\gamma^2}{\chi_\mathrm{A}}\,g_\mathrm{D}(T) \equiv \Omega_\mathrm{A}^2. \tag{8.20b}$$

One of the transverse oscillator frequencies is seen to be zero, which is generally true for any state with $\boldsymbol{k}$-independent preferred direction $\boldsymbol{d}$. In this case the transverse eigenfrequency ω_- is zero and carries zero spectral weight. Hence there is a *unique* transverse resonant frequency $\omega_\perp^{\mathrm{A}}$ given by

$$(\omega_\perp^{\mathrm{A}})^2 = \omega_{\mathrm{L}}^2 + \Omega_{\mathrm{A}}^2. \tag{8.21}$$

The longitudinal resonant frequency is seen to be simply related to the transverse resonance shift by

$$\omega_\parallel^{\mathrm{A}} = \Omega_{\mathrm{A}}. \tag{8.22}$$

The same relation would hold for any A-phase-like state with axially symmetric $\boldsymbol{d}(\boldsymbol{k}) \cdot \boldsymbol{d}^*(\boldsymbol{k})$ in $\boldsymbol{k}$ space.

Turning to the B phase, we recall that a magnetic field orients the spin–orbit rotation axis $\hat{\boldsymbol{n}}$ along the field axis. Accordingly, the only relevant angular dependence of the dipole energy is that on rotations about the z axis. Differentiating the expression (6.107) for Δf_{D},

$$\Delta f_{\mathrm{D}}^{\mathrm{B}} = \tfrac{8}{5} g_{\mathrm{D}}(T)(\cos\theta_z^{\mathrm{s}} + \tfrac{1}{4})^2, \tag{8.23}$$

as required in (8.14), and putting $\cos\theta_z^{\mathrm{s}} = -\frac{1}{4}$, one finds

$$\Omega_{xx} = \Omega_{yy} = 0, \tag{8.24a}$$

$$\Omega_{zz}^2 = 3\frac{\gamma^2}{\chi_{\mathrm{B}}} g_{\mathrm{D}}(T) \equiv \Omega_{\mathrm{B}}^2. \tag{8.24b}$$

There is no transverse frequency shift in the BW state (at least in the so-called "Leggett configuration", where $\hat{\boldsymbol{n}} \parallel \boldsymbol{H}_0$), but there is a longitudinal resonance at frequency

$$\omega_\parallel^{\mathrm{B}} = \Omega_{\mathrm{B}}. \tag{8.25}$$

The behaviour described by (8.21), (8.22) and (8.25) is very well observed in linear CW NMR experiments in the A and B phases of ^{3}He (see Figs. 4.4 and 4.6). In particular, the relation of the longitudinal resonant frequencies in both phases,

$$\frac{\omega_\parallel^{\mathrm{A}}}{\omega_\parallel^{\mathrm{B}}} = \left(\frac{4}{15}\right)^{1/2} \frac{\Delta_0^{\mathrm{A}}}{\Delta^{\mathrm{B}}} \left(\frac{\chi^{\mathrm{B}}}{\chi^{\mathrm{A}}}\right)^{1/2}, \tag{8.26}$$

has been verified near the A–B transition, where both phases may exist, if only in metastable states (see Fig. 4.6) (Ahonen *et al.* 1975c, Osheroff *et al.* 1975, Webb *et al.* 1975a). The ratio of the gap parameters appearing in (8.26) may be extracted from measurements of the specific-heat discontinuity (see (5.14)).

8.2.2 Effect of the magnetic field on NMR

Strictly speaking, the results for the resonant frequencies of the A and B phases given by (8.20) and (8.24) are only correct in the limit of zero magnetic field, since the order-parameter structure itself changes under the influence of a magnetic field (see Section 5.4). As a consequence, the dipole energy will change too, which gives rise to a modification of the results derived above.

In Section 6.3.1 the dipole energy in the A_1 and A_2 phases was shown to be given by (6.105). In the limit of zero magnetic field one has $\Delta_\uparrow = \Delta_\downarrow$ and the last term in (6.105) vanishes, which brings us back to (6.104), and hence (8.19). To determine the minimum of the dipole energy (6.105), we consider first a fixed direction of $\hat{\boldsymbol{d}}$ and $\hat{\boldsymbol{d}} \cdot \hat{\boldsymbol{l}} = \cos\theta$. The last term in (6.105) is minimal for $\hat{\boldsymbol{e}}$ in the plane of $\hat{\boldsymbol{d}}$ and $\hat{\boldsymbol{l}}$ (note that $\hat{\boldsymbol{e}} \cdot \hat{\boldsymbol{d}} = 0$), i.e. $\hat{\boldsymbol{e}} \cdot \hat{\boldsymbol{l}} = \sin\theta$. In the A_2 phase the prefactor of $(\hat{\boldsymbol{d}} \cdot \hat{\boldsymbol{l}})^2$ is larger than that of $(\hat{\boldsymbol{e}} \cdot \hat{\boldsymbol{l}})^2$. It follows that the minimum of Δf_D is given by $\hat{\boldsymbol{d}} \parallel \hat{\boldsymbol{l}}$, with $\hat{\boldsymbol{e}}$ arbitrary. Since $\hat{\boldsymbol{d}} \times \hat{\boldsymbol{e}}$ is along the Cooper-pair spin polarization, $\hat{\boldsymbol{d}}$ and $\hat{\boldsymbol{e}}$ are oriented perpendicular to the applied magnetic field in equilibrium.

In the A_1 phase, on the other hand, where the prefactors of $(\hat{\boldsymbol{d}} \cdot \hat{\boldsymbol{l}})^2$ and $(\hat{\boldsymbol{e}} \cdot \hat{\boldsymbol{l}})^2$ in (6.105) are equal, the dependence on the angle θ drops out and the minimum of f_D is assumed for any planar configuration of $\hat{\boldsymbol{d}}$, $\hat{\boldsymbol{e}}$ and $\hat{\boldsymbol{l}}$, i.e. for $\hat{\boldsymbol{l}}$ anywhere in the plane spanned by $\hat{\boldsymbol{d}}$ and $\hat{\boldsymbol{e}}$. In general, the dipole energy depends on the three angles θ^s_μ, which may be chosen to parametrize the unit vectors $\hat{\boldsymbol{d}}$ and $\hat{\boldsymbol{e}}$ as

$$\hat{\boldsymbol{d}} = (\cos\theta^s_y \cos\theta^s_z, \cos\theta^s_y \sin\theta^s_z, \sin\theta^s_y), \tag{8.27}$$

$$\begin{aligned}\hat{\boldsymbol{e}} = (&\sin\theta^s_x \sin\theta^s_y \cos\theta^s_z - \cos\theta^s_x \sin\theta^s_z, \sin\theta^s_x \sin\theta^s_y \sin\theta^s_z \\ &+ \cos\theta^s_x \cos\theta^s_z, -\sin\theta^s_x \cos\theta^s_y).\end{aligned} \tag{8.28}$$

The angles θ^s_μ describe rotations about the μ axis. Taking $\hat{\boldsymbol{l}}$ along the x axis, one finds the general expression for Δf_D:

$$\begin{aligned}\Delta f_D^{A_{1,2}} = &-\tfrac{3}{5}\lambda_D N_F[\Delta_+^2 \cos^2\theta^s_y \cos^2\theta^s_z \\ &+ \Delta_-^2(\sin\theta^s_x \sin\theta^s_y \cos\theta^s_z - \cos\theta^s_x \sin\theta^s_z)^2].\end{aligned} \tag{8.29}$$

The minimum of Δf_D is at $\theta^s_x = \theta^s_y = \theta^s_z = 0$. For small deviations from the equilibrium, one may expand Δf_D in θ^s_μ as

$$\Delta f_D^{A_{1,2}} = \tfrac{3}{5}\lambda_D N_F[\tfrac{1}{4}(\Delta_{\uparrow\uparrow} + \Delta_{\downarrow\downarrow})^2(\theta^s_y)^2 + \Delta_{\uparrow\uparrow}\Delta_{\downarrow\downarrow}(\theta^s_z)^2]. \tag{8.30}$$

It follows from (8.14) that the longitudinal resonant frequency in the A_2 phase is given by (Osheroff and Anderson 1974, Takagi 1975b)

$$\omega_\parallel^2 = \tfrac{6}{5}\lambda_D N_F \gamma^2 \chi^{-1} \Delta_{\uparrow\uparrow}\Delta_{\downarrow\downarrow}. \tag{8.31}$$

There is no longitudinal resonance in the A_1 phase in this approximation for lack of a restoring force on motions of $\hat{\boldsymbol{d}}$ and $\hat{\boldsymbol{e}}$ in the plane perpendicular to the magnetic field (infinitely degenerate minimum). However, the dipole coupling leads to a minute admixture of the opposite-spin pairing component in the A_1-phase order parameter, which causes a finite, but very small, longitudinal resonant frequency after all (Monien and Tewordt 1985).

The transverse resonant shift is given by

$$\omega_\perp^2 = \omega_L^2 + \tfrac{3}{10}\,\lambda_D N_F \frac{\gamma^2}{\chi}(\Delta_{\uparrow\uparrow} + \Delta_{\downarrow\downarrow})^2. \tag{8.32}$$

These predictions have been verified experimentally (Osheroff and Anderson 1974). The specific dependence of $\omega_\parallel^2$ and $\omega_\perp^2$ on $\Delta_{\uparrow\uparrow}$ and $\Delta_{\downarrow\downarrow}$ offers the possibility of measuring the combinations of strong-coupling parameters β_{24} and β_5 that enter $\Delta_{\uparrow\uparrow}$ and $\Delta_{\downarrow\downarrow}$ (see Section 5.4).

In the B_2 phase the order parameter, as given by (5.66b), is characterized by a preferred direction in spin space, which under equilibrium conditions coincides with the direction of the magnetic field $\hat{\boldsymbol{H}}$. The dipole energy density of the B_2 state has been derived in Section 6.3.1 (see (6.108)–(6.110)). For small deviations from equilibrium (defined by $\hat{\boldsymbol{n}} = \hat{\boldsymbol{H}},\ \theta = \theta_0$) one may expand (6.109) as

$$\Delta f_D^{B_2} = \tfrac{1}{5}\lambda_D N_F\left\{8(\Delta_\perp^2 - \tfrac{1}{2}\Delta_\parallel^2)(\theta_z - \theta_0)^2 - \frac{[f_1(\theta_0) + 2f_2(\theta_0)](\theta_x^2 + \theta_y^2)}{2(1-\cos\theta_0)}\right\}, \tag{8.33}$$

where f_1, f_2 and θ_0 are given in (6.110) and (6.112) and the relation $\delta\hat{n}_{x,y}^2 = \tfrac{1}{2}(1-\cos\theta_0)^{-1}\theta_{x,y}^2$ has been used.

The longitudinal resonance frequency follows from (8.14) as

$$\omega_\parallel^2 = \tfrac{1}{5}\lambda_D N_F \gamma^2 \chi^{-1}(H)(16\Delta_\perp^2 - \Delta_\parallel^2), \tag{8.34}$$

where $\chi(H)$ is the largest eigenvalue of the anisotropic spin susceptibility in the B_2 phase.

In equilibrium the axis $\hat{\boldsymbol{n}}$ is oriented along the magnetic field by the susceptibility anisotropy, i.e. $(\hat{\boldsymbol{n}}\cdot\hat{\boldsymbol{z}}) = 1$. The minimum of the dipole energy is located at the shifted Leggett angle $\theta_L(H)$ defined by (6.112). Thus, as the B_2 transition is approached from well inside the B_2 phase, the Leggett angle decreases from 104° to 90°.

Since $\Delta f_D^{B_2}$ depends on the orientation of $\hat{\boldsymbol{n}}$, there is a transverse resonance shift (Leggett 1974a). The transverse components of the tensor of eigenfrequencies $\Omega_{\mu\nu}^2$ are given by

$$\Omega_{xx}^2 = \Omega_{yy}^2 = \tfrac{2}{5}\lambda_D N_F \gamma^2 \chi^{-1}(\Delta_\perp^2 - \Delta_\parallel^2), \tag{8.35}$$

implying the existence of two elliptically polarized modes with frequency

$$\omega_t^2 = \tfrac{1}{2}\{\omega_L^2 + 2\Omega_{xx} \pm [(\omega_L^2 + 4\Omega_{xx}^2)\omega_L^2]^{1/2}\}. \tag{8.36}$$

In the limit of large fields ($\omega_L \gg \Omega_{xx}$) this reduces to (Greaves 1976; see also Hasegawa 1983)

$$\omega_t^2 = \begin{cases} \omega_L^2 + 2\Omega_{xx}^2 & (+), \qquad (8.37) \\ 0 & (-). \qquad (8.38) \end{cases}$$

From the general result (8.17) on the spectral weight of these resonances, we conclude that the resonance at $\omega_t^2 \approx 0$ carries no weight. The resonant shift $\Delta\omega_t = \omega_t - \omega_L$ of the Larmor frequency in (8.37) has been observed experimentally (Osheroff and Brinkman 1977). It may in principle be used to determine yet another combination of free-energy parameters, β_{345}.

8.3 NONLINEAR NMR PHENOMENA IN UNIFORM TEXTURES

Even in the simple case of spatially uniform spin motion, the nonlinear Leggett equations (8.8a,b) are complex enough to defy general solution. In the following we shall therefore address the A phase and the B phase separately, and we shall discuss a few special cases that allow an analytical solution while still revealing the underlying physics.

8.3.1 The Leggett equations for the A and B phases

Using (8.9b) and (3.68b) to determine $\boldsymbol{R}_\mathrm{D}$, the Leggett equations in the A phase take the form

$$\dot{\boldsymbol{S}} = \gamma \boldsymbol{S} \times \boldsymbol{H} + \Omega_\mathrm{A}^2 \frac{\chi}{\gamma^2} (\hat{\boldsymbol{d}} \times \hat{\boldsymbol{l}})(\hat{\boldsymbol{d}} \cdot \hat{\boldsymbol{l}}), \tag{8.39a}$$

$$\dot{\hat{\boldsymbol{d}}} = \gamma \hat{\boldsymbol{d}} \times \left(\boldsymbol{H} - \frac{\gamma}{\chi} \boldsymbol{S}\right), \tag{8.39b}$$

where $\hat{\boldsymbol{d}}$ and $\hat{\boldsymbol{l}}$ are the preferred directions in spin and orbital spaces and Ω_A is the characteristic frequency of the A phase, defined in (8.20). It follows from (8.39) that $\boldsymbol{S} \cdot \hat{\boldsymbol{d}}$ is a constant of the motion.

The nonlinear spin dynamics of the B phase is more complicated than that of the A phase owing to the spin–orbit correlation of the order parameter $d_\mu(\boldsymbol{k}) = \Delta R_{\mu j} \hat{k}_j$. For most purposes, it is convenient to parametrize the rotation matrix by the unit vector $\hat{\boldsymbol{n}}$ specifying the rotation axis and the rotation angle θ. Alternatively, one may employ a representation in terms of Euler angles. The rotation matrix in the axis/angle variables is given by (6.106).

It is straightforward to write the Leggett equations (8.8a,b) in terms of $\hat{\boldsymbol{n}}$ and θ, using the general expression for the dipole torque valid for unitary states,

$$(R_\mathrm{D})_\mu = -\tfrac{6}{5} g_\mathrm{D}(T) \sum_{\beta\lambda\nu} \epsilon_{\mu\nu\lambda} (A_{\nu\lambda} A_{\beta\beta}^* + A_{\nu\beta} A_{\beta\lambda}^* + A_{\nu\beta} A_{\lambda\beta}^*), \tag{8.40}$$

(6.8) and the relations below (6.106). The result is

$$\dot{\boldsymbol{S}} = \gamma \boldsymbol{S} \times \boldsymbol{H} + \tfrac{4}{15} \Omega_\mathrm{B}^2 \frac{\chi}{\gamma^2} \sin\theta (1 + 4\cos\theta) \hat{\boldsymbol{n}}, \tag{8.41a}$$

$$\dot{\theta} = \frac{\gamma^2}{\chi} \hat{\boldsymbol{n}} \cdot \boldsymbol{S}', \tag{8.41b}$$

$$\dot{\hat{\boldsymbol{n}}} = -\frac{1}{2}\frac{\gamma^2}{\chi} [\hat{\boldsymbol{n}} \times \boldsymbol{S}' + \cot \tfrac{1}{2}\theta\, \hat{\boldsymbol{n}} \times (\hat{\boldsymbol{n}} \times \boldsymbol{S}')], \tag{8.41c}$$

where $\boldsymbol{S}' = \boldsymbol{S} - \gamma^{-1}\chi\boldsymbol{H}$ is the deviation of $\boldsymbol{S}$ from instantaneous equilibrium. It should be noted that the dipole torque is along $\hat{\boldsymbol{n}}$, (8.41a), and that the spin–orbit rotation angle is driven by the projection of $\boldsymbol{S}'$ on $\hat{\boldsymbol{n}}$, (8.41b).

8.3.2 Spin dynamics in zero magnetic field

In the absence of a magnetic field there are no external forces acting on the spin system. If we neglect dissipation for the moment, this implies that the total energy is conserved, i.e. is a constant of motion. The spin motion is then in many ways similar to that of a spinning top in classical mechanics. We shall see that the present problem presents a nice example of the usefulness of differential geometry. It provides a simple mathematical framework within which it is easy to show that the spin dynamics in zero field of both phases can be solved exactly (Vollhardt 1984b).

We shall focus attention on the motion of $\boldsymbol{S}$, i.e. its space curve, when some parameter t is changed. Here t can be, for example, the time. There is, however, another parameter—the so-called "arclength" of a segment of the space curve—which is even more suitable for such a discussion and which leads to the so-called "natural parametrization". The arclength $s(t)$ of a segment is given by (see e.g. Struik 1961)

$$s(t) = \int_{t_0}^{t} \mathrm{d}t' \left|\frac{\mathrm{d}\boldsymbol{S}}{\mathrm{d}t'}\right|, \tag{8.42}$$

and hence $\dot{s}(t) = |\dot{\boldsymbol{S}}|$. We can then take $\boldsymbol{S}$ to be a function of $s(t)$, rather than of t itself, because the connection between s and t is given uniquely by (8.42). In this way, the characterization of the space curve is considerably simplified. For example, the derivative of $\boldsymbol{S}$ with respect to s, denoted by $\boldsymbol{S}' = \mathrm{d}\boldsymbol{S}/\mathrm{d}s$, i.e. the tangent vector to the space curve, is given by

$$\boldsymbol{S}' = \frac{\dot{\boldsymbol{S}}}{\dot{s}} \tag{8.43a}$$

$$= \frac{\dot{\boldsymbol{S}}}{|\dot{\boldsymbol{S}}|} \equiv \hat{\boldsymbol{t}}, \tag{8.43b}$$

so $\boldsymbol{S}' \equiv \hat{\boldsymbol{t}}$ is automatically a unit vector. Furthermore, the second derivative of $\boldsymbol{S}$,

$$\boldsymbol{S}'' = \hat{\boldsymbol{t}}' \equiv \boldsymbol{n}_0 \tag{8.44}$$

defines the "normal" $\boldsymbol{n}_0$ to the curve. It is directed perpendicular to the curve, i.e. to $\hat{\boldsymbol{t}}$, (this is a consequence of $|\hat{\boldsymbol{t}}|^2=1$). Its magnitude $|\boldsymbol{n}_0|$ is called the "curvature" of the curve:

$$\boldsymbol{n}_0=\kappa\hat{\boldsymbol{n}}_0, \tag{8.45}$$

where κ^{-1}, the radius of curvature, is the radius of the circle fitting the curve at that point. The plane passing through the tangent $\hat{\boldsymbol{t}}$ and the normal $\boldsymbol{n}_0$ is called the osculating plane, because it fits perfectly to the curve at every point. A third vector can then be defined by

$$\hat{\boldsymbol{b}}=\hat{\boldsymbol{t}}\times\hat{\boldsymbol{n}}_0, \tag{8.46}$$

which is perpendicular to this plane and is called the "binormal". The three unit vectors form an orthonormal triad that is a new system of reference moving along the curve. Any change of $\hat{\boldsymbol{b}}$ during this motion (which must be parallel to $\hat{\boldsymbol{n}}_0$) describes a tilting of the osculating plane and defines the "torsion" τ of the curve.

$$\hat{\boldsymbol{b}}'=-\tau\hat{\boldsymbol{n}}_0. \tag{8.47}$$

It is a measure of how strongly the curve winds out of the osculating plane. Using (8.43)–(8.47), one easily finds that τ is given by

$$\tau=\frac{\boldsymbol{S}'\cdot(\boldsymbol{S}''\times\boldsymbol{S}''')}{\kappa^2}. \tag{8.48}$$

The A phase

In the case of the A phase one obtains from (8.39a)

$$\hat{\boldsymbol{t}}=\frac{\hat{\boldsymbol{d}}\times\hat{\boldsymbol{l}}}{\sin\theta}, \tag{8.49}$$

$$\hat{\boldsymbol{n}}_0=\sin\theta\left[\frac{\cos\theta}{\boldsymbol{S}\cdot\hat{\boldsymbol{l}}}\hat{\boldsymbol{t}}\times(\hat{\boldsymbol{t}}\times\boldsymbol{S})+\hat{\boldsymbol{d}}\right], \tag{8.50}$$

where θ is the angle between $\hat{\boldsymbol{d}}$ and $\hat{\boldsymbol{l}}$. The binormal $\hat{\boldsymbol{b}}$ is then found to be

$$\hat{\boldsymbol{b}}=\hat{\boldsymbol{t}}\times\hat{\boldsymbol{n}}_0=\hat{\boldsymbol{l}}, \tag{8.51}$$

i.e. it is given by the preferred direction $\hat{\boldsymbol{l}}$. Let us assume $\hat{\boldsymbol{l}}$ to be time-independent, i.e. to be a constant vector in space. Then the torsion (8.48) of the curve described by $\boldsymbol{S}$, being the measure of the change of $\hat{\boldsymbol{b}}$, is identially zero. In other words, $\boldsymbol{S}$ moves in such a way that the projection of $\boldsymbol{S}$ on $\hat{\boldsymbol{b}}=\hat{\boldsymbol{l}}$ is a constant of motion (because $\mathrm{d}(\boldsymbol{S}\cdot\hat{\boldsymbol{b}})/\mathrm{d}s=\boldsymbol{S}\cdot\hat{\boldsymbol{b}}'=0$):

$$\boldsymbol{S}\cdot\hat{\boldsymbol{l}}\equiv C_{\mathrm{A}}=\text{const.} \tag{8.52}$$

This is also evident from (8.39a), where $\dot{\boldsymbol{S}}\cdot\hat{\boldsymbol{l}}=0$ for $\mathbf{H}=0$. Thus there exists a constant vector

$$\boldsymbol{C}_{\mathrm{A}}=(\boldsymbol{S}\cdot\hat{\boldsymbol{l}})\hat{\boldsymbol{l}} \tag{8.53}$$

of magnitude C_A and direction $\hat{\boldsymbol{l}}$, which enables us to find an exact solution of the spin dynamics of the A phase for $\boldsymbol{H}=0$. For this, we make use of the vector identity

$$S^2 = (\boldsymbol{S} \cdot \hat{\boldsymbol{t}})^2 + (\boldsymbol{S} \times \hat{\boldsymbol{t}})^2 \tag{8.54}$$

to express $\boldsymbol{S}$ in terms of the angle θ only. Taking the derivative of $\hat{\boldsymbol{d}} \cdot \hat{\boldsymbol{l}} = \cos\theta$, one finds $\dot{\theta} = -\gamma^2\chi^{-1}(\boldsymbol{S} \cdot \hat{\boldsymbol{t}})$; furthermore, $\boldsymbol{S} \times \hat{\boldsymbol{t}} = C_A \hat{d}/\sin\theta$, which yields

$$S^2 = \left(\frac{\chi}{\gamma^2}\right)^2 \dot{\theta}^2 + \left(\frac{C_A}{\sin\theta}\right)^2. \tag{8.55}$$

The total energy, as determined by (8.1) with $H=0$, is a constant of motion and may then also be written in terms of a single variable only:

$$E_A = \frac{\chi}{2\gamma^2}\left(\dot{\theta}^2 + \frac{\tilde{C}_A^2}{\sin^2\theta} + \Omega_A^2 \sin^2\theta\right), \tag{8.56}$$

where $\tilde{C}_A = (\gamma^2/\chi)C_A$. We are then lead to the equation of motion for θ:

$$\ddot{\theta} - \tilde{C}_A^2 \frac{\cos\theta}{\sin^3\theta} + \tfrac{1}{2}\Omega_A^2 \sin 2\theta = 0. \tag{8.57}$$

The problem of solving the five coupled differential equations in (8.39) has thus been reduced to the solution of a single equation.

Once this equation has been solved, every information about the motion of $\boldsymbol{S}$ and $\hat{\boldsymbol{d}}$ can be obtained by elementary techniques. In particular, $\boldsymbol{S}(t)$ is specified by (i) its magnitude (8.55), (ii) its projection on $\hat{\boldsymbol{b}} = \hat{\boldsymbol{l}}$, $\boldsymbol{S} \cdot \hat{\boldsymbol{b}} = C_A$, and (iii) its projection into the plane perpendicular to $\hat{\boldsymbol{b}}$, where it precesses with the angular velocity

$$\dot{\varphi}_S = \left|\frac{\mathrm{d}}{\mathrm{d}t}\frac{\boldsymbol{S} \times \hat{\boldsymbol{b}}}{|\boldsymbol{S} \times \hat{\boldsymbol{b}}|}\right| \tag{8.58a}$$

$$= \Omega_A^2 \frac{\chi}{\gamma^2} C_A^2 \frac{\cos^2\theta}{S^2 - C_A^2}. \tag{8.58b}$$

The corresponding precession frequency of $\hat{\boldsymbol{d}}$ around $\hat{\boldsymbol{l}}$ is given by $\dot{\varphi}_{\hat{d}} = \tilde{C}_A/\sin^2\theta$. It is interesting to note that (8.56) and (8.57) are very similar to the corresponding expressions for a spinning top, where C_A represents the vertical component of the angular momentum, which is a constant of motion.

Equation (8.57) describes a periodic motion of $\hat{\boldsymbol{d}}$ (and therefore $\boldsymbol{S}$) and can easily be solved analytically by an elliptic integral. The solution depends on the magnitude of C_A. In a turn-off experiment, when an external field $\boldsymbol{H}$ is switched off, C_A can be determined from (8.55), putting $\theta=0$. For this, the initial configuration of $\hat{\boldsymbol{d}}$ and $\boldsymbol{S}$ relative to $\hat{\boldsymbol{l}}$ and the external field $\boldsymbol{H}$ (see Fig. 8.1) has to be determined first. (In fields $H \lessdot \Omega_A/\gamma$ the spin $\boldsymbol{S}$ is not aligned

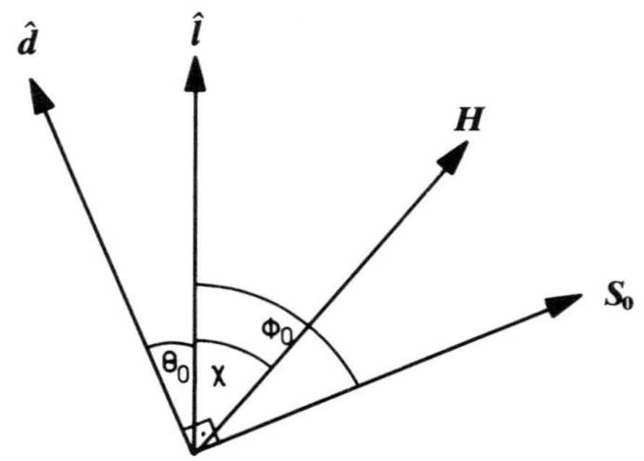

Figure 8.1 Configuration of $\hat{\boldsymbol{d}}$ and $\boldsymbol{S}$ relative to $\hat{\boldsymbol{l}}$ and a fixed external magnetic field $\boldsymbol{H}$.

along $\boldsymbol{H}$ because the dipole coupling favours $\boldsymbol{S} \perp \hat{\boldsymbol{l}}$ such that $\hat{\boldsymbol{d}}$ is not perpendicular to $\boldsymbol{H}$.) Note that once the system has reached equilibrium in the presence of the magnetic field, $\hat{\boldsymbol{d}}$ and $\boldsymbol{S}$ are stationary, lying in the plane of $\hat{\boldsymbol{l}}$ and $\boldsymbol{H}$. This initial condition results solely from the exclusion from (8.27) of relaxation effects that damp the motion of $\hat{\boldsymbol{d}}$ and $\boldsymbol{S}$, which were originally precessing under the influence of $\boldsymbol{H}$. Minimizing the energy with respect to $|\boldsymbol{S}|$ and ϕ_0, the angle between $\boldsymbol{S}$ and $\hat{\boldsymbol{l}}$, yields

$$S = \frac{\chi}{\gamma} H \cos(\phi_0 - \chi), \tag{8.59}$$

where χ is the angle between $\boldsymbol{H}$ and $\hat{\boldsymbol{l}}$, and ϕ_0 $(=\frac{1}{2}\pi - \theta_0)$ is determined by

$$\tan 2\phi_0 = \frac{\sin 2\chi}{\cos 2\chi - (\Omega_A/\gamma H)^2}. \tag{8.60}$$

When the field $\boldsymbol{H}$ is suddenly switched off, $\tilde{C}_A$ is given by

$$\tilde{C}_A = \Omega_A z \cos\phi_0, \tag{8.61}$$

where

$$z = \frac{\omega_L}{\Omega_A} \cos(\phi_0 - \chi) \tag{8.62}$$

and $\omega_L = \gamma H$.

The resulting periodic motion of $\hat{\boldsymbol{d}}$ and $\boldsymbol{S}$ is determined by (8.57) and depends on whether $z \lessgtr 1$. Starting from θ_0, θ initially increases in both cases. However, for $z < 1$, θ cannot exceed a maximum value $\bar{\theta} \equiv \sin^{-1} z$, while, for $z > 1$, one has $\theta_0 \leqslant \theta \leqslant \pi - \theta_0$. The periods are found to be

$$T_r = \frac{4}{\Omega_A \sin\phi_0} K\left(\frac{(z^2 - \cos^2\phi_0)^{1/2}}{\sin\phi_0}\right) \qquad (z \leqslant 1), \tag{8.63a}$$

$$T_r = \frac{4}{\Omega_A (z^2 - \cos^2\phi_0)^{1/2}} K\left(\frac{\sin\phi_0}{(z^2 - \cos^2\phi_0)^{1/2}}\right) \qquad (z \geqslant 1), \tag{8.63b}$$

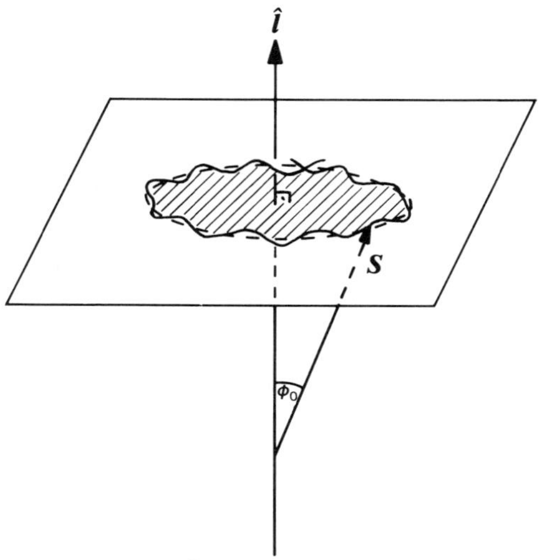

Figure 8.2 Precession of $\boldsymbol{S}$ about $\hat{\boldsymbol{l}}$ and simultaneous nutation, after turn-off of a magnetic field $H \gg \Omega_A/\gamma$.

where $K(x)$ is the complete elliptic integral. At $z = 1$ the energy due to field turn-off is just sufficient to push $\hat{\boldsymbol{d}}$ perpendicular to $\hat{\boldsymbol{l}}$. It takes infinite time to attain this (metastable) position, i.e. the period is infinite.One expects that for $z \approx 1$ relaxation effects become very important because of the long timescales involved.

The resulting motion of the spin vector for $\gamma H \gg \Omega_A$ is shown schematically in Fig. 8.2: $\boldsymbol{S}$ precesses about $\hat{\boldsymbol{l}}$ with an instantaneous angular velocity $\dot{\varphi}_S$, (8.58b), while nutating at the same time with twice the frequency of the θ motion. This nutation is due to the finite spin–orbit coupling, which wants to have $\boldsymbol{S}$ and $\hat{\boldsymbol{l}}$ perpendicular: the spin $\boldsymbol{S}$ is no longer a conserved quantity. As the projection of $\boldsymbol{S}$ on $\hat{\boldsymbol{l}}$ and the total energy are constants of motion, the component of $\boldsymbol{S}$ perpendicular to $\hat{\boldsymbol{l}}$ will therefore change. Its oscillation is a consequence of the periodic tunnelling of Cooper pairs between the two subsystems with spin projection $S_z = \pm 1$ (for a discussion see Section 8.3.3). Note that the precession frequency of $\boldsymbol{S}$ about $\hat{\boldsymbol{l}}$ is quite different from that of the θ motion. It is mainly determined by the size of $\gamma H/\Omega_a$ and the initial direction of $\boldsymbol{H}$ relative to $\hat{\boldsymbol{l}}$. The two frequencies are generally unrelated.

The B phase

In spite of the somewhat more complicated nature of the spin dynamics in the B phase, (8.41), the field-free case can also be solved exactly, as was first shown by Maki and Ebisawa (1976a) using an Euler-angle approach (see also Golo 1981b). Within the present framework, we again study the space curve of $\boldsymbol{S}$ and determine the moving triad $(\hat{\boldsymbol{t}}, \hat{\boldsymbol{n}}_0, \hat{\boldsymbol{b}})$ by taking the derivative with respect to the arclength $s(t)$. We obtain from (8.41a)

$$\hat{\boldsymbol{t}} \equiv \boldsymbol{S}' = \hat{\boldsymbol{n}}, \tag{8.64a}$$

$$\hat{\boldsymbol{n}}_0 \equiv \boldsymbol{S}'' = \frac{\dot{\hat{\boldsymbol{n}}}}{\dot{s}}, \tag{8.64b}$$

$$\boldsymbol{S}''' = \frac{\dot{s}\ddot{\hat{\boldsymbol{n}}} - \ddot{s}\dot{\hat{\boldsymbol{n}}}}{\dot{s}^2}, \tag{8.64c}$$

where $\dot{s} = \frac{4}{15}\,\Omega_B^2 \chi\gamma^{-2} \sin\theta\,(1 + 4\cos\theta)$. Interestingly, the (unit) tangent vector to $\boldsymbol{S}$ is given by the direction $\hat{\boldsymbol{n}}$. The binormal $\hat{\boldsymbol{b}}$ is then given by

$$\hat{\boldsymbol{b}} = \frac{\hat{\boldsymbol{n}} \times \dot{\hat{\boldsymbol{n}}}}{|\dot{\hat{\boldsymbol{n}}}|}. \tag{8.65}$$

Calculating the torsion τ of the curve via (8.47) or (8.48), one finds that τ is identically zero at all times. This implies that $\dot{\hat{\boldsymbol{b}}} = 0$, i.e. the binormal (8.65) is a constant vector in space, and hence the osculating plane is constant too! As mentioned before, zero torsion of $\boldsymbol{S}$ means that $\boldsymbol{S}$ rotates about $\hat{\boldsymbol{b}}$ in such a manner that its projection on $\hat{\boldsymbol{b}}$ is constant:

$$\boldsymbol{S} \cdot \hat{\boldsymbol{b}} = |\hat{\boldsymbol{n}} \times \boldsymbol{S}| \sin \tfrac{1}{2}\theta \tag{8.66a}$$

$$\equiv C_B = \text{const}, \tag{8.66b}$$

where we have used (8.41b) to obtain (8.66a). At the same time, $\hat{\boldsymbol{n}}$ also rotates about $\hat{\boldsymbol{b}}$ but with zero projection on $\hat{\boldsymbol{b}}$ $(\hat{\boldsymbol{n}} \cdot \hat{\boldsymbol{b}} = 0)$: its motion is planar, taking place in the plane perpendicular to $\hat{\boldsymbol{b}}$. We have thus identified three constants of motion. They can be combined into a constant vector

$$\boldsymbol{C}_B = (\boldsymbol{S} \cdot \hat{\boldsymbol{b}})\hat{\boldsymbol{b}}. \tag{8.67}$$

This is essentially the constant vector $\boldsymbol{A}$ $(= -(\gamma^2/\chi)\boldsymbol{C}_B)$ introduced by Brinkman and Cross (1978), whose square A^2 is identical with the corresponding constant of motion found by Maki and Ebisawa (1976a). Making use of the vector identity (8.54) and of (8.66a), we can again express S^2 in terms of one variable only—the rotation angle θ:

$$S^2 = \left(\frac{\chi}{\gamma^2}\right)^2 \dot{\theta}^2 + \left(\frac{C_B}{\sin \frac{1}{2}\theta}\right)^2. \tag{8.68}$$

We note that the expressions (8.55) and (8.68) for S are almost identical! The total energy may now be expressed in terms of θ and $\dot{\theta}$ only, as

$$E = \frac{\chi}{2\gamma^2}\left[\dot{\theta}^2 + \frac{\tilde{C}_B^2}{\sin^2 \frac{1}{2}\theta} + \tfrac{16}{15}\,\Omega_B^2(\tfrac{1}{4} + \cos\theta)^2\right], \tag{8.69}$$

where $\tilde{C}_B = (\gamma^2/\chi)C_B$.

The equation of motion for θ follows immediately as

$$\ddot{\theta} - \tfrac{1}{2}\tilde{C}_{\mathrm{B}}^2 \frac{\cos\frac{1}{2}\theta}{\sin^3\frac{1}{2}\theta} - \tfrac{4}{15}\,\Omega_{\mathrm{B}}^2 \sin\theta(1 + 4\cos\theta) = 0. \tag{8.70}$$

Again, the problem of having to solve a set of (here, six) differential equations has been reduced to the solution of a single one. Once this has been done, S can easily be determined as described below (8.57) in the case of the A phase. Here the frequency of precession of $\boldsymbol{S}$ about $\hat{\boldsymbol{b}}$ is given by

$$\dot{\varphi}_S = \tfrac{8}{15}\,\Omega_{\mathrm{B}}^2 \frac{\chi}{\gamma^2} C_{\mathrm{B}} \frac{\cos^2\frac{1}{2}\theta\,(1 + 4\cos\theta)}{S^2 - C_{\mathrm{B}}^2}, \tag{8.71a}$$

and that of $\hat{\boldsymbol{n}}$ by

$$\dot{\varphi}_{\hat{\boldsymbol{n}}} = \frac{1}{2}\frac{\tilde{C}_{\mathrm{B}}}{\sin^2\frac{1}{2}\theta}. \tag{8.71b}$$

8.3.3 Nonlinear spin dynamics in the A phase

In the presence of an external magnetic field the equations describing the spin dynamics of superfluid ^{3}He are considerably more complicated than those for the field-free case. So far, only certain limiting situations have been solved analytically (see Maki 1975c). One such case is the purely longitudinal spin motion, $\boldsymbol{S} = (0, 0, S_z)$. The vector $\hat{\boldsymbol{d}}$ is then confined to lie in the (x, y) plane, $\hat{\boldsymbol{d}} = (\cos\psi, \sin\psi, 0)$. Taking $\hat{\boldsymbol{l}}$ along the x axis, the minimum of the dipole energy is located at $\psi = 0$. Eliminating S_z, one finds for ψ the equation of motion of a driven mathematical pendulum:

$$\ddot{\psi} + \tfrac{1}{2}\Omega_{\mathrm{A}}^2 \sin 2\psi = -\gamma\dot{H}_z. \tag{8.72}$$

As suggested by Leggett (1975a), (8.72) may be considered as a Josephson oscillator equation. For this, we imagine the A phase to consist of two subsystems, the spin-up and spin-down particles, described by order parameters $\Delta_\uparrow \propto \mathrm{e}^{\mathrm{i}\psi}$ and $\Delta_\downarrow \propto \mathrm{e}^{-\mathrm{i}\psi}$. The dipole interaction introduces a weak coupling between the subsystems, allowing pairs to tunnel from one to the other. The associated change in magnetic energy provides a restoring force for the pair current, giving rise to oscillations in the relative particle number $n_\uparrow - n_\downarrow$, and hence the spin S_z (see Maki and Tsuneto 1975).

The "pendulum" may be excited by means of an impulse—a sudden change of magnitude of the magnetic field by ΔH. The subsequent "ringing" of magnetization is a special case ($\phi_0 = \frac{1}{2}\pi$) of the spin dynamics in the field-free situation. The oscillations of $\boldsymbol{S}$ and, at half the frequencies, of $\hat{\boldsymbol{d}}$, occurs at a frequency $\omega_{\mathrm{r}} = 2\pi/T_{\mathrm{r}}$, where T_{r} is given by (8.63a,b) with $z = \omega_{\mathrm{L}}/\Omega_{\mathrm{a}}$, $\omega_{\mathrm{L}} = \gamma\,\Delta H$ (Maki and Tsuneto 1974b):

$$T_{\mathrm{r}} = \begin{cases} 4\Omega_{\mathrm{A}}^{-1} K(\omega_{\mathrm{L}}/\Omega_{\mathrm{A}}) & (\omega_{\mathrm{L}} < \Omega_A), \\ 4\omega_{\mathrm{L}}^{-1} K(\Omega_{\mathrm{A}}/\omega_{\mathrm{L}}) & (\Omega_A < \omega_{\mathrm{L}}). \end{cases} \tag{8.73}$$

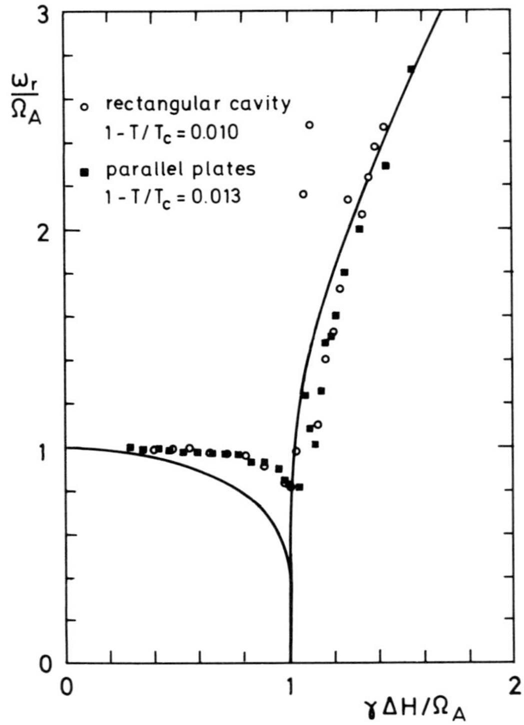

Figure 8.3 Nonlinear parallel ringing in ^{3}He-A at $P = 22.0$ bar in two different experimental geometries. (After Webb *et al.* (1975b).) Solid curve: theoretical results by Maki and Tsuneto (1974b).

In the limit $\omega \to 0$ the ringing frequency ω_r tends to Ω_A, the longitudinal resonant frequency discussed in Section 8.2. For increasing ω_L, i.e. larger amplitude of the pendulum oscillations, ω_r decreases until it approaches zero at $\omega_L = \Omega_A$. This is the point where the impulse is just right to push the pendulum into its (unstable) maximum position. Beyond this point ω_r increases with ω_L, finally following the law $\omega_r = 2\omega_L$ for $\omega_L \gg \Omega_A$. In the pendulum picture the latter case corresponds to the rotating pendulum. This behaviour has been observed experimentally (Webb *et al.* 1975b, 1977), except for the region close to the "frequency dip" at $\omega_L = \Omega_A$, where the system is very sensitive to relaxation effects as well as to perturbations such as walls and heat currents (see Fig. 8.3).

A second case allowing for an analytic solution is the precession of the magnetization about a strong static magnetic field $\boldsymbol{H}_0$ (i.e. $\gamma H_0 \gg \Omega_A$) after a large RF pulse perpendicular to $\boldsymbol{H}_0$ has tipped the magnetization by an angle θ. The dipole interaction may then be considered as a perturbation. In the frame of reference rotating with frequency $\gamma H_0 \equiv \omega_L$ about the direction of the magnetic field, the spin is a constant vector $\boldsymbol{S}' = S(\sin\theta, 0, \cos\theta)$ as long as the dipole interaction is neglected. The vector $\hat{\boldsymbol{d}}$ in the rotating

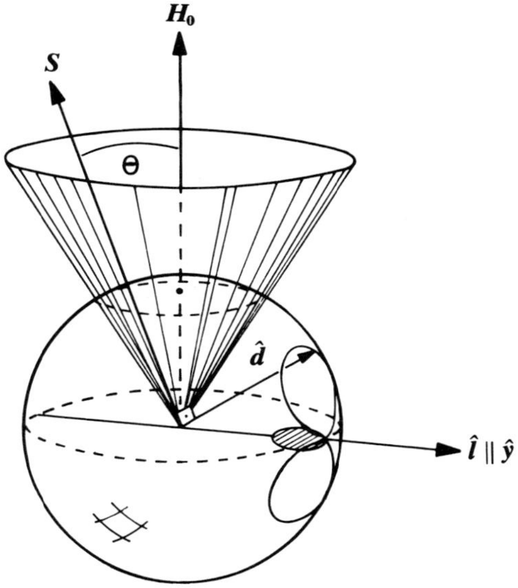

Figure 8.4 Precession of the spin S and corresponding motion of the vector $\hat{d}$ after a spin-tipping pulse with $\theta \approx 30°$.

frame precesses about this fixed vector S' as described by

$$\dot{\hat{d}}' = -\frac{\gamma^2}{\chi}\hat{d}' \times S'. \tag{8.74}$$

Solving (8.63), one finds $\hat{d}$ in the laboratory frame as

$$\hat{d}(t) = \begin{pmatrix} (1-\cos\theta)\sin\omega_L t\cos\omega_L t \\ \cos\theta\sin^2\omega_L t + \cos^2\omega_L t \\ \sin\theta\sin\omega_L t \end{pmatrix}, \tag{8.75}$$

representing a motion where $\hat{d}$ traverses a "figure-of-eight" shaped orbit on the surface of the unit sphere (see Fig. 8.4). The projection of the two branches onto the (x,y)-plane is a circle of radius $\sin^2\frac{1}{2}\theta$, given by the equation

$$x^2 + (y - \cos^2\tfrac{1}{2}\theta)^2 = \sin^4\tfrac{1}{2}\theta, \tag{8.76}$$

which touches the unit sphere at the point $\hat{l}$. Here we have taken the vector as pointing along the y axis, such that it is perpendicular to both the static field H_0 and the RF field that tipped the magnetization. Substituting $\hat{d}(t)$ from (8.75) into the expression for the dipole torque, and neglecting higher harmonics of the Larmor frequency ω_L, which are small by factors Ω_A/ω_L, one finds

$$R_D(t) = \frac{\Omega_A^2}{4\omega_L}(1 + 3\cos\theta)S_y(t)\hat{x}. \tag{8.77}$$

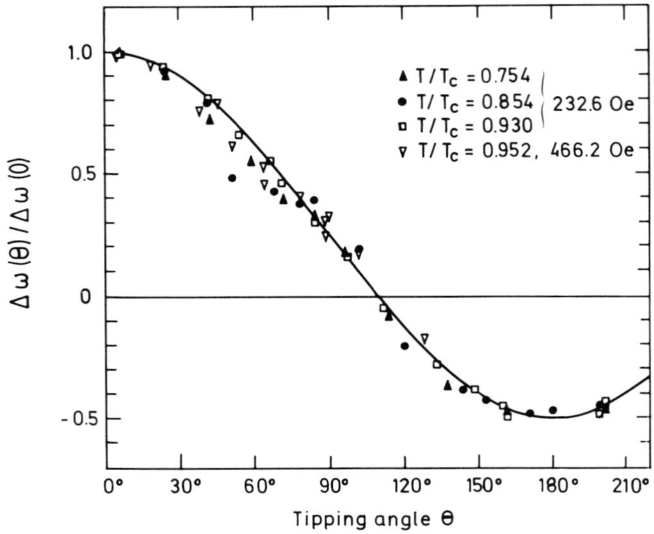

Figure 8.5 Frequency shift $\Delta\omega(\theta) = \omega_\perp - \omega_L$ in ^{3}He-A as a function of the pulse-tipping angle θ. (After Osheroff and Corrucini (1975a).) Solid curve: theroetical result by Brinkman and Smith (1975a) as given by (8.78).

The dipole torque $\boldsymbol{R}_D(t)$ gives rise to a frequency shift of the free-induction signal (Brinkman and Smith 1975a) that depends on the tipping angle and is given by

$$\omega_\perp = \omega_L + \frac{\Omega_A^2}{8\omega_L}(1 + 3\cos\theta). \tag{8.78}$$

For small tipping angles, this result reduces to the transverse resonant frequency in the linear regime (8.21), expanded in the high-field limit. As the tipping angle increases, the frequency shift decreases and eventually becomes negative beyond a critical angle $\theta_c = \cos^{-1}(-\frac{1}{3}) \approx 109°$. This behaviour is due to the fact that for increasing θ, the vector $\boldsymbol{d}$ oscillates with larger and larger amplitude, thereby going through configurations where the dipole torque is weaker or even reverses sign. The frequency shift given by (8.78) has been observed experimentally by Osheroff and Corrucini (1975a,b) (see Fig. 8.5).

As shown by Poluéktov (1984), the Leggett equations may also be solved analytically in the case when the vector $\hat{\boldsymbol{l}}$ is aligned with a constant external magnetic field. Such a configuration may be realized in a parallel-plate geometry (see Section 7.9).

An alternative and in a way more systematic approach to the nonlinear spin dynamics of superfluid ^{3}He has been developed by Fomin (1976, 1978a,b,c, 1979a,b, 1980, 1984a,b, 1985a,b, 1986, 1987a,b) (see also Fomin *et al.* (1982) and the review by Fomin (1981b)). Many of the results for the

A phase discussed so far, as well as those for the B phase to be discussed below, may be derived using this approach. It is based on the observation that the motion of the spin variables may be decomposed into a fast precession and a slow motion of the axis (or the angle) of precession. One may then average over the fast motion to obtain a much simpler set of equations for the slow variables. The theory has also been generalized to include spatial variations and relaxation effects. In particular, it has been shown (Fomin 1984b, 1985b) that even a small inhomogeneity of the magnetic field can have a strong influence on the precession of magnetization in the B phase. It leads to a splitting into two homogeneous domains in the case of a linear magnetic field gradient. While in one domain the magnetization is at rest, it precesses uniformly at a relative angle $\theta_L = 104°$ in the other. This effect has been observed by Bun'kov *et al.* (1984a) and Borovik-Romanov *et al.* (1984b, 1985). The energetic degeneracy of the two possible orientations of $\hat{\boldsymbol{n}}$ is lifted by the motion of the domain wall (Podd'yakova and Fomin 1988).

Instability of homogeneous precession of the magnetization is also expected in the A phase (Fomin 1979b, 1984a, Fomin and Shopova 1989). The predicted time dependence in the linear stage of the development of this instability has been observed by Borovik-Romanov *et al.* (1984a).

The nonlinear spin dynamics in the A_1 phase has been discussed in detail by de Vegvar (1984), employing the adiabatic method of Fomin.

Nonlinear driven longitudinal spin resonance has been investigated both numerically (Katayama *et al.* 1980a,b,c, 1983b, Ooiwa *et al.* 1981, Yamaguchi *et al.* 1982, Yamaguchi 1983) and within an analytical stability analysis (Fishman and Folk 1984a). It has been found that behaviour familiar from other nonlinear systems, such as stable and unstable periodic motion as well as chaotic motion, may occur (Ugulava 1984, Montgomery 1985).

8.3.4 Nonlinear spin dynamics in the B phase

Let us now discuss the solution of (8.41), describing the spin dynamics in the B phase in a finite magnetic field. Again we concentrate on a few simple cases. As in the A phase, there is parallel ringing after a sudden change in the static magnetic field $\boldsymbol{H}_0$, i.e. a longitudinal oscillation of $\boldsymbol{S}$ driven by the dipole potential. If the magnetic field determines the direction of $\hat{\boldsymbol{n}}$ ($\hat{\boldsymbol{n}} \parallel \boldsymbol{H}$), one can find the equation of motion for the angle θ by taking the time derivative of (8.41b) and eliminating S_z with the aid of (8.41c):

$$\ddot{\theta} - \tfrac{4}{15}\,\Omega_B^2 \sin\theta\,(1 + 4\cos\theta) = -\gamma \dot{H}_z. \tag{8.79}$$

For small oscillations about the minimum of the dipole energy at $\theta = \theta_L = \cos^{-1}(-\tfrac{1}{4})$, (8.79) reduces to the linear longitudinal resonance equation (8.10a). In contrast with the A phase, the dipole energy in the BW state has two inequivalent maxima at $\theta = 0$ and $\theta = \pi$ (see Fig. 6.6). Consequently, there will be two values of the change in magnetic field ΔH for which the ringing frequency approaches zero. These values are obtained by requiring that the kinetic energy $E_{\mathrm{kin}} = \frac{1}{2}\chi(\Delta H)^2$ provided by the field change be equal to the difference in dipole energy between the minimum and maximum positions: $\Delta E_{\mathrm{D}} = \frac{8}{15}\,\Omega_{\mathrm{B}}^2(\chi/\gamma^2)(\frac{1}{4} + \cos\theta)^2$. With this, one finds $\gamma\,\Delta H = (\frac{3}{5})^{1/2}\,\Omega_{\mathrm{B}}$ and $\gamma\,\Delta H = (\frac{5}{3})^{1/2}\Omega_{\mathrm{B}}$ (Maki and Hu 1975a). For values of ΔH large compared with $\Omega_{\mathrm{B}}/\gamma$, "rotating-pendulum" behaviour with frequency $\omega_{\mathrm{r}} = 2\gamma\,\Delta H$ occurs as in the A phase. These predictions are in qualitative agreement with experiment (Webb *et al.* 1975b, 1977).

As a second example, we consider the "wall-pinned configuration", where $\hat{\boldsymbol{n}}$ is oriented perpendicular to $\boldsymbol{H}$ under the influence of, say, closely spaced walls (Brinkman 1974, Maki and Hu 1975b; for other configurations see Maki and Ebisawa 1975a). The equation of motion (8.70) for θ, obtained for the field-free case, allows discussion of this more complicated situation. In a field-turn-off experiment, with $\hat{\boldsymbol{n}}\cdot\boldsymbol{H} = 0$ and $\theta = \theta_{\mathrm{L}}$ initially, the energy term involving $\tilde{C}_{\mathrm{B}}$ in (8.69) must be equal to the kinetic energy gained through the field turn-off, i.e. $\tilde{C}_{\mathrm{B}}^2 = \frac{5}{8}(\gamma\,\Delta H)^2$. For small fields, $\gamma\,\Delta H \ll \Omega_{\mathrm{B}}$, the dipole potential dominates in (8.70), and we may put $\theta(t) \approx \theta_{\mathrm{L}}$ for all times. In this case $\boldsymbol{S}$ is stationary and $\hat{\boldsymbol{n}}$ precesses in the plane perpendicular to the axis $\boldsymbol{C}_{\mathrm{B}}$ with frequency $\dot{\varphi}_{\hat{\boldsymbol{n}}}(\theta_{\mathrm{L}}) = \frac{1}{2}\tilde{C}_{\mathrm{B}}/\sin^2\frac{1}{2}\theta_{\mathrm{L}} = (\frac{2}{5})^{1/2}\gamma\,\Delta H$. Note that this frequency is independent of the dipole coupling owing to the special initial condition. The precession of $\hat{\boldsymbol{n}}$ couples back to $\boldsymbol{S}$ if initially θ is not precisely equal to θ_{L}, giving rise to a transverse ringing signal. Wall-pinned ringing of magnetization has been observed experimentally (Webb *et al.* 1975c, 1977). Obviously, the ringing mode discussed above is completely distinct from the oscillations of the vector $\hat{\boldsymbol{d}}$ about the minimum of the dipole well. It may rather be compared to the free precession of $\boldsymbol{S}$ and $\hat{\boldsymbol{n}}$, subject to the constraint $\hat{\boldsymbol{n}}\cdot\boldsymbol{S} = 0$. It *appears* as if the dipole interaction is irrelevant in the latter case because the solution does not explicitly depend on it. This is not so, however: the dipole force is actually responsible for the constraint $\hat{\boldsymbol{n}}\cdot\boldsymbol{S} = 0$.

Finally, we consider the precession of the magnetization tipped by a large angle in a strong magnetic field, $\omega_{\mathrm{L}} \gg \Omega_{\mathrm{B}}$. Under these conditions, the total spin $\boldsymbol{S}$ is approximately a constant vector in the frame rotating with the Larmor frequency, which we take to lie in the (x, z) plane. It is convenient to work in a coordinate system in which the x and z axes are along $\hat{\boldsymbol{S}} - \hat{\boldsymbol{H}}$ and $\hat{\boldsymbol{S}} + \hat{\boldsymbol{H}}$ respectively, where $\hat{\boldsymbol{S}}$ and $\hat{\boldsymbol{H}}$ are unit vectors in the direction of $\boldsymbol{S}$ and $\boldsymbol{H}$. The equations of motion for the spin–orbit rotation angle θ and the vector $\hat{\boldsymbol{n}}$ then take the forms (Brinkman and Smith 1975b):

$$\dot{\theta} = 2\omega_{\mathrm{L}}\sin(\tfrac{1}{2}\phi)\hat{n}_x, \tag{8.80a}$$

$$\dot{\hat{n}}_x = \omega_{\mathrm{L}}[-\cos(\tfrac{1}{2}\phi)\hat{n}_y + \cot(\tfrac{1}{2}\theta)\sin(\tfrac{1}{2}\phi)(1 - \hat{n}_x^2)], \tag{8.80b}$$

$$\dot{\hat{n}}_y = \omega_{\mathrm{L}}[\cos(\tfrac{1}{2}\phi)\hat{n}_x - \cot(\tfrac{1}{2}\theta)\sin(\tfrac{1}{2}\phi)\hat{n}_x\hat{n}_y], \tag{8.80c}$$

$$\dot{\hat{n}}_z = -\omega_{\mathrm{L}}\cot(\tfrac{1}{2}\theta)\sin(\tfrac{1}{2}\phi)\hat{n}_x\hat{n}_z. \tag{8.80d}$$

On a long timescale, we may expect the physical system to be in a state of minimum dipole energy, i.e. $\theta = \theta_L$, and hence $\dot{\theta} = 0$. According to (8.80b), this implies $\hat{n}_x = 0$, and consequently $\dot{\hat{n}}_y = \dot{\hat{n}}_z = 0$ from (8.80c,d). The requirement that the right-hand side of (8.80b) vanish determines $\hat{n}_y$ and $\hat{n}_z$ as $\hat{n}_y = \cot \frac{1}{2}\theta_L \tan \frac{1}{2}\phi$ and $\hat{n}_z = (1 - \hat{n}_y^2)^{1/2}$. Since the dipole energy is a miminum for this solution, no dipole torque is exerted and hence no frequency shift occurs. There is a problem, however: the expression for $\hat{n}_y$ becomes greater than unity for tipping angles $\phi > \theta_L$. Thus the above solution is acceptable only for $\phi < \theta_L$.

For larger tipping angles, one has to explicitly take into account the existence of relaxation processes, which tend to drive the spin–orbit angle θ towards its equilibrium value θ_L. This may be incorporated by adding a term $-(\theta - \theta_L)/\tau$ to the right-hand side of (8.80a). Here τ is a spin relaxation time whose significance and magnitude will be discussed in Section 8.6. With this modification, the quasistationary solution of (8.70) discussed above remains meaningful even in the regime $\phi > \theta_L$: the vector $\hat{\boldsymbol{n}}$ may now escape into the x direction, while staying perpendicular to the z axis.

A solution is then found with $\hat{n}_z = 0$, $\hat{n}_x = (\theta - \theta_L)/(2\omega_L \tau \sin \frac{1}{2}\phi)$ and $\hat{n}_y = \cot \frac{1}{2}\phi \tan \frac{1}{2}\theta$. The stationary value θ_0 of the spin–orbit angle (which has to satisfy $\theta_0 < \phi$) is obtained from the normalization condition $\hat{n}_x^2 + \hat{n}_y^2 = 1$. The angle θ_0 increases with $\omega_L \tau$ from $\theta_0 \approx \theta_L$ for $\omega_L \tau \ll 1$ to $\theta_0 \approx \phi$ for $\omega_L \tau \gg 1$. It should be noted that in general this solution is not consistent with the initial assumptions of a time-independent tipping angle. However, in the limit of weak relaxation, when $\omega_L \tau \gg 1$, one can neglect the relaxation of the z component of the magnetization, because $\hat{\boldsymbol{n}}$ and therefore the dipole torque are parallel to the y axis and hence perpendicular to the magnetic field.

In this case the dipole torque gives rise to a transverse resonant shift of magnitude (Brinkman and Smith 1975b)

$$\omega_\perp - \omega_L = -\frac{4}{15} \frac{\Omega_B^2}{\omega_L} (1 + 4 \cos \phi). \tag{8.81}$$

Such a frequency shift has indeed been observed (Osheroff and Corruccini, 1975b), although there are deviations from the behaviour predicted by (8.81) near $\phi = \pi$ (see Fig. 8.6). Further nonlinear effects have been discussed by Liu and Brinkman (1978).

It has been suggested that in longitudinal driven NMR in the B phase, nonlinear phenomena such as period doubling, intermittent chaos and crisis should occur (Katayama *et al.* 1983a,b, Yamaguchi 1983, Montgomery 1985). Note, however, that driven longitudinal spin resonance has in fact been shown to be unstable with respect to transverse fluctuations (Fishman and Folk 1984b, 1985).

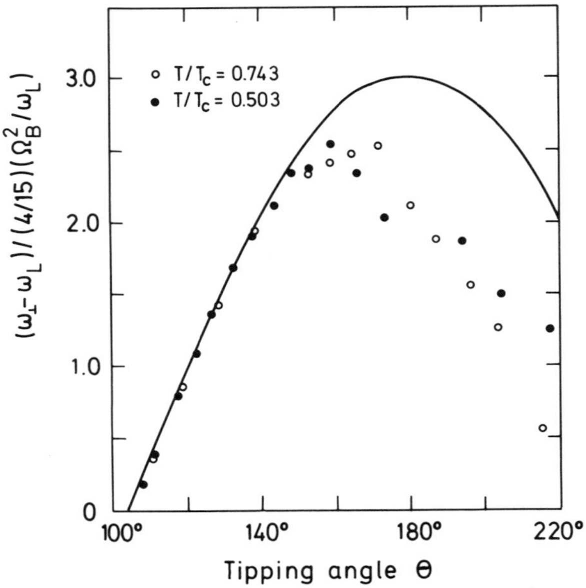

Figure 8.6 Frequency shift $\Delta\omega = \omega_\perp - \omega_L$ as a function of the pulse-tipping angle θ in ^{3}He-B. (After Osheroff and Corrucini (1975b).) Solid curve: theroetical result by Brinkman and Smith (1975b) as given by (8.81).

8.4 TEXTURE-INDUCED MAGNETIC RESONANCE PHENOMENA

So far, we have concentrated on the spin dynamics of uniform textures, in particular of those textures where the orientation of the preferred directions was determined by an external magnetic field or the dipole interaction. As discussed in Chapter 7, one of the salient features of superfluid ^{3}He is the existence of textures, both smoothly varying and strongly localized, of the order-parameter field. Nuclear magnetic resonance experiments provide a powerful tool for studying the textures related to the spin part of the order parameters. Some of the simpler results will be discussed in this section.

In calculating the NMR shifts induced by a given texture, one may distinguish between two limiting situations. If the spatial variation of the texture occurs on a scale ξ_L long compared with the dipole coherence length $\xi_D \ll \xi_L$, one effectively has a model of uncoupled "local oscillators", where each resonator corresponds to a subsystem of extension larger than ξ_D but smaller than ξ_L. In this case the contributions to the NMR power absorption simply add up. The NMR lineshape is then determined by the distribution of subsystems with a given resonance shift. In the opposite limit, $\xi_L \lesssim \xi_D$, the different subsystems are coupled owing to the stiffness of the texture. In this case the local-oscillator picture is clearly insufficient and a more general description is required. For weak excitations the corresponding theory turns out to be equivalent to solving a Schrödinger-type equation where the kinetic energy describes the stiffness of the texture and the potential energy is due to the locally varying dipole forces in the equilibrium texture. The case of strong excitations, where a *non*linear treatment is required, is naturally even more complex and so far has received not much attention.

We begin by considering the linear spin dynamics in spatially uniform textures, whose orientation is not only determined by a magnetic field but also by additional effects like the container walls or a counterflow $\boldsymbol{v}_s - \boldsymbol{v}_n$, i.e. a heat current.

8.4.1 Tilted uniform textures

Suppose that the $\hat{\boldsymbol{l}}$-vector field in the A phase is tilted by one of the abovementioned effects so that the angle ψ enclosed by $\hat{\boldsymbol{l}}$ and the magnetic field $\boldsymbol{H}$ is not $\frac{1}{2}\pi$ as required by the dipole interaction in equilibrium. In this case the the tensor of eigenfrequencies $\Omega^2_{\mu\nu}$ is no longer diagonal in the "laboratory frame", where $\boldsymbol{H}$ is along the z axis, but rather is rotated by an angle $\frac{1}{2}\pi - \psi$, so that

$$\tilde{\Omega}^2_{\mu\nu} = (\boldsymbol{R}\Omega^2\boldsymbol{R}^{-1})_{\mu\nu} = \Omega^2_{\mathrm{A}}\begin{pmatrix} \cos^2\psi & 0 & \sin\psi\cos\psi \\ 0 & 1 & 0 \\ \sin\psi\cos\phi & 0 & \sin^2\psi \end{pmatrix}. \tag{8.82}$$

where $\boldsymbol{R}$ is a rotation matrix describing a rotation by $\frac{1}{2}\pi - \psi$ about $\hat{\boldsymbol{y}}$. Here it is assumed that $\hat{\boldsymbol{l}}$ lies in the (x, z) plane. The resonant frequencies follow from (8.82) as zeros of the determinant $|\omega\delta_{\mu\nu} - \gamma\sum_\lambda \epsilon_{\mu\nu\lambda}H_\lambda - \tilde{\Omega}_{\mu\nu}|$ and are given by (Takagi 1975a, Fetter 1976a)

$$\omega^2_{\pm} = \tfrac{1}{2}\omega^2_{\mathrm{L}} + \Omega^2_{\mathrm{A}} \pm \tfrac{1}{2}\omega_{\mathrm{L}}(\omega^2_{\mathrm{L}} + 4\Omega^2_{\mathrm{A}}\cos^2\psi)^{1/2}. \tag{8.83}$$

The associated modes are elliptically polarized mixed motions of $\boldsymbol{S}$. In the limit of high magnetic fields, $\omega_{\mathrm{L}} \gg \Omega_{\mathrm{A}}$, and/or $\psi \approx \frac{1}{2}\pi$, the two solutions reduce to transverse (+) and longitudinal (−) excitations,

$$\omega^2_{\mathrm{t}} \equiv \omega^2_{+} \approx \omega^2_{\mathrm{L}} + \Omega^2_{\mathrm{A}}(1 + \cos^2\psi), \tag{8.84a}$$

$$\omega^2_{\ell} \equiv \omega^2_{-} \approx \Omega^2_{\mathrm{A}}\sin^2\psi. \tag{8.84b}$$

The third eigenfrequency is zero and does not carry spectral weight. The transverse resonance shift (8.83) caused by the wall orientation of $\hat{\boldsymbol{l}}$ has been observed experimentally (Ahonen *et al.* 1975a,b 1976b). The possibility of the excitation to spin waves with wave vector $|\boldsymbol{q}| \neq 0$ in tilted uniform textures and other dynamic effects have been investigated by Gongadze *et al.* (1980).

In the B phase a rotation of the $\hat{\boldsymbol{n}}$-vector field away from the equilibrium orientation along $\boldsymbol{H} \parallel \hat{\boldsymbol{z}}$ by the angle ψ (about the y axis, say) yields the eigenfrequency matrix

$$\tilde{\Omega}^2_{\mu\nu} = \Omega^2_{\rm B}\begin{pmatrix} \sin^2\psi & 0 & \sin\psi\cos\phi \\ 0 & 0 & 0 \\ \sin\psi\cos\psi & 0 & \cos^2\psi \end{pmatrix} \tag{8.85}$$

and two magnetic resonance frequencies (Osheroff *et al.* 1975)

$$\omega^2_{\pm} = \tfrac{1}{2}(\omega^2_{\rm L} + \Omega^2_{\rm B}) \pm \tfrac{1}{2}[(\omega^2_{\rm L} + \Omega^2_{\rm B})^2 - 4\omega^2_{\rm L}\Omega^2_B\cos^2\psi]^{1/2}. \tag{8.86}$$

For high fields and/or small ψ, ω_+ and ω_- can again be identified with the transverse and longitudinal spin motions respectively.

Experimental studies of B-phase textures have been made in a number of geometries, including (i) cylinders (Osheroff and Brinkman 1974, Osheroff 1974, Brinkman *et al.* 1974b, Ahonon *et al.* 1974c), (ii) powders (Ahonen *et al.* 1974a) and (iii) slabs (Osheroff *et al.* 1975, Ahonen *et al.* 1975a,c Engelsberg *et al.* 1977, Freeman *et al.* 1988).

8.4.2 NMR shifts induced by superflow

The above results for NMR in uniform tilted textures are now used to investigate the effect of superflow on the NMR frequencies. For this, the idealized assumption is made that the resulting textures may be considered as locally uniform, so that the results of the above discussion can be applied. The effect of the coupling of different regions of the sample by the gradient terms in the free energy will be discussed later.

In the A phase the effect of a weak superflow is to orient the vector $\hat{\boldsymbol{l}}$ parallel to itself (see (6.120) and the discussion in Section 7.10). For orientations of the flow direction $\boldsymbol{v}_{\rm s}$ other than perpendicular to an applied external magnetic field $\boldsymbol{H}$ (e.g. $\hat{\boldsymbol{v}}_{\rm s}\cdot\hat{\boldsymbol{H}} = \cos\chi$), there is then a competition between both orienting effects, resulting in an intermediate orientation of $\hat{\boldsymbol{l}}$ with respect to $\boldsymbol{H}$ (i.e. $\hat{\boldsymbol{l}}\cdot\hat{\boldsymbol{H}} = \cos\psi$). When the dipole interaction is the dominating influence, i.e. for a weak magnetic field and weak flow, the texture may be assumed to be dipole-locked, i.e. $\hat{\boldsymbol{l}} \parallel \hat{\boldsymbol{d}}$, and one finds (Fetter 1975b, 1976a)

$$\sin^2\psi = \frac{1}{2}\left[1 + \frac{1 - \eta\cos 2\chi}{(1 - 2\eta\cos 2\chi + \eta^2)^{1/2}}\right], \tag{8.87}$$

where

$$\eta = \frac{(\rho_{\rm s\parallel} - \rho_{\rm s\perp})v_{\rm s}^2}{H^2\,\Delta\chi}.$$

Here it has been assumed that $\hat{\boldsymbol{l}}$ lies in the plane of $\boldsymbol{H}$ and $\boldsymbol{v}_{\rm s}$, which may not be entirely correct in view of the torque exerted on $\hat{\boldsymbol{l}}$ by $\boldsymbol{v}_{\rm s}$ (see the discussion of the helical texture in Section 7.10). For perpendicular flow ($\chi = \tfrac{1}{2}\pi$), one finds $\psi = \tfrac{1}{2}\pi$. By contrast, in the case of parallel flow ($\chi = 0$)

(8.87) yields $\psi = \frac{1}{2}\pi$ for small flow values and $\eta < 1$, while ψ changes discontinuously to 0 for $\eta > 1$. A smooth transition from $\psi = \frac{1}{2}\pi$ for $\eta \to 0$ to $\psi = 0$ for $\eta \to \infty$ is found for intermediate values of χ. The corresponding NMR frequency shifts induced by the superflow follow directly on substituting (8.87) into (8.83).

The limit of large magnetic fields has been discussed by Takagi (1975a). The theoretical predictions are in qualitative agreement with experiment (Flint *et al.* 1978).

In the B phase the vector $\hat{\boldsymbol{n}}$ is oriented by the combined effect of magnetic field and superflow, as given by (6.125). In the limit of weak magnetic field and superflow the orientation of $\hat{n}$ is found to be given by an expression identical with (8.87) except that $\sin^2 \psi$ is replaced by $\cos^2 \psi$ and η by 2η (Fetter 1976a). The NMR frequency shifts follow from (8.83). The discontinuous behaviour at $2\eta = 1$ that is again found for the maximally competing case of perpendicular orientations of $\boldsymbol{H}$ and $\boldsymbol{v}_s$ has been observed experimentally in the NMR spectrum (Hutchins *et al.* 1985), along with other qualitative features predicted by theory.

8.4.3 General theory of texture-induced NMR shifts in the A phase

In order to describe the spin response of a nonuniform $\hat{\boldsymbol{d}}$ texture, one has to take into account the bending energy associated with the texture. The latter acts as an additional restoring force on $\hat{\boldsymbol{d}}$, when $\hat{\boldsymbol{d}}$ is displaced from equilibrium as in the case of an NMR experiment. The equation of motion for $\hat{\boldsymbol{d}}$ may be derived by different, but equivalent, methods: either by a Lagrangian formulation or by a Hamiltonian approach. Both formulations have been used extensively in the literature and will therefore be described separately.

Lagrangian approach

The Lagrangian method was first formulated by Maki (1975b) for the B phase and was later adapted to the A phase by Maki and Kumar (1977b), Maki (1980) and Vollhardt *et al.* (1981). Below, we shall present a general formulation of the spin dynamics in the A phase, valid for arbitrarily strong magnetic fields.

We start from the Lagrangian

$$L = \int \mathrm{d}^3 r\, \mathscr{L}, \tag{8.88}$$

where

$$\mathscr{L} = T - V \tag{8.89}$$

is determined by the kinetic energy density T associated with the $\hat{\boldsymbol{d}}$ motion and

$$V\{\hat{\boldsymbol{d}}\} = f_{\text{bend}} + \Delta f_{\text{D}} \tag{8.90}$$

is the additional potential felt by $\hat{\boldsymbol{d}}$ owing to the bending of the texture, (7.20) and the dipolar coupling to $\hat{\boldsymbol{l}}$, (6.104) (if necessary, other orientational energy densities may be added to V). The kinetic energy of the $\hat{\boldsymbol{d}}$ motion is written as

$$T = \frac{1}{2\gamma^2}(\boldsymbol{\omega} - \boldsymbol{\omega}_{\mathrm{L}})\boldsymbol{\chi}(\boldsymbol{\omega} - \boldsymbol{\omega}_{\mathrm{L}}). \tag{8.91}$$

Here $\boldsymbol{\chi}$ is the full anisotropic spin-susceptibility tensor (3.102):

$$\chi_{\mu\nu} = \chi_{\mathrm{N}}(\delta_{\mu\nu} - a\hat{d}_\mu \hat{d}_\nu), \tag{8.92a}$$

with

$$a = \Delta\chi/\chi_{\mathrm{N}} \tag{8.92b}$$

as the prefactor of the anisotropy term, and the external magnetic field $\boldsymbol{H}$ is assumed to be along the z axis such that $\boldsymbol{\omega}_{\mathrm{L}} = \omega_{\mathrm{L}}\hat{\boldsymbol{z}}$. Furthermore, the local spin $\boldsymbol{S}$, i.e. the induced local magnetization $\boldsymbol{M} = \gamma\boldsymbol{S}$, is represented in terms of the local rotation axis $\boldsymbol{\omega}$ as

$$\boldsymbol{S} = \frac{1}{\gamma^2}\boldsymbol{\chi}\boldsymbol{\omega}. \tag{8.93}$$

$\boldsymbol{\omega}$ is expressed in terms of the Euler angles (α, β, γ) as (cf. the angular velocity in classical mechanics)

$$\left.\begin{aligned} \omega_x &= \gamma_t \cos\alpha \sin\beta - \beta_t \sin\alpha, \\ \omega_y &= \gamma_t \sin\alpha \sin\beta + \beta_t \cos\alpha, \\ \omega_z &= \gamma_t \cos\alpha + \alpha_t, \end{aligned}\right\} \tag{8.94}$$

where $\gamma_t = \partial\gamma/\partial t$ etc. The kinetic-energy density then takes the form

$$\begin{aligned} T = \frac{1}{2}\frac{\chi_{\mathrm{N}}}{\gamma^2}\{&\alpha_t^2 + \beta_t^2 + \gamma_t^2 - 2\alpha_t\omega_{\mathrm{L}} + 2(\alpha_t - \omega_{\mathrm{L}})\gamma_t \cos\beta \\ &- a[\gamma_t + (\alpha_t - \omega_{\mathrm{L}})\cos\beta]^2\}. \end{aligned} \tag{8.95}$$

We must now express $\hat{\boldsymbol{d}}$ in terms of α, β and γ. To this end, we generate the local vector $\hat{\boldsymbol{d}}$ by applying the local rotation matrix $\mathbf{R}(\alpha, \beta, \gamma)$, (7.60), to the unit vector in the direction of the magnetic field $\boldsymbol{H}$, i.e. to $\hat{\boldsymbol{z}}$. This yields

$$\hat{\boldsymbol{d}} = \mathbf{R}\hat{\boldsymbol{z}} \tag{8.96a}$$

$$= (\cos\alpha \sin\beta, \sin\alpha \sin\beta, \cos\beta). \tag{8.96b}$$

Comparing this with the usual parametrization of $\hat{\boldsymbol{d}}$ given by (7.149), we identify

$$\alpha = \psi, \quad \beta = \theta. \tag{8.97}$$

We may therefore replace α and β in (8.95) by ψ and θ. The angle γ, describing rotations *about* $\hat{\boldsymbol{d}}$, does not enter in (8.96b), and is clearly seen to be irrelevant since $\hat{\boldsymbol{d}}$ is a polar vector. In particular, γ does not appear in the potential V of the Lagrangian density $\mathscr{L}$, (8.89), where $V = V\{\psi; \theta\}$ is only a functional of the angles ψ and θ.

The equation of motion for $\hat{\boldsymbol{d}}$ is now given by the general Euler–Lagrange equations

$$\frac{\partial}{\partial t}\frac{\partial \mathscr{L}}{\partial \dot{\xi}}-\frac{\delta \mathscr{L}}{\delta \xi}=0, \tag{8.98a}$$

with

$$\frac{\delta \mathscr{L}}{\delta \xi_i}\equiv\frac{\partial \mathscr{L}}{\partial \xi_i}-\boldsymbol{\nabla}\cdot\frac{\partial \mathscr{L}}{\partial(\boldsymbol{\nabla}\xi_i)} \tag{8.98b}$$

and $\xi_i=\psi,\ \theta,\ \gamma$. Since $\mathscr{L}$ depends on γ only through γ_t in T, γ is a cyclic variable. The Euler–Lagrange equation (8.98) for γ yields a constant of motion

$$P\equiv\gamma_t+(\psi_t-\omega_{\mathrm{L}})\cos\theta, \tag{8.99}$$

which may be determined from the equilibrium configuration $\psi=\psi_0$, $\theta=\theta_0$ with $\psi_t=\theta_t=\gamma_t=0$ as

$$P=-\omega_{\mathrm{L}}\cos\theta_0. \tag{8.100}$$

Eliminating γ_t from T, (8.95), the appropriate new kinetic energy is given by the Legendre transform (see the discussion leading to (7.230))

$$\begin{aligned}T^*&=T-\gamma_t\frac{\partial T}{\partial \gamma_t}\\&=\frac{1}{2}\frac{\chi_{\mathrm{N}}}{\gamma^2}[(\psi_t-\omega_{\mathrm{L}})^2\sin^2\theta+\theta_t^2+2(1-a)P(\psi_t-\omega_{\mathrm{L}})\cos\theta]+\text{const.}\end{aligned} \tag{8.101}$$

The correct Lagrangian, now being a functional of ψ and θ only, is then given by (8.89) with T replaced by T^*, i.e. the spin motion is described completely in terms of ψ and θ.

Introducing fluctuations around the equilibrium position of $\hat{\boldsymbol{d}}$, i.e. longitudinal fluctuations

$$f=\psi-\psi_0 \tag{8.102a}$$

and transverse fluctuations

$$g=\theta-\theta_0, \tag{8.102b}$$

both T^* and V are expressed in terms of f and g. The linearized equations of motion for f and g are then found as (Vollhardt *et al.* 1981)

$$f_{tt}\sin^2\theta_0-\tfrac{1}{2}g_t\omega_{\mathrm{L}}(1+a)\sin 2\theta_0=\frac{\gamma^2}{\chi_{\mathrm{N}}}\frac{\delta V\{f;g\}}{\delta f}, \tag{8.103a}$$

$$g_{tt}+g\omega_{\mathrm{L}}^2[1-(1+a)\cos^2\theta_0]+\tfrac{1}{2}f_t\omega_{\mathrm{L}}(1+a)\sin 2\theta_0=\frac{\gamma^2}{\chi_{\mathrm{N}}}\frac{\delta V^*\{f;g\}}{\delta g}. \tag{8.103b}$$

where

$$\frac{\delta V^*}{\delta g}=\frac{\delta V}{\delta g}+\tfrac{1}{2}\,\Delta\chi\,H^2\sin 2\theta_0. \tag{8.103c}$$

The second term on the right-hand side of (8.103c) is a force term, which is independent of the amplitude of the fluctuation. It corresponds to $[-\partial\Delta f_H(\theta_0+g)/\partial g]_{g=0}$, where $\Delta f_H(\theta)=\frac{1}{2}\Delta\chi H^2\cos^2\theta$ is the magnetic orientation energy of the vector $\hat{\boldsymbol{d}}$, (6.115). This constant term is cancelled by contributions originating from $\delta V/\delta g$ since θ_0, the equilibrium value of the angle θ of the vector $\hat{\boldsymbol{d}}$, is determined by the Euler–Lagrange equation

$$\frac{\delta}{\delta\theta}(V+\Delta f_H)=\frac{\delta V}{\delta g}\bigg|_{g=0}-\frac{\partial(\Delta f_H)}{\partial g}\bigg|_{g=0}=0, \tag{8.103d}$$

where again $\theta=\theta_0+g$.

In general the two equations (8.103a,b) for the longitudinal and transverse fluctuations are coupled. This implies that in a longitudinal NMR experiment transverse fluctuations are excited, and vice versa. However, in the case of a strong external magnetic field $\boldsymbol{H}$ the vector $\hat{\boldsymbol{d}}$ is forced into the plane perpendicular to $\boldsymbol{H}$, so that $\theta_0\rightarrow\frac{1}{2}\pi$; thereby the constant of motion P, (8.100), is made to vanish. Then the two equations reduce to

$$f_{tt}=\frac{\gamma^2}{\chi_{\rm N}}\frac{\delta V}{\delta f}, \tag{8.104a}$$

$$g_{tt}-\omega_{\rm L}g=\frac{\delta V^*}{\delta g}, \tag{8.104b}$$

i.e. they decouple. Stability of the equilibrium $\hat{\boldsymbol{d}}$ texture implies that the potentials V and V^* are quadratic in f and g (see the discussion below (8.103b)). We may therefore express $\delta V/\delta f$ and $\delta V^*/\delta g$ as

$$\frac{\delta V}{\delta f}\equiv-(K_1\Delta_0^2\xi_{\rm D}^{-2})\Lambda^f f, \tag{8.105a}$$

$$\frac{\delta V^*}{\delta g}\equiv-(K_1\Delta_0^2\xi_{\rm D}^{-2})\Lambda^g g, \tag{8.105b}$$

where Λ^f and Λ^g are dimensionless differential operators acting on f and g respectively. Using $K_1\Delta_0^2=(\chi_{\rm N}/\gamma^2)\Omega_{\rm A}^2\xi_{\rm D}^2$ (see (7.18a), (7.27), (8.20b)), (8.104) then takes the form

$$f_{tt}=-\Omega_{\rm A}^2\Lambda^f f, \tag{8.105c}$$

$$g_{tt}-\omega_{\rm L}^2 g=-\Omega_{\rm A}^2\Lambda^g g. \tag{8.105d}$$

In a typical NMR experiment the fluctuations f and g are excited as periodic oscillations

$$f=\mathrm{e}^{-\mathrm{i}\omega_\ell t}\tilde{f}(\boldsymbol{r}), \tag{8.106a}$$

$$g=\mathrm{e}^{-\mathrm{i}\omega_{\rm t}t}\tilde{g}(\boldsymbol{r}). \tag{8.106b}$$

The longitudinal and transverse resonant frequencies ω_ℓ and $\omega_{\rm t}$ are then determined by the eigenvalue equations obtained from (8.105c,d),

$$\Lambda^f f_k = \lambda_{\ell,k} f_k, \tag{8.107a}$$

$$\Lambda^g g_k = \lambda_{t,k} g_k, \tag{8.107b}$$

with the eigenvalues

$$\lambda_{\ell,k} = \left(\frac{\omega_{\ell,k}}{\Omega_A}\right)^2, \quad \lambda_{t,k} = \frac{\omega_{t,k}^2 - \omega_L^2}{\Omega_A^2}. \tag{8.108a}$$

Here the subscript k denotes the kth spin-wave mode. In most NMR investigations of nonuniform textures only the lowest mode ($k=0$) is detected. In the following we shall drop the subscript k in the eigenvalues $\lambda_{\ell,k}$ and $\lambda_{t,k}$ and resonant frequencies $\omega_{\ell,k}$ and $\omega_{t,k}$, unless stated otherwise, and write

$$\lambda_\ell = \left(\frac{\omega_\ell}{\Omega_A}\right)^2, \quad \lambda_t = \frac{\omega_t^2 - \omega_L^2}{\Omega_A^2}. \tag{8.108b}$$

In the bulk liquid $\lambda_\ell = \lambda_t = 1$. However, in the presence of a nonuniform texture the eigenvalues decrease and thus determine the shift of the NMR signal from the bulk values.

In general, the structure of the differential operators Λ^f and Λ^g is too complicated to allow an exact analytic solution of (8.107). In this case a variational approach, employing variational eigenfunctions f and g, allows a convenient approximation of the lowest eigenmode. The eigenvalues are then obtained by

$$\lambda_\ell \leqslant \frac{\int d^3r\, f^* \Lambda^f f}{\int d^3r\, |f|^2}, \quad \lambda_t \leqslant \frac{\int d^2r\, g^* \Lambda^g g}{\int d^3r\, |g|^2}. \tag{8.109}$$

Hamiltonian approach

The spin dynamics of the A phase with nonuniform textures may equally be derived within a Hamiltonian approach, in which the Leggett equations (8.8a,b) are generalized to account for nonuniform textures (Brinkman and Cross 1978, Buchholtz 1978, Bromley 1980, Fetter 1983, Vulovic *et al.* 1984, Ullah 1988). This is achieved by adding the spin part of the bending free-energy density (7.20) to the effective Hamiltonian (8.1).

Using the definition of the dipole healing length ξ_D given in (7.27) and the expression (8.20) for the longitudinal resonant frequency Ω_A, this effective Hamiltonian takes the form

$$H_{\rm eff} = \tfrac{1}{2}\gamma^2 \boldsymbol{S}\chi^{-1}\boldsymbol{S} - \gamma \boldsymbol{S}\cdot\boldsymbol{H}$$
$$- \tfrac{1}{2}\Omega_A^2 \frac{\chi}{\gamma^2}\left\{(\hat{\boldsymbol{d}}\cdot\hat{\boldsymbol{l}})^2 - \xi_D^2 \bar{K}_{\rm sp\parallel} \sum_\mu [(\hat{\boldsymbol{l}}\cdot\nabla)\hat{d}_\mu]^2 - \xi_D^2 \bar{K}_{\rm sp\perp} \sum_{i\mu} [(\hat{\boldsymbol{l}}\times\nabla)_i \hat{d}_\mu]^2\right\}, \tag{8.110}$$

where $\bar{K}_{\mathrm{sp}\parallel,\perp}$ are dimensionless constants of order unity, given in terms of the components of the superfluid spin density tensor by $\bar{K}_{\mathrm{sp}\parallel,\perp} = (\hbar/2m)^2\rho_{\mathrm{sp}\parallel,\perp}(\gamma^2/\Omega_{\mathrm{A}}^2\chi\xi_{\mathrm{D}}^2)$. By employing (8.6), we then find that the equation of motion for the spin density $\boldsymbol{S}(\boldsymbol{r})$, given by (8.8a) for the uniform situation, acquires the following gradient terms:

$$\dot{\boldsymbol{S}} = \gamma\boldsymbol{S}\times\boldsymbol{H} + \tfrac{1}{2}\Omega_{\mathrm{A}}^2\frac{\chi}{\gamma^2}\{(\hat{\boldsymbol{d}}\times\hat{\boldsymbol{l}})(\hat{\boldsymbol{d}}\cdot\hat{\boldsymbol{l}}) + \xi_{\mathrm{D}}^2\bar{K}_{\mathrm{sp}\parallel}[\hat{\boldsymbol{d}}\times(\hat{\boldsymbol{l}}\cdot\nabla)^2\hat{\boldsymbol{d}}] + \xi_{\mathrm{D}}^2\bar{K}_{\mathrm{sp}\perp}[\hat{\boldsymbol{d}}\times(\hat{\boldsymbol{l}}\times\nabla)^2\hat{\boldsymbol{d}}]\}. \quad (8.111)$$

The equation of motion for $\hat{\boldsymbol{d}}$, (8.8b), remains unchanged, up to terms of the order of the dipole coupling λ_{D}.

The linear spin dynamics is obtained by expanding the vector $\hat{\boldsymbol{d}}$ about the equilibrium texture $\hat{\boldsymbol{d}}_0$, $\hat{\boldsymbol{d}} = \hat{\boldsymbol{d}}_0 - \boldsymbol{d}'$, and linearizing in the fluctuating part $\boldsymbol{d}'$. The latter may be expressed as

$$\boldsymbol{d}' = g(\hat{\boldsymbol{d}}_0\times\hat{\boldsymbol{z}}) + f\hat{\boldsymbol{z}}, \quad (8.112)$$

i.e. $\boldsymbol{d}'$ is composed of a transverse part and a longitudinal part relative to $\boldsymbol{H}$, described by the fluctuations g and f respectively, in direct analogy to the discussion above (see (8.102)). The fluctuating part $\boldsymbol{d}'$ will be excited by a weak RF field $\boldsymbol{H}\mathrm{e}^{-\mathrm{i}\omega t} = (\boldsymbol{H}'_\perp + H'_z\hat{\boldsymbol{z}})\mathrm{e}^{-\mathrm{i}\omega t}$ where $\boldsymbol{H}'_\perp = \boldsymbol{H}'\times\hat{\boldsymbol{H}}$ is the component perpendicular to the static field $\boldsymbol{H}\parallel\hat{\boldsymbol{z}}$. The equations of motion for $\boldsymbol{S}$ and $\hat{\boldsymbol{d}}$, (8.111) and (8.8b), are now combined by eliminating $\boldsymbol{S}$. This leads to two coupled linear inhomogeneous equations for f and g.

In the case of a strong static external magnetic field with $\omega_{\mathrm{L}} \gg \Omega_{\mathrm{A}}$ the equilibrium vector $\hat{\boldsymbol{d}}_0$ is confined to the (x, y) plane ($\theta_0 = \frac{1}{2}\pi$)

$$\hat{\boldsymbol{d}}_0 = \cos\psi_0\,\hat{\boldsymbol{x}} + \sin\psi_0\,\hat{\boldsymbol{y}}, \quad (8.113)$$

in which case the two equations decouple (cf. the discussion below (8.103d)). In particular, the equation for g, relevant in the case of transverse NMR, becomes (Fetter 1983, 1986)

$$(\mathrm{D} + V_{\mathrm{t}})g = \frac{1}{\Omega_{\mathrm{A}}^2}\{[\omega_{\mathrm{t}}^2 - (\omega_{\mathrm{L}}^2 + \Omega_{\mathrm{A}}^2)]g - \mathrm{i}\omega_{\mathrm{t}}\gamma H'_\perp - \omega_{\mathrm{L}}\gamma H'_z\}, \quad (8.114)$$

where D is a differential operator defined by

$$\mathrm{D} = -\xi_{\mathrm{D}}^2[\bar{K}_{\mathrm{sp}\perp}\nabla^2 + (\bar{K}_{\mathrm{sp}\parallel} - \bar{K}_{\mathrm{sp}\perp})(\nabla\cdot\hat{\boldsymbol{l}})(\hat{\boldsymbol{l}}\cdot\nabla)] \quad (8.115)$$

and

$$V_{\mathrm{t}} = -\{(\hat{\boldsymbol{l}}\times\hat{\boldsymbol{d}}_0)^2 + l_z^2 + \xi_{\mathrm{D}}^2[\bar{K}_{\mathrm{sp}\perp}(\hat{\boldsymbol{l}}\times\nabla\psi)^2 + \bar{K}_{\mathrm{sp}\parallel}(\hat{\boldsymbol{l}}\cdot\nabla\psi)^2]\} \quad (8.116)$$

Equation (8.114) resembles the Schrödinger equation for a particle moving in an attractive potential V_{t}. The general solution may be expressed in terms of the eigenstates $\psi_{\mathrm{t},k}$ and the ("energy") eigenvalues $E_{\mathrm{t},k}$ of the homogeneous problem:

$$(\mathrm{D} + V_{\mathrm{t}})\psi_{\mathrm{t},k} = E_{\mathrm{t},k}\psi_{\mathrm{t},k}. \quad (8.117)$$

Note that the "local-oscillator" model is obtained from (8.117) in the limit where ξ_D is much smaller than the spatial variation of the texture, such that $D \to 0$. The eigenvalues $E_{t,k}$ describe the frequency *shift* of the kth transverse spin-wave mode,

$$\omega_{t,k}^2 = \omega_L^2 + (1 + E_{t,k})\Omega_A^2, \tag{8.118}$$

relative to the resonance of the uniform bulk liquid, for which $E_t = 0$. Note that $E_{t,k}$ is related to the eigenvalue $\lambda_{t,k}$, introduced in (8.107b) and (8.108b), by

$$E_{t,k} = \lambda_{t,k} - 1. \tag{8.119}$$

Since the potential V_t is attractive, the lowest eigenvalues $E_{t,k}$ are negative, corresponding to bound states. The corresponding frequency shift is negative too.

A further simplification of (8.117) is achieved by assuming $\hat{\boldsymbol{d}}_0$, (8.113), to be uniform and along $\hat{\boldsymbol{x}}$, say (Fetter 1983, 1986). In the case of a strong magnetic field $\boldsymbol{H} \parallel \hat{\boldsymbol{z}}$ considered here, this approximation can be expected to be obeyed rather well throughout most of the superfluid, except for regions of extension $r \leqslant \max(\xi_D, \xi_H)$ where $\hat{\boldsymbol{l}}$ may deviate from its asymptotic uniform configuration (for example in the soft core of the high-field vortex (Volovik and Hakonen 1981) in the rotating superfluid). Then the potential V_t, (8.116), reduces to

$$V_t = -(l_y^2 + 2l_z^2). \tag{8.120}$$

A convenient approximation for the lowest eigenvalue ($k = 0$) is provided by the variational expression

$$\lambda_t \leqslant \frac{\xi_D^2 \int d^3r\, [\bar{K}_{sp\perp} |\boldsymbol{\nabla}\Psi|^2 + (\bar{K}_{sp\parallel} - \bar{K}_{sp\perp}) |\hat{\boldsymbol{l}} \cdot \boldsymbol{\nabla}\Psi|^2 + \xi_D^{-2} V_t |\Psi|^2]}{\int d^3r\, |\Psi|^2}, \tag{8.121}$$

where the test functions Ψ are normalizable functions. The higher eigenstates are usually difficult to detect.

The longitudinal shifts may be derived similarly (Vulovic *et al.* 1984). The homogeneous effective Schrödinger equation has the form

$$(D + V_\ell)\psi_{\ell,k} = E_{\ell,k}\psi_{\ell,k}, \tag{8.122}$$

where now (with the approximation $\hat{\boldsymbol{d}}_0 = \hat{\boldsymbol{x}}$)

$$V_\ell = -(2l_y^2 + l_z^2) \tag{8.123}$$

and $E_{\ell,k}$ is related to $\lambda_{\ell,k}$, introduced in (8.107a) and (8.108a), by

$$E_{\ell,k} = \lambda_{\ell,k} - 1. \tag{8.124}$$

In the uniform bulk liquid the resonance is unshifted, so that $E_\ell = 0$. The theory described above will now be applied to determine the NMR response of several characteristic non-uniform textures in the A phase.

Parallel-plate geometry

The $\hat{\boldsymbol{l}}$ texture in a slab extending from $z=-\frac{1}{2}L$ to $z=\frac{1}{2}L$, subjected to a magnetic field in the z direction normal to the walls, has been discussed in Section 7.9 In particular, for a strong magnetic field, $\hat{\boldsymbol{d}}$ lies in the plane of the slab, say along $\hat{\boldsymbol{x}}$. The resulting nontrivial $\hat{\boldsymbol{l}}$ texture for $L > \xi_D$ (see Fig. 7.30) is planar, i.e. parallel to the (x, z) plane. The detailed form of the texture has been determined approximately by a variational method (Bruinsma and Maki 1979a), i.e. assuming the form

$$\sin\beta = k\,\mathrm{sn}\,(\eta z \mid k), \tag{8.125}$$

where $\beta(z)$ is the angle made by $\hat{\boldsymbol{l}}$ with the surface normal, sn is a Jacobian elliptic function (Abramowitz and Stegun 1965) with $k=\sin\beta(0)$, and η is a variational parameter; $\beta(0)$ and η are related by the condition $K(k)=0$, where $K(k)$ is a complete elliptic integral.

To determine the NMR shifts caused by the texture, we have to obtain the operators Λ^f and Λ^g in (8.107). This is done by expanding (7.183) in terms of fluctuations f and g of the vector

$$\hat{\boldsymbol{d}} = (\sin(\theta_0+g)\cos(\psi_0+f), \sin(\theta_0+g)\sin(\psi_0+f), \cos(\theta_0+g)) \tag{8.126a}$$

$$= \hat{\boldsymbol{d}}_0 - (\tfrac{1}{2}(f^2+g^2), f, -g) \tag{8.126b}$$

around the equilibrium texture $\hat{\boldsymbol{d}}_0=\hat{\boldsymbol{x}}$, i.e. $\theta_0=\frac{1}{2}\pi$, $\psi_0=0$. The operators Λ^f and Λ^g are thus found as

$$\Lambda^f = -\xi_D^2\frac{\mathrm{d}}{\mathrm{d}z}\left[(1+\sin^2\beta)\frac{\mathrm{d}}{\mathrm{d}z}\right]+\sin^2\beta, \tag{8.127a}$$

$$\Lambda^g = -\xi_D^2\frac{\mathrm{d}}{\mathrm{d}z}\left[(1+\sin^2\beta)\frac{\mathrm{d}}{\mathrm{d}z}\right]-\cos 2\beta. \tag{8.127b}$$

Even without a detailed treatment, the qualitative behaviour of the frequency eigenvalues λ_ℓ and λ_t, (8.108b), may easily be understood: for large plate separation ($L \gg \xi_D$), the NMR frequencies approach their bulk values, i.e. $\lambda_\ell=\lambda_t=1$. As L is decreased, the eigenvalues decrease, too. At $L\approx\xi_D$ the $\hat{\boldsymbol{l}}$ texture resembles the so-called "bending $\hat{\boldsymbol{l}}$ soliton" (Maki and Kumar 1977d), for which $\lambda_\ell=0.116$ and $\lambda_t=0.687$. For even smaller L ($L \ll \xi_D$), when the uniform configuration $\hat{\boldsymbol{l}}\perp\hat{\boldsymbol{d}}$ is assumed, the eigenvalues approach the limiting values $\lambda_\ell=0$ and $\lambda_t=-1$. The above problem has also been considered by Fetter (1977), who used a different approximation for calculating the $\hat{\boldsymbol{l}}$ texture and who determined the NMR response in the local-oscillator approximation, i.e. where the local spin precession is assumed to be given by decoupled local oscillators (see the beginning of Section 8.4).

A detailed discussion of the NMR response of superfluid ^{3}He confined in very thin slabs (width of the order of the coherence length $\xi(T)$) has been presented by Ullah (1988), who finds that NMR can be a very sensitive probe of the order parameter in such a geometry.

Cylindrical geometry

The energetically most favourable $\hat{\boldsymbol{l}}$ textures in a long narrow cylinder have been discussed in Section 7.9. There it was found that, for radii $R \gtrsim 10\,\xi(T)$, the Mermin–Ho (MH) texture, parametrized by (7.191), is stable. Only for very narrow cylinders ($R \lesssim 10\xi$) is a disgyration-type texture of $\hat{\boldsymbol{l}}$, (7.197) or (7.199), energetically favoured. Their NMR signals have been calculated by Bruinsma and Maki (1978), as will be discussed below (see also Fishman 1978).

In the following we restrict the discussion to the MH texture in the dipole-locked regime. The corresponding bending free energy, in terms of the angle χ made by $\hat{\boldsymbol{l}} = \hat{\boldsymbol{d}}$ with the cylinder axis, was derived in (7.192). The exact solution $\chi(\rho)$ is given by (7.193), where $\rho = \alpha r/R$. Here r is the distance from the cylinder axis and $\alpha = [2/(3-5^{1/2})]^2$ in weak coupling. The azimuthal angle is denoted by φ. The operators Λ^f and Λ^g in (8.107) for this case are given by

$$\Lambda^f = -\alpha^2\left(\frac{\xi_D}{R}\right)^2\left\{\frac{1}{r\sin^2\chi}\frac{\partial}{\partial r}\left[r(1+\cos^2\chi)\sin^2\chi\frac{\partial}{\partial r}\right]+\frac{2}{r^2}\frac{\partial^2}{\partial\varphi^2}\right\}+1, \tag{8.128a}$$

$$\Lambda^g = -\alpha^2\left(\frac{\xi_D}{R}\right)^2\left\{\frac{1}{r}\frac{\partial}{\partial r}\left[r(1+\cos^2\chi)\frac{\partial}{\partial r}\right]+\frac{2}{r^2}\left[\frac{\partial^2}{\partial\varphi^2}+\cos 2\chi\right]\right\}+1. \tag{8.128b}$$

For symmetry reasons, the longitudinal fluctuation f must be radially symmetric, whereas the transverse fluctuation depends on φ according to $g(\rho,\varphi) = \tilde{g}(\rho)e^{\pm i\varphi}$. The boundary conditions on f and g at the surface of the cylinder, where $\chi = \frac{1}{2}$, may be obtained by integrating (8.107) over an infinitesimal interval about the surface, with the result that

$$\left.\frac{\partial f}{\partial\rho}\right|_{r=R} = \left.\frac{\partial g}{\partial\rho}\right|_{r=R} = 0. \tag{8.129}$$

The lowest eigenvalue of the equation for f is obtained for $f = \text{const}$, i.e.

$$\lambda_\ell = 1. \tag{8.130a}$$

The lowest eigenvalue of the operator Λ^g may be determined variationally and is found as

$$\lambda_t = 1 + \eta\left(\frac{\xi_D}{R}\right)^2, \tag{8.130b}$$

with $\eta = 4.14$. The corresponding values for the radial disgyration (7.197) and the "double-half" disgyration (7.199) are also given by (8.130a,b) with $\eta = 5.56$ and $\eta = 2.44$ respectively.

While the preceding derivation holds for arbitrary values of ξ_D/R because of the scale invariance of the MH texture, the assumption of dipole-locking has to be abandoned for $\xi_D/R \lesssim 1$ or for a layer of thickness ξ_D near the wall of the cylinder. One may estimate the latter contribution for a wide cylinder ($R > \xi_D$) by noting that the maximum angle between $\hat{\boldsymbol{l}}$ and $\hat{\boldsymbol{d}}$ at the surface is of order ξ_D/R. Therefore the dipole energy density in this layer is of order $(\xi_D/R)^2$ and the contribution to the eigenvalue obtained by integrating over the cross-section is of order $(\xi_D/R)^3$.

In the opposite limit ($R \ll \xi_D$) the $\hat{\boldsymbol{d}}$ texture is uniform. One may then argue that the longitudinal resonance frequency for a radially symmetric $\hat{\boldsymbol{l}}$ texture is *zero* (Gould 1979). Indeed, for a spatially independent fluctuation f, the operator Λ^f, (8.105a), is determined solely by the potential provided by the dipole interaction (6.104), so that

$$\Lambda^f = \lambda_\ell = l_x^2 - l_y^2, \tag{8.131}$$

where l_x and l_y are the components of $\hat{\boldsymbol{l}}$ in the plane perpendicular to the magnetic field, with $\hat{\boldsymbol{d}} \parallel \hat{\boldsymbol{x}}$. In the local-oscillator model the *average* of this frequency shift over the cross-section of the cylinder therefore vanishes. On the other hand, the transverse shift for cylinders with $R = 1\ \mu\text{m}$ is found as $\lambda_t = 0.20$ (Bruinsma and Maki 1979d).

Experimental NMR studies on narrow cylinders have been performed by Saunders *et al.* (1978) and Gould and Lee (1978), with different results: while the first group observed temperature-independent frequency ratios and a longitudinal resonance, the second did not. It has been suggested that the discrepancy between the two measurements might be due to the presence of persistent currents in the cylinders of the first group (Truscott 1979).

NMR response of the helical texture

As discussed in Section 7.10, the uniform $\hat{\boldsymbol{l}}$ and $\hat{\boldsymbol{d}}$ texture is unstable against the formation of a helical texture within a certain range of superfluid currents and magnetic fields. Apart from the sound-attenuation experiments described in this context, NMR experiments should provide a unique probe to this texture, since the chiral symmetry of the direction of superflow is broken. Therefore the transverse resonant frequency splits into two branches depending on the polarization of the applied RF field.

The NMR frequencies of the helical texture in an open system and their dependence on the external superflow and magnetic field $\boldsymbol{H} \parallel \boldsymbol{v}_s$ have been calculated by Vollhardt *et al.* (1979, 1981). In an open geometry the wave vector k of the helix may adjust so as to minimize the free energy. The case where k is not free to change (e.g. in a toroidal geometry) has been addressed by Bromley (1980), who derived the NMR frequencies in terms of the orientations of $\hat{\boldsymbol{l}}$ and $\hat{\boldsymbol{d}}$.

In the helical texture the equilibrium $\hat{\boldsymbol{d}}$ vector is characterized by the angles $\theta_0 = \text{const}$ and $\psi_0 = kz$ (see (7.235)); in the following the subscript zero on θ_0 and ϕ_0 will be dropped. As discussed by Vollhardt *et al.* (1979, 1981), the equations of motion for the fluctuations f and g around ψ and θ are given by (8.103a,b). The right-hand side of (8.103a,b), with $V = \mathcal{G}$ given by (7.232), is obtained as

$$\frac{\gamma^2}{\chi_N}\frac{\delta V}{\delta f} = \Omega_A^2[f_{ss}\sin^2\theta\,(1+\sin^2\beta) - f\sin\theta\sin\beta\cos(\theta-\beta) + g_s\bar{k}\sin 2\theta\,(1+\sin^2\beta)], \tag{8.132a}$$

$$\frac{\gamma^2}{\chi_N}\frac{\delta V^*}{\delta g} = \Omega_A^2\{g_{ss}(1+\sin^2\beta) - g[\bar{k}^2\cos 2\theta\,(1+\sin^2\beta) + \cos 2(\theta-\beta)] - f_s\bar{k}\sin 2\theta\,(1+\sin^2\beta)\}, \tag{8.132b}$$

with $\bar{k} = k\xi_D$ and $s = z/\xi_D$. A spin-wave solution of the form

$$f,\ g \propto e^{i(qz-\omega t)} \tag{8.133}$$

is determined by the dispersion relation

$$(\omega^2 - L_{11}\Omega_A^2)\{\omega^2 - \omega_L^2[1-(1+a)\cos^2\theta] - L_{22}\Omega_A^2\} - [\omega\omega_L(1+a) - L_{12}\Omega_A^2]^2\cos^2\theta = 0, \tag{8.134}$$

where ($\bar{q} = q\xi_D$)

$$\left.\begin{aligned} L_{11} &= \frac{\sin\beta\cos(\beta-\theta)}{\sin\theta} + (1+\sin^2\beta)\bar{q}^2, \\ L_{22} &= \frac{\sin 2\beta}{\sin 2\theta} + h^2\cos 2\theta + (1+\sin^2\beta)\bar{q}^2, \\ L_{12} &= 2\bar{k}\bar{q}(1+\sin^2\beta). \end{aligned}\right\} \tag{8.135}$$

An equation similar to (8.134) was derived by Bromley (1980) for flow in a torus, where k is quantized by the boundary conditions, so that $\mathscr{G}$ cannot be minimized with respect to k.

In the vicinity of the phase boundary between the uniform texture (region I in Fig. 7.35) and the helical texture (region II), the angles β and θ are small and (8.134) can be solved exactly:

$$\omega = [\tfrac{1}{4}\,\omega_L^2(1+a)^2 + \Omega_A^2(r_\ell + \bar{q}^2 \mp 2\bar{k}\bar{q})]^{1/2} \pm \tfrac{1}{2}\omega_L(1+a), \tag{8.136}$$

where $r_\ell = 1 + \bar{k}^2 - h^2$ is the ratio of β and θ. Making use of the fact that the longitudinal RF field couples to the $q = 0$ mode, while the transverse RF field couples to $q = \pm k$, one obtains from (8.136) two longitudinal resonant frequencies

$$\omega_{\ell,\pm} = [r_\ell\Omega_A^2 + \tfrac{1}{4}\omega_L^2(1+a)^2]^{1/2} \pm \tfrac{1}{2}\omega_L(1+a) \tag{8.137}$$

and four transverse resonant frequencies

$$\left.\begin{aligned} \omega_{t_1,\pm} &= [r_t\Omega_A^2 + \tfrac{1}{4}\omega_L^2(1+a)^2]^{1/2} \pm \tfrac{1}{2}\omega_L(1+a), \\ \omega_{t_2,\pm} &= [\Omega_A^2 + \tfrac{1}{4}\omega_L^2(1-a)^2]^{1/2} \pm \tfrac{1}{2}\omega_L(1+a), \end{aligned}\right\} \tag{8.138}$$

i.e. two for each helicity; here $r_t = r_\ell + 3\bar{k}^2$. To calculate the six resonant frequencies in the full stability region of the helix, (8.134) has to be solved (Vollhardt *et al.* 1981). The resulting frequencies give a distinct signature of the helical texture as a function of magnetic field for given superflow.

The NMR response of the elliptically distorted helix, which may be stabilized if the magnetic field is perpendicular to the flow ($\boldsymbol{H} \perp \boldsymbol{v}_s$, see the discussion in Section 7.10), has been determined by Lin-Liu *et al.* (1979) along the phase boundary to the uniform texture.

8.4.4 Dynamics of solitons in the A phase

The term "soliton" was introduced into the field of superfluid ^{3}He by Maki and Tsuneto (1975) following their investigation of the nonlinear longitudinal NMR in the A phase. In this problem the relative phase $\phi = 2\psi$ between the up- and down-spin population oscillates ("internal Josephson effect", see (8.72)); here ψ is the angle between $\hat{\boldsymbol{d}}$ and $\hat{\boldsymbol{l}}$ in the plane perpendicular to $\boldsymbol{H}$. The equation of motion for ϕ may be derived from the Lagrangian L, (8.88), given in terms of (8.101) and (7.22d), (6.104) with $\hat{\boldsymbol{l}} \parallel \hat{\boldsymbol{x}}$. For a purely one-dimensional variation of ϕ (along $\hat{\boldsymbol{x}}$, say) and neglecting dissipation, the Euler–Lagrange equation for $\phi(x, t)$ is

$$\frac{\partial^2 \phi}{\partial t^2} - c_{\text{sp}}^2 \frac{\partial^2 \phi}{\partial x^2} = -\Omega_{\text{A}}^2 \sin \phi, \tag{8.139}$$

where $c_{\text{sp}} = \Omega_{\text{A}} \xi_{\text{D}}$ is the spin-wave velocity; see (11.20). This is the socalled "sine–Gordon" equation, which occurs in many areas in physics (see e.g. Whitham 1974, Maki 1982). Equation (8.139) has several periodic solutions, one of which represents a "soliton" (also called "kink", "hump" etc. in other areas of physics), i.e. a spatially confined solution

$$\tan \tfrac{1}{4}\phi = \mathrm{e}^{-u} \tag{8.140}$$

with

$$u = \frac{1}{\Omega_{\text{A}}} \frac{x - vt}{(c_{\text{sp}}^2 - v^2)^{1/2}}, \quad v < c_{\text{sp}}, \tag{8.141}$$

where v is an arbitrary velocity.

This describes a single peak moving in the x direction. There are infinitely many solutions to (8.139), since $\phi \to \phi + 2\pi N$, $N \in \mathbb{Z}$, also solves (8.139). The properties of solitons, in particular their topologicical stability, have been discussed in detail in Section 7.8.

The spin dynamics of ^{3}He-A is therefore seen to allow spatially confined magnetic excitations (Maki and Tsuneto 1975, Maki and Ebisawa 1976c, Maki and Kumar 1976a,b). These are propagating domain walls in the $\hat{\boldsymbol{d}}$ field where the orientation of $\hat{\boldsymbol{d}}$ changes from $\hat{\boldsymbol{d}} \parallel \hat{\boldsymbol{l}}$ to $-\hat{\boldsymbol{d}} \parallel \hat{\boldsymbol{l}}$ as shown in Fig. 7.22(d). Although the other domain walls discussed in Section 7.8, where both $\hat{\boldsymbol{d}}$ and $\hat{\boldsymbol{l}}$ participate in the reorientation, do not obey (8.139) and hence are not solitons in the strict mathematical sense, they are nevertheless often referred to as "solitons" too, since they are also spatially confined topologically stable solutions of a nonlinear differential equation.

Creation and propagation of solitons

One expects that pure $\hat{\boldsymbol{d}}$ solitons are created in a typical field turn-off or pulse experiment (Maki and Kumar 1976b, Kurkijärvi 1978): in such an experiment the vector $\hat{\boldsymbol{d}}$ is made to rotate in the plane perpendicular to the static magnetic field *inside* the magnetic coil, while it remains static outside. The $\hat{\boldsymbol{d}}$ texture inside the coil is thereby given a finite winding number relative to the outside, as a consequence of which closely packed soliton-like excitations are created. The excitations will propagate with a velocity smaller than, but of the order of, the spin-wave velocity (see (8.141))—a result that has been verified experimentally by a "time-of-flight" experiment for solitons (Gould *et al.* 1980). At temperatures that are not too low the solitons are slowed down by interaction with thermal excitations. This effect may be taken into account by adding a spin-diffusion term to the equation of motion (Maki and Ebisawa 1976a). Once a soliton has come to rest after an estimated time of about 10^{-4} s, it starts to locally distort the $\hat{\boldsymbol{l}}$ texture (which was previously assumed to be uniform) via the dipolar coupling. This is energetically favourable since the $\hat{\boldsymbol{d}}$ texture is "stiffer" than the $\hat{\boldsymbol{l}}$ texture (see (7.21a)) and hence the $\hat{\boldsymbol{d}}$ texture wants to reduce its bending energy at the expense of the $\hat{\boldsymbol{l}}$ texture. It has been proposed that the array of high-density $\hat{\boldsymbol{d}}$ solitons is eventually transformed into a regular one-dimensional lattice of high-density *composite* solitons, (7.163), which finally expands into a dilute "soliton gas" owing to mutual repulsion of the domain walls (Maki and Kumar 1977a,b, 1978, Maki 1977b). In the composite soliton, which has a considerably lower energy than the $\hat{\boldsymbol{d}}$ soliton, (7.163), most of the reorientation from $\hat{\boldsymbol{d}} \parallel \hat{\boldsymbol{l}}$ to $\hat{\boldsymbol{d}} \parallel -\hat{\boldsymbol{l}}$ is done by the vector $\hat{\boldsymbol{l}}$. This reorientation is acted against by surface orientation effects and the presence of other solitons. These effects are difficult to estimate and therefore the explicit path for the decay of the $\hat{\boldsymbol{d}}$ soliton into the composite soliton remains an open question.

8.4.5 NMR signature of solitons in the A phase

As in Section 7.8, we assume a static magnetic field $\boldsymbol{H} \parallel \hat{z}$ to confine the equilibrium configuration of $\hat{\boldsymbol{d}}$ and $\hat{\boldsymbol{l}}$ to the (x, y) plane (see below (7.149)). We introduce small fluctuations f and g around the equilibrium angles of the soliton, $f = \psi - \psi_0$ and $g = \theta - \theta_0$, where f and g describe in-plane and out-of-plane oscillations, respectively. The weak-coupling expressions for the operators Λ^f and Λ^g, (8.105), valid for all temperatures and including Fermi-liquid corrections, have been derived by Maki and Kumar (1978) and Maki (1986a) on the basis of the free energy (7.153) with (7.20) and (7.23) as

$$\Lambda^f = -2\xi_D^2 \frac{d}{ds}\left[(1-\bar{a}^2)\frac{d}{ds}\right] + 1 - 2\sin^2(\alpha-\psi), \tag{8.142a}$$

$$\Lambda^g = -2\xi_D^2 \left\{\frac{d}{ds}\left[(1-\bar{a}^2)\frac{d}{ds}\right] + (1-\bar{a}^2)\psi_s^2\right\} + \cos^2(\alpha-\psi), \tag{8.142b}$$

where $s = \hat{\boldsymbol{k}} \cdot \boldsymbol{r}$, with $\hat{\boldsymbol{k}}$ the surface normal to the domain wall and $\bar{a}^2 = (1-\lambda)(\hat{k}_x \cos\alpha + \hat{k}_y \sin\alpha)^2$ with $\lambda = \rho_{sp\parallel}/\rho_{sp\perp}$.

For the pure $\hat{d}$ soliton, where $\alpha = 0$ and ψ is given by (7.157) with $\xi_D'^2 = 2[1-[1-\lambda)\hat{k}_x^2]\xi_D^2$ so that $\psi_s = (\xi_D')^{-1}\sin^2\psi$, (8.142a,b) further reduce to

$$\Lambda^f = \Lambda^g = -\xi_D'^2 \frac{d^2}{ds^2} + 1 - 2\,\mathrm{sech}^2\left(\frac{s}{\xi_D'}\right). \tag{8.143}$$

Hence the operators Λ^f and Λ^g are found to be *identical* in this case. The corresponding eigenvalue equations (8.107) have only one bound state, with the eigenvalues λ_ℓ and λ_t and the eigenfunctions f and g given by

$$\lambda_\ell = \lambda_t = 0, \tag{8.144a}$$

$$f = g \propto \mathrm{sech}^2\left(\frac{s}{\xi_d'}\right). \tag{8.144b}$$

Hence the $\hat{\boldsymbol{d}}$ soliton is seen to produce *no* longitudinal magnetic resonance, and the transverse response occurs at the Larmor frequency. This is a consequence of the assumed uniformity of the $\hat{\boldsymbol{l}}$ texture, over which the $\hat{\boldsymbol{d}}$ soliton may then slide freely.

For the twist composite soliton (7.162b), where the surface normal $\hat{\boldsymbol{k}}$ is along the external magnetic field $\boldsymbol{H}$ ($\bar{a} = 0$) and which has the lowest energy of all domain walls, the operators Λ^f and Λ^g are found from (8.142a,b) as

$$\Lambda^f = -2\xi_D^2 \frac{d^2}{dz^2} + 1 - 2\,\mathrm{sech}^2\,\eta z, \tag{8.145a}$$

$$\Lambda^g = -2\xi_D^2 \frac{d^2}{dz^2} + 1 - [1 + \tfrac{1}{2}(\eta\xi_D)^{-2}]\,\mathrm{sech}^2\,\eta z. \tag{8.145b}$$

Here we have made use of the general analytic solution for the twist composite soliton, valid at all temperatures (Maki and Kumar 1978),

$$u = \alpha + \frac{1}{\kappa}\psi = \text{const}, \tag{8.146a}$$

$$v = \alpha - \psi = 2\tan^{-1}(e^{\pm\eta z}), \tag{8.146b}$$

with

$$\eta = \left(\frac{1+\kappa}{2\kappa}\right)^{1/2} \xi_D^{-1}, \tag{8.147}$$

$$\kappa = \frac{1}{12}\frac{\rho_{s\parallel}^0}{\rho_{s\perp}^0}\left(1 - \frac{\frac{1}{3}F_1^a}{1+\frac{1}{3}F_1^a}\frac{\rho_{s\perp}^0}{\rho}\right)\left[4 + \frac{\rho_{s\perp}^0}{\rho_{s\parallel}^0} + \frac{F_1^s}{1+\frac{1}{3}F_1^s(1-\rho_{s\parallel}^0/\rho)}\frac{\rho_{s\parallel}^0}{\rho}\right]. \quad (8.148)$$

In the Ginzburg–Landau regime one has $\kappa \approx \frac{1}{4}$ and the solution (7.162b) is recovered.

Again the operators (8.142a,b) are found to have only a single bound state, characterized by the exact eigenfunctions (Maki 1986a)

$$f(z) \propto A_\nu \operatorname{sech}^\nu \eta z, \quad (8.149a)$$

$$g(z) \propto A_\mu \operatorname{sech}^\mu \eta z \quad (8.149b)$$

with eigenvalues

$$\lambda_\ell = \frac{1}{2\kappa}\{[(1+\kappa)(1+9\kappa)]^{1/2} - 1 - 3\kappa\}, \quad (8.150a)$$

$$\lambda_t = \frac{1}{1+\kappa}, \quad (8.150b)$$

where

$$\nu = \frac{1}{2}\left[\left(\frac{1+9\kappa}{1+\kappa}\right)^{1/2} - 1\right], \quad (8.151a)$$

$$\mu = \frac{\kappa}{1+\kappa}, \quad (8.151b)$$

$$A_\alpha = \left[\frac{\eta}{B(\frac{1}{2}, \alpha)}\right]^{1/2}, \quad (8.152)$$

with $B(x, y)$ the beta function. In particular, in the Ginzburg–Landau regime ($T \lesssim T_c$) the eigenvalues are given by

$$\lambda_\ell = \tfrac{1}{2}[(65)^{1/2} - 7] \approx 0.531, \quad (8.153a)$$

$$\lambda_t = \tfrac{4}{5}. \quad (8.153b)$$

With decreasing temperature, the parameter κ increases owing to the large Landau parameter F_1^s; at $T = 0$ it would be $\kappa = \frac{5}{12}(1+\frac{1}{5}F_1^s)/(1+\frac{1}{3}F_1^a) \approx 1$–2. The eigenvalues $\lambda_{\ell,t}$ therefore decrease with decreasing temperature. The temperature dependence of the longitudinal and transverse NMR modes of this soliton in the immediate vicinity of the transition into the A_1 phase has been investigated by Kharadze and Smith (1978).

The intensities of the satellite resonances are proportional to the squares of the integrated bound spin-wave functions. Hence the intensity corresponds roughly to that of the total volume of the domain wall, being given by the cross-section $\sigma(\hat{\boldsymbol{k}})$ times the thickness η^{-1}.

For other composite solitons, the eigenvalues for the satellite frequencies have to be obtained numerically. In particular, the NMR satellites of the splay composite soliton, with surface normal perpendicular to $\boldsymbol{H}$, are found to be characterized by $\lambda_\ell = 0.40$ and $\lambda_t = 0.67$ in the Ginzburg–Landau regime.

As discussed by Maki and Kumar (1978), the NMR satellites observed by Avenel *et al.* (1975) and Gould and Lee (1976) (see also Kokko *et al.* 1978a) may be interpreted as the signature of composite solitons, namely of the twist composite soliton in the case of the longitudinal resonance experiment and of the splay composite soliton in the transverse case. The existence of two different kinds of solitons in these two experimental situations is indeed made plausible by the fact that Gould and Lee (1976) rotated the stationary magnetic field, rather than the AC field, by 90° to perform the two experiments, thereby stabilizing the twist and splay composite solitons respectively as discussed below (7.163).

The soliton lattice

In a field-turn-off or tipping experiment a large number of domain walls in the $\hat{\boldsymbol{d}}$-vector field are expected to arise. A possible scenario for the decay of these $\hat{\boldsymbol{d}}$ solitons into an array of composite solitons has been discussed above. Assuming this lattice of domain walls to be one-dimensional, the spatial configuration can be calculated (see (7.165) for the twist composite soliton lattice). Since the soliton lattice is expected to expand as a result of the mutual repulsion of the solitons, the dependence of the frequencies of the NMR satellites and their intensities on the spacing of the domain walls gives important information (Bruinsma and Maki 1979c).

The operators Λ^f and Λ^g for this case are still determined by (8.142a,b), the equilibrium configuration being given by (7.165) for the twist composite soliton lattice. In the case of the period $L_0/\xi_D \to \infty$, the eigenvalue equation for the longitudinal resonance reduces to (8.145a), with λ_ℓ given by (8.150a). For $L_0/\xi_D \ll 1$, when $v = \alpha - \psi$ has the form (7.168), it becomes a Mathieu equation, which picks out the spin-wave mode with zero wave vector; in this limit $\lambda_\ell = 0$. For intermediate values of L_0/ξ_D, a variational ansatz for $f(z)$ may be employed to calculate λ_ℓ. A calculation of the intensity of the signal shows that for densely packed domain walls ($L_0/\xi_D \leqslant 1$), the signal is completely exhausted by the bound state. For $L_0 \geqslant 10\xi_D$, the intensity of the satellite peak begins to decrease. There is then an infinite series of scattering states contributing to the longitudinal resonance, all of which reduce to the Leggett frequency Ω_A in the limit $L_0/\xi_D \to \infty$. Similar behaviour is found for the transverse resonance, where λ_t increases from $\lambda_t = 0.5$ at $L_0/\xi_D \ll 1$ to the value given by (8.153b) for $L_0/\xi_D \gg 1$. Again, for $L_0 \geqslant 10\xi_D$, the intensity of the satellite peak begins to decline and the scattering states start to dominate.

A numerical analysis of a lattice of splay composite solitons yields a qualitatively similar result: λ_ℓ increases from 0.08 at small L_0 to the asymptotic value 0.40 as $L_0 \to \infty$, reaching the latter value at about $L_0 = 10\xi_D$. The same is true for $0.55 \leqslant \lambda_t \leqslant 0.67$.

Therefore one finds that both the resonant frequencies and the intensities of the satellite change with soliton separation L_0. A similar effect occurs when the spatial extension of a single soliton changes, for example during the process of soliton formation (Vollhardt and Kumar 1978).

As discussed by Bruinsma and Maki (1979c) and Nakahara and Maki (1983), the above results for the soliton lattice seem to explain the findings of the field-tipping experiment of Gould *et al.* (1980). Immediately after the magnetization has been tipped by a large angle, a transverse resonance with negative shift, consistent with a rotation of $\hat{\boldsymbol{d}}$ in the (x, y) plane (see (8.78)), was observed in this experiment. While this resonance disappeared on a timescale of order 10^{-2} s, a *new* resonance at frequency $\omega \approx \omega_L + 0.54\Omega_A^2/2\omega_L$ appeared with increasing intensity, which soon afterwards exhausted the total absorption. The frequency of this satellite was then found to increase with time, arriving at an asymptotic frequency consistent with a single splay composite soliton on a timescale of order seconds or minutes. This time evolution of the satellite frequency and the intensity is indeed what one would expect if the nonequilibrium precession of $\hat{\boldsymbol{d}}$ transformed into a dense splay composite soliton lattice, which then expanded in space until the separation was large enough for the concept of separate solitons to apply. In a similar experiment by Giannetta *et al.* (1981) the comparison with theory suggests that an isolated composite soliton has been observed (Kharadze and Maki 1982, Nakahara and Maki 1983). This may be a result of the time lapse between the large-angle tipping and the subsequent NMR measurements in this experiment, during which the soliton lattice had perhaps already expanded.

To investigate the creation, propagation and metamorphosis of solitons, a detailed "time-of-flight" experiment between two collinear coils was performed by Bartolac *et al.* (1981) (see also Bartolac 1980). In order to control the texture, these authors confined the liquid to a stack of sheets each separated by 17 μm from the next. The vector $\boldsymbol{l}$ was thereby expected to be fixed perpendicular to the surfaces, so that the $\hat{\boldsymbol{d}}$ solitons created by the tipping could not transfer their winding to the $\hat{\boldsymbol{l}}$ texture, i.e. could not decay. Since $\hat{\boldsymbol{d}}$ solitons do not produce any NMR shifts ($\lambda_\ell = \lambda_t = 0$), one would expect the evolution of the satellite frequencies to be drastically different from that in the bulk case. However, the satellite frequencies measured in this experiment were found to be essentially the *same* as in the unrestricted geometry. This clearly indicates that they involve a strongly nonuniform $\hat{\boldsymbol{l}}$ texture, although this seems to be in conflict with the orientational effect due to the walls in the slabs. Perhaps the spacing of 17 μm between the plates was still too wide to clamp $\hat{\boldsymbol{l}}$ rigidly, since it is about twice as large as ξ_D (see Section 7.9.3). Alternatively, the $\hat{\boldsymbol{l}}$ texture in the slabs could have been a priori nonuniform, i.e. there could have existed regions with $\hat{\boldsymbol{l}}$ pointing up or down, separated by a two-dimensional domain wall, which were formed when the sample was cooled into the A phase. The boundaries would then have provided a potential for the vector $\hat{\boldsymbol{d}}$ to give rise to NMR satellites. In this case an average value of $\lambda_t = 0.5$ might result

(Sonin 1983b). Nevertheless, the dynamics of the soliton propagation and the time dependence of the NMR signal still await an explanation.

Solitons in the presence of superflow

Here we restrict our discussion to the case $\boldsymbol{H} \parallel \boldsymbol{v}_s \parallel \hat{\boldsymbol{z}}$ with surface normal $\hat{\boldsymbol{k}} \parallel \hat{\boldsymbol{z}}$, which corresponds to the pure twist composite soliton in the limit of zero external superflow (see Section 7.10). Since the flow induces a finite component of $\hat{\boldsymbol{l}}$ along $\boldsymbol{H}$—expressed by the equilibrium configuration (7.252) with $\beta \neq \frac{1}{2}\pi$—the operators Λ^f and Λ^g take slightly more complicated forms than in (8.142a,b). In the Ginzburg–Landau regime an expansion of the generalized free energy (7.232) in terms of the fluctuations f and g yields

$$\Lambda^f = -\xi_D^2 \frac{d}{dz}\left[(1+\sin^2\beta)\frac{d}{dz}\right] + \sin^2\beta\,[1 - 2\sin^2(\alpha-\psi)], \tag{8.154a}$$

$$\Lambda^g = -\xi_D^2 \left\{\frac{d}{dz}\left[(1+\sin^2\beta)\frac{d}{dz}\right] + (1+\sin^2\beta)\psi_z^2\right\} + \sin^2\beta\cos^2(\alpha-\psi) - \cos^2\beta. \tag{8.154b}$$

The corresponding eigenvalue equations cannot be solved in closed form, but may be treated variationally (Vollhardt and Maki 1979b). For increasing superflow p, λ_ℓ is found to stay almost constant at the value given by (8.153a), while λ_t decreases slightly from its value 0.8 at $p=0$. At the critical current $p_{\perp 1}=0.45$, where the twist solution becomes unstable with respect to a purely planar, bendlike $\hat{\boldsymbol{l}}$ texture ($\alpha=\psi=0$), λ_ℓ jumps upwards to a value of about 0.7, while λ_t drops abruptly to a very small value ($\lambda_t \approx 0.1$). Further increase of the superflow with $p < p_{\perp 2} = 0.625$ brings little change in λ_ℓ and λ_t.

Flow-induced soliton lattice

For superflow $p > p_{\perp 2}$, the uniform texture was found to become unstable against the spontaneous formation of a flow-induced domain-wall lattice (see Section 7.10). The operators Λ^f and Λ^g are then obtained from (8.154a,b), with $\alpha=\psi=0$, as

$$\Lambda^f = -\xi_D^2 \frac{d}{dz}\left[(1+\sin^2\beta)\frac{d}{dz}\right] + \sin^2\beta, \tag{8.155a}$$

$$\Lambda^g = -\xi_D^2 \frac{d}{dz}\left[(1+\sin^2\beta)\frac{d}{dz}\right] - 1 + 2\sin^2\beta. \tag{8.155b}$$

The lowest eigenvalues for λ_ℓ and λ_t may be obtained by explicit integration of (8.155a,b) (Vollhardt and Maki 1979a). Assuming that one deals with a *uniform* texture for $p < p_{\perp 2}$, characterized by $\lambda_\ell = \lambda_t = 1$, the eigenvalues are found to drop sharply (i.e. continuously but with infinite slope) at $p_{\perp 2}$, with λ_t becoming *negative,* thus indicating the appearance of the soliton lattice. Then both λ_ℓ and λ_t decrease monotonically. At $p = 2^{1/2}$, where the uniform texture becomes unstable, the eigenvalues have reached $\lambda_\ell = 0.43$ and $\lambda_t = -0.15$.

8.4.6 General theory of texture-induced NMR shifts in the B phase

The general spin dynamics for the B phase in the presence of nonuniform textures may be formulated by the same methods discussed in detail in the case of the A phase. Such a general approach was first derived by Maki (1975b) within a Lagrangian formulation (see below (8.88)). In the following we shall discuss the equivalent Hamiltonian approach along the lines presented by Fetter (1986).

To obtain the equation of motion for the spin $\boldsymbol{S}$ in the presence of a nonuniform texture, the bending energy has to be included. The linearized equation, valid for small deviations from equilibrium, may conveniently be obtained by parametrizing the order parameter in terms of θ^s_μ, the three angles of rotation about the Cartesian axes in spin space

$$A_{\mu j} = \sum_\nu R_{\mu\nu}(\boldsymbol{\theta}^s) A^0_{\nu j}. \tag{8.156}$$

Here $A^0_{\nu j}$ describes the equilibrium texture and the rotation matrix $\boldsymbol{R}$ is expanded as

$$R_{\mu\nu}(\boldsymbol{\theta}^s) = \delta_{\mu\nu} - \sum_\lambda \epsilon_{\mu\nu\lambda}\theta^s_\lambda + \tfrac{1}{2}[\theta^s_\mu\theta^s_\nu - \delta_{\mu\nu}(\boldsymbol{\theta}^s)^2] + \ldots . \tag{8.157}$$

(In the following the superscript "s" on θ^s_μ will be dropped.) The angular variables θ_μ are taken as operators representing collective quantum variables. They are canonically conjugate to the spin density operators $S_\mu(\boldsymbol{r})$, the generators of local rotations, i.e.

$$[\theta_\mu(\boldsymbol{r}), S_\nu(\boldsymbol{r}')] = i\hbar\delta_{\mu\nu}\delta(\boldsymbol{r}-\boldsymbol{r}'), \tag{8.158}$$

while they commute with one another.

Written in terms of the variables θ_μ, the dipole energy density (6.107) takes the form (Brinkman and Smith 1974)

$$\Delta f_D = \frac{1}{2}\frac{\chi}{\gamma^2}\Omega_B^2(\hat{\boldsymbol{n}}\cdot\boldsymbol{\theta})^2, \tag{8.159}$$

where the definition of Ω_B in (8.24) has been employed. Substituting (8.156) into the general expression (7.17) for the gradient energy in the Ginzburg–Landau regime (where, again, $d_{\mu j} = 3^{1/2}\Delta(T)A_{\mu j}$), two different contributions are seen to arise: those involving gradients of θ_μ and of the equilibrium texture $A^0_{\mu j}$ respectively. Assuming the equilibrium $\hat{\boldsymbol{n}}$ texture to vary on the scale of ξ_H (which is much longer than the scale for variations in θ_μ, given by ξ_D) and that the texture is dipole-locked, i.e. that the spin–orbit rotation angle is fixed at $\theta = \theta_L$, we may neglect gradients of $A^0_{\mu j}$. The resulting bending energy expression is given by (Smith *et al.* 1977, Brinkman and Cross 1978, Buchholtz 1978)

$$f_{\text{bend}} = \frac{1}{2}\frac{\chi}{\gamma^2}\,\Omega_B^2\xi_D^2\left\{\bar{K}\sum_{\mu\nu}(\nabla_\mu\theta_\nu)^2 - \bar{K}'\left[\sum_{\mu\nu}R^L_{\mu\nu}(\nabla_\mu\theta_\nu)\right]^2\right\}, \qquad (8.160)$$

where $\bar{K}' = \bar{K}_2 + \bar{K}_3$ and $\bar{K} = 2\bar{K}_1 + \bar{K}'$ and the dimensionless parameters $\bar{K}_1$, $\bar{K}_2$ and $\bar{K}_3$ are defined in terms of the coefficients K_j in (7.17) as $\bar{K}_j = K_j\Delta^2(\gamma^2/\chi\Omega_B^2\xi_D^2)$. In the weak-coupling Ginzburg–Landau limit $\bar{K} = \frac{8}{5}$ and $\bar{K}' = \frac{4}{5}$. The tensor quantity $R^L_{\mu\nu}$ is the rotation matrix characterizing the B phase at the dipole-locked rotation angle $\theta = \theta_L$. It contains the information about the equilibrium texture (see (6.106)):

$$R^L_{\mu\nu}(\boldsymbol{r}) = \frac{1}{4}\left[\delta_{\mu\nu} - 5\hat{n}_\mu\hat{n}_\nu + (15)^{1/2}\sum_\lambda \epsilon_{\mu\nu\lambda}\hat{n}_\lambda\right]. \qquad (8.161)$$

The equation of motion for the spin density $\boldsymbol{S}(\boldsymbol{r})$ is derived by calculating the commutator of $\boldsymbol{S}$ with the total effective Hamiltonian

$$\mathrm{i}\hbar\dot{\boldsymbol{S}} = [\boldsymbol{S}, \tfrac{1}{2}\gamma^2\boldsymbol{S}\cdot\boldsymbol{\chi}^{-1}\cdot\boldsymbol{S} - \gamma\boldsymbol{H}\cdot\boldsymbol{S} + f_{\text{bend}} + \Delta f_D] \qquad (8.162)$$

using (8.158). The result is

$$\dot{\boldsymbol{S}} = \gamma\boldsymbol{S}\times\boldsymbol{H} + \frac{\chi}{\gamma^2}\,\Omega_B^2\xi_D^2\,\{-\bar{K}\,\nabla^2\boldsymbol{\theta} + \bar{K}'(\boldsymbol{R}^L\boldsymbol{\nabla})[(\boldsymbol{R}^L\boldsymbol{\nabla})\cdot\boldsymbol{\theta}]\}$$
$$+ \frac{\chi}{\gamma^2}\,\Omega_B^2\,(\hat{\boldsymbol{n}}\cdot\boldsymbol{\theta})\hat{\boldsymbol{n}}. \qquad (8.163)$$

The equation of motion for the angular variables θ_μ is not affected by the bending energy (to order λ_D) and is still given by (8.8b), which translates into

$$\dot{\boldsymbol{\theta}} = -\gamma\left(\boldsymbol{H} - \frac{\gamma}{\chi}\boldsymbol{S}\right). \qquad (8.164)$$

We mention in passing that a particular limit of (8.163) and (8.164), corresponding to the so-called "double sine–Gordon equation", has been studied as an example of a nonintegrable partial differential equation yielding soliton solutions (Kitchenside *et al.* 1978, 1979a,b).

In the situation considered here, a weak RF field $\boldsymbol{H}'\mathrm{e}^{-\mathrm{i}\omega t}$ in the presence of a static uniform field $\boldsymbol{H}_0$ along $\hat{\boldsymbol{z}}$ excites oscillations of the texture about its equilibrium configuration. The dynamical equation for the μth component of $\boldsymbol{\theta}$, θ_μ, is obtained by taking the time derivative of (8.164) and eliminating $\boldsymbol{S}$ by means of (8.162), yielding

$$\Omega_B^2\sum_\nu(\mathrm{D}_{\mu\nu} + \hat{n}_\mu\hat{n}_\nu - \delta_{\mu\nu})\theta_\nu - (\omega^2 - \Omega_B^2)\theta_\mu + \mathrm{i}\omega\omega_L(\boldsymbol{\theta}\times\hat{\boldsymbol{z}})_\mu = \mathrm{i}\omega\gamma H'_\mu. \qquad (8.165)$$

Here θ_μ and $\hat{n}_\mu$ are functions of $\boldsymbol{r}$ and $\mathrm{D}_{\mu\nu}$ is a tensor differential operator with components

$$\mathrm{D}_{\mu\nu} = \xi_{\mathrm{D}}^2\Big[-\delta_{\mu\nu}\bar{K}\nabla^2 + \bar{K}'\sum_{\lambda\lambda'}(R^{\mathrm{L}}_{\mu\lambda}\nabla_\lambda)(R^{\mathrm{L}}_{\nu\lambda'}\nabla_{\lambda'})\Big]. \tag{8.166}$$

The last term on the left-hand side of (8.165) may be diagonalized by projecting onto the spherical unit vectors $\hat{\eta}_\pm = 2^{-1/2}(\hat{\boldsymbol{x}} \pm \mathrm{i}\hat{\boldsymbol{y}})$, $\hat{\eta}_0 = \hat{z}$ as $\theta_m = \boldsymbol{\theta}\cdot\hat{\boldsymbol{\eta}}_m$, $\mathrm{D}_{mm'} = \hat{\boldsymbol{\eta}}_m\boldsymbol{D}\hat{\boldsymbol{\eta}}^*_{m'}$. Here $m = +, -, 0$ correspond to the polarization of the vector field. The remaining off-diagonal terms in $\mathrm{D}_{mm'}$ are small for sufficiently large magnetic fields, such that $(\Omega_{\mathrm{B}}/\omega_{\mathrm{L}})^2 \ll 1$ (Theodorakis and Fetter 1983). In this case (8.165) separates into three inhomogeneous differential equations of the form

$$(\mathrm{D}_m + V_m)\theta_m = \frac{1}{\Omega_{\mathrm{B}}^2}[(\omega^2 - m\omega\omega_{\mathrm{L}} - \Omega_{\mathrm{B}}^2)\theta_m + \mathrm{i}\omega\gamma H'_m]. \tag{8.167}$$

with $\mathrm{D}_m \equiv \mathrm{D}_{mm}$ and $V_m = -1 + |\hat{\boldsymbol{n}}_m|^2$. The boundary conditions on θ_m at the surface of the sample are given by the requirement that there be no spin supercurrent through the surface (Brinkman and Cross 1978). This gives rise to the Neumann boundary condition on θ_m

$$\hat{\boldsymbol{s}}\cdot\boldsymbol{\nabla}\theta_m = 0, \tag{8.168}$$

where $\hat{\boldsymbol{s}}(\boldsymbol{r})$ is the surface normal.

Equation (8.167) is seen to be the analogue to (8.114) valid in the A phase, and indeed has the same structure: the homogeneous part of the differential equation (8.167) for θ_m has the form of a Schrödinger equation, with D_m and V_m playing the roles of the kinetic and potential energy respectively. As in the A phase (see (8.117), (8.122)) the solution of the inhomogeneous equation may be expressed in terms of the eigenfunctions $\psi_{m,k}$ and the ("energy") eigenvalues $E_{m,k}$ of the homogeneous problem

$$(\mathrm{D}_m + V_m)\psi_{m,k} = E_{m,k}\psi_{m,k}, \tag{8.169}$$

subject to the same boundary conditions (8.168). Assuming the eigenfunctions $\psi_{m,k}$ to form a complete set of orthonormal functions, the solution $\theta_m(\boldsymbol{r})$ may be expanded in terms of $\psi_{m,k}$ as

$$\theta_m(\boldsymbol{r}) = \sum_k c_{m,k}\psi_{m,k}. \tag{8.170}$$

Substituting (8.170) into (8.167), the coefficients $c_{m,k}$ are obtained by multiplication of the equation by $\psi_{m,k'}$ and integration over $\boldsymbol{r}$:

$$c_{m,k} = \mathrm{i}\omega\gamma\frac{\int \mathrm{d}^3r\,\psi^*_{m,k}(\boldsymbol{r})H'_m(\boldsymbol{r})}{(\omega_{m,k} - \omega - \mathrm{i}0)(\omega_{m,k} + \omega - m\omega_{\mathrm{L}} + \mathrm{i}0)}. \tag{8.171}$$

As usual, an infinitesimal imaginary part has been added to the frequency $(\omega \to \omega + \mathrm{i}0)$, expressing the fact that the perturbing RF field has been switched on adiabatically slowly, so that causality is guaranteed. As expected, the magnitude of $c_{m,k}$ (and thereby the contribution of $\psi_{m,k}$ to θ_m) is determined by the overlap of the kth mode with the RF field, the resonance being strong if the external frequency ω is near the eigenfrequency $\omega_{m,k}$. The latter is itself determined by the eigenvalue $E_{m,k}$ via

$$\omega_{m,k}^2 - m\omega_{m,k}\omega_{\mathrm{L}} - \Omega_{\mathrm{B}}^2 = E_{m,k}\Omega_{\mathrm{B}}^2. \tag{8.172}$$

The excitations correspond to spin waves.

The quantity observed in an NMR experiment is the rate of power absorption. With $\boldsymbol{M} = \gamma\boldsymbol{S}$ and using (8.164), the latter is given by

$$\frac{\mathrm{d}U}{\mathrm{d}t} = \tfrac{1}{2}\omega \int \mathrm{d}^3r \,\mathrm{Im}\, \boldsymbol{H}'^* \cdot \boldsymbol{M}(\boldsymbol{r}) \tag{8.173a}$$

$$= -\frac{1}{2}\frac{\chi}{\gamma}\,\omega\,\mathrm{Im}\left[\mathrm{i}\omega H_m'^* \int \mathrm{d}^3r\, \theta_m(\boldsymbol{r})\right], \tag{8.173b}$$

where the RF field has been assumed to be uniform. By expressing $\theta_m(\boldsymbol{r})$ in terms of (8.171) and making use of $(x - \mathrm{i}0)^{-1} = \mathscr{P}(1/x) + \mathrm{i}\pi\delta(x)$, where $\mathscr{P}$ is the principal-value integral, (8.173) reduces to

$$\frac{\mathrm{d}U}{\mathrm{d}t} \approx \tfrac{1}{2}\pi\chi\omega^3\, |H_m'|^2 \sum_k \frac{\delta(\omega - m\omega_{\mathrm{L}} + \omega_{m,k})}{2\omega - m\omega_{\mathrm{L}}} \left|\int \mathrm{d}^3r\, \psi_{m,k}(\boldsymbol{r})\right|^2, \tag{8.174}$$

where the RF field has been assumed to be weak. As expected, the main resonance line ($k = 0$) is shifted and additional satellite peaks appear at frequencies corresponding to the excited-state frequencies $\omega_{m,k}$.

The lowest eigenvalue $E_{m,0}$ may be conveniently determined by employing a variational method, which makes use of the Hermiticity and the existence of a lower bound of the spectrum of the operator $\mathrm{D}_m + V_m$. A trial function Ψ, satisfying the boundary condition (8.168), therefore yields

$$E_{m,0} \leqslant \frac{\int \mathrm{d}^3r\, \Psi^*(\mathrm{D}_m + V_m)\Psi}{\int \mathrm{d}^3r\, |\Psi|^2}. \tag{8.175}$$

For longitudinal NMR ($m = 0$) in the uniform bulk liquid, where $\mathrm{D}_0 = V_0 \equiv 0$, the eigenvalue E_0 is zero. This is in direct analogy with the A phase, where the corresponding eigenvalue E_ℓ, (8.124), is also zero for the uniform bulk liquid.

Parallel-plate geometry

As a first example, let us discuss the NMR signal from the B phase in a parallel-plate geometry as considered by Smith *et al.* (1977) (see also

Lin-Liu and Maki 1978), with a strong static magnetic field $\boldsymbol{H} \parallel \hat{\boldsymbol{x}}$ parallel to the plates. The equilibrium $\hat{\boldsymbol{n}}$ texture in this geometry has been discussed in Section 7.9. In this case the magnetization and the variables θ_μ depend only on $|z|$, the coordinate along the normal to the plates ($|z| \leqslant \frac{1}{2}L$). In the centre of the slab, i.e. for small values of $|z|$, $\hat{\boldsymbol{n}}$ tends to lie along the direction of the magnetic field. The orientation changes into that favoured by the magnetic surface energy, i.e. $\hat{\boldsymbol{n}} = 5^{-1/2}(1, 3^{1/2}, 1)$ in a layer of thickness ξ_H at the surfaces.

For transverse NMR ($m = +$), the eigenvalue equation (8.169) reduces to an ordinary differential equation, given by

$$(-\bar{K} + \bar{K}' \, |R^{\mathrm{L}}_{+z}|^2)\xi_{\mathrm{D}}^2 \frac{\mathrm{d}^2\psi(z)}{\mathrm{d}z^2} + \hat{n}_x^2 = E_{+,k}\psi(z), \tag{8.176}$$

where $R^{\mathrm{L}}_{+z} = \hat{\boldsymbol{\eta}}_+ \boldsymbol{R}^{\mathrm{L}} \hat{\boldsymbol{z}}$ and here $\hat{\boldsymbol{\eta}}_+ = 2^{-1/2}(\hat{\boldsymbol{y}} + \mathrm{i}\hat{\boldsymbol{z}})$. The absolute square of R^{L}_{+z} is calculated as

$$|R^{\mathrm{L}}_{+z}|^2 = 1 + 5[5\hat{n}_\perp^2\hat{n}_z^2 - 2\hat{n}_z^2 + 3\hat{n}_x^2 - 2(15)^{1/2}\hat{n}_x\hat{n}_y\hat{n}_z], \tag{8.177}$$

where $\hat{n}_\perp^2 = \hat{n}_y^2 + \hat{n}_z^2$.

For large plate separation ($L > \xi_H$), the $\hat{\boldsymbol{n}}$ texture is uniformly oriented along $\boldsymbol{H}$ over most of the sample. In this situation the solution of (8.176) is characterized by periodic oscillations of the magnetization. The detection of excited states in an NMR experiment is then hardly possible since the integrated magnetization averages to zero. By contrast, for smaller plate separation ($L \leqslant \xi_H$), $\hat{n}_x^2 = (\hat{\boldsymbol{n}} \cdot \hat{\boldsymbol{H}})^2$ varies quadratically with z (see (7.176b)). If one now approximates $|R^{\mathrm{L}}_{+z}|^2$ in (8.176)) by a constant then (8.176) corresponds to the equation for a harmonic oscillator. The eigenvalues are therefore evenly spaced. The intensity of the excited modes given by (8.174) decreases as k^{-1}. A detailed numerical calculation of both the texture and the eigenstates is found to give excellent agreement with experiment (Osheroff 1977, Osheroff *et al.* 1977; see also the earlier work of Osheroff *et al.* 1975). This also shows that the qualitative treatment discussed above already includes the essential physics. The agreement may be taken as a proof that spin waves exist in ^{3}He-B.

Cylindrical geometry

As a second example, let us discuss the NMR signal from the B phase in a long circular cylinder of radius R with a magnetic field along the cylinder axis. For $R > \xi_{\mathrm{s}}$, the equilibrium texture in this geometry (see (7.182)) may be written as

$$\hat{n}_\pm = 2^{-1/2}(\hat{n}_x \pm \mathrm{i}\hat{n}_y) \tag{8.178a}$$

$$= 2^{-1/2} \sin\beta \, \mathrm{e}^{\pm\mathrm{i}(\alpha+\varphi)}, \tag{8.178b}$$

$$\hat{n}_0 = \hat{n}_z = \cos\beta. \tag{8.178c}$$

It follows that the effective attractive potential $V_m = -1 + |n_m|^2$ is axisymmetric and given by $V_\pm = -1 + \frac{1}{2}\sin^2\beta$ and $V_0 = -\sin^2\beta$ respectively.

Eigenstates of (8.167) in an axisymmetric potential can be classified by their azimuthal angular dependence $e^{il\varphi}$. In the usual case of an axisymmetric RF field only the $l=0$ modes are excited.

For the transverse case ($m=+$), the eigenvalue equation (8.169) reduces to the following ordinary linear second-order differential equation:

$$(-\bar{K}+\bar{K}'\,|R^{\mathrm{L}}_{+r}|^2)\xi_{\mathrm{D}}^2\left[\frac{1}{r}\frac{\mathrm{d}}{\mathrm{d}r}\left(r\frac{\mathrm{d}}{\mathrm{d}r}\right)\right]\psi_{+,k}-(1-\tfrac{1}{2}\sin^2\beta)\psi_{+,k}=E_{+,k}\psi_{+,k}, \tag{8.179}$$

where $R^{\mathrm{L}}_{+\mathrm{r}}=\hat{\boldsymbol{\eta}}_{+}\boldsymbol{R}^{\mathrm{L}}\hat{\boldsymbol{r}}$ (Hakonen and Volovik 1982, Maki and Nakahara 1983, Fetter and Theodorakis 1983, Jacobsen and Smith 1983). The potential $-1+\frac{1}{2}\sin^2\beta$ has a quadratic dependence on r near the centre, with very small curvature, and rises to its boundary value of $-\frac{3}{5}$ only within a surface layer of thickness $\xi_H \ll R$. As a first crude approximation, one may therefore replace the potential by a harmonic one, with $\frac{1}{2}\sin^2\beta=(r/L_1)^2$, and $|R^{\mathrm{L}}_{+r}|^2$ by its value at $r=0$, $|R^{\mathrm{L}}_{+r}|^2=\frac{1}{2}$. This leaves us with the "Schrödinger equation" for a *two*-dimensional harmonic oscillator, a standard problem in quantum mechanics. The only characteristic length of the problem is given by $r_1(\xi_{\mathrm{D}}L_1)^{1/2}$, i.e. the eigenfunctions vary as r/r_1. The eigenvalues are evenly spaced and given by

$$E_{+,k}=-1+2(\bar{K}-\tfrac{1}{2}\bar{K}')^{1/2}\frac{\xi_{\mathrm{D}}}{L_1}(k+1). \tag{8.180}$$

The corresponding NMR frequencies are then determined by (8.172). Multiplets with odd k do not contain s-wave states ($l=0$) and therefore do not couple. Within the approximation discussed here, the intensities of the satellite resonances are found to be independent of k. The ratio of the intensity of the kth normal mode, relative to the uniform-texture signal from the area πR^2, is given by

$$\frac{I_k}{I_{\mathrm{uniform}}}=\left(\frac{96}{5}\right)^{1/2}\frac{\xi_{\mathrm{D}}L_1}{R^2}. \tag{8.181}$$

NMR experiments in stationary cylinders do indeed show a set of nearly equally spaced absorption peaks above the Larmor frequency, but the intensity falls off as a function of increasing k (Hakonen *et al.* 1983a). These deviations have been explained by a numerical solution of (8.179), using the exact equilibrium $\hat{\boldsymbol{n}}$ texture for various stationary cylinders (Jacobsen and Smith 1983).

For small magnetic fields ($H<H_{\mathrm{S}}$) and radii $R<10\xi_{\mathrm{S}}$, the "in-plane" texture (7.177), (7.178) is stable. This situation has been studied theoretically as well as experimentally by Brinkman *et al.* (1974b).

8.5 NMR IN ROTATING SUPERFLUID ^{3}He

The NMR response of the rotating liquid can be calculated by the same method described above for the stationary liquid. As discussed in Section 7.6 the rotation generates a lattice of vortices parallel to the rotation axis. The effect of the vortices, as far as NMR is concerned, is to change the equilibrium texture. This in turn modifies the operator D and the potential V in the "Schrödinger equations" (8.117), (8.122) and (8.169) and thereby changes the eigenvalues and eigenstates (see Fetter 1986, Salomaa and Volovik 1987). As a result, the pattern of resonance lines is expected to vary as a function of angular velocity Ω, both in position and intensity (see Figs. 4.10 and 4.11).

8.5.1 The A phase

The NMR response in the A phase is governed by the differential equations (8.117) and (8.122) for the fluctuations of the vector $\hat{\boldsymbol{d}}$. For a given $\hat{\boldsymbol{l}}$ texture, the "kinetic-energy" operator D and the potential V are fixed. The spin-wave frequencies follow from the eigenvalues of the operator $\mathrm{D}+V$, whereas calculation of the intensity requires knowledge of the eigenstates as well.

Let us now discuss the case of transverse NMR. In a relatively strong external magnetic field $\boldsymbol{H} \parallel \hat{\boldsymbol{z}}$ the $\hat{\boldsymbol{d}}$ texture is confined to lie in the plane perpendicular to $\boldsymbol{H}$ and is essentially uniform throughout the sample (say, $\hat{\boldsymbol{d}} \parallel \hat{\boldsymbol{x}}$) to minimize the bending energy. Likewise, the $\hat{\boldsymbol{l}}$ texture is also essentially uniform, being oriented along $\hat{\boldsymbol{d}}$. Deviations of $\hat{\boldsymbol{l}}$ from uniformity only occur on a lengthscale ξ_{D} around each vortex core, which is much less than the separation between vortices (see Section 7.6). Therefore the potential $V_{\mathrm{t}} = -(l_y + 2l_z)$ in (8.117) consists of a regular array of attractive cylindrical potentials with radii of the order of ξ_{D} for both the nonsingular doubly quantized vortices and the singular singly quantized vortices (Seppälä and Volovik 1983, Vulovic *et al.* 1984). The low-lying states of this potential are well localized within a single potential well. It is therefore sufficient to look for the bound spin-wave states of a single vortex core. A glance at $\mathrm{D}+V_{\mathrm{t}}$ shows that the potential has a range ξ_{D} and unit depth, which allows only a small number of bound states. Note that the potential V_{t} is not cylindrically symmetric, precluding a decomposition in angular-momentum functions. A numerical evaluation of the $\hat{\boldsymbol{l}}$ texture and the eigenvalue equations (8.117) for the ground state at $\Omega = 1$ rad/s yields the eigenvalues $E_{\mathrm{t},0} = -0.5$ and $E_{\mathrm{t},0} = -0.02$ and the intensities 0.047 and 0.56 relative to the signal of the bulk A phase for the NMR satellites of the nonsingular and the singular vortex lattices respectively (Seppälä and Volovik 1983; see also Maki 1983a,b,c). Both $E_{\mathrm{t},0}$ and the relative intensities are found to decrease with decreasing temperature (Vulovic *et al.* 1984).

NMR experiments in the rotating A phase, at 29.3 bar pressure and for a range of temperatures, show a satellite peak (see Fig. 4.10), which may be interpreted as a spin-wave resonance located at the vortex cores (Seppälä *et al.* 1984, Hakonen *et al.* 1985b). One finds that the eigenvalue $E_{t,0}$ derived from the observed NMR satellite shift is independent of the relative orientation of the magnetic field with respect to the rotation axis, and decreases nearly linearly with decreasing temperature, from $E_{t,0} \approx -0.4$ at T_c to $E_{t,0} \approx -0.7$ at $T/T_c = 0.7$. The strength of the satellite absorption peak varies linearly with the rotation speed and hence with the number of vortices in the system, which proves that the satellites are associated with vortices. The relative intensity at $\Omega = 1$ rad/s is 0.06 and increases with the tipping angle, in agreement with theory (Maki and Zotos 1985b). These observations point to the presence of doubly quantized nonsingular vortices, even though theoretical estimates indicate that singly quantized singular vortices should be lower in energy in the high-field regime (Seppälä and Volovik 1983, Vulovic *et al.* 1984). It is interesting to note that ion-mobility experiments do indeed suggest the presence of singular vortices at magnetic fields $H \gtrsim 4$ G (Simola *et al.* 1987); see also Section 7.5. The effect of broadening of the NMR line due to spin-wave scattering on vortices with integer quanta of circulation has been considered by Fomin and Kamenskii (1982).

Experimental results on longitudinal NMR shifts in the rotating superfluid, theoretically described by (8.122), are not available so far.

8.5.2 The B phase

In contrast with the A phase, in the B phase only the large-scale textures are of relevance for NMR properties. This is because the characteristic lengths are much longer than in the A phase. Vortices affect the $\hat{\boldsymbol{n}}$ texture mainly through the change in the magnetic orientation energy. As discussed in Section 7.5, there are two contributions to the magnetic energy of vortices: the first, quantified by the parameter λ, is due to the suceptibility anisotropy present in the cores of most vortices (see (7.120)); the second, described by the parameter H_M, is due to the gyromagnetic effect (see (7.125)). Let us recall that both λ and H_M are positive and proportional to the angular velocity Ω of the rotating container. The combined effect of both terms is to weaken the bulk magnetic orientation energy by a factor $1 - \lambda \mp H_M/H$ (assuming the angle $\beta \ll 1$) and to increase the magnetic healing length ξ_H as given by (7.127). This in turn leads to a narrow central region of the texture (where $\beta \approx 0$) and to a more extended surface region (where $\beta(r)$ approaches its boundary value), i.e. the curvature of the parabolic potential $V_+ = -1 + \frac{1}{2}\sin^2\beta$ in (8.179) increases. Thus the spacing of the transverse spin-wave modes in (8.180) should be expected to vary with the rotation speed of the container—an effect first predicted by

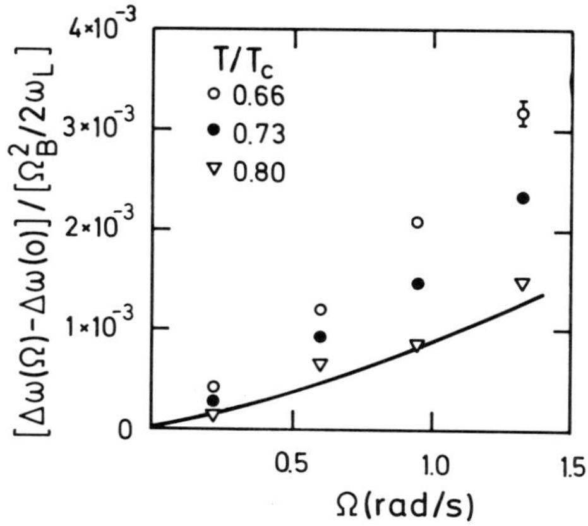

Figure 8.7 Frequency shift of the lowest spin-wave resonance in the B phase as a function of rotation speed for various values of temperature. The full line is the theoretical result of Hakonen and Volovik (1982). (After Hakonen *et al.* (1983a).)

Hakonen and Volovik (1982) and Maki and Nakahara (1983). This is indeed observed experimentally, as shown in Figs. 4.11 and 8.7, where the splitting is found to increase linearly with Ω (Hakonen *et al.* 1983a). Furthermore, the NMR frequency shift is seen to depend on the relative orientation of the applied magnetic field with respect to the rotation axis (Hakonen *et al.* 1983c).

A particularly interesting experimental finding was the discovery of a discontinuity in the spin-wave resonant frequencies at a temperature $T_v \approx 0.6_c$, as shown in Fig. 8.8 (Ikkala *et al.* 1982, Hakonen *et al.* 1983a,c). The transition temperature T_v is found to be independent of angular velocity Ω, magnetic field H and the relative orientation of the two, and to depend only on pressure. In the low-temperature phase the gyromagnetic effect is clearly seen, i.e. the frequency shift depends on the sign of $\boldsymbol{\Omega} \cdot \boldsymbol{H}$. The location of the phase boundary $T_v(P)$ between the two vortex states is shown in Fig. 8.9. In persistent-current experiments an abrupt change has also been observed in the critical velocity, which occurs approximately along the same line in the P–T phase diagram (Pekola *et al.* 1984c). Indeed, it has been argued that the two transitions have the same origin, namely a change of the vortex core structure (Pekola *et al.* 1984a). A different interpretation of the transition found in the flow experiment has been proposed by Thuneberg (1987b), who argues that the transitions in the two experiments are qualitatively different. Ruling out a vortex core transition or surface-related effects (Thuneberg 1986b), Thuneberg concludes that the transition is due to a new type of vortex, involving spin superflow rather than mass superflow around the line defect.

As discussed above, rotation of the sample influences the $\hat{\boldsymbol{n}}$ texture and thereby changes the NMR frequencies. An even stronger effect is observed

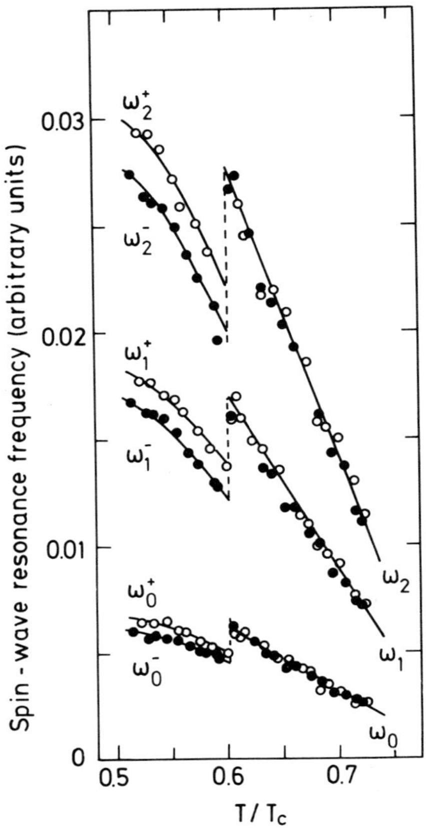

Figure 8.8 Resonant frequencies of the three lowest spin-wave modes at 29.3 bar in rotating ^{3}He-B ($\Omega = 0.6$ rad/s) as a function of temperature, for magnetic field orientation parallel (○) and antiparallel (●) to the rotation axis $\boldsymbol{\Omega}$. (After Hakonen *et al.* (1983a).)

if the magnetic field is tilted with respect to the rotation axis, because in this case $\hat{\boldsymbol{n}}$ is not parallel to $\boldsymbol{H}$ even in the bulk liquid (Bun'kov *et al.* 1983a,b, 1984b, Hakonen *et al.* 1985a). A numerical analysis of the spin-wave eigenvalue equation for several vortex cores has been performed by Hakonen *et al.* (1983c) in order to determine the vortex parameters λ and H_M as functions of temperature. As can be seen from Fig. 8.10, λ is found to increase with decreasing temperature both above and below the vortex core transition and makes a jump of relative size 10–30%, which depends on the tilt angle of the magnetic field with respect to the rotation axis. In Fig. 8.11 the parameter H_M is plotted versus temperature. Above the transition, H_M is seen to be very small, indicating that—if at all—there is only a minute gyromagnetic effect associated with the vortex cores in this case. A calculation of λ and H_M, based on the assumption that in the

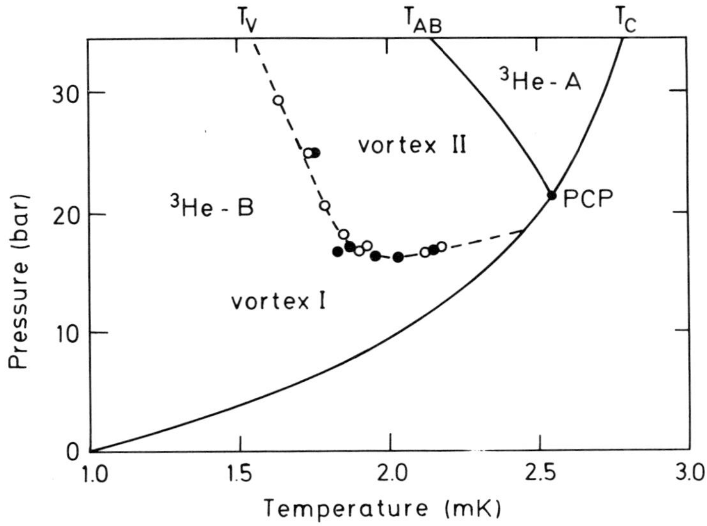

Figure 8.9 Phase diagram of rotating liquid ^{3}He in a magnetic field H (○, $H = 284$ G; ●, $H = 568$ G), indicating a discontinuous transition betwen two vortex states in the B phase. (After Hakonen *et al.* (1983a) and Pekola *et al.* (1984a).)

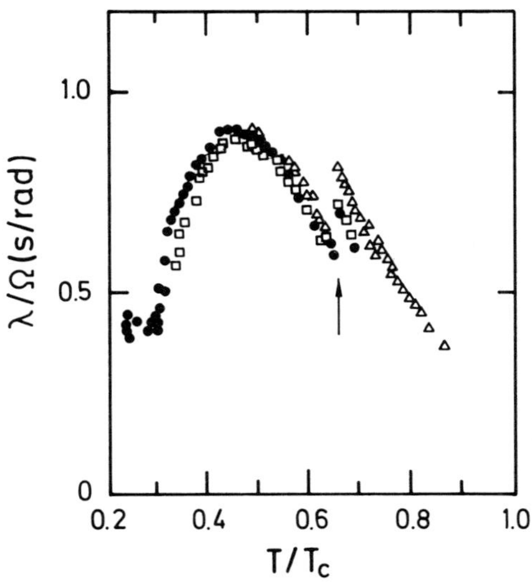

Figure 8.10 Vortex parameter λ characterizing the orientational effect of axisymmetric vortices on the bulk texture in ^{3}He-B. The values are deduced from the analysis of NMR frequency shifts for various magnitudes of the magnetic field $\boldsymbol{H}$ and orientations relative to the rotation axis $\boldsymbol{\Omega}$. The arrow indicates the position of the discontinuity at the vortex transition temperature. (After Salomaa and Volovik (1987).)

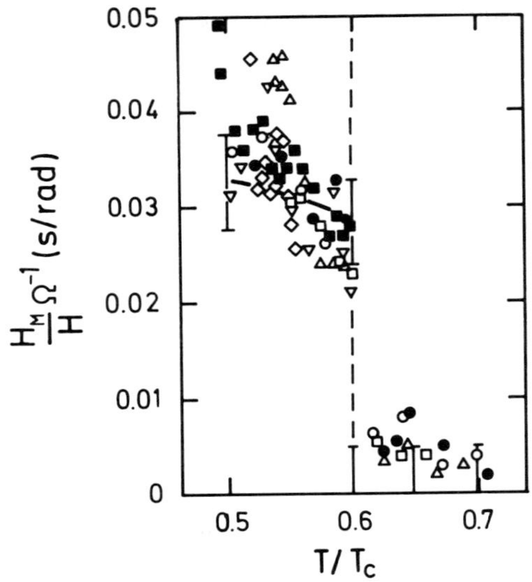

Figure 8.11 Parameter H_M describing the gyromagnetic effect of the ferromagnetic vortex cores in ^{3}He-B. The values are extracted from the analysis of NMR frequency shifts for various orientations of the magnetic field $\boldsymbol{H}$ with respect to the rotation axis $\boldsymbol{\Omega}$. The solid line is obtained from a comparison of the calculated NMR frequencies with the measured spectrum in the case $\boldsymbol{H} \parallel \Omega$. (After Hakonen *et al.* (1983c) and Salomaa and Volovik (1987).)

high-temperature/high-pressure phase the vortex cores are of the double-core type whereas at low temperature the axisymmetric v-type vortex is stable (Thuneberg 1986a, 1987a, Salomaa and Volovik 1986), is in qualitative agreement with the data.

8.6 SPIN-RELAXATION PHENOMENA

Since the total spin of a strictly isolated system of individual spins is conserved, spin relaxation can only occur by coupling to other spins outside the system or by coupling to an orbital angular momentum within the system itself. In this way, angular momentum is transferred to a heat bath, where it is dissipated. As a consequence, the spin relaxes to its equilibrium state. In liquid ^{3}He, spin relaxation in the bulk takes place via the spin–orbit forces given by the dipole–dipole coupling of the ^{3}He nuclei. In the normal state the spin-relaxation rate due to this coupling is proportional to the square of the dipole coupling constant (see Abragam 1961). In the limit of low temperatures the relaxation times become very long ($\propto T^{-2}$), i.e. of order 10^5 s at 1–10 mK (Vollhardt and Wölfle 1981a,b). Experimentally, one usually observes shorter relaxation times (of about 10^2 s) due to the presence of a surface relaxation mechanism, which dominates the bulk mechanism at low T. It therefore came as a surprise when NMR experiments in the superfluid phases appeared to indicate a much faster spin-relaxation time, of order 10^{-3} s.

In fact, as pointed out by Leggett and Takagi (1975) (see also Combescot and Ebisawa 1974, Combescot 1975a, 1976, Maki and Ebisawa 1976d, Ambegaokar 1975, Bhattacharyya *et al.* 1975, Ambegaokar and Levy 1976), the increased relaxation rate may be understood on the basis of Leggett's theory of spin dynamics, if one allows both the condensate and the quasiparticle systems to have their own dynamics.

Following Leggett and Takagi (1975, 1977), we define the partial spin polarizations of the Cooper-pair condensate $\boldsymbol{S}^{\mathrm{p}}$ and of the quasiparticle system (normal-fluid component) $\boldsymbol{S}^{\mathrm{q}}$, which add up to the total spin polarization $\boldsymbol{S}$:

$$\boldsymbol{S} = \boldsymbol{S}^{\mathrm{p}} + \boldsymbol{S}^{\mathrm{q}}. \tag{8.182}$$

Consider a change in the applied external magnetic field. In the hydrodynamic regime the spins of the pair condensate will adjust to the new field within a time period of order $\hbar/\Delta$, i.e. essentially instantaneously compared with the quasiparticle relaxation time. We may then define an "instantaneous" value $\boldsymbol{H}_{\mathrm{inst}}$ of the effective magnetic field acting on the Cooper-pair spins by

$$\boldsymbol{S}^{\mathrm{p}} = \frac{1}{\gamma}\chi^{\mathrm{p}0}\boldsymbol{H}_{\mathrm{inst}}. \tag{8.183}$$

Here $\chi^{\mathrm{p}0}$ is the relevant eigenvalue of the partial spin susceptibility of the condensate, given by (10.53) in Chapter 10, and the superscript zero on the susceptibility implies the absence of Fermi-liquid corrections. The "instantaneous" magnetic field $\boldsymbol{H}_{\mathrm{inst}}$ is given by the sum of $\boldsymbol{H}_{\mathrm{eff}}$, the external magnetic field corrected by Fermi-liquid interaction effects,

$$\boldsymbol{H}_{\mathrm{eff}} = \boldsymbol{H} - \frac{\gamma}{\chi_{\mathrm{N}}^{0}} F_0^{\mathrm{a}}\boldsymbol{S} \tag{8.184}$$

(where we neglect the Landau parameters F_l^{a} for $l \geqslant 2$), and the magnetic field equivalent to the difference in the chemical potentials of the Cooper pairs with spins up and down respectively (see (10.141b). The dipole interaction induces a weak coupling of the Cooper-pair subsystems of different spin orientation and thus provides a mechanism for spin exchange between the subsystems. In equilibrium the respective chemical potentials must therefore be equal. The field driving the motion of the vector $\boldsymbol{d}$ is given precisely by this chemical potential difference, i.e. by $\boldsymbol{H}_{\mathrm{eff}} - \boldsymbol{H}_{\mathrm{inst}}$ rather than by $\boldsymbol{H} - (\gamma/\chi)\boldsymbol{S}$ as in (8.8b). Consequently, the equation of motion for the vector $\boldsymbol{d}$, (8.8b), has to be modified as follows:

$$\dot{\boldsymbol{d}}(\boldsymbol{k}) = \boldsymbol{d}(\boldsymbol{k}) \times \gamma(\boldsymbol{H}_{\mathrm{eff}} - \boldsymbol{H}_{\mathrm{inst}}) \tag{8.185a}$$

$$= \boldsymbol{d}(\boldsymbol{k}) \times \gamma\left(\boldsymbol{H} - \frac{\gamma}{\chi_{\mathrm{N}}^{0}} F_0^{\mathrm{a}}\boldsymbol{S} - \frac{\gamma}{\chi^{\mathrm{p}0}}\boldsymbol{S}^{\mathrm{p}}\right). \tag{8.185b}$$

In contrast with the Cooper-pair spin $\boldsymbol{S}^{\mathrm{p}}$, the quasiparticle spin $\boldsymbol{S}^{\mathrm{q}}$ does *not* follow the field $\boldsymbol{H}_{\mathrm{inst}}$ instantaneously. Rather $\boldsymbol{S}^{\mathrm{q}}$ relaxes to

$$\boldsymbol{S}^{\mathrm{q}}_{\mathrm{eq}} = \chi^{\mathrm{q}0}\boldsymbol{H}_{\mathrm{inst}} \tag{8.186}$$

through the action of collision processes. (This is the value it would attain if $\boldsymbol{H}_{\mathrm{inst}}$ were itself time-independent.) Here

$$\chi^{\mathrm{q}0} = \chi^{0} - \chi^{\mathrm{p}0} \tag{8.187}$$

is the partial susceptibility of the quasiparticle system. The timescale for this adjustment to occur is hence given by a quasiparticle collision time τ_{CE} (Combescot and Ebisawa 1974). Taking into account the precession of $\boldsymbol{S}^{\mathrm{q}}$ in the screened magnetic field $\boldsymbol{H}_{\mathrm{eff}}$, we may then write the equation of motion for $\boldsymbol{S}^{\mathrm{q}}$ as

$$\dot{\boldsymbol{S}}^{\mathrm{q}} = \boldsymbol{S}^{\mathrm{q}} \times \boldsymbol{H}_{\mathrm{eff}} - \frac{1}{\tau_{\mathrm{CE}}}(\boldsymbol{S}^{\mathrm{q}} - \boldsymbol{S}^{\mathrm{q}}_{\mathrm{eq}}). \tag{8.188}$$

As one approaches the normal state from below T_{c}, the quasiparticle spin $\boldsymbol{S}^{\mathrm{q}}$ becomes equal to the total spin $\boldsymbol{S}$, which is a conserved quantity. It follows that τ_{CE} has to diverge in the limit $T \to T_{\mathrm{c}}$. From microscopic theory, one indeed finds $\tau_{\mathrm{CE}} \propto (1 - T/T_{\mathrm{c}})^{-1/2}$. (For a detailed derivation of τ_{CE} in the B phase, see (10.149); the same qualitative behaviour is also found in the A phase.) In the usual experimental situation the value of the instantaneous field $\boldsymbol{H}_{\mathrm{inst}}$ is not an externally controlled variable. Rather, $\boldsymbol{H}_{\mathrm{inst}}$ is determined by the total spin $\boldsymbol{S}$ and the actual value of $\boldsymbol{S}^{\mathrm{q}}$, as seen by inverting (8.183) and using (8.182):

$$\boldsymbol{H}_{\mathrm{inst}} = \frac{\gamma}{\chi^{\mathrm{p}0}}(\boldsymbol{S} - \boldsymbol{S}^{\mathrm{q}}). \tag{8.189}$$

With the aid of (8.189), the local equilibrium value of $\boldsymbol{S}^{\mathrm{q}}$, given by (8.186), can be expressed in terms of $\boldsymbol{S}$ and $\boldsymbol{S}^{\mathrm{q}}$. Substituting this expression for $\boldsymbol{S}^{\mathrm{q}}_{\mathrm{eq}}$ into (8.188), the equation of motion for $\boldsymbol{S}^{\mathrm{q}}$ may be rewritten in the form

$$\dot{\boldsymbol{S}}^{\mathrm{q}} = \boldsymbol{S}^{\mathrm{q}} \times \boldsymbol{H}_{\mathrm{eff}} - \frac{1}{\tau_{\mathrm{LT}}}(\boldsymbol{S}^{\mathrm{q}} - \boldsymbol{S}^{\mathrm{q}}_{0}), \tag{8.190}$$

which describes the relaxation of $\boldsymbol{S}^{\mathrm{q}}$ to the equilibrium value

$$\boldsymbol{S}^{\mathrm{q}}_{0} = \frac{\chi^{\mathrm{q}0}}{\chi^{0}}\boldsymbol{S}. \tag{8.191}$$

It simply expresses the fact that the effective magnetic fields responsible for the total spin polarization $\boldsymbol{S}$ and the quasiparticle spin polarization $\boldsymbol{S}^{\mathrm{q}}$ must be equal in equilibrium. The relaxation time τ_{LT} was first introduced by Leggett and Takagi (1975) and is related to τ_{CE} by (see also (10.152))

$$\tau_{\mathrm{LT}} = \frac{\chi^{\mathrm{p}0}}{\chi^{0}}\tau_{\mathrm{CE}}. \tag{8.192}$$

Since χ^{p0} vanishes linearly in $\Delta(T)$ for $T \to T_c$ (see (10.153) for the B-phase result), τ_{LT} stays *finite* in this limit. In fact, it may be shown from microscopic theory that τ_{LT} becomes equal to the normal-state relaxation time τ_N^0 in this case (see (10.154)).

The equations of motion for $\boldsymbol{S}$, $\boldsymbol{d}(\boldsymbol{k})$ and $\boldsymbol{S}^q$, (8.8a), (8.185b) and (8.190), together with the constitutive equations (8.182), provide a complete description of spin relaxation in superfluid ^{3}He. It may be inferred from the derivations given above that they should be valid as long as the superfluid component is in local equilibrium, i.e. for frequencies $\omega \ll \Delta/\hbar$, independently of the dynamics of the quasiparticles. On the other hand, the microscopic derivation presented in Section 10.3 shows that this is not strictly true. Indeed, the quasiparticle dynamics are different in the hydrodynamic ($\omega\tau \ll 1$) and collisionless regimes ($\omega\tau \gg 1$). Therefore one should expect the relaxation rates τ_{CE}^{-1} and τ_{LT}^{-1} to be frequency-dependent. This is borne out by microscopic theory. In fact, the frequency dependence of the relaxation rate as calculated in (10.148b) turns out to be weak. Therefore the equation of motion for $\boldsymbol{S}^q$, (8.190), with a frequency-independent relaxation time τ_{LT}, is expected to be a reasonable approximation even for $\omega\tau > 1$.

The derivation of the equation of motion presented above can actually be formulated in a more precise way (Leggett and Takagi 1977). In fact, the reactive part of the equations of motion for $\boldsymbol{S}$, $\boldsymbol{d}$ and $\boldsymbol{S}^q$ may be derived from the effective Hamiltonian

$$H = f_D\{\boldsymbol{d}(\boldsymbol{k})\} - \gamma \boldsymbol{S}\cdot\boldsymbol{H} + \frac{1}{2}\frac{\gamma^2}{\chi_N^0} F_0^a \boldsymbol{S}^2 + \frac{1}{2}\frac{\gamma^2}{\chi^{p0}}(\boldsymbol{S}^p)^2 + \frac{1}{2}\frac{\gamma^2}{\chi^{q0}}(\boldsymbol{S}^q)^2 \qquad (8.193)$$

by assuming that $\boldsymbol{S}^p$ and $\boldsymbol{S}^q$ commute and the components of $\boldsymbol{S}^p$ and $\boldsymbol{S}^q$ obey spin commutation relations.

Let us now turn to the solution of (8.190). In the hydrodynamic regime, $\omega\tau_{LT} \ll 1$, the equation of motion for $\boldsymbol{S}^q$ can be approximated by

$$\boldsymbol{S}^q - \boldsymbol{S}_0^q \approx \tau_{LT}(\gamma \boldsymbol{S}_0^q \times \boldsymbol{H}_{eff} - \dot{\boldsymbol{S}}_0^q) \qquad (8.194a)$$

$$= -\tau_{LT}\frac{\chi^{q0}}{\chi^0}\boldsymbol{R}_D, \qquad (8.194b)$$

since $\boldsymbol{S}^q$ approaches the equilibrium value $\boldsymbol{S}_0^q$ in the limit $\omega\tau_{LT} \to 0$. Equation (8.194b) follows by expanding $\boldsymbol{S}_0^q$ in terms of $\boldsymbol{S}$, using (8.191) and employing the equation of motion (8.8a) for $\boldsymbol{S}$. According to (8.194b), the dipole torque is effectively driving the quasiparticle spin polarization out of equilibrium, although it only acts on the Cooper-pair spins! Substituting (8.194b) into (8.185b), one obtains the equation of motion for $\boldsymbol{d}(\boldsymbol{k})$ including relaxation effects to lowest order in $\omega\tau$:

$$\dot{\boldsymbol{d}}(\boldsymbol{k}) = \boldsymbol{d}(\boldsymbol{k}) \times \gamma\left(\boldsymbol{H} - \frac{\gamma}{\chi}\boldsymbol{S}'\right), \qquad (8.195)$$

where

$$\boldsymbol{S}' = \boldsymbol{S} + \tau_{\mathrm{R}}\boldsymbol{R}_{\mathrm{D}} \tag{8.196a}$$

$$= \boldsymbol{S} + \tau_{\mathrm{R}}(\dot{\boldsymbol{S}} - \gamma \boldsymbol{S} \times \boldsymbol{H}). \tag{8.196b}$$

Here we have defined an effective relaxation time τ_{R} by

$$\tau_{\mathrm{R}} = \frac{\chi}{\chi^0}\frac{\chi^{\mathrm{q}0}}{\chi^{\mathrm{p}0}}\tau_{\mathrm{LT}}. \tag{8.197}$$

In the limit $T \to T_{\mathrm{c}}$, τ_{R} diverges as $(1 - T/T_{\mathrm{c}})^{-1/2}$ (see Fig. 10.5), but this divergence in (8.196) is over-compensated by the temperature dependence of $|\boldsymbol{R}_{\mathrm{D}}| \propto 1 - T/T_{\mathrm{c}}$. At low temperatures the rapid increase of $\tau_{\mathrm{LT}} \propto \tau$ (see (10.150)) for the B phase is reduced by the decrease of $\chi^{\mathrm{q}0}$ so that τ_{R} is a considerably weaker function of temperature than τ_{LT}.

We may now use (8.195) to calculate the NMR resonance linewidths in the linear regime. As can be seen from (8.8b), the only change that has to be made in the oscillator equation (8.10) is to replace $\delta\boldsymbol{S}$ by $\delta\boldsymbol{S}'$ according to (8.196). This leads to the Fourier-transformed equation

$$[\omega^2 - (1 - \mathrm{i}\omega\tau_R)\boldsymbol{\Omega}^2]\,\delta\boldsymbol{S} - \mathrm{i}\gamma(\omega + \mathrm{i}\tau_{\mathrm{R}}\boldsymbol{\Omega}^2)\,\delta\boldsymbol{S} \times \boldsymbol{H}_0 = -\gamma\boldsymbol{S}_0 \times \delta\boldsymbol{H}. \tag{8.198}$$

For longitudinal spin motion, one finds the resonance condition

$$\omega^2 = (1 - \mathrm{i}\omega\tau_{\mathrm{R}})\Omega_{\parallel}^2, \tag{8.199}$$

where $\Omega_{\parallel}^2$ is given by Ω_{A}^2 and Ω_{B}^2 for the A and B phases respectively. The linewidth of the longitudinal resonance may be extracted from (8.199) (taking into account that $\Omega_{\parallel}\tau_{\mathrm{R}} \ll 1$) as

$$\Gamma_{\parallel} = \Omega_{\parallel}^2\tau_{\mathrm{R}}. \tag{8.200}$$

Note that $\Gamma_{\parallel}$ is predicted to vanish $\propto(1 - T/T_{\mathrm{c}})^{1/2}$ as $T \to T_{\mathrm{c}}$, which is equivalent to a finite relative linewidth $\Gamma_{\parallel}/\Omega_{\parallel}$. In the B phase at low temperatures (8.200) yields the temperature dependence $\Gamma_{\parallel} \propto (T/T_{\mathrm{c}})^3$, in qualitative agreement with experiment (Bloyet *et al.* 1979). A more precise expression can be derived using (10.150).

The width of the transverse resonance in the A phase is found from (8.198) as (Combescot and Ebisawa 1974)

$$\Gamma_{\perp}^{\mathrm{A}} = \Gamma_{\parallel}^{\mathrm{A}}\frac{\Omega_{\mathrm{A}}^2}{\omega_{\mathrm{L}}^2 + \Omega_{\mathrm{A}}^2}, \tag{8.201}$$

i.e. the resonance narrows $\propto H^{-2}$ at high fields ($\omega_{\mathrm{L}} \gg \Omega_{\mathrm{A}}$). Note that in deriving (8.201) only the weaker condition $(\omega - \omega_{\mathrm{L}})\tau_{\mathrm{R}} \ll 1$ had to be satisfied. In the B phase, on the other hand, the linewidth is zero in the present approximation, at least as long as there is no transverse resonance shift. As can be seen from (8.194b), in the absence of the corresponding component of the dipole torque the quasiparticle spin system is not disturbed by the motion of the Cooper-pair spins, and no dissipation can occur. In fact, since the dipole torque $\boldsymbol{R}_D$ (and hence the dissipative torque given by (8.196)) is directed along $\hat{\boldsymbol{n}}$, dissipative terms appear only in the equation of motion for θ, (8.41b), but not in the equation of motion for $\hat{\boldsymbol{n}}$, (8.41c). However, as discussed below (8.34), there is a small transverse resonance shift $\Delta\omega_t$ even in the B phase for perfect orientation of the axis $\hat{\boldsymbol{n}}$ along the magnetic field, due to the slight distortion of the gap parameter by a magnetic field. This residual dipole torque will give rise to a small line broadening $\Gamma_{\perp}^{B} \propto \Gamma_{\parallel}^{B}(\Delta\omega_t/\omega_L)^2$.

The above theoretical results on the CW resonance linewidths are in partial agreement with experiment. This has been discussed in detail by Lee and Richardson (1978). (See Avenel *et al* (1975), Gully *et al.* (1976), Sager *et al.* (1977, 1978b) and Webb (1978b) for measurements of the width of the longitudinal resonance in the A phase and Bozler *et al.* (1974) and Gully *et al.* (1975) for measurements of the transverse linewidth, as well as Corruccini and Osheroff (1975b) for measurements of the relaxation time T_1.) In particular, for $T \to T_c$, the longitudinal linewidth in the A phase is found to *increase* rather than decrease as predicted by theory. This may possibly be attributed to spin supercurrents flowing from outside into the interior of the magnetic coil, thereby destroying the magnetization in the coil (Vuorio 1974, 1976, Sonin 1979). Another possibility is that the assumption of a fixed $\hat{\boldsymbol{l}}$-vector orientation, which is implicitly contained in the above derivation, fails close to T_c.

Leggett and Takagi (1977) have derived a very useful expression for the total energy change by intrinsic spin relaxation, which allows one to calculate the global effect of dissipation even for complicated spin motions. Starting from the expression for the total energy (8.1), the time derivative of the energy may be calculated in the hydrodynamic approximation (8.196) as

$$\frac{\mathrm{d}E}{\mathrm{d}t} = -\frac{\gamma^2}{\chi}\tau_R \boldsymbol{R}_D^2. \tag{8.202}$$

As a first application of (8.202), we calculate the recovery time of the longitudinal magnetization after a sudden change in the longitudinal magnetization, as happens for example in a field-turn-off experiment (Combescot 1976, Maki and Ebisawa 1976d). The time rate of change of the total energy as given by (8.202) is determined by the average value of $\boldsymbol{R}_D^2$. For a large turnoff field ΔH along $\hat{z}$, such that $\gamma\,\Delta H \gg \Omega_{\parallel}$, the longitudinal NMR response has been shown in Section 8.3 to be equivalent to that of a mathematical pendulum rotating with a frequency corresponding to the initial impulse. (This analogy is exact in the case of the A phase, see (8.72); in the case of the B phase, (8.79), the existence of two inequivalent minima suggest the model of *two* pendula, coupled in such a way that their angles are always in the proportion 1:2). Immediately after the pulse, $\hat{\boldsymbol{d}}$ therefore rotates in the (x, y) plane with a constant angular velocity, the motion being described by some angle θ (e.g. the relative angle to $\hat{\boldsymbol{l}}$ in the case of the A phase). In other words, the angle θ runs through all possible values with roughly equal probability. One may therefore average $\boldsymbol{R}_{\mathrm{D}}^2$ over all values of θ. In the A phase, where, according to (8.39a), $|\boldsymbol{R}_{\mathrm{D}}| = (\chi/2\gamma^2)\Omega_{\mathrm{A}}^2 \sin 2\theta$, the average comes out as

$$\langle \boldsymbol{R}_{\mathrm{D}}^2 \rangle_\theta = \frac{1}{2}\left(\frac{\chi}{2\gamma^2}\Omega_{\mathrm{A}}^2\right)^2, \tag{8.203}$$

so that the average energy dissipation (8.202) is given by

$$\left\langle \frac{\mathrm{d}E}{\mathrm{d}t} \right\rangle = -\frac{1}{8}\frac{\chi}{\gamma^2}\Gamma_{\parallel}^{\mathrm{A}}\Omega_{\mathrm{A}}^2. \tag{8.204}$$

Hence one obtains a *constant* time rate of change of the total energy, i.e. a *linear* decrease of energy with time. Relating the energy change (8.202) to the initial energy $\frac{1}{2}\chi(\Delta H)^2$, the recovery time $T_{\mathrm{rec},\parallel}^{\mathrm{A}}$ is found from (8.204) as

$$T_{\mathrm{rec},\parallel}^{\mathrm{A}} = 4\left(\frac{\gamma\,\Delta H}{\Omega_{\mathrm{A}}}\right)^2 \frac{1}{\Gamma_{\parallel}^{\mathrm{A}}}. \tag{8.205}$$

An analogous derivation for the B phase, where (see (8.41a)) $|\boldsymbol{R}_{\mathrm{D}}| = \frac{4}{15}(\chi/\gamma^2)\Omega_{\mathrm{B}}^2 \sin\theta\,(1 + 4\cos\theta)$, so that $\langle \boldsymbol{R}_{\mathrm{D}}^2\rangle_\theta = \frac{8}{45}(\chi\Omega_{\mathrm{B}}^2/\gamma^2)^2$, yields

$$T_{\mathrm{rec},\parallel}^{\mathrm{B}} = \frac{45}{16}\left(\frac{\gamma\,\Delta H}{\Omega_{\mathrm{B}}}\right)^2 \frac{1}{\Gamma_{\parallel}^{\mathrm{B}}}. \tag{8.206}$$

As a second example, the recovery time for the magnetization after a large-angle *tipping* pulse may be calculated from (8.202), if during the motion of $\boldsymbol{d}$ the whole unit sphere is essentially covered. This is the case in the A phase, where a fixed reference direction, given by the vector $\hat{\boldsymbol{l}}$, forces $\hat{\boldsymbol{d}}$ to perform large-amplitude oscillations. Averaging $\boldsymbol{R}_{\mathrm{D}}^2$ over the unit sphere, and considering that the total energy dissipated is given by $\chi(1-\cos\phi)H_0^2$, where H_0 is the applied static field and ϕ is the tipping angle, one finds a recovery time

$$T_{\mathrm{rec},\perp}^{\mathrm{A}}(\phi) = \frac{15}{2}\left(\frac{\gamma H_0}{\Omega_{\mathrm{A}}}\right)^2 (1-\cos\phi)\frac{1}{\Gamma_{\parallel}^{\mathrm{A}}}. \tag{8.207}$$

The above predictions on the recovery of magnetization in the A phase are in qualitative agreement with experiment (Corruccini and Osheroff 1975b). So the linear recovery in time as well as an increase of T_{rec} with magnetic field are clearly observed. The deviations are probably again due to the effect of spin currents.

In the B phase the situation is more complicated (Brinkman and Smith 1975b). As we have seen, there exists a kind of motion where the system is almost in the minimum of the dipole well, i.e. $\theta \approx \theta_{\mathrm{L}}$, whereas $\boldsymbol{S}$ and $\hat{\boldsymbol{n}}$ precess about a common axis. The relaxation during the time interval in which the latter mode of motion is approached is described roughly by a recovery time of the order of (8.207). The relaxation in the long-lived mode, however, must be calculated by taking the time dependence of $\boldsymbol{R}_{\mathrm{D}}^2$ explicitly into account. The above theory is in qualitative agreement with the observed spin-relaxation behaviour in the B phase after a large tipping pulse (Ishikawa *et al.* 1989).

As an application of (8.202) to a problem involving the *long-lived motion* of the collective variables in the B phase, we now consider the damping of the wall-pinned ringing mode (Leggett 1975b, Maki and Ebisawa 1976d, Leggett and Takagi 1977, Takagi 1977, Soda 1977a, Cross 1978). As discussed at the end of Section 8.3, this particular mode is excited by switching off a small magnetic field $\Delta H \ll \Omega_{\mathrm{B}}/\hbar$ in a configuration with $\hat{\boldsymbol{n}}$ initially perpendicular to the magnetic field. This mode is caused by a large-amplitude precession of $\hat{\boldsymbol{n}}$ about the axis $\boldsymbol{C}_{\mathrm{B}}$ with frequency $\omega(0) = (\frac{2}{5})^{1/2}\gamma\,\Delta H$, which excites a small-amplitude precession of $\boldsymbol{S}$ with the same frequency. Under these circumstances, the system always stays close to the minimum of the dipole energy, i.e. $\theta \approx \theta_{\mathrm{L}}$. Any function of θ may therefore be approximated well by expanding it to lowest order in $\theta - \theta_{\mathrm{L}}$. Substituting such an expansion of the dipole torque $\boldsymbol{R}_{\mathrm{D}}$ (see (8.41a)) into (8.202), one may express the energy relaxation as

$$\frac{\mathrm{d}E}{\mathrm{d}t} = -\frac{\chi}{\gamma^2}\tau_{\mathrm{R}}\Omega_{\mathrm{B}}^4(\theta - \theta_{\mathrm{L}})^2. \tag{8.208}$$

Substituting the precession frequency $\omega(t)$ $(=\dot{\varphi}_{\hat{\boldsymbol{n}}})$ as given by (8.71b) into (8.70), eliminating $\tilde{C}_{\mathrm{B}}$ and dropping the small term $\ddot{\theta}$, $\theta - \theta_{\mathrm{L}}$ may be related to $\omega(t)$ by

$$\theta(t) - \theta_{\mathrm{L}} = \left(\frac{15}{16}\right)^{1/2}\left[\frac{\omega(t)}{\Omega_{\mathrm{B}}}\right]^2. \tag{8.209}$$

In this simple problem the total energy may hence be calculated *explicitly* in terms of $\omega(t)$. From the general expression (8.69) for $E(t)$, one obtains the simple relation

$$E(t) = \frac{5}{4}\frac{\chi}{\gamma^2}\,\omega^2(t). \tag{8.210}$$

Here we have again neglected the terms $\dot{\theta}^2$ and $(\theta - \theta_{\mathrm{L}})^2$, which are small, i.e. of order $(\omega/\Omega_{\mathrm{B}})^2$. Combining (8.208), (8.209) and (8.210), one is led to the following differential equation for $\omega(t)$:

$$\frac{\mathrm{d}}{\mathrm{d}t}\,\omega^2(t) = -\tfrac{3}{4}\,\tau_{\mathrm{R}}\omega^4(t), \tag{8.211}$$

with the solution

$$\omega(t) = \frac{\omega(0)}{[1 + \frac{3}{4}\tau_{\mathrm{R}}\omega^2(0)t]^{1/2}}. \tag{8.212}$$

The precession frequency is hence found to decrease with time as energy is dissipated and taken out of the mode. The relaxation time for this is given by $[\omega^2(0)\tau_R]^{-1}$, which under suitable experimental conditions can be much longer than the longitudinal relaxation time given by (8.206). The ringing down of the wall-pinned mode as predicted by (8.212) has been observed experimentally (Wheatley 1978; see also Webb 1978a,c). It provides a method for measuring the intrinsic spin relaxation time τ_{LT}. Numerical solutions to the Leggett–Takagi equations have been obtained by Katayama *et al.* (1980a,b,c, 1983a,b), Katayama (1981) and Ooiwa (1982).

The effect of dissipation on the nonlinear spin dynamics of ^{3}He-B discussed in Section 8.3 has been studied by Golo (1981a), Golo and Leman (1982a,b, 1983), Fomin (1983), Golo *et al.* (1983) and Kesaev and Ugulava (1984); for a review see Golo and Leman (1989). It is found that the interplay between the Leggett–Takagi dissipative mechanism and the nonlinearity (particularly where the nonlinear ringing frequency vanishes) leads to interesting effects. In particular, a new type of propagating magnetic excitation in the B phase, i.e. a spatial modulation of nonlinear ringing of the magnetization, has been proposed by Golo (1984). The nonlinear interaction of the latter mode with acoustic modes has been studied with the aim of finding a new way of detecting second sound (Golo and Kats 1986).

The existence of surface spin waves at a boundary characterized by a diffuse scattering law and the corresponding contribution to the NMR absorption have been discussed by Fishman (1979).

The homogeneous spin dynamics of the A_1 phase in the nonlinear regime, including spin relaxation, has been discussed by de Vegvar (1984) within the adiabatic approximation. The intrinsic superfluid relaxation is seen to be dominated by normal T_1 processes, leading to magnetization recoveries as in the A (i.e. A_2) phase.

Finally, we comment briefly on the effect of spin diffusion, which provides another mechanism for dissipation in spatially nonuniform situations. The corresponding terms in the equations of motion for the spin density are considered in Chapter 9 (equations (9.29)–(9.31)), and the spin-diffusion coefficient is calculated within kinetic theory in Chapter 10 (equations 10.138a–d)).

The combined effect of the Leggett–Tagaki relaxation mechanism and spin diffusion has been discussed by Fomin (1984b, 1985b), who thereby explained the duration of the long-lived induction signal in the B phase measured by Borovik-Romanov *et al.* (1984b). Spin-relaxation processes in ^{3}He-A have been studied in particular by Bun'kov *et al.* (1985) using a pulsed NMR method. They found that at low tilt angles the relaxation was indeed due mainly to the Leggett–Takagi mechanism, while for tilt angles $\gtrsim 40°$ an instability occurred in the homogeneous spin precession leading to large-amplitude spin waves. Their findings are in agreement with Fomin's earlier theoretical treatment of the problem of fast spin relaxation (Fomin 1979b, 1981a, 1984a).

Early attempts to determine the spin-diffusion coefficient experimentally, using the spin-echo method, failed (Corruccini and Osheroff 1978). In later experiments, spin echoes were found in both the A and B phases (Eska *et al.* 1981). These echoes were found to have a rather peculiar structure: while a single echo could be detected in the B phase at sufficiently small tipping angles, a sequence of *multiple* spin echoes was observed in the A phase and in the B phase at large tipping angles. Multiple spin echoes generally appear if the resonant frequency, or the spin-diffusion coefficient, vary with the spin-tipping angle. The structure of these echoes has been explained on the basis of the theory presented in this chapter for the case without dissipation (Hasegawa 1982), and including spin relaxation but neglecting spin diffusion (Fomin *et al.* 1984). We mention in passing that multiple spin echoes have also been seen experimentally in normal liquid ^{3}He (Einzel *et al.* 1984a).

Relaxation of spin superflow in the A_1 phase has been investigated experimentally by Lu *et al.* (1989). The discontinuity in the relaxation time as a function of temperature observed by these authors may be explained by the effect of the A_1–A_2 interface present in the experiment (Grabinski 1989).

FURTHER READING

General reviews and books

Abragam A and Goldman M 1982 *Nuclear Magnetism: Order and Disorder* (Clarendon Press, Oxford)
Anderson P W and Brinkman W F 1978 in *The Physics of Liquid and Solid Helium,* Part II, ed. K H Bennemann and J B Ketterson (Wiley, New York), p. 177
Lee D M and Richardson R C 1978 in *The Physics of Liquid and Solid Helium,* Part II, ed. K H Bennemann and J B Ketterson (Wiley, New York), p. 287
Leggett A J 1975 *Rev. Mod. Phys.* **47** 331
Wheatley J C 1975 *Rev. Mod. Phys.* **47** 415

Specialized reviews and articles

Brinkman W F and Cross M C 1978 in *Progress in Low Temperature Physics,* Vol. VIIA, ed. D F Brewer (North-Holland, Amsterdam), p. 105
Fetter A L 1986 in *Progress in Low Temperature physics,* Vol. X, ed. D F Brewer (North-Holland, Amsterdam), p. 1
Fomin I A 1981 *Soviet Scientific Reviews,* Section A: *Physics Reviews,* Vol. 3 (Harwood Academic Publishers, Chur), p. 275
Giannetta R W, Smith E N and Lee D M 1981 *J. Low Temp. Phys.* **45** 295
Golo V L and Leman A A 1989 in *Helium 3,* ed. W P Halperin and L P Pitaevskii (North-Holland, Amsterdam)
Leggett A J 1974 *Ann. Phys.* (*NY*) **85** 11
Leggett A J 1978 in *Quantum Liquids,* ed. J Ruvalds and T Regge (North-Holland, Amsterdam), p. 167
Leggett A J and Takagi S 1977 *Ann. Phys.* (*NY*) **106** 79
Maki K 1986 in *Solitons,* ed. S E Trullinger, V. E. Zakharov and V. L. Pokrovskii (North-Holland, Amsterdam), p. 435
Salomaa M M and Volovik G E 1987 *Rev. Mod. Phys.* **59** 533

9

Hydrodynamic Theory

The dynamics of any macroscopic system in the limit of low frequencies and long wavelengths can be completely described in terms of a few so-called "hydrodynamic" variables. These variables owe their existence to the behaviour of the system under continuous symmetry transformations (see Chapter 6). Consequently, the structure of the equations of motion obeyed by these variables—the hydrodynamic equations—is uniquely determined by the symmetry properties of the system.

Let us consider a nonequilibrium initial state of a system. Most degrees of freedom will relax towards equilibrium within a time period of the order of a microscopic relaxation time τ. In the case of ^{3}He, τ is essentially the quasiparticle collision time, which has been discussed in Chapter 2 for the normal state. The corresponding collision time for the superfluid state will be investigated in Chapter 10. This collision time is of order 10^{-7} s near T_c, increases at lower temperatures and eventually diverges for $T \rightarrow 0$. Now, the important point is that a small number of macroscopic degrees of freedom are not affected at all by microscopic relaxation processes—either (i) because there is a conservation law rendering collision processes ineffective, or (ii) because there is a spontaneously broken continuous symmetry. In the latter case the equilibrium state is energetically degenerate with respect to changes of the symmetry variable. Hence a uniform change of this variable is unobservable. The set of variables consisting of these two types, namely the conserved quantities and those describing the broken symmetry, is sufficient to completely determine the state of the system in the hydrodynamic regime.

The existence of conserved quantities is intimately related to the continuous symmetries under which the system is invariant. Thus gauge symmetry, translational symmetry in time, translational symmetry in space, and rotational symmetry in position space and in spin space (neglecting the weak spin–orbit coupling) give rise to the conservation of particle number (or mass), energy, momentum, angular momentum and spin. The corre-

Table 9.1 Hydrodynamic variables and associated symmetries.

Variable	Symmetry
mass density $\rho(\boldsymbol{r}, t)$	gauge
momentum density $\boldsymbol{g}(\boldsymbol{r}, t)$	space translation
angular-momentum density $\boldsymbol{L}(\boldsymbol{r}, t)$	orbital rotation
energy density $\epsilon(\boldsymbol{r}, t)$	time translation
spin density $\boldsymbol{S}(\boldsymbol{r}, t)$	spin rotation
phase variable ϕ	broken gauge
orbital rotation angles θ_j^{o}	broken orbital rotation
spin rotation angles θ_μ^{s}	broken spin rotation

sponding hydrodynamic variables are listed in Table 9.1 (top) together with the variables associated with potentially broken symmetries in the superfluid phases of ^{3}He (bottom).

9.1 GENERAL PRINCIPLES

The hydrodynamic variables are related to each other by virtue of the general principles of thermodynamics. To see this, one should bear in mind that a change in the energy of the total system may be caused by changes in (i) the heat content $T\,\mathrm{d}S$, (ii) the mechanical volume energy $P\,\mathrm{d}V$, (iii) the energy of linear motion $\boldsymbol{v}\cdot\mathrm{d}\boldsymbol{P}$ (where $\boldsymbol{P}$ is the total momentum and $\boldsymbol{v}$ is the velocity of the system), (iv) the energy of rotational motion $\boldsymbol{\Omega}\cdot\mathrm{d}\boldsymbol{L}_{\mathrm{tot}}$ (where $\boldsymbol{L}_{\mathrm{tot}}$ is the angular momentum and $\boldsymbol{\Omega}$ is the angular velocity of the system), (v) the mass $\mathrm{d}M$ and (vi) the magnetization $\gamma\,\mathrm{d}\boldsymbol{S}$. Obviously the system's energy does not depend on a symmetry variable itself (provided the symmetry holds strictly), but it does depend on the gradient of a symmetry variable (here we assume that the gradient is uniform).

9.1.1 Thermodynamic identities

This leaves us with the following thermodynamic identity for the whole system:

$$\mathrm{d}E = T\,\mathrm{d}S - P\,\mathrm{d}V + \mu\,\mathrm{d}M + \boldsymbol{v}\cdot\mathrm{d}\boldsymbol{P} + \boldsymbol{\Omega}\cdot\mathrm{d}\boldsymbol{L}_{\mathrm{tot}} + \sum_\nu \omega_\nu^{\mathrm{s}}\,\mathrm{d}S_\nu + \boldsymbol{J}^\phi\cdot\mathrm{d}\boldsymbol{\nabla}\phi, \tag{9.1}$$

where μ is the chemical potential per unit mass, the ω_ν^{s} are angular velocity components conjugate to the total spin and j^ϕ is a "current" associated with $\boldsymbol{\nabla}\phi$. The last term in (9.1) stands for all terms induced by the spontaneously broken symmetries. In the context of hydrodynamics, we are obviously interested in local rather than global thermodynamic identities. Let us therefore change to a local description by introducing the densities of energy $\epsilon = E/V$, entropy $s = S/V$, mass $\rho = M/V$, momentum $\boldsymbol{g} = \boldsymbol{P}/V$, angular momentum $\boldsymbol{L} = \boldsymbol{L}_{\text{tot}}/V$ and current $\boldsymbol{j}^{\phi} = \boldsymbol{J}^{\phi}/V$ induced by symmetry breaking. Applying the identity (9.1) to a volume element of fixed size small enough that spatial dependences of the thermodynamic variables can be neglected (but large enough that local equilibrium is established), one derives the local thermodynamic identity

$$d\epsilon = T\,ds + \mu\,d\rho + \boldsymbol{v} \cdot d\boldsymbol{g} + \boldsymbol{\Omega} \cdot d\boldsymbol{L} + \boldsymbol{\omega}^{s} \cdot d\boldsymbol{S} + \boldsymbol{j}^{\phi} \cdot d\nabla\phi. \tag{9.2}$$

This identity is valid for sufficiently slow processes, i.e. in the hydrodynamic regime.

Let us next consider an infinitesimal change in volume, dV. Then it follows from (9.1) that the pressure P may be completely expressed in terms of the conserved densities:

$$P = -\epsilon + Ts + \mu\rho + \boldsymbol{v} \cdot \boldsymbol{g} + \boldsymbol{\Omega} \cdot \boldsymbol{L}. \tag{9.3}$$

Combining the two identities, one finds for the change in pressure

$$dP = s\,dT + \rho\,d\mu + \boldsymbol{g} \cdot d\boldsymbol{v} + \boldsymbol{L} \cdot d\boldsymbol{\Omega} + \boldsymbol{S} \cdot d\boldsymbol{\omega}^{s} - \boldsymbol{j}^{\phi} \cdot d\nabla\phi. \tag{9.4}$$

The thermodynamic identities (9.2) and (9.4), when combined with the equations of motion for the hydrodynamic variables and the second law of thermodynamics, suffice to determine the structure of the hydrodynamic equations. More precisely, the thermodynamic identities serve to relate changes in the hydrodynamic variables to the change in entropy density. At this point it is important to remember that the entropy density is subject to constraints following from the second law of thermodynamics, which states that the total entropy of a closed system never decreases. This severely restricts the form of the equation describing the time evolution of the entropy density, and therefore the equations of motion for the hydrodynamic variables.

9.1.2 Standard procedure for deriving hydrodynamic equations

The derivation of the hydrodynamic equations proceeds according to the following general rules (Khalatnikov 1965, Liu 1982).

(i) As a first step, one has to identify the conserved densities and symmetry variables of the system in question. This turns out to be somewhat subtle in the case of superfluid ^{3}He. While in both the A and B phases several symmetries are simultaneously broken, this does not automatically imply that an *independent* dynamical variable has to be added to the set of hydrodynamical variables for each broken symmetry. In fact, as first pointed out by Liu and Cross (1979), when two symmetry operations affect the order parameter in the same way, only *one* independent variable associated with the relative symmetry transformations of the two respective symmetries need be included in the set of hydrodynamic variables. For example, in the A phase a gauge transformation can be undone by a corresponding rotation in orbital space about the preferred direction (see Section 6.1 for a detailed discussion). The dynamical variable accounting for both symmetry transformations is the difference of the phase angle ϕ and the rotation angle θ_z^{o} (assuming $\hat{\boldsymbol{l}}$ to lie along the z direction). Similarly, in the B phase the change in the order parameter under an arbitrary rotation in orbital space may be compensated for by a corresponding rotation in spin space. Therefore, instead of the six rotation angles θ_i^{o} and θ_i^{s}, only three relative angles θ_i^{so} are allowed as independent hydrodynamic variables.

(ii) Next, the appropriate conservation laws and equations of motion for the symmetry variables are listed. These equations have the following general structure (see e.g. (2.21)–(2.24)): the time derivative of the hydrodynamic variable is equal to the negative of the divergence of the corresponding current. The currents are expressed in terms of gradients of the hydrodynamic variables, the so-called "thermodynamic forces" or "thermodynamically conjugate fields". The behaviour of the currents under transformations to linearly moving (Galilean) and rotating frames as well as under time reversal provides constraints on the form of these currents. In this way, some of the coefficients in the expressions for the currents may be determined.

(iii) It is now necessary to employ the thermodynamic identity (9.2) and to express the change in entropy density ds in terms of differentials of the conserved quantities and the symmetry variables. By interpreting ds as an entropy change in time, this may be converted into an equation for the time derivative of the entropy density in terms of the time derivatives of the hydrodynamic variables.

(iv) Using the conservation laws, one may now replace the time derivative of each hydrodynamic variable by the divergence of the corresponding current. This yields the time derivative of the entropy in terms of spatial derivatives of the currents. To lowest order in the gradients, i.e. on the level of local equilibrium, entropy is conserved. In general, however, there is a finite positive entropy production. With this in mind, the time evolution of the entropy density may be written as

$$\frac{\mathrm{d}s}{\mathrm{d}t} + \boldsymbol{\nabla} \cdot \boldsymbol{j}_s = \frac{R}{T}, \tag{9.5}$$

where $\boldsymbol{j}_s$ is the entropy current density and R is the heat-production function. The latter is a positive-definite quadratic form (up to surface terms) in the gradients of the thermodynamically conjugate variables

$(\partial s/\partial H_i)$:

$$R = \sum_{ij} \nabla\left(\frac{\partial s}{\partial H_i}\right) L_{ij} \cdot \nabla\left(\frac{\partial s}{\partial H_j}\right), \tag{9.6}$$

where H_i is any one of the hydrodynamic variables.

Note that only the symmetric part of the matrix of transport coefficients L_{ij} contributes to the dissipation function R (the Onsager relations). In addition, the heat production R is an even function under time reversal. On the other hand, the two terms on the left-hand side of (9.5), representing the conservation of entropy, must have identical time-reversal properties. As the first term is the time derivative of the scalar equilibrium quantity s, it is obviously odd under time reversal; consequently the entropy current density has to be odd, too, as one might expect. The terms in the equation of motion for any hydrodynamic variable H_i may be classified as "reactive" or "dissipative" respectively, according to whether they have the same or the opposite time-reversal property as $\partial H_i/\partial t$. Writing $\partial s/\partial t$ as in (9.5) puts sufficient constraints on the possible functional structure of the currents to determine their precise form.

Before we turn to the detailed derivation of the hydrodynamic equations for the superfluid phases of ^{3}He, we first consider the equations of motion for the symmetry variables in general, neglecting dissipation.

9.1.3 Equations of motion for the symmetry variables

The spontaneous breakdown of a continuous symmetry gives rise to a modification of the hydrodynamic theory in two ways (Liu 1982). First, the new symmetry variable leads to a new equation of motion, which has to be included in the set of hydrodynamic equations. Secondly, the new hydrodynamic variable enters the already existing hydrodynamic equations. In particular, there appears a dissipationless current proportional to the gradient of the symmetry variable in the conservation law related to the symmetry that is broken.

In local equilibrium, the equation of motion of the symmetry variable, ϕ say, is given by equating the material derivative to a generalized force

$$\frac{\mathrm{d}\phi}{\mathrm{d}t} = \frac{\partial \phi}{\partial t} + \boldsymbol{v} \cdot \nabla \phi = \frac{\partial E}{\partial G}, \tag{9.7}$$

where G is the thermodynamic variable corresponding to the generator of the symmetry transformation in question.

Substituting (9.7) into the entropy-production equation (9.5), one is led to a term $\boldsymbol{j}^{\phi} \cdot \nabla(\partial E/\partial G)$, which is reactive. As such, it must be part of the divergence of the entropy current density on the left-hand side of (9.5). However, this term is not a divergence of a current itself. Consequently, there must exist an additional term of the form $(\partial E/\partial G)\nabla\cdot\boldsymbol{j}^{\phi}$, which completes it to give a divergence. The prefactor $\partial E/\partial G$ appears in the thermodynamic density (9.2) in front of the differential $\mathrm{d}G$, indicating that this additional term comes from the conservation law for G. In fact, it is generated precisely by an additional contribution $-\boldsymbol{j}^{\phi}$ to the current density of G. The ensuing contribution to the entropy equation (9.5) combines to give an additional entropy current density $\boldsymbol{j}^{\phi}(\partial E/\partial G)$.

As a first example, let us take the broken gauge invariance characteristic of a superfluid. The corresponding symmetry variable is the phase variable ϕ. The generator of gauge transformations in a pair-correlated system is $\frac{1}{2}N$, where N is the particle number operator. Hence the equation of motion for ϕ, neglecting dissipative terms, is

$$\frac{\mathrm{d}\phi}{\mathrm{d}t}=-\frac{2m}{\hbar}\mu, \tag{9.8}$$

with $\mu=m^{-1}\partial E/\partial N$ being the chemical potential per unit mass (the change in sign, $\phi\to-\phi$, is conventional). In this case a nondissipative current appears in the conservation law associated with gauge symmetry, i.e. in the continuity equation (2.21). It is the well-known superfluid mass current.

Similarly, the broken rotational symmetry in spin space as encountered for instance in an antiferromagnetic normal fluid provides three additional hydrodynamic variables; namely the three rotation angles $\theta^{\mathrm{s}}_{\mu}$ about the three axes in spin space. The fact that these variables are globally ill-defined, because of the noncommutativity of three-dimensional rotations, has to be kept in mind when dealing with second- or higher-order derivatives, as will be explained later. The generators of rotations in spin space are the three components of the spin operator S_{μ}. According to the general prescription formulated above, the equations of motion for the symmetry variables $\theta^{\mathrm{s}}_{\mu}$ are then given by

$$\frac{\mathrm{d}\theta^{\mathrm{s}}_{\mu}}{\mathrm{d}t}=\frac{\partial E}{\partial S_{\mu}}\equiv\omega^{\mathrm{s}}_{\mu}. \tag{9.9}$$

From the magnetic-energy expression

$$E_M=\frac{\gamma^2}{2\chi}\boldsymbol{S}^2-\gamma\boldsymbol{S}\cdot\boldsymbol{H}, \tag{9.10}$$

one finds the field driving the motion of $\theta^{\mathrm{s}}_{\mu}$ as

$$\omega^{\mathrm{s}}_{\mu}=\gamma\left(\frac{\gamma S_{\mu}}{\chi}-H_{\mu}\right). \tag{9.11}$$

This necessitates the addition of a nondissipative current contribution

$$\boldsymbol{j}^{M}_{\mu}=\frac{-\delta E}{\delta(\nabla\theta^{\mathrm{s}}_{\mu})}$$

to the spin conservation equation.

Finally, the broken rotation symmetry in orbital space occurring, for example, in a biaxial nematic liquid cyrstal is specified by three rotation angles θ_j^{o}. The generator of rotations is the angular-momentum operator $\boldsymbol{L}$. However, as mentioned above, angular momentum and momentum are closely related. A local angular-momentum density $\boldsymbol{L}$ contributes to the momentum density as

$$\boldsymbol{g} = \boldsymbol{p} + \tfrac{1}{2}\boldsymbol{\nabla} \times \boldsymbol{L}, \tag{9.12}$$

where $\boldsymbol{p}$ is the linear-momentum density. The term $\frac{1}{2}\boldsymbol{\nabla} \times \boldsymbol{L}$ is akin to the magnetization current in electrodynamics. Differentiating the energy of motion $E = \int \mathrm{d}^3 \boldsymbol{r}\, \boldsymbol{v} \cdot \boldsymbol{g}$ with respect to the angular-momentum density $\boldsymbol{L}$, one finds

$$\frac{\mathrm{d}\theta_j^{\text{o}}}{\mathrm{d}t} = \frac{\delta E}{\delta L_j} \equiv \Omega_j, \tag{9.13}$$

where

$$\boldsymbol{\Omega} = \tfrac{1}{2}\boldsymbol{\nabla} \times \boldsymbol{v}$$

is the vorticity. According to the general rule described above, we expect the appearance of a dissipation-free current in the conservation law for angular momentum. However, the angular-momentum conservation law is *not* an independent hydrodynamic equation. Rather, the close connection of $\boldsymbol{g}$ and $\boldsymbol{L}$ expressed by (9.12) leads to a nondissipative current of the form $\frac{1}{2}\sum_i \boldsymbol{\nabla} \times (\boldsymbol{\nabla}_i\, \delta E/\delta(\nabla_i \boldsymbol{\theta}^{\text{o}}))$ in the momentum conservation law.

9.2 HYDRODYNAMIC EQUATIONS FOR THE B PHASE

The order parameter of the BW state as given in (6.23) is the product of a phase factor $\mathrm{e}^{\mathrm{i}\phi}$ and a real orthogonal matrix $R_{\mu j}$ describing a rotation of spin space relative to orbital space. It follows that there are four additional symmetry variables: (i) the phase variable $\phi(\boldsymbol{r}, t)$ associated with broken gauge symmetry; and (ii) the three rotation angles $\theta_j^{\text{so}}(\boldsymbol{r}, t)$ associated with the broken relative spin–orbit rotation symmetry. In the following we shall only work with the differentials

$$\mathrm{d}\theta_j^{\text{so}} = \mathrm{d}\theta_j^{\text{o}} - \sum_{\mu} R_{\mu j}\, \mathrm{d}\theta_\mu^{\text{s}} \tag{9.14}$$

(where $R_{\mu j}$ describes the local order-parameter structure), because the variables θ_j^{so} are globally ill-defined.

Even when working with infinitesimal changes in the rotation angles, we have to be careful if we are dealing with operations involving two successive infinitesimal rotations about different axes. In general, we have to observe the commutation rule (Liu and Cross 1979)

$$(\delta\nabla - \nabla\delta)\boldsymbol{\theta}^{\text{so}} = \delta\boldsymbol{\theta}^{\text{so}} \times \nabla\boldsymbol{\theta}^{\text{so}}, \tag{9.15}$$

where δ and ∇ denote two arbitrary differentiations. In order to demonstrate the validity of (9.15), consider two successive infinitesimal rotations described by rotation angles $\mathrm{d}\theta_1$ and $\mathrm{d}\theta_2$ respectively, such that $\boldsymbol{r}\rightarrow \boldsymbol{r}+\boldsymbol{r}\times \mathrm{d}\boldsymbol{\theta}_1$ etc., where $\mathrm{d}\boldsymbol{\theta}_i=\mathrm{d}\theta_i\,\hat{\boldsymbol{n}}_i$ and $\hat{\boldsymbol{n}}_i$ is a unit vector in the direction of the rotation axis. The difference in the resulting vectors after carrying out the successive rotations in the order $\mathrm{d}\theta_1\,\mathrm{d}\theta_2$ and in the order $\mathrm{d}\theta_2\,\mathrm{d}\theta_1$ is just $\boldsymbol{r}\times(\mathrm{d}\boldsymbol{\theta}_1\times \mathrm{d}\boldsymbol{\theta}_2)$, which agrees with (9.15).

9.2.1 The thermodynamic identity and equilibrium conditions

Recalling that a gradient in the phase variable ϕ gives rise to a supercurrent (see Chapter 7), it is useful to eliminate $\boldsymbol{\nabla}\phi$ in favour of the superfluid velocity

$$\boldsymbol{v}_\mathrm{s}=\frac{\hbar}{2m}\boldsymbol{\nabla}\phi. \tag{9.16}$$

The thermodynamic identity for the B phase (omitting the angular-momentum term for the moment) then takes the form (Graham and Pleiner 1976, Liu and Cross 1978)

$$\begin{aligned}T\,\mathrm{d}s=\mathrm{d}\epsilon-\mu\,\mathrm{d}\rho-\boldsymbol{v}_\mathrm{n}\cdot\mathrm{d}\boldsymbol{g}-\boldsymbol{g}_\mathrm{s}\cdot\mathrm{d}\boldsymbol{v}_\mathrm{s}-\sum_\mu \omega^\mathrm{s}_\mu\,\mathrm{d}S_\mu\\ -\sum_{ij}\frac{\partial f_\mathrm{bend}}{\partial(\nabla_j\theta^\mathrm{so}_i)}\mathrm{d}(\nabla_j\theta^\mathrm{so}_i)-\sum_i\frac{\partial f}{\partial\theta^\mathrm{so}_i}\mathrm{d}\theta^\mathrm{so}_i,\end{aligned} \tag{9.17}$$

where f_bend is the bending free-energy density of the B phase (7.25b). Here we have denoted the velocity conjugate to the momentum density by $\boldsymbol{v}_\mathrm{n}$, because it has to be identified with the so-called normal-fluid velocity, as will become clear below. The variable $\boldsymbol{g}_\mathrm{s}$ has the dimensions of a momentum density, too; i.e. it must be given by the product of a mass density times a velocity. In the frame moving with velocity $\boldsymbol{v}_\mathrm{n}$, the only other velocity vector left is $\boldsymbol{v}_\mathrm{s}-\boldsymbol{v}_\mathrm{n}$, and therefore

$$\boldsymbol{g}_\mathrm{s}=\rho_\mathrm{s}(\boldsymbol{v}_\mathrm{s}-\boldsymbol{v}_\mathrm{n}). \tag{9.18}$$

Going back to the laboratory frame, the momentum density is given by

$$\boldsymbol{g}=\rho\boldsymbol{v}_\mathrm{n}+\boldsymbol{g}_\mathrm{s}=\rho_\mathrm{n}\boldsymbol{v}_\mathrm{n}+\rho_\mathrm{s}\boldsymbol{v}_\mathrm{s}, \tag{9.19}$$

where $\rho_\mathrm{n}=\rho-\rho_\mathrm{s}$ and $\boldsymbol{v}_\mathrm{n}$ are the normal-fluid density and velocity discussed in Section 3.5. The contribution $\rho_\mathrm{s}\boldsymbol{v}_\mathrm{s}$ is precisely the dissipationless current mentioned after (9.8).

The field conjugate to $\nabla_j\theta^\mathrm{so}_i$ (in 9.17) may be calculated by functionally differentiating the bending free-energy density for the BW state, (7.25b), with respect to $\nabla_j\theta^\mathrm{so}_i$:

$$\frac{\partial f_\mathrm{bend}}{\partial(\nabla_l\theta^\mathrm{so}_k)}=2\left(\frac{\hbar}{2m}\right)^2[a\nabla_l\theta^\mathrm{so}_k+b\nabla_k\theta^\mathrm{so}_l+c\boldsymbol{\nabla}\cdot\boldsymbol{\theta}^\mathrm{so}]. \tag{9.20}$$

9.2.2 Hydrodynamic equations for the symmetry variables

We have so far assumed the system to be invariant under rotations in spin–orbit space. However, as elaborated in Section 6.3, there is a small spin–orbit interaction—the dipole force—which acts as a kind of anisotropy field on the preferred directions. This effect is contained in the last term of the thermodynamic identity (9.17). Noting that the dipole energy density f_D, (6.100), may be expressed in terms of the spin–orbit variables $\hat{\boldsymbol{n}}$ and $\boldsymbol{\theta}^{so}$ as given by (8.159), the conjugate field $\partial f_D/\partial\boldsymbol{\theta}^{so} = \Omega_B^2(\chi/\gamma^2)(\hat{\boldsymbol{n}}\cdot\boldsymbol{\theta}^{so})\hat{\boldsymbol{n}}$ is nothing but the dipole torque $\boldsymbol{R}_D$ discussed in Section 8.1 (Brinkman and Smith 1974).

From the requirement that the entropy should be a maximum in equilibrium, it follows via (9.17) that the temperature T, the chemical potential μ and the normal velocity $\boldsymbol{v}_n$ as well as the spin angular velocity $\boldsymbol{\omega}^s$ are constant in equilibrium. The vanishing of the variation of s with the phase variable ϕ implies $\boldsymbol{\nabla}\cdot\boldsymbol{g}_s = 0$.

The equilibrium condition on the spin–orbit angles θ_j^{so} is obtained by taking the functional derivative of the total free energy $F_{bend} = \int d^3r\, f_{bend}$ with respect to θ_j^{so}:

$$\frac{\delta f_{bend}}{\delta\theta_j^{so}} \equiv \psi_j = \sum_k \nabla_k\left[\frac{\partial f_{bend}}{\partial(\nabla_j\theta_k^{so})} - \sum_{il}\epsilon_{jil}\theta_i^{so}\frac{\partial f_{bend}}{\partial(\nabla_l\theta_k^{so})}\right] + \varphi n_j, \tag{9.21}$$

where

$$\varphi = \hat{\boldsymbol{n}}\cdot\frac{\partial f_D}{\partial\boldsymbol{\theta}^{so}} = \Omega_B^2\frac{\chi}{\gamma^2}\hat{\boldsymbol{n}}\cdot\boldsymbol{\theta}^{so}.$$

The second term on the right-hand-side of (9.21) comes from the commutation relation (9.15), which has to be applied when a partial integration is performed. In equilibrium the free energy is minimal and hence $\psi_j \equiv 0$.

The thermodynamic forces driving the system towards equilibrium are given by the gradients of those quantities that are spatially uniform in equilibrium, i.e. by $\boldsymbol{\nabla}T$, $\boldsymbol{\nabla}\mu$ (or equivalently $\boldsymbol{\nabla}P$), $\nabla_i v_{nj}$, $\boldsymbol{\nabla}g_s$, $\boldsymbol{\nabla}\omega_\mu^s$ and ω_j. In general, the currents in the hydrodynamic equations can be represented by linear combinations involving all thermodynamic forces. However, only some of the thermodynamic forces are allowed by symmetry for each given case.

With the aid of (9.8), the equation of motion for the superfluid velocity $\boldsymbol{v}_s$ is obtained as

$$\frac{d\boldsymbol{v}_s}{dt} + \boldsymbol{\nabla}(\mu + \mu^D) = 0. \tag{9.22}$$

The dissipative part of the chemical potential is a scalar of first order in the gradients. The only quantities of this type that may be formed by the conjugate fields are the divergences of the normal velocity $\boldsymbol{v}_n$ and the superfluid momentum density $\boldsymbol{g}_\mathrm{s}$, leading to the expression

$$\mu^\mathrm{D} = -\zeta_1 \boldsymbol{\nabla} \cdot \boldsymbol{v}_\mathrm{n} - \zeta_3 \boldsymbol{\nabla} \cdot \boldsymbol{g}_\mathrm{s}, \tag{9.23}$$

where ζ_1 and ζ_3 are bulk viscosity coefficients. The expression for μ^D is identical to the one for an isotropic superfluid (Khalatnikov 1965).

Similarly, the equation of motion for the spin–orbit angles θ_i^so follows by combining (9.9) and (9.13) as (Graham and Pleiner 1976, Liu and Cross 1978)

$$\frac{\mathrm{d}\theta_j^\mathrm{so}}{\mathrm{d}t} + \sum_\mu R_{\mu j}\omega_\mu^\mathrm{s} - \tfrac{1}{2}(\boldsymbol{\nabla} \times \boldsymbol{v}_\mathrm{n})_j + \Omega_j^\mathrm{D} = 0. \tag{9.24}$$

The dissipative part Ω_j^D again must be given by a linear combination of the thermodynamic forces. The only force that has the symmetry of a vector in orbit space is ψ_j as defined in (9.21), and hence

$$\Omega_j^\mathrm{D} = -\nu\psi_j, \tag{9.25}$$

where ν is a kind of spin–orbit viscosity.

9.2.3 Hydrodynamic equations for the conserved quantities

The remaining hydrodynamic equations are given by the conservation laws.

(i) The continuity equation has the usual form (see (2.21)), with the momentum density given by (9.16).

(ii) The momentum conservation equation is given explicitly by

$$\frac{\mathrm{d}g_i}{\mathrm{d}t} + \nabla_i P - \tfrac{1}{2}(\boldsymbol{\nabla} \times \boldsymbol{\psi})_i + \sum_k \nabla_k(\Pi_{ik}^\mathrm{e} + \Pi_{ik}^\mathrm{D}) = 0. \tag{9.26}$$

Here the term $-\frac{1}{2}(\boldsymbol{\nabla} \times \boldsymbol{\psi})$ is due to the reactive current in the angular-momentum conservation law, which appears as a consequence of the broken spin–orbit rotational symmetry, as explained below (9.13). The contribution Π_{ik}^e to the stress tensor is similar to the so-called Erikson stress tensor for nematic liquid crystals: it describes the stress of an anisotropic medium, i.e.

$$\Pi_{ik}^\mathrm{e} = \sum_j \frac{\partial f_\mathrm{bend}}{\partial(\nabla_k \theta_j^\mathrm{so})} \nabla_i \theta_j^\mathrm{so} + g_{\mathrm{s},k} v_{\mathrm{s},i}. \tag{9.27}$$

Technically speaking, this contribution may be traced back to the last term in the expression for the change in pressure $\mathrm{d}P$, as given by (9.4). Note that the first term on the right-hand side of (9.27) involves θ^so and hence induces a coupling of mechanical and magnetic degrees of freedom. This effect might be observable in a torsional-oscillator experiment, where the shear flow would give rise to an NMR signal. In addition, the symmetry of the dissipative part of the stress tensor Π_{ij}^D allows one to add a term proportional to

$\nabla \cdot \boldsymbol{g}_s$ to the expression for the normal fluid so that in the B phase it has the general form

$$\Pi^{\mathrm{D}}_{ik} = -\eta(\nabla_i v_{\mathrm{n},k} + \nabla_k v_{\mathrm{n},i} - \tfrac{2}{3}\delta_{ik}\nabla \cdot \boldsymbol{v}_\mathrm{n}) - (\zeta_1 \nabla \cdot \boldsymbol{g}_\mathrm{s} + \zeta_2 \nabla \cdot \boldsymbol{v}_\mathrm{n})\delta_{ik}. \tag{9.28}$$

Of the three coefficients η, ζ_1 and ζ_2, the bulk viscosities ζ_1 and ζ_2 will be shown to be order $(T/T_\mathrm{F})^2$ small compared with the shear viscosity η, and may therefore be neglected. Note that Onsager's symmetry principle requires that the coefficient of $\nabla \cdot \boldsymbol{g}_\mathrm{s}$ in (9.28) be equal to the coefficient of $\nabla \cdot \boldsymbol{v}_\mathrm{n}$ in (9.23).

(iii) The equation of motion for the spin density $\boldsymbol{S}$ is obtained as

$$\frac{\mathrm{d}S_\mu}{\mathrm{d}t} - \gamma(\boldsymbol{S} \times \boldsymbol{H})_\mu + \sum_{ij} \nabla_i \left[\frac{\partial f_{\mathrm{bend}}}{\partial(\nabla_i \theta_j^{\mathrm{so}})} R_{\mu j} + \alpha \sum_k \epsilon_{ijk} R_{\mu j} \nabla_k T + \delta_{ij} X^{\mathrm{D}}_{\mu i} \right] = \sum_k R_{\mu k}\varphi_k, \tag{9.29}$$

where the first term in the square brackets is the reactive spin supercurrent, i.e. the symmetry current discussed after (9.7), and the second term is a thermally driven reactive spin current. The dissipative spin current is due entirely to spin diffusion:

$$X^{\mathrm{D}}_{\mu i} = -\sum_{\nu j} D_{\mu i \nu j} \nabla_j \omega^{\mathrm{s}}_\nu. \tag{9.30}$$

The spin-diffusion coefficient is a fourth-rank tensor, which may be characterized by three independent components D_1, D_2 and D_3:

$$D_{\mu i \nu j} = D_1 \delta_{ij}\delta_{\mu\nu} + D_2 R_{\mu i} R_{\nu j} + D_3 R_{\mu j} R_{\nu i}. \tag{9.31}$$

Employing the above hydrodynamic equations and the thermodynamic identities (9.17), the entropy-production equation is found to be

$$\frac{\mathrm{d}s}{\mathrm{d}t} + \sum_i \nabla_i \left[f^{\mathrm{D}}_i + \alpha \sum_{j,k,\mu} \epsilon_{ijk} R_{\mu j} \nabla_k \omega^{\mathrm{s}}_\mu \right] = \frac{R}{T}, \tag{9.32}$$

where f^{D}_i is the usual entropy current

$$f^{\mathrm{D}}_i = -\frac{\kappa}{T}\nabla_i T = \frac{1}{T} j_{Qi}, \tag{9.33}$$

with $\boldsymbol{j}_Q$ the heat current.

The entropy-production function may be cast into the form

$$\begin{aligned} R = {} & \frac{\kappa}{T}(\nabla T)^2 + \frac{1}{\zeta_2}(\zeta_2 \nabla \cdot \boldsymbol{v}_\mathrm{n} + \zeta_1 \nabla \cdot \boldsymbol{g}_\mathrm{s})^2 + \frac{1}{\zeta_2}(\zeta_2\zeta_3 - \zeta_1^2)(\nabla \cdot \boldsymbol{g}_\mathrm{s})^2 \\ & + \tfrac{1}{2}\eta \sum_{ij} (\nabla_i v_{\mathrm{n},j} + \nabla_j v_{\mathrm{n},i} - \tfrac{2}{3}\delta_{ij}\nabla \cdot \boldsymbol{v}_\mathrm{n})^2 + \nu \psi^2 \\ & + (D_1 + D_3)\sum_{ij} B^{\mathrm{s}}_{ij} B^{\mathrm{s}}_{ji} + D_2 \Big(\sum_i B^{\mathrm{s}}_{ii} \Big)^2 + (D_1 - D_3) \sum_{ij} B^{\mathrm{a}}_{ij} B^{\mathrm{a}}_{ji}, \end{aligned} \tag{9.34}$$

where

$$B^{s,a}_{\alpha\beta} = \tfrac{1}{2} \sum_i (R_{\alpha i} \nabla_i \omega^s_\beta \pm R_{\beta i} \nabla_i \omega^s_\alpha).$$

Positivity of R requires

(a) the transport coefficients κ, ζ_2, ζ_3, η, v and D_1 to be positive;

(b) $\zeta_2 \zeta_3 > \zeta_1^2$; (9.35)

(c) $D_1 > |D_3|$; $D_2 > -\tfrac{1}{3}(D_1 + D_3)$. (9.36)

9.3 HYDRODYNAMIC EQUATIONS FOR THE A PHASE

The hydrodynamics of the A phase has been a source of considerable controversy. It is only recently that a consistent picture has begun to emerge. Following the first derivation of the linearized hydrodynamic equations by straightforward application of the rules laid down in Section 9.1 (Graham 1974, Graham and Pleiner 1975, 1976), several amendments and refinements have been proposed (Liu 1975, 1976, 1977, 1978, Volovik and Mineev 1976c, 1981, Liu and Cross 1979, Mermin and Ho 1976). In particular, the role of an intrinsic orbital angular-momentum density associated with the Cooper pairs has attracted much interest (see also Section 7.3). This is partly because microscopic derivations of the hydrodynamic equations revealed the existence of terms that could be interpreted as being due to an intrinsic angular momentum (Combescot and Dombre 1980, Combescot 1978b, 1980a, Mermin and Muzikar 1980, Nagai 1980, Ishikawa *et al.* 1980). While microscopic theories are of importance in providing quantitative information about the coefficients in the hydrodynamic equations, they are usually model-dependent. The microscopic approach consequently cannot be expected to provide the most general form of the hydrodynamic equations admitted by symmetry. Following Miyake *et al.* (1981), Liu (1985) and Hall and Hook (1986) (see also Hu and Saslow 1977, Ho 1977, Lhuillier 1977, Nagai 1981a,b, Sonin 1984b), we shall therefore attempt to derive the hydrodynamics of the A phase phenomenologically, assuming this state to possess an intrinsic orbital angular momentum of density $\boldsymbol{l}(\boldsymbol{r}, t)$. Note that an intrinsic angular momentum is necessarily of quantum-mechanical origin. Alternative but similar derivations of the hydrodynamics of ^{3}He-A have been considered by Khalatnikov and Lebedev (1977a, 1977b, 1980), Lebedev and Khalatnikov (1977) and Gu (1982a, b). Early discussions of the orbital dynamics of ^{3}He-A include those of Hall (1975, 1976, 1977, 1978), Hall and Hook (1977), Pleiner (1977b) and Brand and Pleiner (1981a). A hydrodynamic description of ^{3}He-A in terms of quantum fields, where in addition to the ordinary (bosonic) variables anticommuting (fermionic) variables are also introduced ("superhydrodynamics"), has been formulated by Andreev and Kagan (1987).

9.3.1 The concept of an intrinsic angular momentum

What are the changes in the hydrodynamic equations if there is a finite angular-momentum density $\boldsymbol{t}$? As mentioned in Section 9.1, in the absence of an intrinsic angular momentum the conservation of orbital angular momentum $\boldsymbol{L}$ does not give rise to an additional independent conservation law. In fact, the angular-momentum conservation law can be derived from the momentum conservation law by employing the definition of the orbital angular-momentum density $\boldsymbol{L}=\boldsymbol{r}\times\boldsymbol{g}$. Note that angular momentum is only conserved if the stress tensor is symmetric. In the presence of an intrinsic angular momentum, the angular-momentum density with respect to any given centre is given by

$$\boldsymbol{L}=\boldsymbol{r}\times\boldsymbol{g}+\boldsymbol{t}. \tag{9.37}$$

Here $\boldsymbol{t}$ is defined to be the intrinsic part of $\boldsymbol{L}$, i.e. $\boldsymbol{t}$ is independent of the size of the system, whereas the position-dependent part $\boldsymbol{r}\times\boldsymbol{g}$ scales with the size R of the system as R^2.

The thermodynamic equilibrium state of such a system is found by maximizing the entropy as given in (9.4). The resulting state is characterized by a uniform temperature T, chemical potential μ, angular velocity $\boldsymbol{\Omega}$ and uniform linear velocity $\boldsymbol{u}$. However, because of the appearance of $\boldsymbol{g}$ in the expression for $\boldsymbol{L}$, (9.37), the total velocity $\boldsymbol{v}_{\mathrm{n}}$ obtained by taking the derivative of ϵ with respect to $\boldsymbol{g}$ at fixed $\boldsymbol{t}$ is nonuniform. In fact, it is given by a superposition of linear and rotational motion:

$$\boldsymbol{v}_{\mathrm{n}}=\boldsymbol{u}+\boldsymbol{\Omega}\times\boldsymbol{r}. \tag{9.38}$$

Equation (9.38) may be used to express the angular velocity in terms of the velocity $\boldsymbol{v}_{\mathrm{n}}$ as

$$\boldsymbol{\Omega}=\tfrac{1}{2}\nabla\times\boldsymbol{v}_{\mathrm{n}}. \tag{9.39}$$

This relation carries over to a local-equilibrium situation, i.e. into the hydrodynamic regime. It implies that $\boldsymbol{\Omega}$ is not an independent variable. Furthermore, the thermodynamic force $\nabla_i\Omega_j$ is a second-order spatial derivative of v_{nk} and may only give rise to dissipative terms of higher order in the gradients. Hence additional transport coefficients connected with the internal angular momentum do not appear.

There are, however, new reactive terms in the momentum conservation equation. Since $\boldsymbol{r}\times\boldsymbol{g}$ is no longer a conserved angular-momentum density, the stress tensor is no longer symmetric. Rather, a term matching the antisymmetric part of the stress tensor appears in the equation of motion for $\boldsymbol{t}$, so that the conservation of $\boldsymbol{L}=\boldsymbol{r}\times\boldsymbol{g}+\boldsymbol{t}$ is guaranteed. It is now convenient to define a new momentum density $\boldsymbol{j}$ (Martin *et al.* 1972) by

$$\boldsymbol{j}=\boldsymbol{g}+\tfrac{1}{2}\nabla\times\boldsymbol{t}, \tag{9.40}$$

with the properties (i) $\int d^3r\,\boldsymbol{j}=\int d^3r\,\boldsymbol{g}$ up to surface terms and (ii) $\int d^3r\,\boldsymbol{r}\times\boldsymbol{j}=\int d^3r\,(\boldsymbol{r}\times\boldsymbol{g}+\boldsymbol{t})$. Since $\boldsymbol{r}\times\boldsymbol{j}$ can be seen to be identical with the angular-momentum density $\boldsymbol{L}$, the stress tensor in the equation of motion for $\boldsymbol{j}$ must be symmetric. This property facilitates the task of collecting all possible terms in the stress tensor allowed by symmetry.

The intrinsic angular momentum $\boldsymbol{t}$ partly originates from the orbital angular momentum of a Cooper pair, which in the A phase is directed along the vector $\hat{\boldsymbol{l}}$. This part, $\boldsymbol{t}_0$, points along $\hat{\boldsymbol{l}}$, i.e.

$$\boldsymbol{t}_0 = t_0 \hat{\boldsymbol{l}}. \tag{9.41}$$

The angular-momentum density $\boldsymbol{t}_0$ is a nonhydrodynamical variable, and may be viewed in analogy to the superfluid density ρ_s, for example.

To understand the significance of $\boldsymbol{t}_0$, we consider a uniformly rotating container filled with ^{3}He-A. In equilibrium the normal fluid will move rigidly with the walls of the container. There will be a change in the free energy proportional to the angular velocity $\boldsymbol{\Omega}$. For a large system, this change will be dominated by the macroscopic contribution to the angular-momentum density, $\boldsymbol{r} \times \boldsymbol{g}$, which scales with the size of the system as $R^2 \times$ volume. However, there will also be a change in free energy induced by the intrinsic angular momentum. It has not so far been possible to experimentally determine $\boldsymbol{t}_0$ (but see Hall 1985, Liu 1985).

For finite vorticity $\boldsymbol{\Omega}$, there is also an induced intrinsic angular momentum

$$\boldsymbol{t}_{\text{ind}} = \boldsymbol{\theta}\boldsymbol{\Omega}, \tag{9.42}$$

where $\boldsymbol{\theta}$ is a moment-of-inertia tensor, which has to be positive for stability reasons. The total intrinsic angular momentum is the sum of the two contributions (9.41) and (9.42):

$$\boldsymbol{t} = \boldsymbol{t}_0 + \boldsymbol{t}_{\text{ind}}. \tag{9.43}$$

The induced angular momentum $\boldsymbol{t}_{\text{ind}}$ should be expected to exist even in the normal phase, but its value would be rather small—of order $\hbar\Omega/k_F v_F$ per particle, where Ω is the angular velocity of the system.

9.3.2 Equations of motion for the symmetry variables

We are now in a position to embark on the derivation of the hydrodynamic equations for the A phase. As discussed in detail in Chapter 6, in this phase the rotational symmetries in orbital and spin spaces and the gauge symmetry are broken, i.e. all the symmetries are separately broken. However, it has already been pointed out that gauge transformations and rotations about the vector $\hat{\boldsymbol{l}}$ are indistinguishable as far as the transformation properties of the order parameter are concerned. In other words, only the relative gauge–orbit symmetry (GOS) is a relevant dynamical symmetry. The dynamical symmetry variables are therefore the three GOS angles θ_i^{go}. Their differentials are related to the infinitesimal changes in the orbital rotation angles $\delta\theta_i^{\rm o}$ and the phase variable $\delta\phi^{\rm o}$ by (Liu and Cross 1979)

$$\delta\theta_i^{\rm go} = \delta\theta_i^{\rm o} - l_i\,\delta\phi^{\rm o}. \tag{9.44}$$

Note that only one of the angles $\theta_i^{\rm go}$, namely the one describing rotations about $\hat{\boldsymbol{l}}$, is a relative symmetry variable, corresponding to the $U(1)$ symmetry discussed in Chapter 6. The notation used for the infinitesimal changes in $\theta_i^{\rm o}$, $\phi^{\rm o}$ and $\theta_i^{\rm go}$ is meant to indicate that they are not differentials in the usual sense owing to the noncommutativity of three-dimensional rotations.

We shall not consider the spin degrees of freedom of the order parameter explicitly, because for frequencies much lower than the longitudinal NMR resonance frequency $\Omega_{\rm L}$, the vector $\hat{\boldsymbol{d}}$ will be locked in the direction along $\hat{\boldsymbol{l}}$ by virtue of the dipole force ("dipole-locked regime").

On the other hand, for frequencies of order $\Omega_{\rm L}$ or higher, the $\hat{\boldsymbol{l}}$-vector field cannot follow, and may be considered as being in the equilibrium state, since its characteristic frequencies (orbital viscosity, see below) are much lower.

It is convenient to represent $\delta\theta_i^{\rm go}$ in terms of changes in a phase variable ϕ and the vector $\hat{\boldsymbol{l}}$ by defining

$$\left.\begin{aligned} \delta\phi &= -\hat{\boldsymbol{l}}\cdot\delta\boldsymbol{\theta}^{\rm go},\\ \delta\hat{\boldsymbol{l}} &= \delta\boldsymbol{\theta}^{\rm go}\times\hat{\boldsymbol{l}}. \end{aligned}\right\} \tag{9.45}$$

Furthermore, a superfluid velocity variable $\boldsymbol{v}_{\rm s}$ may be introduced in the usual way (see (7.10))

$$\boldsymbol{v}_{\rm s} = \frac{\hbar}{2m}\frac{\delta\phi}{\delta\boldsymbol{r}} = -\frac{\hbar}{2m}\sum_i l_i\boldsymbol{\nabla}\theta_i^{\rm go}. \tag{9.46}$$

In a nonuniform texture the superfluid velocity thus defined will not be irrotational. This follows from the noncommutativity of three-dimensional rotations, which gives rise to the noncommutativity of two (or more) successive derivatives of $\theta_i^{\rm go}$, exactly in the same way as found for the spin–orbit angle $\theta^{\rm so}$ in the B phase. Indeed, by employing the relation (9.15) as applied to $\theta_i^{\rm go}$ and making use of $\mathrm{d}\boldsymbol{\theta} = \hat{\boldsymbol{l}}\times\mathrm{d}\hat{\boldsymbol{l}}$, we find for the curl of $\boldsymbol{v}_{\rm s}$ the so-called Mermin–Ho relation (Mermin and Ho 1976)

$$\begin{aligned} (\boldsymbol{\nabla}\times\boldsymbol{v}_{\rm s})_i &= -\frac{\hbar}{2m}\sum_{jkm}\epsilon_{ijk}(\nabla_j l_m\nabla_k\theta_m^{\rm go})\\ &= \frac{\hbar}{4m}\sum_{jk}\epsilon_{ijk}\hat{\boldsymbol{l}}\cdot(\nabla_j\hat{\boldsymbol{l}}\times\nabla_k\hat{\boldsymbol{l}}). \end{aligned} \tag{9.47}$$

which has already been derived in (7.53).

We may now write down the reactive terms in the equations of motion for the symmetry variables. They are obtained by taking the time derivative of (9.44) and substituting for $\mathrm{d}\theta_i^{\rm o}/\mathrm{d}t$ and $\mathrm{d}\phi^0/\mathrm{d}t$ from (9.13) and (9.8). One

finds

$$\frac{\mathrm{d}\theta_i^{\mathrm{go}}}{\mathrm{d}t} = \Omega_i + \hat{l}_i \frac{2m\mu}{\hbar}. \tag{9.48}$$

In terms of ϕ and $\hat{\boldsymbol{l}}$, (9.48) takes the form

$$\frac{\mathrm{d}\phi}{\mathrm{d}t} + \hat{\boldsymbol{l}} \cdot \boldsymbol{\Omega} + \frac{2m\mu}{\hbar} = 0, \tag{9.49}$$

$$\frac{\mathrm{d}\hat{\boldsymbol{l}}}{\mathrm{d}t} = \boldsymbol{\Omega} \times \hat{\boldsymbol{l}}. \tag{9.50}$$

Note that the chemical potential in (9.48) is the one at fixed $\boldsymbol{t}$, i.e. $\mu = \mu_{\boldsymbol{t}}$. This is related to the chemical potential at fixed $\boldsymbol{\Omega}$, $\mu_{\boldsymbol{\Omega}}$, by

$$\mu_{\boldsymbol{t}} = \mu_{\boldsymbol{\Omega}} - \boldsymbol{\Omega} \cdot \left(\frac{\partial \boldsymbol{t}}{\partial \rho}\right)_{\boldsymbol{\Omega}}, \tag{9.51}$$

as will become clear later (see (9.60)). The chemical potential at fixed $\boldsymbol{\Omega}$ differs from its value at rest, μ_0, by the kinetic energy per particle, i.e.

$$\mu_{\boldsymbol{\Omega}} = \mu_0 + O(\Omega^2), \tag{9.52}$$

in analogy to the dependence of μ on the linear velocity $\boldsymbol{v}_{\mathrm{n}}$ given by (9.56). To first order in $\boldsymbol{\Omega}$, one therefore finds

$$\frac{\mathrm{d}\phi}{\mathrm{d}t} + \left(1 - \frac{\partial \rho_{\mathrm{L}}}{\partial \rho}\right)\hat{\boldsymbol{l}} \cdot \boldsymbol{\Omega} + \frac{2m\mu_0}{\hbar} = 0, \tag{9.53}$$

where $\rho_{\mathrm{L}} = (2m/\hbar)t_0$ is an effective mass density associated with the intrinsic angular momentum. It follows that a finite vorticity $\boldsymbol{\Omega} = \frac{1}{2}(\boldsymbol{\nabla} \times \boldsymbol{v}_{\mathrm{n}})$, due for example to a mechanical rotation of the sample, will wind up the phase and will eventually lead to a supercurrent. This is the remarkable gauge-wheel effect (Liu and Cross 1979, Ho and Mermin 1980, Miyake and Usui 1980b, Combescot 1982a), which offers a direct link between macroscopic mechanical motion and quantum-mechanical phase. It is seen that the prefactor of the $\hat{\boldsymbol{l}} \cdot \boldsymbol{\Omega}$ term in (9.53) depends on the density variation of the intrinsic angular momentum. Microscopic derivations of the hydrodynamics (Nagai 1980, Combescot 1980a,b, 1981a,b) give a strongly temperature-dependent coefficient of the $\hat{\boldsymbol{l}} \cdot \boldsymbol{\Omega}$ term, which even vanishes for $T \rightarrow 0$. As will be shown below, the latter property may actually be derived if one assumes the existence of an intrinsic angular momentum.

9.3.3 Thermodynamic identities and equilibrium conditions

In order to proceed with the derivation of the full hydrodynamic equations, we now establish the local thermodynamic relations. Let us consider the entropy density s as a function of the energy density ϵ, the mass density ρ, the momentum density $\boldsymbol{g}$, the superfluid velocity $\boldsymbol{v}_s$ and the $\hat{\boldsymbol{l}}$ field, as well as $\nabla_i l_j$ and the internal angular-momentum density $\boldsymbol{t}$. From the Galilean invariance of s, it follows that

$$s(\epsilon, \rho, \boldsymbol{g}, \boldsymbol{v}_s, \hat{\boldsymbol{l}}, \nabla_i l_j, \boldsymbol{t}) = s(\epsilon + \boldsymbol{g}\cdot\boldsymbol{u} + \tfrac{1}{2}\rho u^2, \rho, \boldsymbol{g} + \rho\boldsymbol{u}, \boldsymbol{v}_s + \boldsymbol{u}, \hat{\boldsymbol{l}}, \nabla_i l_j, \boldsymbol{t}), \tag{9.54}$$

where $-\boldsymbol{u}$ is the velocity of the moving frame. Observing that in local equilibrium the temperature T, chemical potential μ and normal velocity $\boldsymbol{v}_n$ are related to the partial derivatives of s by

$$\frac{\partial s}{\partial \epsilon} = \frac{1}{T}, \quad \frac{\partial s}{\partial \rho} = -\frac{\mu}{T}, \quad \frac{\partial s}{\partial \boldsymbol{g}} = -\frac{\boldsymbol{v}_n}{T},$$

we find that the chemical potential transforms according to

$$\mu' = \mu - \boldsymbol{v}_n\cdot\boldsymbol{u} - \tfrac{1}{2}u^2, \tag{9.55}$$

so that μ may be expressed in terms of the chemical potential μ_0 in the $\boldsymbol{v}_n = 0$ frame as

$$\mu = \mu_0 - \tfrac{1}{2}v_n^2. \tag{9.56}$$

Putting $\boldsymbol{u} = -\boldsymbol{v}_s$ in (9.54) and taking the derivative with respect to $\boldsymbol{v}_s$, we find furthermore

$$-T\frac{\partial s}{\partial \boldsymbol{v}_s} = \boldsymbol{g} - \rho\boldsymbol{v}_n \equiv \boldsymbol{g}_s. \tag{9.57}$$

The rotational invariance of the entropy density implies

$$s(\epsilon, \rho, \boldsymbol{g}, \boldsymbol{v}_s, \hat{\boldsymbol{l}}, \nabla_i l_j, \boldsymbol{t}) = s(\epsilon, \rho, \boldsymbol{g} + \delta\boldsymbol{\omega}\times\boldsymbol{g}, \boldsymbol{v}_s + \delta\boldsymbol{\omega}\times\boldsymbol{v}_s, \hat{\boldsymbol{l}} + \delta\boldsymbol{\omega}\times\hat{\boldsymbol{l}},$$
$$\nabla_i l_j + (\delta\boldsymbol{\omega}\times\boldsymbol{\nabla})_i l_j + \nabla_i(\delta\boldsymbol{\omega}\times\hat{\boldsymbol{l}})_j, \boldsymbol{t} + \delta\boldsymbol{\omega}\times\boldsymbol{t}) \tag{9.58}$$

for infinitesimal rotation angles $\delta\boldsymbol{\omega}$. To first order in $\delta\boldsymbol{\omega}$, (9.58) provides us with the relation

$$\hat{\boldsymbol{l}}\times\left(\frac{\partial s}{\partial \hat{\boldsymbol{l}}}\right)_{\boldsymbol{t}} + \sum_i \left[\boldsymbol{\nabla} l_i \times \frac{\partial s}{\partial(\boldsymbol{\nabla} l_i)} + \nabla_i \hat{\boldsymbol{l}} \times \frac{\partial s}{\partial(\nabla_i \hat{\boldsymbol{l}})}\right] + \frac{1}{T}\boldsymbol{g}_s\times(\boldsymbol{v}_s - \boldsymbol{v}_n) + \frac{1}{T}\boldsymbol{\Omega}\times\boldsymbol{t} = 0. \tag{9.59}$$

Here the derivatives of s have to be taken at fixed $\boldsymbol{t}$.

The local thermodynamic identity following from (9.54) takes the form

$$T\,\mathrm{d}s = \mathrm{d}\epsilon - \mu\,\mathrm{d}\rho - \boldsymbol{v}_n\cdot\mathrm{d}\boldsymbol{g} - \boldsymbol{g}_s\cdot\mathrm{d}\boldsymbol{v}_s - \frac{\partial f_{\text{bend}}}{\partial\hat{\boldsymbol{l}}}\cdot\mathrm{d}\hat{\boldsymbol{l}} - \sum_{ij}\frac{\partial f_{\text{bend}}}{\partial(\nabla_i l_j)}\mathrm{d}(\nabla_i l_j) - \boldsymbol{\Omega}\cdot\mathrm{d}\boldsymbol{t}, \tag{9.60}$$

where f_{bend} is the bending free-energy density (7.20).

The pressure is defined by considering a change in volume with the mass, energy, momentum and angular momentum, as well as the symmetry variables, held fixed:

$$P \equiv T\frac{\partial S}{\partial V} = Ts - \epsilon + \rho\mu + \boldsymbol{v}_\mathrm{n}\cdot\boldsymbol{g} + \boldsymbol{\Omega}\cdot\boldsymbol{t}. \tag{9.61}$$

On combining (9.60) and (9.61), we are led to the Duhem–Gibbs relation

$$\begin{aligned} \mathrm{d}P &= s\,\mathrm{d}T + \rho\,\mathrm{d}\mu + \boldsymbol{g}\cdot\mathrm{d}\boldsymbol{v}_\mathrm{n} - \boldsymbol{g}_\mathrm{s}\cdot\mathrm{d}\boldsymbol{v}_\mathrm{s} \\ &\quad + T\frac{\partial s}{\partial\hat{\boldsymbol{l}}}\cdot\mathrm{d}\hat{\boldsymbol{l}} + T\sum_{ij}\frac{\partial s}{\partial(\nabla_i l_j)}\mathrm{d}(\nabla_i l_j) + \boldsymbol{t}\cdot\mathrm{d}\boldsymbol{\Omega}. \end{aligned} \tag{9.62}$$

The appearance of the last term in (9.62) is one of the observable consequences of an intrinsic angular-momentum density $\boldsymbol{t}$: the pressure should increase or decrease in a rotating vessel depending on the relative orientation of $\hat{\boldsymbol{l}}$ and $\boldsymbol{\Omega}$. Even if one assumes $|\boldsymbol{t}| \approx (\hbar/m)\rho$, which is probably an upper limit on $|\boldsymbol{t}|$, the ensuing pressure difference is quite small, namely of order 10^{-7} Pa for a rotational speed 1 rad/s.

The conditions for total equilibrium at temperature T can be obtained by minimizing the free-energy density $f = \epsilon - Ts = f(T, \rho, \boldsymbol{g}, \boldsymbol{v}_\mathrm{s}, \nabla_i l_j, \hat{\boldsymbol{l}}, t)$ subject to the constraint $\hat{\boldsymbol{l}}\cdot\hat{\boldsymbol{l}} = 1$ and the Mermin–Ho relation (9.47). The condition on the normal velocity $\boldsymbol{v}_\mathrm{n}$ is given by (9.38), i.e. $\boldsymbol{\nabla}\cdot\boldsymbol{v}_\mathrm{n} = 0$. In order to guarantee stationary mass and energy density, the divergence of $\boldsymbol{g}_\mathrm{s}$ must vanish: $\boldsymbol{\nabla}\cdot\boldsymbol{g}_\mathrm{s} = 0$.

The $\hat{\boldsymbol{l}}$ texture in equilibrium is determined by requiring that the change in the free energy with respect to a local change in $\hat{\boldsymbol{l}}$ vanish. In other words, the local torque on the texture must be zero:

$$\hat{\boldsymbol{l}} \times \boldsymbol{\psi} = 0, \tag{9.63a}$$

where

$$\psi_i = \frac{\partial f_\mathrm{bend}}{\partial l_i} - \boldsymbol{\nabla}\cdot\frac{\partial f_\mathrm{bend}}{\partial(\boldsymbol{\nabla} l_i)} + \frac{\hbar}{2m}[\hat{\boldsymbol{l}}\times(\boldsymbol{g}_\mathrm{s}\cdot\boldsymbol{\nabla})\hat{\boldsymbol{l}}]_i. \tag{9.63b}$$

Here the derivatives are understood to be taken at fixed $\boldsymbol{t}$. The last term in (9.63b) derives from the dependence of f_bend on the superfluid velocity $\boldsymbol{v}_\mathrm{s}$. As $\boldsymbol{v}_\mathrm{s}$ is defined in terms of an orbital rotation about $\hat{\boldsymbol{l}}$ (see (9.45)), a change in $\hat{\boldsymbol{l}}$ naturally leads to a change in $\boldsymbol{v}_\mathrm{s}$. More explicitly,

$$\frac{\partial v_{\mathrm{s},j}}{\partial l_i} = \frac{\hbar}{2m}\sum_k \frac{\partial}{\partial l_i}(l_k \nabla_j \theta_k^{\mathrm{go}}) = \frac{\hbar}{2m}\nabla_j \theta_i^{\mathrm{o}} = \frac{\hbar}{2m}(\hat{\boldsymbol{l}}\times\nabla_j\hat{\boldsymbol{l}})_i, \tag{9.64}$$

where in the last equality use has been made of $\mathrm{d}\boldsymbol{\theta}^\mathrm{o} = \hat{\boldsymbol{l}}\times\mathrm{d}\hat{\boldsymbol{l}}$.

From the equilibrium conditions as stated above, we may conclude that the thermodynamic forces driving the system towards equilibrium are given by

(i) the temperature gradient $\boldsymbol{\nabla}T$;

(ii) the gradients of the normal velocity, of which three independent combinations may be formed:

(a) the shear flow

$$V_{ij} = \tfrac{1}{2}(\nabla_i v_{n,j} + \nabla_j v_{n,i}) - \tfrac{1}{3}\delta_{ij}\nabla \cdot \boldsymbol{v}_n, \tag{9.65}$$

(b) the bulk flow $\nabla \cdot \boldsymbol{v}_n$, and
(c) the superfluid bulk flow $\nabla \cdot \boldsymbol{g}_s$;

(iii) the torque acting on $\hat{\boldsymbol{l}}$, $\hat{\boldsymbol{l}} \times \boldsymbol{\psi}$.

The spatial derivatives of the vorticity, which have to be taken into account as thermodynamic forces in rotating equilibrium, like $\nabla \cdot \boldsymbol{\Omega}$, are one order higher in the gradients in the nonrotating case and will be discarded.

9.3.4 Entropy production

In full generality, the equations of motion for the order parameter fields $\hat{\boldsymbol{l}}$ and $\boldsymbol{v}_s$ have the form (Ho 1977)

$$\frac{\partial \hat{\boldsymbol{l}}}{\partial t} = \boldsymbol{\omega} \times \hat{\boldsymbol{l}}, \tag{9.66}$$

$$\frac{\partial v_{s,j}}{\partial t} = -\nabla_j\left(\mu + \frac{\hbar}{2m}\hat{\boldsymbol{l}} \cdot \boldsymbol{\omega}\right) + \frac{\hbar}{2m}\hat{\boldsymbol{l}} \cdot \left(\frac{\partial \hat{\boldsymbol{l}}}{\partial t} \times \nabla_j \hat{\boldsymbol{l}}\right), \tag{9.67}$$

where $\boldsymbol{\omega} = \boldsymbol{\Omega}$ to lowest order in the gradients. The form of the $\boldsymbol{\omega}$-dependent terms is determined by the requirement that the order parameter rotate as a rigid structure. The last term in (9.67) comes from the noncommutativity of the time derivative in (9.49) and the spatial derivative in the definition (9.46) of $\boldsymbol{v}_s$: in order to derive (9.66) from (9.49), the two derivatives have to be interchanged, which introduces a term $\partial \hat{\boldsymbol{l}}/\partial t \times \nabla \hat{\boldsymbol{l}}$ via the relation (9.15). Equation (9.67) is usually referred to as a "Josephson equation".

The conservation laws for mass, momentum, energy and entropy have their usual form. In particular, the mass current, being identical with the momentum density (a conserved quantity), can be expressed in terms of the two local equilibrium quantities $\rho \boldsymbol{v}_n$ and $\boldsymbol{g}_s$, which have the correct behaviour under Galilean transformations:

$$\boldsymbol{g} = \rho \boldsymbol{v}_n + \boldsymbol{g}_s. \tag{9.68}$$

The conservation law for angular momentum may be stated as an equation of motion for the intrinsic angular-momentum density:

$$\frac{\partial t_i}{\partial t} = -\sum_{jk} \epsilon_{ijk} \Pi_{jk} - \sum_j \nabla_j \tilde{B}_{ji}, \tag{9.69}$$

where $\tilde{B}_{ji}$ is an, as yet undetermined, angular-momentum current density. In writing (9.69), we have split off the term $-\sum_{jk} \epsilon_{ijk} \Pi_{jk}$, representing the antisymmetric part of the stress tensor. This term is necessary to guarantee conservation of the total angular momentum $\boldsymbol{L}=\boldsymbol{r}\times\boldsymbol{g}+\boldsymbol{t}$. The entropy-production equation may now be derived from the local thermodynamic identity (9.60) by substituting the conservation laws (2.21)–(2.23) and the equations of motion for $\hat{\boldsymbol{l}}$, (9.66), and $\boldsymbol{v}_s$, (9.67):

$$T\frac{\partial s}{\partial T}=-\boldsymbol{\nabla}\cdot\boldsymbol{j}_\epsilon+\mu\boldsymbol{\nabla}\cdot\boldsymbol{g}+\sum_{ij}v_{n,j}\nabla_i\Pi_{ij}$$
$$+\boldsymbol{g}_s\cdot\left(\boldsymbol{\nabla}\mu+\frac{\hbar}{2m}\sum_i l_i\boldsymbol{\nabla}\omega_i\right)-\frac{\partial f_{\text{bend}}}{\partial\hat{\boldsymbol{l}}}\cdot(\boldsymbol{\omega}\times\hat{\boldsymbol{l}})$$
$$-\sum_{ij}\frac{\partial f_{\text{bend}}}{\partial(\nabla_i l_j)}\nabla_i(\boldsymbol{\omega}\times\hat{\boldsymbol{l}})_j+\sum_{ij}\Omega_i\Big(\sum_k\epsilon_{ijk}\,\Pi_{jk}+\nabla_j\tilde{B}_{ji}\Big). \quad (9.70)$$

Note that the term involving the time derivative of $\hat{\boldsymbol{l}}$ in (9.67) has been written as $\hat{\boldsymbol{l}}\cdot(\partial\hat{\boldsymbol{l}}/\partial t\times\nabla_j\hat{\boldsymbol{l}})=\nabla_j(\hat{\boldsymbol{l}}\cdot\boldsymbol{\omega})-\hat{\boldsymbol{l}}\cdot(\nabla_j\boldsymbol{\omega})$, using (9.15).

Our next goal is to cast the right-hand side of (9.70) into the form of a sum of a divergence and terms proportional to the thermodynamic forces. This is achieved by

(i) splitting off the pressure term $P\delta_{ij}$ from the stress tensor Π_{ij} and making use of the Duhem–Gibbs relation (9.62);

(ii) expressing the gradient of the normal velocity as

$$\nabla_i v_{n,j}=V_{ij}+\tfrac{1}{3}\delta_{ij}\boldsymbol{\nabla}\cdot\boldsymbol{v}_n+\epsilon_{ijk}\Omega_k; \quad (9.71a)$$

(iii) employing the Mermin–Ho relation (9.47) to write

$$\boldsymbol{g}_s\cdot(\boldsymbol{v}_n\cdot\boldsymbol{\nabla})\boldsymbol{v}_s=\boldsymbol{v}_n\cdot(\boldsymbol{g}_s\cdot\boldsymbol{\nabla})\boldsymbol{v}_s+\frac{\hbar}{2m}\hat{\boldsymbol{l}}\cdot[(\boldsymbol{v}_n\cdot\boldsymbol{\nabla})\hat{\boldsymbol{l}}\times(\boldsymbol{g}_s\cdot\boldsymbol{\nabla})\hat{\boldsymbol{l}}]; \quad (9.71b)$$

(iv) making use of the following equality derived from the identity (9.59):

$$-\sum_i\boldsymbol{\nabla}l_i\times\frac{\partial f_{\text{bend}}}{\partial(\boldsymbol{\nabla}l_i)}+\boldsymbol{g}_s\times\boldsymbol{v}_s^0=\hat{\boldsymbol{l}}\times\boldsymbol{\psi}-\frac{\hbar}{2m}(\boldsymbol{g}_s\cdot\boldsymbol{\nabla})\hat{\boldsymbol{l}}$$
$$+\sum_i\nabla_i\left[\hat{\boldsymbol{l}}\times\frac{\partial f_{\text{bend}}}{\partial(\nabla_i\hat{\boldsymbol{l}})}\right]-\boldsymbol{\Omega}\times\boldsymbol{t}; \quad (9.71c)$$

(v) writing all terms containing spatial derivatives other than the thermodynamic forces as a full divergence minus the terms that were added to make it a full divergence.

The result of this reformulation is

$$\frac{\partial s}{\partial t}+\boldsymbol{\nabla}\cdot\left(s\boldsymbol{v}_n+\frac{\boldsymbol{j}_Q}{T}\right)=-\frac{1}{T}\left[\frac{1}{T}\boldsymbol{j}_Q\cdot\boldsymbol{\nabla}T+\sum_{ij}P_{ij}V_{ij}+\frac{1}{3}\sum_i P_{ii}\boldsymbol{\nabla}\cdot\boldsymbol{v}_n\right.$$
$$\left.+\frac{\hbar}{2m}[\hat{\boldsymbol{l}}\cdot(\boldsymbol{\omega}_0-\boldsymbol{\Omega})]\boldsymbol{\nabla}\cdot\boldsymbol{g}+(\boldsymbol{\omega}-\boldsymbol{\Omega})\cdot(\hat{\boldsymbol{l}}\times\boldsymbol{\psi})+\sum_{ij}B_{ij}\nabla_i\Omega_j\right], \quad (9.72)$$

where P_{ij} (a generalized pressure) and B_{ij} (connected to the orbital angular-momentum current) are determined below.

Here the heat current density $\boldsymbol{j}_Q$ is given by

$$\boldsymbol{j}_Q = \boldsymbol{j}_\epsilon^0 - \boldsymbol{g}_s\left(\mu_0 + \frac{\hbar}{2m}\hat{\boldsymbol{l}}\cdot\boldsymbol{\omega}_0\right) - T\sum_i \frac{\partial s}{\partial(\boldsymbol{\nabla} l_i)}(\boldsymbol{\omega}_0 \times \hat{\boldsymbol{l}})_i - \boldsymbol{B\Omega}. \tag{9.73}$$

The index 0 indicates a quantity in the frame of reference where $\boldsymbol{v}_n = 0$. The transformation laws are given by (9.55) and by

$$\hat{\boldsymbol{l}}\cdot\boldsymbol{\omega} = \hat{\boldsymbol{l}}\cdot\boldsymbol{\omega}_0 + \frac{2m}{\hbar}\boldsymbol{v}_n\cdot\boldsymbol{v}_s, \tag{9.74}$$

$$\hat{\boldsymbol{l}}\times\boldsymbol{\omega} = \hat{\boldsymbol{l}}\times\boldsymbol{\omega}_0 + (\boldsymbol{v}_n\cdot\boldsymbol{\nabla})\hat{\boldsymbol{l}}, \tag{9.75}$$

$$\boldsymbol{j}_\epsilon = \boldsymbol{j}_\epsilon^0 + \epsilon_0\boldsymbol{v}_n + \boldsymbol{\Pi}\boldsymbol{v}_n - \tfrac{1}{2}v_n^2\boldsymbol{g}, \tag{9.76}$$

where

$$\epsilon_0 = \epsilon - \boldsymbol{v}_n\cdot\boldsymbol{g}_s - \tfrac{1}{2}\rho v_n^2, \tag{9.77}$$

$$\Pi_{ij} = \Pi_{ij}^0 + g_{s,i}v_{n,j} + v_{n,i}g_{s,j} + \rho v_{n,i}v_{n,j}. \tag{9.78}$$

The Galilean transformation properties of the energy density ϵ, the energy current density $\boldsymbol{j}_\epsilon$ and the stress tensor Π_{ij} are the same as for an isotropic superfluid (Khalatnikov 1965), with, however, nonsymmetric Π_{ij}.

The quantities P_{ij} and B_{ij} are found as

$$P_{ij} = \Pi_{ij}^0 - P\delta_{ij} - g_{s,i}(\boldsymbol{v}_s - \boldsymbol{v}_n)_j + T\frac{\partial s}{\partial(\nabla_i\hat{\boldsymbol{l}})}\cdot\nabla_j\hat{\boldsymbol{l}}, \tag{9.79}$$

$$B_{ij} = \tilde{B}_{ij} - v_{n,i}t_j - \frac{\hbar}{2m}g_{s,i}\hat{l}_j - \left[\hat{\boldsymbol{l}}\times T\frac{\partial s}{\partial(\nabla_i\hat{\boldsymbol{l}})}\right]_j. \tag{9.80}$$

9.3.5 Derivation of the hydrodynamic currents

We now write the currents as linear combinations of the thermodynamic forces. Since the equilibrium state is characterized by the $\hat{\boldsymbol{l}}$ texture, the expansion coefficients will in general depend on $\hat{\boldsymbol{l}}$. It is now important to realize that $\hat{\boldsymbol{l}}$ is proportional to the angular momentum of the Cooper pairs. Consequently, $\hat{\boldsymbol{l}}$ changes sign under time reversal. This property provides us with a certain freedom to add factors of $\hat{l}_i$ to the entropy production that are odd (reactive) or even (dissipative) under time reversal.

(i) In the case of the stress tensor we have to allow for two reactive terms in P_{ij}, proportional to $\hat{\boldsymbol{l}}\times\boldsymbol{\psi}$ and to V_{ij}, (9.65), respectively. Adding up all terms that are allowed by symmetry, the stress tensor takes the following form (with $(\boldsymbol{V}\times\hat{\boldsymbol{l}})_{ij}\equiv\epsilon_{jkl}\hat{l}_k V_{il}$):

$$\begin{aligned}\Pi_{ij} = P\delta_{ij} &+ \frac{\partial f_{\text{bend}}}{\partial(\nabla_i\hat{\boldsymbol{l}})}\cdot\nabla_j\hat{\boldsymbol{l}} + g_{\text{s},i}v_{\text{s},j} + v_{\text{n},i}g_j + \tfrac{1}{2}\gamma_0(\psi_i^\perp\hat{l}_j + \psi_j^\perp\hat{l}_i)\\ &+ \gamma_\perp(\boldsymbol{V}\times\hat{\boldsymbol{l}})^{\text{s}}_{ij} + \gamma_\parallel[(\hat{\boldsymbol{l}}\times\boldsymbol{V}\hat{\boldsymbol{l}})_i l_j]^{\text{s}} + \beta_0^{\text{R}}(\hat{l}_i\psi_j - \hat{l}_j\psi_i)\\ &+ \varepsilon_{ijk}\hat{l}_k(\beta_1^{\text{R}}\nabla\cdot\boldsymbol{g}_{\text{s}} + \beta_2^{\text{R}}\nabla\cdot\boldsymbol{v}_{\text{n}} + \beta_3^{\text{R}}\hat{\boldsymbol{l}}\boldsymbol{V}\hat{\boldsymbol{l}}) + \beta_4^{\text{R}}\sum_{lk}\epsilon_{ijl}V_{lk}\hat{l}_k + \Pi_{ij}^{\text{D}},\end{aligned} \tag{9.81}$$

where γ_0, $\gamma_\perp$, $\gamma_\parallel$ and β_i^{R} $(i=1,\ldots,4)$ are reactive transport coefficients and $\psi_i^\perp=\psi_i-\hat{l}_i(\hat{\boldsymbol{l}}\cdot\boldsymbol{\psi})$.

The dissipative part of the stress tensor, Π_{ij}^{D}, contains the usual terms proportional to V_{ij}, $\nabla\cdot\boldsymbol{v}_{\text{n}}$ and $\nabla\cdot\boldsymbol{g}_{\text{s}}$, and in addition a term involving the textural force $\hat{\boldsymbol{l}}\times\boldsymbol{\psi}$. By taking cross-products of $\hat{\boldsymbol{l}}$ with $\boldsymbol{V}\hat{\boldsymbol{l}}$ or $\hat{\boldsymbol{l}}\times\boldsymbol{\psi}$, one can also construct antisymmetric terms. Introducing transport coefficients in a way that is as close as possible to the usual notation, one may write Π_{ij}^{D} as

$$\begin{aligned}\Pi_{ij}^{\text{D}} = &-(2\eta_{\text{L}}^\parallel\hat{l}_i\hat{l}_j - \eta_{\text{L}}^\perp\delta_{ij}^\perp)(\hat{\boldsymbol{l}}\boldsymbol{V}\hat{\boldsymbol{l}}) - 2\eta^\perp\sum_{kl}\delta_{ik}^\perp V_{kl}\delta_{lj}^\perp\\ &-2\eta^\parallel\Big[l_i\sum_{kl}l_k V_{kl}\delta_{lj}^\perp + (i\leftrightarrow j)\Big] - (\zeta_2^\perp\delta_{ij}^\perp + \zeta_2^\parallel\hat{l}_i\hat{l}_j)\nabla\cdot\boldsymbol{v}_{\text{n}}\\ &-(\zeta_1^\perp\delta_{ij}^\perp + \zeta_1^\parallel\hat{l}_i\hat{l}_j)\nabla\cdot\boldsymbol{g}_{\text{s}} - \tfrac{1}{2}\lambda_1[(\hat{\boldsymbol{l}}\times\boldsymbol{\psi})_i\hat{l}_j + (\hat{\boldsymbol{l}}\times\boldsymbol{\psi})_j\hat{l}_i]\\ &-\frac{1}{2}\sum_k\epsilon_{ijk}[-\lambda_2(\hat{\boldsymbol{l}}\times\mathbf{V}\hat{\boldsymbol{l}})_k + \lambda_3[\hat{\boldsymbol{l}}\times(\hat{\boldsymbol{l}}\times\boldsymbol{\psi})]_k],\end{aligned} \tag{9.82}$$

where $\delta_{ij}^\perp\equiv\delta_{ij}-\hat{l}_i\hat{l}_j$, and $\eta^{\parallel,\perp}$, $\eta_{\text{L}}^{\parallel,\perp}$, λ_1, λ_2, λ_3, $\zeta_1^{\parallel,\perp}$ and $\zeta_2^{\parallel,\perp}$ are the eleven transport coefficients characterizing the stress tensor.

The shear-viscosity coefficients $\eta^{\parallel,\perp}$ and $\eta_{\text{L}}^\parallel$ are such that they reduce to the usual shear-viscosity coefficient in the limit $T\to T_{\text{c}}$, while $\eta_{\text{L}}^\perp$ tends to zero in this limit. The same is true for the bulk-viscosity coefficients $\zeta_1^{\parallel,\perp}$ and $\zeta_2^{\parallel,\perp}$. The antisymmetric terms of Π_{ij}^{D} do not contribute to the entropy production.

(ii) In the heat current density $\boldsymbol{j}^Q$ a reactive term appears that is linear in $\hat{\boldsymbol{l}}$ (odd time-reversal symmetry). It is directed perpendicularly to ∇T (no dissipation). In the dissipative part there is only the usual diffusive heat current if one drops terms involving gradients in $\boldsymbol{\Omega}$. The total heat current is then given by

$$\boldsymbol{j}_{\text{Q}} = \gamma(\hat{\boldsymbol{l}}\times\nabla T) - \kappa_\perp\nabla T - (\kappa_\parallel-\kappa_\perp)[\hat{\boldsymbol{l}}(\hat{\boldsymbol{l}}\cdot\nabla T)]. \tag{9.83}$$

Here $\kappa_\parallel$ and $\kappa_\perp$ are the coefficients of the thermal conductivity for heat flow parallel and perpendicular to $\hat{\boldsymbol{l}}$ respectively.

(iii) In the Josephson equation (9.67) symmetry does not allow for reactive terms besides those already discussed in the context of (9.48) and (9.52), i.e. one has $\hat{\boldsymbol{l}}\cdot\boldsymbol{\omega}_0=\hat{\boldsymbol{l}}\cdot\boldsymbol{\Omega}$ in (9.66). There are, however, dissipative terms that have to be added to the chemical potential, which are proportional to $\boldsymbol{\nabla}\cdot\boldsymbol{v}_{\rm n}$ and $\boldsymbol{\nabla}\cdot\boldsymbol{g}_{\rm s}$. In addition there is a term proportional to $\hat{\boldsymbol{l}}\boldsymbol{V}\hat{\boldsymbol{l}}$, the component of V_{ij} parallel to $\hat{\boldsymbol{l}}$; one then obtains the equation of motion for $\boldsymbol{v}_{\rm s}$:

$$\frac{\partial v_{{\rm s},j}}{\partial t}+\nabla_j\left[\mu_0+\boldsymbol{v}_{\rm n}\cdot\boldsymbol{v}_{\rm s}+\frac{\hbar}{2m}\left(1+\frac{\partial\rho_{\rm L}}{\partial\rho}\right)\hat{\boldsymbol{l}}\cdot\boldsymbol{\Omega}+\mu^{\rm D}\right]=\frac{\hbar}{2m}\hat{\boldsymbol{l}}\cdot\left(\frac{\partial\hat{\boldsymbol{l}}}{\partial t}\times\nabla_j\hat{\boldsymbol{l}}\right),\tag{9.84}$$

where

$$\mu^{\rm D}=-\zeta_3\boldsymbol{\nabla}\cdot\boldsymbol{g}_{\rm s}-\zeta_4\boldsymbol{\nabla}\cdot\boldsymbol{v}_{\rm n}-\zeta_5\hat{\boldsymbol{l}}\boldsymbol{V}\hat{\boldsymbol{l}}.\tag{9.85}$$

The coefficients ζ_4 and ζ_5 are related to the transport coefficients in the stress tensor (9.82) by Onsager's symmetry relations in the following way:

$$\zeta_4=\tfrac{1}{3}(\zeta_1^{\parallel}+2\zeta_1^{\perp}),\tag{9.86}$$

$$\zeta_5=\zeta_1^{\parallel}-\zeta_1^{\perp}.\tag{9.87}$$

(iv) Equation for $\hat{\boldsymbol{l}}$: in the thermodynamic flux $\boldsymbol{\omega}_0-\boldsymbol{\Omega}$ multiplying the $\hat{\boldsymbol{l}}$-texture torque $\hat{\boldsymbol{l}}\times\boldsymbol{\psi}$ in (9.72) symmetry allows for terms driven by $\hat{\boldsymbol{l}}\times(\hat{\boldsymbol{l}}\times\boldsymbol{\psi})$ and by $\hat{\boldsymbol{l}}\times\boldsymbol{V}\hat{\boldsymbol{l}}$ on the reactive level. The first term yields a vanishing entropy production, while the second term is cancelled by the term with coefficient γ_0 in the stress tensor given by (9.81). In the dissipative part $\boldsymbol{\omega}^{\rm D}$ only $\hat{\boldsymbol{l}}\times\boldsymbol{\psi}$ and $\boldsymbol{V}\hat{\boldsymbol{l}}$ are possible thermodynamic forces. The equation of motion for the vector $\hat{\boldsymbol{l}}$ consequently takes the form

$$\frac{\partial\hat{\boldsymbol{l}}}{\partial t}+(\boldsymbol{v}_{\rm n}\cdot\boldsymbol{\nabla})\hat{\boldsymbol{l}}=\boldsymbol{\Omega}\times\hat{\boldsymbol{l}}-\gamma_0(\boldsymbol{V}\hat{\boldsymbol{l}})^{\perp}+\alpha\hat{\boldsymbol{l}}\times\boldsymbol{\psi}-\lambda_1\hat{\boldsymbol{l}}\boldsymbol{V}\times\hat{\boldsymbol{l}}-\lambda_4\boldsymbol{\psi}^{\perp},\tag{9.88}$$

where the superscript $\perp$ denotes the component perpendicular to $\hat{\boldsymbol{l}}$ and γ_0, α, λ_1 and λ_4 are transport coefficients. Note that $1/\lambda_4\equiv\mu_l$ is known as the "orbital" viscosity coefficient (see (10.166). In deriving (9.88), use has again been made of Onsager's symmetry principle in identifying the coefficient of the term $\hat{\boldsymbol{l}}\boldsymbol{V}\times\hat{\boldsymbol{l}}$ with the coefficient λ_1 appearing in the equation for the stress tensor, (9.82).

9.3.6 Further consequences of an intrinsic orbital angular momentum

The coefficients $\beta_i^{\rm R}$ (which should not be confused with the strong-coupling parameters $\{\beta_i\}$ in the expression for the free energy (5.4)) in the stress tensor (9.81) may actually be related directly to the intrinsic angular-momentum density in the event that the induced angular momentum $\boldsymbol{t}_{\rm ind}$ of (9.42) can be neglected (Hall and Hook 1986). In this case $\boldsymbol{t}=t_0\hat{\boldsymbol{l}}$ is no longer an independent variable, but is completely determined by $\hat{\boldsymbol{l}}$ and the two scalar densities ρ and s, i.e. $t_0=t_0(\rho,s)$. If one now writes the antisymmetric part of the stress tensor in terms of derivatives of $\boldsymbol{t}$ as

$$\begin{aligned}\sum_{j,k}\epsilon_{ijk}\Pi_{jk}=&-\left(\frac{\partial}{\partial t}+\boldsymbol{v}_{\rm n}\cdot\boldsymbol{\nabla}\right)(t_0 l_i)+\sum_j\nabla_j B_{ji}\\&-\sum_j\nabla_j\left\{\frac{\hbar}{2m}g_{{\rm s},j}l_i+\left[\hat{\boldsymbol{l}}\times T\frac{\partial s}{\partial(\nabla_j\hat{\boldsymbol{l}})}\right]_i\right\},\end{aligned}\tag{9.89}$$

using (9.69) and (9.80), one may convert the derivatives of t_0 into derivatives of ρ and s, that is to say, into already known quantities.

With the aid of the equations of motion for ρ, s and $\hat{\boldsymbol{l}}$ and the identity (9.71), $\sum_{j,k} \epsilon_{ijk}\Pi_{jk}$ can be shown to be given by the following linear combination of forces:

$$\sum_{j,k} \epsilon_{ijk}\Pi_{jk} = -(1+t_0\alpha)(\hat{\boldsymbol{l}} \times \boldsymbol{\psi})_i + t_0\gamma_0(\boldsymbol{V}\hat{\boldsymbol{l}})_i^{\perp} + \hat{l}_i(\alpha_1 \nabla \cdot \boldsymbol{v}_\mathrm{n} + \alpha_2 \nabla \cdot \boldsymbol{g}_\mathrm{s})$$

$$+ \sum_j \left[\nabla l_j \times T \frac{\partial s}{\partial(\nabla l_j)}\right]_i + [\boldsymbol{g}_\mathrm{s} \times (\boldsymbol{v}_\mathrm{s} - \boldsymbol{v}_\mathrm{n})]_i, \tag{9.90}$$

where

$$\left.\begin{aligned} \alpha_1 &= \rho \frac{\partial t_0}{\partial \rho} + s \frac{\partial t_0}{\partial s} - t_0, \\ \alpha_2 &= \frac{\partial t_0}{\partial \rho} - \frac{\hbar}{2m}. \end{aligned}\right\} \tag{9.91}$$

By comparison with the general expression for Π_{ij} given in (9.81), the coefficients of the antisymmetric terms are identified as follows

$$\left.\begin{aligned} &\beta_0^\mathrm{R} = -\tfrac{1}{2}(1+t_0\alpha), \\ &\beta_1^\mathrm{R} = \tfrac{1}{2}\alpha_2, \quad \beta_2^\mathrm{R} = \tfrac{1}{2}\alpha_1, \\ &\beta_3^\mathrm{R} = -\beta_4^\mathrm{R} = -t_0\gamma_0. \end{aligned}\right\} \tag{9.92}$$

The term $\nabla_j B_{ji}$ has been neglected here, since B_{ji} is the flux conjugate to $\nabla_j \Omega_i$, as can be seen from (9.72), and hence is proportional to the thermodynamic forces itself.

Let us finally comment on the superfluid current density $\boldsymbol{g}_\mathrm{s}$ as defined by (9.57). It is not identical with what is usually calculated in Ginzburg–Landau theory (see (7.35)). The difference is that $\boldsymbol{g}_\mathrm{s}$, and hence $\boldsymbol{g}$ as defined in this chapter, is given by the derivative of the free energy with respect to $\boldsymbol{v}_\mathrm{s}^0 = \boldsymbol{v}_\mathrm{s} - \boldsymbol{v}_\mathrm{n}$ at fixed internal angular-momentum density $\boldsymbol{t}$, whereas in a static microscopic calculation the derivative is taken assuming $\boldsymbol{v}_\mathrm{n} \equiv 0$. This difference appears only in the coefficient $C_\perp$ of the $(\nabla \times \hat{\boldsymbol{l}})_\perp$ term in (7.39).

The additional contribution to $C_\perp$ arises because in the microscopic derivation of hydrodynamic theory one finds a term in the momentum density in addition to the terms discussed in (7.39), which is of the form (Nagai 1980, 1981a, Combescot and Dombre 1980, Combescot 1980a, Mermin and Muzikar 1980)

$$\Delta \boldsymbol{g} = -a(T)\hat{\boldsymbol{l}} \times \nabla \rho \tag{9.93}$$

where $a(T)$ may be related to the superfluid mass density $\rho_{\mathrm{s}\perp}$ and the coefficient $C_\perp$ (in the calculation where $\boldsymbol{v}_\mathrm{n} \equiv 0$), given by (7.22a) and (7.22b) respectively, as

$$a(T) = \frac{2}{3\rho} \frac{\hbar}{2m} (\rho_{\mathrm{s}\perp} + C_\perp) \tag{9.94}$$

As mentioned at the beginning of this subsection, we have adopted the point of view put forward by Mermin and Muzikar (1980) that (9.94) is nothing but a contribution to $\boldsymbol{g}$ from an assumed intrinsic angular-momentum density

$$\boldsymbol{t} = 2\rho a(T)\hat{\boldsymbol{l}}. \tag{9.95}$$

The contribution (9.93) to the current is due to additional terms in the free-energy density

$$\Delta f_{\text{bend}} = -\tfrac{1}{2}(\boldsymbol{v}_{\text{s}} - \boldsymbol{v}_{\text{n}}) \cdot (\hat{\boldsymbol{l}} \times \boldsymbol{\nabla} t_0). \tag{9.96}$$

On performing a partial integration, this may be cast into the form

$$\Delta f_{\text{bend}} = -\tfrac{1}{2}t_0(\boldsymbol{v}_{\text{s}} - \boldsymbol{v}_{\text{n}}) \cdot (\boldsymbol{\nabla} \times \hat{\boldsymbol{l}}) - t_0\hat{\boldsymbol{l}} \cdot \boldsymbol{\Omega} + \text{surface terms}. \tag{9.97}$$

Here the second term on the right-hand side is interpreted as the energy of a spinning top with angular momentum $t_0\hat{\boldsymbol{l}}$ in a system rotating with angular frequency $\boldsymbol{\Omega}$. We note that the first term has the same structure as the term multiplied by $C_\perp$ in (7.39). This allows one to combine the two to yield a single term with coefficient $\bar{C}_\perp$, where

$$\bar{C}_\perp = C_\perp - \tfrac{1}{2}t_0. \tag{9.98}$$

Note that the momentum density $\boldsymbol{g}$ in the frame where $\boldsymbol{v}_{\text{n}} = 0$ is given by differentiating f_{bend} with respect to $\boldsymbol{v}_{\text{n}}$ at fixed $\boldsymbol{\Omega} = \tfrac{1}{2}\boldsymbol{\nabla} \times \boldsymbol{v}_{\text{n}}$:

$$\boldsymbol{g} = -\left(\frac{\partial f_{\text{bend}}}{\partial \boldsymbol{v}_{\text{n}}}\right)_{\boldsymbol{\Omega}}. \tag{9.99}$$

The unconstrained functional derivative of f_{bend}, which is usually taken in microscopic theories, gives an extra term $\Delta\boldsymbol{g} = \tfrac{1}{2}\boldsymbol{\nabla} \times \boldsymbol{t}$ from differentiating $\boldsymbol{\Omega}$ in the second term of (9.97), so that $-\partial f_{\text{bend}}/\partial \boldsymbol{v}_{\text{n}} = \boldsymbol{j}$, the current defined by (9.40).

9.4 HYDRODYNAMICS FOR FINITE MAGNETIC FIELD

In a magnetic field the nature of pairing changes as a result of the Zeeman energy of the Cooper-pair spins. The resulting distortion of the order parameter has been discussed in Section 5.4. In small magnetic fields and if one is not too close to the transition, the order-parameter structure may be approximated by its zero-field form. This approximation has already been used in most parts of Chapter 8 and also in the derivation of the spin hydrodynamics for the B phase in Section 9.2. In this section we want to explore some of the qualitative differences in the hydrodynamics brought about by the modification in symmetry of the order parameter due to a magnetic field.

9.4.1 The A phase in a magnetic field

A particularly drastic difference exists between the order parameters of the A phase and the A_1 phase given by (3.68) and (5.59). In this case a change in the phase of the $|\uparrow\uparrow\rangle$ Cooper-pair wave function can be undone not only by a corresponding rotation about the axis $\hat{\boldsymbol{l}}$ in orbital space but equally well by a rotation about the axis $\hat{\boldsymbol{f}}$ in spin space (see Chapter 6). Here one has a spontaneously broken relative spin–orbit–gauge symmetry (SOGS) as compared with the broken gauge–orbit symmetry in the A phase in zero magnetic field (Liu 1979a). In the A_2 phase SOGS is broken for both spin components, $|\uparrow\uparrow\rangle$ and $|\downarrow\downarrow\rangle$, of the condensate.

The proper dynamical variable characterizing the broken gauge symmetry is therefore not the phase variable $\phi^0_{\uparrow,\downarrow}$ itself, but rather the variable describing the difference of the phase variable and the rotation angles about $\hat{\boldsymbol{l}}$ and $\hat{\boldsymbol{f}}$ in orbital space and spin space respectively. For a small change in this variable, one therefore has

$$\left.\begin{aligned} \mathrm{d}\phi_\uparrow &= \mathrm{d}\phi^0_\uparrow - \hat{\boldsymbol{l}}\cdot \mathrm{d}\boldsymbol{\theta}^{\mathrm{o}} - \hat{\boldsymbol{f}}\cdot \mathrm{d}\boldsymbol{\theta}^{\mathrm{s}}, \\ \mathrm{d}\phi_\downarrow &= \mathrm{d}\phi^0_\downarrow - \hat{\boldsymbol{l}}\cdot \mathrm{d}\boldsymbol{\theta}^{\mathrm{o}} + \hat{\boldsymbol{f}}\cdot \mathrm{d}\boldsymbol{\theta}^{\mathrm{s}}. \end{aligned}\right\} \tag{9.100}$$

The sign of the last term in the expressions for $\mathrm{d}\phi_\uparrow$ and $\mathrm{d}\phi_\downarrow$ is ∓ 1 according to the different axes of spin polarization, $\pm\hat{\boldsymbol{f}}$, in the two cases.

In the A_2 phase it is physically meaningful to also introduce the total phase ϕ and the relative phase of the two spin populations, ϕ_{sp}:

$$\left.\begin{aligned} \mathrm{d}\phi &= \tfrac{1}{2}(\mathrm{d}\phi_\uparrow + \mathrm{d}\phi_\downarrow), \\ \mathrm{d}\phi_{\mathrm{sp}} &= \tfrac{1}{2}(\mathrm{d}\phi_\uparrow - \mathrm{d}\phi_\downarrow). \end{aligned}\right\} \tag{9.101}$$

Employing the general relations (9.8), (9.9) and (9.13), the equations of motion for $\phi_{\uparrow,\downarrow}$ are obtained as

$$\frac{\mathrm{d}\phi_{\uparrow,\downarrow}}{\mathrm{d}t} = -\frac{2m}{\hbar}\mu_{\uparrow,\downarrow} - \hat{\boldsymbol{l}}\cdot\boldsymbol{\Omega} \mp \hat{\boldsymbol{f}}\cdot\boldsymbol{\omega}^{\mathrm{s}}, \tag{9.102}$$

where $\mu_{\uparrow,\downarrow} = m^{-1}(\partial\epsilon/\partial n_{\uparrow,\downarrow})_{\iota,\boldsymbol{S},\dots}$ are the chemical potentials of the $(\uparrow,\downarrow)$ spin species and the term involving $\boldsymbol{\omega}^{\mathrm{s}}$ (see (9.9)) describes the restoring force in a situation where the spin $\boldsymbol{S}$ is out of equilibrium. For the A_1 phase one has only one of the equations (9.102), say that for $\phi_\uparrow$. The last term in (9.102) implies that a supercurrent (see (9.105)) is generated by a gradient in the applied magnetic field (magnetic fountain effect). This has been observed experimentally by Ruel and Kojima (1985), although unexpected nonlinear behaviour has been found (Ruel and Kojima 1986).

The equation of motion for the total phase ϕ has the same form as in the A phase:

$$\frac{\mathrm{d}\phi}{\mathrm{d}t} = -\frac{2m}{\hbar}\mu - \hat{\boldsymbol{l}}\cdot\boldsymbol{\Omega}, \tag{9.103}$$

with $\mu = \tfrac{1}{2}(\mu_\uparrow + \mu_\downarrow)$.

A new equation appears here, namely that for the spin phase:

$$\frac{\mathrm{d}\phi_{\mathrm{sp}}}{\mathrm{d}t} = -\hat{\boldsymbol{f}} \cdot \boldsymbol{\omega}_{\mathrm{s}} - \frac{m}{\hbar}(\mu_{\uparrow} - \mu_{\downarrow}). \tag{9.104}$$

Apart from the last term, this is the longitudinal part of Leggett's equation (8.27b) for the vector $\hat{\boldsymbol{d}}$. But this last term vanishes in local equilibrium, when $\mu_{\uparrow} = \mu_{\downarrow}$.

The hydrodynamics of the A_2 phase may be looked upon as a three-fluid dynamics (Liu 1980, Gurgenishvili and Kharadze 1980; see also Baramidze *et al.* 1983), the three interpenetrating fluids being the two superfluid components of the up- and down-spin populations and the normal component. In the following we consider only the case of uniform $\hat{\boldsymbol{l}}$- and $\hat{\boldsymbol{f}}$-vector fields. The theory for the general case has not yet been completely worked out (see, however, Pleiner and Graham 1976, Pleiner 1977a,b, Saslow and Hu 1981, Pleiner and Brand 1981, Brand and Pleiner 1981a,b, 1982, Dombre and Combescot 1982a).

We start by defining superfluid velocities $\boldsymbol{v}_{\uparrow,\downarrow}$ for the $\uparrow$, $\downarrow$ subsystems:

$$\boldsymbol{v}_{\uparrow,\downarrow} = \frac{\hbar}{2m} \boldsymbol{\nabla}\phi_{\uparrow,\downarrow}. \tag{9.105}$$

The superfluid velocities $\boldsymbol{v}_{\mathrm{s}}$ and $\boldsymbol{v}_{\mathrm{sp}}$ driven by the gradients in the total phase ϕ and the spin phase ϕ_{sp} are defined by

$$\left.\begin{aligned} \boldsymbol{v}_{\mathrm{s}} &= \tfrac{1}{2}(\boldsymbol{v}_{\uparrow} + \boldsymbol{v}_{\downarrow}) = \frac{\hbar}{2m}\boldsymbol{\nabla}\phi = \frac{\hbar}{2m}\sum_i \hat{m}_i \boldsymbol{\nabla}\hat{n}_i, \\ \boldsymbol{v}_{\mathrm{sp}} &= \tfrac{1}{2}(\boldsymbol{v}_{\uparrow} - \boldsymbol{v}_{\downarrow}) = \frac{\hbar}{2m}\boldsymbol{\nabla}\phi_{\mathrm{sp}} = \frac{\hbar}{2m}\hat{d}_{\mu}\boldsymbol{\nabla}\hat{e}_{\mu}. \end{aligned}\right\}$$

It follows from the last equation that $\boldsymbol{\nabla} \times \boldsymbol{v}_{\mathrm{s}}$ and $\boldsymbol{\nabla} \times \boldsymbol{v}_{\mathrm{sp}}$ satisfy Mermin–Ho relations (9.47) involving the $\hat{\boldsymbol{l}}$ and $\hat{\boldsymbol{f}}$ fields respectively. Note that $\boldsymbol{v}_{\mathrm{sp}}$ does not transform like a velocity under Galilean transformations; in fact it does not change at all!

The mass (spin) currents in the normal-fluid rest frame ($\boldsymbol{v}_{\mathrm{n}} = 0$) are obtained by adding (subtracting) the contributions from the $\uparrow$, $\downarrow$ subsystems:

$$\left.\begin{aligned} \boldsymbol{g}_{\mathrm{s}} &= \rho_{\uparrow}(\boldsymbol{v}_{\uparrow} - \boldsymbol{v}_{\mathrm{n}}) + \rho_{\downarrow}(\boldsymbol{v}_{\downarrow} - \boldsymbol{v}_{\mathrm{n}}), \\ \boldsymbol{j}_{\mathrm{sp}} &= \frac{\hbar}{2m}[\rho_{\uparrow}(\boldsymbol{v}_{\uparrow} - \boldsymbol{v}_{\mathrm{n}}) - \rho_{\downarrow}(\boldsymbol{v}_{\downarrow} - \boldsymbol{v}_{\mathrm{n}})]. \end{aligned}\right\} \tag{9.107}$$

Here $\rho_{\uparrow,\downarrow}$ are uniaxial tensors oriented along $\hat{\boldsymbol{l}}$. It is useful to rewrite (9.107) in terms of $\boldsymbol{v}_{\mathrm{s}}$ and $\boldsymbol{v}_{\mathrm{sp}}$:

$$\left.\begin{aligned} \boldsymbol{g}_{\mathrm{s}} &= \rho_{\mathrm{s}}(\boldsymbol{v}_{\mathrm{s}} - \boldsymbol{v}_{\mathrm{n}}) + \rho_{\uparrow\downarrow}\boldsymbol{v}_{\mathrm{sp}}, \\ \boldsymbol{j}_{\mathrm{sp}} &= \frac{\hbar,}{2m}[\rho_{\mathrm{sp}}\boldsymbol{v}_{\mathrm{sp}} + \rho_{\uparrow\downarrow}(\boldsymbol{v}_{\mathrm{s}} - \boldsymbol{v}_{\mathrm{n}})], \end{aligned}\right\} \tag{9.108}$$

where

$$\rho_s = \rho_{sp} = \rho_\uparrow + \rho_\downarrow, \tag{9.109a}$$

$$\rho_{\uparrow\downarrow} = \rho_\uparrow - \rho_\downarrow. \tag{9.109b}$$

The equality of ρ_s and ρ_{sp} only holds if the two subsystems are strictly decoupled. In the real liquid, however, the exchange part of the Fermi-liquid interaction provides a considerable coupling, and this causes ρ_{sp}, (7.23c), to deviate substantially from ρ_s, except near T_c. The difference between $\rho_\uparrow$ and $\rho_\downarrow$ is due entirely to the different magnitudes of the gap $\Delta_{\uparrow,\downarrow}$ (see Section 5.4), and may be calculated within weak-coupling theory.

It turns out that $\rho_{\uparrow\downarrow}$ is sizeable only very close to the A_2 transition and, of course, in the A_1 phase, where $\rho_{\uparrow\downarrow} = -\rho_\uparrow$. It is interesting to note that in the region where $\rho_{\uparrow\downarrow}$ is comparable to ρ_s, $\boldsymbol{v}_s$ no longer has the meaning of the superfluid velocity of mass flow (see (9.108)). Rather, the characteristic velocity changes from $\boldsymbol{v}_s$ to $\boldsymbol{v}_\uparrow$ as the A_1 phase is approached.

We shall not reproduce the equations of motion for the transverse parts of the orbital and spin symmetry variables $\hat{\boldsymbol{l}}$ and $\hat{\boldsymbol{f}}$, or the conservation laws (except for spin conservation), since they are similar to the equations for the A phase discussed in the previous section. The longitudinal component of the spin conservation law on the reactive level is given by

$$\frac{\partial S_z}{\partial t} + \boldsymbol{\nabla} \cdot (S_z \boldsymbol{v}_n + \boldsymbol{j}^{sp}) = \frac{\chi}{\gamma^2} \Omega_{\parallel}^2 \phi_{sp}, \tag{9.110}$$

where the dipole-torque term has been included on the right-hand side.

9.4.2 The B phase in a magnetic field

Just as in the case of the A phase, it is also useful here to consider the spin components separately. There are three spin components in this case, which have the orbital structure given in (5.66). Each spin component is invariant under a particular combination of a gauge transformation $e^{i\varphi_{\uparrow\uparrow}}$, $e^{i\varphi_{\downarrow\downarrow}}$ and an orbital as well as a spin rotation. These degeneracies are eliminated from the dynamics by introduction of the following phase variables for the three condensates:

$$\left.\begin{aligned} d\phi_{\uparrow\uparrow} &= d\varphi_{\uparrow\uparrow} + \hat{\boldsymbol{l}} \cdot d\boldsymbol{\theta}^o - \hat{\boldsymbol{f}} \cdot d\boldsymbol{\theta}^s, \\ d\phi_{\downarrow\downarrow} &= d\varphi_{\downarrow\downarrow} - \hat{\boldsymbol{l}} \cdot d\boldsymbol{\theta}^o + \hat{\boldsymbol{f}} \cdot d\boldsymbol{\theta}^s, \\ d\phi_{\uparrow\downarrow} &= \tfrac{1}{2}(d\varphi_{\uparrow\uparrow} + d\varphi_{\downarrow\downarrow}). \end{aligned}\right\} \tag{9.111}$$

the last equation takes into account that the phase relation between the $\uparrow\downarrow$ and $\uparrow\uparrow$ or $\downarrow\downarrow$ components of this real-valued order parameter has to be preserved (see (5.66)). It also follows that $d\phi_{\uparrow\downarrow} = \frac{1}{2}(d\phi_{\uparrow\uparrow} + d\phi_{\downarrow\downarrow})$. Hence there exist two independent phase variables here, which may conveniently be chosen as the total phase ϕ and the spin phase ϕ_{sp}:

$$d\phi = \tfrac{1}{2}(d\phi_{\uparrow\uparrow} + d\phi_{\downarrow\downarrow}) \equiv d\phi_{\uparrow\downarrow}, \tag{9.112a}$$

$$d\phi_{sp} = \tfrac{1}{2}(d\phi_{\uparrow\uparrow} - d\phi_{\downarrow\downarrow}). \tag{9.112b}$$

The corresponding Josephson equations are given by

$$\frac{d\phi}{dt} = -\frac{2m}{\hbar}\mu, \tag{9.113a}$$

$$\frac{d\phi_{sp}}{dt} = \hat{\boldsymbol{l}} \cdot \boldsymbol{\Omega} - \hat{\boldsymbol{f}} \cdot \boldsymbol{\omega}^{s} - \frac{m}{\hbar}(\mu_{\uparrow} - \mu_{\downarrow}), \tag{9.113b}$$

where $\mu_{\uparrow,\downarrow} = m^{-1}\,\partial\epsilon/\partial n_{\uparrow,\downarrow}$ is the chemical potential of the $\uparrow$, $\downarrow$ populations. The term $\hat{\boldsymbol{l}} \cdot \boldsymbol{\Omega}$ in the equation for ϕ_{sp} implies that there is a gauge-wheel effect for the spin-phase variable rather than for the total phase variable as is the case in the A phase. In equilibrium, a finite excess magnetization ($\boldsymbol{\omega}^{s} \neq 0$) causes the sample to rotate ("Einstein–de Haas effect"). Equivalently, a solid-body rotation of the sample ($\boldsymbol{\Omega} \neq 0$) causes a magnetization $\boldsymbol{\omega}^{s}$ ("Barnett effect"; see Combescot 1980a, Volovik and Mineev 1984).

The last term of (9.113b) would again be zero in local equilibrium were it not for the fact that we have to allow for an intrinsic orbital angular momentum density related to $\boldsymbol{S}$. According to microscopic theory (Combescot 1980a), a finite spin polarization $\boldsymbol{S} = S\hat{\boldsymbol{f}}$ gives rise to an intrinsic angular-momentum density

$$\boldsymbol{t} = -\beta_{s}S\hat{\boldsymbol{l}}, \tag{9.114}$$

where β_{s} is a coefficient that varies between 1 and 0 as one goes from $T = 0$ to T_{c}.

We now recall that the chemical potentials $\mu_{\uparrow,\downarrow}$ in (9.113) are defined at fixed $\boldsymbol{t}$. However, in a rotating vessel the quantity characterizing the equilibrium state is the vorticity $\boldsymbol{\Omega}$, not $\boldsymbol{t}$, and only the chemical potentials $\mu_{\uparrow,\downarrow}(\boldsymbol{t})$ at fixed $\boldsymbol{\Omega}$ are equal. If we then express $\mu_{\uparrow,\downarrow}(\boldsymbol{t})$ in terms of $\mu_{\uparrow,\downarrow}(\boldsymbol{\Omega})$ by

$$\mu_{\uparrow,\downarrow}(\boldsymbol{t}) = \mu_{\uparrow,\downarrow}(\boldsymbol{\Omega}) - \boldsymbol{\Omega} \cdot \frac{\partial \boldsymbol{t}}{\partial \rho_{\uparrow,\downarrow}}, \tag{9.115}$$

using $(\partial E_{\uparrow,\downarrow}/\partial \boldsymbol{t})_{\rho} = \boldsymbol{\Omega}$, and take the derivative of $\boldsymbol{t}$, assuming $\boldsymbol{t} = t_{0}(S)\hat{\boldsymbol{l}}$, as

$$\frac{\partial \boldsymbol{t}}{\partial \rho_{\uparrow,\downarrow}} = -\beta_{s}\hat{\boldsymbol{l}}\frac{\partial S}{\partial \rho_{\uparrow,\downarrow}} = \mp\frac{\hbar}{2m}\beta_{s}\hat{\boldsymbol{l}}, \tag{9.116}$$

we are led to the Josephson–Leggett equation

$$\frac{d\phi_{sp}}{dt} = -\hat{\boldsymbol{f}} \cdot \left(\frac{\gamma^{2}}{\chi}\boldsymbol{S} - \gamma\boldsymbol{H}\right) + (1 - \beta_{s})\hat{\boldsymbol{l}} \cdot \boldsymbol{\Omega}. \tag{9.117}$$

The Barnett effect is seen to disappear at $T = 0$, when $\partial t_{0}/\partial S = 1$.

FURTHER READING

Review articles and books

Brinkman W F and Cross M C 1978 in *Progress in Low Temperature Physics,* Vol. VIIIA, ed. D F Brewer (North-Holland, Amsterdam), p. 105

Forster D 1975 *Hydrodynamic Fluctuations, Broken Symmetry and Correlation Functions* (Benjamin, Reading, Massachusetts)

Hall H E and Hook J R 1986 in *Progress in Low Temperature Physics,* Vol. IX, ed. D F Brewer (North-Holland, Amsterdam), p. 143

Khalatnikov I M 1965 *An Introduction to the Theory of Superfluidity* (Benjamin, New York)

Putterman S J 1974 *Superfluid Hydrodynamics* (North-Holland, Amsterdam)

Articles

Ho T L 1977 in *Quantum Fluids and Solids,* ed. S B Trickey, E D Adams and J W Dufty (Plenum, New York), p. 97

Hu C R and Saslow W 1977 *Phys. Rev. Lett.* **38** 605

Liu M 1982 *Physica* **109/110 B + C** 1615 (Proceedings of the 16th International Conference on Low Temperature Physics, LT-16)

Mermin N D and Muzikar P 1980 *Phys. Rev.* **B21** 980

Miyake K, Takagi H and Usui T 1981 *Prog. Theor. Phys.* **65** 1115

Volovik G and Mineev V P 1981 *Zh. Eksp. Teor. Fiz.* **81** 989 [*Sov. Phys. JETP* **54** 524 (1981)]

10

Transport Properties

The last two chapters dealt with the phenomenological theory of dynamical properties (spin dynamics and hydrodynamics) of superfluid ^{3}He. This theory, however, is still incomplete since it contains a number of as yet unknown coefficients. Furthermore, the hydrodynamic theory is restricted to the regime of small frequencies ($\omega\tau \ll 1$) and wavenumbers ($qv_F\tau \ll 1$). In this and the next chapter we therefore want to address these questions within the framework of a semimicroscopic theory. At present such a theory has only been explicitly worked out for the case of linear response and within the weak-coupling model.

Quite generally, the dynamics of a system possessing long-range order is determined (i) by the structure of the elementary excitations and their interactions and (ii) by the laws of motion governing the order parameter (which, in a sense, describe the ground-state correlations). In the case of a Fermi liquid the most important elementary excitations are the single-particle excitations, the so-called Bogoliubov quasiparticles, which are fermions. At very low temperatures ($T < 0.1T_c$) gapless elementary excitations of the Bose type, like phonons, magnons (i.e. those that do not require a minimum energy) will eventually become important. In view of the temperature range that is experimentally accessible at present and in the foreseeable future, we shall confine our discussion to Fermi excitations.

One may distinguish four dynamical regimes in a pair-correlated Fermi liquid:

(i) *The hydrodynamic regime.* In the case of very slow disturbances, the fluid of thermal excitations and the Cooper-pair condensate are both in a state of local thermodynamic equilibrium, and hence are completely described by a few collective variables only. This is the hydrodynamic regime considered in Chapter 9. For the thermal excitations to be in local equilibrium, the frequency ω and the wavenumber q of the disturbance are subject to the conditions ω, $v_F q \ll 1/\tau$, where τ is a typical quasiparticle collision time and v_F is the Fermi velocity. The corresponding conditions for the pair condensate are ω, $v_F q \ll \Delta/\hbar$, where Δ is the averaged temperature-dependent energy gap. These latter conditions are much less stringent, except for a very narrow temperature range below the transition, where $\tau\Delta \lesssim \hbar$.

(ii) *The macroscopic regime.* For somewhat larger frequencies and wave vectors such that $\tau^{-1} \ll \omega$, $v_F q \ll \Delta/\hbar$, the pair condensate is still in local equilibrium, but the quasiparticle fluid is not. In this "collisionless" regime the description of the quasiparticle fluid requires the specification of infinitely many variables—namely the components of a distribution function in momentum space. This is so because in this case quasiparticles from different portions of the Fermi surface may respond quite differently to an applied field.

(iii) *The microscopic regime.* In the case of rapidly varying disturbances such that $\omega \gtrsim \Delta/\hbar$, $v_F q \gtrsim \Delta/\hbar$, the fluid of thermal excitations and the Cooper pairs are no longer well separated, because the external field may induce transitions between the subsystems. Then the pair condensate is no longer characterized by a few symmetry variables, but rather by a full distribution function in momentum space.

(iv) Finally, for ω, $v_F q \gg \Delta/\hbar$, that part of the available phase space where pair-correlation effects matter constitutes only a small portion of the total phase space accessible. Under these conditions the system reacts as a normal Fermi system.

In the present chapter we shall concentrate on the hydrodynamic region (case (i)). However, rather than deriving the complete hydrodynamic equations from microscopic theory (a task that would, to some extent, duplicate the derivations given in Chapter 9), we shall confine ourselves to the calculation of only some of the more important transport coefficients. Indeed, the phenomenological equations established in Chapter 9 would be of little value without at least some information on the phenomenological coefficients involved. These coefficients fall into one of two categories: (i) the generalized susceptibilities and (ii) the transport coefficients. We have already calculated most of the susceptibilities in Chapters 3 and 7, for example the normal and superfluid densities ρ_n and ρ_s, as well as similar coefficients for the A phase: the supercurrent, the specific heat and the spin susceptibility. The susceptibilities are equilibrium quantities and as such are given as derivatives of the free energy. On the other hand, as exemplified in Chapter 2 for the normal Fermi liquid, the transport coefficients are much harder to calculate, because they characterize properties of a *non-equilibrium* state of the system.

In setting up a theory of the dynamical properties of superfluid ^{3}He, we may again take advantage of the fact that the transition temperature to the superfluid state is much lower than the Fermi temperature. Consequently, the energy spectrum of the liquid is modified only in a small shell of order T_c about the Fermi surface. This of course does not imply that the change in the physical quantities is small, as we have seen. We may, however, hope that a kinetic theory of interacting quasiparticles might provide a unified and intuitively appealing framework for describing such diverse dynamical phenomena as stationary transport properties and high-frequency collective modes. This expectation is nourished by the splendid success of Landau's kinetic theory of normal Fermi liquids. Compared with a completely microscopic theory, such a theory has the advantage of being much more tractable, since all the simplifying features of the problem have been incorporated at an early stage. The starting point of a kinetic theory is the kinetic equation, a generalized Boltzmann equation for the quasiparticle distribution functions. In the pair-correlated state this has to be supplemented by the equation of motion for the order parameter.

As shown in Section 10.1, kinetic equations in the microscopic regime (case (iii)) may be derived from rather elementary considerations, although their mathematical structure is somewhat more complex. We shall consider these full equations in Chapter 11, where we study collective modes in the microscopic regime. In the macroscopic regime the kinetic equations can be considerably simplified by applying a Bogoliubov transformation and a set of gauge transformations. One arrives at a Landau–Boltzmann equation for Bogoliubov quasiparticles, with the hydrodynamic condensate variables appearing as external driving forces. The collision integral requires special consideration and will be discussed in a separate section.

10.1 KINETIC EQUATIONS

In the framework of BCS mean-field theory the equilibrium state of a pair-correlated Fermi system is specified by two momentum-dependent distribution functions $\mathbf{n}_k$ and $\mathbf{F}_k$, which are 2×2 matrices in spin space, defined by their elements

$$n_{k\alpha\beta} = \langle a^{+}_{k\beta} a_{k\alpha} \rangle, \tag{10.1a}$$

$$F_{k\alpha\beta} = \langle a_{-k\beta} a_{k\alpha} \rangle. \tag{10.1b}$$

Note that $n_{k\alpha\beta}$ is diagonal in α, β except for situations involving transverse spin motion. Under the action of external disturbances, these distribution functions will deviate from their equilibrium values, i.e. $n_{k\alpha\beta}(\boldsymbol{r}, t)$ and $F_{k\alpha\beta}(\boldsymbol{r}, t)$ will in general become functions of space and time. As long as the scales of the disturbance in space and time are larger than the Fermi wavelength $\lambda_F = 2\pi/k_F$ and the inverse Fermi frequency $\hbar/\epsilon_F$ respectively, the wave nature of the quasiparticles is not essential and a local quasiclassical description is meaningful. It is now convenient to introduce yet another 2×2 matrix representation, this time in *particle–hole space,* even though this introduces a certain redundancy. Following Nambu (1960), we may then combine the two distribution functions $n_{k\alpha\beta}$ and $F_{k\alpha\beta}$ into a 4×4 matrix distribution function

$$\mathcal{N}_k = \begin{pmatrix} \mathbf{n}_k & \mathbf{F}_k \\ \mathbf{F}_k^{+} & \mathbf{1} - \mathbf{n}_{-k} \end{pmatrix} \tag{10.2a}$$

($\mathbf{1}$ is the 2×2 unit matrix), which may also be expressed as

$$\mathcal{N}_{\boldsymbol{k}} = \begin{pmatrix} \langle a^+_{\boldsymbol{k}\beta} a_{\boldsymbol{k}\alpha}\rangle & \langle a_{-\boldsymbol{k}\beta} a_{\boldsymbol{k}\alpha}\rangle \\ \langle a^+_{\boldsymbol{k}\beta} a^+_{-\boldsymbol{k}\alpha}\rangle & \langle a_{-\boldsymbol{k}\beta} a^+_{-\boldsymbol{k}\alpha}\rangle \end{pmatrix}. \tag{10.2b}$$

Consequently $\mathbf{n}_{\boldsymbol{k}}$ and $\mathbf{F}_{\boldsymbol{k}}$ are referred to as diagonal and off-diagonal distribution functions. Employing the results of Chapter 3 (see (3.43)), the equilibrium distribution function $\mathcal{N}^0_{\boldsymbol{k}}$ takes the form

$$\mathcal{N}^0_{\boldsymbol{k}} = \tfrac{1}{2}\left[\mathit{1} - \boldsymbol{\varepsilon}^0_{\boldsymbol{k}} \frac{1}{E_{\boldsymbol{k}}} \tanh \frac{E_{\boldsymbol{k}}}{2k_{\mathrm{B}}T}\right] \tag{10.3a}$$

($\mathit{1}$ is the 4×4 unit matrix), where we have defined the 4×4 equilibrium energy matrix

$$\boldsymbol{\varepsilon}^0_{\boldsymbol{k}} = \begin{pmatrix} \xi_{\boldsymbol{k}\alpha}\delta_{\alpha\beta} & \Delta_{\boldsymbol{k}\alpha\beta} \\ \Delta^+_{\boldsymbol{k}\alpha\beta} & -\xi_{-\boldsymbol{k}\alpha}\delta_{\alpha\beta} \end{pmatrix}. \tag{10.3b}$$

For simplicity, we consider only the case of zero magnetic field ($\xi_{\boldsymbol{k}\alpha} = \xi_{\boldsymbol{k}}$) for the moment. For unitary states, the square of the energy matrix is just given by $\boldsymbol{\varepsilon}^0_{\boldsymbol{k}}\boldsymbol{\varepsilon}^0_{\boldsymbol{k}} = E^2_{\boldsymbol{k}}\mathit{1}$. Using this relation, the equilibrium distribution function may be interpreted as a Fermi function $f(x) = (\mathrm{e}^{x/k_{\mathrm{B}}T}+1)^{-1}$ of the matrix energy

$$\mathcal{N}^0_{\boldsymbol{k}} = f(\boldsymbol{\varepsilon}^0_{\boldsymbol{k}}), \tag{10.3c}$$

defined by its power series. In the following we present the derivation of a matrix kinetic equation along the lines formulated for usual superconductors by Betbeder-Matibet and Nozières (1969) and for superfluid ^{3}He by Wölfle (1976a). An alternative but equivalent formulation has been given on the basis of the so-called "quasiclassical Green's-function technique" (Eckern 1981, Serene and Rainer 1983).

10.1.1 Matrix kinetic equation

In the presence of a disturbance with wave vector $\boldsymbol{q}$ the distribution function $\mathcal{N}$ acquires nondiagonal matrix elements $\langle \boldsymbol{k}+\tfrac{1}{2}\boldsymbol{q}|\,\mathcal{N}\,|\boldsymbol{k}-\tfrac{1}{2}\boldsymbol{q}\rangle$ in momentum space, since, for example, the expectation value $\langle a^+_{\boldsymbol{k}+\boldsymbol{q}/2} a_{\boldsymbol{k}-\boldsymbol{q}/2}\rangle \neq 0$. Also, the energy matrix will be nondiagonal in momentum space. The energy matrix may actually be considered as an effective Hamiltonian in the restricted space of states spanned by single-particle excitations only. The time evolution of the matrix distribution function may then be written in shorthand notation as

$$\mathrm{i}\hbar \frac{\partial}{\partial t}\mathcal{N} = [\mathcal{N}, \boldsymbol{\varepsilon}], \tag{10.4}$$

where $\mathcal{N}$ is a kind of density matrix of the subspace considered. $\mathcal{N}$ and $\boldsymbol{\varepsilon}$ are matrices in particle–hole and spin spaces as well as in momentum space.

In the case of weak disturbances we may linearize the kinetic equation in terms of the deviations of $\mathcal{N}$ and $\boldsymbol{\varepsilon}$ from their equilibrium values. Defining

$$\langle \boldsymbol{k}+\tfrac{1}{2}\boldsymbol{q}|\,\mathcal{N}\,|\boldsymbol{k}-\tfrac{1}{2}\boldsymbol{q}\rangle = \delta_{\boldsymbol{q},0}\mathcal{N}^0_{\boldsymbol{k}} + \delta\mathcal{N}_{\boldsymbol{k}}(\boldsymbol{q}, t), \tag{10.5a}$$

$$\langle \boldsymbol{k}+\tfrac{1}{2}\boldsymbol{q}|\,\boldsymbol{\varepsilon}\,|\boldsymbol{k}-\tfrac{1}{2}\boldsymbol{q}\rangle = \delta_{\boldsymbol{q},0}\boldsymbol{\varepsilon}^0_{\boldsymbol{k}} + \delta\boldsymbol{\varepsilon}_{\boldsymbol{k}}(\boldsymbol{q}, t), \tag{10.5b}$$

one obtains the linearized matrix kinetic equation ($\boldsymbol{k}\pm\frac{1}{2}\boldsymbol{q}\equiv\boldsymbol{k}_\pm$)

$$i\hbar\frac{\partial}{\partial t}\delta\mathcal{N}_{\boldsymbol{k}}(\boldsymbol{q}, t) - (\boldsymbol{\varepsilon}^0_{\boldsymbol{k}_+}\delta\mathcal{N}_{\boldsymbol{k}} - \delta\mathcal{N}_{\boldsymbol{k}}\,\boldsymbol{\varepsilon}^0_{\boldsymbol{k}_-}) + (\mathcal{N}^0_{\boldsymbol{k}_+}\delta\boldsymbol{\varepsilon}_{\boldsymbol{k}} - \delta\boldsymbol{\varepsilon}_{\boldsymbol{k}}\,\mathcal{N}^0_{\boldsymbol{k}_-}) = \delta\mathcal{I}_{\boldsymbol{k}}. \tag{10.6}$$

This equation is reminiscent of the Landau–Boltzmann equation for normal Fermi liquids. The time rate of change of the distribution function is seen to be given by (i) the free motion of the system (the second and third terms on the left-hand side of (10.6), corresponding to the streaming of particles in real space in a normal system), and (ii) the change induced by forces applied either externally or generated by the mean fields in the system (the last two terms on the left-hand side of (10.6)). A third source of changes of the distribution function is due to the collisions among the quasiparticles. A corresponding term has been included on the right-hand side of (10.6). A derivation of the collision term will be presented in Section 10.2.

The change in the energy matrix caused by the application of external fields is given by the sum of a direct term and a mean-field term induced by the change in the distribution function, as in Fermi-liquid theory. The latter contribution can be expressed quite generally in terms of a linear and local functional relationship between $\delta\boldsymbol{\varepsilon}_{\boldsymbol{k}}$ and $\delta\mathcal{N}_{\boldsymbol{k}}$ involving a generalized Fermi-liquid interaction $\tilde{f}_{\boldsymbol{kk}'}$:

$$\delta\boldsymbol{\varepsilon}_{\boldsymbol{k}}(\boldsymbol{r}, t) = \delta\boldsymbol{\varepsilon}^{\text{ext}}_{\boldsymbol{k}} + \sum_{\boldsymbol{k}}\tilde{f}_{\boldsymbol{kk}'}\,\delta\mathcal{N}_{\boldsymbol{k}'}(\boldsymbol{r}, t). \tag{10.7}$$

However, the relation (10.7) can be simplified if one observes that cross-terms connecting the diagonal energy $\delta\xi_{\boldsymbol{k}}$ with the off-diagonal distribution $\delta F_{\boldsymbol{k}}$ (and vice versa) are necessarily proportional to the gap Δ. On the other hand, the reference energy for the Fermi-liquid interactions is given by the Fermi energy. We may therefore assume that the off-diagonal parts of the generalized Fermi-liquid interactions $\tilde{f}_{\boldsymbol{kk}'}$ are order T_c/T_F small. The same holds for the changes in the diagonal parts of $\tilde{f}$ induced by the pair correlations (for a justification from microscopic theory, see Leggett 1965a, 1966). Hence, to lowest order in T_c/T_F, the changes in the quasiparticle energy matrix components are given by

$$\delta\xi_{\boldsymbol{k}\alpha\beta} = \sum_{\boldsymbol{k}'}\sum_{\alpha'\beta'} f_{\boldsymbol{k}\alpha\beta,\boldsymbol{k}'\alpha'\beta'}\,\delta n_{\boldsymbol{k}'\alpha'\beta'} + \delta\varepsilon^{\text{ext}}_{\boldsymbol{k}\alpha\beta} \tag{10.8}$$

and

$$\delta\Delta_{\boldsymbol{k}\alpha\beta} = \sum_{\boldsymbol{k}'} V_{\boldsymbol{kk}'}\,\delta F_{\boldsymbol{k}'\alpha\beta} \tag{10.9}$$

for the diagonal and off-diagonal components respectively. Here $f_{k\alpha\beta,k'\alpha'\beta'}$ is the normal-state Fermi-liquid interaction (2.6) and $V_{kk'}$ is the pair interaction.

On making the same assumptions about the $\boldsymbol{k}$ dependence of $f_{kk'}$ as discussed before (2.6), the first terms in a Legendre expansion of the diagonal energy change $\delta\xi_{k\alpha\beta}$ may be written as

$$\delta\xi_{k\alpha\beta} = \delta_{\alpha\beta}N_F^{-1}F_0^s\,\delta n + (\tfrac{1}{2}\hbar N_F)^{-1}F_0^a \sum_\mu (\sigma_\mu)_{\alpha\beta}S_\mu$$
$$+\,\delta_{\alpha\beta}\rho^{-1}\frac{\frac{1}{3}F_1^s}{1+\frac{1}{3}F_1^s}\hbar\boldsymbol{k}\cdot\boldsymbol{g} + S_0^{-1}\frac{\frac{1}{3}F_1^a}{1+\frac{1}{3}F_1^a}\hbar\boldsymbol{k}\sum_\mu (\sigma_\mu)_{\alpha\beta}\boldsymbol{j}_\mu^\sigma. \qquad (10.10)$$

Here δn, S_μ, $\boldsymbol{g}$ and $\boldsymbol{j}_\mu^\sigma$ are the changes in particle density, spin density, momentum density and spin current density respectively, which will be defined in Section 10.1.2; $S_0 = \frac{1}{2}\hbar n$ is the maximum spin polarization.

The matrix kinetic equation (10.6), supplemented by the relations (10.8)–(10.10), will be solved in Chapter 11. In the present chapter the matrix kinetic equations will be cast into a form suitable for the calculation of transport properties. First, however, we need to know how the physically observable properties are actually obtained from the matrix distribution function.

10.1.2 Conserved quantities and conservation laws

The expectation values of conserved quantities can be expressed in terms of the diagonal distribution function $\mathbf{n}_k$, which is defined as the average of the general single-particle operator product $a^+_{k+q/2,\beta}a_{k-q/2,\alpha}$ in a given nonequilibrium state. Consequently, the deviations from equilibrium of the conserved densities of particle number, spin and momentum are given by

$$\delta n = \tfrac{1}{2}\sum_k \mathrm{tr}_4\,(\boldsymbol{\tau}_0\,\delta\mathcal{N}_k) = \sum_k \mathrm{tr}_\sigma\,(\delta\mathbf{n}_k), \qquad (10.11a)$$

$$S_\mu = \tfrac{1}{4}\hbar\sum_k \mathrm{tr}_4\,(\boldsymbol{\tau}_\mu\,\delta\mathcal{N}_k) = \tfrac{1}{2}\hbar\sum_k \mathrm{tr}_\sigma\,(\boldsymbol{\sigma}_\mu\,\delta\mathbf{n}_k), \quad \mu = 1, 2, 3, \qquad (10.11b)$$

$$\boldsymbol{g} = \tfrac{1}{2}\sum_k \hbar\boldsymbol{k}\,\mathrm{tr}_4\,(\delta\mathcal{N}_k) = \sum_k \hbar\boldsymbol{k}\,\mathrm{tr}_\sigma\,(\delta\mathbf{n}_k), \qquad (10.11c)$$

where "tr_4" and "tr_σ" denote the traces on 4×4 particle–hole matrix quantities and on 2×2 spin matrix quantities respectively. In addition, the energy density contains an off-diagonal contribution:

$$\delta\varepsilon = \tfrac{1}{2}\sum_k \mathrm{tr}_4\,[(\boldsymbol{\varepsilon}_k^0 + \mu\boldsymbol{1})\,\delta\mathcal{N}_k]. \qquad (10.11d)$$

The 4×4 matrices $\boldsymbol{\tau}_\mu$, $\mu = 0, 1, 2, 3$, introduced in (10.11a,b) are defined as direct products of the unit matrix $\mathbf{1}$ $(\mu = 0)$ or the Pauli matrices $\boldsymbol{\sigma}_\mu$ $(\mu = 1, 2, 3)$ in spin space with the Pauli matrix $\boldsymbol{\sigma}_3$ in particle–hole space:

$$\boldsymbol{\tau}_0 = \begin{pmatrix} \mathbf{1} & 0 \\ 0 & -\mathbf{1} \end{pmatrix}, \quad \boldsymbol{\tau}_\mu = \begin{pmatrix} \boldsymbol{\sigma}_\mu & 0 \\ 0 & -\boldsymbol{\sigma}_\mu^{\mathrm{T}} \end{pmatrix}, \tag{10.12}$$

where superscript T denotes the transposed matrix. The conservation laws for particle number, spin, momentum and energy may now be derived by multiplying the kinetic equations (10.6) by $\boldsymbol{\tau}_0$, $\frac{1}{2}\hbar\boldsymbol{\tau}_\mu$, $\hbar\boldsymbol{k}$ and $\boldsymbol{\varepsilon}_{\boldsymbol{k}}^0 + \mu\mathbf{1}$, then taking the 4-trace and finally summing over $\boldsymbol{k}$. In this way, the particle-number conservation equation, for example, takes the form

$$\omega\,\delta n - \tfrac{1}{2}\sum_{\boldsymbol{k}} \mathrm{tr}_4\,[(\boldsymbol{\tau}_0\boldsymbol{\varepsilon}_{\boldsymbol{k}_+}^0 - \boldsymbol{\varepsilon}_{\boldsymbol{k}_-}^0\boldsymbol{\tau}_0)\,\delta\mathcal{N}_{\boldsymbol{k}} - (\boldsymbol{\tau}_0\mathcal{N}_{\boldsymbol{k}_+}^0 - \mathcal{N}_{\boldsymbol{k}_-}^0\boldsymbol{\tau}_0)\,\delta\boldsymbol{\varepsilon}_{\boldsymbol{k}}] = 0, \tag{10.13}$$

where use has been made of the cyclic invariance of the trace and a Fourier transform with respect to time has been taken. The collision integral drops out because of the particle-number-conserving nature of collisions. The trace expression cancels for a spatially homogeneous system $(\boldsymbol{q} = 0)$ by virtue of the equilibrium gap equation. On the other hand, the terms linear in $\boldsymbol{q}$ are interpreted as the divergence of the current density $\boldsymbol{j}$:

$$\boldsymbol{j} = \sum_{\boldsymbol{k}} \boldsymbol{v}_{\boldsymbol{k}}\,\mathrm{tr}_\sigma\,(\delta\mathbf{n}'_{\boldsymbol{k}}). \tag{10.14}$$

Here $\boldsymbol{v}_{\boldsymbol{k}} = \hbar\boldsymbol{k}/m^*$ is the normal-state quasiparticle velocity, and

$$\delta\mathbf{n}'_{\boldsymbol{k}} = \delta\mathbf{n}_{\boldsymbol{k}} + \phi'_{\boldsymbol{k}}\,\delta\boldsymbol{\varepsilon}_{\boldsymbol{k}}, \tag{10.15a}$$

where

$$\phi'_{\boldsymbol{k}} = \frac{\partial}{\partial\xi_{\boldsymbol{k}}}\left(\frac{\xi_{\boldsymbol{k}}}{2E_{\boldsymbol{k}}}\,t_{\boldsymbol{k}}\right) \tag{10.15b}$$

and $t_{\boldsymbol{k}} = \tanh\,(E_{\boldsymbol{k}}/2k_{\mathrm{B}}T)$.

Note that, owing to Galilean invariance, the momentum density is equal to the bare mass times the current density:

$$\boldsymbol{g} = m\boldsymbol{j}. \tag{10.16}$$

The equality of (10.16) and (10.11c) is guaranteed by the effective-mass relation (2.16), which continues to hold even in the superfluid state. In analogy, the spin current density $\boldsymbol{j}_\mu^\sigma$, the momentum current density Π_{ij} and the energy current density $\boldsymbol{j}^\varepsilon$ are found to be given by

$$\boldsymbol{j}_\mu^\sigma = \tfrac{1}{2}\hbar\sum_{\boldsymbol{k}} \boldsymbol{v}_{\boldsymbol{k}}\,\mathrm{tr}_\sigma\,(\boldsymbol{\sigma}_\mu\,\delta\boldsymbol{n}'_{\boldsymbol{k}}), \tag{10.17a}$$

$$\Pi_{ij} = \sum_{\boldsymbol{k}} \hbar k_i v_{\boldsymbol{k},j}\,\mathrm{tr}_\sigma\,(\delta\boldsymbol{n}'_{\boldsymbol{k}}), \tag{10.17b}$$

$$\boldsymbol{j}_\varepsilon = \frac{\mu}{m}\boldsymbol{g} + \sum_{\boldsymbol{k}} \xi_{\boldsymbol{k}}\boldsymbol{v}_{\boldsymbol{k}}\,\mathrm{tr}_\sigma\,(\delta\boldsymbol{n}'_{\boldsymbol{k}}). \tag{10.17c}$$

Evaluating the $\boldsymbol{k}$ sum in the usual way, namely by taking integrals over the energy $\xi_{\boldsymbol{k}}$ and the angular variables, the spin current, for example, may be written as

$$\boldsymbol{j}_{\mu}^{\sigma}=\tfrac{1}{2}\hbar(1+\tfrac{1}{3}F_1^{\mathrm{a}})\sum_{\boldsymbol{k}}\boldsymbol{v}_{\boldsymbol{k}}\,\delta n_{\boldsymbol{k},\mu}. \tag{10.18a}$$

Here we have made use of the integral

$$\int_{-\infty}^{\infty}\mathrm{d}\xi_{\boldsymbol{k}}\,\phi'_{\boldsymbol{k}}=\int_{-\infty}^{\infty}\mathrm{d}\xi_{\boldsymbol{k}}\left(\frac{|\Delta_{\boldsymbol{k}}|^2}{2E_{\boldsymbol{k}}^3}t_{\boldsymbol{k}}+\frac{\xi_{\boldsymbol{k}}^2}{2E_{\boldsymbol{k}}^2}\frac{\partial t_{\boldsymbol{k}}}{\partial E_{\boldsymbol{k}}}\right)=1 \tag{10.18b}$$

and introduced the notation

$$\delta n_{\boldsymbol{k},\mu}=\mathrm{tr}_{\sigma}\,(\boldsymbol{\sigma}_{\mu}\,\delta\mathbf{n}_{\boldsymbol{k}}). \tag{10.18c}$$

The last term of (10.10) has in fact already been expressed in terms of the spin current via (10.18a).

10.1.3 Kinetic equation for Bogoliubov quasiparticles

As already mentioned, the matrix kinetic equation may be simplified in the low-frequency macroscopic regime, where the pair condensate is in a local equilibrium state. In this regime transitions between the normal liquid and the condensate rarely occur. The Bogoliubov quasiparticles (BQP) may then be considered to be stable entities. Under these circumstances, a description in terms of a distribution function for BQP rather than by a matrix distribution function appears promising. This can be achieved by performing a particular Bogoliubov transformation on the matrix distribution function. In the present matrix notation the Bogoliubov transformation is defined as the unitary transformation diagonalizing the energy matrix:

$$\mathcal{U}_{\boldsymbol{k}}\boldsymbol{\varepsilon}_{\boldsymbol{k}}^0\mathcal{U}_{\boldsymbol{k}}^{+}=E_{\boldsymbol{k}}\boldsymbol{\tau}_0, \tag{10.19}$$

where the components of $\mathcal{U}_{\boldsymbol{k}}$,

$$\mathcal{U}_{\boldsymbol{k}}=\begin{pmatrix}u_{\boldsymbol{k}\alpha\beta} & v_{\boldsymbol{k}\alpha\beta}\\ -v_{\boldsymbol{k}\alpha\beta} & u_{\boldsymbol{k}\alpha\beta}\end{pmatrix}, \tag{10.20}$$

have been calculated in (3.25). There we found

$$u_{\boldsymbol{k}\alpha\beta}=\delta_{\alpha\beta}\left[\frac{1}{2}\left(1+\frac{\xi_{\boldsymbol{k}}}{E_{\boldsymbol{k}}}\right)\right]^{1/2}, \tag{10.21a}$$

$$v_{\boldsymbol{k}\alpha\beta}=\frac{\Delta_{\boldsymbol{k}\alpha\beta}}{|\Delta_{\boldsymbol{k}}|}\left[\frac{1}{2}\left(1-\frac{\xi_{\boldsymbol{k}}}{E_{\boldsymbol{k}}}\right)\right]^{1/2}. \tag{10.21b}$$

The Bogoliubov transformation of the equilibrium distribution function is immediately obtained as

$$\mathscr{U}_k \mathcal{N}_k^0 \mathscr{U}_k^+ = f(E_k \boldsymbol{\tau}_0) = \begin{pmatrix} f(E_k)\mathbf{1} & 0 \\ 0 & f(-E_k)\mathbf{1} \end{pmatrix}, \tag{10.22}$$

where $f(E)$ is the Fermi function.

We are now in a position to diagonalize the matrices multiplying $\delta\mathcal{N}$ and $\delta\boldsymbol{\varepsilon}$ in the matrix kinetic equation. For this, we multiply (10.6) from the left by $\mathscr{U}_{k_+}$ and from the right by $\mathscr{U}_{k_-}^+$, which yields the result

$$\omega\, \delta\tilde{\mathcal{N}}_k - E_{k_+}\boldsymbol{\tau}_0\, \delta\tilde{\mathcal{N}}_k + \delta\tilde{\mathcal{N}}_k\, \boldsymbol{\tau}_0 E_{k_-} + f(\boldsymbol{\tau}_0 E_{k_+})\, \delta\tilde{\mathscr{E}}_k - \delta\tilde{E}_k\, f(\boldsymbol{\tau}_0 E_{k_-}) = \mathscr{U}_{k_+}\, \delta\mathscr{I}_k\, \mathscr{U}_{k_-}^+. \tag{10.23}$$

Here $\delta\tilde{\mathcal{N}}_k$ and $\delta\tilde{\mathscr{E}}_k$ are the Bogoliubov-transformed matrix distribution function and energy change respectively. They are defined by

$$\delta\tilde{\mathcal{N}}_k = \mathscr{U}_{k_+}\, \delta\mathcal{N}_k\, \mathscr{U}_{k_-}^+ \equiv \begin{pmatrix} \delta f_{k\alpha\beta} & \delta h_{k\alpha\beta} \\ \delta h_{k\alpha\beta}^+ & -\delta f_{-k\alpha\beta}^{\mathrm{T}} \end{pmatrix}, \tag{10.24a}$$

$$\delta\tilde{\mathscr{E}}_k = \mathscr{U}_{k_+}\, \delta\boldsymbol{\varepsilon}_k\, \mathscr{U}_{k_-}^+ \equiv \begin{pmatrix} \delta E_{k\alpha\beta} & \delta\Gamma_{k\alpha\beta} \\ \delta\Gamma_{k\alpha\beta}^+ & -\delta E_{-k\alpha\beta}^{\mathrm{T}} \end{pmatrix}. \tag{10.24b}$$

An expansion of (10.23) in small q leads to a Boltzmann equation of the new (diagonal!) distribution $\delta\mathbf{f}_k$:

$$\omega\, \delta\mathbf{f}_k - (\boldsymbol{q}\cdot\boldsymbol{\nabla}_k E_k)(\delta\mathbf{f}_k - f_k'\, \delta\mathbf{E}_k) = (\mathscr{U}\, \delta\mathscr{I}\, \mathscr{U}^+)_{++}, \tag{10.25a}$$

where a suffix $++$, $+-$ etc. denotes components in particle–hole space; i.e. $++$ is the first diagonal element while $+-$ is the upper off-diagonal element, etc. Furthermore,

$$f_k' = \frac{\partial f_k}{\partial E_k} = -\frac{1}{2}\frac{\partial t_k}{\partial E_k}, \tag{10.25b}$$

$$\boldsymbol{\nabla}_k E_k = \frac{\xi_k}{E_k}\frac{\hbar \boldsymbol{k}}{m^*}. \tag{10.25c}$$

In (10.25c) a contribution proportional to $\boldsymbol{\nabla}_k \boldsymbol{\Delta}_k$ has been neglected, since it is order T_c/T_F small. Note that $\boldsymbol{\nabla}_k E_k$, the group velocity of BQP, vanishes at the Fermi energy $\xi_k = 0$. In (10.25a) the change in the BQP energy is found as

$$\delta\mathbf{E}_k = \tfrac{1}{2}\left(1 + \frac{\xi_k}{E_k}\right)\delta\xi_k - \tfrac{1}{2}\left(1 - \frac{\xi_k}{E_k}\right)\frac{1}{|\Delta_k|^2}\boldsymbol{\Delta}_k\, \delta\xi_{-k}^{\mathrm{T}}\, \boldsymbol{\Delta}_k^+ + \frac{1}{2E_k}(\delta\boldsymbol{\Delta}_k\, \boldsymbol{\Delta}_k^+ + \boldsymbol{\Delta}_k\, \delta\boldsymbol{\Delta}_k^+). \tag{10.26}$$

As we shall see later, $\delta\boldsymbol{E}_k$ is related to $\delta\mathbf{f}_k$ in a more complicated way than in normal Fermi-liquid theory, because $\delta\xi_k$ and $\delta\boldsymbol{\Delta}_k$ are defined in terms of $\delta\mathbf{n}_k$ and $\delta\mathbf{F}_k$ rather than $\delta\mathbf{f}_k$ and $\delta\mathbf{h}_k$ appearing in (10.24a).

Indeed, the kinetic equation for the new off-diagonal distribution function $\delta\mathbf{h}_k$ is found to take quite a different form:

$$(\omega - 2E_k)\, \delta\mathbf{h}_k - t_k\, \delta\boldsymbol{\Gamma}_k = (\mathscr{U}\, \delta\mathscr{I}\, \mathscr{U}^+)_{+-}. \tag{10.27}$$

For frequencies ω much less than the average gap frequency $\Delta/\hbar$ (and also for collision rates $\tau^{-1} < \Delta/\hbar$), (10.27) may be solved for $\delta\mathbf{h}_k$ to give

$$\delta\mathbf{h}_k = -\frac{t_k}{2E_k}\,\delta\mathbf{\Gamma}_k + O(q^2, \tau^{-1}). \tag{10.28}$$

Equation (10.28) must be interpreted as the local equilibrium value of $\delta\mathbf{h}_k$. The new off-diagonal energy change is calculated as

$$\delta\mathbf{\Gamma}_k = -(\delta\xi_k\,\mathbf{\Delta}_k + \mathbf{\Delta}_k\,\delta\xi^{\mathrm{T}}_{-k})\frac{1}{2E_k} - \frac{1}{2\,|\Delta_k|^2}\mathbf{\Delta}_k(\delta\mathbf{\Delta}_k^+\,\mathbf{\Delta}_k - \mathbf{\Delta}_k^+\,\delta\mathbf{\Delta}_k)$$
$$+\frac{\xi_k}{E_k}\frac{1}{2\,|\Delta_k|^2}\mathbf{\Delta}_k(\delta\mathbf{\Delta}_k^+\,\mathbf{\Delta}_k + \mathbf{\Delta}_k^+\,\delta\mathbf{\Delta}_k) + O(q^2). \tag{10.29}$$

In order to close the system of equations, one also needs to express the old distribution functions $\delta\mathbf{n}_k$ and $\delta\mathbf{F}_k$, entering the expressions for $\delta\xi_k$ and $\delta\mathbf{\Delta}_k$, in terms of the new ones. This is easily obtained from the reverse Bogoliubov transformation. One finds

$$\delta\mathbf{n}_k = \tfrac{1}{2}\left(1 + \frac{\xi_k}{E_k}\right)\delta\mathbf{f}_k - \tfrac{1}{2}\left(1 - \frac{\xi_k}{E_k}\right)\frac{1}{|\Delta_k|^2}\mathbf{\Delta}_k\,\delta\mathbf{f}^{\mathrm{T}}_{-k}\,\mathbf{\Delta}_k^+$$
$$-\frac{t_k}{4E_k^3}(|\Delta_k|^2\,\delta\xi_k + \mathbf{\Delta}_k\,\delta\xi^{\mathrm{T}}_{-k}\,\mathbf{\Delta}_k^+)$$
$$+\frac{t_k}{4E_k^2}\frac{\xi_k}{E_k}(\mathbf{\Delta}_k\,\delta\mathbf{\Delta}_k^+ + \delta\mathbf{\Delta}_k\,\mathbf{\Delta}_k^+) + O(q), \tag{10.30a}$$

$$\delta\mathbf{F}_k = \frac{1}{2E_k}(\delta\mathbf{f}_k\mathbf{\Delta}_k + \mathbf{\Delta}_k\,\delta\mathbf{f}^{\mathrm{T}}_{-k}) + \frac{t_k}{4E_k^2}(\delta\xi_k\,\mathbf{\Delta}_k + \mathbf{\Delta}_k\,\delta\xi^{\mathrm{T}}_{-k}) - \frac{t_k}{2E_k}\frac{\xi_k}{E_k}\,\delta\mathbf{\Delta}_k. \tag{10.30b}$$

In principle, one would now have to eliminate $\delta\mathbf{h}_k$ via (10.28) in favour of $\delta\mathbf{f}_k$ and $\delta\mathbf{\Delta}_k$ in all expressions. This is a rather cumbersome procedure, because it turns out that all the expressions have to be evaluated at least to order q. After all, we have to remember that the hydrodynamic variables associated with the order parameter are given by gradients of the angles specifying the rotation or the gauge of the gap parameter and not by the angles themselves.

Fortunately, there is a way to circumvent this laborious task. The clue is that the change in the order parameter in the hydrodynamic limit is describable in terms of gauge transformations of the order parameter. By performing the gauge transformations directly on the matrix distribution function, one is able to generate the terms coupling the condensate to the quasiparticle system in a natural way. The price that one has to pay for this is that one loses the equation of motion for the order parameter. This, however, is easily remedied: as we have seen before, every hydrodynamic variable of the condensate corresponds to a broken symmetry. On the other hand, every one of these symmetries gives rise to a conservation law in which the particular hydrodynamic variable enters. Thus, by enforcing the conservation law, one is able to fix the value of the hydrodynamic variable.

Let us demonstrate this procedure for one particular example—the phase of the order parameter.

10.1.4 Gauge transformation of the kinetic equation

A change in the phase of the gap parameter

$$\Delta_{\boldsymbol{k}} = e^{i\delta\phi}\Delta^{g}_{\boldsymbol{k}} \tag{10.31}$$

is accomplished by a unitary transformation of the energy matrix

$$\boldsymbol{\varepsilon}_{\boldsymbol{k}} = \mathcal{U}_{\phi}\boldsymbol{\varepsilon}^{g}_{\boldsymbol{k}}\mathcal{U}^{+}_{\phi} = \boldsymbol{\varepsilon}^{g}_{\boldsymbol{k}} + \delta\boldsymbol{\varepsilon}^{\phi}_{\boldsymbol{k}}, \tag{10.32a}$$

where

$$\mathcal{U}_{\phi} = \exp\left(\tfrac{1}{2}i\boldsymbol{\tau}_0\,\delta\phi\right), \tag{10.32b}$$

$$\delta\boldsymbol{\varepsilon}^{\phi}_{\boldsymbol{k}} = \tfrac{1}{2}i\,\delta\phi\,(\boldsymbol{\tau}_0\boldsymbol{\varepsilon}^{g}_{\boldsymbol{k}} - \boldsymbol{\varepsilon}^{g}_{\boldsymbol{k}}\boldsymbol{\tau}_0). \tag{10.32c}$$

Here the superscript ϕ indicates the *fixed* gauge. Because a phase change $\delta\phi$ in $\Delta_{\boldsymbol{k}}$ carries over to the nondiagonal distribution function $\mathbf{F}_{\boldsymbol{k}}$ as well, the matrix distribution function also transforms as

$$\mathcal{N}_{\boldsymbol{k}} = \mathcal{U}_{\phi}\mathcal{N}^{g}_{\boldsymbol{k}}\mathcal{U}^{+}_{\phi}. \tag{10.33}$$

We may now choose the phase change $\delta\phi(\boldsymbol{r}, t)$ at any point in space and time such that $\boldsymbol{\varepsilon}_{\boldsymbol{k}}$ has constant phase. In other words, we may "gauge away" the phase. For spatially varying phase $\delta\phi(\boldsymbol{r}, t)$, or equivalently for its Fourier transform $\delta\phi(\boldsymbol{q}, \omega)$, this is accomplished by substituting

$$\begin{aligned}\delta\mathcal{N}_{\boldsymbol{k}}(\boldsymbol{q}, \omega) &= \delta(\mathcal{U}_{\phi}\mathcal{N}^{g}_{\boldsymbol{k}}\mathcal{U}^{+}_{\phi})\\ &= \delta\mathcal{N}^{g}_{\boldsymbol{k}} + \tfrac{1}{2}i\,\delta\phi\,(\boldsymbol{\tau}_0\mathcal{N}^{0}_{\boldsymbol{k}_-} - \mathcal{N}^{0}_{\boldsymbol{k}_+}\boldsymbol{\tau}_0)\end{aligned} \tag{10.34}$$

into the matrix kinetic equation (10.6).

After rearranging terms, the kinetic equation for the gauge-invariant distribution function $\delta\mathcal{N}^{g}_{\boldsymbol{k}}$ is found to have essentially the same form as before. The only difference is due to the appearance of a number of additional terms in the change in energy:

$$\delta\boldsymbol{\varepsilon}_{\boldsymbol{k}} \rightarrow \delta\boldsymbol{\varepsilon}^{\text{tot}}_{\boldsymbol{k}} = \delta\boldsymbol{\varepsilon}_{\boldsymbol{k}} - \tfrac{1}{2}i\hbar\omega\,\delta\phi\,\boldsymbol{\tau}_0 + \tfrac{1}{2}i\,\delta\phi\,(\boldsymbol{\varepsilon}^{0}_{\boldsymbol{k}_+}\boldsymbol{\tau}_0 - \boldsymbol{\tau}_0\boldsymbol{\varepsilon}^{0}_{\boldsymbol{k}_-}). \tag{10.35}$$

For $\boldsymbol{q} = 0$, the last term is exactly equal to the change in $\delta\boldsymbol{\varepsilon}$ under a change of phase by $-\delta\phi$ as described by (10.32c). (Note that (10.34) is equivalent to a phase transformation of $\delta\mathcal{N}_{\boldsymbol{k}}$ by $-\delta\phi$.)

We therefore define the gauge-invariant energy change by

$$\delta\boldsymbol{\varepsilon}^{g}_{\boldsymbol{k}} = \delta\boldsymbol{\varepsilon}_{\boldsymbol{k}} - \delta\boldsymbol{\varepsilon}^{-\phi}_{\boldsymbol{k}}. \tag{10.36}$$

Furthermore, we introduce the notation

$$\delta\mu = \tfrac{1}{2}i\hbar\omega\,\delta\phi, \tag{10.37a}$$

$$\boldsymbol{v}_s = \frac{\hbar}{2m}i\boldsymbol{q}\,\delta\phi, \tag{10.37b}$$

which is borrowed from the phenomenological theory that relates the time derivative of the phase to the chemical potential μ and the gradient of the phase to the superfluid velocity $\boldsymbol{v}_s$ (see (9.8) and (9.16)). Note that in this chapter μ is the chemical potential per particle. This identification can be verified later.

Expansion of the last term in (10.35) to first order in q gives an expression for the change in the energy matrix in the gauge-invariant formulation:

$$\delta\boldsymbol{\varepsilon}_{\boldsymbol{k}}^{\text{tot}} = \delta\boldsymbol{\varepsilon}_{\boldsymbol{k}}^{\text{g}} - \delta\mu\,\boldsymbol{\tau}_0 + \frac{m}{m^*}\hbar(\boldsymbol{k}\cdot\boldsymbol{v}_s)\boldsymbol{1}. \tag{10.38}$$

The degrees of freedom of the order parameter that relate to rotations in spin space or orbital space may be treated in an analogous way (see Combescot 1974a). For example, the rotation of the order parameter in spin space can be compensated by a non-Abelian gauge transformation $\mathbf{U}_\theta = \exp(\frac{1}{2}\mathrm{i}\sum_\mu \boldsymbol{\sigma}_\mu\,\delta\theta_\mu^{\text{s}})$, where $\delta\theta_\mu^{\text{s}}$ are the rotation angles in spin space and $\boldsymbol{\sigma}_\mu$ are the three Pauli matrices. This allows an interpretation of the time and the space derivatives of $\delta\theta_\mu^{\text{s}}$ as an effective magnetic field $\boldsymbol{H}^{\text{g}}$,

$$\mu_0 H_\mu^{\text{g}} = \tfrac{1}{2}\mathrm{i}\hbar\omega\,\delta\theta_\mu^{\text{s}}, \tag{10.39a}$$

and a spin superfluid velocity,

$$\boldsymbol{v}_{\text{sp},\mu} = -\frac{\hbar}{2m}\mathrm{i}\boldsymbol{q}\,\delta\theta_\mu^{\text{s}}, \tag{10.39b}$$

respectively. In this way, one generates additional gauge-dependent terms in the energy change,

$$\delta\boldsymbol{\varepsilon}_{\boldsymbol{k}}^{\text{spin}} = \sum_\mu\left[-\mu_0 H_\mu^{\text{g}}\boldsymbol{\tau}_\mu + \frac{m}{m^*}\hbar(\boldsymbol{k}\cdot\boldsymbol{v}_{\text{sp},\mu})\boldsymbol{\tau}_\mu\boldsymbol{\tau}_0\right], \tag{10.40}$$

which have to be added to the right-hand side of (10.38).

It remains to express the observable quantities, i.e. the densities and currents, in terms of the gauge-invariant distribution function $\delta\mathcal{N}_{\boldsymbol{k}}^{\text{g}}$. For this, we first rewrite the diagonal elements of $\delta\mathcal{N}_{\boldsymbol{k}}$ in terms of $\delta\mathcal{N}_{\boldsymbol{k}}^{\text{g}}$ (see (10.34)):

$$\delta n_{\boldsymbol{k}\alpha\beta} = \delta n_{\boldsymbol{k}\alpha\beta}^{\text{g}} + m\left[\delta_{\alpha\beta}\boldsymbol{v}_s + \sum_\mu (\sigma_\mu)_{\alpha\beta}\boldsymbol{v}_{\text{sp},\mu}\right]\cdot\boldsymbol{v}_{\boldsymbol{k}}\phi_{\boldsymbol{k}}'. \tag{10.41}$$

Equation (10.41) is then inserted into the expressions for the conserved densities. One finds that in the equations (10.11a, b, d) for δn, S_μ and $\delta\varepsilon$ the distribution function $\delta\mathbf{n}_{\boldsymbol{k}}$ is simply replaced by $\delta\mathbf{n}_{\boldsymbol{k}}^{\text{g}}$. However, (10.11c) for the momentum density acquires an additional term such that

$$\boldsymbol{g} = \sum_{\boldsymbol{k}\alpha}\hbar\boldsymbol{k}\,\delta n_{\boldsymbol{k}\alpha\alpha}^{\text{g}} + \rho\boldsymbol{v}_s. \tag{10.42}$$

The momentum density $\boldsymbol{g}$ is hence seen to be given by the sum of the momentum density in the rest frame of the superfluid, described by $\delta n_{\boldsymbol{k}}^{\text{g}}$, and the contribution from a Galilean transformation to the system with $\boldsymbol{v}_s \neq 0$.

Since the energy change induced by the Fermi-liquid interaction involves the ungauged distribution function $\delta\mathbf{n}_{\boldsymbol{k}}$, this part contains gauge-dependent terms too. Inserting (10.41) into the definition of $\delta\boldsymbol{\varepsilon}_{\boldsymbol{k}}$, (10.7), and combining all the gauge terms, one finds for the diagonal part of $\delta\boldsymbol{\varepsilon}_{\boldsymbol{k}}^{\text{tot}}$,

$$\delta\boldsymbol{\xi}_{\boldsymbol{k}}^{\text{tot}} = \sum_{\boldsymbol{k}'}\left[\mathbf{1}f_{\boldsymbol{k}\boldsymbol{k}'}^{\text{s}}\,\text{tr}_\sigma\,(\delta\mathbf{n}_{\boldsymbol{k}'}^{\text{g}}) + \sum_\mu f_{\boldsymbol{k}\boldsymbol{k}'}^{\text{a}}\boldsymbol{\sigma}_\mu\,\text{tr}_\sigma\,(\boldsymbol{\sigma}_\mu\,\delta\mathbf{n}_{\boldsymbol{k}'}^{\text{g}})\right]$$
$$-\,\delta\mu\,\mathbf{1} - \mu_0\sum_\mu H_\mu^{\text{g}}\boldsymbol{\sigma}_\mu + \hbar(\boldsymbol{k}\cdot\boldsymbol{v}_{\text{s}})\mathbf{1} + \frac{m}{m_{\text{s}}}\sum_\mu \hbar(\boldsymbol{k}\cdot\boldsymbol{v}_{\text{sp},\mu})\boldsymbol{\sigma}_\mu. \quad (10.43)$$

The effect of replacing the ungauged distribution function by the gauged one in (10.43) is to remove the factor m/m^* in front of $\boldsymbol{k}\cdot\boldsymbol{v}_{\text{s}}$ (in compliance with Galilean invariance) and to renormalize the prefactor of the spin supercurrent term (last term in (10.43)). Here we have defined a "spin" effective mass by

$$\frac{m_{\text{s}}}{m} = \frac{1+\frac{1}{3}F_1^{\text{s}}}{1+\frac{1}{3}F_1^{\text{a}}}. \quad (10.44)$$

The Bogoliubov-transformed kinetic equation for the gauge-transformed distribution function $\delta\mathbf{f}_{\boldsymbol{k}}^{\text{g}}$ still has the form (10.25a). However, the change in the BQP energy $\delta\mathbf{E}_{\boldsymbol{k}}^{\text{g}}$ is now given by (10.26) with $\delta\boldsymbol{\xi}_{\boldsymbol{k}}$ replaced by $\delta\boldsymbol{\xi}_{\boldsymbol{k}}^{\text{tot}}$.

10.1.5 Gauge-invariant densities and currents

In order to calculate the transport properties from the kinetic equation (10.25a), we need to know how to express the currents in the hydrodynamic equations in terms of the gauge-invariant BQP distribution function $\delta\mathbf{f}_{\boldsymbol{k}}^{\text{g}}$. For this, we have to gauge-transform (10.30), which amounts to replacing $\delta\mathbf{n}_{\boldsymbol{k}}$ by $\delta\mathbf{n}_{\boldsymbol{k}}^{\text{g}}$ and $\delta\mathbf{f}_{\boldsymbol{k}}$ by $\delta\mathbf{f}_{\boldsymbol{k}}^{\text{g}}$ as well as $\delta\boldsymbol{\varepsilon}_{\boldsymbol{k}}$ by $\delta\boldsymbol{\varepsilon}_{\boldsymbol{k}}^{\text{tot}}$. Then we insert this gauge-transformed equation as well as (10.41) into the expressions (10.11)–(10.18). In this way, we find for the particle-number density

$$\delta n = \sum_{\boldsymbol{k}}\text{tr}_\sigma\left[\frac{\xi_{\boldsymbol{k}}}{E_{\boldsymbol{k}}}\,\delta\mathbf{f}_{\boldsymbol{k}}^{\text{g}} - \frac{t_{\boldsymbol{k}}}{2E_{\boldsymbol{k}}^3}\,|\Delta_{\boldsymbol{k}}|^2\,\delta\boldsymbol{\xi}_{\boldsymbol{k}}^{\text{g}}\right] + \left(\frac{\partial n}{\partial\mu}\right)_P^0\delta\mu, \quad (10.45\text{a})$$

where

$$\left(\frac{\partial n}{\partial\mu}\right)_{\text{P}}^0 = \sum_{\boldsymbol{k}\alpha}\frac{t_{\boldsymbol{k}}}{2E_{\boldsymbol{k}}^3}\,|\Delta_{\boldsymbol{k}}|^2. \quad (10.45\text{b})$$

As is apparent from (10.45a), the quantity $(\partial n/\partial u)_P^0$ relates a change in the chemical potential to a change in density at fixed BQP distribution (i.e. $\delta\mathbf{f}_{\boldsymbol{k}}=0$) and in the absence of Fermi-liquid interaction effects. Thus it only describes a part of the total density response and is therefore proportional to the so-called "bare partial compressibility" κ_{P}^0. An equivalent (and very useful) form of (10.45) is given by

$$\delta n = \frac{\partial n}{\partial \mu}\left[N_F^{-1}\sum_{\boldsymbol{k}}\frac{\xi_{\boldsymbol{k}}}{E_{\boldsymbol{k}}}\operatorname{tr}_\sigma(\delta\mathbf{f}_{\boldsymbol{k}}^{\mathrm{g}} - f'_{\boldsymbol{k}}\,\delta\mathbf{E}_{\boldsymbol{k}}^{\mathrm{g}}) + \delta\mu\right], \tag{10.46}$$

where $\partial n/\partial \mu$ is the normal-state compressibility (see (2.19)).

The momentum density $\boldsymbol{g}$ is obtained from its definition (10.11c) by using (10.18b), (10.30) and (10.41):

$$\boldsymbol{g} = \sum_{\boldsymbol{k}} \hbar\boldsymbol{k}\operatorname{tr}_\sigma(\delta\mathbf{f}_{\boldsymbol{k}}^{\mathrm{g}}) + \rho\boldsymbol{v}_{\mathrm{s}}. \tag{10.47}$$

Similarly, the momentum current density is found to be given by

$$\Pi_{ij} = \sum_{\boldsymbol{k}} \hbar k_i v_{\boldsymbol{k}j}\operatorname{tr}_\sigma\left[\frac{\xi_{\boldsymbol{k}}}{E_{\boldsymbol{k}}}(\delta\mathbf{f}_{\boldsymbol{k}}^{\mathrm{g}} - f'_{\boldsymbol{k}}\,\delta\mathbf{E}_{\boldsymbol{k}}^{\mathrm{g}})\right] + \delta_{ij} n\,\delta\mu, \tag{10.48}$$

where (10.18b) has again been used.

The energy density and the energy current density are obtained as

$$\delta\varepsilon = \sum_{\boldsymbol{k}} E_{\boldsymbol{k}}\operatorname{tr}_\sigma(\delta\mathbf{f}_{\boldsymbol{k}}^{\mathrm{g}}) + \mu\,\delta n, \tag{10.49}$$

$$\begin{aligned}\boldsymbol{j}_\varepsilon &= \sum_{\boldsymbol{k}} \boldsymbol{v}_{\boldsymbol{k}}\xi_{\boldsymbol{k}}\operatorname{tr}_\sigma(\delta\mathbf{f}_{\boldsymbol{k}}^{\mathrm{g}} + \phi'_{\boldsymbol{k}}\,\delta\xi_{\boldsymbol{k}}) + \frac{1}{m}\mu\boldsymbol{g} \\ &\equiv \boldsymbol{j}_Q + \frac{1}{m}\mu\boldsymbol{g}.\end{aligned} \tag{10.50a}$$

Note that in a system with strict particle–hole symmetry the term involving $\delta\xi_{\boldsymbol{k}}$ is zero. The first term on the right-hand side of (10.50a) may be interpreted as the heat current. It consists of a sum of a convective and a diffusive term

$$\boldsymbol{j}_Q = Ts\boldsymbol{v}_{\mathrm{n}} + \sum_{\boldsymbol{k}} \boldsymbol{v}_{\boldsymbol{k}}\xi_{\boldsymbol{k}}\operatorname{tr}_\sigma[\delta\mathbf{f}_{\boldsymbol{k}}^{\mathrm{g}} + f'_{\boldsymbol{k}}(\hbar\boldsymbol{k}\cdot\boldsymbol{v}_{\mathrm{n}})\mathbf{1}], \tag{10.50b}$$

where $\boldsymbol{v}_{\mathrm{n}}$ is the local drift velocity of BQP, i.e. the normal-fluid velocity.

The corresponding expressions for the spin density $\boldsymbol{S}$ and the spin current $\boldsymbol{j}^\sigma$ are slightly more complicated because of the spin structure of the order parameter. By inserting (10.30) into the definition (10.11b), one finds

$$S_\mu = \frac{\hbar}{2}\sum_{\boldsymbol{k}\nu}\left\{s_{\boldsymbol{k},\mu\nu}\,\delta f_{\boldsymbol{k},\nu} - \frac{t_{\boldsymbol{k}}}{2E_{\boldsymbol{k}}}\frac{|\Delta_{\boldsymbol{k}}|^2}{E_{\boldsymbol{k}}^2}[\delta_{\mu\nu} - \Lambda_{\mu\nu}(\boldsymbol{k})](\delta\xi_{\boldsymbol{k},\nu} - \mu_0 H_\nu^{\mathrm{g}})\right\}, \tag{10.51}$$

where a notation analogous to (10.18c) has been employed. In the presence of an external magnetic field H_ν^{g} is replaced by $H_\nu^{\mathrm{g}} + H_\nu^{\mathrm{ext}}$. Here $s_{\boldsymbol{k},\mu\nu}$ is the BQP spin weight factor, which specifies the contribution to the spin component S_μ of a BQP with momentum $\boldsymbol{k}$ and spin along the ν direction. It is given by

$$s_{\boldsymbol{k},\mu\nu} = \frac{\xi_{\boldsymbol{k}}}{E_{\boldsymbol{k}}}\delta_{\mu\nu} + \left(1 - \frac{\xi_{\boldsymbol{k}}}{E_{\boldsymbol{k}}}\right)\Lambda_{\mu\nu}(\boldsymbol{k}), \tag{10.52a}$$

where $\Lambda_{\mu\nu}(\boldsymbol{k})$ is the projector

$$\Lambda_{\mu\nu}(\boldsymbol{k}) = \frac{d_\mu^*(\boldsymbol{k})d_\nu(\boldsymbol{k})}{\boldsymbol{d}^*\cdot\boldsymbol{d}}. \tag{10.52b}$$

In deriving (10.52a), use has been made of the relation

$$\frac{1}{2}\frac{1}{|\Delta_{\boldsymbol{k}}|^2}\operatorname{tr}_\sigma\,[\boldsymbol{\sigma}_\mu\boldsymbol{\Delta}_{\boldsymbol{k}}(\boldsymbol{\sigma}_\nu)^{\mathrm{T}}\boldsymbol{\Delta}_{\boldsymbol{k}}^{+}]=\delta_{\mu\nu}-2\Lambda_{\mu\nu}. \tag{10.52c}$$

In analogy to the partial compressibility κ_{P}^{0}, we may define a Cooper-pair partial spin-susceptibility tensor by

$$\chi_{\mu\nu}^{\mathrm{P0}}=\mu_0^2\sum_{\boldsymbol{k}}\frac{t_{\boldsymbol{k}}}{2E_{\boldsymbol{k}}}\frac{|\Delta_{\boldsymbol{k}}|^2}{E_{\boldsymbol{k}}^2}[\delta_{\mu\nu}-\Lambda_{\mu\nu}(\boldsymbol{k})]. \tag{10.53}$$

Equation (10.51) allows one to calculate higher-order Fermi-liquid corrections to the thermodynamic spin susceptibility, i.e. those beyond the usual molecular-field correction considered in (10.10). To this end, $\delta f_{\boldsymbol{k},\mu}$ is replaced by its local-equilibrium form

$$\delta f_{\boldsymbol{k},\mu}=f'_{\boldsymbol{k}}\,\delta E_{\boldsymbol{k},\mu}, \tag{10.54}$$

where, using (10.26), the BQP energy change is derived as

$$\delta E_{\boldsymbol{k},\mu}=\sum_{\nu}s_{\boldsymbol{k},\mu\nu}(\delta\xi_{\boldsymbol{k},\nu}-\mu_0H_\nu). \tag{10.55}$$

Here $H_\nu=H_\nu^{\mathrm{ext}}+H_\nu^{\mathrm{g}}$, with $\boldsymbol{H}^{\mathrm{ext}}$ the applied external magnetic field. In the B phase only the $l=2$ Fermi-liquid correction enters besides the $l=0$ one, owing to the fact that only the weight factor $s_{\boldsymbol{k},\mu\nu}$ and the projector $\Lambda_{\mu\nu}(\boldsymbol{k})$ have an angular dependence. The resulting spin susceptibility is given by (Czerwonko 1967)

$$\chi^{\mathrm{B}}\equiv\frac{\gamma S_\mu}{H_\mu}=\chi_{\mathrm{N}}^{0}\frac{\tilde{\chi}_0}{1+F_0^{\mathrm{a}}\tilde{\chi}_0},\qquad \mu=1,2,3, \tag{10.56a}$$

where

$$\tilde{\chi}_0=\tfrac{2}{3}+\tfrac{1}{3}Y_0+\tfrac{2}{15}F_2^{\mathrm{a}}(1-Y_0)^2 \tag{10.56b}$$

and Y_0 is the Yosida function defined in (3.101c).

The spin current $\boldsymbol{j}_\mu^\sigma$, defined in (10.17a), may be written in terms of the gauge-invariant distribution function as

$$\boldsymbol{j}_\mu^\sigma=\frac{\hbar}{2}\sum_{\boldsymbol{k},\nu}\frac{\boldsymbol{k}}{m_{\mathrm{s}}}\left[p_{\boldsymbol{k},\mu\nu}\,\delta f_{\boldsymbol{k},\nu}^{\mathrm{g}}-\frac{t_{\boldsymbol{k}}}{2E_{\boldsymbol{k}}^3}|\Delta_{\boldsymbol{k}}|^2\,\Lambda_{\mu\nu}\left(\delta\xi_{\boldsymbol{k},\nu}+\frac{m}{m^*}\hbar\boldsymbol{k}\cdot\boldsymbol{v}_{\mathrm{sp},\nu}\right)\right]$$
$$+\frac{m}{m_{\mathrm{s}}}S_0\boldsymbol{v}_{\mathrm{sp},\mu}. \tag{10.57}$$

In (10.57) we have defined $p_{\boldsymbol{k},\mu\nu}$, the contribution of a BQP of spin component ν and momentum $\boldsymbol{k}$ to the spin current with spin orientation μ given by

$$p_{\boldsymbol{k},\mu\nu}=\delta_{\mu\nu}-\left(1-\frac{\xi_{\boldsymbol{k}}}{E_{\boldsymbol{k}}}\right)\Lambda_{\mu\nu}(\boldsymbol{k}). \tag{10.58}$$

In local equilibrium the BQP distribution function is again given by (10.54); however, this time only the part of $\delta E_{\boldsymbol{k},\mu}$ that is odd in $\boldsymbol{k}$ contributes, i.e.

$$\delta E_{\boldsymbol{k},\mu} = \sum_{\nu} p_{\boldsymbol{k},\mu\nu}\left(\delta\xi_{\boldsymbol{k},\nu} + \frac{m}{m^*}\hbar\boldsymbol{k}\cdot\boldsymbol{v}_{\mathrm{sp},\nu}\right). \tag{10.59}$$

This may be substituted into (10.57). With the aid of (10.18b), it yields the local-equilibrium expression for the spin current:

$$\boldsymbol{j}_{\mu}^{\sigma,\mathrm{loc}} = \frac{m}{m^*}\sum_{\lambda\nu}\left[\mathbf{1} - \tfrac{1}{3}F_1^{\mathrm{a}}\frac{m_{\mathrm{s}}}{m^* S_0}\boldsymbol{\rho}_0^{\mathrm{sp}}\right]^{-1}_{\mu\lambda}\boldsymbol{\rho}^{\mathrm{sp}}_{0,\lambda\nu}\boldsymbol{v}_{\mathrm{sp},\nu} \equiv \sum_{\nu}\boldsymbol{\rho}^{\mathrm{sp}}_{\mu\nu}\boldsymbol{v}_{\mathrm{sp},\nu}. \tag{10.60}$$

Here the spin superfluid density in the absence of Fermi-liquid corrections is given by

$$(\rho_0^{\mathrm{sp}})^{ij}_{\mu\nu} = 3S_0\langle\hat{k}_i\hat{k}_j[1 - Y_0(\hat{\boldsymbol{k}};T)][\delta_{\mu\nu} - \Lambda_{\mu\nu}(\hat{\boldsymbol{k}})]\rangle_{\boldsymbol{k}}. \tag{10.61}$$

Note that $\boldsymbol{j}_{\mu}^{\sigma,\mathrm{loc}}$ is equal to the spin supercurrent $\boldsymbol{j}_{\mu}^{\mathrm{sp}}$ defined in (7.16).

10.2 COLLISION INTEGRAL

10.2.1 Matrix operator for binary collision processes

The time rate of change of the matrix distribution function caused by binary quasiparticle collision processes is given by the golden-rule expression involving the scattering amplitude T:

$$\begin{aligned}
I^{\gamma_1}_{\boldsymbol{k}_1 ss'} = &-\frac{1}{3!}\frac{1}{2}\sum_{\boldsymbol{k}_2\boldsymbol{k}_3\boldsymbol{k}_4}\sum_{\substack{s_1\ldots s_4\\ s_1'\ldots s_4'}}\sum_{\gamma_2,\gamma_3,\gamma_4=\pm1}\langle\boldsymbol{k}_1 s_1,\,\boldsymbol{k}_2 s_2|\,T\,|k_3 s_3,\,\boldsymbol{k}_4 s_4\rangle\\
&\times\langle\boldsymbol{k}_3 s_3',\,\boldsymbol{k}_4 s_4'|\,T\,|\boldsymbol{k}_1 s_1',\,\boldsymbol{k}_2 s_2'\rangle(2\pi)^4\,\delta^3\Big(\sum_i{}'\,\boldsymbol{k}_i\Big)\,\delta\Big(\sum_i{}'\,\gamma_i E_{\boldsymbol{k}_i}\Big)\\
&\times\Big\{\delta_{ss_1}[N^{\gamma_1}_{\boldsymbol{k}_1 s_1' s'}N^{\gamma_2}_{\boldsymbol{k}_2 s_2' s_2}(\mathcal{P}^{\gamma_3}_{\boldsymbol{k}_3} - \mathcal{N}^{\gamma_3}_{\boldsymbol{k}_3})_{s_3 s_3'}(\mathcal{P}^{\gamma_4}_{\boldsymbol{k}_4} - \mathcal{N}^{\gamma_4}_{\boldsymbol{k}_4})_{s_4 s_4'}\\
&-(\mathcal{P}^{\gamma_1}_{\boldsymbol{k}_1} - \mathcal{N}^{\gamma_1}_{\boldsymbol{k}_1})_{s_1' s'}(\mathcal{P}^{\gamma_2}_{\boldsymbol{k}_2} - \mathcal{N}^{\gamma_2}_{\boldsymbol{k}_2})_{s_2' s_2}N^{\gamma_3}_{\boldsymbol{k}_3, s_3 s_3'}N^{\gamma_4}_{\boldsymbol{k}_4, s_4 s_4'}] + \begin{pmatrix} s\leftrightarrow s_1'\\ s'\leftrightarrow s_1\end{pmatrix}\Big\}.
\end{aligned} \tag{10.62}$$

To simplify the notation, the particle–hole index ρ and spin index α have been combined into a single 4-component index $s = (\rho, \alpha)$. Since the components of the distribution function belonging to the energy eigenvalues $+E_{\boldsymbol{k}}$ and $-E_{\boldsymbol{k}}$ have to be treated separately, an additional index $\gamma = \pm 1$ has been introduced in (10.62). Only in this way can we take energy conservation during collision into account, as expressed by the delta function with argument $\sum_i' \gamma_i E_{\boldsymbol{k}_i} = \gamma_1 E_{\boldsymbol{k}_1} + \gamma_2 E_{\boldsymbol{k}_2} - \gamma_3 E_{\boldsymbol{k}_3} - \gamma_4 E_{\boldsymbol{k}_4}$. This generalization leads to a further technical complication, since we now have to introduce a quantity $\mathscr{P}_k^\gamma$, which is the probability matrix for the state $|\boldsymbol{k}\rangle$ (with energy $\gamma E_{\boldsymbol{k}}$) to be occupied or empty. This is not equal to unity, since the quasiparticle may be in state $|\boldsymbol{k}\rangle$ with energy $+E_{\boldsymbol{k}}$ *or* $-E_{\boldsymbol{k}}$. Of course, the sum of the spectral weight factors $\mathscr{P}_k^\gamma$ adds up to unity $\sum_\gamma \mathscr{P}_k^\gamma = \boldsymbol{1}$.

The conservation of momentum is incorporated by the delta function with argument $\sum_i' \boldsymbol{k}_i = \boldsymbol{k}_1 + \boldsymbol{k}_2 - \boldsymbol{k}_3 - \boldsymbol{k}_4$. Of course, the collisions also conserve the total particle number and spin. This implies that the scattering amplitude $\langle \boldsymbol{k}_1 s_1,\ \boldsymbol{k}_2 s_2 |\ T\ | \boldsymbol{k}_3 s_3,\ \boldsymbol{k}_4 s_4 \rangle$ is nonzero only for $\rho_1 + \rho_2 = \rho_3 + \rho_4$ and $\rho_1\alpha_1 + \rho_2\alpha_2 = \rho_3\alpha_3 + \rho_4\alpha_4$, where ρ_i is the particle–hole index and α_i is the spin index of state $|\boldsymbol{k}_i s_i\rangle$.

The scattering amplitude represents the sum of interaction processes involving momentum transfer over the entire Fermi surface. The pair correlations, on the other hand, are effective only in a small region of momentum space near the Fermi surface. Hence to order T_c/T_F the scattering amplitude T is given by the scattering amplitude t in the normal state. More precisely, the components of T may be identified with

$$\left.\begin{aligned} T_{++++} &= T_{----} = t(1, 2; 3, 4), \\ T_{+-+-} &= T_{-+-+} = t(1, -4; 3, -2), \\ T_{+--+} &= T_{-++-} = t(1, -3; -2, 4), \end{aligned}\right\}$$

where the subscripts +, − refer to the particle–hole indices ρ_i, and we have used the abbreviations $1 \equiv (\boldsymbol{k}_1, \alpha_1)$, $-2 \equiv (-\boldsymbol{k}_2, -\alpha_2)$, etc. The factor 1/3! in (10.62) accounts for the fact that all 3! permutations of (2, 3, 4) in the sums in (10.62) are equivalent and should be counted only once. Finally, the matrix structure of $\mathscr{N}_k^\gamma$ introduces an asymmetry, for example into the first term in the curly brackets in (10.62), which is removed by taking half of the symmetric combination (for a derivation from microscopic theory, see Wölfle 1976a). The collision integral has been investigated, in particular in the limits $T \to T_c$ and $T \to 0$, by Pethick *et al.* (1975, 1976, 1977).

Next, we derive explicit expressions for the linearized collision integral $\delta\mathscr{I}$ in (10.6) using (10.62). The equilibrium distribution function $\mathscr{N}_k^{\gamma 0}$, describing the distribution of quasiparticles with energy $\gamma E_{\boldsymbol{k}}$, must be proportional to $f(\gamma E_{\boldsymbol{k}})$. It is then easy to decompose the complete equilibrium distribution function $\mathscr{N}_k^0$, (10.3a), into components

$$\mathscr{N}_k^0 = \sum_{\gamma=\pm 1} \mathscr{N}_k^{\gamma 0}, \tag{10.64a}$$

with

$$\mathscr{N}_k^{\gamma 0} = \frac{1}{2}\left(\boldsymbol{1} + \frac{\gamma \boldsymbol{\varepsilon}_k^0}{E_k}\right) f(\gamma E_k). \tag{10.64b}$$

The linearization of $\mathscr{N}_k^\gamma$ is done about the equilibrium value $\mathscr{N}_k^{\gamma 0}$:

$$\mathscr{N}_k^\gamma = \mathscr{N}_k^{\gamma 0} + \delta\mathscr{N}_k^\gamma. \tag{10.65}$$

On the other hand, the distribution of quasiholes with energy γE_k is proportional to $1-f(\gamma E_{\boldsymbol{k}})$ and hence

$$\mathscr{P}_{\boldsymbol{k}}^{\gamma 0}-\mathcal{N}_{\boldsymbol{k}}^{\gamma 0}=\tfrac{1}{2}\left(\boldsymbol{1}+\frac{\gamma \boldsymbol{\varepsilon}_{\boldsymbol{k}}^{0}}{E_{\boldsymbol{k}}}\right)[1-f(\gamma E_{\boldsymbol{k}})]. \tag{10.66a}$$

From (10.64b) and (10.66a), the equilibrium spectral weight function $\mathscr{P}_{\boldsymbol{k}}^{\gamma 0}$ is found as

$$\mathscr{P}_{\boldsymbol{k}}^{\gamma 0}=\tfrac{1}{2}\left(\boldsymbol{1}+\frac{\gamma \boldsymbol{\varepsilon}_{\boldsymbol{k}}^{0}}{E_{\boldsymbol{k}}}\right), \quad \gamma=\pm 1. \tag{10.66b}$$

Note that $\mathscr{P}_{\boldsymbol{k}}^{\gamma 0}$ is a projector, i.e. $(\mathscr{P}_{\boldsymbol{k}}^{\gamma 0})^2=\mathscr{P}_{\boldsymbol{k}}^{\gamma 0}$, and $\mathscr{P}_{\boldsymbol{k}}^{+0}$ and $\mathscr{P}_{\boldsymbol{k}}^{-0}$ are orthogonal, $\mathscr{P}_{\boldsymbol{k}}^{+0}\mathscr{P}_{\boldsymbol{k}}^{-0}=\boldsymbol{0}$.

Upon substituting the equilibrium distribution functions $\mathcal{N}_{\boldsymbol{k}}^{\gamma 0}$, (10.64b), and $\mathscr{P}_{\boldsymbol{k}}^{\gamma 0}$, (10.66b), the collision integral (10.62) is found to be zero. This follows from the fact that, owing to energy conservation, the combinations of Fermi factors in the out- and in-scattering parts, $f_1 f_2(1-f_3)(1-f_4)$ and $(1-f_1)(1-f_2)f_3 f_4$, have exactly the same value. Consequently, the two parts of the collision integral are found to cancel each other. This is true not only in global equilibrium, but also for any local equilibrium distribution functions, i.e. distribution functions of the form (10.64b) and (10.66a), where $\boldsymbol{\varepsilon}_{\boldsymbol{k}}$ and $E_{\boldsymbol{k}}$ depend on space and time.

The linearized collision integral takes the form

$$\begin{aligned}\delta I_{\boldsymbol{k}_1 ss'}^{\gamma_1}=&-\frac{1}{3!}\frac{1}{2}\sum_{\boldsymbol{k}_2\boldsymbol{k}_3\boldsymbol{k}_4}\sum_{\substack{s_1\ldots s_4\\ s'_1\ldots s'_4}}\sum_{\gamma_2,\gamma_3,\gamma_4=\pm 1}\langle \boldsymbol{k}_1 s_1,\boldsymbol{k}_2 s_2|\,T\,|\boldsymbol{k}_3 s_3,\boldsymbol{k}_4 s_4\rangle\\ &\times\langle \boldsymbol{k}_3 s'_3,\boldsymbol{k}_4 s'_4|\,T\,|\boldsymbol{k}_1 s'_1,\boldsymbol{k}_2 s'_2\rangle(2\pi)^4\delta^3\Big({\sum_i}'\boldsymbol{k}_i\Big)\delta^3\Big({\sum_i}'\gamma_i E_{\boldsymbol{k}_i}\Big)\\ &\times f_{\boldsymbol{k}_1}^{\gamma_1}f_{\boldsymbol{k}_2}^{\gamma_2}(1-f_{\boldsymbol{k}_3}^{\gamma_3})(1-f_{\boldsymbol{k}_4}^{\gamma_4})\Big\{P_{\boldsymbol{k}_1 s'_1 s'}^{\gamma_1 0}P_{\boldsymbol{k}_2 s'_2 s_2}^{\gamma_2 0}P_{\boldsymbol{k}_3 s_3 s'_3}^{\gamma_3 0}P_{\boldsymbol{k}_4 s_4 s'_4}^{\gamma_4 0}\\ &\times[\delta\phi_{\boldsymbol{k}_1 s'_1 s_1}^{\gamma_1}+\delta\phi_{\boldsymbol{k}_2 s'_2 s_2}^{\gamma_2}-\delta\phi_{\boldsymbol{k}_3 s_3 s'_3}^{\gamma_3}-\delta\phi_{\boldsymbol{k}_4 s_4 s'_4}^{\gamma_4}]+\begin{pmatrix}s\leftrightarrow s'_1\\ s'\leftrightarrow s_1\end{pmatrix}\Big\},\end{aligned} \tag{10.67}$$

where $\delta\phi$ is a combination of linearized distribution functions,

$$\delta\phi_{\boldsymbol{k}ss'}^{\gamma}=[f_{\boldsymbol{k}}^{\gamma}(1-f_{\boldsymbol{k}}^{\gamma})P_{\boldsymbol{k}ss'}^{\gamma}]^{-1}[\delta N_{\boldsymbol{k}ss'}^{\gamma}-f_{\boldsymbol{k}}^{\gamma}\,\delta P_{\boldsymbol{k}ss'}^{\gamma}-P_{\boldsymbol{k}ss'}^{\gamma}f'_{\boldsymbol{k}}\gamma\,\delta E_{\boldsymbol{k}}], \tag{10.68}$$

and $f_{\boldsymbol{k}}^{\gamma}=f(\gamma E_{\boldsymbol{k}})$ is the Fermi function. The last term in (10.68), which is proportional to the change in energy $\delta E_{\boldsymbol{k}}$, comes from the linearization of the energy variables in the energy delta function in (10.62). Because the collision integral is zero in local equilibrium, a change in energy $\delta E_{\boldsymbol{k}}$ in the delta function is equivalent to a change $-\delta E_{\boldsymbol{k}}$ in the corresponding equilibrium distribution function. This leads to the additional change in the distribution function given by the last terms of (10.68).

The linearized collision integral (10.67) is seen to act on four distribution functions, $\delta\mathcal{N}_{\boldsymbol{k}}^{\gamma}$ and $\delta\mathscr{P}_{\boldsymbol{k}}^{\gamma}$, $\gamma=\pm 1$, instead of the one function $\delta\mathcal{N}_{\boldsymbol{k}}$ we started from originally. However, these functions are not independent. This may be seen by looking at the kinetic equations for these four functions, which are obtained by decomposing the kinetic equation (10.6) for $\delta\mathcal{N}_k$ into its $\gamma = \pm 1$ parts. To do this, one simply replaces $\mathcal{N}_k^0$ by $\sum_\gamma \mathcal{N}_k^{\gamma 0}$. The equations for $\delta\mathcal{N}_k^\gamma$ are then given by the kinetic equations for $\delta\mathcal{N}_k$ with $\mathcal{N}_k^0$ replaced by $\mathcal{N}_k^{\gamma 0}$ and the full collision integral replaced by the component $\delta\mathcal{I}_k^\gamma$. In order to derive the equation for $\delta\mathcal{P}_k^\gamma$, one starts from the matrix kinetic equation for the quasihole distribution function $\delta\mathcal{P}_k^\gamma - \delta\mathcal{N}_k^\gamma$. This equation is given by the normal kinetic equation with $\mathcal{N}_k^0$ replaced by $\mathcal{P}_k^{\gamma 0} - \mathcal{N}_k^{\gamma 0}$. The collision integral has the opposite sign from the one in the quasiparticle equation because adding a quasiparticle to a quasiparticle state is equivalent to removing a quasihole from the same state. The sum of the two kinetic equations for quasiparticles and quasiholes with energy γE_k yields the kinetic equation for $\delta\mathcal{P}_k^\gamma$:

$$\omega\, \delta\mathcal{P}_k^\gamma - \boldsymbol{\varepsilon}_{k_+}\, \delta\mathcal{P}_k^\gamma + \delta\mathcal{P}_k^\gamma\, \boldsymbol{\varepsilon}_{k_-} = \delta\boldsymbol{\varepsilon}_k\, \mathcal{P}_{k_-}^{\gamma 0} - \mathcal{P}_{k_+}^{\gamma 0}\, \delta\boldsymbol{\varepsilon}_k. \tag{10.69}$$

Note that here the collision integral has dropped out. On inserting $\mathcal{P}_k^{\gamma 0}$ into (10.69), the right-hand side of (10.69) is seen to be proportional to γ, and hence

$$\delta\mathcal{P}_k^\gamma = \gamma\, \delta\mathcal{P}_k. \tag{10.70}$$

From the kinetic equations for $\delta\mathcal{N}_k^\gamma$, $\delta\mathcal{N}_k$ and $\delta\mathcal{P}_k^\gamma$, it is straightforward to derive $\delta\mathcal{N}_k^\gamma$ in terms of $\delta\mathcal{N}_k$ and $\delta\mathcal{P}_k$ alone:

$$\delta\mathcal{N}_k^\gamma = \tfrac{1}{2}(\mathcal{P}_{k_+}^\gamma\, \delta\mathcal{N}_k + \delta\mathcal{N}_k\, \mathcal{P}_{k_-}^\gamma) + \tfrac{1}{2}\gamma(\delta\mathcal{P}_k\, \mathcal{N}_{k_+} + \mathcal{N}_{k_-}\, \delta\mathcal{P}_k). \tag{10.71}$$

It follows that only two of the four distribution functions, $\delta\mathcal{N}_k$ and $\delta\mathcal{P}_k$, are independent. While $\delta\mathcal{N}_k$ describes the distribution of quasiparticles and Cooper pairs in momentum space, corresponding to both branches $\pm E_k$ of the excitation spectrum, the additional function $\delta\mathcal{P}_k$ determines the relative spectral weight of the two excitation branches.

When substituting (10.71) into the collision integral (10.67) the small momentum changes $\pm\frac{1}{2}\boldsymbol{q}$ due to the external field can be neglected, as they only introduce correction terms of order q/k_{F} into the collision integral.

10.2.2 Conservation properties of the collision integral

The conservation of mass, momentum, energy and spin is reflected in corresponding integral properties of $\delta\mathcal{I}$. One finds that for any distribution function $\delta\mathcal{N}_k$,

$$\sum_{k,\gamma=\pm 1} \mathrm{tr}_4\,(\delta\mathcal{I}_k^\gamma\, \boldsymbol{\tau}_0) = 0, \tag{10.72a}$$

$$\sum_{k,\gamma=\pm 1} \boldsymbol{k}\, \mathrm{tr}_4\,(\delta\mathcal{I}_k^\gamma) = 0, \tag{10.72b}$$

$$\sum_{k,\gamma=\pm 1} \mathrm{tr}_4\,[\delta\mathcal{I}_k^\gamma\,(\boldsymbol{\varepsilon}_k^0 + \mu\boldsymbol{\tau}_0)] = 0, \tag{10.72c}$$

$$\sum_{k,\gamma=\pm 1} \mathrm{tr}_4\,(\delta\mathcal{I}_k^\gamma\, \boldsymbol{\tau}_\nu) = 0, \quad \nu = 1, 2, 3. \tag{10.72d}$$

10.2.3 Collision integral for Bogoliubov quasiparticles

In the macroscopic regime, i.e. when the condensate is in local equilibrium, the matrix collision integral (10.67) assumes a simple form. Let us first perform the Bogoliubov transformation (10.20) on the distribution function $P^{\gamma}_{kss'}\,\delta\phi^{\gamma}_{kss'}$ appearing in the linearized collision integral (10.67). The result is

$$P^{\gamma}_{kss'}\,\delta\phi^{\gamma}_{kss'} \to [f^{\gamma}_{k}(1-f^{\gamma}_{k})]^{-1}\frac{1}{2}\begin{pmatrix}(1+\gamma)\delta\bar{\mathbf{f}}_{k} & \delta\bar{\mathbf{h}}_{k} \\ \delta\bar{\mathbf{h}}^{+}_{k} & -(1-\gamma)\delta\bar{\mathbf{f}}_{-k}\end{pmatrix}, \tag{10.73}$$

where

$$\delta\bar{\mathbf{f}}_{k} = \delta\mathbf{f}_{k} - f'_{k}\,\delta\mathbf{E}_{k}, \tag{10.74a}$$

$$\delta\bar{\mathbf{h}}_{k} = \delta\mathbf{h}_{k} + \frac{t_{k}}{2E_{k}}\,\delta\mathbf{\Gamma}_{k}. \tag{10.74b}$$

However, according to the discussion below (10.27), $\delta\bar{\mathbf{h}}_{k}$ may be put equal to zero in the macroscopic regime, so that the transformed matrix distribution function $P^{\gamma}_{kss'}\,\delta\phi^{\gamma}_{k}$ becomes diagonal. Also, only one of the diagonal elements is nonzero for given γ. In deriving (10.73), small terms of order q and ω have been neglected and the solution of (10.69) for $\delta\mathcal{P}^{\gamma}_{k}$ (more precisely its Bogoliubov transform), i.e.

$$\delta\bar{\mathcal{P}}_{k} = \frac{1}{2E_{k}}\begin{pmatrix}0 & \delta\mathbf{\Gamma}_{k} \\ \delta\mathbf{\Gamma}^{+}_{k} & 0\end{pmatrix} + O\left(\frac{v_{\mathrm{F}}q}{\Delta}, \frac{\omega}{\Delta}\right) \tag{10.75}$$

has been employed.

The collision integral of the scalar kinetic equation (10.67) for the distribution function of Bogoliubov quasiparticles may now be written as the sum of an out-scattering and an in-scattering contribution (Einzel and Wölfle 1978, Einzel 1984):

$$\delta\bar{\mathbf{I}}_{k_1} = (\mathcal{U}\,\delta\mathcal{I}\,\mathcal{U}^{+})_{++} = -\frac{\mathrm{i}}{\tau_{k_1}}\,\delta\bar{\mathbf{f}}_{k_1} + \delta\bar{\mathbf{I}}^{\mathrm{in}}_{k_1}. \tag{10.76}$$

Here the Bogoliubov quasiparticle relaxation rate is obtained from the Bogoliubov transformation of the term proportional to $\delta\phi_{k}$ in (10.67). Employing the fact that at low T all momenta lie on the Fermi surface, the energy and angular dependences may be separated (see the discussion for the normal state after (2.35)). Owing to the anisotropy of the condensate, the relaxation rate is also anisotropic in general, i.e. it depends on the vector $\boldsymbol{k}_1$. Furthermore, the orientation of the momentum vectors is specified by three angles instead of two, as in the normal phase: (i) the angle θ enclosed by $\boldsymbol{k}_1$ and $\boldsymbol{k}_2$, (ii) the angle ϕ between the planes spanned by $(\boldsymbol{k}_1, \boldsymbol{k}_2)$ and $(\boldsymbol{k}_3, \boldsymbol{k}_4)$, and (iii) the angle ϕ_2 describing rotations about $\boldsymbol{k}_1$.

The result of these reformulations is ($\xi_i \equiv \xi_{k_i}$, $E_i \equiv E_{k_i}$, $f_i \equiv f_{k_i}$)

$$\frac{1}{\tau_{k_1}} = \frac{\pi}{32\hbar\varepsilon_F}(1 + e^{-E_k/k_BT})\int_0^\infty d\xi_2\, d\xi_3\, d\xi_4 \sum_{\substack{\mu_2,\mu_3,\mu_4 \\ =\pm 1}} \langle \delta(E_1 + \mu_2 E_2 - \mu_3 E_3 - \mu_4 E_4)$$
$$\times f_2(1 - f_3)(1 - f_4)\Omega(\boldsymbol{k}_1, \boldsymbol{k}_2, \boldsymbol{k}_3, \boldsymbol{k}_4)\rangle_a. \quad (10.77a)$$

Here we have defined a generalized transition probability by

$$\Omega(\boldsymbol{k}_1, \boldsymbol{k}_2, \boldsymbol{k}_3, \boldsymbol{k}_4) = W - W_{AB}[\boldsymbol{D}_1^* \cdot \boldsymbol{D}_3 + \boldsymbol{D}_2^* \cdot \boldsymbol{D}_4] - W_2(\boldsymbol{D}_1^* \cdot \boldsymbol{D}_2^*)(\boldsymbol{D}_3 \cdot \boldsymbol{D}_4)$$
$$+ 2W_3(\boldsymbol{D}_1^* \cdot \boldsymbol{D}_3)(\boldsymbol{D}_2^* \cdot \boldsymbol{D}_4) + 2W_4(\boldsymbol{D}_1^* \cdot \boldsymbol{D}_4)(\boldsymbol{D}_2^* \cdot \boldsymbol{D}_3), \quad (10.77b)$$

where the quantities W_j are combinations of normal-state quasiparticle scattering amplitudes given by (2.39) and (2.49) and

$$W_{AB}(\theta, \phi) = (A^0 + A^1)(B^0 + B^1) + 4A^1B^1. \quad (10.78)$$

The quantities $A^S(\theta, \phi)$ and $B^S(\theta, \phi)$ are the singlet ($S = 0$) and triplet ($S = 1$) components of the quasiparticle scattering amplitude introduced in Chapter 2 (see (2.36)):

$$A = N_F a(\boldsymbol{k}_1\alpha_1, \boldsymbol{k}_2\alpha_2; \boldsymbol{k}_3\alpha_3, \boldsymbol{k}_4\alpha_4), \quad (10.79a)$$
$$B = N_F a(\boldsymbol{k}_1\alpha_1, -\boldsymbol{k}_4 - \alpha_4; \boldsymbol{k}_3\alpha_3, -\boldsymbol{k}_2 - \alpha_2), \quad (10.79b)$$

where $\alpha_i = \pm 1$ are spin indices. The vector quantity $\boldsymbol{D}_j$ is the normalized equilibrium gap parameter,

$$\boldsymbol{D}_j = \mu_j \boldsymbol{d}(\boldsymbol{k}_j)/E_{k_j}, \quad (10.80)$$

and the angular brackets in (10.77a) denote the average,

$$\langle \Omega \rangle_a \equiv \int_0^1 d(\cos \tfrac{1}{2}\theta) \int_0^{2\pi} \frac{d\phi}{2\pi} \int_0^{2\pi} \frac{d\phi_2}{2\pi}\, \Omega. \quad (10.81)$$

The terms involving $\boldsymbol{D}_j$ in (10.77b) come from processes in which a quasiparticle transforms into a hole by condensing into a Cooper pair (or vice versa). The quasiparticle or quasihole character of a momentum label $\boldsymbol{k}_j$ may therefore be different for either of the two scattering amplitudes, which leads to the asymmetric combinations in (10.78).

The in-scattering part $\delta\tilde{\mathbf{I}}^{\text{in}}$ has a very similar structure. We only give the result for the spin-symmetric part:

$$\delta\tilde{\mathbf{I}}_{k_1}^{\text{in},s} = i\frac{\pi}{32\hbar\varepsilon_F}\int_0^\infty d\xi_2\, d\xi_3\, d\xi_4 \sum_{\substack{\mu_3,\mu_3,\mu_4 \\ =\pm 1}} \langle \delta(E_1 + \mu_2 E_2 - \mu_3 E_3 - \mu_3 E_3 - \mu_4 E_4)$$
$$\times f_1 f_2(1 - f_3)(1 - f_4)[\Omega(\boldsymbol{k}_1, \boldsymbol{k}_2, \boldsymbol{k}_3, \boldsymbol{k}_4)(\delta\phi_3 + \delta\phi_4 - \delta\phi_2)]\rangle_a, \quad (10.82a)$$

where Ω is defined in (10.77b). The distribution function appears in the combination

$$\delta\phi_j = \frac{1}{2}\frac{(1 + \mu_j)\,\delta\bar{f}_{k_j} - (1 - \mu_j)\,\delta\bar{f}_{-k_j}}{f_{k_j}(1 - f_{k_j})}, \quad \mu_j = \pm 1. \quad (10.82b)$$

Equation (10.82a) is valid in the case that $\delta f_{\boldsymbol{k}}$ is an even function of $\xi_{\boldsymbol{k}}$. In the opposite case the quantity in square brackets has to be replaced by

$$\frac{\xi_{\boldsymbol{k}_j}}{E_{\boldsymbol{k}_1}}\left[W(\delta\psi_3+\delta\psi_4-\delta\psi_2)-W_{AB}(\boldsymbol{D}_2^*\cdot\boldsymbol{D}_4)\,\delta\psi_3\right] \tag{10.83a}$$

with W, W_{AB} and $\boldsymbol{D}_j$ defined in (2.39) and (10.78) and

$$\delta\psi_j=\frac{1}{2}\frac{\xi_{\boldsymbol{k}_j}}{E_{\boldsymbol{k}_j}}\frac{(1+\mu_j)\,\delta\bar{f}_{\boldsymbol{k}_j}+(1-\mu_j)\,\delta\bar{f}_{-\boldsymbol{k}_j}}{f_{\boldsymbol{k}_j}(1-f_{\boldsymbol{k}_j})}. \tag{10.83b}$$

The complete collision integral for the spin-antisymmetric case has been derived by Einzel (1984).

The collision integral (10.76) does not conserve Bogoliubov quasiparticle number and spin. This is due to the fact that BQP may be generated or annihilated during a collision process. However, momentum and energy are still conserved. This gives rise to the following integral properties of $\delta\tilde{\mathbf{I}}$:

$$\sum_{\boldsymbol{k}}\boldsymbol{k}\,\mathrm{tr}_\sigma\,(\delta\tilde{\mathbf{I}}_{\boldsymbol{k}})=0, \tag{10.84a}$$

$$\sum_{\boldsymbol{k}}E_{\boldsymbol{k}}\,\mathrm{tr}_\sigma\,(\delta\tilde{\mathbf{I}}_{\boldsymbol{k}})=0. \tag{10.84b}$$

At the same time, the collision integral vanishes for any distribution functions of the form

$$\delta f^{\mathrm{loc}}_{\boldsymbol{k}\alpha\beta}=f'_{\boldsymbol{k}}\left(\delta E_{\boldsymbol{k}}-\hbar\boldsymbol{k}\cdot\boldsymbol{v}_{\mathrm{n}}-E_{\boldsymbol{k}}\frac{\delta T}{T}\right)\delta_{\alpha\beta}, \tag{10.85}$$

where $\boldsymbol{v}_{\mathrm{n}}$ and δT must be interpreted as the local velocity and a change in local temperature of the BQP. This implies that collisions do not alter a distribution of the form (10.85). Consequently, the distribution function (10.85) must correspond to a local-equilibrium state of the BQP system.

10.2.4 Relaxation-time approximation

The collision integral for Bogoliubov quasiparticles derived in Section 10.2.3 is complicated enough to defy any general solution. It is therefore necessary to introduce an approximation that simplifies the collision operator sufficiently to allow an exact solution for a large class of transport problems. At the same time, this approximation must be able to describe the essential physical properties of the system under consideration. This task is achieved by the following relaxation-time approximation (RTA):

$$\delta\tilde{\mathbf{I}}_{\boldsymbol{k}}=-\frac{\mathrm{i}}{\tau_{\boldsymbol{k}}}\left[\delta\bar{\mathbf{f}}_{\boldsymbol{k}}-4\pi\sum_{l=0}^{\infty}\sum_{m=-l}^{l}\sum_{\sigma}\sum_{r=\pm1}\lambda^{\sigma}_{lr}\sum_{\boldsymbol{k}'}\mathsf{P}^{\sigma}_{lmr}(\boldsymbol{k},\boldsymbol{k}')\,\delta\bar{\mathbf{f}}_{\boldsymbol{k}'}\right]. \tag{10.86a}$$

Here the sum on $\sigma = \mathrm{s, a}$ extends over spin-symmetric and -antisymmetric components, and the sum on $r = t(-1)^l = \pm 1$ runs over even $(t = +1)$ and odd $(t = -1)$ functions in ξ. For the spin-symmetric case the projectors P are defined by

$$\mathsf{P}^{\sigma}_{lmr}(\boldsymbol{k}, \boldsymbol{k}') = Q^r_l(\xi)\frac{1}{\tau_{\boldsymbol{k}}} f'_{\boldsymbol{k}} Y_{lm}(\hat{\boldsymbol{k}})\left[\sum_{\boldsymbol{k}_2} (3E_{\boldsymbol{k}_2})^{1-r}\frac{1}{\tau_{\boldsymbol{k}_2}} f'_{\boldsymbol{k}_2} |Y_{lm}(\hat{\boldsymbol{k}}_2)|^2\right]^{-1} \times Q^r_l(\xi')\frac{1}{\tau_{\boldsymbol{k}'}} Y_{lm}(\hat{\boldsymbol{k}}')\mathsf{P}^{\sigma}_t, \tag{10.86b}$$

where

$$Q^r_l(\xi) = \frac{1}{2}\left[1 + \frac{\xi}{E} r(-1)^l\left(1 - \frac{\xi}{E}\right)\right]E^{(1-r)/2}, \tag{10.86c}$$

and $Y_{lm}(\hat{\boldsymbol{k}})$ are spherical harmonics. The operators P^{σ}_t project onto the space of functions $\delta\bar{\mathbf{f}}_{\boldsymbol{k}}$ with spin symmetry σ (=s, a) and particle–hole symmetry $t = r(-1)^l$.

It is easy to verify that the above relaxation-time approximation guarantees the two conservation properties (10.84) for momentum and energy. In the limit $\Delta \to 0$, i.e. $T \to T_c$, (10.86) reduces to a relaxation-time approximation for the normal-state collision integral, which yields results for the transport properties within one per cent of the exact results (Wölfle 1978a,b). In the opposite limit, i.e. for $T \to 0$, only quasiparticles with energies close to the gap edge are excited, such that $\xi_{\boldsymbol{k}} \ll \langle \Delta \rangle$, where $\langle \Delta \rangle$ is the average energy gap. In this limit the in-scattering part of the spin-symmetric collision integral vanishes more rapidly than the out-scattering term owing to factors of $\xi^2 \approx k_B T \Delta$. This property is also shared by the RTA collision integral (10.86). On these grounds, we can expect the RTA to be a reasonable representation to the collision integral.

10.2.5 Bogoliubov-quasiparticle relaxation rate

The main features of the collision integral (10.76), such as its anisotropy and its energy and temperature dependences, are contained in the first term, the "out-scattering term". This term describes relaxation of the distribution function $\delta\mathbf{f}_{\boldsymbol{k}}$ with a rate $1/\tau_{\boldsymbol{k}}$, which we shall refer to as the "BQP relaxation rate". Actually, the BQP quasiparticle lifetime is twice the relaxation time $\tau_{\boldsymbol{k}}$, because $1/\tau_{\boldsymbol{k}}$ is the sum of the decay rates for a particle and a hole, which are equal.

Let us discuss the relaxation rate $1/\tau_{\boldsymbol{k}}$ for the A and B phases in detail, in order to estimate the relevance of collision processes.

The B phase

In the B phase the dependence on energy and angular variables can be completely separated because of the isotropic excitation spectrum. One

finds

$$\frac{1}{\tau_{\boldsymbol{k}}}=\frac{1}{\tau_{\mathrm{N}}^{0}}\left[I_{\boldsymbol{k}}^{(0)}-\gamma_0(I_{\boldsymbol{k}}^{(1)}+I_{\boldsymbol{k}}^{(2)})+\delta_0 I_{\boldsymbol{k}}^{(3)}\right], \tag{10.87a}$$

where τ_{N}^{0} is the quasiparticle relaxation time at the Fermi level in the normal phase given by (2.38). The quantities $I_{\boldsymbol{k}}^{(n)}$ are dimensionless energy- and temperature-dependent functions defined by

$$\begin{aligned} I_{\boldsymbol{k}_1}^{(n)} &= \int_0^\infty \mathrm{d}\xi_2\,\mathrm{d}\xi_3\,\mathrm{d}\xi_4 \sum_{\substack{\mu_2,\mu_3,\mu_4\\ =\pm 1}} f_1 f_2(1-f_3)(1-f_4) \\ &\quad\times \delta(E_1+\mu_2E_2-\mu_3E_3-\mu_4E_4)K^{(n)} \\ &\equiv \int_0^\infty \mathrm{d}\xi_2\, J_n(\xi_1,\,\xi_2), \end{aligned} \tag{10.87b}$$

with

$$\left.\begin{aligned} K^{(0)} &= 1, \\ K^{(1)} &= \frac{1}{E_1}\frac{\mu_3}{E_3}\Delta^2, \\ K^{(2)} &= \frac{\mu_2}{E_2}\frac{\mu_4}{E_4}\Delta^2, \\ K^{(3)} &= \frac{1}{E_1}\frac{\mu_2}{E_2}\frac{\mu_3}{E_3}\frac{\mu_4}{E_4}\Delta^4. \end{aligned}\right\} \tag{10.87c}$$

The dimensionless parameters γ_l and δ_0 depend only on the quasiparticle scattering amplitude in the normal phase and are given by

$$\gamma_l=\frac{\langle W_{AB}c_3P_l(c_3)\rangle_{\mathrm{a}}}{\langle W\rangle_{\mathrm{a}}}, \quad l=0,\,1,\,\ldots, \tag{10.88a}$$

$$\delta_0=\frac{\langle -W_2\cos^2\theta+W_3c_3^2+W_4c_4^2\rangle_{\mathrm{a}}}{\langle W\rangle_{\mathrm{a}}}, \tag{10.88b}$$

where

$$c_{3,4}=\hat{\boldsymbol{k}}_1\cdot\hat{\boldsymbol{k}}_{3,4}=\cos^2\tfrac{1}{2}\theta\pm\sin^2\tfrac{1}{2}\theta\cos\phi=\hat{\boldsymbol{k}}_2\cdot\hat{\boldsymbol{k}}_{3,4}, \tag{10.88c}$$

and $P_l(z)$ are Legendre polynomials. For later purposes we have defined γ_l for $l\neq 0$ as well. The cross-sections W, $W_j(\theta,\,\phi)$ and W_{AB} have been given in (2.39), (2.49) and (10.78) respectively, and the angular brackets denote an angular average as defined in (10.81).

It is worth mentioning that the collision integral (10.87) describes three different types of collision processes:

(i) the usual process of scattering of two quasiparticles into two final momentum states (e.g. for $\mu_2=\mu_3=\mu_4=1$);

(ii) the decay of the initial quasiparticle into three final quasiparticles (e.g. for $-\mu_2=\mu_3=\mu_4=1$);

(iii) the coalescence of three quasiparticles into one final quasiparticle (e.g. for $\mu_2 = -\mu_3 = \mu_4 = 1$).

The latter two processes are possible because the quasiparticle number is not conserved. Two quasiparticles may combine into a Cooper pair, or conversely a Cooper pair may split up into two quasiparticles during a collision process.

In the limit $T \to T_c$ the functions $I^{(n)}$ for $n = 1, 2, 3$ vanish owing to the factors of Δ^2 and Δ^4 respectively, whereas

$$I^{(0)}_{\boldsymbol{k}_1} = 1 + \left(\frac{\xi_{\boldsymbol{k}_1}}{\pi k_B T}\right)^2 \quad \text{as} \quad T \to T_c. \tag{10.89}$$

The result (2.37) for the normal-state relaxation rate is recovered in this limit.

In the opposite limit of low temperatures, all functions $K^{(n)}$ tend to zero as $e^{-E_{\boldsymbol{k}}/k_B T}$. The quasiparticle relaxation rate in this limit is found as

$$\lim_{T\to 0} \frac{1}{\tau_{\boldsymbol{k}}} \equiv \frac{1}{\tau} = \frac{1}{\tau^0_N(T)} \frac{3}{(2\pi)^{1/2}} W_0 \left(\frac{\Delta}{k_B T}\right)^{3/2} e^{-\Delta/k_B T}\left(1 + \alpha_\tau \frac{k_B T}{\Delta}\right) \quad \text{as } T \to 0, \tag{10.90a}$$

where

$$W_0 = 1 - \tfrac{2}{3}\gamma_0 + \delta_0, \tag{10.90b}$$

$$\alpha_\tau = \tfrac{3}{4} - \frac{\delta_0 - \frac{1}{3}\gamma_0}{W_0}. \tag{10.90c}$$

The first finite-temperature correction in (10.90a), proportional to $k_B T/\Delta(0)$, as well as that in (10.112a), (10.125a) and (10.131a) (see below), has been calculated exactly by Einzel (1984).

The rate $\tau_{\boldsymbol{k}}^{-1}$ is seen to be proportional to the density of excitations (taking into account the T^2 dependence of $(\tau^0_N)^{-1}$):

$$n_{ex} = \sum_{\boldsymbol{k}} f_{\boldsymbol{k}} \approx \tfrac{3}{2}(2\pi)^{1/2} n \frac{(\Delta k_B T)^{1/2}}{E_F} e^{-\Delta/k_B T} \quad \text{as } T \to 0. \tag{10.91}$$

This behaviour, reminiscent of that of a classical gas, is obtained because the occupation numbers $f_{\boldsymbol{k}}$ are very small and the excitation spectrum is quadratic at low temperatures. The Fermi blocking effect is of little importance in this case.

Given the relaxation time, which may be interpreted as the mean time between successive collisions, one can derive the mean free path l of BQP, which is defined as the product of the velocity v_{av} and the mean collision time τ_{av}. As the group velocity of BQP depends strongly on energy (and in the A phase also on direction in $\boldsymbol{k}$ space), we shall use the r.m.s. average over the thermally excited states for the definition of v_{av}:

$$v_{av} = \left\{\left[\sum_{\boldsymbol{k}} (\nabla_{\boldsymbol{k}} E_{\boldsymbol{k}})^2 f'_{\boldsymbol{k}}\right] \Big/ \left[\sum_{\boldsymbol{k}} f'_{\boldsymbol{k}}\right]\right\}^{1/2}. \tag{10.92a}$$

The mean collision time τ_{av} will be defined by the inverse of the average relaxation rate

$$\tau_{av} = \left\{ \left[\sum_{k} \frac{1}{\tau_k} f'_k \right] \Big/ \left[\sum_{k} f'_k \right] \right\}^{-1}, \tag{10.92b}$$

so that the mean free path is given by

$$l = v_{av} \tau_{av}. \tag{10.92c}$$

This yields a low-temperature limiting behaviour

$$l = \frac{(2\pi)^{1/2}}{3} \left(\frac{T_c}{\Delta(0)} \right)^2 \frac{1}{W_0} [v_F \tau_N^0(T_c)] e^{\Delta/T}. \tag{10.93}$$

The normalized mean free path is seen to increase exponentially for $T \to 0$ (see Fig. 10.1). It should be noted that this normalized mean free path $l(T)/l(T_c)$ depends only weakly on pressure. It is of the order of centimetres for $T \approx 0.2 T_c$. In this temperature regime, and even more so at still lower T, the collisions of quasiparticles with the wall of the container dominate the collisions *between* the quasiparticles. This is the so-called "Knudsen

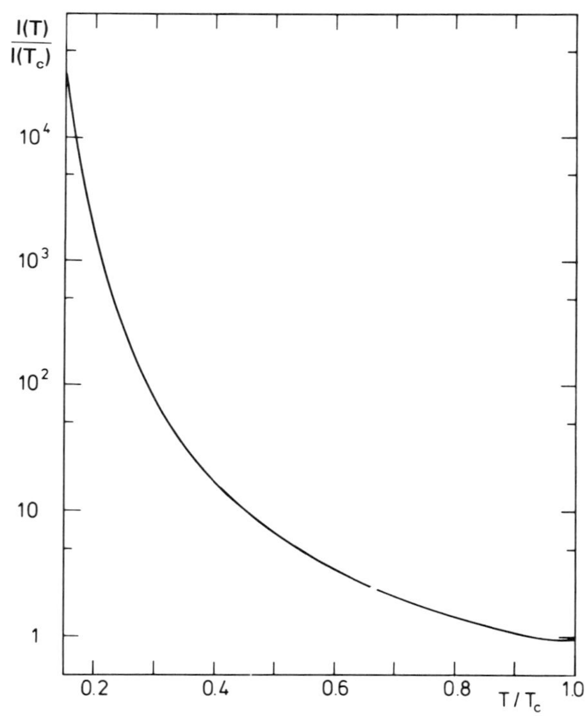

Figure 10.1 Mean free path $l(T)$ of Bogoliubov quasiparticles in the B phase versus temperature.

regime", or "ballistic quasiparticle regime", some aspects of which will be considered in Section 10.4.

The A phase

The evaluation of the BQP relaxation rate in the A phase is more cumbersome as energy and angular integrations do not separate in this case.

At low temperatures, where the A phase may exist in the presence of a magnetic field, only quasiparticles with momenta in the neighbourhood of the two nodes of the gap, i.e. along $\pm\hat{\boldsymbol{l}}$, where $\Delta_{\boldsymbol{k}} \ll k_B T$, will be thermally excited. These quasiparticles will in turn collide mainly with quasiparticles of the same kind, i.e. with momenta $\boldsymbol{k} \approx \hat{\boldsymbol{l}} k_F$. The fraction of momenta $\boldsymbol{k}$ inside the region considered is approximately equal to the area on the Fermi surface where the condition

$$\theta_{\boldsymbol{k}} < \frac{k_B T}{\Delta_0(T)} \tag{10.94}$$

is satisfied ($\theta_{\boldsymbol{k}}$ is the angle between $\boldsymbol{k}$ and $\hat{\boldsymbol{l}}$) divided by the total area, i.e. $[k_B T/\Delta_0(T)]^2$. In the case where both initial momenta satisfy the condition (10.94), momentum conservation requires the final momenta to satisfy this condition, too. Since all quasiparticles taking part in a collision process of this type have energy gaps much less than the thermal energy, their contribution to the relaxation rate is essentially the same as in the normal state. The relaxation rate of BQP with momenta $\boldsymbol{k}_{\parallel}$ parallel to $\hat{\boldsymbol{l}}$ is hence found approximately as

$$\frac{1}{\tau_{\boldsymbol{k}\parallel}} \approx \frac{1}{\tau_N^0(T)} \left[\frac{k_B T}{\Delta_0(0)}\right]^2 \approx \frac{1}{\tau_N^0(T_c)} \left(\frac{T}{T_c}\right)^4. \tag{10.95a}$$

Since the group velocity of these BQP is equal to that of normal quasiparticles, i.e. it is given approximately by v_F, their mean free path in the direction $\hat{\boldsymbol{l}}$ is roughly

$$l = v_F \tau_{\boldsymbol{k}\parallel}. \tag{10.95b}$$

10.3 TRANSPORT COEFFICIENTS

As a first application of the kinetic theory formulated in the preceding sections, we shall now calculate the transport coefficients that have been introduced phenomenologically in the hydrodynamic equations of Chapter 9. Quite generally, the transport coefficients may be expressed as limiting values of the various current correlation functions for frequency ω and wavenumber $\boldsymbol{q}$ tending to zero (for a formal discussion see Brand *et al.* 1979, Dörfle *et al.* 1980). The correlation functions, or equivalently the expectation values of the currents, may be conveniently calculated from the quasiparticle distribution function, as will be discussed below. This amounts to solving the kinetic equation for slowly varying disturbances, i.e. those with frequency ω and wavenumber q small in comparison with the relaxation rate τ^{-1} and the inverse mean free path l^{-1} respectively. Of course, the conditions for the condensate to be in local equilibrium (ω, $v_F q \ll \Delta/\hbar$) have to be satisfied as well.

The scalar kinetic equation (10.25) may then be solved by expanding it in powers of ω and q (the Chapman–Enskog procedure well known from the transport theory of classical gases). We shall follow the discussion given by Einzel and Wölfle (1978), Wölfle and Einzel (1978) and Einzel (1984). In the first step we write the distribution function $\delta\mathbf{f}_k$ (in the following we drop the index "g" referring to the gauge-transformed quantity) as a sum of the local equilibrium distribution function $\delta\mathbf{f}_k^{\text{loc}}$ and the deviation $\delta\mathbf{f}'_k$:

$$\delta\mathbf{f}_k = \delta\mathbf{f}_k^{\text{loc}} + \delta\mathbf{f}'_k. \tag{10.96}$$

The correction term $\delta\mathbf{f}'_k$ is order-(ω, q) small, since in a time-independent ($\omega = 0$) and spatially uniform ($q = 0$) state the system is in local equilibrium, i.e. $\delta\mathbf{f}'_k \equiv 0$. Actually, $\delta\mathbf{f}'_k$ describes the *approach* to local equilibrium. We shall show that the transport coefficients can be expressed in terms of $\delta\mathbf{f}'_k$.

We now insert $\delta\mathbf{f}_k$, as given by (10.96), into the kinetic equation (10.25) and expand to first order in ω and q. Since the "streaming terms" on the left-hand side are already proportional to ω and q, we may drop $\delta\mathbf{f}'_k$ there. On the right-hand side, i.e. in the collision integral, the local-equilibrium part $\delta\mathbf{f}_k^{\text{loc}}$ drops out. (Recall that by definition a local-equilibrium state is not changed by collisions.) We are then led to the following inhomogeneous integral equation for $\delta\mathbf{f}'_k$:

$$\omega\, \delta\mathbf{f}_k^{\text{loc}} - (\boldsymbol{\nabla}_k E_k \cdot \boldsymbol{q})(\delta\mathbf{f}_k^{\text{loc}} - f'_k\, \delta\mathbf{E}_k) = \delta\tilde{\mathbf{I}}\{\delta\mathbf{f}'_k\}. \tag{10.97}$$

This equation describes time-independent transport phenomena involving the stationary flow of mass, momentum, energy and spin as well as time-dependent phenomena in the macroscopic regime. None the less, even for stationary transport problems, the first term on the left-hand side is not necessarily zero, because it contains terms proportional to the time derivative of the density or spin density, which by the continuity equation are equivalent to the divergence of the respective currents. Using (10.10), (10.26), (10.55), (10.59) and (10.85), one may write

$$\begin{aligned}\omega\, \delta f^{\text{loc}}_{k\alpha\beta} &\approx \omega f'_k\, \delta E^{\text{tot}}_{k\alpha\beta}\\ &\approx \omega f'_k\Bigg[\frac{\xi_k}{E_k}(f_0^{\text{s}}\,\delta n - \delta\mu)\delta_{\alpha\beta} + \sum_{\mu\nu}(\sigma_\mu)_{\alpha\beta} s_{k,\mu\nu}\left(\frac{2}{\hbar} f_0^{\text{a}} S_\nu - \mu_0 H_\nu^{\text{g}}\right)\\ &\quad + \sum_{\mu\nu}(\sigma_\mu)_{\alpha\beta} p_{k,\mu\nu}\left(S_0^{-1}\frac{\frac{1}{3}F_1^{\text{a}}}{1+\frac{1}{3}F_1^{\text{a}}}\hbar \boldsymbol{k}\cdot \boldsymbol{j}_\nu^\sigma + \frac{m}{m^*}\hbar\boldsymbol{k}\cdot \boldsymbol{v}_{\text{sp},\nu}\right)\Bigg].\end{aligned} \tag{10.98}$$

In a state of local equilibrium, changes in density and chemical potential are related by a thermodynamic derivative, i.e.

$$\delta\mu = \frac{\partial\mu}{\partial n}\,\delta n = N_F^{-1}(1+F_0^s)\,\delta n. \tag{10.99a}$$

Likewise, the effective gauge field $\boldsymbol{H}^g$ is related to the spin polarization $\boldsymbol{S}$ by the magnetic-susceptibility tensor $\boldsymbol{\chi}$, which may be replaced by its largest eigenvalue χ when the magnetic field is the dominant orienting force:

$$\begin{aligned} H_\mu^g &= \frac{1}{\chi}(\gamma S_\mu - \chi H_\mu^{ext}) \\ &= \frac{\gamma}{\chi}\,\omega_\mu^s. \end{aligned} \tag{10.99b}$$

Here ω_μ^s is the thermodynamic field introduced in (9.9). The last term in (10.98), which is proportional to $\omega\boldsymbol{v}_{sp,\mu}$, may be expressed in terms of $q\omega_\mu^s$ using (10.39b).

Exploiting the conservation law for mass and spin in order to replace $\omega\,\delta n$ by $m^{-1}\boldsymbol{q}\cdot\boldsymbol{g}$ and ωS_μ by $\boldsymbol{q}\cdot\boldsymbol{j}_\mu^\sigma$, the transport equation (10.97) may finally be written explicitly as

$$\begin{aligned} f_k'\Big\{\mathbf{1}\frac{\xi_k}{E_k}\Big[&-\frac{1}{3\rho}k_F v_F\boldsymbol{q}\cdot\boldsymbol{g}_s + \tfrac{1}{3}(\boldsymbol{k}\cdot\boldsymbol{v}_k - k_F v_F)\boldsymbol{q}\cdot\boldsymbol{v}_n \\ &-i\boldsymbol{k}\boldsymbol{V}\boldsymbol{v}_k + E_k\boldsymbol{v}_k\cdot\boldsymbol{q}\frac{\delta T}{T}\Big] + \sum_{\mu\nu\lambda}\boldsymbol{\sigma}_\mu s_{k,\mu\nu}\Big(\frac{\mu_0\gamma}{\chi}\Big)_{\nu\lambda}\boldsymbol{q}\cdot\boldsymbol{j}_\lambda^\sigma \\ &-\tfrac{1}{2}\sum_{\mu\nu\lambda\lambda'}\boldsymbol{\sigma}_\mu p_{k,\mu\nu}\boldsymbol{k}\Big(\tfrac{1}{3}F_1^a\frac{m_s}{m^*}S_0^{-1}\boldsymbol{\rho}_{\nu\lambda}^{sp} + \frac{m}{m^*}\delta_{\nu\lambda}\mathbf{1}\Big)\boldsymbol{q}\Big(\frac{\mu_0\gamma}{\chi}\Big)_{\lambda\lambda'}\omega_{\lambda'}^s\Big\} \\ = \delta\tilde{\mathbf{I}}\{\delta\mathbf{f}_k'\}&, \end{aligned} \tag{10.100a}$$

where the tensor $\boldsymbol{V}$ is defined by (see (9.65))

$$V_{ij} = \tfrac{1}{2}\mathrm{i}(q_i v_{nj} + q_j v_{ni} - \tfrac{2}{3}\delta_{ij}\boldsymbol{q}\cdot\boldsymbol{v}_n). \tag{10.100b}$$

Equation (10.100a) is an inhomogeneous integral equation for $\delta\mathbf{f}_k'$, which cannot be solved in general terms because of the complexity of the collision operator in a superfluid Fermi liquid. Exact results can be found for some of the transport coefficients close to T_c and in the limit of low temperatures (Pethick *et al.* 1975, 1976, 1977, Bhattacharyya *et al.* 1977b, Einzel 1984). This is due to the fact that for some of the transport quantities the leading correction term to the normal-state result just below T_c is of kinematic origin and may be calculated from the normal-state kinetic equation with the BQP group velocity inserted in place of the normal QP velocity. At low T the back-scattering part of the collision integral, i.e. the one involving integrations, vanishes in some cases more rapidly than the out-scattering part, so that the simple relaxation time approximation becomes exact. In general, the integral equation (10.100) can only be solved approximately, for example by assuming the kernel to be separable (Wölfle and Einzel 1978, Einzel 1984) or by employing a variational method (Ono *et al.* 1977, 1982, Hara *et al.* 1980, Ono and Hara 1981, Hara 1981). For a somewhat different formulation based on the Green's-function technique, see Houghton and Maki (1977) and Dörfle *et al.* (1980).

10.3.1 Transport coefficients of the B phase

In the B phase the task of solving (10.100a) is greatly simplified by the isotropy of the energy spectrum. This allows one to separate the dependences on energy and angles, as in the normal state. The mathematical problem may then be reduced to the solution of integral equations in the energy variable. The angular dependence is easily obtained from the left-hand side in the kinetic equation (10.100a) for each particular case.

Shear viscosity

In response to shear flow, characterized by $q_i v_{nj} \neq 0$, $i \neq j$, the spin-symmetric distribution function, hereinafter denoted by $\delta f'_{\boldsymbol{k}}$, assumes an angular dependence $(\nabla_{\boldsymbol{k}} E_{\boldsymbol{k}})_i k_j$. Separating out this behaviour as well as the main dependence on energy, $\delta f'_{\boldsymbol{k}}$ may be written in terms of a dimensionless function $\psi_\eta(\xi)$:

$$\delta f'_{\boldsymbol{k}} = \mathrm{i} \sum_{ij} \tau_{\mathrm{N}}^0 f'_{\boldsymbol{k}} (\nabla_{\boldsymbol{k}} E_{\boldsymbol{k}})_i k_j q_i v_{nj} \psi_\eta(\xi_{\boldsymbol{k}}). \tag{10.101}$$

Substituting this into the collision integral (10.76), (10.82a), (10.87a), ψ_η is found to satisfy the integral equation

$$\psi_\eta(\xi_1) = \bar{\tau}(\xi_1)\left[1 + \int_0^\infty \mathrm{d}\xi_2\, S_2(\xi_1, \xi_2)\left(\frac{\xi_2}{E_2}\right)^2 \psi_\eta(\xi_2)\right], \tag{10.102a}$$

where

$$\bar{\tau}(\xi) = \tau_{\boldsymbol{k}}/\tau_{\mathrm{N}}^0 \tag{10.102b}$$

is the BQP relaxation time (10.87a) divided by the normal-state relaxation rate τ_{N}^0, (2.38). The integral operator S_2 in (10.102a) is a special case of a more general operator S_l (i.e. the one for $l = 2$):

$$S_l(\xi_1, \xi_2) = \lambda_{l+}^{\mathrm{s}} J_0(\xi_1, \xi_2) - \gamma_l J_2(\xi_1, \xi_2), \tag{10.103}$$

which is determined by the functions J_0 and J_2 defined in (10.87b) and by the parameters $\lambda_{l+}^{\mathrm{s}}$ and γ_l introduced in (2.51)–(2.53) and (10.88a) respectively. This follows from specializing (10.83a) to the B phase.

To proceed, we equate the off-diagonal components of the stress tensor (10.48) and their hydrodynamic expression (9.28):

$$\Pi_{ij} = \sum_{\boldsymbol{k}} (\nabla_{\boldsymbol{k}} E_{\boldsymbol{k}})_i k_j\, \delta f'_{\boldsymbol{k}} = -\mathrm{i}\eta q_i v_{nj}, \quad i \neq j. \tag{10.104}$$

Substituting for $\delta f'_{\boldsymbol{k}}$ from (10.101), the following expression for the shear-viscosity coefficient is found:

$$\eta = \tau_{\mathrm{N}}^{0} \sum_{\boldsymbol{k}\alpha} (\nabla_{\boldsymbol{k}} E_{\boldsymbol{k}})_i^2 \hbar^2 k_j^2 (-f'_{\boldsymbol{k}}) \psi_\eta(\xi_{\boldsymbol{k}}) = \eta_0 \left\langle \frac{\xi^2}{E^2} \psi_\eta \right\rangle_{\xi}. \tag{10.105a}$$

Here

$$\eta_0 = \tfrac{1}{5} n m^* v_{\mathrm{F}}^2 \tau_{\mathrm{N}}^{0}, \tag{10.105b}$$

and we have defined the energy average

$$\left\langle \frac{\xi^2}{E^2} \psi_\eta \right\rangle_{\xi} \equiv \int_{-\infty}^{\infty} \mathrm{d}\xi_{\boldsymbol{k}}\, (-f'_{\boldsymbol{k}}) \frac{\xi_{\boldsymbol{k}}^2}{E_{\boldsymbol{k}}^2} \psi_\eta(\xi_{\boldsymbol{k}}). \tag{10.105c}$$

Note that (10.105a) does not depend on i or j for symmetry reasons. In the immediate vicinity of T_{c} the temperature variation of η is dominated by the change in the quasiparticle excitation spectrum, which introduces a factor ξ/E into the expression for the group velocity (10.25c) (Shumeiko 1972). The ensuing factor of $(\xi/E)^2$, for example in the energy average in (10.105a), gives rise to a square-root dependence of the change of η in the superfluid state: $\eta(T) - \eta(T_{\mathrm{c}}) \propto -(T_{\mathrm{c}} - T)^{1/2}$. This may be seen from the limiting behaviour of the ξ integral over the expression

$$\lim_{\Delta \to 0} \left(\frac{\xi^2}{E^2}\right) = \lim_{\Delta \to 0} \left(1 - \frac{\Delta^2}{\xi^2 + \Delta^2}\right) = 1 - \pi \Delta \delta(\xi) \tag{10.106}$$

if we recall that $\Delta \propto (T_{\mathrm{c}} - T)^{1/2}$ for $T \to T_{\mathrm{c}}$. The integral equation (10.102a) for $\psi_\eta(\xi)$ may be solved exactly to first order in Δ, using the eigenfunctions of the transport problem in the normal state, as discussed in Section 2.3 (Pethick *et al.* 1976). In fact, the only correction linear in Δ (and hence nonanalytic in the "natural variable" Δ^2) comes from replacing the singular term ξ_2^2/E_2^2 in (10.102a) by (10.106). The relaxation time $\bar{\tau}(\xi_2)$ as well as the kernel $S_2(\xi_1, \xi_2)$ possess analytic expansions in Δ^2, as may be seen by inspection, and to order Δ may therefore be approximated by their normal-state expressions. Introducing a function

$$\phi(x) = \frac{1}{\pi^2} \psi_\eta(\xi) \operatorname{sech}^2 \tfrac{1}{2}x, \tag{10.107}$$

where $x = \xi/k_{\mathrm{B}}T$, the integral equation (10.102a) takes the form

$$(\pi^2 + x^2)\phi(x) - \lambda_2 \int \mathrm{d}x'\, K(x, x')\phi(x') = \operatorname{sech}^2 \tfrac{1}{2}x - \pi\lambda_2 \frac{\Delta}{T} K(x, 0)\phi(0) \tag{10.108}$$

(here and in the following we write λ_2 as shorthand for $\lambda_{2+}^{\mathrm{s}}$). To first order in Δ, the quantity $\phi(0)$ is approximated well by its $\Delta = 0$ value. Using the eigenfunction expansion (2.61) of the integral kernel $K(x, x')$, (10.108) is easily solved. The result is conveniently expressed as

$$\phi(x) = \phi_0(x) + \delta\phi(x), \tag{10.109a}$$

where

$$\phi_0(x) = \sum_{n \text{ odd}} \psi_n(x)(1 - \lambda_2 k_n)^{-1} \int dx' \, \psi_n(x') \tag{10.109b}$$

is the solution in the normal state and

$$\int_{-\infty}^{\infty} \delta\phi(x) \, dx = \frac{\pi\Delta}{T}[1 - \pi^2 \phi_0(0)]\phi_0(0). \tag{10.109c}$$

The shear-viscosity coefficient is obtained by inserting (10.109) into (10.105a), using the relation (10.106):

$$\eta = \eta_N - \frac{\pi^5}{4} \eta_0[\phi_0(0)]^2 \frac{\Delta}{k_B T} + O(\Delta^2), \tag{10.110a}$$

where η_N is the normal-state result (2.66). The quantity $\phi_0(0)$ may be obtained as a rapidly converging series whose first few terms are

$$\phi_0(0) = \frac{1}{\pi^2}\left(1 + \frac{3}{4}\frac{\lambda_2}{1 - \lambda_2} + \frac{7}{64}\frac{\lambda_2}{6 - \lambda_2} + \ldots\right). \tag{10.110b}$$

Turning now to the limit of low temperatures, we observe that the integral term in (10.102a) vanishes somewhat faster than the term on the left-hand side owing to the extra factor of $\xi_2^2/E_2^2 \propto T/\Delta$ in the limit $T \to 0$. For $T/T_c \ll 1$, we may therefore neglect the integral term. The function ψ is then given by

$$\psi_\eta(\xi) = \bar{\tau}(\xi)[1 + O(T/T_c)]. \tag{10.111}$$

Substituting this into (10.105a) and using the low-temperature result for τ given by (10.90), one finds a remarkable result: the viscosity coefficient tends to a *finite* value as $T \to 0$ (Pethick *et al.* 1975, Einzel 1980, 1984):

$$\lim_{T \to 0} \eta(T) = c_\eta \eta(T_c)\left(1 + \alpha_\eta \frac{k_B T}{\Delta}\right), \tag{10.112a}$$

where

$$c_\eta = \frac{2\pi}{3}\left[\frac{k_B T_c}{\Delta(0)}\right]^2 \frac{1}{W_0 f_\eta(\lambda_2)}, \tag{10.112b}$$

$$\alpha_\eta = 4\alpha_\tau - \frac{\frac{5}{3}\lambda_2 - \gamma_2}{W_0} \tag{10.112c}$$

and W_0, $f_\eta(\lambda_2)$ and α_τ have been defined by (10.90b), (2.68) and (10.90c) respectively. Estimates of the coefficient c_η using realistic values of $\Delta(0)$, W_0 and λ_2 yield values of 0.1–0.2, depending on pressure.

The behaviour of η at intermediate temperatures may be calculated either by solving the integral equation (10.102a) approximately using a variational method, or by approximating the integral kernel in (10.102a) by a separable kernel as given by (10.86), which then allows an exact solution of the integral equation. In the present problem the symmetry is specified by $l=2$ and $t=+1$ (even function in ξ), and the problem is symmetric in spin space. The integral kernel in (10.102a) may therefore be approximated by

$$S_l^{\mathrm{RTA}}(\xi_1, \xi_2) = \lambda_l \frac{1}{\bar{\tau}(\xi_1)} \frac{1}{\langle 1/\bar{\tau}(\xi)\rangle_\xi} \frac{1}{\bar{\tau}(\xi_2)}, \tag{10.113}$$

where $\bar{\tau}(\xi)$ is the normalized relaxation time given by (10.102b) and the energy average $\langle\ \rangle_\xi$ is defined by (10.105c). The integral equation (10.102a) with the separable kernel (10.113) is readily solved. The viscosity coefficient calculated from (10.105) using this solution is then obtained as

$$\eta = \eta_0\left[\left\langle \frac{\xi^2}{E^2}\bar{\tau}(\xi)\right\rangle_\xi + \lambda_2 \frac{\langle \xi^2/E^2\rangle_\xi^2}{\langle 1/\bar{\tau}(\xi)\rangle_\xi - \lambda_2\langle(\xi^2/E^2)/\bar{\tau}\rangle_\xi}\right]. \tag{10.114}$$

Note that the result (10.114) becomes exact in the low-temperature limit and yields the correct behaviour near T_c as given by (10.110), with $\phi_0(0)$ approximated by the first two terms in the parentheses on the right-hand side of (10.110b).

A numerical evaluation of (10.114) with parameter values appropriate for pressures of 0, 10, 20 and 30 bar but neglecting surface effects is shown in Fig. 10.2. The initial drop in η below T_c is well described theoretically, whereas the low-temperature plateau is not borne out by experiment. Qualitatively, the reason for this can easily be understood: it is due to the deviations from the hydrodynamic behaviour associated with the increasing mean free path at low T. The latter may eventually become larger than the characteristic dimension of the sample used in the experiment. Consequently, scattering processes of the quasiparticles at the wall become more important than those between the quasiparticles. This effect is not included in the theory described above and will be further discussed at the end of this chapter.

Bulk viscosity

The hydrodynamic equations for the B phase derived in Chapter 9 contain three coefficients of the bulk (or "second") viscosity: ζ_1, ζ_2 and ζ_3. These are dissipative coefficients multiplying $\nabla\cdot\boldsymbol{v}_\mathrm{n}$ and $\nabla\cdot\boldsymbol{g}_\mathrm{s}$ in the expressions for the stress tensor and the dissipative part of the chemical potential (9.28) and (9.23). On the other hand, the trace of the dissipative part of the stress tensor is obtained from (10.48) by inserting the deviation from local equilibrium of the distribution function:

$$\begin{aligned}\tfrac{1}{3}\sum_i \Pi_{ii}^{\mathrm{D}} &= \tfrac{1}{3}\sum_{\boldsymbol{k}\alpha} \boldsymbol{k}\cdot\boldsymbol{v}_{\boldsymbol{k}} \frac{\xi_{\boldsymbol{k}}}{E_{\boldsymbol{k}}}\,\delta f'_{\boldsymbol{k}} + nm\,\delta\mu^{\mathrm{D}} \\ &= -\zeta_1\nabla\cdot\boldsymbol{g}_\mathrm{s} - \zeta_2\nabla\cdot\boldsymbol{v}_\mathrm{n}.\end{aligned} \tag{10.115}$$

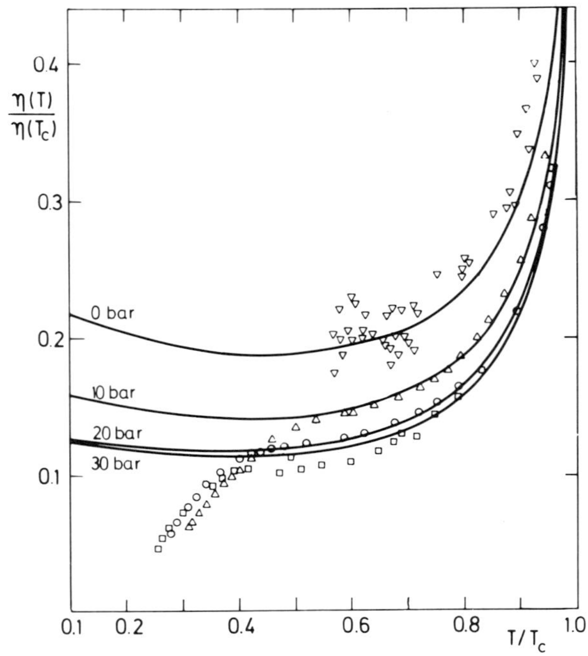

Figure 10.2 Shear-viscosity coefficient $\eta(T)$ for ^{3}He-B versus temperature at pressures $P = 0$, 10, 20, 30 bar as calculated from (10.114) (after Einzel 1984); neglecting surface effects. Experimental values are from vibrating-wire measurements of Carless *et al.* (1983a,b): ▽, 0.1 bar; △, 9.89 bar; ○, 19.89 bar; □, 29.34 bar.

The dissipative part of the chemical potential, i.e. $\delta\mu^{\mathrm{D}} = \delta\mu^{\mathrm{loc}} - \delta\mu^{\mathrm{loc}}$, where $\delta\mu^{\mathrm{loc}}$ is the local equilibrium value, can be determined with the aid of (10.46), which expresses the density δn in terms of $\delta\mu$ and $\delta\bar{f}_{\boldsymbol{k}}$. By definition, the local-equilibrium value of the density change δn^{loc} is equal to the exact value δn, which implies that the dissipative part on the right-hand side of (10.46) has to vanish, and hence

$$\delta\mu^{\mathrm{D}} = -\frac{1}{m} N_{\mathrm{F}}^{-1} \sum_{\boldsymbol{k}\alpha} \frac{\xi_{\boldsymbol{k}}}{E_{\boldsymbol{k}}} \delta f'_{\boldsymbol{k}} = -\zeta_1 \boldsymbol{\nabla}\cdot\boldsymbol{v}_{\mathrm{n}} - \zeta_3 \boldsymbol{\nabla}\cdot\boldsymbol{g}_{\mathrm{s}}. \tag{10.116}$$

Here the factor $1/m$ accounts for the fact that $\delta\mu^{\mathrm{D}}$ is the chemical potential per unit mass.

It is clear that the two terms in (10.115) may be combined into a single expression:

$$\tfrac{1}{3}\sum_i \Pi_{ii}^{\mathrm{D}} = \tfrac{1}{3}\hbar \sum_{\boldsymbol{k}\alpha} (\boldsymbol{k}\cdot\boldsymbol{v}_{\boldsymbol{k}} - k_{\mathrm{F}} v_{\mathrm{F}}) \frac{\xi_{\boldsymbol{k}}}{E_{\boldsymbol{k}}} \delta f'_{\boldsymbol{k}}. \tag{10.117}$$

Now the quantity in parentheses vanishes at the Fermi energy. Therefore only particle–hole-*asymmetric* terms contribute to the viscosity coefficients ζ_1 and ζ_2. Since each particle–hole-asymmetric term is order-T_c/T_F small and since one needs two of these to get a nonvanishing result for the energy integration in (10.117), ζ_1 and ζ_2 are at least order-$(T_c/T_F)^2$ smaller than, for example, the shear viscosity. The coefficient ζ_2 multiplying $\boldsymbol{\nabla}\cdot\boldsymbol{v}_n$ acquires yet another factor of $(T_c/T_F)^2$ from the driving term in (10.100a), which is of the same structure as the term multiplying $\delta f'_{\boldsymbol{k}}$ in (10.117) (Shumeiko 1972).

The remaining coefficient ζ_3 describes the friction arising from a normal–superfluid counterflow with nonvanishing divergence, which exerts a force on the superflow. It is again convenient to split off the dependences contained in the driving term by defining

$$\delta f'_{\boldsymbol{k}} = \mathrm{i}\tau_N^0 \frac{k_F v_F}{3\rho} f'_{\boldsymbol{k}} \frac{\xi_{\boldsymbol{k}}}{E_{\boldsymbol{k}}} \boldsymbol{q}\cdot\boldsymbol{g}_s \psi_\zeta(\xi_{\boldsymbol{k}}). \tag{10.118}$$

The dimensionless function ψ_ζ is found to satisfy the integral equation

$$\psi_\zeta(\xi_1) = \bar{\tau}(\xi_1)\left[1 + \int_0^\infty \mathrm{d}\xi_2\, S_0(\xi_1, \xi_2) \frac{\xi_2^2}{E_2^2} \psi_\zeta(\xi_2)\right], \tag{10.119}$$

which is rather similar to the corresponding equation (10.102a) in the case of the shear viscosity except that the kernel is now given by S_0 (see (10.103)). From the experience with the shear-viscosity calculation, we expect to be able to calculate ζ_3 exactly near T_c and for $T\to 0$.

The general result for ζ_3 is obtained by substituting (10.119) into (10.116):

$$\zeta_3 = \zeta_3^0 \left\langle \frac{\xi^2}{E^2} \psi_\zeta(\xi) \right\rangle_\xi. \tag{10.120a}$$

Here

$$\zeta_3^0 = \frac{k_F}{3\rho} v_F \tau_N^0 = \frac{5}{3} \frac{1}{mn^2} \eta_0 \tag{10.120b}$$

sets the scale and the angular brackets denote the energy average defined by (10.105c).

In the limit $\Delta\to 0$ there is again a nonanalytic contribution from the factor of ξ_2^2/E_2^2 in (10.119). Replacing $\bar{\tau}$ and S_0 by their normal-state limits and introducing a new function

$$\phi(x) = \frac{1}{\pi^2} \psi_\zeta(\xi)\, \mathrm{sech}^2\, \tfrac{1}{2}x, \tag{10.121}$$

where $x = \xi/k_B T$, one arrives at the following integral equation:

$$(\pi^2 + x^2)\phi(x) - \int \mathrm{d}x'\, K(x, x')\left[1 - \pi \frac{\Delta}{k_B T} \delta(x')\right]\phi(x') = \mathrm{sech}^2\, \tfrac{1}{2}x. \tag{10.122}$$

However, this time one is not allowed to treat the term proportional to Δ in perturbation theory, because the coefficient λ_{0+}^{s} appearing in the definition (10.103) of S_0 is unity and is thus equal to the eigenvalue $k_1 = 1$ of the first of the eigenfunctions $\psi_n(x)$ introduced in (2.56). This simply expresses the fact that particle number is conserved in the normal state ($\lambda_{0+}^{s} = 1!$) but ceases to be conserved in the superfluid state, where the number of quasiparticles is no longer equal to the number of atoms in the system.

As one approaches T_c from below, the coefficient of the eigenfunction

$$\psi_1(x) = \frac{3^{1/2}}{4\pi} \operatorname{sech}^2 \tfrac{1}{2}x \tag{10.123}$$

in the expansion of $\phi(x)$ in terms of eigenfunctions $\psi_n(x)$ diverges as Δ^{-1}. The leading term in the bulk viscosity ζ_3 is easily found by employing the eigenfunction representation (2.61) of the kernel $K(x, x')$; it is given by Wölfle 1977b)

$$\zeta_3 = \zeta_3^0 \left[\frac{4}{\pi} \frac{k_B T}{\Delta} + \text{const} \right] \quad \text{as } T \to T_c. \tag{10.124}$$

In other words, ζ_3 diverges as $(1 - T/T_c)^{-1/2}$ as $T \to T_c$. The corresponding term in the Josephson equation (9.22), $-\zeta_3 \nabla \cdot \boldsymbol{g}_s$, still vanishes as it should, because $\boldsymbol{g}_s = \rho_s(\boldsymbol{v}_s - \boldsymbol{v}_n)$ approaches zero as Δ^2, and therefore overcompensates the divergence of ζ_3.

In the opposite limit of low temperatures the integral term in (10.119) again may be neglected compared with the out-scattering term. For the same reason as discussed in the context of shear viscosity, ζ_3 is found to tend to a constant:

$$\zeta_3(T) = c_\zeta \zeta_3^0(T_c) \left(1 + \alpha_\zeta \frac{k_B T}{\Delta} \right) \quad \text{as } T \to 0, \tag{10.125a}$$

where

$$c_\zeta = \frac{2\pi}{3} \frac{1}{W_0} \left(\frac{k_B T_c}{\Delta(0)} \right)^2, \tag{10.125b}$$

$$\alpha_\zeta = 4\alpha_\tau - \frac{\frac{5}{3} - \gamma_0}{W_0}. \tag{10.125c}$$

At intermediate temperatures we may approximate the integral kernel in (10.119) by the relaxation-time expression (10.113) with $\lambda_{0+}^{2} = 1$. The resulting expression for ζ_3 is found to be

$$\zeta_3 = \zeta_3^0 \left[\left\langle \frac{\xi^2}{E^2} \bar{\tau}(\xi) \right\rangle_\xi + \frac{\langle \xi^2/E^2 \rangle_\xi^2}{\langle (\Delta^2/E^2)/\bar{\tau}(\xi) \rangle_\xi} \right]. \tag{10.126}$$

Equation (10.126) tends to the correct limiting values for $T \to T_c$ and $T \to 0$. In Fig. 10.3 the result of a numerical evaluation of ζ_3 according to (10.126) is shown versus temperature.

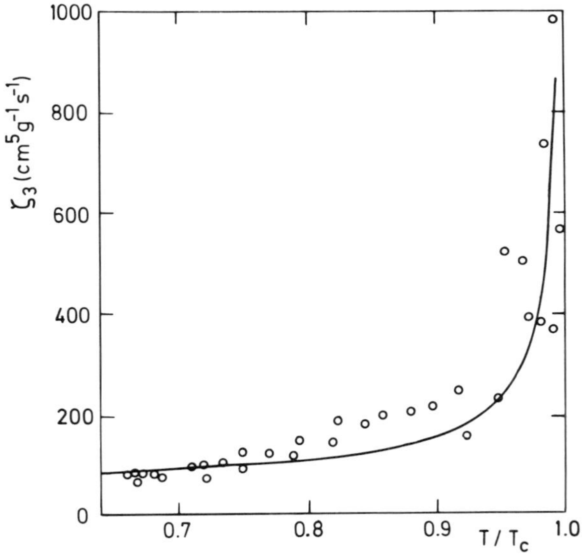

Figure 10.3 Second viscosity coefficient $\zeta_3(T)$ for ^{3}He-B versus temperature at $P = 0$ as calculated from (10.126) (after Einzel 1984). Experimental values (○) are from the vibrating-wire measurements at 1.28 bar of Carless *et al.* (1983a,b).

The coefficient ζ_3 is not easily accessible experimentally. In principle, it determines the attenuation of fourth sound, as will be shown in Section 11.1, but in practice this contribution is very small compared with the one from mean-free-path effects. It has recently become possible to extract ζ_3 from vibrating-wire experiments (Carless *et al.* 1983a,b). A comparison of the data with theory is shown in Fig. 10.3.

Thermal conductivity

We begin by splitting off the $\boldsymbol{k}$ dependence of the driving term in (10.100a) from the distribution function:

$$\delta f'_{\boldsymbol{k}} = \mathrm{i}\tau_{\mathrm{N}}^{0} f'_{\boldsymbol{k}} \frac{\xi_{\boldsymbol{k}}}{E_{\boldsymbol{k}}} \frac{E_{\boldsymbol{k}}}{T} (\boldsymbol{v}_{\boldsymbol{k}} \cdot \boldsymbol{q})\, \delta T\, \psi_{\kappa}(\xi). \tag{10.127}$$

It can be seen that $\delta f'_{\boldsymbol{k}}$ is odd in ξ as in the case of η and ζ_3, but is also odd in $\boldsymbol{k}$, in contrast with the previous cases. The function $\psi_{\kappa}(\xi)$ is found to satisfy the integral equation

$$\psi_{\kappa}(\xi_1) = \bar{\tau}(\xi_1)\left[1 + \frac{1}{E_1}\int_0^{\infty} \mathrm{d}\xi_2\, S_1^{(-)}(\xi_1, \xi_2)\frac{\xi_2^2}{E_2}\psi_{\kappa}(\xi_2)\right], \tag{10.128}$$

where one factor of E^{-1} has now been shifted outside the integral, owing to the extra factor of E in (10.127). Taking into account the oddness of $\delta f_{\boldsymbol{k}}$ with respect to $\boldsymbol{k} \to -\boldsymbol{k}$, the kernel $S_1^{(-)}$ is obtained from (10.82) and (10.83) as

$$S_1^{(-)}(\xi_1, \xi_2) = \lambda_{1-}^{\mathrm{s}} J_0^{(-)}(\xi_1, \xi_2) - \gamma_1 J_2^{(-)}(\xi_1, \xi_2). \tag{10.129}$$

Here $J_l^{(-)}$ is defined by (10.87b) with, however, an additional factor of $-\mu_2$ in the μ_2 sum. The parameters $\lambda_{1-}^{\mathrm{s}}$ and γ_1 are given by (2.52) and (10.88a) respectively.

Inserting (10.127) into the expression (10.50b) for the heat current and comparing with (9.33), one obtains the thermal conductivity as

$$\kappa = \kappa_0(T) \frac{3}{(\pi k_{\mathrm{B}} T)^2} \langle \xi^2 \psi_\kappa \rangle_\xi, \tag{10.130a}$$

with

$$\kappa_0(T) = \tfrac{1}{2} C_{\mathrm{N}}(T) v_{\mathrm{F}}^2 \tau_{\mathrm{N}}^0(T). \tag{10.130b}$$

The thermal conductivity varies smoothly near T_{c}, because the factor ξ_2^2/E_2^2 (giving rise to the nonanalytic behaviour of η and ζ_3) has been replaced by the nonsingular term ξ_2^2/E_2. Nevertheless, the slope of κ changes discontinuously at T_{c}.

In the limit of low temperatures the integral term in (10.128) is again negligible owing to the factor ξ_2^2. The various factors of T from the relaxation time $\bar{\tau}$ and the energy averages (accidentally) happen to combine to a T^{-1} law as in the normal state. The complete result is

$$\kappa(T) = c_\kappa \kappa_0(T)\left(1 + \alpha_\kappa \frac{k_{\mathrm{B}} T}{\Delta}\right) \quad \text{as } T \to 0, \tag{10.131a}$$

where

$$c_\kappa = \frac{2}{\pi} \frac{1}{W_0}, \tag{10.131b}$$

$$\alpha_\kappa = 4\alpha_\tau - 3 - \frac{\lambda_{1-}^{\mathrm{s}} - \frac{5}{3}\gamma_1}{W_0}, \tag{10.131c}$$

with W_0 given by (10.90b)

The relaxation-time approximation for the kernel $S_1^{(-)}$ should take into account the fact that $J_0^{(-)}(\xi_1, \xi_2)$ vanishes for $\xi_1 = 0$ and $\xi_2 = 0$ in the normal state. This property is readily incorporated by introducing factors E_1 and E_2 such that

$$S_1^{(-)}(\xi_1, \xi_2)\,\big|_{\mathrm{RTA}} = \tfrac{1}{3}\lambda_{1-}^{\mathrm{s}} \frac{E_1}{\bar{\tau}(\xi_1)} \frac{1}{\langle E^2/\bar{\tau}(\xi) \rangle_\xi} \frac{E_2}{\bar{\tau}(\xi_2)}. \tag{10.132}$$

With this approximation, excellent results are obtained in the normal state.

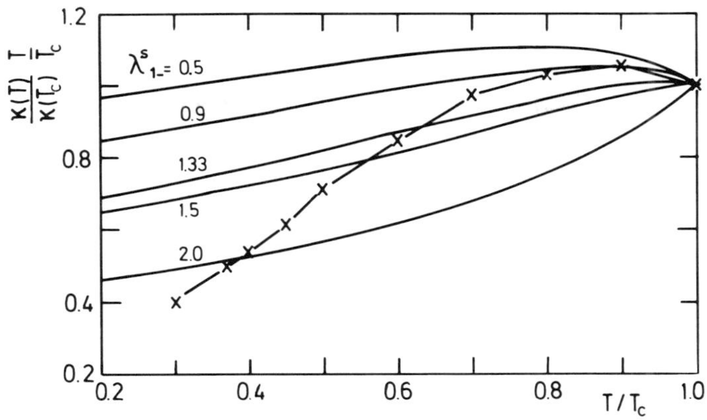

Figure 10.4 The scaled thermal conductivity $\kappa(T)$ of ^{3}He-B versus temperature at $P = 20$ bar according to (10.133) using λ_{1-}^s values from 0.5 to 2.0 (after Einzel 1984). Experimental data points (×) are from heat-pulse measurements at 20 bar by Wellard (1982) (see Hall and Hook 1986).

For the B phase, a simple calculation yields

$$\kappa(T) = \kappa_0 \frac{3}{(\pi k_B T)^2}\left[\langle \xi^2 \bar{\tau}(\xi)\rangle_\xi + \tfrac{1}{3}\lambda_{1-}^s \frac{\langle \xi^2\rangle_\xi^2}{\langle E^2/\bar{\tau}(\xi)\rangle_\xi - \tfrac{1}{3}\lambda_{1-}^s\langle \xi^2/\bar{\tau}(\xi)\rangle_\xi}\right]. \tag{10.133}$$

At low temperatures (10.133) reduces to the exact behaviour (10.131). A numerical evaluation of (10.133) shows that $T\kappa(T)$ is roughly constant (see Fig. 10.4). This is essentially borne out by experiment. However, at low temperatures the experimental curve (Wellard 1982, Wellard *et al.* 1982) falls below the theoretical results. This may again be caused by finite-mean-free-path effects. Additional scattering processes involving phonons and, at high pressure, fluctuations of the solid phase have also been considered but have been found to be small (Geilikman and Chechetkin 1975).

Spin diffusion

The spin-diffusion coefficient describes the dissipative spin current, which is driven by a gradient in ω_μ, the deviation of the spin density from its equilibrium value (see (9.29) and (9.30)). Therefore the last term in (10.100a) is the relevant driving term. For simplicity, we assume the Landau parameter F_1^a to be equal to zero. When we discussed the spin density and spin current density in (10.51) and (10.57), we already noticed that the coherence factors here are rather complicated. Indeed, the spin-antisymmetric component of the collision integral is so complex that we shall not attempt to discuss it here. A complete treatment has been given by Einzel (1980, 1981). Here we only consider the simplest relaxation-time approximation, where the in-scattering part is completely neglected. The distribution function is then given by (see Combescot 1975c)

$$\delta f'_{k\alpha\beta} = \mathrm{i}\frac{\mu_0\gamma}{\chi}\sum_{\mu\nu} \tau_k f'_k(\boldsymbol{v}_k \cdot \boldsymbol{q})(\sigma_\mu)_{\alpha\beta}\, p_{k,\mu\nu}\omega^{\mathrm{s}}_\nu. \tag{10.134}$$

Inserting this into the definition of the dissipative spin current,

$$\begin{aligned}(j^\sigma_{\mu i})^{\mathrm{D}} &= \tfrac{1}{2}\hbar\sum_{k\nu} v_{ki}p_{k,\mu\nu}\,\mathrm{tr}_\sigma\,(\boldsymbol{\sigma}_\nu\,\delta\mathbf{f}'_k)\\ &= -\sum_{\nu j} D_{\mu i\nu j}\,\nabla_j\omega^{\mathrm{s}}_\nu,\end{aligned} \tag{10.135}$$

one finds the spin-diffusion coefficient

$$D_{\mu i\nu j} = \frac{\mu_0^2}{\chi}2\sum_k (-f'_k)v_{ki}v_{kj}(\boldsymbol{p}^2_k)_{\mu\nu}\tau_k. \tag{10.136}$$

Using the definition (10.58) of $p_{k,\mu\nu}$ and the projector property of $\Lambda_{\mu\nu}$, $\Lambda^2 = \Lambda$, one has

$$(\boldsymbol{p}^2_k)_{\mu\nu} = \delta_{\mu\nu} - \sum_{ij}\frac{\Delta^2}{E^2_k}R_{\mu i}\hat{k}_i R_{\nu j}\hat{k}_j, \tag{10.137}$$

where $R_{\mu i}$ is the rotation matrix characterizing the B phase. As anticipated in (9.31), it follows that the diffusion tensor consists of three independent components:

$$D_{\mu i\nu j} = D_1\delta_{\mu\nu}\delta_{ij} + D_2R_{\mu i}R_{\nu j} + D_3R_{\mu j}R_{\nu i}, \tag{10.138a}$$

with

$$D_1 = D_0\left\langle \bar{\tau}(\xi)\left(1 - \frac{1}{5}\frac{\Delta^2}{E^2}\right)\right\rangle_\xi, \tag{10.138b}$$

$$D_2 = D_3 = -\tfrac{1}{5}D_0\left\langle \bar{\tau}(\xi)\frac{\Delta^2}{E^2}\right\rangle_\xi, \tag{10.138c}$$

$$D_0 = \tfrac{1}{3}v_{\mathrm{F}}^2\tau^0_{\mathrm{N}}\frac{\chi^0_{\mathrm{N}}}{\chi}. \tag{10.138d}$$

We already know that the factors ξ^2/E^2 produce nonanalytic dependences proportional to Δ near T_{c}. Hence $D_{2,3}$ vanishes as Δ for $T \to T_{\mathrm{c}}$, whereas D_1 tends to the normal-state coefficient D^σ. Just below T_{c}, D_1 falls, which is similar to the behaviour of the viscosity η, but the effect is smaller here, owing to the factor of $\frac{1}{5}$ in front of ξ^2/E^2.

In the limit of low temperatures the spin-diffusion coefficients of (10.136) are found as

$$D_1 = -4D_{2,3} = \frac{8\pi}{15} D_0 \frac{1}{W_0} \frac{k_B T}{\Delta(0)}. \tag{10.139}$$

Since $D_0(T)$ is proportional to τ_N^0, (2.38), whose low-temperature dependence is given by $\tau_N^0(T) = \tau_N^0(T_c)(T_c/T)^2$, the spin-diffusion coefficients are seen to diverge as T^{-1} for $T \to 0$. The qualitative behaviour near T_c and as $T \to 0$ shown by (10.136) is in agreement with a more refined calculation (Einzel 1981).

The spin-diffusion coefficient is difficult to measure quantitatively, because the dipole torque and spin supercurrents tend to mask the effects of spin diffusion (see the discussion of spin-echo experiments at the end of Section 8.6).

10.3.2 Intrinsic spin relaxation

In this subsection we discuss situations in which, by some means, normal and superfluid components are driven out of equilibrium with respect to each other (see Section 8.6). For example, we might envisage increasing the density of the normal component (instantaneously) by injecting hot quasiparticles into the system. This type of experiment has been carried out on superconductors, where a suitable device is available—the tunnel junction. One expects the system to relax towards equilibrium within some intrinsic relaxation time. In superfluid ^{3}He the equivalent of a tunnel junction is not available, but there is none the less a way to prepare a nonequilibrium state of this kind for the spin densities, as discussed phenomenologically in Section 8.6. There we argued that the dipole torque acts only on the Cooper-pair spins, which adjust to this field within a time of order $\hbar/\Delta$. The quasiparticle spins, on the other hand, relax towards the mutual equilibrium state via quasiparticle-number-nonconserving collisions in which quasiparticles are converted into Cooper pairs and vice versa. The corresponding intrinsic spin-relaxation rate is typically of the same order as the quasiparticle collision time and hence much longer than $\hbar/\Delta$, except very close to T_c.

In the expression (10.51) for the spin density S_μ one can identify the quasiparticle contribution as

$$S_\mu^q = \tfrac{1}{2}\hbar \sum_{k\nu} s_{k,\mu\nu}\, \delta f_{k,\nu}, \tag{10.140}$$

where $s_{k,\mu\nu}$ is the BQP spin weight factor defined in (10.52a) and a notation as introduced in (10.18c) is used.

The remaining terms in (10.51) describe the polarization of the Cooper pairs by the gauge field H_ν^g, the external field H_ν^{ext} and the Fermi-liquid molecular field. In the approximation where the Landau parameters F_l^a with

$l \geq 2$ are neglected, the Cooper-pair spin density is given by

$$S^{\mathrm{p}}_{\mu} = \gamma^{-1} \sum_{\nu} \chi^{\mathrm{p0}}_{\mu\nu} H^{\mathrm{eff}}_{\nu}, \tag{10.141a}$$

where χ^{p0} is the Cooper-pair partial susceptibility defined in (10.53) and

$$H^{\mathrm{eff}}_{\nu} = H^{\mathrm{ext}}_{\nu} + H^{\mathrm{g}}_{\nu} - \frac{\gamma}{\chi^0_{\mathrm{N}}} F^{\mathrm{a}}_0 S_{\nu} \tag{10.141b}$$

is the total effective magnetic field acting on the Cooper-pair spins. As usual, this field is screened by the total spin polarization S_{ν} via the Landau parameter F^{a}_0.

The effective field H^{eff}_{μ} may be expressed in terms of the spin densities by inverting (10.141a):

$$H^{\mathrm{eff}}_{\mu} = \gamma \sum_{\nu} (\chi^{\mathrm{p0}})^{-1}_{\mu\nu} (S_{\nu} - S^{\mathrm{q}}_{\nu}), \tag{10.141c}$$

where we have defined the quasiparticle spin density $S^{\mathrm{q}}_{\mu} \equiv S_{\mu} - S^{\mathrm{p}}_{\mu}$.

The quasiparticle spin density in a local-equilibrium state, described by the distribution function

$$\begin{aligned} \delta f^{\mathrm{loc}}_{\boldsymbol{k},\mu} &= f'_{\boldsymbol{k}}(\delta \varepsilon_{\boldsymbol{k},\mu} - \mu_0 H^{\mathrm{ext}}_{\mu} - \mu_0 H^{\mathrm{g}}_{\mu}) \\ &\approx -f'_{\boldsymbol{k}} \mu_0 H^{\mathrm{eff}}_{\mu}, \end{aligned} \tag{10.142}$$

is found by substituting (10.142) into (10.140a):

$$S^{\mathrm{q,loc}}_{\mu} = \gamma^{-1} \sum_{\nu} \chi^{\mathrm{q0}}_{\mu\nu} H^{\mathrm{eff}}_{\nu}. \tag{10.143a}$$

Here we have defined the quasiparticle partial spin susceptibility by

$$\chi^{\mathrm{q0}}_{\mu\nu} = \mu_0^2 \sum_{\boldsymbol{k}} (-f'_{\boldsymbol{k}}) (\boldsymbol{s}^2_{\boldsymbol{k}})_{\mu\nu}, \tag{10.143b}$$

where

$$(\boldsymbol{s}^2_{\boldsymbol{k}})_{\mu\nu} = \left(\frac{\xi_{\boldsymbol{k}}}{E_{\boldsymbol{k}}}\right)^2 \delta_{\mu\nu} + \frac{|\Delta_{\boldsymbol{k}}|^2}{E^2_{\boldsymbol{k}}} \Lambda_{\mu\nu}, \tag{10.143c}$$

with $\Lambda_{\mu\nu}$ given by (10.52b). Of course, the two partial susceptibilities add up to the usual spin susceptibility χ^0, discussed in (3.102)–(3.104):

$$\chi^0_{\mu\nu} = \chi^{\mathrm{q0}}_{\mu\nu} + \chi^{\mathrm{p0}}_{\mu\nu}. \tag{10.144}$$

In order to calculate the relaxation time τ_{CE} introduced in (8.188) from the kinetic equation, we need an approximation for the relevant part of the collision integral. To neglect the in-scattering part completely, as we did in the case of the spin-diffusion coefficient, is not allowed here, because spin is a conserved quantity in the normal state. As one approaches T_{c} from below, the Cooper-pair density S^{p}_{μ} vanishes and $S^{\mathrm{q}}_{\mu} \rightarrow S_{\mu}$. However, in the absence of spin-conserving interactions, the spin-relaxation time is infinite in the normal state, and one therefore expects τ_{CE} to diverge for $T \rightarrow T_{\mathrm{c}}$. This behaviour is brought about by a cancellation of the in-scattering and out-scattering parts of the collision integral. A simple relaxation-time approximation along the lines of (10.68a), which keeps much of the structure of the exact collision integral, is given by

$$\delta\tilde{I}_{\mathbf{k},\mu} = -\frac{\mathrm{i}}{\tau_{\boldsymbol{k}}}\left[\delta f'_{\boldsymbol{k},\nu} - \sum_{\nu}\left(f'_{\boldsymbol{k}} s_{\boldsymbol{k},\mu\nu}\frac{\sum_{\boldsymbol{k}'\lambda} s_{\boldsymbol{k}',\nu\lambda}\,\delta f'_{\boldsymbol{k}',\lambda}/\tau_{\boldsymbol{k}'}}{\sum_{\boldsymbol{k}''} f'_{\boldsymbol{k}''}/\tau_{\boldsymbol{k}''}}\right)\right]. \tag{10.145}$$

The sum over $\boldsymbol{k}'$ in the last term of (10.145) may be related to S^{q}_{μ}. Multiplying the kinetic equation (10.25a) in the limit $q=0$ (we consider spatially homogeneous relaxation processes here) by $\sum_{\mu} s_{\boldsymbol{k},\mu\nu}\,\mathrm{tr}_{\sigma}\,(\boldsymbol{\sigma}^{\mu}\ldots)$ and summing over $\boldsymbol{k}$, one finds

$$\sum_{\boldsymbol{k}\nu}\left(\frac{s_{\boldsymbol{k},\mu\nu}\,\delta f'_{\boldsymbol{k},\nu}}{\tau_{\boldsymbol{k}}}\right) = \mathrm{i}\omega S^{\mathrm{q}}_{\mu}\frac{\sum_{\boldsymbol{k}'} f'_{\boldsymbol{k}'}/\tau_{\boldsymbol{k}'}}{\sum_{\boldsymbol{k}'}(1-s^2_{\boldsymbol{k}'})f'_{\boldsymbol{k}'}/\tau_{\boldsymbol{k}'}}. \tag{10.146}$$

In deriving (10.146), we have replaced $(\boldsymbol{s}^2_{\boldsymbol{k}})_{\mu\nu}$ by

$$(\boldsymbol{s}^2_{\boldsymbol{k}})_{\mu\nu} \equiv \delta_{\mu\nu}s^2_{\boldsymbol{k}} = \delta_{\mu\nu}\times\begin{cases}\left(1-\dfrac{2}{3}\dfrac{\Delta^2}{E^2}\right) & \text{(B phase)},\\[2ex] \left(1-\dfrac{\Delta^2_{\boldsymbol{k}}}{E^2_{\boldsymbol{k}}}\right) & \text{(A phase)}.\end{cases} \tag{10.147}$$

This is an exact result for the B phase. In the case of the A phase we assume that S^{q}_{μ} is approximately parallel to S^{p}_{μ}, i.e. perpendicular to $\hat{\boldsymbol{d}}$. Then the term proportional to $\Lambda_{\mu\nu}$ in (10.143) does not contribute to $\mathbf{s}^2_{\boldsymbol{k}}$.

The equation of motion for S^{q}_{μ} is obtained by substitution of the local-equilibrium solution (10.142) into $\delta\mathbf{f}'_{\boldsymbol{k}} = \delta\mathbf{f}_{\boldsymbol{k}} - \delta\mathbf{f}^{\mathrm{loc}}_{\boldsymbol{k}}$ in the first term of the collision integral, then solving for $\delta\mathbf{f}_{\boldsymbol{k}}$ and finally integrating as before. The resulting expression is so simple that we need not specialize to the hydrodynamic limit $\omega\tau \ll 1$, but may keep the ω dependence, valid for $\hbar\omega < \Delta$, as a by-product:

$$\omega S^{\mathrm{q}}_{\mu} = -\frac{\mathrm{i}}{\tau_{\mathrm{eff}}(\omega)}(S^{\mathrm{q}}_{\mu} - S^{\mathrm{q,loc}}_{\mu}). \tag{10.148a}$$

The complex-valued relaxation time $\tau_{\mathrm{eff}}(\omega)$ is found as

$$\tau_{\mathrm{eff}}(\omega) = \frac{\sum_{\boldsymbol{k}} s^2_{\boldsymbol{k}} f'_{\boldsymbol{k}}}{\sum_{\boldsymbol{k}}(1-s^2_{\boldsymbol{k}})f'_{\boldsymbol{k}}/\tau_{\boldsymbol{k}}} + \frac{\sum_{\boldsymbol{k}} s^2_{\boldsymbol{k}} f'_{\boldsymbol{k}}\tau_{\boldsymbol{k}}/(\mathrm{i}+\omega\tau_{\boldsymbol{k}})}{\sum_{\boldsymbol{k}} s^2_{\boldsymbol{k}} f'_{\boldsymbol{k}}/(\mathrm{i}+\omega\tau_{\boldsymbol{k}})}. \tag{10.148b}$$

In the hydrodynamic limit $\omega\tau_{\boldsymbol{k}} \ll 1$, τ_{eff} may be identified with the relaxation time introduced phenomenologically by Combescot and Ebisawa

(1974). By taking the limit $\omega \to 0$ of (10.148b), we find the explicit expression for the B phase:

$$\tau_{\mathrm{CE}}^{\mathrm{B}} = \frac{3\langle 1 - \frac{2}{3}\Delta^2/E^2 \rangle_\xi}{2\langle (\Delta^2/E^2)/\tau(E) \rangle_\xi} + \frac{\langle (1 - \frac{2}{3}\Delta^2/E^2)\tau(E) \rangle_\xi}{\langle 1 - \frac{2}{3}\Delta^2/E^2 \rangle_\xi}. \tag{10.148c}$$

The first term in (10.148c) clearly diverges for $\Delta \to 0$. Using (10.106), the singular term is easily evaluated as

$$\tau_{\mathrm{CE}}^{\mathrm{B}}(T) = \frac{6}{\pi} \tau_{\mathrm{N}}^{0} \frac{k_{\mathrm{B}} T_{\mathrm{c}}}{\Delta(T)} \quad \text{as } T \to T_{\mathrm{c}}. \tag{10.149}$$

According to (10.149), the intrinsic spin-relaxation time at fixed effective field H^{eff} diverges as $(1 - T/T_{\mathrm{c}})^{-1/2}$ for $T \to T_{\mathrm{c}}$. The result (10.149) agrees with an exact calculation (Bhattacharyya *et al.* 1975, 1977b; see also Combescot 1975a). In the limit of low temperatures, τ_{CE} is found from (10.148c) as

$$\tau_{\mathrm{CE}}^{\mathrm{B}} = C_{\mathrm{CE}} \tau \quad \text{as } T \to 0, \tag{10.150}$$

i.e. τ_{CE} is essentially given by the BQP relaxation time (10.90a). From (10.148c), the coefficient C_{CE} follows as $\frac{3}{2}$, whereas an exact treatment of the collision integral yields a somewhat different combination of angular averages of the scattering amplitude than that appearing in τ (Bagnuls *et al.* 1979, Dombre *et al.* 1980, Einzel 1984). Similar expressions may in principle also be derived for the A phase.

In most experiments the spin relaxation does not proceed at fixed effective H^{eff}, but rather at fixed total spin density S_μ. It is a simple matter to rewrite the relaxation equation (10.148a) under these conditions by expressing H_μ^{eff} in terms of S_μ and S_μ^{q} with the aid of (10.141c):

$$\omega S_\mu^{\mathrm{q}} = -\frac{\mathrm{i}}{\tau_{\mathrm{LT}}} [S_\mu^{\mathrm{q}} - (S_0^{\mathrm{q}})_\mu] \tag{10.151a}$$

The new equilibrium quasiparticle spin density $(S_0^{\mathrm{q}})_\mu$ is then given by

$$\boldsymbol{S}_0^{\mathrm{q}} = \frac{\chi^{\mathrm{q}0}}{\chi^0} \boldsymbol{S}, \tag{10.151b}$$

where $\chi^{\mathrm{q}0}$ and χ^0 are the relevant eigenvalues of the tensors $\boldsymbol{\chi}^{\mathrm{q}0}$ and $\boldsymbol{\chi}^0$. The new relaxation time τ_{LT} was introduced by Leggett and Takagi (1975). It is related to τ_{CE} by

$$\tau_{\mathrm{LT}} = \frac{\chi^{\mathrm{p}0}}{\chi^0} \tau_{\mathrm{CE}}. \tag{10.152}$$

The ratio of susceptibilities appearing in (10.152) vanishes as Δ for $T \to T_{\mathrm{c}}$ and therefore cancels the divergence of τ_{CE}. More precisely, from (10.53), one finds for the B phase

$$\frac{\chi^{\mathrm{p}0}}{\chi^0} = \frac{2}{3} \int_{-\infty}^{\infty} \mathrm{d}\xi_{\boldsymbol{k}} \frac{t_{\boldsymbol{k}}}{2E_{\boldsymbol{k}}} \frac{\Delta^2}{E_{\boldsymbol{k}}^2} = \frac{\pi}{6} \frac{\Delta}{k_{\mathrm{B}} T_{\mathrm{c}}} \quad \text{as } T \to T_{\mathrm{c}}. \tag{10.153}$$

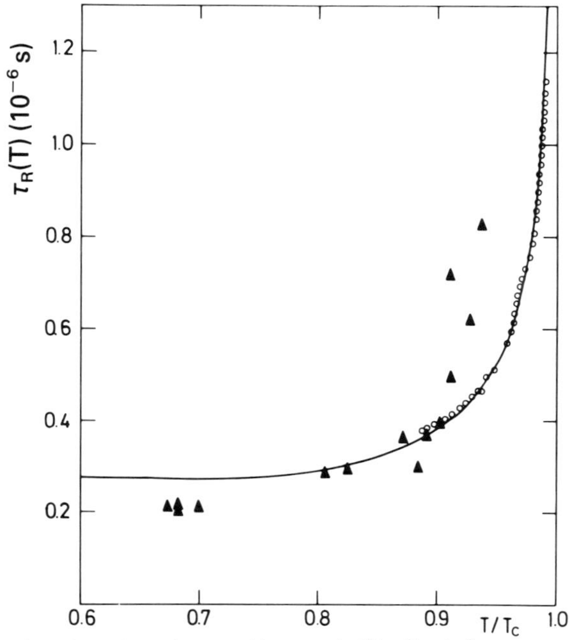

Figure 10.5 Spin-relaxation time $\tau_R(t) = \tau_{LT}(\chi^{q0}/\chi^{p0})\chi/\chi^0$ defined in (8.197) versus temperature at $P = 20$ bar (after Einzel 1981). Experimental data are from Webb *et al.* (1975a) (○) and Eska *et al.* (1982) (▲).

Consequently, τ_{LT} is found to become equal to the quasiparticle relaxation time at the Fermi energy in the normal state

$$\tau_{LT} = \tau_N^0 \quad \text{as } T \to T_c. \tag{10.154}$$

Equation (10.154) is also valid for the A phase (Bhattacharyya *et al.* 1975, 1977b). This important result implies that the normal-state quantity τ_N^0, which is not experimentally accessible in the normal state, may be measured in a spin-relaxation experiment in the superfluid state close to T_c (see Section 8.6). In the limit of low temperatures one has $\chi^{p0}/\chi^0 \to 1$ and consequently $\tau_{LT} \to \tau_{CE}$. A comparison of τ_{LT} as calculated from (10.148) and experiment is given in Fig. 10.5.

10.3.3 Transport coefficients of the A phase

The calculation of the transport properties of ^{3}He-A is considerably more complicated than for the B phase. Technically speaking, the anisotropy of the gap $\Delta_{\boldsymbol{k}}$, i.e. the explicit $\boldsymbol{k}$ dependence of $\Delta_{\boldsymbol{k}}$, forbids a separation of the energy dependence and the angular dependence. Consequently, analytic results may only be obtained in particular cases, such as the first-order corrections in Δ to some of the transport-coefficients near T_c. Indeed, the results for these corrections obtained in the preceding section for the B phase may immediately be taken over by simply replacing the isotropic B-phase energy gap Δ by the angular average $\langle \Delta_{\boldsymbol{k}} \rangle_{\hat{\boldsymbol{k}}}$ in the A phase.

Besides the additional mathematical complications, the anisotropy of the A phase in orbital space also leads to many more transport coefficients than in the B phase. In the limit of low temperatures these behave in quite different ways, depending on the relative orientation of the orbital axis $\boldsymbol{l}$ to the direction of transport or to the spatial orientation of the quantity transported (e.g. momentum). In the following we shall discuss several important transport coefficients, employing the simplest possible relaxation-time approximation.

Viscosity

Let us first consider the shear-viscosity coefficients $\eta^{\parallel}$, $\eta^{\perp}$, $\eta_{\mathrm{L}}^{\perp}$ and the bulk-viscosity coefficients $\zeta_1^{\parallel,\perp}$, $\zeta_2^{\parallel,\perp}$ appearing in the expression for the stress tensor Π_{ij}, (9.82). From the general definition of Π_{ij} as obtained in (10.48) and the expression for the dissipative part of the chemical-potential change (10.116), it follows that the dissipative component of Π_{ij} is given by

$$\Pi_{ij}^{\mathrm{D}} = \sum_{\boldsymbol{k}\alpha} \frac{\xi_{\boldsymbol{k}}}{E_{\boldsymbol{k}}} \hbar(v_{\boldsymbol{k}i}k_j - \tfrac{1}{3}\delta_{ij}v_{\mathrm{F}}k_{\mathrm{F}})\, \delta f'_{\boldsymbol{k}}. \tag{10.155}$$

The change in the spin-symmetric BQP distribution $\delta f'_{\boldsymbol{k}}$ due to velocity fields $\boldsymbol{v}_{\mathrm{n}}$ and $\boldsymbol{v}_{\mathrm{s}}$ is immediately obtained from (10.100a) in the approximation where the collision integral is assumed as

$$\delta \tilde{I}_{\boldsymbol{k}} = -\frac{\mathrm{i}}{\tau_{\boldsymbol{k}}}\, \delta f'_{\boldsymbol{k}},$$

with

$$\delta f'_{\boldsymbol{k}} = f'_{\boldsymbol{k}} \tau_{\boldsymbol{k}} \frac{\xi_{\boldsymbol{k}}}{E_{\boldsymbol{k}}} \left[\boldsymbol{k} \cdot \boldsymbol{V} \cdot \boldsymbol{v}_{\boldsymbol{k}} + \tfrac{1}{3}\mathrm{i}(\boldsymbol{k} \cdot \boldsymbol{v}_{\boldsymbol{k}} - k_{\mathrm{F}}v_{\mathrm{F}})\boldsymbol{q} \cdot \boldsymbol{v}_{\mathrm{n}} - \frac{\mathrm{i}}{3} \frac{k_{\mathrm{F}}v_{\mathrm{F}}}{\rho} \boldsymbol{q} \cdot \boldsymbol{g}_{\mathrm{s}} \right]. \tag{10.156}$$

It follows from (10.155) that the trace of $\Pi_{\mathrm{ij}}^{\mathrm{D}}$ is of order $(T_c/T_{\mathrm{F}})^2$, i.e. it is small owing to the factor of $\boldsymbol{v}_{\boldsymbol{k}} \cdot \boldsymbol{k} - v_{\mathrm{F}}k_{\mathrm{F}}$. This gives rise to the following relations among the viscosity coefficients in (9.82) (for the first relation see Combescot 1975):

$$\eta^{\parallel} - \eta_{\mathrm{L}}^{\perp} - \eta^{\perp} = O\left[\left(\frac{T}{T_{\mathrm{F}}}\right)^2\right], \tag{10.157a}$$

$$2\zeta_i^{\perp} + \zeta_i^{\parallel} = O\left[\left(\frac{T}{T_{\mathrm{F}}}\right)^2\right] \quad (i = 1, 2). \tag{10.157b}$$

For all practical purposes, one may put the right-hand sides in (10.157) equal to zero.

Substituting (10.156) into (10.155) and defining a function

$$\Lambda_{\boldsymbol{k}} = (-f'_{\boldsymbol{k}})\tau_{\boldsymbol{k}}\left(\frac{\xi_{\boldsymbol{k}}}{E_{\boldsymbol{k}}}\right)^2, \tag{10.158}$$

one finds the following result for the transport coefficients in the simple relaxation-time approximation (see also Shahzamanian 1975):

$$\eta^{\perp} = \sum_{\boldsymbol{k}\alpha} \Lambda_{\boldsymbol{k}} v_{\boldsymbol{k},x}^2 k_y^2, \tag{10.159a}$$

$$\eta^{\parallel} = \sum_{\boldsymbol{k}\alpha} \Lambda_{\boldsymbol{k}} v_{\boldsymbol{k},z}^2 k_y^2, \tag{10.159b}$$

$$\eta_{\mathrm{L}}^{\parallel} = (v_{\mathrm{F}}k_{\mathrm{F}})^2 \sum_{\boldsymbol{k}\alpha} \Lambda_{\boldsymbol{k}}(\hat{k}_z^2 - \tfrac{1}{3})^2, \tag{10.159c}$$

$$\zeta_1^{\parallel} = \frac{k_{\mathrm{F}}v_{\mathrm{F}}}{3\rho} \sum_{\boldsymbol{k}\alpha} \Lambda_{\boldsymbol{k}}(\hat{k}_z^2 - \tfrac{1}{3}), \tag{10.159d}$$

where the z axis has been chosen to be along $\boldsymbol{l}$. The coefficients $\zeta_2^{\parallel,\perp}$ are order $(T/T_{\mathrm{F}})^2$ small.

The transport coefficients ζ_3, ζ_4 and ζ_5 in the Josephson equation (9.84), defined in (9.85), are calculated similarly by evaluating the expression (10.116) for the dissipative contribution to the chemical potential. As ζ_4 and ζ_5 are related to $\zeta_1^{\parallel,\perp}$ by (9.86) and (9.87), we need only consider ζ_3, which is obtained by inserting (10.156) into (10.116):

$$\zeta_3 = \left(\frac{\hbar k_{\mathrm{F}}v_{\mathrm{F}}}{3\rho}\right)^2 \sum_{\boldsymbol{k}\alpha} \Lambda_{\boldsymbol{k}}. \tag{10.160}$$

In the limit $T \to T_{\mathrm{c}}$ one finds $\eta^{\parallel} = \eta^{\perp} = \eta_{\mathrm{L}}^{\parallel} = \eta_0(T_{\mathrm{c}})$, with η_0 defined in (10.105b), and $\zeta_1^{\parallel,\perp} = \zeta_2^{\parallel,\perp} = 0$. The coefficient ζ_3 diverges like Δ_0^{-1}, as given by (10.124) with Δ replaced by $\langle \Delta_{\hat{k}} \rangle_{\hat{k}}$.

At low temperatures Bogoliubov quasiparticles tend to be concentrated in two small areas on the Fermi surface of relative extension $k_{\mathrm{B}}T/\Delta_0$ around the nodes of the gap (chosen to be along $\pm\hat{z}$). Hence the momentum sum in (10.158)–(10.160) gives rise to a factor $(T/T_{\mathrm{c}})^2$, whereas each factor of k_x^2 or k_y^2 under the $\boldsymbol{k}$ sum contributes an additional factor of $(T/T_{\mathrm{c}})^2$. Together with the leading temperature dependence of the BQP relaxation time given in (10.95a), one finds in the limit $T \to 0$,

$$\eta^{\perp}(T) \sim \left(\frac{T}{T_{\mathrm{c}}}\right)^2 \eta_0(T_{\mathrm{c}}), \tag{10.161a}$$

$$\eta^{\parallel}(T) \sim \mathrm{const} \times \eta_0(T_{\mathrm{c}}), \tag{10.161b}$$

$$\eta_{\mathrm{L}}^{\parallel}(T) \sim \left(\frac{T_{\mathrm{c}}}{\mathrm{T}}\right)^2 \eta_0(T_{\mathrm{c}}), \tag{10.161c}$$

$$\zeta_1^{\parallel} \sim \frac{\hbar k_{\mathrm{F}}v_{\mathrm{F}}}{3\rho} \eta_{\mathrm{L}}^{\parallel}, \tag{10.161d}$$

$$\zeta_3 \sim \left(\frac{\hbar k_{\mathrm{F}}v_{\mathrm{F}}}{3\rho}\right)^2 \eta_{\mathrm{L}}^{\parallel}. \tag{10.161e}$$

The coefficients $\eta_L^{\parallel}$, $\zeta_1^{\parallel}$ and ζ_3 are seen to diverge as T^{-2} at low temperatures.

Measurements of the various viscosity coefficients are made difficult by the existence of nonuniform $\hat{\boldsymbol{l}}$ textures in any given sample. Nevertheless, it has been possible to determine the coefficients $\eta^{\parallel}$ and $\eta^{\perp}$ quantitatively (and the remaining coefficients at least qualitatively) by means of a torsional-oscillator experiment, in which the frequency shift and the damping are measured as functions of temperature. In order to relate these data to the transport coefficients, a theoretical analysis is required that involves the calculation of the possible $\hat{\boldsymbol{l}}$ textures in a slab geometry (Hook 1989, Hook *et al.* 1989).

Thermal conductivity

According to (9.83), the diffusive thermal transport is characterized by the thermal-conductivity tensor κ_{ij}, which in the A phase is completely determined by the two eigenvalues $\kappa^{\parallel}$ and $\kappa^{\perp}$ for transport parallel and perpendicular to $\hat{\boldsymbol{l}}$. The diffusive heat current in the kinetic theory is given by (10.50b) as

$$\boldsymbol{j}_Q^{\mathrm{D}} = \sum_{\boldsymbol{k}\alpha} \boldsymbol{v}_{\boldsymbol{k}} \xi_{\boldsymbol{k}}\, \delta f'_{\boldsymbol{k}}, \tag{10.162}$$

where particle–hole symmetry has been employed. It follows that the thermal-conductivity coefficients are given by

$$\kappa^{\parallel,\perp} = \frac{1}{k_{\mathrm{B}}T} \sum_{\boldsymbol{k}\alpha} (-f'_{\boldsymbol{k}}) \tau_{\boldsymbol{k}} \xi_{\boldsymbol{k}}^2 (v_{\boldsymbol{k}}^{\parallel,\perp})^2, \tag{10.163}$$

where $v_{\boldsymbol{k}}^{\parallel,\perp}$ is a component of the velocity vector $\hbar\boldsymbol{k}/m^*$ parallel or perpendicular to $\boldsymbol{l}$.

At low temperatures, using (10.95a), $\kappa^{\parallel,\perp}$ are found to obey the power laws (note that $\langle \xi_{\boldsymbol{k}} \rangle \sim T$):

$$\left.\begin{aligned} \kappa^{\parallel}(T) &\sim \frac{T_{\mathrm{c}}}{T} \kappa_{\mathrm{N}}(T_{\mathrm{c}}), \\ \kappa^{\perp}(T) &\sim \frac{T}{T_{\mathrm{c}}} \kappa_{\mathrm{N}}(T_{\mathrm{c}}). \end{aligned}\right\} \tag{10.164}$$

Orbital relaxation

In the equation of motion for $\hat{\boldsymbol{l}}$, (9.87), four transport coefficients have been introduced. Among these, the coefficient λ_4 relating the bending force $\boldsymbol{\psi}$ to $\partial_t \hat{\boldsymbol{l}}$ is of particular importance (Brinkman and Cross 1978). It describes the effect of viscous forces, caused by the dependence of the quasiparticle energy $E_{\boldsymbol{k}}$ on $\hat{\boldsymbol{l}}$. The remaining coefficients appear to be small (Nagai 1980). An infinitesimal rotation of $\hat{\boldsymbol{l}}$ changes the energy of the quasiparticle system by the amount

$$\delta E_{\mathrm{n}} = \sum_{k\alpha} \delta \hat{\boldsymbol{l}} \cdot \frac{\partial E_{\boldsymbol{k}}}{\partial \boldsymbol{l}} \delta f_{\boldsymbol{k}}, \tag{10.165}$$

where $\delta f_{\boldsymbol{k}}$ is the deviation of the quasiparticle distribution from equilibrium (corresponding to a quasiparticle energy spectrum with gap axis along the instantaneous direction of $\hat{\boldsymbol{l}}$). Since the quasiparticle system needs a finite time to adjust to a new gap orientation $\hat{\boldsymbol{l}}$, the quasiparticle distribution always lags behind the motion of $\hat{\boldsymbol{l}}$, i.e. $\delta f_{\boldsymbol{k}} \propto \partial_t \hat{\boldsymbol{l}}$. This "normal-locking" effect (Leggett and Takagi 1976, 1978) produces a viscous force (Cross and Anderson 1975; see also Combescot 1975d, 1978a)

$$\frac{\delta E_{\mathrm{n}}}{\delta \hat{\boldsymbol{l}}} \equiv \psi_{\mathrm{visc}} = -\frac{1}{\lambda_4} \dot{\hat{\boldsymbol{l}}}, \tag{10.166}$$

where $\mu_l = 1/\lambda_4$ is called the "orbital viscosity".

According to (10.165), this force may be expressed in terms of the quasiparticle distribution function $\delta f_{\boldsymbol{k}}$. Employing the simple relaxation-time approximation for the Boltzmann equation

$$\partial_t \, \delta f_{\boldsymbol{k}} = -\frac{1}{\tau_{\boldsymbol{k}}} \delta f_{\boldsymbol{k}}, \tag{10.167}$$

and replacing $\delta f_{\boldsymbol{k}}$ on the left-hand side by the local-equilibrium expression

$$\partial_t \, \delta f_{\boldsymbol{k}}^{\mathrm{loc}} = f'_{\boldsymbol{k}} \frac{\partial E_{\boldsymbol{k}}}{\partial \hat{\boldsymbol{l}}} \cdot \partial_t \hat{\boldsymbol{l}}, \tag{10.168}$$

one finds as an approximate distribution function in the hydrodynamic regime

$$\delta f_{\boldsymbol{k}} = -\tau f'_{\boldsymbol{k}} \frac{\partial E_{\boldsymbol{k}}}{\partial \hat{\boldsymbol{l}}} \cdot \partial_t \hat{\boldsymbol{l}}. \tag{10.169}$$

On combining (10.165), (10.166) and (10.169) and using $\partial E_{\boldsymbol{k}}/\partial \hat{\boldsymbol{l}} = -\hat{\boldsymbol{k}}(\hat{\boldsymbol{k}} \cdot \hat{\boldsymbol{l}})\Delta_0^2/E_{\boldsymbol{k}}$, one finally obtains the transport coefficient μ_l in the form

$$\mu_l = \tfrac{1}{2}\Delta_0^2 \tau \sum_{\boldsymbol{k}} f'_{\boldsymbol{k}} (\hat{\boldsymbol{k}} \cdot \hat{\boldsymbol{l}})^2 \left(\frac{\Delta_{\boldsymbol{k}}}{E_{\boldsymbol{k}}}\right)^2. \tag{10.170}$$

The limiting temperature dependence is easily found to be given by

$$\frac{1}{\mu_l} = [\tau_{\mathrm{N}}^0(T_{\mathrm{c}}) N_{\mathrm{F}} (k_{\mathrm{B}} T_{\mathrm{c}})^2]^{-1} \times \begin{cases} a_{\mathrm{c}} \left[\dfrac{k_{\mathrm{B}} T_{\mathrm{c}}}{\Delta_0(T)}\right]^3 & (T \to T_{\mathrm{c}}), \\ a_0 \left(\dfrac{T}{T_{\mathrm{c}}}\right)^2 & (T \to 0). \end{cases} \tag{10.171}$$

As shown by Pethick and Smith (1977), the coefficient a_{c} may be calculated exactly, yielding $a_{\mathrm{c}} = 128/\pi^2$.

The motion of $\hat{l}$ is dominated by the viscous interaction with the normal fluid. This is confirmed by experiments (Paulson *et al.* 1976c) in which the relaxation of $\hat{l}$, following a sudden change in the orientational force exerted on $\hat{l}$ by a magnetic field via the dipole interaction, has been studied. The observed relaxation rate is in good agreement with the theoretical estimate (10.170). The limiting value of μ_l as $T \to T_c$ provides yet another measurement of the quasiparticle relaxation rate $1/\tau_N^0$ at T_c. Long-lived oscillations of the $\hat{l}$ field with typical periods of the order of seconds, rather than pure relaxation, have also been observed (Wheatley 1978). These are probably to be explained as nonlinear periodic solutions of the hydrodynamic equations.

10.4 FLOW IN RESTRICTED GEOMETRIES

We have seen in Section 10.2 that the mean free path l of Bogoliubov quasiparticles may become very long at low temperatures. When l exceeds the dimensions of the sample cell or some other characteristic dimension like the size of an object immersed in the liquid (e.g. a vibrating wire or a torsion pendulum), the state of the quasiparticle system is determined by the interaction of BQP with the confining boundaries rather than by the collision processes within the liquid. It goes without saying that the flow behaviour expected on the basis of bulk hydrodynamic theory will be altered considerably in such a situation. A theoretical interpretation of flow experiments thus requires the consideration of quasiparticle–boundary scattering in addition to interquasiparticle scattering in the bulk. At present very little is known about the interaction of quasiparticles with a solid wall. A convenient model of a rough surface is provided by the assumption of *diffuse* scattering, whereby quasiparticles leaving the wall are assumed to be in local equilibrium with the wall. At a free surface of helium the assumption of *specular* scattering is probably more adequate.

10.4.1 Slip correction to hydrodynamics

Bulk hydrodynamic theory is valid in the limit $l/R \to 0$, where R is a characteristic dimension of the boundary. The boundary condition on the fluid velocity field in this limit is given by the "sticking" condition: the tangential velocity of the normal-fluid component at the wall must be equal to the velocity of the wall. In contrast, the superfluid velocity and the component of the normal velocity perpendicular to the boundary need not satisfy the sticking boundary condition. In fact, a normal–superfluid counterflow $\boldsymbol{g}_s$ perpendicular to the boundary (at zero mass flow) is possible—provided that heat may be transferred through the boundary into the adjoining solid.

A first-order correction in l/R to bulk hydrodynamics may be obtained by allowing for a modified boundary condition, i.e. fluid slip at the boundaries.

In order to calculate the amount of fluid slip, it is sufficient to consider the simplest situation (Couette flow), i.e. shear flow parallel to an infinitely extended plane wall. In the usual bulk hydrodynamics—valid in a sample sufficiently far from the wall that collision processes in the liquid dominate over surface scattering—the velocity profile will be linear. This is a consequence of the hydrodynamic equations. The linear behaviour, however, need not extrapolate to zero velocity at the wall (assuming the wall to be at rest). Rather, there will be a finite intercept of the linear extrapolation to the velocity profile at the wall, i.e. to some extent the fluid slips along the wall, depending on the nature of the quasiparticle scattering at the wall. The slip effect may conveniently be characterized by the so-called slip length ζ, which is defined as the distance behind the wall at which the velocity extrapolates to zero (see Fig. 10.6). The slip length is usually of the order of the mean free path. Although the actual velocity field does not remain linear close to the wall (as assumed by this extrapolation), but falls somewhat below linearity, the effect of fluid slip on the bulk properties is described correctly by the slip length. Namely, the (transverse) velocity profile $u(z)$ in the bulk shifts by the constant amount

$$u_0 = \zeta\left(\frac{\partial u}{\partial z}\right)_\infty. \tag{10.172}$$

This effect may easily be incorporated into the hydrodynamic description, by replacing the usual boundary condition $u(0) = u_{\text{wall}}$ by the slip boundary condition

$$u(0) = \zeta\left(\frac{\partial u}{\partial z}\right)_{z=0} + u_{\text{wall}}. \tag{10.173}$$

Examples of the application of this boundary condition will be discussed below.

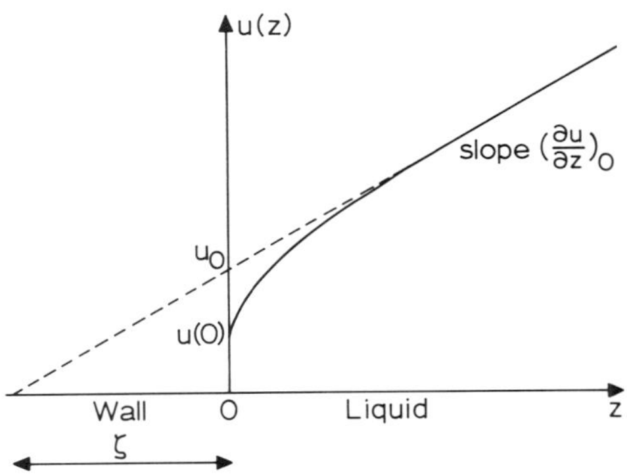

Figure 10.6 The velocity profile $u(z)$ near a wall (schematic). The extrapolated velocity u_0 and the slip length ζ are indicated.

The slip length may be calculated within the kinetic theory discussed in this chapter by adding a boundary condition on the distribution function $\delta f_{\boldsymbol{k}}$ to the kinetic equation (10.25). In general, the boundary condition connects the distribution functions for particles approaching and leaving the wall, $\delta f_{\boldsymbol{k}}^{(-)}$ and $\delta f_{\boldsymbol{k}}^{(+)}$ respectively:

$$(\hat{\boldsymbol{n}} \cdot \boldsymbol{V}_{\boldsymbol{k}})\, \delta f_{\boldsymbol{k}}^{(+)} = \sum_{\substack{\boldsymbol{k}' \\ (\hat{\boldsymbol{n}} \cdot \boldsymbol{V}_{\boldsymbol{k}'} < 0)}} (\boldsymbol{k}|\, R\, |\boldsymbol{k}')(\hat{\boldsymbol{n}} \cdot \boldsymbol{V}_{\boldsymbol{k}'})\, \delta f_{\boldsymbol{k}}^{(-)}. \tag{10.174}$$

Here $\boldsymbol{V}_{\boldsymbol{k}} = \boldsymbol{\nabla}_{\boldsymbol{k}} E_{\boldsymbol{k}}$ is the BQP velocity, $\hat{\boldsymbol{n}}$ is the surface normal, and the $\boldsymbol{k}'$ sum is restricted to states with $\hat{\boldsymbol{n}} \cdot \boldsymbol{V}_{\boldsymbol{k}'} < 0$. The transition probability $(\boldsymbol{k}|\, R\, |\boldsymbol{k}')$ satisfies the detailed-balance relation

$$|\hat{\boldsymbol{n}} \cdot \boldsymbol{V}_{\boldsymbol{k}'}|\, (\boldsymbol{k}|\, R\, |\boldsymbol{k}') = |\hat{\boldsymbol{n}} \cdot \boldsymbol{V}_{\boldsymbol{k}}|\, (-\boldsymbol{k}'|\, R\, |-\boldsymbol{k}), \tag{10.175}$$

as well as the sum rule valid for elastic processes

$$\sum_{\substack{\boldsymbol{k} \\ (\hat{\boldsymbol{n}} \cdot \boldsymbol{V}_{\boldsymbol{k}} > 0)}} (\boldsymbol{k}|\, R\, |\boldsymbol{k}') = 1, \tag{10.176}$$

which ensures quasiparticle conservation.

For the normal Fermi liquid as well as for an isotropic superfluid (B phase), the slip length ζ may be calculated exactly by assuming diffuse boundary scattering and employing a relaxation-time approximation of the collision integral that is consistent with momentum conservation (Einzel *et al.* 1983). For the normal Fermi liquid, one finds $\zeta = 0.579l$, where $l = v_{\mathrm{F}}\tau$. For more general types of scattering, variational methods are useful, allowing one to obtain bounds on the slip length.

For example, a lower bound on the slip length, which is within one per cent of the exact solution for diffuse scattering, is given by

$$\zeta > \frac{(L_2 - K_{21})(L_2 - K_{12}) + (L_1 + K_{11})(L_3 - K_{22})}{2L_2(L_1 + K_{11})}, \tag{10.177}$$

where

$$L_n = \hbar^2 \sum_{\boldsymbol{k}}{}' (-f'_{\boldsymbol{k}}) k_x^2 V_{\boldsymbol{k},z} (\tau V_{\boldsymbol{k},z})^{n-1},$$

$$K_{mn} = \hbar^2 \sum_{\boldsymbol{k}}{}' \sum_{\boldsymbol{k}'}{}' (-f'_{\boldsymbol{k}'}) k_x k'_x V_{\boldsymbol{k}',z} (\tau V_{\boldsymbol{k},z})^{m-1} (\tau V_{\boldsymbol{k}',z})^{n-1} (\boldsymbol{k}|\, R\, |-\boldsymbol{k}'),$$

and the prime on the summation sign indicates $V_{\boldsymbol{k},z} > 0$ or $V_{\boldsymbol{k}',z} > 0$.

10.4.2 Andreev reflection

At a boundary the order parameter is distorted within a layer of thickness $\xi(T)$, the coherence length. This leads to a modification of the scattering of BQP off a boundary. In conventional superconductors this effect may occur at a normal–superconducting interface and will lead to important changes in the transport properties, as first discussed by Andreev (1964). Indeed, the thermal conductivity of a superconducting metal in the intermediate state is dominated by the scattering of quasiparticles off the interfaces between normal and superconducting regions, at which the gap parameter tends to zero within a distance $\xi(T)$. A similar effect occurs at a boundary in superfluid ^{3}He, where the gap parameter varies in space.

The gap distortion induces an interaction of the BQP with the boundary. This effect, called Andreev reflection, acts in addition to the microscopic scattering process right at the wall. It is similar to the reflection of a quantum-mechanical particle by a potential threshold in the case of a particle approaching the threshold from the top of the step. The Bogoliubov quasiparticle is turned back by means of an elastic transition from the particle-like excitation branch with $\xi_{\mathbf{k}} > 0$ to the hole-like excitation branch with $\xi_{\mathbf{k}} < 0$ (or vice versa), thereby reversing its group velocity $\nabla_{\mathbf{k}} E_{\mathbf{k}} = (\xi_{\mathbf{k}}/E_{\mathbf{k}})\mathbf{k}/m^*$. On the other hand, its momentum is hardly affected at all in this process and is changed by only a negligible amount $\hbar\,\Delta k \approx \hbar/\xi$, where ξ is the coherence length. Note that, since there is essentially no momentum transfer to the wall in an Andreev reflection process, there is no contribution to friction.

In order to discuss the importance of Andreev reflection, we need to know the actual distortion of the gap. In an ordinary superconductor the gap function does not change much at the boundary of the superconductor with an insulator. This is a consequence of the pairing of time-reversed states and of the isotropy of the gap parameter in an s-wave superconductor (Anderson 1960). In a p-wave superfluid, however, the situation is quite different, because scattering of the two partners of a Cooper pair by the boundary carries the quasiparticles into all directions on the Fermi sphere and thereby tends to wash out the intrinsic anisotropy of the gap parameter associated with any non-s-wave pairing. Hence boundary scattering in the p-wave superfluid ^{3}He acts as a pair-breaking mechanism, as does impurity scattering in general.

Indeed, it has been shown that specular scattering of quasiparticles from a plane boundary leads to complete suppression of the orbital components of the gap parameter parallel to the boundary normal (Ambegaokar *et al.* 1974). This is the reason for the strong orientation of the vector $\hat{\boldsymbol{l}}$ in the A phase perpendicular to a boundary surface (see Section 6.3.6). Intuitively, such a result is quite plausible. Using a classical picture of two bound particles orbiting about their centre of mass, it is clear that the orbital plane will tend to orient itself parallel to a nearby surface in order to avoid bumping into the wall (Leggett 1975a). Even for the more realistic case of diffuse boundary scattering, one finds a considerable, although not complete, suppression of the parallel component (Buchholtz and Rainer 1979, Buchholtz and Zwicknagl 1981, Zhang *et al.* 1985, 1987, 1988, Buchholtz 1986; see also Fal'ko, 1985). The region of suppressed order parameter near a wall forms a potential well in which bound states of quasiparticles may exist (Privorotskii 1977). An experimental determination of the order-parameter healing length has been reported by Ichikawa *et al.* (1987); see also Chainer *et al.* (1984).

Given the gap profile, the probability for Andreev reflection, $R(E)$, may be expressed in terms of the usual quasiparticle amplitudes $u_{\boldsymbol{k}}$ and $v_{\boldsymbol{k}}$, defined in (3.18), which determine the reflected and incident currents. Using (3.16), it may be readily shown that the stationary "quasiclassical Schrödinger equations" for $u_{\boldsymbol{k}}$ and $v_{\boldsymbol{k}}$, known as the Andreev equations, take the form (Andreev 1964)

$$\left.\begin{aligned}\left(E_{\boldsymbol{k}} - \mathrm{i}v_{\mathrm{F}}\frac{\mathrm{d}}{\mathrm{d}z}\right)u_{\boldsymbol{k}}(z) &= \Delta(z)v_{\boldsymbol{k}}(z),\\ \left(E_{\boldsymbol{k}} + \mathrm{i}v_{\mathrm{F}}\frac{\mathrm{d}}{\mathrm{d}z}\right)v_{\boldsymbol{k}}(z) &= \Delta(z)u_{\boldsymbol{k}}(z),\end{aligned}\right\} \tag{10.178}$$

where z is the distance from the surface and $\Delta(z)$ is a given gap profile. In the case of diffuse scattering at the wall the transmitted quasiparticles are necessarily scattered into different $\boldsymbol{k}$ states, which means that they are no longer able to interfere constructively. Such a process may be modelled by assuming the quasiparticles to propagate from the threshold to infinity. The differential equations (10.178) may then be solved easily. At the gap edge, i.e. for $\xi_{\boldsymbol{k}} \to 0$, total reflection is found to occur, with corrections linear in $\xi_{\boldsymbol{k}}$. A useful interpolation expression is given by

$$R_{\mathrm{A}}(E) = \frac{E - \xi}{E - \alpha\xi}, \tag{10.179a}$$

with a coefficient $\alpha < 1$ depending on the details of $\Delta(z)$ (Kieselmann and Rainer 1983, Einzel *et al.* 1984b). A comprehensive treatment of Andreev scattering using the quasiclassical Green's-function formalism has been presented by Kurkijärvi and Rainer (1989).

The transition probability in the general boundary condition (10.174), which takes account of Andreev reflection, is thus given by

$$(\boldsymbol{k}|\, R\, |\boldsymbol{k}') = N_{\mathrm{F}}^{-1} R_{\mathrm{A}}(E_{\boldsymbol{k}})\delta(\hat{\boldsymbol{k}} - \hat{\boldsymbol{k}}')\delta(\xi_{\boldsymbol{k}} + \xi_{\boldsymbol{k}'}) . \tag{10.179b}$$

The expression (10.179) for the transition probability $(\boldsymbol{k}|\, R\, |\boldsymbol{k}')$ may conveniently be substituted into the general formulae for the slip length obtained by variational methods, and into similar expressions for the integrated Poiseuille flow, the surface impedance and other quantities. One finds that Andreev reflection leads to an increased slip length and a reduced effective viscosity.

Andreev reflection also occurs in *bulk* ^{3}He-A with nonuniform $\hat{\boldsymbol{l}}$ texture. Greaves and Leggett (1983a,b) predicted that if a quasiparticle arrives at a point where $E < |\Delta_{\hat{\boldsymbol{k}}}|$ because $\hat{\boldsymbol{l}}$ has changed its direction, it will be Andreev-reflected. Hence its momentum is virtually unchanged as it propagates across a texture.

Furthermore, the small momentum transfer occurring in the Andreev reflection process at the phase boundary between the bulk A and B phases (Yip 1985, Yip and Leggett 1986, Leggett and Yip 1989; see also Palmeri 1989) may explain the surprisingly high propagation velocity v_{AB} of the A–B interface observed experimentally (Buchanan *et al.* 1986). A somewhat similar theory of the dynamics of the interface was developed by Kopnin (1987), while Markelov (1987) used a hydrodynamic approach. A different aspect of the moving A–B interface has been studied by Salomaa (1988), who finds that the moving superfluid A–B interface undergoes a phase transition associated with a change in the discrete symmetry and topology of the energy gap (see also Salomaa and Volovik 1988). Such a transition should lead to a discontinuity in $\mathrm{d}v_{AB}/\mathrm{d}T$, i.e. in the slope of v_{AB}. This effect has not yet been seen.

10.4.3 Poiseuille flow

The simplest example of boundary effects on fluid flow is provided by stationary flow through a channel of constant cross-section. For example, in a parallel-plate geometry (of width d) the total flux is given in terms of the pressure gradient $\partial P/\partial x$ along the channel by

$$G = \frac{\partial P}{\partial x}\frac{d^3}{12\eta_{\mathrm{eff}}}. \qquad (10.180)$$

Here η_{eff} is an effective-viscosity coefficient; it reduces to the usual shear viscosity η in the limit $l/d \to 0$. The first correction in the so-called "Knudsen number" $Kn \equiv l_\eta/d$ may be calculated by employing the slip boundary condition (10.173). Here l_η is the viscous mean free path, defined by $l_\eta = l/[1 - \lambda_2\langle \xi^2/E^2\rangle_\xi/\langle 1\rangle_\xi]$, with l and $\lambda_2 \equiv \lambda^{\mathrm{s}}_{2+}$ defined by (10.92c) and (2.51) respectively. The corrected expression is given by

$$\frac{\eta}{\eta_{\mathrm{eff}}} = 1 + 6\frac{\zeta}{d}. \qquad (10.181)$$

For general Knudsen number, the total flow has been calculated using a variational solution of the kinetic boundary-value problem for normal and superfluid Fermi liquids assuming diffuse scattering (Højgaard Jensen *et al.* 1980; for the case of cylindrical geometry, see also Topsoe and Højgaard Jensen 1984). As expected, the scaled integrated flow G/d^2 is found to decrease with decreasing plate separation d, reach the so-called Knudsen minimum and rise again in the Knudsen regime ($d/l_\eta \ll 1$). At $d/l_\eta \lesssim 2$ the slip approximation deviates significantly from the full result for general Knudsen number. Experimentally, the flow of normal ^{3}He through a cylindrical tube was studied by Eisenstein *et al.* (1980a), who found reasonable agreement with the slip theory. Extending earlier experiments by Parpia *et al.* (1978) and Archie *et al.* (1981), Parpia and Rhodes (1983) measured the damping and frequency shift of a torsional oscillator with parallel-plate geometry, which also relates back to Poiseuille flow at the low frequencies used. In the normal state they found a slip length that at zero pressure is 40% smaller than the value calculated for diffuse surface scattering whereas at melting pressure good agreement with experiment is obtained (see Einzel and Parpia 1987a). They also found a minimum in the integrated flow, which was, however, displaced towards a higher value of d/l when compared with the theoretical curve. Deviations from hydrodynamic flow have also been observed in experiments on spin transport along narrow capillaries (Sachrajda *et al.* 1984).

In the superfluid the effective viscosity η_{eff} falls off rapidly with decreasing temperature as shown in Fig. 10.7 (Einzel and Parpia 1987a,b). Figure 10.7 also shows the dependence of η_{eff} on the Knudsen number $Kn = l_\eta/d$. This behaviour can be explained quantitatively at high pressures and semiquantitatively at lower pressures by assuming diffuse and Andreev scattering at the boundaries (Einzel and Parpia 1987a,b, Zhang and Kurkijärvi 1988). The effect of the additional scattering of the quasiparticles due to friction between the normal and superfluid components in rotating superfluid ^{3}He has been investigated by Sonin (1981). In narrow channels and at high flow

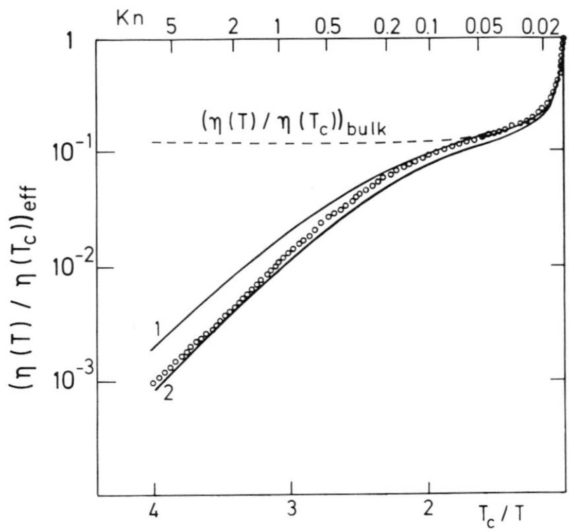

Figure 10.7 Effective-viscosity coefficient $\eta_{\text{eff}}(T)$, introduced in (10.180), versus T_c/T and Knudsen number $Kn = l_\eta/d$ at $P = 20$ bar. The full curves are the theoretical results for diffuse scattering only (curve 1) and for diffuse scattering plus Andreev scattering from a steplike order-parameter profile near the surface (curve 2). The open circles represent the data points. Also shown (dashed line) is the theoretical result when surface effects are neglected. (After Einzel and Parpia (1987a).)

rates critical velocities associated with pair breaking have been observed in the B phase (Eisenstein *et al.* 1979, Maninnen and Pekola 1982); see Sections 7.11 and 7.12).

10.4.4 Vibrating-wire experiments

The first experiments to determine the shear viscosity in superfluid ^{3}He were performed by measuring the response of a vibrating wire immersed in the liquid, i.e. the damping and resonant-frequency shift caused by the liquid (Alvesalo *et al.* 1974, 1975). The method has since been used extensively to study normal and superfluid ^{3}He (Carless *et al.* 1983a,b, Hall 1983) as well as dilute mixtures of ^{3}He in ^{4}He (Guénault *et al.* 1983a). The physics of a vibrating wire is characterized by three typical lengths: the radius R of the (cylindrical) wire, the viscous penetration depth $\delta = (2\eta/\rho_n\omega)^{1/2}$ (given by the wave vector for shear flow) and the mean free path l. In the limit $l \ll R$, hydrodynamics is valid. For this situation, the force per unit length exerted on the wire, including the slip correction of first order in l/R, has been calculated by Højgaard Jensen *et al.* (1980). A solution of the problem including the Knudsen regime ($l \ll R$) has not yet been given, although a variational theory may be formulated (Beyer 1986). The most recent measurements show that the vibrating-wire technique provides a sensitive probe of the quasiparticle dynamics even in the ballistic regime (Guénault *et al.* 1986, Coates *et al.* 1986). Extreme nonlinear damping by the quasiparticle gap was observed in these experiments (Carney *et al.* 1989). Distortion of the order parameter by a moving surface has also been detected in these experiments (Guénault *et al.* 1983b, Castelijns *et al.* 1986). This was inferred from the observation of a sharp critical velocity $v_{s,c}$, above which high dissipation sets in. Comparison with the bulk value $v_{s,c}^B$ of the depairing critical velocity showed $v_{s,c} \approx \frac{1}{2} v_{s,c}^B$, suggestive of a 50% depression of the gap parameter near the surface.

10.4.5 Sound propagation and other flow problems

When sound propagates in restricted geometries, its velocity and attenuation are affected by the interaction of the particles in the medium with the walls. In a normal fluid the friction of the sound wave at the walls of a channel of width d leads to a complete damping of sound for $d \ll \delta$, where δ is the viscous penetration depth. By contrast, in a superfluid system the superfluid component is still able to flow (provided that $d > \xi$, the coherence length) and may carry a density wave, so-called fourth sound (see Section 11.1). Since the viscous penetration depth increases rapidly with decreasing temperature, a transition from first sound to fourth sound may be achieved by simply cooling a given resonator. In the transition region the sound attenuation is expected to go through a maximum, while the sound velocity drops by a factor $(\rho_s/\rho)^{1/2}$ (Højgaard Jensen *et al.* 1983). The transition is particularly sensitive to the liquid–boundary interaction. It is plausible that in the regime $d/l \ll 1$ the sound dispersion relation is governed by the stationary Poiseuille flow properties (see (11.10)). On the other hand, for $d/l \gg 1$, only a thin layer at the walls of the cavity is disturbed by the boundary scattering, and the momentum transfer to the wall is described by the transverse surface impedance $Z_\perp = \Pi_{xz}/u_x^{\text{wall}}$, as described by (11.3).

The surface impedance is perhaps the physical quantity that reflects the properties of the boundary surface most directly. The transverse surface impedance has been calculated from an exact solution of the model kinetic equation for the B phase by applying a diffuse-scattering boundary condition (Einzel *et al.* 1983; see also Ebisawa 1980) and using a variational approach for a general scattering law (Einzel *et al.* 1987). Within the slip approximation (valid for $\omega\tau \ll 1$), the transverse surface impedance is readily found from the hydrodynamic equations as

$$Z_\perp \equiv \frac{\Pi_{xz}}{u_x^{\text{wall}}} = (1-\mathrm{i})\frac{\eta}{\delta}\left(1 - \frac{1-\mathrm{i}}{\delta}\zeta\right), \tag{10.182}$$

where $\delta = (2\eta/\rho_n\omega)^{1/2}$ is the viscous penetration depth and ζ is the slip length. The surface impedance in the A phase has been discussed by Hook and Zimmermann (1981).

Another flow situation in which boundary effects are very important occurs in a so-called U-tube oscillation experiment. There the two vertical columns of a U-shaped vessel are connected by a narrow channel, such that only superflow connects the two parts. When the level of liquid in one of the columns is raised by means of an electric field applied to the liquid, and if the field is then switched off suddenly, one might expect oscillations to occur about the equilibrium position of the liquid levels in both columns. However, experimentally, only a slow relaxation back to equilibrium is observed (Eisenstein and Packard 1982, Eisenstein 1983). This finding may be explained by taking into account superfluid-to-normal-fluid interconversion near the free surfaces and the friction of the normal flow at the walls of the vertical tubes (Brand and Cross 1982). From the hydrodynamic equations supplemented by the usual boundary conditions on the normal-fluid velocity field, $\boldsymbol{v}_n = \boldsymbol{v}_{\text{surface}}$ at the free surface and $\boldsymbol{v}_n = 0$ at the walls of the channel (plate separation d), one finds a pressure change associated with the velocity $\dot{x}$ of the moving free surface:

$$\delta P = \rho(12\eta\zeta_3)^{1/2}\frac{\dot{x}}{d}. \tag{10.183}$$

At lower temperatures, $T \ll T_c$, the mean free path is no longer small in comparison with the diameter of the tube. In this case the viscosity η must be replaced by the effective viscosity coefficient η_{eff} relevant for Poiseuille flow (Einzel and Parpia 1987a).

The boundary condition on the normal-fluid velocity component perpendicular to a boundary surface depends in principle on the properties of the superfluid itself, as well as on the nature of the surface. Normal-fluid flow carries entropy, and therefore the physical quantity that should determine $\boldsymbol{v}_n$ at a boundary is the heat current $\boldsymbol{j}Q$ through the boundary. As discussed in Chapter 9, the heat current in a superfluid has two additive contributions: a convective one and a diffusive one, i.e. $\boldsymbol{j}_Q = Ts\boldsymbol{v}_n - \kappa\boldsymbol{\nabla}T$. Thus, in order to determine the normal-fluid velocity from a given heat current through the boundary, one needs to know the temperature gradient at the boundary. The general hydrodynamic boundary conditions for an isotropic superfluid relating $\boldsymbol{j}_Q$ and the superfluid flow $\boldsymbol{g}_s$ to the temperature discontinuity ΔT and $\mathrm{d}\boldsymbol{v}_n/\mathrm{d}x$ at the boundary (surface normal parallel to $\hat{\boldsymbol{x}}$) are given by

$$\begin{pmatrix} \boldsymbol{j}_Q \\ \boldsymbol{g}_s \end{pmatrix} = \begin{pmatrix} aT & c\alpha T/\rho \\ c & b\alpha T/\rho \end{pmatrix} \begin{pmatrix} \Delta T\hat{\boldsymbol{x}} \\ \mathrm{d}\boldsymbol{v}_n/\mathrm{d}x \end{pmatrix}, \tag{10.184}$$

where

$$\alpha = \tfrac{4}{3}\eta - 2\zeta_1\rho + \zeta_2 + \zeta_3\rho^2, \tag{10.185}$$

and η, ζ_1, ζ_2 and ζ_3 are the viscosity coefficients of an isotropic superfluid defined in (9.28) and (9.23) (Grabinski and Liu 1987). For a description of a general boundary condition, three so-called "surface Onsager coefficients" a, b and c are required. The coefficients have been calculated from kinetic theory for the case of diffuse scattering at the boundary in ^{3}He-B by Sun *et al.* (1989), who find that the coefficient a dominates. This implies that, to a good approximation, $\boldsymbol{v}_n = 0$ and $\boldsymbol{j}_Q = -\kappa\boldsymbol{\nabla}T$ at the boundary and justifies the choice of boundary condition for the above U-tube oscillation experiment. By contrast, in superfluid ^{4}He one has $\boldsymbol{j}_Q = Ts\boldsymbol{v}_n$ and $\boldsymbol{\nabla}T = 0$ at the boundary. This drastic difference in behaviour is associated with the characteristic length λ governing the temperature variations in a superfluid in the presence of a stationary heat flow. From (9.22), (9.26) and (9.32), one finds that the spatial variation of the temperature and the normal-fluid velocity component along the direction of heat flow $\hat{\boldsymbol{x}}$ are given by (Saslow 1971)

$$T - T_\infty = T_s^- \mathrm{e}^{x/\lambda} + T_s^+ \mathrm{e}^{-x/\lambda}, \tag{10.186}$$

$$sv_n = \frac{\kappa}{T\lambda}(T - T_\infty) + \frac{j_Q}{T}; \tag{10.187}$$

i.e. there is a temperature jump at the boundary, followed by an exponential relaxation extending over a distance λ. The characteristic length λ can be expressed in terms of the transport coefficients as

$$\lambda = \left(\frac{\alpha\kappa}{s^2 T}\right)^{1/2}, \tag{10.188}$$

with α defined above. For superfluid ^{3}He, λ is of order $(T_F/T_c)l$, i.e. much longer than the mean free path l, whereas for superfluid ^{4}He, $\lambda \approx l$.

The temperature jump at the boundary in the case of ^{3}He is found to be negligibly small, and hence the so-called "Kapitza resistance" is also small. (In the case of ^{4}He the temperature jump and the exponential relaxation may be combined into a single effective temperature jump.) The heat resistance associated with a plane boundary is then given by $R_T \equiv \Delta T_W / j_Q = s^{-1}(\alpha/\kappa T)^{1/2}$, where ΔT_W is the difference in the temperatures of the superfluid far from the boundary and at the (solid) boundary. We note that the mean free path drops out of the ratio α/κ. In the B phase, where $s \propto e^{-\Delta/k_B T}$, the thermal boundary resistance is seen to increase exponentially at low temperatures, in agreement with experiment (Parpia 1985, Castelijns *et al.* 1985).

If the channel connecting the two reservoirs in a U-tube experiment is sufficiently narrow that only tunnelling of Cooper pairs is possible, one should expect the analogue of the Josephson effect to occur, as found for superconducting weak links (Josephson 1965, Anderson 1967). Josephson flow oscillations across an orifice connecting two reservoirs of ^{3}He-B may be induced by changing the pressure on one side. The optimal conditions for such an experiment, i.e. the shape and dimensions of the orifice, have been investigated theoretically by Monien and Tewordt (1986). These authors numerically solved the corresponding Ginzburg–Landau boundary-value problem for the order parameter and found the optimal length of the channel to be given by about seven Ginzburg–Landau coherence lengths. The Josephson flow through an aperture of radius R much less than the coherence length has been discussed by Kopnin (1986a).

An unambiguous observation of Josephson oscillations in superfluid ^{4}He was first reported by Avenel and Varoquaux (1985, 1986), who were able to identify individual phase-slip events. They were then also successful in observing near-ideal Josephson effects in ^{3}He-B at zero pressure (Avenel and Varoquaux 1987; for a discussion of a Josephson experiment see also Lounasmaa *et al.* 1983). These findings gave final proof for the actual existence of this important phenomenon peculiar to superfluids. Phase slippage has in fact also been measured in the *spin* supercurrent (Borovik-Romanov *et al.* 1987a,b, 1988). Hence one may expect that the corresponding Josephson effect is also observable (Markelov 1988).

10.4.6 Superfluidity in ^{3}He films

Following the first discussion of the effect of boundaries on the superfluidity of a ^{3}He film of thickness comparable to the coherence length by Ambegaokar *et al.* (1974), the variation of the transition temperature with the channel width between the plates was calculated in more detail by Kjäldman *et al.* (1978) (see also Tešanović and Valls 1986a,b). Thermodynamic stability of various phases in narrow channels had been studied earlier by Barton and Moore (1975b). As already mentioned, a realistic calculation of the effect of boundaries on the superfluid state requires formidable numerical effort, except for the simplest case of specular scattering and the somewhat more realistic case of diffuse scattering by the confining walls. An extensive treatment of the dynamical properties of superfluid ^{3}He films using the quasiclassical Green's-function approach has been given by Kopnin (1986b). The superfluid transition temperature of a film may even be higher than that of the bulk liquid owing to enhanced pair attraction in the vicinity of the surface. This hypothesis is not implausible, since the first few layers of ^{3}He at a solid surface are known to be nearly ferromagnetic (Ahonen *et al.* 1976a, 1977a, 1978a, Bozler *et al.* 1978, Godfrin *et al.* 1978). Theoretical models that account for the observed behaviour have been developed by Béal-Monod and Doniach (1977) and Béal-Monod and Mills (1978, 1980). In later experiments on superfluid ^{3}He films of 0.3 μm thickness confined between a stack of mylar sheets, it was found that the A phase is stabilized down to zero pressure and down to the lowest temperatures of about 0.5 mK (Freeman *et al.* 1988). Superflow in adsorbed films of thickness <0.5 μm has been studied by Davis *et al.* (1988). These authors determined the suppression of T_c with decreasing film thickness and found the critical current to scale with temperature as $(1 - T/T_c)^{3/2}$.

So far we have considered ^{3}He films with thicknesses of the order of the coherence length. These are still three-dimensional systems. A truly two-dimensional ^{3}He film, i.e. a monolayer or less of ^{3}He, is expected to form on the free surface of a dilute solution of ^{3}He in ^{4}He (Andreev 1966, Zinov'eva and Boldarev 1969). Estimates of the pair interaction indicate a p-wave superfluid transition (Edwards *et al.* 1977). The fascinating properties of such a two-dimensional Fermi liquid have been studied by Stein and Cross (1979), Fujita *et al.* (1980), Brusov and Popov (1981), Ashida and Takagi (1981), Miyake (1983), Tešanović (1984), Korshunov (1985) and Nakahara (1986). It is found that there exist two unitary phases corresponding to the A and B phases of the three-dimensional liquid, which are, however, energetically degenerate in the weak-coupling limit because both phases have isotropic energy gaps. Stein and Cross (1979) point out the possibility of an Ising-like phase transition in the "A" phase in addition to the Berezinski–Kosterlitz–Thouless vortex-binding transition expected for a two-dimensional superfluid (Berezinskii 1970, Kosterlitz and Thouless 1972, 1973, Kosterlitz 1974). Korshunov (1985) has calculated the strong-coupling corrections following Rainer and Serene (1976) and finds that the "A" phase is favoured over the "B" phase if the particle–hole scattering dominates over the particle–particle scattering, and vice versa. He also stresses that the symmetry group of the "A" phase is $U(1)/\mathbb{Z}_2$ because states differing by a rotation in the plane by π are equivalent (see the discussion on the topological classification of defects in the A phase in Section 7.4). As a consequence, the discontinuity in the superfluid density at the vortex-binding transition is four times the conventional value.

As discussed by Volovik and Yakovenko (1989), two-dimensional films of superfluid ^{3}He may exhibit novel properties that are due to the existence of a topological term ("Chern–Simons term") in the action of the corresponding quantum field theory (for a discussion of topological terms see Wilczek and Zee 1983). This term implies that in a film of ^{3}He-A, particle-like solitons in the $\hat{\boldsymbol{d}}$-vector field will obey either Fermi–Dirac or Bose–Einstein statistics, depending on the film thickness. Furthermore, Volovik and Yakovenko (1989) predict a specific quantized Hall effect for the spin current in ^{3}He-A. In the planar phase of the ^{3}He film the nonsingular 4π spin disclination is found to have a fractional fermion charge (equal to the number of particles contained in the defect) of $\frac{1}{2}n$ (with n an integer), while the spin disclination in the A_1 phase has both fractional spin $\frac{1}{4}\hbar n$ and fractional fermion charge $\frac{1}{4}n$. Most interestingly, these "particles" (spin disclination) in the A_1 phase obey "fractional statistics" with parameter $\theta = \frac{1}{2}\pi$, i.e. the state vector of the system changes by a factor $e^{i\pi/2}$ upon interchange of two "particles".

10.4.7 Ion mobility

The motion of ions in liquid ^{3}He is determined by their interaction with the Bogoliubov quasiparticles in the liquid. The order parameter is essentially unchanged because the ion diameter is much smaller than the coherence length. In an applied uniform electric field $\boldsymbol{E}$ the electron acquires an average drift velocity $\boldsymbol{v}$ proportional to the field (for sufficiently small fields):

$$\boldsymbol{v} = \mu_{\text{ion}}\boldsymbol{E}, \tag{10.189}$$

where μ_{ion} is the ion mobility. The mobility depends strongly on the effective mass of the ion. Indeed, there are only two kinds of ion that are not firmly bound at the container walls: (i) positive helium ions, which form a sort of solid "snowball" of mass $M^* \approx 30m$, and (ii) those negative ions consisting of a free electron in a bubble established by the electron's zero-point motion ($M^* \approx 300m$). In both cases the ionic diameter, being of order 10 Å, is much smaller than the quasiparticle mean free path.

The mobility is determined by the force on the ion moving with velocity $\boldsymbol{v}$ due to collisions with the quasiparticles in the liquid. This may be expressed by the Golden-Rule formula (Josephson and Lekner 1969)

$$\mu_{\text{ion}}^{-1} = \frac{e}{v^2}\sum_{\boldsymbol{kk}'} \hbar(\boldsymbol{k}-\boldsymbol{k}')\cdot\boldsymbol{v}W_{\boldsymbol{kk}'}f_{\boldsymbol{k}}(1-f_{\boldsymbol{k}'})S(\boldsymbol{q},\omega_{\boldsymbol{v}}), \tag{10.190}$$

where $W_{\boldsymbol{kk}'}$ is the scattering cross-section for quasiparticles with momentum $\boldsymbol{k}$ to be scattered into state $\boldsymbol{k}'$, $f_{\boldsymbol{k}}$ is the Fermi occupation factor and $\omega_{\boldsymbol{v}}$ is a frequency depending on the velocity $\boldsymbol{v}$ (see below). In contrast with the boundary problems discussed before, the mass of the moving object cannot be taken to be infinite in this case. Rather, the scattering process is inelastic owing to the finite recoil of the ion. This implies that the quasiparticle–ion interaction is retarded. The dependence of the transition probability on the transferred energy is given by the dynamical structure factor $S(\boldsymbol{q}, \omega)$ of the ion in the system moving with velocity $\boldsymbol{v}$, i.e. $\boldsymbol{q} = \boldsymbol{k} - \boldsymbol{k}'$ and $\hbar\omega = \hbar\omega_{\boldsymbol{v}} \equiv E_{\boldsymbol{k}} - E_{\boldsymbol{k}'} + \hbar\boldsymbol{v} \cdot \boldsymbol{q}$. In general, many quasiparticle collisions have to be taken into account within the characteristic recoil time. This is the hydrodynamic regime, in which the ion performs a random diffusive motion. The dynamical structure factor is then given by $S(\boldsymbol{q}, \omega) = \gamma_{\boldsymbol{q}}/(\omega^2 + \gamma_{\boldsymbol{q}}^2)$, where $\gamma_{\boldsymbol{q}} = Dq^2$ describes the diffusional broadening of the spectral function. It is very important to realize that the diffusion constant D is connected with the mobility μ_{ion} by way of the Einstein relation $D = (\mu_{\text{ion}}/e)k_{\text{B}}T$. This makes (10.190) a self-consistent equation for the mobility.

For the electron bubble, one may estimate $\gamma_{\boldsymbol{q}}$ to be much less than the thermal frequency $k_{\text{B}}T/\hbar$. This may be verified by substituting the experimental value for μ_{ion} and using the fact that typical momentum transfers $\hbar q$ are of order $\hbar k_{\text{F}}$. It implies that $S(\boldsymbol{q}, \omega)$ is a sharply peaked function on the scale of the occupation numbers $f_{\boldsymbol{k}}$ and may be replaced by an energy-conserving delta function, so that the scattering is essentially elastic (Fetter and Kurkijärvi 1977a,b, Baym *et al.* 1977a,b). The mobility in the normal state is then found to be temperature-independent, as observed experimentally (Ahonen *et al.* 1976b, 1977b).

For the lighter positive ion, a much higher mobility is measured experimentally (Roach *et al.* 1977, Ahonen *et al.* 1978b, Kokko *et al.* 1978b). In fact, the condition $\hbar\gamma_{\boldsymbol{q}} < k_{\text{B}}T$ for elastic scattering is violated below about 100 mK. A full solution of the self-consistent equation (10.190) yields an increasing mobility at lower temperatures, in agreement with experiment (Bowley 1977, Wölfle *et al.* 1980, Wölfle 1980; see also the diagrammatic theory of Kondo and Soda 1983, Soda 1984).

In the superfluid phases the mobility is expected to rise. Assuming the scattering to be elastic and the transition probability $W_{\boldsymbol{k}\boldsymbol{k}'}$ to be given by its value in the normal state (apart from coherence factors due to the Bogoliubov transformation), the expression (10.190) may easily be evaluated for the B phase; one finds (Bowley 1976; see also Soda 1975, 1977b)

$$\frac{\mu_{\text{ion}}(T)}{\mu_{\text{ion}}^{\text{N}}} = \tfrac{1}{2}(\mathrm{e}^{\Delta/k_{\text{B}}T} + 1). \tag{10.191}$$

The increase in μ_{ion} near T_{c}, proportional to Δ, is clearly seen in the experiment, but the agreement is not quantitative (cf. the experimental references cited above). In fact, the Born approximation is not sufficient to describe the quasiparticle scattering off the ion, because of the diverging density of states $N(E) \propto E/\xi$. An exact calculation of the scattering amplitude for elastic scattering shows that the differential cross-section is strongly peaked in the forward direction (Baym *et al.* 1977a,b, 1979,

Salomaa *et al* 1980a,b, Salomaa 1982). At the same time, the total cross-section remains roughly at its normal-state value, implying that the rate of momentum transfer is reduced (see also Bromley 1981).

The interaction of ions with vortex structures is of considerable interest since ions are a useful probe of both the structure of individual vortices and a vortex lattice. Calculations of the interaction potential of ions and singular vortex cores have been performed by Rainer and Vuorio (1977), Thuneberg *et al.* (1981a,b, 1982) and Mineev and Salomaa (1984) for the o- and v-vortices of the B phase. It appears that ions are trapped by the vortex cores as in the more familiar case of superfluid ^{4}He (Donnelly 1965). An ion-mobility experiment would therefore provide information about the thermal excitations in the vortex core region and hence the local order parameter. There may even exist bound quasiparticle states, giving rise to resonant interaction processes (Thuneberg *et al.* 1981b).

In the A phase the mobility is a tensor quantity characterized by two eigenvalues, $\mu_{\parallel}$ and $\mu_{\perp}$, for motion parallel and perpendicular to the gap axis $\hat{\boldsymbol{l}}$. Since at low temperatures thermally excited quasiparticles are concentrated in the regions on the Fermi surface around $\hat{\boldsymbol{k}} = \pm\hat{\boldsymbol{l}}$, ions travelling along $\hat{\boldsymbol{l}}$ are much more effectively scattered than those moving perpendicular to $\hat{\boldsymbol{l}}$, i.e. $\mu_{\perp} > \mu_{\parallel}$. The difference $\mu_{\perp} - \mu_{\parallel}$ increases with decreasing temperature. This qualitative picture is borne out by detailed calculations (Salomaa *et al.* 1980a,b, Salmelin and Salomaa 1987a,b) as well as by experimental observation (Ahonen *et al.* 1976b, 1977b, 1978b, Roach *et al* 1977).

Ion-mobility experiments are a powerful tool for investigating the $\hat{\boldsymbol{l}}$ textures in the A phase, as the anisotropy of the mobility tensor leads to a focusing effect similar to the focusing of light in a medium with spatially varying index of refraction. Since the mobility is lowest for ions moving along $\hat{\boldsymbol{l}}$, the path of an ion will tend to follow the "potential lines" of $\hat{\boldsymbol{l}}$ in a given texture. This effect has been exploited for the investigations of vortex structures in rotating ^{3}He-A. As discussed in Section 7.6, one of the candidate states, a nonsingular vortex structure, has $\hat{\boldsymbol{l}}$ in the plane perpendicular to the axis far from the vortex core, flaring out along the axis at the vortex-core centre. Ions drifting along the rotation axis are expected to be captured by the vortex cores, where they move more slowly. This effect has been seen (Simola *et al.* 1986), and provides an important clue to the nature of vortex cores in rotating superfluid ^{3}He-A (see the discussion in Section 7.6). In later experiments two types of vortices were identified (Nummila *et al.* 1989).

As discussed by Salmelin *et al.* (1989a), a moving ion will feel the orbital angular momentum of the coherently aligned Cooper pairs in the A phase. This leads to a measurable "internal Magnus effect" on the ion, manifesting itself as an asymmetry of the associated quasiparticle scattering.

Pronounced nonlinear behaviour of the mobility is observed in the B phase at ion velocities above about 5–10 cm/s (Ahonen *et al.* 1976b). This may be associated with pair-breaking processes, which set in at recoil energies $\omega_{\text{rec}} = \hbar \boldsymbol{q} \cdot \boldsymbol{v}$ exceeding the threshold value 2Δ. The corresponding velocity of the ion ("Landau critical velocity") is given by $v_c = \Delta/p_F$, as discussed in Section 7.11.

The nonlinear dependence of μ_{ion} on the drift velocity has been considered by Fetter and Kurkijärvi (1977a,b).

FURTHER READING

Review articles

Brinkman W F and Cross M C 1978 in *Progress in Low Temperature Physics,* Vol. VIIA, ed. D F Brewer (North-Holland, Amsterdam), p. 105

Hall H E and Hook J R 1986 in *Progress in Low Temperature Physics,* Vol. IX, ed. D F Brewer (North-Holland, Amsterdam), p. 143

Kurkijärvi J and Rainer D 1989 in *Helium 3,* ed. W P Halperin and L P Pitaevskii (North-Holland, Amsterdam)

Leggett A J and Yip S 1989 in *Helium 3,* ed. W P Halperin and L P Pitaevskii (North-Holland, Amsterdam)

Salomaa M M 1982 *Contemp. Phys.* **23** 169

Serene J W and Rainer D 1983 *Phys. Rep.* **101** 221

Smith H 1984 *Physica* **126B + C** 267 (Proceedings of the 17th International Conference on Low Temperature Physics, LT-17)

Smith H 1987 in *Progress in Low Temperature Physics,* Vol. XI, ed. D F Brewer (North-Holland, Amsterdam), p. 75

Wölfle P 1978 in *Progress in Low Temperature Physics,* Vol. VIIA, ed. D F Brewer (North-Holland, Amsterdam), p. 191

Wölfle P 1979 *Rep. Prog. Phys.* **42** 269

Articles

Betbeder-Matibet O and Nozières P 1969 *Ann. Phys.* (*NY*) **51** 392

Combescot R 1974 *Phys. Rev.* **A10** 1700

Hara J and Nagai K 1979 *J Low Temp. Phys.* **34** 351

Leggett A J and Takagi S 1977 *Ann. Phys.* (*NY*) **106** 79

Nagai K 1980 *J. Low Temp. Phys.* **38** 677

Pethick C J and Smith H 1977 *Physica* **90B + C**, 107

11

Collective Modes

The dynamic behaviour of a many-body system on a macroscopic scale is governed by its collective modes. These modes involve a weakly damped and consequently long-lived coherent motion of a large fraction of particles in the system. In a normal liquid, for example, there exists one well-defined collective mode: ordinary longitudinal sound. By contrast, superfluid ^{3}He can support a number of different collective modes, some of which are of a type never before encountered in nature. The existence of these modes is intimately related to the peculiar type of macroscopic order present in the superfluid phases of ^{3}He.

Quite generally the collective modes can be grouped into two categories: (i) the so-called gapless modes (or "massless modes", as they are called in the general context of field theory), whose eigenfrequency tends to zero as the wavelength becomes infinitely long ($\boldsymbol{q} = 0$); and (ii) modes with nonzero eigenfrequency, which have a gap in the dispersion relation for $\boldsymbol{q} = 0$ ("massive modes"). According to a theorem of Goldstone (1961), in a system with short-range interactions a gapless collective (so-called "Goldstone") mode appears whenever a continuous symmetry is broken in the ordered state. This may be easily understood if one recalls that in the case of a spontaneously broken continuous symmetry the free energy is degenerate with respect to the spontaneously preferred value of the associated symmetry variable (see Section 6.1). Thus a long-wavelength excitation of the symmetry variable does not involve the order-parameter structure itself and only costs the bending energy of the order-parameter field, which is proportional to the square of the wavenumber q. By contrast, the massive modes do involve the internal structure of the order parameter. In the case of superfluid ^{3}He they may be envisaged as phase-coherent oscillations of the Cooper pairs about the equilibrium configuration. The vibration frequency may be expected to be of the order of the binding energy of a Cooper pair, $\omega \approx \Delta/\hbar$.

In the next section we consider the collective modes in the hydrodynamic regime ($\omega\tau \ll 1$). Since the collision rate τ^{-1} is much less than Δ except for a tiny region close to T_c, and since $\omega\tau \ll 1$, only gapless modes are accessible in this regime. However, gaps may appear in the collective mode spectra owing to symmetry-breaking terms introduced by a magnetic field (reducing rotational symmetry in spin space) and the dipole interaction (lifting spin–orbit symmetry).

In Section 11.2 we present a group-theoretical classification of the order-parameter collective modes, which provides a framework for microscopic calculations of the collective-mode frequencies performed in Section 11.3. In Section 11.4 the question of observability of these modes is addressed. We shall present the detailed theory for the currently most powerful detection method, namely ultrasound propagation.

11.1 HYDRODYNAMIC MODES

The hydrodynamic modes are obtained as eigensolutions of the linearized hydrodynamic equations. Consequently the number of modes equals the number of hydrodynamic variables. Not all of these modes are well defined, in the sense of long-lived wavelike excitations. Some are overdamped or even purely diffusive. Let us begin with the sound modes familiar from isotropic superfluids.

11.1.1 Sound modes

In an ordinary liquid the sound mode is a density wave. In a superfluid the density is not only coupled to the longitudinal component of the normal velocity, but also to the superfluid velocity (which in the B phase, as well as in the A phase with uniform $\hat{\boldsymbol{l}}$ texture, has only a longitudinal component, i.e. is curl-free) and to the temperature. This follows from the hydrodynamic equations for the B phase (2.21), (9.22), (9.26) and (9.32). The two types of eigenmodes of these four equations are given by (i) a simultaneous oscillation of normal and superfluid components ($\boldsymbol{v}_n = \boldsymbol{v}_s$), so-called first sound, and (ii) by a counterphase oscillation with total mass current density $\boldsymbol{g} = \rho_n \boldsymbol{v}_n + \rho_s \boldsymbol{v}_s = 0$, so-called second sound (Wölfle 1973b; see also Saslow 1973, 1976a).

The dispersion law of first sound is found as

$$\omega^2 = c_1^2 q^2 \left(1 - \tfrac{4}{3} \mathrm{i}\omega \frac{\bar{\eta}}{\rho c_1^2}\right). \qquad (11.1a)$$

Here contributions due to second viscosity and the coupling to entropy fluctuations, which are both of higher order in T/T_F, have been neglected. In (11.1a) $c_1 = (\partial P/\partial \rho)^{1/2}$ is the ordinary sound velocity, which may be expressed in terms of Fermi-liquid parameters as

$$c_1^2 = \tfrac{1}{3} v_F^2 (1 + F_0^s)(1 + \tfrac{1}{3} F_1^s), \qquad (11.1b)$$

which is identical with the normal-state result (2.20). The damping parameter $\bar{\eta}$ is equal to the shear viscosity η in the B phase.

In the A phase with uniform $\hat{\boldsymbol{l}}$ texture $\bar{\eta}$ is given by the following combination of viscosity coefficients $\eta^{\perp}$, $\eta^{\parallel}$ and $\eta_{\mathrm{L}}^{\parallel}$ defined in (9.82):

$$\bar{\eta} = \eta_{\mathrm{L}}^{\parallel} \cos^4 \theta + (3\eta^{\parallel} - \eta_{\mathrm{L}}^{\parallel}) \sin^2 \theta \cos^2 \theta + \tfrac{1}{4}(3\eta^{\perp} + \eta_{\mathrm{L}}^{\parallel}) \sin^4 \theta. \quad (11.2)$$

Here θ is the angle between the preferred direction $\hat{\boldsymbol{l}}$ and the sound-propagation direction $\hat{\boldsymbol{q}}$. At low temperatures the sound attenuation α, which is proportional to $\bar{\eta}$, becomes very anisotropic. Using the low-temperature limiting behaviour of $\eta^{\perp}$, $\eta^{\parallel}$ and $\eta_{\mathrm{L}}^{\parallel}$ calculated in (10.161), one finds a maximum $\alpha_{\max} \propto (T_{\mathrm{c}}/T)^2$ for $\hat{\boldsymbol{q}} \parallel \hat{\boldsymbol{l}}$ and a minimum $\alpha_{\min} = \mathrm{const}$ for $\theta = \cos^{-1} 3^{-1/2} \approx 55°$ (Combescot 1975b).

From (11.1), it might appear as if first-sound attenuation measurements would be an ideal probe for determining the shear-viscosity coefficients. Unfortunately, as one goes to lower temperatures, the quasiparticle collision time increases rapidly (see Section 10.2), so that the hydrodynamic regime is limited to rather low frequencies and long wavelengths.

Experimentally, the dissipative losses caused by the interaction of the liquid with the wall dominate the first-sound attenuation even at the lowest frequencies attainable for a given resonator geometry (Eska *et al.* 1980, 1983; see also Kodama and Kojima 1981). In this case the sound dispersion for a parallel-plate resonator of width d is given by (Nagai and Wölfle 1981)

$$\omega^2 = c_1^2 q^2 \left[1 - \mathrm{i}\frac{2Z_{\perp}(\omega)}{\omega \rho d}\right], \quad (11.3)$$

where $Z_{\perp}(\omega)$ is the surface impedance discussed in Section 10.5 (see (10.182) *et seq.*).

The dispersion law of second sound is obtained as

$$\omega^2 = c_2^2 q^2 (1 - \mathrm{i}\omega\tau_2), \quad (11.4)$$

where the velocity c_2 and the relaxation time τ_2 are given by

$$c_2^2 = \frac{s^2 T}{\rho C_V}\overline{\left(\frac{\rho_{\mathrm{s}}}{\rho_{\mathrm{n}}}\right)}, \quad (11.5)$$

$$\tau_2 = \frac{C_V}{s^2 T}\left[\tfrac{4}{3}\bar{\eta} + \rho^2 \zeta_3 + \frac{\rho\bar{\kappa}}{C_V}\Big/\overline{\left(\frac{\rho_{\mathrm{s}}}{\rho_{\mathrm{n}}}\right)}\right]. \quad (11.6)$$

Here s and ρ are the entropy and mass density, C_V is the specific heat, and ρ_{s} and ρ_{n} are the densities of the superfluid and normal component. The bar is used to denote the component of the respective tensor quantity in the direction of sound propagation $\hat{\boldsymbol{q}}$. For example, $\bar{\eta}$ is an average of the shear viscosity as defined by (11.2) for the anisotropic A phase and $\bar{\kappa} = \hat{\boldsymbol{q}}\boldsymbol{\kappa}\hat{\boldsymbol{q}}$ is the relevant component of the thermal-conductivity tensor. In (11.6) ζ_3 is the second-viscosity coefficient.

Second sound is essentially a wave-like excitation of the entropy density. Since the entropy of the Fermi liquid ^{3}He is smaller than that of the Bose liquid ^{4}He by a factor of T/T_{F}, the second-sound velocity turns out to be about three orders of magnitude smaller than the Fermi velocity. The damping, on the other hand, is quite large. Although this makes second sound difficult to observe, it has been detected by Lu and Kojima (1985). In the limit $T \to T_c$ the term in τ_2 involving the thermal conductivity diverges as $(T_c - T)^{-1}$. This divergence is removed in the dispersion law (11.4) by the factor c_2^2, so that (11.4) goes over into the dispersion law for thermal diffusion for $T \geqslant T_c$:

$$\omega = -\mathrm{i} D_{\mathrm{T}} q^2 \tag{11.7}$$

where $D_{\mathrm{T}} = \kappa / C_V$.

Fortunately, there is a more effective way to induce oscillating superflow. Suppose that the normal-fluid fraction is clamped by the scattering of quasiparticles with the surface of a very narrow channel, as for example in a porous medium. The superfluid fraction is barely affected by the confining walls, provided that the channel diameter is greater than the coherence length $\xi(T)$, and thus may move freely. The oscillatory motion of the superfluid in such a confined geometry (superleak) is called fourth sound. The dispersion law of this collective mode is obtained from the continuity equation and the Josephson equation upon putting the normal-fluid velocity equal to zero. Note that the momentum conservation law does not hold here since momentum is continuously transferred from the quasiparticle system to the walls. One finds the linear dispersion law

$$\omega^2 = c_4^2 q^2 (1 - \mathrm{i}\omega\tau_4), \tag{11.8a}$$

where the velocity of fourth sound is given by

$$c_4^2 = c_1^2 \bar{\rho}_{\mathrm{s}} / \rho. \tag{11.8b}$$

Thus the ratio of the velocities of fourth and first sound is given by the square root of the superfluid density fraction, which vanishes as $(1 - T/T_c)^{1/2}$ for $T \to T_c$ and tends to unity for $T \to 0$. Here $\bar{\rho}_{\mathrm{s}}$ is the diagonal component of the superfluid density tensor in the direction of sound propagation $\hat{\boldsymbol{q}}$.

The damping of fourth sound is determined by the relaxation time τ_4. In the B phase and in the uniform A phase τ_4 is given by

$$\tau_4 = \rho \zeta_3 / c_1^2, \tag{11.9}$$

where ζ_3 is the coefficient of second viscosity appearing in the Josephson equations (9.22) and (9.84). Fourth sound has been observed experimentally in a number of experiments (Yanof and Reppy 1974, Kojima *et al.* 1974, 1975, Chainer *et al.* 1980). Measurements of the fourth-sound velocity provide a convenient technique for determining the superfluid density component ρ_{s}. In principle, it should be possible to determine the second-viscosity coefficient from the damping of fourth sound. In practice, however, this intrinsic contribution to the damping is often obscured by the so-called "normal-fluid slip", i.e. the effect of a small residual motion of the normal-fluid component.

This contribution may be related to the effective viscosity describing

Poiseuille flow through narrow channels discussed in Section 10.4 (see (10.178) *et seq.*). For a parallel-plate geometry the dispersion relation of fourth sound including viscous losses at the channel walls for arbitrary ratio of channel width d to mean free path is found as (Højgaard Jensen *et al.* 1983)

$$\omega^2 = c_4^2 q^2\left(1 - \mathrm{i}\omega\tau_4 - \frac{\mathrm{i}}{6}\frac{\rho_\mathrm{n}}{\rho_\mathrm{s}}\frac{d^2}{\delta_\mathrm{eff}^2}\right). \tag{11.10}$$

Here $\delta_\mathrm{eff} = (2\eta_\mathrm{eff}/\rho_\mathrm{n}\omega)^{1/2}$ is the effective viscous penetration depth and η_eff has been discussed in Section 10.3, (10.180). The additional damping contribution deriving from the last term in (11.10) comes close to, but is still somewhat less than, what is observed experimentally (Yanof and Reppy 1974, Chainer *et al.* 1980).

Another possibility of observing a kind of second sound is as entropy waves in the A_1 phase. Since superfluid mass current and spin current are identical in ^{3}He-A_1, a superfluid density wave associated with second sound is equivalent to a spin wave. It is therefore stabilized by the spin stiffness, which is much stronger than the entropy stiffness. This mode will be discussed at the end of Section 11.1.2.

11.1.2 Spin-wave modes

The fact that in superfluid ^{3}He the Cooper pairs carry spin leads to well-defined spin waves in both the A and B phases. Crudely speaking, the spin waves owe their existence to the stiffness of the order parameter against local rotations in spin space. The spin waves are the Goldstone modes associated with the spontaneously broken rotational symmetry in spin space. While this is strictly true for the A phase, where one has a preferred direction $\hat{\boldsymbol{d}}$ in spin space, in the B phase only the relative spin–orbit symmetry breaking is relevant for the dynamics (see Chapters 6 and 9). The spectrum of spin waves has been discussed by Brinkman and Smith (1974), Combescot (1974a,b), Maki (1973, 1975b, 1976b), Maki and Ebisawa (1974a,b,c), Maki and Tsuneto (1974b, 1975), Graham and Pleiner (1975), Pleiner and Graham (1976) and Pleiner (1977a,b).

The equations of motion governing the dynamics of spin waves are the spin conservation law and the Josephson equations for the spin phase variables θ^s and θ^so respectively. Using (8.8), (9.29), (9.110) and (10.60), we may write the equation of motion for the spin-density components S_μ as

$$\frac{\mathrm{d}S_\mu}{\mathrm{d}t} - \gamma(\boldsymbol{S}\times\boldsymbol{H}_0)_\mu + \sum_\nu \boldsymbol{\nabla}\boldsymbol{\rho}^\mathrm{sp}_{\mu\nu}\boldsymbol{v}_{\mathrm{sp},\nu} + \boldsymbol{\nabla}\cdot\boldsymbol{X}^\mathrm{D}_\mu = R_{\mathrm{D},\mu}, \tag{11.11}$$

where $\boldsymbol{\rho}^\mathrm{sp}$ is the superfluid spin-density tensor, $\boldsymbol{R}_\mathrm{D}$ is the dipole torque and $\boldsymbol{X}^\mathrm{D}_\mu$ is the dissipative part of the spin current, for example the spin diffusion current. The equation of motion for the spin superfluid velocity $\boldsymbol{v}_{\mathrm{sp},\mu}$ takes the general form

$$\frac{\mathrm{d}}{\mathrm{d}t}\boldsymbol{v}_{\mathrm{sp},\mu}+\frac{1}{2m}\boldsymbol{\nabla}(\omega^{\mathrm{s}}_{\mu}+\Omega^{\mathrm{D}}_{\mu})=0, \tag{11.12}$$

where $\boldsymbol{\nabla}\Omega^{\mathrm{D}}_{\mu}$ is the dissipative force.

Substituting (11.12) into (11.11) and neglecting dissipative effects, one finds the spin-wave equation of motion

$$\frac{\mathrm{d}^2 S_{\mu}}{\mathrm{d}t^2}-\frac{\gamma^2\hbar}{2m\chi}\boldsymbol{\nabla}\boldsymbol{\rho}^{\mathrm{sp}}_{\mu\nu}\boldsymbol{\nabla}S_{\mu}=\frac{\mathrm{d}}{\mathrm{d}t}R_{\mathrm{D},\mu}. \tag{11.13}$$

The A phase

In the A phase the spin superfluid density tensor is given (see (10.60)) by

$$\boldsymbol{\rho}^{\mathrm{sp}}_{\mu\nu}=(\delta_{\mu\nu}-\hat{d}^{0}_{\mu}\hat{d}^{0}_{\nu})\boldsymbol{\rho}^{\mathrm{sp}}, \tag{11.14}$$

where $\hat{\boldsymbol{d}}^0$ is the equilibrium $\hat{\boldsymbol{d}}$ vector and the components of $\boldsymbol{\rho}^{\mathrm{sp}}$ parallel and perpendicular to $\hat{\boldsymbol{l}}$ are

$$\rho^{\mathrm{sp}}_{\parallel,\perp}=\frac{1+\frac{1}{3}F^{\mathrm{a}}_1}{1+\frac{1}{3}F^{\mathrm{s}}_1}\frac{1-Y_{\parallel,\perp}}{1+\frac{1}{3}F^{\mathrm{a}}_1 Y_{\parallel,\perp}}S_0, \tag{11.15}$$

where $Y_{\parallel,\perp}=3\langle\hat{k}^2_{\parallel,\perp}Y_0(\hat{\boldsymbol{k}};T)\rangle_{\hat{\boldsymbol{k}}}$ and $S_0=\frac{1}{2}\hbar n$ is the saturation spin polarization. Here $\hat{k}_{\parallel,\perp}$ are the components of $\hat{\boldsymbol{k}}$ parallel and perpendicular to $\hat{\boldsymbol{l}}$, and $Y_0(\hat{\boldsymbol{k}};T)$ is the Yosida function defined in (3.88). In the A phase the dipole torque term on the right-hand side of (11.13) takes the form (see Section 8.2)

$$\frac{\mathrm{d}\boldsymbol{R}_{\mathrm{D}}}{\mathrm{d}t}=\Omega^2_{\mathrm{A}}[\delta\boldsymbol{S}-\hat{\boldsymbol{d}}_0(\delta\boldsymbol{S}\cdot\hat{\boldsymbol{d}}_0)], \tag{11.16}$$

where $\hat{\boldsymbol{d}}_0$ is the equilibrium $\hat{\boldsymbol{d}}$ vector (usually oriented along $\hat{\boldsymbol{l}}$) and $\delta\boldsymbol{S}=\boldsymbol{S}-\gamma^{-1}\chi\boldsymbol{H}$.

On Fourier-transforming (11.13), one obtains the dispersion law for spin waves of longitudinal and transverse polarization relative to the magnetic field:

$$\omega^2_{\parallel}=\Omega^2_{\mathrm{A}}+c^2_{\mathrm{sp}}(\hat{\boldsymbol{q}})q^2, \tag{11.17a}$$

$$\omega^2_{\perp}=\omega^2_{L}+\omega^2_{\parallel}. \tag{11.17b}$$

where the spin-wave velocity c_{sp} is given by

$$c_{\mathrm{sp}}(\hat{\boldsymbol{q}})=\left(\frac{\hbar\gamma^2}{2m\chi}\hat{\boldsymbol{q}}\boldsymbol{\rho}^{\mathrm{sp}}\hat{\boldsymbol{q}}\right)^{1/2}. \tag{11.18}$$

The velocity of spin waves depends on the orientation of $\hat{\boldsymbol{l}}$ with respect to the direction of propagation. Nonlinear spin waves in the A phase have been discussed by Poluéktov (1983).

The A_1 and A_2 phases

As already mentioned, the hydrodynamics of the A_1 phase has a peculiar structure because it is a superfluid magnetic liquid. Here one encounters a broken relative gauge and spin rotational symmetry (the simultaneously broken rotational symmetry in orbital space (see Chapter 6) is not important for this discussion). Therefore the two associated Goldstone modes, second sound and longitudinal spin waves, are expected to appear as a single mode, which combines properties of both of the abovementioned modes (Maki 1975a, Liu 1979a,b, 1980, Gongadze *et al.* 1978a,b, Gurgenishvili and Kharadze 1980, Takagi 1981a,b). At the transition to the A_2 phase this mode has to split again into the two ordinary Goldstone modes.

We now wish to calculate the eigenmodes of the system. To do this, we combine the hydrodynamic equations (9.110), (9.108), (9.104) and (9.106). Neglecting the weak coupling between the mass-density oscillations and the entropy and spin modes, the mass current density may be put equal to zero. This condition allows the normal-fluid velocity to be expressed in terms of the superfluid velocities $\boldsymbol{v}_s$ and $\boldsymbol{v}_{sp}$ by

$$\boldsymbol{v}_n = -\boldsymbol{\rho}_n^{-1}(\boldsymbol{\rho}_s\boldsymbol{v}_s + \boldsymbol{\rho}_{\uparrow\downarrow}\boldsymbol{v}_{sp}). \tag{11.19}$$

Thus one finds the equation of motion for the spin component along the preferred direction $\hat{\boldsymbol{f}}$:

$$\frac{\partial^2}{\partial t^2}(\hat{\boldsymbol{f}}\cdot\boldsymbol{S}) + \frac{\chi}{\gamma^2}\Omega_{\parallel}^2\hat{\boldsymbol{f}}\cdot\boldsymbol{\omega}_s = \frac{\hbar}{2m}\{(\boldsymbol{\nabla}\tilde{\boldsymbol{\rho}}_{sp})\boldsymbol{\nabla}(\hat{\boldsymbol{f}}\cdot\boldsymbol{\omega}_s) + \rho[\boldsymbol{\nabla}(\boldsymbol{\rho}_{\uparrow\downarrow}\boldsymbol{\rho}_n^{-1})]\boldsymbol{\nabla}\delta\mu\}. \tag{11.20}$$

Here we have introduced

$$\tilde{\boldsymbol{\rho}}_s = \boldsymbol{\rho}_{sp} + \boldsymbol{\rho}_{\uparrow\downarrow}\boldsymbol{\rho}_n^{-1}\boldsymbol{\rho}_{\uparrow\downarrow}, \tag{11.21}$$

which may be interpreted as an effective superfluid spin density. The angular frequency $\boldsymbol{\omega}_s$ is defined in (9.11). The first term on the right-hand side of (11.20) describes something like the "stiffness" of the spins.

In the absence of $\delta\mu$, the spin-wave frequency is obtained as

$$\omega_{sp}^2 = \frac{\hbar\gamma^2}{2m\chi}\boldsymbol{q}\tilde{\boldsymbol{\rho}}_{sp}\boldsymbol{q} + \Omega_A^2. \tag{11.22}$$

However, the pure spin-wave frequency is *not* an eigenfrequency of (11.20) (although it is almost so). Rather, the spin density couples to the temperature field by virtue of the last term on the right-hand side of (11.20) via

$$\delta\mu = -\rho s\,\delta T, \tag{11.23}$$

where s is the entropy density. An equation of motion for the temperature fluctuation δT may be obtained with the aid of the entropy conservation law

(9.27) and by substituting $\boldsymbol{v}_n$ using (11.21); this yields

$$\frac{\partial}{\partial t}\left(\frac{\partial s}{\partial T}\delta T\right) - s\boldsymbol{\nabla}\cdot[\boldsymbol{\rho}_n^{-1}(\boldsymbol{\rho}_s\boldsymbol{v}_s + \boldsymbol{\rho}_{\uparrow\downarrow}\boldsymbol{v}_{sp})] = 0. \tag{11.24}$$

Now the Josephson equations (9.103) and (9.104) are used to replace the superfluid velocities $\boldsymbol{v}_s$ and $\boldsymbol{v}_{sp}$ in (11.24) in terms of gradients of the chemical potentials $\delta\mu$ and $\hat{\boldsymbol{f}}\cdot\boldsymbol{\omega}_s$. Finally, by combining (11.23) and (11.24) and after performing a Fourier transformation, one is led to the dispersion law

$$(\omega^2 - \omega_{sp}^2)(\omega^2 - \omega_2^2) = \frac{\hbar\gamma^2}{2m\chi}\frac{s^2}{\partial s/\partial T}(\boldsymbol{q}\boldsymbol{\rho}_{\uparrow\downarrow}\boldsymbol{\rho}_n^{-1}\boldsymbol{q})^2. \tag{11.25}$$

Here the usual second-sound frequency has been introduced:

$$\omega_2^2 = \frac{s^2}{\rho\,\partial s/\partial T}\boldsymbol{q}(\boldsymbol{\rho}_s\boldsymbol{\rho}_n^{-1})\boldsymbol{q}. \tag{11.26}$$

Far away from the A_2 phase transition, one has $\rho_{\uparrow\downarrow} \ll \rho_s$ and the right-hand side of (11.25) is small. Hence spin waves and second sound decouple into separate modes with frequencies ω_{sp} and ω_2 respectively. On the other hand, in the A_1 phase one has $\rho_{\uparrow\downarrow} = -\rho_{sp} = -\rho_s$ and $\Omega_A = 0$, whence the right-hand side of (11.25) becomes $\omega_{sp}^2\omega_2^2$. Therefore the relevant eigenfrequency in the A_1 phase is

$$\omega^2 = \omega_{sp}^2 + \omega_2^2. \tag{11.27}$$

The second solution, $\omega^2 = 0$, carries zero weight and may be discarded. The first mode is a combined temperature–spin wave with linear dispersion and velocity approximately equal to the *spin-wave* velocity, because $\omega_{sp}^2 \gg \omega_2^2$. On moving into the A_2 phase, the second eigenfrequency begins to grow until it reaches the value ω_2 not far from the transition. This mode, which was predicted by Liu (1979a), has been observed experimentally by Corruccini and Osheroff (1980), using an oscillating superheat transducer. The efficiency of such transducers in superfluid ^{3}He has been investigated theoretically by Liu and Stern (1982) and Johnson (1982). A different method for detecting spin waves in ^{3}He-A, using the scattering of light from the associated magnetization fluctuations, has been proposed by Golo and Kats (1984). As pointed out by Stern and Liu (1982), the damping of spin waves may be dominated by spin-relaxation processes at the walls of the container rather than by spin diffusion (Nagai 1981b).

The B phase

As in the A phase, we are again interested in determining the eigenmodes of the equation of motion of the spin density, in this case given by (11.11). Here we neglect the dissipation, i.e. the term $\boldsymbol{\nabla}\cdot\boldsymbol{X}_\mu^D$ in (11.11). In the B phase the situation is somewhat more complicated owing to the coupling of spin and orbital spaces. This is, for example, expressed in the specific structure of the superfluid spin density as given by (10.60). However, in the representation where the spin coordinate system is rotated relative to the coordinate system in position space by $R_{\mu j}$ (the rotation matrix characterizing the B phase), the superfluid spin density tensor has the simpler form

$$(\rho^{\rm sp})^{ij}_{\bar\mu\bar\nu} = \rho_1\delta_{ij}\delta_{\bar\mu\bar\nu} + \rho_2\delta_{i\bar\nu}\delta_{j\bar\mu} + \rho_3\delta_{i\bar\mu}\delta_{j\bar\nu}, \tag{11.28}$$

where $\bar\mu$ and $\bar\nu$ label the Cartesian components in the rotated spin space. In the absence of Fermi-liquid corrections, the coefficients ρ_i in (11.28) are obtained from (10.60) as

$$\rho_1^0 = -4\rho_2^0 = -4\rho_3^0 = \tfrac{4}{5}S_0(1-Y_0), \tag{11.29}$$

where Y_0 is the Yosida function defined in (3.88). The respective renormalizations of the ρ_i are then readily found from the general expression (10.60) for $\rho^{\rm sp}$:

$$\rho_1 = |\rho_2^0|\frac{4a_1 + a_2}{a_1^2 - a_2^2}, \tag{11.30}$$

$$\rho_2 = \rho_2^0\frac{a_1 + 4a_2}{a_1^2 - a_2^2}, \tag{11.31}$$

$$\rho_3 = \rho_3^0\frac{1}{a_1 + a_2}, \tag{11.32}$$

where

$$a_1 = 1 + \tfrac{1}{3}F_1^{\rm a}\left(1 + \frac{m_{\rm s}}{m^* S_0}\rho_1^0\right),$$

$$a_2 = \tfrac{1}{3}F_1^{\rm a}\frac{m_{\rm s}}{m^* S_0}\rho_2^0.$$

Substituting (11.21) and the expression for the dipole torque,

$$\frac{\rm d}{{\rm d}t}\boldsymbol{R}_{\rm D} = -\Omega_{\rm B}^2\hat{\boldsymbol{n}}(\hat{\boldsymbol{n}}\cdot\boldsymbol{S}), \tag{11.33}$$

into the spin-wave equation (11.13), one is led to

$$\omega^2\boldsymbol{S} - {\rm i}\omega\gamma(\boldsymbol{S}\times\boldsymbol{H}_0) - \alpha_1 q^2\boldsymbol{S} - \alpha_2\boldsymbol{q}(\boldsymbol{q}\cdot\boldsymbol{S}) = \Omega_{\rm B}^2\hat{\boldsymbol{n}}(\hat{\boldsymbol{n}}\cdot\boldsymbol{S}), \tag{11.34}$$

where the stiffness parameters α_1 and α_2 are given by

$$\alpha_1 = \frac{\hbar\gamma^2}{2m\chi}\rho_1, \tag{11.35}$$

$$\alpha_2 = \frac{\hbar\gamma^2}{2m\chi}(\rho_2 + \rho_3). \tag{11.36}$$

The eigenfrequencies of (11.34) are conveniently calculated in a coordinate system $(\hat{\boldsymbol{x}}, \hat{\boldsymbol{y}}, \hat{\boldsymbol{z}})$ with $\hat{\boldsymbol{n}}$ along $\hat{\boldsymbol{z}}$ and $\hat{\boldsymbol{x}}$ both perpendicular to $\hat{\boldsymbol{n}}$ and the wave-propagation direction $\hat{\boldsymbol{q}}$ (Brinkman and Smith 1974).

In the limit of zero magnetic field the mode spectrum is found as

$$\left.\begin{aligned} \omega_1^2 &= \alpha_1 q^2, \\ \omega_{2,3}^2 &= \tfrac{1}{2}[\tilde{\Omega}_B^2 + 2\alpha_1 q^2 \pm (\tilde{\Omega}_B^4 - 4\alpha_2 \Omega_B^2 q_y^2)^{1/2}], \end{aligned}\right\} \tag{11.37}$$

where $\tilde{\Omega}_B^2 = \Omega_B^2 + \alpha_2 q^2$ (note that $\alpha_2 < 0$). Here ω_1 is the frequency of a spin wave that is linearly polarized in the direction orthogonal to $\hat{\boldsymbol{n}}$ and $\hat{\boldsymbol{q}}$, whereas the remaining two modes are elliptically polarized in the plane spanned by $\hat{\boldsymbol{n}}$ and $\hat{\boldsymbol{q}}$. For large q, (11.34) reduce to the linear dispersion laws

$$\omega_j = c_j^{\text{sp}} q, \quad j = 1, 2, 3, \tag{11.38}$$

with the spin-wave velocities

$$c_1^{\text{sp}} = \alpha_1^{1/2}, \tag{11.39}$$

$$c_{2,3}^{\text{sp}} = (\alpha_1 \pm |\alpha_2|)^{1/2}. \tag{11.40}$$

A static magnetic field $\boldsymbol{H}_0$ introduces further gaps into the excitation spectra. The result for small q (such that $c^{\text{sp}} q \ll \Omega_B$), but arbitrary frequency ω_L compared with Ω_B, is

$$\omega_{1,2}^2 = \tfrac{1}{2}\{\tilde{\omega}_L^2 \pm [\tilde{\omega}_L^4 - 4\alpha_1 q^2(\alpha_1 q^2 + \alpha_2 q_y^2)]^{1/2}\}, \tag{11.41}$$

$$\omega_3^2 = \Omega_B^2 + \alpha_1 q^2 + \alpha_2 q_z^2, \tag{11.42}$$

where $\tilde{\omega}_L^2 = \omega_L^2 + 2\alpha_1 q^2 + \alpha_2 q_y^2$.

Spin waves are difficult to detect experimentally because there is no analogue to a sound transducer or absorber used in sound-wave measurements. Furthermore, an NMR pick-up coil, which might appear to be an ideal detector, only allows one to measure the *integrated* magnetization in the interior of the coil, which averages to zero in the usual limit of wavelengths small compared with the size of the coil. None the less, in the case of inhomogeneous textures and fields, standing spin waves do lead to detectable NMR resonances provided that the wave train is sufficiently modulated in amplitude that it has a net magnetization. As discussed in Section 8.4, spin waves in a nonuniform $\hat{\boldsymbol{n}}$ texture have been investigated by Smith *et al.* (1977).

The spectrum of standing spin waves in a stack of parallel plates has been observed experimentally (Osheroff *et al.* 1977). The results are in close agreement with theory. Spin waves in the collisionless regime are discussed at the end of Section 10.3. The collective modes of the B phase in a strong magnetic field (i.e. such that the distortion of the gap is not negligible) have been discussed by Pleiner and Brand (1983).

11.1.3 Other hydrodynamic modes

The preceeding discussion did not include all hydrodynamic modes. In fact, according to the general classification of hydrodynamic modes presented earlier in this chapter, there must be several other such modes.

In the linear hydrodynamic regime (i.e. where the hydrodynamic variables enter only linearly, as is the case for weak excitations) there is a rigorous correspondence between the number of hydrodynamic variables and the number of possible modes. For example, in Chapter 9 we found that the hydrodynamic properties of the B phase are described by the following set of 12 hydrodynamic variables: $\{\rho, T, S_\mu, \boldsymbol{v}_n, \boldsymbol{v}_s, \boldsymbol{v}_\mu^{sp}; \mu = 1, 2, 3\}$; here the four superfluid velocities $\boldsymbol{v}_s$ and $\boldsymbol{v}_\mu^{sp}$ have each been counted just once because only their longitudinal component enters. As discussed above, the corresponding modes are (i) two first-sound modes, $\omega = \pm c_1 q$ and two second-sound modes $\omega = \pm c_2 q$ associated with the four variables ρ, T, $\boldsymbol{v}_s$ and the longitudinal part of $\boldsymbol{v}_n$, and (ii) six spin–orbit wave modes $\omega = \pm c_s q$ connected with the six variables S_μ and $\boldsymbol{v}_\mu^{sp}$. The two remaining variables are the transverse components of $\boldsymbol{v}_n$, which give rise to two diffusive shear modes with dispersion $\omega = \mathrm{i}Dq^2$.

An analogous classification of modes holds for the A phase. There the $\hat{\boldsymbol{l}}$-vector field provides two more hydrodynamic variables. The associated "orbital-wave" modes attracted a great deal of interest in the early stages of research on superfluid ^{3}He (de Gennes 1973, Anderson 1973, Wölfle 1974, Volovik 1975a, Saslow 1976, Combescot and Combescot 1976a, Cross 1977a; see also Brinkman and Cross 1978, Leggett and Takagi 1978). However, as shown in Section 10.3, the motion of $\hat{\boldsymbol{l}}$ at finite temperatures is severely inhibited by the strong coupling of $\hat{\boldsymbol{l}}$ to the quasiparticle system, giving rise to the so-called "orbital viscosity" (Cross and Anderson 1975). Thus orbital waves are overdamped except in the limit $T \to 0$, where they merge with the so-called "normal-flapping" mode that is discussed in the next section.

11.2 SYMMETRY CLASSIFICATION OF ORDER-PARAMETER MODES IN THE COLLISIONLESS REGIME

In contrast with a BCS superconductor, the Cooper pairs in the superfluid phases of ^{3}He have an internal structure. This in turn allows collective modes of the order parameter. The nature of these modes is determined by the equilibrium order-parameter structure itself. In the A phase, for example, the unit vectors $\hat{\boldsymbol{m}}$ and $\hat{\boldsymbol{n}}$ characterizing the orbital part of the order parameter may oscillate about their equilibrium configuration $\hat{\boldsymbol{m}} \perp \hat{\boldsymbol{n}}$ by vibrating within the plane spanned by these two unit vectors. This is the so-called "clapping mode", the origin of whose name is self-evident (see Fig. 11.2). The frequencies of such order-parameter modes are of the order of the gap frequency (only the "normal flapping mode" is an exception—see below). In addition to these modes, there are Goldstone modes with frequency tending to zero in the limit as the wave vector $\boldsymbol{q} \to 0$. Depending on the symmetry of the equilibrium order parameter, the eigenfrequencies of different Goldstone modes may or may not be equal. Accordingly, the 18 components of the order parameter, which are the dynamical variables of this problem, may be grouped into multiplets. This fact is familiar from atomic and solid-state physics. As in atomic physics, the multiplet structure can essentially be derived on the basis of symmetry considerations. The classification is then used to perform the actual microscopic calculation of the eigenfrequencies in an economic way.

The multiplet structure of the collective-mode spectrum is completely determined by the symmetry group H of the order parameter, i.e. the set of symmetry operations, both continuous and discrete, under which the order parameter is invariant (Volovik and Khazan 1983). The group H is itself a subgroup of the overall symmetry group $G = \mathrm{SO}(3) \times \mathrm{SO}(3) \times \mathrm{U}(1)$ describing the free-energy functional of p-wave superfluidity (see Chapter 6). The degeneracies of the eigenfrequencies (if any) are given by the dimensions of the different representations of the group H. The representations are labelled by the eigenvalues of the Casimir operators of the group H. Although in the cases of interest there exist infinitely many representations, the problem of selecting a particular set of representations is uniquely determined by the dimension of the space of states of dynamical variables. Hence in the case where only p-wave order parameters are considered this space is 18-dimensional.

11.2.1 The B phase

As discussed in Section 6.3, the symmetry group H of the B phase is $\mathrm{SO}(3)_{\boldsymbol{L}+\boldsymbol{S}}$, i.e. the order parameter is invariant under simultaneous rotations in spin space and orbital space. In addition, the order-parameter components are real, apart from an overall phase factor. This implies that $d'_{\mu j}$ and $d''_{\mu j}$, the real and imaginary parts of the *change* in the order parameter, $\delta d_{\mu j}$, are not coupled. Here we have introduced a shorthand notation, i.e. $d'_{\mu j} \equiv \mathrm{Re}\,[\delta d_{\mu j}(\boldsymbol{r}, t)]$, $d''_{\mu j} \equiv \mathrm{Im}\,[\delta d_{\mu j}(\boldsymbol{r}, t)]$. The complete group H is then given by

$$H_{\mathrm{B}} = \mathrm{SO}(3)_{\boldsymbol{L}+\boldsymbol{S}}. \tag{11.43}$$

We are now interested in the Casimir operator of the group H because it commutes with all elements of this symmetry group and because its eigenvalues label the different irreducible representations.

The Casimir operator of $\mathrm{SO}(3)_{\boldsymbol{L}+\boldsymbol{S}}$ is $\boldsymbol{J}^2 = (\boldsymbol{L} + \boldsymbol{S})^2$, with eigenvalues $J(J+1)$, where $J = 0, 1, 2, \ldots$ is the total angular-momentum quantum number of the p-wave Cooper pair. The representations labelled by J are

$(2J+1)$-dimensional. The 18 order-parameter variables may be grouped into 9 real and 9 imaginary components, which fall into a singlet ($J=0$), a triplet ($J=1$) and a quintet ($J=2$) (Maki 1974). The classification scheme is shown in Table 11.1. The different modes listed in column 1 are labelled by three quantum numbers (columns 2 and 3): the angular momentum J, a quantity K taking the values (+) and (−) for the real and imaginary parts respectively, and the magnetic quantum number $m_J = 0, \pm1, \ldots, \pm J$. The transformation properties associated with these quantum numbers lead to the linear combinations of order parameter components shown in column 4. The square of the eigenfrequency in the limit $q \to 0$, as obtained from weak-coupling theory (see below), is shown in column 5, and the coefficient of the $(v_F q)^2$ contribution is shown in column 6. The Fermi-liquid correction factor given in column 7 renormalizes the bare frequency $\omega^2(q)$. The order to which the order parameter couples to spin or density fluctuations is given in column 8 (up to order q^2). Finally, in column 9 the magnitude of the magnetic-field splitting is shown, where ω_L is the Larmor frequency and the g factor is discussed below.

From the table, we can see that there are four Goldstone (i.e. gapless) modes: the imaginary $J=0$ mode and the three real $J=1$ modes (relative spin–orbit rotations), which manifest themselves as second or fourth sound and spin–orbit waves respectively. For the purpose of discussion, we have neglected the dipole interaction. The frequencies of the remaining modes are of order $\Delta(T)$ ("pair vibration modes").

In the presence of a magnetic field the overall symmetry of the free-energy functional of the system is reduced to the group $G = \mathrm{U}(1)_{S_z} \times \mathrm{SO}(3)_{\boldsymbol{L}} \times \mathrm{U}(1)_\phi$ (see Chapter 6), and the subgroup H_B is correspondingly reduced to

$$H_B = \mathrm{U}(1)_{L_z+S_z}. \tag{11.44}$$

The latter group has only one-dimensional representations, so that the degeneracies of the m_J sublevels are completely lifted. For small magnetic fields, when the distortion of the gap itself may be neglected, the splitting is symmetric and linear in the field.

Similarly, in a spatially inhomogeneous situation characterized by a finite $\boldsymbol{q}$ vector, the symmetry group of the system is reduced to $G = \mathrm{SO}(3)_{\boldsymbol{S}} \times \mathrm{U}_L \times \mathrm{U}(1)_\phi$, where U_L is the group of rotations about the direction of $\boldsymbol{q}$ plus the reflection on the plane perpendicular to $\boldsymbol{q}$. Since the rotations do not commute with the reflection operation, the group U_L is non-Abelian. The subgroup H is then reduced to

$$H_B = \mathrm{U}_{L,S}, \tag{11.45}$$

where $\mathrm{U}_{L,S}$ is isomorphic to U_L, but contains relative spin–orbit rotations about the axes $\boldsymbol{q}$ and $\boldsymbol{Rq}$, respectively, plus the reflection operation. The group has one- and two-dimensional representations, such that the eigenvalues of $J_z = \pm m$ states are still degenerate. The situation is similar to the

Table 11.1 Collective modes of the B phase.

Mode	J, K	m_J	Variable	ω^2	Dispersion $\dfrac{\omega^2(q)-\omega^2(0)}{(v_F q)^2}$	Fermi-liquid correction factor	Coupling to observables	Magnetic-field splitting
	0^+	0	$d'_{11}+d'_{22}+d'_{33}$	$4\Delta^2$	$0.24 - i0.3$		Density, $O(\omega/E_F)$	
Spin waves	1^+	0 ±1	$d'_{12}-d'_{21}$ $d'_{31}-d'_{13}\pm i(d'_{32}-d'_{23})$	0	$\frac{1}{5}$ $\frac{2}{5}$	$(1+\frac{2}{3}F_0^a)[1+O(F_1^a)]$	Spin density, $O(q^0)$	$\omega_L m_J g_{1^+}$
Real squashing mode	2^+	0 ±1 ±2	$d'_{11}+d'_{22}-2d'_{33}$ $d'_{31}+d'_{13}\pm i(d'_{32}+d'_{23})$ $d'_{11}-d'_{22}\pm i(d'_{12}+d'_{21})$	$\frac{8}{5}\Delta^2$	≈ 0.44 $\frac{2}{5}(1+\frac{2}{105}\lambda_1)$ $\frac{1}{5}(1-\frac{16}{35}\lambda_1)$	$1+\frac{3}{25}\lambda F_2^a$	Spin density, $O(q^2)$; density, $O(q^2\omega/E_F)$	$\omega_L m_J g_{2^+}$
Sound	0^-	0	$d''_{11}+d''_{22}+d''_{33}$	0	$\frac{1}{3}$	$(1+F_0^s)(1+\frac{1}{3}F_1^s)$	Density, $O(q^0)$	
	1^-	0 ±1	$d''_{12}-d''_{21}$ $d''_{31}-d''_{13}\pm i(d''_{32}-d''_{23})$	$4\Delta^2$	0.11–i0.17 0.35–i0.33		Spin density, $O(\omega/E_F)$	$\omega_L m_J g_{1^-}$
Squashing mode	2^-	0 ±1 ±2	$d''_{11}+d''_{22}-2d''_{33}$ $d''_{31}+d''_{13}\pm i(d''_{32}+d''_{23})$ $d''_{11}-d''_{22}\pm i(d''_{12}+d''_{21})$	$\frac{12}{5}\Delta^2$	$\frac{1}{5}(\frac{7}{3}-\frac{16}{35}\lambda_1)$ ≈ 0.42 $\frac{1}{5}(1+\frac{16}{35}\lambda_1)$	$1+\frac{2}{25}\lambda F_2^s$	Density, $O(q^2)$; spin density, $O(q^2\omega/E_F)$	$\omega_L m_J g_{2^-}$

coupling of an electric field to an atom in atomic physics, leading to doubly degenerate Stark levels.

11.2.2 The A phase

The order parameter of the A phase, (3.68), is given by $d_{\mu j} = \Delta_0 \hat{z}_\mu(\hat{x}_j + \mathrm{i}\hat{y}_j)$ if we take the preferred directions in orbital and spin space, $\hat{\boldsymbol{m}}$, $\hat{\boldsymbol{n}}$ and $\hat{\boldsymbol{d}}$, to lie along the x, y and z axes respectively. The symmetry group of the order parameter in this case is (Volovik and Khazan 1983)

$$H_{\mathrm{A}} = \mathrm{U}^{(1)}_{S_z,\Phi} \times \mathrm{U}^{(2)}_{L_z,\Phi}. \tag{11.46}$$

The group $\mathrm{U}^{(1)}_{S_z,\Phi}$ comprises (i) the group of rotations about the z axis, $\mathrm{U}(1)_{S_z}$, and (ii) the combined symmetry operation P_1, which consists of a rotation by π about an axis perpendicular to z carrying $\hat{\boldsymbol{d}}$ into $-\hat{\boldsymbol{d}}$ and a gauge transformation by π, which compensates for the minus sign. The effect of the gauge transformation is seen to be equivalent to a space inversion. (Note that for a discussion of the bulk free energy of a homogeneous system as in Chapter 6 the symmetry transformation P_1 is irrelevant, so that, instead of $\mathrm{U}^{(1)}_{S_z,\Phi}$ in (11.46), one only needs $\mathrm{U}(1)_{S_z}$.) The operator $P_1 = \exp(\mathrm{i}\pi S_y)\, P_{\mathrm{s}}$, with P_{s} the inversion operator in spin space, does not commute with the operators $\exp(\mathrm{i}\theta^{\mathrm{s}}_z S_z)$ of $\mathrm{U}(1)_{S_z}$ (the components of the spin operator, S_μ, are invariant under space inversion). Hence the group $\mathrm{U}^{(1)}$ must have multidimensional representations. The Casimir operator of $\mathrm{U}^{(1)}$ is S_z^2. For a triplet order parameter, the possible eigenvalues of S_z^2 are 0 and 1. The representation labelled by $S_z = 0$ describes the transformation of the $\mu = z$ component of the order parameter and is therefore one-dimensional. The nontrivial representation, labelled by $|S_z| = 1$, is two-dimensional. It mixes the $\mu = x$, y components of $d_{\mu j}$ under the transformations of the group H_{A}.

The group $\mathrm{U}^{(2)}_{L_z,\Phi}$ of the orbital and gauge degrees of freedom contains (i) the group of gauge–orbit rotations $\mathrm{U}(1)_{L_z+I}$, and (ii) the discrete symmetry operation $P_2 = \exp(\mathrm{i}\pi L_x)\, K$ consisting of a rotation in orbital space about the x axis by π, carrying $\hat{\boldsymbol{x}} + \mathrm{i}\hat{\boldsymbol{y}}$ into $\hat{\boldsymbol{x}} - \mathrm{i}\hat{\boldsymbol{y}}$, and the operation of complex conjugation K, restoring the original order parameter. As in the case of the group $\mathrm{U}^{(1)}_{S_z,\Phi}$, the additional discrete symmetry transformation P_2 is irrelevant for the consideration of homogeneous bulk properties, so that $\mathrm{U}^{(2)}_{L_z,\Phi} \rightarrow \mathrm{U}(1)_{L_z-\Phi}$ in Section 6.2. Since the operator P_2 does not commute with the generator of $\mathrm{U}(1)_{L_z+I}$, i.e. $P_2(L_z - I)P_2 = -(L_z - I)$, the group $\mathrm{U}^{(2)}$ is again non-Abelian and must have multidimensional representations. It follows from the simple transformation property of $L_z - I$ under P_2 stated above that the Casimir operator of $\mathrm{U}^{(2)}$ is given by $(L_z - I)^2$. Since the possible eigenvalues of L_z for a p-wave order parameter are given by 0, ± 1, and the relevant eigenvalue of the generator of gauge transformations $I = -\mathrm{i}\,\partial/\partial\phi$ is

$+1$, there appear three representations labelled by $|L_z - I| = 0, 1, 2$ in this case. The representations have dimension one for $|L_z - I| = 0$ and two for $|L_z - I| = 1, 2$. The two-dimensional representations are a consequence of the operator P_2 mixing real and imaginary parts of a given linear combination of order-parameter variables $d_{\mu j}$.

The representations of the group H are obtained as direct products of the representations of $U^{(1)}$ and $U^{(2)}$, and may be labelled by $(|L_z - I|, |S_z|)$. Except for the singlet representation with $|L_z - I| = 0$, i.e. $(0, 0)$ and $(0, 1)$—which necessarily appears twice, both for the real and imaginary parts of the order parameter—the representation for $|L_z - I| \neq 0$ occurs only once. Accordingly, there are four doublets, labelled $(1, 0)$, $(2, 0)$ and $(0, 1)$, and two quartets, $(1, 1)$ and $(2, 1)$. The multiplet spectrum of collective modes is shown in Table 11.2. As in Table 11.1 for the B phase, the respective modes, their quantization numbers and the associated order-parameter variables are shown in the first four columns. The next two columns contain the weak-coupling results for the eigenfrequency in the temperature limits $T \to T_c$ and $T \to 0$. Next, the coupling to order q^2 is indicated. The set of eigenvectors in column 4 is of course not uniquely determined, since the eigenvectors within each multiplet transform among each other under the group H. In principle, the eigenvectors may be determined by diagonalizing the Casimir operators, but in the present case they may simply be found by inspection.

As is evident from Table 11.2, some of the weak-coupling collective frequencies are equal, although they belong to different multiplets. This implies an additional symmetry of the order parameter in this case. To identify the additional symmetry transformations, we recall that in weak-coupling theory the two Cooper-pair systems of up- and down-spin pairs are independent. Hence the orbital structures of the two types of pairs may be oriented arbitrarily with respect to each other. This is expressed by the more general order-parameter structure

$$d_{\mu j} = \tfrac{1}{2}\Delta_0[(\hat{d}_\mu + \mathrm{i}\hat{e}_\mu)(\hat{m}_{\uparrow j} + \mathrm{i}\hat{n}_{\uparrow j}) + (\hat{d}_\mu - \mathrm{i}\hat{e}_\mu)(\hat{m}_{\downarrow j} + \mathrm{i}\hat{n}_{\downarrow j})], \qquad (11.47)$$

where $\hat{\boldsymbol{d}}$, $\hat{\boldsymbol{e}}$, $\hat{\boldsymbol{m}}_{\uparrow,\downarrow}$ and $\hat{\boldsymbol{n}}_{\uparrow,\downarrow}$ are unit vectors and $\hat{\boldsymbol{d}} \cdot \hat{\boldsymbol{e}} = 0$, $\hat{\boldsymbol{m}}_\uparrow \cdot \hat{\boldsymbol{n}}_\uparrow = 0$, $\hat{\boldsymbol{m}}_\downarrow \cdot \hat{\boldsymbol{n}}_\downarrow = 0$. Strong-coupling corrections in the free energy, however weak, will tend to align the orbital structures of spin-up and spin-down Cooper pairs, such that $\hat{\boldsymbol{m}}_\uparrow = \hat{\boldsymbol{m}}_\downarrow = \hat{\boldsymbol{m}}$, $\hat{\boldsymbol{n}}_\uparrow = \hat{\boldsymbol{n}}_\downarrow = \hat{\boldsymbol{n}}$. In this case the vector $\hat{\boldsymbol{e}}$ drops out of (11.4). Nevertheless, the collective-mode spectrum will exhibit the additional degeneracies associated with the larger symmetry group of the order-parameter structure (11.47). (These degeneracies are, of course, only approximate because of the aforementioned strong-coupling corrections.) Let us now find out what the additional symmetries correspond to mathematically. Using (11.47), it is clear that any rotation about the axis $\hat{\boldsymbol{d}} \times \hat{\boldsymbol{e}}$ in spin space by an angle α may be compensated by a corresponding rotation of $\hat{\boldsymbol{m}}_\uparrow$ and $\hat{\boldsymbol{n}}_\downarrow$ about the axis $\hat{\boldsymbol{m}}_\uparrow \times \hat{\boldsymbol{n}}_\uparrow$ by an angle $-\alpha$ and a

Table 11.2 Collective modes of the A phase.

Mode	$\|L_z - l\|$	$\|S_z\|$	Variable	Weak-coupling frequency ω $T \to T_c$	Weak-coupling frequency ω $T \to 0$	Coupling	P_1	Variables in a magnetic field (field along $\boldsymbol{y}$)
Sound	0	0	d''_{z1}	$\propto v_F q$	$3^{-1/2} v_F q$	Density, $O(q^0)$	1	d''_{z1}
Spin waves		1	d'_{x1}, d'_{y1}			Spin density, $O(q^0)$	1, −1	d'_{x1}, d'_{y1}
None		0	d'_{z1}			Density, $O(\omega/E_F)$	1	d'_{z1}
		1	d''_{x1}, d''_{y1}			Spin density, $O(\omega/E_F)$	1, −1	d''_{x1}, d''_{y1}
Orbital waves	1	0	d'_{z0}	$\omega_{orb} \propto i^{1/2} q^2$	$\propto iq^3$	Density, $O(q^2\omega/E_F)$	1	d'_{z0}
Flapping mode			d''_{z0}			Density, $O(q^2)$		d''_{z0}
Flapping modes		1	d'_{x0}, d'_{y0}	$\omega_{nfl} = (\frac{4}{5})^{1/2}\Delta_0(T)$	$\propto T$	Spin density, $O(q^2)$	1	d'_{x0}, d''_{x0}
			d''_{x0}, d''_{y0}	$\omega_{sfl} = 2\Delta_0(T)$	$1.56\Delta_0(0)$	Spin density, $O(q^2\omega/E_F)$	−1	d'_{y0}, d''_{y0}
Clapping modes	2	0	$d'_{z,-1}$	$1.23\Delta_0(T)$	$1.22\Delta_0(T)$	Density, $O(q^2\omega/E_F)$	1	$d'_{z,-1}$, $d''_{z,-1}$
			$d''_{z,-1}$			Density, $O(q^2)$		
		1	$d'_{x,-1}$, $d'_{y,-1}$			Spin density, $O(q^2)$	1	$d'_{x,-1}$, $d''_{x,-1}$
			$d''_{x,-1}$, $d''_{y,-1}$			Spin density, $O(q^2\omega/E_F)$	−1	$d'_{y,-1}$, $d''_{y,-1}$

simultaneous rotation of $\hat{\boldsymbol{m}}_\downarrow$ and $\hat{\boldsymbol{n}}_\downarrow$ about $\hat{\boldsymbol{m}}_\downarrow \times \hat{\boldsymbol{n}}_\downarrow$ by an angle $+\alpha$. Now if the relative orientation of the triads $(\hat{\boldsymbol{m}}_\uparrow, \hat{\boldsymbol{n}}_\uparrow, \hat{\boldsymbol{m}}_\uparrow \times \hat{\boldsymbol{n}}_\uparrow)$ and $(\hat{\boldsymbol{m}}_\downarrow, \hat{\boldsymbol{n}}_\downarrow, \hat{\boldsymbol{m}}_\downarrow \times \boldsymbol{n}_\downarrow)$ is irrelevant on account of the independence of spin-up and spin-down Cooper pairs then the order parameter (11.47) is invariant under rotations in spin space about any axis perpendicular to $\hat{\boldsymbol{d}}$, in addition to rotations about the axis $\hat{\boldsymbol{d}}$ considered in the strong-coupling case. The symmetry group of the weak-coupling ABM order parameter is then given by

$$H_{\mathrm{A}}^{\mathrm{wc}} = \mathrm{SO}(3)_{\boldsymbol{S}} \times \mathrm{U}^{(2)}_{L_z,\Phi} \tag{11.48}$$

(for a somewhat different discussion, see Volovik and Khazan 1983). Since the Cooper-pair spin is $S=1$, the relevant representation of SO(3) is three-dimensional. Including the orbital part, which has either one-dimensional ($|L_z - I| = 0$) or two-dimensional ($|L_z - I| = 1,\ 2$) representations, one finds the following degeneracies of the collective-mode spectrum (see Table 11.2): there are two threefold- and two sixfold-degenerate states. The four additional Goldstone modes, which may be interpreted as spin–orbit waves, result from the degeneracy of the $S_z = 0$ and $|S_z| = 1$ modes in the $|L_z - I| = 1$ multiplet. These findings are in accordance with the results of the explicit calculations presented in the next section.

In Table 11.2 three different frequencies have been entered for the case $|L_z - I| = 1$. As will be shown later, the transcendental equations determining the eigenfrequencies are nonlinear in ω^2 in this case, and therefore allow more than one solution.

Finally, we want to discuss the influence of a magnetic field on the symmetry. In the presence of a magnetic field $\boldsymbol{H}$ along $\hat{\boldsymbol{y}}$ the rotational symmetry in spin space is reduced to U(1) so that $G = \mathrm{U}(1)_{S_y} \times \mathrm{SO}(3)_{\boldsymbol{L}} \times \mathrm{U}(1)_\Phi$ (we retain the direction of $\hat{\boldsymbol{d}}$ along the z axis; consequently $\hat{\boldsymbol{d}} \cdot \boldsymbol{H} = 0$ in equilibrium). The subgroup H is then reduced to $H = \mathbb{Z}_2 \times \mathrm{U}^{(2)}_{L_z,\Phi}$, where the group $\mathbb{Z}_2$ consists of the two elements P_1 and 1. The modes are then only labelled by the quantum numbers $|L_z - I| = 0,\ 1,\ 2$ and $P_1 = \pm 1$. Since the group $\mathbb{Z}_2$ has only a one-dimensional representation, the degeneracies in the spectrum are lifted to the extent that only singlets and doublets remain. This is shown in detail in Table 11.2.

We do not consider the small effect of residual interactions like the dipole interaction and particle–hole asymmetry here, but they may be readily included in the scheme (Volovik and Khazan 1983).

Having listed the various possibilities for the existence of collective modes allowed by symmetry, we now want to calculate the mode frequencies and the order-parameter response functions explicitly. This can only be done within a model theory. Under the given circumstances, where $T_c \ll T_F$, the obvious model is weak-coupling theory.

11.3 TIME-DEPENDENT MEAN-FIELD THEORY IN THE COLLISIONLESS REGIME

In weak-coupling theory, i.e. to leading order in the small parameter T_c/T_F, the dynamics of a pair-correlated Fermi system in the regime of external frequencies and wave vectors $\hbar\omega \ll E_F$, $|\boldsymbol{q}| \ll k_F$, is completely described by the matrix kinetic equation (10.6) if supplemented by the definitions of the change in the diagonal energy, $\delta\boldsymbol{\xi}_k$, and of the order parameter, $\delta\boldsymbol{\Delta}_k$, defined in (10.8) and (10.9). The mean fields $\delta\boldsymbol{\xi}_k$ and $\delta\boldsymbol{\Delta}_k$ are functions of the position on the Fermi surface, $\hat{\boldsymbol{k}}$, and of spin, but they are independent of the magnitude of the momentum $|\boldsymbol{k}| = k_F$ up to order T_c/T_F. The $|\boldsymbol{k}|$ dependence may therefore be neglected. Nevertheless, in order to keep the notation simple, we adhere to the notation $\boldsymbol{\Delta}_k$, $\delta\boldsymbol{\Delta}_k$, $\delta\boldsymbol{\xi}_k$ (rather than $\boldsymbol{\Delta}(\hat{\boldsymbol{k}})$ etc.), where it is implied that $|\boldsymbol{k}| = k_F$. The neglect of the $|\boldsymbol{k}|$ dependence allows one to cast the matric kinetic equation into a set of coupled integral equations in $\hat{\boldsymbol{k}}$ for the mean fields. In order to achieve this goal we define a four-component vector distribution function by

$$\delta\mathbf{N}_k = \begin{pmatrix} \delta\mathbf{n}_k + \delta\mathbf{n}_{-k} \\ \delta\mathbf{n}_k - \delta\mathbf{n}_{-k} \\ \dfrac{1}{|\Delta_k|}(\delta\mathbf{F}_k\,\boldsymbol{\Delta}_k^+ + \boldsymbol{\Delta}_k\,\delta\mathbf{F}_k^+) \\ \dfrac{1}{|\Delta_k|}(\delta\mathbf{F}_k\,\boldsymbol{\Delta}_k^+ - \boldsymbol{\Delta}_k\,\delta\mathbf{F}_k^+) \end{pmatrix}, \tag{11.49}$$

where each element represents a 2×2 spin matrix. A vector of the quasiparticle energy change, $\delta\boldsymbol{\epsilon}_k$, may be defined in complete analogy with (11.49), i.e. with $\delta\mathbf{n}_k$ and $\delta\mathbf{F}_k$ in (11.49) replaced by $\delta\boldsymbol{\xi}_k$ and $\delta\boldsymbol{\Delta}_k$ respectively. In the nonmagnetic case, when $\delta\mathbf{n}_k$ and $\delta\boldsymbol{\xi}_k$ are proportional to the unit matrix in spin space, the matrix kinetic equation can then be written as a system of four inhomogeneous linear equations for the components of $\delta\mathbf{N}_k$:

$$\boldsymbol{\Omega}_k\delta\mathbf{N}_k = \mathbf{M}_k\delta\boldsymbol{\epsilon}_k. \tag{11.50a}$$

The matrix of coefficients $\boldsymbol{\Omega}_k$ is given by

$$\boldsymbol{\Omega}_k = \begin{pmatrix} \hbar\omega & \xi_- - \xi_+ & 0 & 2\Delta \\ \xi_- - \xi_+ & \hbar\omega & 0 & 0 \\ 0 & 0 & \hbar\omega & -(\xi_+ + \xi_-) \\ 2\Delta & 0 & -(\xi_+ + \xi_-) & \hbar\omega \end{pmatrix}, \tag{11.50b}$$

and similarly

$$\mathbf{M}_k = \begin{pmatrix} 0 & \xi_+\theta_+ - \xi_-\theta_- & 0 & -\Delta(\theta_+ + \theta_-) \\ \xi_+\theta_+ - \xi_-\theta_- & 0 & \Delta(\theta_+ - \theta_-) & 0 \\ 0 & \Delta(\theta_+ - \theta_-) & 0 & \xi_+\theta_+ + \xi_-\theta_- \\ -\Delta(\theta_+ + \theta_-) & 0 & \xi_+\theta_+ + \xi_-\theta_- & 0 \end{pmatrix}. \tag{11.50c}$$

Here we have used the abbreviations

$$\left.\begin{aligned} \Delta &= |\Delta_k|, \\ \xi_\pm &= \xi_{k\pm q/2}, \quad \theta_\pm = \theta_{k\pm q/2}, \\ \theta_k &= \frac{1}{2E_k}\tanh\frac{E_k}{2k_B T}. \end{aligned}\right\} \tag{11.51}$$

Also, we have restricted ourselves to the collisionless regime, i.e. we have dropped the collision integral in (10.6). An extension of the theory including collisions has been given by Wölfle (1978a,b). The system of equations (11.50a) may be readily solved for $\delta\mathbf{N}_k$:

$$\delta\mathbf{N}_k = (\boldsymbol{\Omega}_k)^{-1}\mathbf{M}_k\,\delta\boldsymbol{\epsilon}_k, \tag{11.52}$$

and, consequently, also for $\delta\mathbf{n}_k$ and $\delta\mathbf{F}_k$. Employing the definition of $\delta\boldsymbol{\xi}_k$, (10.8), and the decomposition (11.59), the spin-symmetric component $\delta\xi_k$ is found after some algebraic manipulations and expansions in q as

$$\begin{aligned} \delta\xi_p - \delta\xi_p^{\text{ext}} = \sum_k f_{pk}^{\text{s}}\,\text{tr}_\sigma \Big\{ & \frac{u}{\omega - u}(1-\lambda_k)\,\delta\boldsymbol{\xi}_k - \tfrac{1}{2}\lambda_k(\delta\boldsymbol{\xi}_k + \delta\boldsymbol{\xi}_{-k}) \\ & + \tfrac{1}{4}\hbar(\omega+u)\bar{\lambda}_k(\delta\boldsymbol{\Delta}_k\,\boldsymbol{\Delta}_k^+ - \boldsymbol{\Delta}_k\,\delta\boldsymbol{\Delta}_k^+) \\ & + \frac{1}{\omega}[u\theta_k' + \xi_k(\omega+u)\bar{\lambda}_k](\delta\boldsymbol{\Delta}_k\,\boldsymbol{\Delta}_k^+ + \boldsymbol{\Delta}_k\,\delta\boldsymbol{\Delta}_k^+)\Big\}. \end{aligned} \tag{11.53a}$$

The spin-antisymmetric components $\delta\xi_{p,\mu}$ are given by an analogous equation, where f_{pk}^{s} is replaced by f_{pk}^{a} and $\text{tr}_\sigma\{\cdots\}$ is replaced by $\text{tr}_\sigma\{\boldsymbol{\sigma}_\mu(\cdots)\}$. Furthermore, using the definition of $\delta\mathbf{F}_k$, (10.30b), one obtains

$$\begin{aligned} \delta\boldsymbol{\Delta}_p + \sum_k V_{pk}\theta_k\,\delta\boldsymbol{\Delta}_k = \sum_k V_{pk}\Big\{ & \boldsymbol{\Delta}_k\Big[\hbar(\omega+u)\Big(1+\frac{2\xi_k}{\hbar\omega}\Big)\bar{\lambda}_k + 2\frac{u}{\omega}\theta_k'\Big]\delta\boldsymbol{\xi}_k \\ & + \bar{\lambda}_k\Big[\boldsymbol{\Delta}_k\boldsymbol{\Delta}_k^+\,\delta\boldsymbol{\Delta}_k + \boldsymbol{\Delta}_k\,\delta\boldsymbol{\Delta}_k^+\,\boldsymbol{\Delta}_k \\ & - \tfrac{1}{2}\hbar^2(\omega^2-u^2)\Big(1+\frac{2\xi_k}{\hbar\omega}\Big)\,\delta\boldsymbol{\Delta}_k\Big] + \hbar\frac{u^2}{\omega}\theta_k'\,\delta\boldsymbol{\Delta}_k\Big\}. \end{aligned} \tag{11.53b}$$

Here we have defined

$$u = \frac{\hbar}{m^*} \boldsymbol{k} \cdot \boldsymbol{q}, \tag{11.54a}$$

$$\theta'_{\boldsymbol{k}} = \frac{\partial \theta_{\boldsymbol{k}}}{\partial \xi_{\boldsymbol{k}}}, \tag{11.54b}$$

$$\lambda_{\boldsymbol{k}} = -2 \frac{|\Delta_{\boldsymbol{k}}|^2}{D_{\boldsymbol{k}}} [(\hbar\omega)^2(\theta_+ + \theta_-) + (\theta_+ - \theta_-)(\xi_+^2 - \xi_-^2)] \tag{11.55a}$$

$$\equiv |\Delta_{\boldsymbol{k}}|^2 \bar{\lambda}_{\boldsymbol{k}}, \tag{11.55b}$$

where

$$D_{\boldsymbol{k}} = [(\hbar\omega)^2 - (\xi_+ - \xi_-)^2][(\hbar\omega)^2 - (\xi_+ + \xi_-)^2] - 4(\hbar\omega)^2 |\Delta_{\boldsymbol{k}}|^2. \tag{11.55c}$$

The quantity $\lambda_{\boldsymbol{k}}$ is a kind of Cooper-pair susceptibility (Tsuneto 1960). Its poles at $\hbar\omega = \pm 2E_{\boldsymbol{k}} + O(q)$ describe the excitation of pairs into two Bogoliubov quasiparticles of momenta $\boldsymbol{k}$ and $-\boldsymbol{k}$.

In (11.53a,b) we have retained terms asymmetric with respect to an interchange of particles and holes (i.e. terms odd in $\xi_{\boldsymbol{k}}$). In spite of their numerical smallness, they are nevertheless quite important in some instances, as will be seen later.

The time-dependent mean-field equations (11.53a,b) still contain the complete dynamics of the system in the collisionless, linear-response regime, but they are much easier to solve than the original matrix kinetic equation. Equivalent equations may be derived by Green's-function methods (Serene 1974, Maki 1974, Ebisawa and Maki 1974, Nagai 1975, Tewordt 1974a,b, Serene and Rainer 1983) and by functional-integral methods (Alonso and Popov 1977, Kleinert 1978, Brusov and Popov 1980a,b,c, 1984a,b).

Let us first consider the collective modes of the order parameter, neglecting Fermi-liquid corrections. They are given by the eigensolutions of (11.53b) at fixed diagonal mean field $\delta\xi_{\boldsymbol{k}}$. In the case of a pure spin-triplet order parameter the equation of motion for $\delta\boldsymbol{\Delta}_{\boldsymbol{k}}$ may be rewritten in terms of the vector order-parameter fluctuation $\delta d_\mu(\hat{\boldsymbol{k}}; \boldsymbol{q}, \omega)$, defined as usual by

$$\delta\Delta_{\boldsymbol{k},\alpha\beta}(\boldsymbol{q}, \omega) = \sum_\mu \delta d_\mu(\hat{\boldsymbol{k}}; \boldsymbol{q}, \omega)\, (\boldsymbol{\sigma}_\mu \mathrm{i}\boldsymbol{\sigma}_2)_{\alpha\beta}, \tag{11.56}$$

in the form

$$\begin{aligned}
&\Big\langle Y^*_{ml}(\hat{\boldsymbol{k}}) \Big\{ \bar{\lambda}(\hat{\boldsymbol{k}})[(\hbar\omega)^2 - u^2]\, \delta d_\mu(\hat{\boldsymbol{k}}) \\
&\quad - 2\bar{\lambda}(\hat{\boldsymbol{k}}) \sum_\nu [|d^0_\nu(\hat{\boldsymbol{k}})|^2\, \delta d_\mu(\hat{\boldsymbol{k}}) - [d^0_\nu(\hat{\boldsymbol{k}})]^2\, \delta d^*_\mu(\hat{\boldsymbol{k}}) \\
&\quad + 2 d^0_\mu(\hat{\boldsymbol{k}}) d^0_\nu(\hat{\boldsymbol{k}})\, \delta d^*_\nu(\hat{\boldsymbol{k}})] + 4\Xi(\hat{\boldsymbol{k}})\, \delta d_\mu(\hat{\boldsymbol{k}}) \Big\} \Big\rangle_{\hat{k}} \\
&= 2\Big\langle Y^*_{lm}(\hat{\boldsymbol{k}}) \bar{\lambda}(\hat{\boldsymbol{k}}) \Big[(\hbar\omega + u)\Big(1 + 2\lambda_{\mathrm{A}}(\hat{\boldsymbol{k}}) \frac{\Delta_{\boldsymbol{k}}}{\hbar\omega}\Big) + 2\theta_{\mathrm{A}}(\hat{\boldsymbol{k}}) u \frac{\Delta_{\boldsymbol{k}}}{\hbar\omega} \Big] \\
&\quad \times \Big[d^0_\mu(\hat{\boldsymbol{k}})\, \delta\xi_{\boldsymbol{k}} - \mathrm{i} \sum_{\nu\lambda} \epsilon_{\mu\nu\lambda} d^0_\nu(\hat{\boldsymbol{k}})\, \delta\xi_{\hat{\boldsymbol{k}},\lambda} \Big] \Big\rangle_{\hat{\boldsymbol{k}}}.
\end{aligned} \tag{11.57}$$

Here $d_\mu^0(\hat{k})$ is the equilibrium order parameter, and we have allowed for pairing in an arbitrary odd-l state. The pair-interaction constant has been expressed in terms of the gap parameter by means of the equilibrium gap equations, leading to the term

$$\Xi(\hat{\boldsymbol{k}}) = \int \mathrm{d}\xi_{\boldsymbol{k}} \left[\theta_{\boldsymbol{k}} - \frac{\langle \theta_{\boldsymbol{k}} \, |d^0(\hat{\boldsymbol{k}})|^2 \rangle_{\hat{\boldsymbol{k}}}}{\langle |d^0(\hat{\boldsymbol{k}})|^2 \rangle_{\hat{\boldsymbol{k}}}} \right]. \tag{11.58}$$

Furthermore, the diagonal energy has been expanded in spin matrices as

$$\delta\xi_{\boldsymbol{k},\alpha\beta} = \delta\xi_{\boldsymbol{k}} \, \delta_{\alpha\beta} + \sum_{\mu=1}^{3} \delta\xi_{\boldsymbol{k},\mu} (\sigma_\mu)_{\alpha\beta}. \tag{11.59}$$

The angular brackets denote the average over the Fermi surface. The quantity $d_\mu^*(\hat{\boldsymbol{k}})$ is the complex conjugate of $d_\mu(\hat{\boldsymbol{k}}; -\boldsymbol{q}, -\omega)$. We have kept the particle–hole-asymmetric terms in $\delta\xi_{\boldsymbol{k}}$, because they can lead to a small but important coupling of $\delta\Delta_{\boldsymbol{k}}$ to $\delta\xi_{\boldsymbol{k}}$, but have neglected corresponding terms in $\delta\Delta_{\boldsymbol{k}}$, since they cause only a negligible frequency shift.

The quantities $\lambda(\hat{\boldsymbol{k}})$ and $\bar{\lambda}(\hat{\boldsymbol{k}})$ are defined by

$$\begin{aligned} \lambda(\hat{\boldsymbol{k}}) &= \int \mathrm{d}\xi_{\boldsymbol{k}} \, \lambda_{\boldsymbol{k}} \\ &\equiv |\Delta_{\boldsymbol{k}}|^2 \, \bar{\lambda}(\hat{\boldsymbol{k}}), \end{aligned} \tag{11.60}$$

and the particle–hole-asymmetry functions $\lambda_{\mathrm{A}}(\hat{\boldsymbol{k}})$ and $\theta_{\mathrm{A}}(\hat{\boldsymbol{k}})$ are given by

$$\lambda_{\mathrm{A}}(\hat{\boldsymbol{k}}) = \frac{1}{\lambda(\hat{\boldsymbol{k}})} \int \mathrm{d}\xi_{\boldsymbol{k}} \frac{\xi_{\boldsymbol{k}}}{|\Delta_{\boldsymbol{k}}|} \lambda_{\boldsymbol{k}}, \tag{11.61a}$$

$$\theta_{\mathrm{A}}(\hat{\boldsymbol{k}}) = \frac{1}{\lambda(\hat{\boldsymbol{k}})} \int \mathrm{d}\xi_{\boldsymbol{k}} \, |\Delta_{\boldsymbol{k}}| \frac{\partial \theta_{\boldsymbol{k}}}{\partial \xi_{\boldsymbol{k}}}. \tag{11.61b}$$

In the limit $\boldsymbol{q} \to 0$ the function $\lambda(\hat{\boldsymbol{k}})$ has the following limiting values.

(i) For $T \to T_{\mathrm{c}}$, the main contribution is given by

$$\lambda(\hat{\boldsymbol{k}}) = \pi \frac{\Delta_{\boldsymbol{k}}}{\hbar\omega} \frac{\tanh(\hbar\omega/4k_{\mathrm{B}}T)}{[4 - (\hbar\omega/|\Delta_{\boldsymbol{k}}|)^2]^{1/2}}, \tag{11.62a}$$

indicating that $\lambda(\hat{\boldsymbol{k}})$ takes on imaginary or real values according to whether $\omega > 2|\Delta_{\boldsymbol{k}}|$ or vice versa. For fixed ω, and $\Delta_{\boldsymbol{k}} \to 0$, $\lambda(\hat{\boldsymbol{k}})$ tends to zero as

$$\lambda(\hat{\boldsymbol{k}}) \to \frac{\pi |\Delta_{\boldsymbol{k}}|}{4 k_{\mathrm{B}} T} \quad \text{as } T \to T_{\mathrm{c}} \tag{11.62b}$$

(assuming $\hbar\omega \ll 4k_{\mathrm{B}}T$).

(ii) At $T=0$ one finds

$$\lambda(\hat{\boldsymbol{k}})=\frac{1}{4}\frac{1}{[y(y-1)]^{1/2}}\ln\frac{2y-1-2[y(y-1)]^{1/2}}{2y-1+2[y(y-1)]^{1/2}}, \qquad (11.62c)$$

where $y=(\hbar\omega/2\Delta_{\boldsymbol{k}})^2$. It follows that at $\omega=0$ one has $\lambda(\hat{\boldsymbol{k}})=1$.

11.3.1 Collective modes in the B phase

To calculate the collective modes in the B phase, we begin by substituting the corresponding equilibrium gap parameter $d^0_\mu(\hat{\boldsymbol{k}})=\Delta\hat{k}_\mu$ into (11.57) (remember that spin space is rotated relative to orbital space by the action of the dipole force; this is, however, irrelevant for the present calculations as long as we neglect the tiny effect of the dipole interaction on the mode frequencies). Defining the real part and the imaginary part (up to a factor i) of the fluctuation variables by

$$\tfrac{1}{2}[\delta d_\mu(\hat{\boldsymbol{k}};\boldsymbol{q},\omega)\pm\delta d^*_\mu(\hat{\boldsymbol{k}};-\boldsymbol{q},-\omega)]=\begin{cases}\sum_j d'_{\mu j}\hat{k}_j & \text{for } (+),\\ \sum_j d''_{\mu j}\hat{k}_j & \text{for } (-),\end{cases} \qquad (11.63)$$

we are then led to the set of equations

$$(\hbar\omega)^2 d'_{\mu j}-\tfrac{4}{5}\Delta^2\Big[\delta_{\mu j}\sum_\nu d'_{\nu\nu}+d'_{\mu j}+d'_{j\mu}\Big]$$
$$=2i\hbar\omega\Delta\sum_\nu\epsilon_{\mu j\nu}\langle\delta\xi_{\boldsymbol{k},\nu}\rangle_{\boldsymbol{k}}+4\lambda_{\mathrm{A}}\Delta^2\delta_{\mu j}\langle\delta\xi_{\boldsymbol{k}}\rangle_{\hat{\boldsymbol{k}}}, \qquad (11.64a)$$

$$[(\hbar\omega)^2-4\Delta^2]d''_{\mu j}+\tfrac{4}{5}\Delta^2\Big[\delta_{\mu j}\sum_\nu d''_{\nu\nu}+d''_{\mu j}+d''_{j\mu}\Big]$$
$$=2\hbar\omega\,\delta_{\mu j}\Delta\langle\delta\xi_{\boldsymbol{k}}\rangle_{\hat{\boldsymbol{k}}}+4i\lambda_{\mathrm{A}}\Delta^2\sum_\nu\epsilon_{\mu j\nu}\langle\delta\xi_{\boldsymbol{k},\nu}\rangle_{\hat{\boldsymbol{k}}}. \qquad (11.64b)$$

For simplicity, we have assumed $\boldsymbol{q}=0$. The function $\lambda_{\boldsymbol{k}}$ is isotropic for $\boldsymbol{q}\to 0$ and has dropped out except for the particle–hole-asymmetric terms, where it appears in the parameter λ_{A} defined in (11.61a). Note that, owing to the isotropy of the B phase, λ_{A} does not depend on $\hat{\boldsymbol{k}}$. Similarly, the quantity Ξ has disappeared because of the isotropy of the quasiparticle energy. Since (11.64a,b) are linear in ω^2, there exists exactly one eigenmode for each variable. The set of equations (11.64a,b) is diagonalized by constructing the linear combinations of order-parameter components listed in Table 11.1. This immediately yields the 18 eigenfrequencies also given in Table 11.1 (see Maki 1974, 1976b, Serene 1974, Nagai 1975, Tewordt *et al.* 1975b, Tewordt and Einzel 1976, Jacak 1977, Chechetkin 1976b,c,d, Hasegawa and Namaizawa 1982).

(*i*) *Goldstone modes.* There are altogether four Goldstone modes with $\omega \propto q$. One of them (with quantum number $(J, K) = 0^-$) is associated with $\sum_\nu d''_{\nu\nu}$, the variable corresponding to the total phase of the order parameter. This is the well-known Anderson–Bogoliubov mode of isotropic neutral superfluids (Anderson 1958b, Bogoliubov *et al.* 1959). It may be excited via its coupling to density fluctuations.

The other three gapless modes (with $J = 1$, $K = +1$, i.e. 1^+ in the classification of Table 11.1) are associated with the variables $\sum_{\mu,j} \epsilon_{\nu\mu j} d'_{\mu j}$ ($\nu = 1, 2, 3$). They correspond to relative rotations of spin and orbital variables and may be identified as spin–orbit waves that couple to spin-density fluctuations.

(*ii*) *Pair-vibration modes.* The $J = 2$ modes are found to have eigenfrequencies $\hbar\omega_{2^+} = (\frac{8}{5})^{1/2}\Delta$ (real order parameter, i.e. $K = +1$) and $\hbar\omega_{2^-} = (\frac{12}{5})^{1/2}\Delta$ (imaginary order parameter, i.e. $K = -1$) in the limit $\boldsymbol{q} \to 0$. Each of these modes is fivefold degenerate, in compliance with the symmetry classification. The $J = 2$ modes of the B phase were first calculated by Vdovin (1963), i.e. long before the experimental discovery of the superfluidity of ^{3}He, but the results later fell into oblivion. Of these ten modes, the 2^- modes couple to inhomogeneous (i.e. finite-q) density fluctuations, such as sound waves, while the 2^+ modes couple to spin fluctuations with finite q. In fact, the latter also couple to sound waves through particle–hole-asymmetric terms, but the coupling is much smaller than that of the 2^- modes (Koch and Wölfle 1981). The 2^- modes and 2^+ modes have been dubbed the "(imaginary) squashing modes" and "real squashing modes" respectively, because they describe an oscillatory, quadrupolar mode in which the spherical gap parameter is squashed in its imaginary and real parts.

It is interesting to note that the sum of the squares of the frequencies of the $2^\pm$ modes is equal to the square of the pair-breaking frequency 2Δ, i.e. $(\hbar\omega_{J^-})^2 + (\hbar\omega_{J^+})^2 = (2\Delta)^2$. This is a consequence of a supersymmetry of the effective Hamiltonian (Nambu 1985), which relates the energy gap of the basic fermionic constituents and the frequencies of the bosonic collective modes for each angular-momentum channel.

The eigenfrequencies of the four remaining modes 0^+ and 1^- are exactly at the pair-breaking edge, i.e. $\omega = 2\Delta$. They couple to homogeneous density or spin-density variations through particle–hole-asymmetric terms only.

Finite wave vector, $|\boldsymbol{q}| \neq 0$

The generalization of (11.64) to nonzero wave vectors is straightforward (Wölfle 1977a, Brusov and Popov 1980a,b, Combescot 1982b). The variables decouple in the (J, J_z) representation, as noted before on the basis of symmetry considerations. At large values of q such that $\hbar v_F q \gg \Delta$, all the

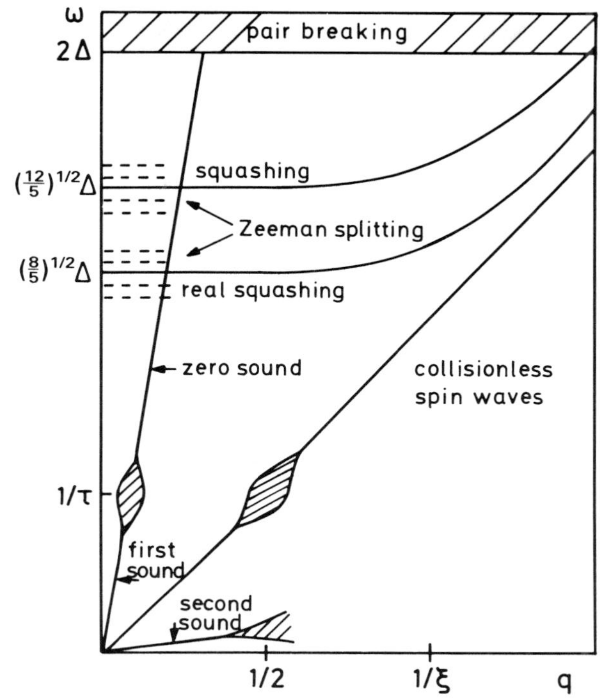

Figure 11.1 Collective modes in the B phase (schematic). The transition from the hydrodynamic to the collisionless regime is indicated by a broadening of the excitation branch. The dashed lines indicate the splitting in a magnetic field. The collective-mode frequencies at $q=0$ indicated in the figure are the weak-coupling results.

modes possess linear dispersion. The leading q^2 correction for small q is listed in Table 11.1. The numerical coefficient in some cases involves the quantity λ_1, which follows from expanding $\lambda(\hat{\boldsymbol{k}})$ to first order in q^2, and is defined by

$$\lambda_1[\omega(\boldsymbol{q}=0)] = \left[\frac{1}{\lambda(\hat{\boldsymbol{k}})}\frac{\Delta^2}{(\hbar\omega)^2}\int \mathrm{d}\xi_{\boldsymbol{k}}\frac{4\Delta^2}{4E_{\boldsymbol{k}}^2-(\hbar\omega)^2}\left[\frac{4\xi_{\boldsymbol{k}}^2-(\hbar\omega)^2}{4E_{\boldsymbol{k}}^2-(\hbar\omega)^2}\theta_{\boldsymbol{k}}+\frac{\xi_{\boldsymbol{k}}^2}{E_{\boldsymbol{k}}}\frac{\mathrm{d}\theta_{\boldsymbol{k}}}{\mathrm{d}E_{\boldsymbol{k}}}\right]\right]_{\boldsymbol{q}=0}. \tag{11.65}$$

The frequency in the argument of λ_1 is the mode frequency at $\boldsymbol{q}=0$. A schematic plot of the collective mode frequencies as functions of q is shown in Fig. 11.1.

Effect of a magnetic field

An external magnetic field may be incorporated into the description by allowing for the Zeeman splitting in the normal quasiparticle energies: $\xi_{\boldsymbol{k}s}=\xi_{\boldsymbol{k}}\pm s\mu_0 H$, $s=\pm 1$, where μ_0 is the nuclear magnetic moment. The equations then become quite complicated, so we do not reproduce them here (see Tewordt and Schopohl 1979b, Schopohl and Tewordt 1981). The magnetic field induces a spin-singlet component of the order-parameter fluctuations, which is responsible for a *linear* rather than quadratic splitting of the mode. Such a splitting has already been calculated for the spin-wave spectrum, i.e. the 1^+ mode, in (11.37).

For small magnetic fields the linear splitting of the $J = 2$ collective-mode frequencies is given by

$$\omega_{2^\pm}^{(m_J)}(T, H) = \omega_{2^\pm}(T) + m_J g_{2^\pm}(T)\omega_L(T, H), \tag{11.66}$$

where $m_J = 0, \pm 1, \pm 2$. The g factors may be expressed in terms of the Yosida function Y_0 defined in (3.101c) and the function λ as

$$g_{2^\pm}(T) = \frac{1}{12}\left[1 \pm \frac{1 - Y_0}{\lambda(\omega_{2^\pm})}\right], \tag{11.67a}$$

and

$$\omega_L(T, H) = \frac{\gamma H}{1 + F_0^a(\frac{2}{3} + \frac{1}{3}Y_0)} \tag{11.67b}$$

is the renormalized Fermi-liquid Larmor frequency. The splitting of the real squashing mode ($J = 2^+$) has been observed experimentally (Avenel *et al.* 1980) and is in full agreement with the theoretical prediction by Tewordt and Schopohl (1979b). The splitting of the squashing mode is harder to observe because of the stronger coupling of the sound to the collective mode (Meisel *et al.* 1983c). It has nevertheless been detected in measurements of the group velocity (Movshovich *et al.* 1988) by employing the theoretical analysis of this experimental situation developed by Varoquaux *et al.* (1986). At higher fields the splitting deviates substantially from linear behaviour (Shivaram *et al.* 1983, 1986) because the equilibrium gap parameter is itself distorted by the magnetic field (Schopohl *et al.* 1983, Brusov and Popov 1984a,b,c, Fishman and Sauls 1986, 1988b). The possibility of inducing transitions between the sublevels by an RF field has been discussed by Dombre and Combescot (1982b). The collective-mode frequencies are also affected by superflow. The effect is somewhat similar to that of a magnetic field (Kleinert 1980a, Sauls and Serene 1984, Brusov 1985a,b).

11.3.2 Collective modes in the A phase

In the axial state the equilibrium gap parameter is given by

$$d_\mu^0(\hat{\boldsymbol{k}}) = (\tfrac{8}{3}\pi)^{1/2}\Delta_0 \hat{d}_\mu^0 Y_{11}(\hat{\boldsymbol{k}}). \tag{11.68}$$

We shall take $\hat{d}_\mu^0$ to lie along the z axis. Since $d_\mu^0(\hat{\boldsymbol{k}})$ is proportional to Y_{11}, it is convenient to expand the order-parameter fluctuations in terms of spherical harmonics Y_{1m}:

$$\delta d_\mu(\hat{\boldsymbol{k}}) = \sum_{m=0,\pm 1} \delta d_{\mu m}\, Y_{1m}(\hat{\boldsymbol{k}}). \tag{11.69}$$

Furthermore, we introduce real and imaginary parts of $\delta d_{\mu m}$, $d'_{\mu m}$ and $d''_{\mu m}$, in analogy with (11.63). Substituting (11.69) into (11.57), we find in the limit $\boldsymbol{q}\to 0$ the following set of equations:

$$\left[\left(\frac{\hbar\omega}{\Delta_0}\right)^2 \bar{\lambda}_m - 2\lambda_m + 4\Xi_m\right] d'_{\mu m} + 2\delta_{m1}\lambda_m\left(d'_{\mu m} - 2\hat{d}^0_\mu \sum_\nu \hat{d}^0_\nu d'_{\nu m}\right)$$
$$= -2\mathrm{i}\frac{\hbar\omega}{\Delta_0}\delta_{m1}\lambda_1 \sum_{\nu\lambda} \epsilon_{\mu\nu\lambda}\hat{d}^0_\nu \langle \delta\xi_{\boldsymbol{k},\lambda}\rangle_{\hat{\boldsymbol{k}}}, \tag{11.70a}$$

$$\left[\left(\frac{\hbar\omega}{\Delta_0}\right)^2 \bar{\lambda}_m - 2\lambda_m + 4\Xi_m\right] d''_{\mu m} - 2\delta_{m1}\lambda_m\left(d''_{\mu m} - 2\hat{d}^0_\mu \sum_\nu \hat{d}^0_\nu d''_{\nu m}\right)$$
$$= 2\frac{\hbar\omega}{\Delta_0}\delta_{m1}\lambda_1 \hat{d}^0_\mu \langle \delta\xi_{\boldsymbol{k}}\rangle_{\hat{\boldsymbol{k}}}. \tag{11.70b}$$

Here we have defined

$$\left.\begin{aligned} \lambda_m &= \int \mathrm{d}\Omega_{\hat{\boldsymbol{k}}}\, |Y_{1m}(\hat{\boldsymbol{k}})|^2\, \lambda(\hat{\boldsymbol{k}}), \\ \bar{\lambda}_m &= \int \mathrm{d}\Omega_{\hat{\boldsymbol{k}}}\, |Y_{1m}(\hat{\boldsymbol{k}})|^2 \frac{\lambda(\hat{\boldsymbol{k}})}{|\Delta(\hat{\boldsymbol{k}})|^2}\Delta_0^2, \\ \Xi_m &= \int \mathrm{d}\Omega_{\hat{\boldsymbol{k}}}\, |Y_{1m}(\hat{\boldsymbol{k}})|^2 \Xi(\hat{\boldsymbol{k}}). \end{aligned}\right\} \tag{11.70c}$$

In deriving (11.70a,b), we have exploited the axial symmetry of the functions $\lambda(\hat{\boldsymbol{k}})$ and $\Xi(\hat{\boldsymbol{k}})$ about $\hat{\boldsymbol{l}}$ for $q\to 0$.

In contrast with the B phase, different angular averages of $\lambda(\hat{\boldsymbol{k}})$ are involved in (11.70a,b,c), so that $\lambda_{\boldsymbol{k}}$ does not simply cancel as in (11.64a,b). A further difficulty arises from the frequency dependence of $\lambda(\hat{\boldsymbol{k}})$, as defined by (11.55). The eigenvalue equations, obtained by putting the right-hand sides of (11.70a,b) equal to zero, are now nonlinear in ω^2 and may have more than one solution.

As usual, a nontrivial solution of a system of linear homogeneous equations exists if the determinant of coefficients vanishes. A numerical analysis of the eigenvalue problem for (11.70) yields well-defined zeros of the real part of the determinant at real frequencies (Wölfle 1976c, 1977a, Tewordt and Schopohl 1979a). However, the imaginary part of the determinant does *not* vanish at these frequencies (in contrast with what happens in the B phase). This is a consequence of pair-breaking processes, which may occur at any frequency owing to the two nodes in the energy gap in the axial state. Consequently, one might expect a zero of the determinant to exist somewhere in the complex frequency plane. But this is not the case, as will be discussed below. In fact, it may be shown that—strictly speaking—the determinant does not have any zeros in the complex frequency plane, except for $\omega = 0$ (Dombre and Combescot 1981). Hence in the A phase there are no well-defined modes at all—there are only "resonances". Nevertheless, we shall call these resonances "collective modes".

The limit $|\boldsymbol{q}| = 0$

Following Table 11.2, we group the collective modes according to the label $m = 0, \pm 1$ of the order-parameter fluctuations $d'_{\mu m}$ and $d''_{\mu m}$ (at least in weak-coupling theory).

(*i*) $m = 1$. For $m = \pm 1$, the angular average of $\Xi(\hat{\boldsymbol{k}})$ defined in (11.70c) vanishes, i.e. $\Xi_{\pm 1} = 0$. It turns out that in this case there is at most one collective mode per variable.

The order-parameter variables d'_{x1}, d'_{y1} and d''_{z1} are associated with rotations in spin space about the x and y axes and the gauge–orbit symmetry respectively. Their eigenfrequencies are zero (they are Goldstone modes) and the corresponding dispersion laws are linear in q, as will be discussed below. The remaining variables d'_{z1}, d''_{x1} and d''_{y1} do not even show any resonance because the real part of the eigenvalue equation

$$\left(\frac{\hbar\omega}{\Delta_0}\right)^2 \bar{\lambda}_1 - 4\lambda_1 = 0 \tag{11.71}$$

does not have a solution for real ω (at least in the absence of Fermi-liquid corrections). Note the simplicity of (11.71), which is a consequence of $\Xi_{m=\pm 1} = 0$.

(*ii*) $m = -1$. The six variables $d'_{\mu,-1}$, $d''_{\mu,-1}$ ($\mu = x, y, z$) are found to oscillate at a frequency

$$\hbar\omega_{\text{cl}} = \left(\frac{2\,\text{Re}\,\lambda_1}{\text{Re}\,\bar{\lambda}_1}\right)^{1/2} \Delta_0 \approx 1.23\Delta_0(T). \tag{11.72}$$

Hence they correspond to pair vibration modes.

Since the motion of the order-parameter structure, i.e. the motion of the unit vectors $\hat{\boldsymbol{m}}$ and $\hat{\boldsymbol{n}}$, resembles the clapping of hands, as explained in the beginning of Section 11.2, this mode has been called the "clapping mode" (see Fig. 11.2). It is remarkable that ω_{cl} scales with the energy gap $\Delta_0(T)$ at all temperatures within one per cent accuracy (see Fig. 11.3) in spite of the presence of the λ factors in (11.72).

Among the six modes, only $d''_{z,-1}$ couples to sound waves (neglecting the coupling due to particle–hole-asymmetric terms), whereas $d'_{x,-1}$ and $d'_{y,-1}$ couple to spin waves.

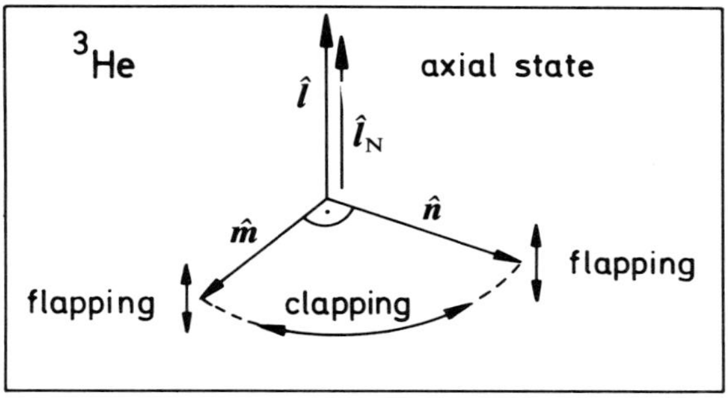

Figure 11.2 A pictorial representation of the order-parameter modes in the A phase.

(*iii*) $m=0$. The case $m=0$ plays a special role because here there exist *three* resonances for each of the variables $d'_{\mu,0}$ and $d''_{\mu,0}$. This is due to the nonzero value of Ξ_0, given by

$$\Xi_0 = -\frac{1}{2}\int \mathrm{d}\Omega_{\hat{k}}\, |Y_{10}(\hat{\boldsymbol{k}})\Delta(\hat{\boldsymbol{k}})|^2 \int \mathrm{d}\xi_{\boldsymbol{k}} \frac{1}{E_{\boldsymbol{k}}}\frac{\partial \theta_{\boldsymbol{k}}}{\partial E_{\boldsymbol{k}}} \tag{11.73}$$

(here a partial integration with respect to $\cos\theta$ has been performed to bring Ξ_0 into the form (11.73)). A small admixture to the gap parameter proportional to $Y_{10}(\hat{\boldsymbol{k}})$, as given by these variables, causes the preferred directions $\hat{\boldsymbol{m}}$ and $\hat{\boldsymbol{n}}$ to oscillate out of plane. This motion may be visualized as a flapping of wings (see Fig. 11.2)—hence the name "flapping mode" (Wölfle 1975b). The $m=0$ eigenvalue equation is given by

$$\left(\frac{\hbar\omega}{\Delta_0}\right)^2 \bar{\lambda}_0 - 2\lambda_0 + 4\Xi_0 = 0. \tag{11.74}$$

More-or-less stable oscillations are found to occur at two different nonzero eigenfrequencies. In addition, there is a solution at zero frequency corresponding to a diffusive mode (see below).

Let us discuss the solution of (11.74) in more detail. There is one solution with a strongly temperature-dependent eigenfrequency:

$$\frac{\hbar\omega_{\mathrm{nfl}}}{\Delta_0} \approx \frac{T}{T_{\mathrm{c}}}. \tag{11.75}$$

More precisely, in the limits $T\to T_{\mathrm{c}}$ and $T\to 0$ the analytic results are

$$\frac{\hbar\omega_{\mathrm{nfl}}}{\Delta_0} = \left(\tfrac{4}{5}\right)^{1/2}\frac{T}{T_{\mathrm{c}}}\left[1-\frac{28}{3\pi^4}\zeta(3)\frac{\Delta_0(T)}{k_{\mathrm{B}}T}\right] \quad \text{as } T\to T_{\mathrm{c}}, \tag{11.76a}$$

$$\hbar\omega_{\mathrm{nfl}} = \left(\tfrac{2}{3}\right)^{1/2}\pi k_{\mathrm{B}}T\left(\ln\frac{T_{\mathrm{c}}}{T}+\mathrm{const}\right)^{-1/2} \quad \text{as } T\to 0. \tag{11.76b}$$

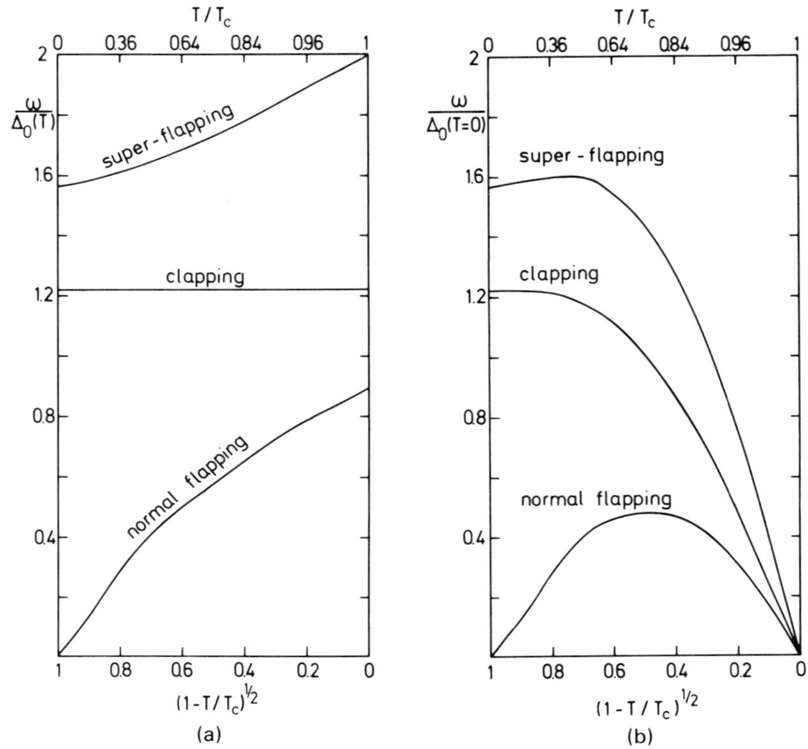

Figure 11.3 Collective-mode frequencies of the A phase versus temperature in units (a) of the temperature-dependent gap $\Delta_0(T)$ and (b) of the gap at $T = 0$, $\Delta_0(0)$.

The complete numerical result is shown in Fig. 11.3. The physical origin of this mode may easily be visualized. It arises from the fact that the flapping of the vectors $\hat{\boldsymbol{m}}$ and $\hat{\boldsymbol{n}}$ is equivalent to an oscillation of the vector $\hat{\boldsymbol{l}}$ about its equilibrium direction (remember that $\hat{\boldsymbol{l}} = \hat{\boldsymbol{m}} \times \hat{\boldsymbol{n}}$). Since $\hat{\boldsymbol{l}}$ specifies the axis of the energy gap, a motion of $\hat{\boldsymbol{l}}$ will change the quasiparticle spectrum and hence the equilibrium quasiparticle distribution $f^0_{\boldsymbol{k}}(\hat{\boldsymbol{l}})$. The quasiparticles then have to adjust to the new equilibrium state by means of collision processes. Hence the relaxation to equilibrium will take a characteristic time τ_l, which is of the order of the quasiparticle collision time τ. As one may expect, τ_l is closely related to the orbital-viscosity coefficient λ_4 calculated in (10.170). Now, if the oscillation of $\hat{\boldsymbol{l}}$ is too rapid for the quasiparticles to follow, the quasiparticle distribution will not be changed much. In this case there is a well-defined direction $\hat{\boldsymbol{l}}_{\mathrm{N}}$ given by the time averaged axis of the energy gap, about which the superfluid component of the system oscillates. The restoring force that maintains this oscillation is provided by the potential well formed by the energy of the quasiparticle system. Therefore the oscillation frequency will decrease with temperature and will eventually tend to zero as $T \to 0$ in the same way as the density of quasiparticles goes to zero in this limit. Because of the important role played by the normal-fluid component in the mechanism of this mode, it has been called the "normal-flapping mode". It is interesting to note that this mode is exceptional since it owes its existence to the thermal excitations in the superfluid and hence is not a property of the superfluid condensate itself.

In contrast with the "normal-flapping mode", the frequency of the other flapping mode scales approximately with the gap, similarly to the clapping mode. It may therefore be viewed as a ground-state property of the superfluid as in the case of the squashing modes in the B phase. Hence we shall refer to this mode as the "super-flapping" mode. The ratio of its eigenfrequency to $\Delta_0(T)$, i.e. $\hbar\omega_{\text{sfl}}/\Delta_0(T)$, is found to decrease from 2 at T_c to approximately 1.56 at $T = 0$ (see Fig. 11.3). The leading correction for $T \to T_c$ is given by

$$\frac{\hbar\omega_{\text{sfl}}}{\Delta_0(T)} = 2\left[1 - \frac{28}{5\pi^4}\,\zeta(3)\,\frac{\Delta_0(T)}{k_B T}\right] \quad \text{as } T \to T_c. \tag{11.77}$$

Pair-breaking effects

The anisotropy of the energy gap in the A phase, $|\Delta(\hat{\boldsymbol{k}})| = \Delta_0 \sin\theta$, implies that the minimum energy required to break a Cooper pair into two quasiparticles varies from zero to $2\Delta_0$, depending on the position on the Fermi surface. Hence pair breaking via absorption of collective-mode quanta may occur at arbitrarily low frequencies. However, at low frequencies ω this effect is reduced because of the limited phase space available. Mathematically, the pair-breaking process is described by the vanishing of the energy denominator in the expression for $\lambda_{\boldsymbol{k}}$ given by (11.55), which gives rise to an imaginary part of $\lambda(\hat{\boldsymbol{k}})$, as evident from (11.62a,c).

To determine the frequency eigenvalues, we have so far omitted the imaginary part, because it is relatively small as long as $\hbar\omega < 2\Delta_0$ and has little influence on the real part of the eigenfrequency. However, one has to keep in mind that the collective modes discussed above are damped by pair-creation processes. Clearly, the higher the frequency, the more damping will occur. We defer a discussion of the damping until later, because in cases of practical interest (e.g. coupling of collective modes to sound waves) the broadening of the absorption peaks caused by the collective modes is also influenced by the imaginary part of the coupling function. Therefore it is more sensible to discuss directly the physically observable quantity (e.g. sound absorption) rather than collective-mode frequencies themselves.

There is yet another problem associated with pair breaking, which has already been briefly mentioned. Even though the small imaginary part in

(11.72) and (11.74) induced by pair-breaking processes may not be of great quantitative importance, it is conceptually relevant. Namely, it implies that, strictly speaking, there are no collective poles in the complex frequency plane corresponding to the clapping and flapping modes. These modes are only peculiar resonances on the real frequency axis, and are not related to a complex pole. The argument runs as follows: In the absence of collision effects, the left-hand sides of (11.71) and (11.74) are functions of ω^2, expressing the fact that time-reversal invariance is strictly valid in this limit. Then, for any given zero of, say, (11.71), at frequency $\omega_+ = \omega_{\text{cl}} - \mathrm{i}\Gamma$ (assuming $\Gamma > 0$), there is a corresponding zero at $\omega_- = -\omega_{\text{cl}} + \mathrm{i}\Gamma$, which lies in the upper half-plane. However, any collective pole in the upper plane violates causality and causes an instability of the system. Hence there can be no complex zeros of (11.71) and (11.74) other than $\omega = 0$. An investigation of the functions on the left-hand sides of (11.71) and (11.74) (Dombre and Combescot 1981) has shown that this is indeed true.

Finite wave vector, $|\boldsymbol{q}| \neq 0$

Even at $|\boldsymbol{q}| \neq 0$ the different spin components of the order parameter decouple. This is a consequence of the fact that the spin and orbital degrees of freedom separate. On the other hand, the respective orbital components become mixed. The eigenvalue problem thus reduces to a set of six 3×3 matrix equations. In the following we only consider the special case $\hat{\boldsymbol{q}} \parallel \hat{\boldsymbol{l}}$ for simplicity, where the 3×3 matrices assume diagonal form.

The dispersion of the Goldstone (i.e. gapless) modes with $m = 1$ associated with the variables d''_{z1} and d'_{x1}, d'_{y1} is given by the single expression

$$\omega^2 = (v_{\text{F}} q)^2 \lim_{\boldsymbol{q} \to 0} \frac{\langle (\hat{\boldsymbol{k}} \cdot \hat{\boldsymbol{q}})^2 \lambda(\hat{\boldsymbol{k}}) \rangle_{\hat{\boldsymbol{k}}}}{\langle \lambda(\hat{\boldsymbol{k}}) \rangle_{\hat{\boldsymbol{k}}}}. \tag{11.78}$$

In the limit $T \to 0$ this reduces to

$$\omega = \pm 3^{-1/2} v_{\text{F}} q. \tag{11.79}$$

A further gapless mode, with $m = 0$, is obtained for the "flapping variables" $d'_{\mu,0}$ and $d''_{\mu,0}$ as a solution of (11.74) in the limit ω, $\boldsymbol{q} \to 0$. The λ functions in (11.74) are highly singular in this limit and have to be analysed with great care. One finds

$$\omega = \alpha (v_{\text{F}} q)^2 \quad \text{as } T \to T_{\text{c}}, \tag{11.80}$$

where

$$\alpha \approx \hbar (1 - \mathrm{i}) \left[\frac{21}{20\pi^3} \frac{1}{\Delta_0(T) k_{\text{B}} T} \right]^{1/2},$$

while in the low-temperature limit

$$\omega = -\mathrm{i} \frac{3}{\pi^3} \left(\frac{\hbar}{k_{\text{B}} T} \right)^2 \ln \left(\frac{\Delta_0}{k_{\text{B}} T} \right) (v_{\text{F}} q)^3 \quad \text{as } T \to 0. \tag{11.81}$$

This mode is the so-called "orbital wave", i.e. the Goldstone mode associated with the breakdown of rotational symmetry in orbital space. It can be seen that this mode is overdamped at all temperatures because of its strong coupling to the quasiparticle system (see the discussion on orbital relaxation in Section 10.3).

The dispersion of the normal-flapping mode follows from (11.74) as

$$\omega_{\mathrm{nfl}}^2(\boldsymbol{q}) = \omega_{\mathrm{nfl}}^2(\boldsymbol{q}=0) + v_{\mathrm{F}}^2(\boldsymbol{q}\cdot\hat{\boldsymbol{l}})^2. \tag{11.82}$$

where (11.82) now holds for arbitrary directions of $\hat{\boldsymbol{q}}$ and $\hat{\boldsymbol{l}}$. At $T=0$, where $\omega_{\mathrm{nfl}}\,(\boldsymbol{q}=0)=0$, the dispersion of the normal-flapping mode comes out to be linear. This behaviour is changed into a quadratic behaviour owing to particle–hole-asymmetric terms in the very-low-frequency regime $\hbar\omega \ll k_{\mathrm{B}}T_{\mathrm{c}}(T_{\mathrm{c}}/T_{\mathrm{F}})$ (Combescot and Combescot 1976a, Cross 1977a).

The collective modes discussed above are all listed in Table 11.2. The effects of the Fermi-liquid interaction, i.e. Fermi-liquid corrections, have been calculated (Ashida and Nagai 1984), but are not included in the table for lack of space. The effects of the dipole interaction on the collective modes have been explored by Tewordt *et al.* (1977).

Effect of a magnetic field

In contrast with the rather complicated situation in the B phase, the effect of a magnetic field on collective modes in the A phase may easily be obtained. This is true at least in weak-coupling theory, where the two spin species are uncoupled, and follows from the fact that spin and orbital structures are separated in the A phase. Therefore the preceding results remain valid for the individual spin-up and spin-down systems. One simply has to take the expressions obtained so far and insert the appropriate gap functions $\Delta_{\uparrow}(T)$ or $\Delta_{\downarrow}(T)$. As a result, one finds a mode splitting (e.g. for the clapping mode) proportional to $\Delta_{\uparrow}-\Delta_{\downarrow}$. There is only one mode of each type in the A_1 phase. A second mode appears below the transition to the A_2 phase (Koch 1980, Schopohl *et al.* 1985).

Another interesting effect involves the motion of the vector $\hat{\boldsymbol{l}}$, i.e. the so-called "flapping modes". If strong-coupling effects are taken into account, the relative orientation of the $\hat{\boldsymbol{l}}$ vectors $\hat{\boldsymbol{l}}_{\uparrow}$ and $\hat{\boldsymbol{l}}_{\downarrow}$ of the two spin subsystems is no longer undetermined, but acquires a fixed value. This gives rise to a further collective mode consisting of relative oscillations of $\hat{\boldsymbol{l}}_{\uparrow}$ and $\hat{\boldsymbol{l}}_{\downarrow}$ (Wartak 1978, 1980a,b).

11.3.3 Effect of residual interactions

Corrections to the collective-mode spectra of both the B and A phases are not only due to Fermi-liquid effects but are also induced by higher-l

components in the pair interaction, i.e. $V_{\boldsymbol{k}-\boldsymbol{k}'} = \sum_l (2l+1)V_l P_l(\hat{\boldsymbol{k}} \cdot \boldsymbol{k}')$ for $|\xi_{\boldsymbol{k}}|, |\xi_{\boldsymbol{k}'}| < \epsilon_c$ (see (3.46a)). In the B phase a finite value of the $l = 3$ and $l = 5$ pair-interaction parameters V_3 and V_5 leads to corrections to the mode frequencies of the order of a few per cent. They are comparable to those induced by the Landau parameters $F_2^{s,a}$ (Sauls and Serene 1981c, Fishman and Sauls 1986). In the case that V_3 and V_5 are sufficiently close to V_1 there will even appear a new mode in the gap (i.e. for $\hbar\omega < 2\Delta$) with total angular momentum $J = 4$. The effect of a small f-wave interaction component V_3 on the A-phase collective modes has also been considered (Wojtanowski and Wölfle 1986a,b, 1987, Wojtanowski 1986, Sauls 1986, Hirashima 1987). The effect of an attractive s-wave pairing constant on a p-wave state has been discussed by Chervonko (1976).

Another type of residual interaction effect is due to collisions among the quasiparticles. These tend to destroy the coherence of the collective motion. This is particularly important in the B phase, where the squashing modes are not damped by pair-breaking processes. As a rule of thumb, the imaginary part of the mode frequency generated in this way is of the order of the quasiparticle relaxation rate τ^{-1} multiplied by a factor proportional to the density of excitations:

$$\omega_{\text{coll}}^{\text{B}} = \frac{\gamma_0}{\hbar} \Delta(T) + \frac{\text{i}}{\tau_{\text{coll}}}. \tag{11.83a}$$

Here $\gamma_0 = (\frac{12}{5})^{1/2}$ or $(\frac{8}{5})^{1/2}$ and the relaxation rate τ_{coll}^{-1} is found as (Wölfle 1975a, 1976b)

$$\frac{1}{\tau_{\text{coll}}} = \frac{1}{\tau} \sum_{\boldsymbol{k}} (-f'_{\boldsymbol{k}}) \frac{\Delta^2}{E_{\boldsymbol{k}}^2 - (\frac{1}{2}\hbar\omega_{\text{coll}})^2}. \tag{11.83b}$$

In the A phase the collision-induced broadening adds to the pair-breaking-induced broadening (Wölfle and Koch 1978).

A strong-coupling theory of the collective modes has not yet been worked out, although it is in principle known how to calculate the leading correction in T_c/T_F (Serene and Rainer 1978, 1983); see, however, Wartak (1978, 1980a,b, 1981, 1982). Judging from the smallness of the strong-coupling corrections left in the thermodynamic properties once the strong-coupling gap parameter is inserted in the weak-coupling theory, one may anticipate that genuine strong-coupling corrections will be small.

11.3.4 Observability of collective modes

Among the set of collective modes found in each of the phases, only the Goldstone mode associated with broken gauge symmetry (which is essentially a sound wave) may be excited and detected directly by experiment (Wheatley 1978). Besides that, it has also been possible to observe standing spin waves using a gradient magnetic field (Masuhara *et al.* 1984). These Goldstone modes are in turn found to couple to some of the order-parameter pair-vibration modes. This allows for an indirect detection of the latter modes via absorption and dispersion effects on the Goldstone modes. In particular, sound waves are very well suited for this purpose since the sound velocity is high compared with the Fermi velocity, implying that the sound-excitation branch will intersect the pair-vibration branch (see Fig. 11.1).

This is not necessarily the case with the spin-wave branch. So far, collective modes have only been seen in sound-propagation experiments. In fact, sound waves may be used as a spectroscopic probe to investigate the pair-vibration modes in superfluid ^{3}He, much as electromagnetic waves are employed to study the electronic-excitation spectra of atoms and molecules. In the next section we shall therefore give a relatively detailed account of the theory of sound propagation in the collisionless regime. A general discussion of the symmetry aspects of the coupling of collective modes to observable quantities has been given by Serene (1983).

11.4 COLLISIONLESS SOUND

The detailed discussion of order-parameter modes in the preceding section has equipped us to calculate the physically observable effects of p-wave pair correlations on the high-frequency dynamics of ^{3}He. Our starting point will be the matrix kinetic equation (10.6) and the conservation laws for particle number and momentum (2.21) and (2.22), together with the definition of the momentum current density Π_{ij} given in (10.17b). Combining (2.21) and 2.22), we find at once the Fourier-transformed equation of motion for the mass-density fluctuations $\delta\rho(\boldsymbol{q}, \omega)$ in terms of the momentum current density:

$$\omega^2\,\delta\rho + \omega\boldsymbol{q}\boldsymbol{\Pi}\boldsymbol{q} = 0. \tag{11.84}$$

The second term in (11.84) may be rewritten by substituting the expression for $\boldsymbol{\Pi}$, (10.17b); this yields

$$\boldsymbol{q}\boldsymbol{\Pi}\boldsymbol{q} = \frac{\hbar^2}{m^*}\sum_{\boldsymbol{k}}(\boldsymbol{k}\cdot\boldsymbol{q})^2(\delta n_{\boldsymbol{k}} + \phi'_{\boldsymbol{k}}\,\delta\xi_{\boldsymbol{k}}). \tag{11.85}$$

Using $(\hat{\boldsymbol{k}}\cdot\hat{\boldsymbol{q}})^2 = \frac{2}{3}P_2(\hat{\boldsymbol{k}}\cdot\hat{\boldsymbol{q}}) + \frac{1}{3}$, where P_2 is the $l = 2$ Legendre polynomial, and employing the integral property (10.18b) of $\phi'_{\boldsymbol{k}}$, this may be cast into the form

$$\boldsymbol{q}\boldsymbol{\Pi}\boldsymbol{q} = \frac{1}{3}\frac{m^*}{m}(v_{\mathrm{F}}q)^2\Big[(1 + F_0^{\mathrm{s}})\,\delta\rho + 2m(1 + \tfrac{1}{5}F_2^{\mathrm{s}})\sum_{\boldsymbol{k}}P_2(\hat{\boldsymbol{k}}\cdot\hat{\boldsymbol{q}})\,\delta n_{\boldsymbol{k}}\Big]. \tag{11.86}$$

It follows from (11.84) that the dispersion law for sound waves is

$$\omega^2 = c_1^2q^2[1 + 2Z\xi(\omega)]. \tag{11.87}$$

where

$$\xi(\omega) = \frac{5}{2} \frac{\sum_{\boldsymbol{k}} P_2(\hat{\boldsymbol{k}} \cdot \hat{\boldsymbol{q}})\, \delta n_{\boldsymbol{k}}}{\sum_{\boldsymbol{k}} \delta n_{\boldsymbol{k}}}, \tag{11.88}$$

$$Z = \frac{2}{5} \frac{1 + \frac{1}{5} F_2^{\mathrm{s}}}{1 + F_0^{\mathrm{s}}}. \tag{11.89a}$$

In fact, if small terms of order $(v_{\mathrm{F}}/c_1)^4$ are neglected, the coefficient Z is given by the relative difference between the velocities of zero sound c_0 and first sound c_1 in the normal state:

$$Z \approx \frac{c_0 - c_1}{c_1}. \tag{11.89b}$$

The response function $\xi(\omega)$ may be expressed in terms of $\delta\xi_{\boldsymbol{k}}$ and $\delta\Delta_{\boldsymbol{k}}$ by taking the $l = 2$ component of the diagonal mean-field equation (11.53a):

$$\begin{aligned}\sum_{\boldsymbol{k}} P_2(\hat{\boldsymbol{k}} \cdot \hat{\boldsymbol{q}})\, \delta n_{\boldsymbol{k}} = N_{\mathrm{F}} \Big\langle P_2(\hat{\boldsymbol{k}} \cdot \hat{\boldsymbol{q}}) \Big\{ &\frac{u}{\omega - u} [1 - \lambda(\hat{\boldsymbol{k}})]\, \delta\xi_{\boldsymbol{k}} - \lambda(\hat{\boldsymbol{k}})\, \delta\xi_{\boldsymbol{k}} \\ &+ \tfrac{1}{4}\hbar\omega\bar{\lambda}(\hat{\boldsymbol{k}})(\delta\Delta_{\boldsymbol{k}}\, \Delta_{\boldsymbol{k}}^{+} - \Delta_{\boldsymbol{k}}\, \delta\Delta_{\boldsymbol{k}}^{+}) \\ &+ \lambda_{\mathrm{A}}(\hat{\boldsymbol{k}})\bar{\lambda}(\boldsymbol{k})(\delta\Delta_{\boldsymbol{k}}\, \Delta_{\boldsymbol{k}}^{+} + \Delta_{\boldsymbol{k}}\, \delta\Delta_{\boldsymbol{k}}^{+}) \Big\} \Big\rangle_{\hat{\boldsymbol{k}}}. \end{aligned} \tag{11.90}$$

We now take advantage of the fact that the Fermi velocity is much smaller than the sound velocity. This allows us to calculate $\delta\xi_{\boldsymbol{k}}$ and $\delta\Delta_{\boldsymbol{k}}$ in an expansion in powers of $u/\omega = (\hat{\boldsymbol{k}} \cdot \hat{\boldsymbol{q}}) v_{\mathrm{F}}/c_1$, which immediately yields an expansion in powers of v_{F}/c_1.

The sound attenuation $\alpha(\omega)$ and the sound velocity $c(\omega)$ may be expressed in terms of the imaginary and real parts of $\xi(\omega)$ respectively as

$$\left.\begin{aligned} \alpha(\omega) &= -qZ \operatorname{Im} \xi(\omega), \\ c(\omega) &= c_1[1 + Z \operatorname{Re} \xi(\omega)]. \end{aligned}\right\} \tag{11.91}$$

11.4.1 Phenomenological model

Before we embark on the microscopic calculation of the response function $\xi(\omega)$ (see Section 11.4.2), let us have a look at a simple phenomenological model (Wölfle 1978c, 1979). While simple, it nevertheless contains the basic physics and is able to reproduce the main features of the results obtained from microscopic theory. Furthermore, it not only applies to the collisionless regime but even covers the hydrodynamic regime. It will enable us to express the momentum current density $\boldsymbol{\Pi}$ in terms of the density ρ so that (11.84) becomes a self-contained equation of motion for the density.

The model is based on the assumption that the longitudinal momentum current density $\hat{\boldsymbol{q}}\boldsymbol{\Pi}\hat{\boldsymbol{q}}$ may be decomposed into (i) a local-equilibrium term given by the pressure P, (ii) a viscoelastic term Π^{n}, and (iii) a contribution Π^{s} from the internal vibrations of the order parameter:

$$\hat{\boldsymbol{q}}\boldsymbol{\Pi}\hat{\boldsymbol{q}} = P + \Pi^{\mathrm{n}} + \Pi^{\mathrm{s}}. \tag{11.92}$$

While at low frequencies the term Π^{n} accounts for normal-fluid friction, at high frequencies ($\omega > \tau^{-1}$) it describes the elastic behaviour of the medium due to the Fermi-liquid interaction (see the discussion of collective modes in the normal state in Section 2.3). At low frequencies the normal-fluid friction term Π^{n} is simply given by the hydrodynamic expression

$$\begin{aligned} \Pi^{\mathrm{hyd}} &= -\mathrm{i}\frac{4}{3}\frac{\eta}{\rho}\boldsymbol{q}\cdot\boldsymbol{g} \\ &= -\mathrm{i}\frac{4}{3}\frac{\eta}{\rho}\,\omega\,\delta\rho, \end{aligned} \tag{11.93}$$

where in the second equality the continuity equation (2.21) has been used. At higher frequencies, $\omega \gtrsim \tau^{-1}$, the collisions are no longer effective enough to establish complete local equilibrium. Rather, the normal component Π^{n} relaxes towards the local-equilibrium value at a rate given by the viscous-relaxation time τ_η:

$$\omega\Pi^{\mathrm{n}} = -\frac{\mathrm{i}}{\tau_\eta}(\Pi^{\mathrm{n}} - \Pi^{\mathrm{hyd}}). \tag{11.94}$$

We omit the usual divergence of a current term in (11.94) because it gives rise to correction terms of order v_{F}/c_1 only.

At still higher frequencies, $\hbar\omega \gtrsim \Delta$, the excitation of pair vibration modes takes place, which in turn generate a contribution to the longitudinal stress proportional to the time derivative of the collective variables $\delta\Delta_l^{\mathrm{coll}}$, where l labels the different collective modes:

$$\Pi^{\mathrm{s}} = \frac{-\mathrm{i}\omega}{\Delta}\sum_l \Lambda_l(\hat{\boldsymbol{q}})\,\delta\Delta_l^{\mathrm{coll}}. \tag{11.95}$$

Here $\delta\Delta_l^{\mathrm{coll}}$ represents any one of the order-parameter components $d'_{\mu j}$, $d''_{\mu j}$ or combinations thereof that we have associated with a finite frequency at $\boldsymbol{q} = 0$. The $\Lambda_l(\boldsymbol{q})$ in (11.95) are dimensionless anisotropic coupling parameters, which depend on internally preferred directions (vector $\hat{\boldsymbol{l}}$ in the A phase) or an external field (magnetic field in the B phase).

We have already discussed the equations of motion for the collective variables $\delta\Delta_l^{\mathrm{coll}}$ in the absence of collisions. In the case of the B phase these were found to be given by harmonic-oscillator equations. If collisions are included in a phenomenological way, the equations may be expected to change into ones that describe damped harmonic oscillators

$$\left(\omega^2 + \mathrm{i}\frac{\omega}{\tau_l} - \omega_l^2\right)\delta\Delta_l^{\mathrm{coll}} = \mathrm{i}\Lambda_l'\Delta\frac{E_{\mathrm{F}}}{\hbar\rho}\boldsymbol{q}\cdot\boldsymbol{g}. \tag{11.96}$$

Here τ_l is a relaxation time, which, by comparison with (11.83), may be identified with $2\tau_{\mathrm{coll}}$, and ω_l are the collective-mode frequencies, which in principle contain an imaginary part originating from pair-breaking processes. The force term on the right-hand side of (11.96) is proportional to the divergence of the mass current, i.e. $\boldsymbol{q}\cdot\boldsymbol{g}$. It must be linear in Δ, since $\delta\Delta^{\mathrm{coll}}$ is expected to scale approximately with the equilibrium gap. For dimensional reasons, a further factor with the dimension of a frequency must appear on the right-hand side of (11.96). Since this cannot be the external frequency itself, the gap frequency or the relaxation rate, it must be given by the characteristic frequency of the normal state, $E_{\mathrm{F}}/\hbar$. The dimensionless coupling constants Λ_l' and Λ_l will be of order unity except close to T_{c} or if the coupling is due to particle–hole-asymmetric terms only.

Combining (11.84) and (11.92)–(11.96), one is led to the following general structure of the sound dispersion relation:

$$\omega^2 = c_1^2 q^2\left[1 - \frac{4}{3}\frac{\eta}{\rho c_1^2}\frac{\mathrm{i}\omega}{1-\mathrm{i}\omega\tau_\eta} - \frac{1}{2}\left(\frac{v_{\mathrm{F}}}{c_1}\right)^2\sum_l \Lambda_l\Lambda_l'\frac{\omega^2}{\omega^2 + \mathrm{i}\omega/\tau_l - \omega_l^2}\right] \tag{11.97}$$

(cf. (2.76) for the normal state). This expression contains the main features of sound propagation in superfluid ^{3}He. The first correction term, involving the shear-viscosity coefficient η, describes the transition from ordinary hydrodynamic first sound to collisionless zero sound at $\omega\tau_\eta \approx 1$. Substituting the result for η as obtained from simple kinetic theory, $\eta = \frac{1}{5}\rho_{\mathrm{n}}\bar{v}^2\tau_\eta$, one finds that in the limit $T\to 0$ the zero-sound velocity c_0 approaches the first-sound velocity roughly as $c_0 - c_1 \propto \rho_{\mathrm{n}}$, with ρ_{n} the normal-fluid density.

The last term accounts for the excitation of collective modes, which are damped both by collision processes and by pair breaking (contained in the imaginary part of ω_l). Incoherent pair-breaking processes give rise to an imaginary part of the coupling functions Λ_l. It is clear that Λ_l vanishes proportionally to Δ as $T\to T_{\mathrm{c}}$. A similar model description may be employed to analyse the acoustic impedance (Ketterson *et al.* 1983a).

11.4.2 Sound propagation in the B phase

Longitudinal sound is characterized by a single vector, i.e. the propagation direction $\hat{\boldsymbol{q}}$. The relevant fluctuations of the order parameter in this case are described by two components d_0 and d_1, defined by

$$\delta d_{\mu j}(\boldsymbol{q},\omega) \pm \delta d_{\mu j}^*(-\boldsymbol{q},-\omega) = \begin{cases} d_0'\delta_{\mu j} + \frac{1}{2}d_1'(3\hat{q}_\mu\hat{q}_j - \delta_{\mu j}) & \text{for } (+), \\ d_0''\delta_{\mu j} + \frac{1}{2}d_1''(3\hat{q}_\mu\hat{q}_j - \delta_{\mu j}) & \text{for } (-). \end{cases} \tag{11.98}$$

For $\hat{\boldsymbol{q}}$ along $\hat{\boldsymbol{z}}$, the components of the order parameter appearing in (11.98) are given by $d_0^{\prime,\prime\prime} = \frac{2}{3}\,\mathrm{Tr}\, d_{\mu j}^{\prime,\prime\prime}$ and $d_1^{\prime,\prime\prime} = \frac{2}{3}(2d_{zz}^{\prime,\prime\prime} - d_{xx}^{\prime,\prime\prime} - d_{yy}^{\prime,\prime\prime})$. Remember that the real parts couple to density waves only via particle–hole-asymmetric terms. In the limit $\boldsymbol{q}\to 0$ the solution of (11.64) yields the order-parameter fluctuation

$$\left.\begin{aligned} d'_{\mu j} &= \lambda_{\mathrm{A}} \frac{8\Delta^2}{(\hbar\omega)^2 - 4\Delta^2} \langle \delta\xi_{\boldsymbol{k}} \rangle \delta_{\mu j}, \\ d''_{\mu j} &= \frac{4\Delta}{\hbar\omega} \langle \delta\xi_{\boldsymbol{k}} \rangle \delta_{\mu j}, \end{aligned}\right\} \tag{11.99}$$

where $\lambda_{\mathrm{A}} \ll 1$ is the particle–hole-asymmetry parameter. Substituting the $\boldsymbol{q}=0$ solution into the q^2 terms on the left-hand side of (11.57), one obtains, to first order in q^2,

$$\left.\begin{aligned} d'_1 &= \frac{32}{15}\frac{1}{N_{\mathrm{F}}}\frac{\Delta^2}{(\hbar\omega)^2 - \frac{8}{5}\Delta^2}\left[\lambda_{\mathrm{A}}\frac{(\hbar\omega)^2}{(\hbar\omega)^2 - 4\Delta^2}\left(\frac{v_{\mathrm{F}}q}{\omega}\right)^2 F_0^{\mathrm{s}} \right. \\ &\quad \left. + (\lambda_{\mathrm{A}} + \theta_{\mathrm{A}})\frac{m}{m^*}F_1^{\mathrm{s}}\right]\delta n, \\ d''_1 &= \frac{16}{5}\frac{1}{N_{\mathrm{F}}}\frac{\hbar\omega\Delta}{(\hbar\omega)^2 - \frac{12}{5}\Delta^2}\delta n, \end{aligned}\right\} \tag{11.100}$$

where δn is the particle number-density fluctuation caused by the sound wave. In deriving (11.100) we have used (10.10) and the continuity equation to express $\delta\xi_{\boldsymbol{k}}$ as

$$\delta\xi_{\boldsymbol{k}} = \frac{1}{N_{\mathrm{F}}}\left[F_0^{\mathrm{s}} + \frac{F_1^{\mathrm{s}}}{1 + \frac{1}{3}F_1^{\mathrm{s}}}\frac{\omega}{v_{\mathrm{F}}q}\hat{\boldsymbol{k}}\cdot\hat{\boldsymbol{q}}\right]\delta n. \tag{11.101}$$

Also, we have replaced $(v_{\mathrm{F}}q)^2$ by $\omega^2/[\frac{1}{3}F_0^{\mathrm{s}}(1 + \frac{1}{3}F_1^{\mathrm{s}})]$, using the zeroth-order sound dispersion relation. The result (11.100) may now be substituted into (11.90), using the relation

$$\tfrac{1}{2}\,\mathrm{tr}_\sigma\,[\delta\boldsymbol{\Delta}_{\boldsymbol{k}}\,\boldsymbol{\Delta}_{\boldsymbol{k}}^{+} \pm \boldsymbol{\Delta}_{\boldsymbol{k}}\,\delta\boldsymbol{\Delta}_{\boldsymbol{k}}^{+}] = \Delta \times \begin{cases} [d'_0 + P_2(\hat{\boldsymbol{k}}\cdot\hat{\boldsymbol{q}})d'_1] & \text{for } (+), \\ [d''_0 + P_2(\hat{\boldsymbol{k}}\cdot\hat{\boldsymbol{q}})d''_1] & \text{for } (-), \end{cases} \tag{11.102}$$

which finally yields

$$\xi(\omega) = 1 - \lambda\left[\frac{3}{5}\frac{(\hbar\omega)^2 - 4\Delta^2}{(\hbar\omega)^2 - \frac{12}{5}\Delta^2} - \tfrac{16}{5}\bar{\lambda}_{\mathrm{A}}^2\frac{\Delta^2}{(\hbar\omega)^2 - \frac{8}{5}\Delta^2}\right]. \tag{11.103}$$

The coupling constant to the real squashing mode is given by

$$\bar{\lambda}_{\mathrm{A}}^2 = \frac{1}{3}\frac{F_1^{\mathrm{s}}(\lambda_{\mathrm{A}} + \theta_{\mathrm{A}}) - 2\lambda_{\mathrm{A}}}{1 + \frac{1}{3}F_1^{\mathrm{s}}}\lambda_{\mathrm{A}} \approx O\left[\left(\frac{T_{\mathrm{c}}}{T_{\mathrm{F}}}\right)^2\right]. \tag{11.104}$$

Note that in the limit $\hbar\omega \ll \Delta$ one finds $\xi = 1 - \lambda$, which tends to zero as

$T \rightarrow 0$. In other words, the sound mode approaches the Anderson–Bogoliubov mode in this limit.

The frequency variable ω in (11.103) contains a small imaginary part in order to guarantee the correct causal behaviour of the response function ξ. A more general theory based on the matrix kinetic equation including the collision integral (Wölfle 1978a,b) shows that the main effect of collisions is to broaden the infinitely sharp collective resonances in (11.103). In addition,

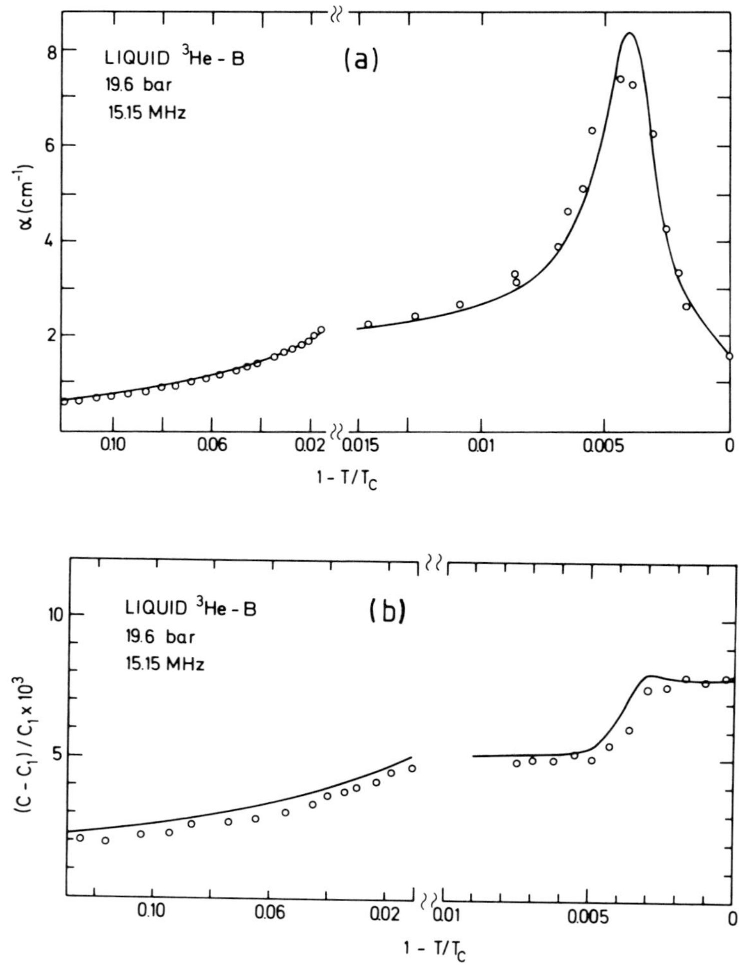

Figure 11.4 (a) Attenuation and (b) velocity shift of 15 MHz sound in ^{3}He-B versus temperature. Experimental data after Paulson *et al.* (1973b); theoretical curve after Wölfle (1976b).

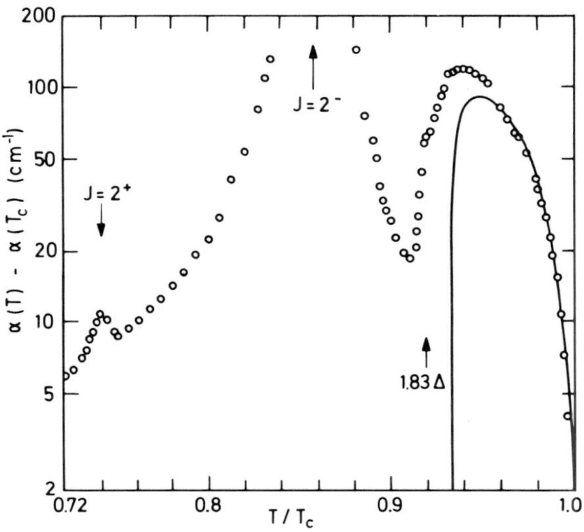

Figure 11.5 Attenuation of 44.2 MHz sound in the B phase at 2.44 bar pressure. (After Saunders *et al.* (1983).) The contributions of the pair-breaking continuum (solid curve) calculated from (11.103) and the positions of the collective-mode peaks are indicated. The additional structure at $\hbar\omega = 1.83\Delta_{BCS}$ is as yet unexplained.

there is a collision-induced background. Both effects may be crudely described by replacing ω in (11.103) by $\tilde{\omega} = \omega + i/\tau_{coll}$, where τ_{coll} is the relaxation time discussed in (11.83).

The result of an evaluation of the sound attenuation (11.91a) and velocity change (11.91b) including collision effects in a more rigorous fashion (Wölfle 1976b) is shown in Fig. 11.4, together with the experimental data of Paulson *et al.* (1973b). A comparison with data of Ihas *et al.* (1985), who used 5 MHz sound, appears to indicate additional absorption at lower temperatures. Data at still higher sound frequencies are shown in Fig. 11.5. Three main contributions to the attenuation are clearly discernable, although the centre portion of the second peak could not be measured owing to the high absorption level. The round peak just below T_c is caused by pair-breaking processes, while the other broad peak at a still lower temperature T_{sq}, defined by $\hbar\omega = (\frac{12}{5})^{1/2}\Delta(T_{sq})$, is due to absorption into a squashing mode (Meisel *et al.* 1983a). Finally, the real squashing mode leads to a sharp peak at T_{rsq}, the temperature where the resonance condition for the real squashing mode, $\hbar\omega = (\frac{8}{5})^{1/2}\Delta(T_{rsq})$, is satisfied (Koch and Wölfle 1981). In spite of the small coupling of the latter mode to sound (due to particle–hole-asymmetric terms) it was detected experimentally by Giannetta *et al.* (1980) and Mast *et al.* (1980). The sound velocity $c_p(\omega) = \omega/q$, i.e. the phase velocity, also shows considerable structure. This structure is seen to be even more pronounced in the group velocity of sound $c_g(\omega) = \partial\omega/\partial q$, in particular in the neighbourhood of the collective modes, where values as low as one-third of the phase velocity have been detected (Calder *et al.* 1980; Halperin 1982). Accurate measurements of the collective-mode frequencies at low temperature have been performed using a dynamic pressure sweeping technique (Adenwalla *et al.* 1989). The propagation of sound pulses at frequencies near the real squashing mode and at low temperatures, where the damping is negligibly small, leads to interesting beat patterns, even in the linear regime (Avenel *et al.* 1983, Varoquaux *et al.* 1986).

The collective mode with quantum number $J = 1^-$ and frequency $\omega = 2\Delta/\hbar$ (see Table 11.1) is barely observable in zero magnetic field (Daniels *et al.* 1983), but becomes clearly observable in a sufficiently high magnetic field (Saunders *et al.* 1983 Ling *et al.* 1987). That is, the field has to be sufficiently high that the frequency (split by the magnetic field into three substates) is moved below the (likewise shifted) pair-breaking threshold. The observed magnetic-field dependence is in qualitative agreement with theory (Schopohl *et al.* 1983; Schopohl and Tewordt 1984). In general, the experimental sound data are described well by theory. This is also true for the magnetic-field splitting of the $J = 2^{\pm}$ modes mentioned in Section 11.4.1, although there are still some unexplained features, which may perhaps be related to textures (Ketterson *et al.* 1983b). In particular, the central peak in the fivefold-split attenuation peak structure is observed to split again into two peaks at higher magnetic fields (Shivaram *et al.* 1982). This effect has been explained on the basis of a local resonance shift of the collective mode induced by the $\hat{\boldsymbol{n}}$ texture (Volovik 1984c, Volovik and Khazan 1984, Mineev 1985c, Fishman and Sauls 1988a,b).

In fact, the agreement between theory and experiment is so good that one can even extract values of the quasiparticle collision time τ (Wölfle 1978a), the Landau parameters $F_2^{s,a}$ (Engel and Ihas 1985) and the ratio of the pair-interaction constants V_3/V_1 by fitting the above theory to the data (Sauls and Serene 1982, Fishman and Sauls 1986, Meisel *et al.* 1987). A contribution from the squashing mode to the longitudinal surface impedance has been detected experimentally (Avenel *et al.* 1981a). While the above derivations made use of an expansion in the small parameter $(v_F/c_1)^2$, expressions for the complete density and current response functions have been derived by Czerwonko (1980a,b, 1982, 1985).

The effect of vortices on sound propagation in the rotating B phase has been investigated in a recent experiment (Salmelin *et al.* 1989b). The five collective-mode peaks, associated with the real squashing mode in a magnetic field, are found to depend strongly on the angular velocity of rotation and the preparation of the system, in qualitative agreement with theory (Salomaa and Volovik 1989).

11.4.3 Sound propagation in the A phase

The situation in the A phase is slightly more complicated owing to the orbital anisotropy of the liquid. On the other hand, spin and orbital degrees of freedom are decoupled here, if we neglect the tiny spin–orbit interaction. Furthermore, since all the collective modes couple to density fluctuations even without the aid of particle–hole-asymmetric terms, the latter may be discarded. The relevant combinations of order-parameter variables are then d''_{zm}, $m = 0, \pm 1$, as defined in (11.69).

Following the procedure described in Section 11.4.2, one finds to order $(v_F q/\omega)^2$

$$\left.\begin{aligned} d''_{z1} &= 4\frac{\Delta_0}{\hbar\omega}\frac{\left\langle\left[1+\frac{3}{F_0^s}\left(\frac{1}{2}\sin^2\beta\sin^2\theta+\cos^2\beta\cos^2\theta\right)\right]\lambda(\hat{\boldsymbol{k}})\,\delta\xi_{\boldsymbol{k}}\right\rangle_{\hat{\boldsymbol{k}}}}{\langle\lambda(\hat{\boldsymbol{k}})\rangle_{\hat{\boldsymbol{k}}}},\\ d''_{z,0} &= 2^{3/2}3\frac{\Delta_0}{\hbar\omega}\frac{1}{F_0^s}\frac{(\hbar\omega)^2\sin\beta\cos\beta\,\langle\cos^2\theta\,\lambda(\hat{\boldsymbol{k}})\,\delta\xi_{\hat{\boldsymbol{k}}}\rangle_{\hat{\boldsymbol{k}}}}{\langle(\hbar\omega)^2\lambda(\hat{\boldsymbol{k}})\cot\theta-2\Delta_0^2[\lambda(\hat{\boldsymbol{k}})+\Xi(\hat{\boldsymbol{k}})]\cos^2\theta\rangle_{\hat{\boldsymbol{k}}}},\\ d''_{z,-1} &= 3\frac{\Delta_0}{\hbar\omega}\frac{1}{F_0^s}\frac{(\hbar\omega)^2\sin^2\beta\,\langle\sin^2\theta\,\lambda(\hat{\boldsymbol{k}})\,\delta\xi_{\hat{\boldsymbol{k}}}\rangle_{\hat{\boldsymbol{k}}}}{\langle\lambda(\hat{\boldsymbol{k}})[(\hbar\omega)^2-2\Delta_0^2\sin^2\theta]\rangle_{\hat{\boldsymbol{k}}}}, \end{aligned}\right\} \tag{11.105}$$

where $\cos\beta = \hat{\boldsymbol{q}}\cdot\hat{\boldsymbol{l}}$, $\cos\theta = \hat{\boldsymbol{k}}\cdot\hat{\boldsymbol{l}}$ and $\hat{\boldsymbol{l}}$ is the axis of the gap. We want to assume a uniform $\hat{\boldsymbol{l}}$ configuration. As discussed in Section 11.3, $d''_{z,0}$ and $d''_{z,-1}$ possess resonances as functions of ω, i.e. the flapping and clapping modes.

Substituting the results (11.105) into the expression for $\xi(\omega)$, (11.88), one finds (Wölfle 1973a)

$$\xi(\omega) = \xi_{\parallel}\cos^4\beta + 2\xi_c\sin^2\beta\cos^2\beta + \xi_{\perp}\sin^4\beta, \tag{11.106}$$

where

$$\left.\begin{aligned} \xi_{\parallel} &= 1-\frac{45}{4}\left[\langle\lambda(\hat{\boldsymbol{k}})\cos^4\theta\rangle_{\hat{\boldsymbol{k}}}-\frac{\langle\lambda(\hat{\boldsymbol{k}})\cos^2\theta\rangle_{\hat{\boldsymbol{k}}}^2}{\langle\lambda(\hat{\boldsymbol{k}})\rangle_{\hat{\boldsymbol{k}}}}\right],\\ \xi_c &= 1-\frac{45}{8}\left[3\langle\lambda(\hat{\boldsymbol{k}})\sin^2\theta\cos^2\theta\rangle_{\hat{\boldsymbol{k}}}\right.\\ &\quad-\frac{\langle\lambda(\hat{\boldsymbol{k}})\cos^2\theta\rangle_{\hat{\boldsymbol{k}}}\langle\lambda(\hat{\boldsymbol{k}})\sin^2\theta\rangle_{\hat{\boldsymbol{k}}}}{\langle\lambda(\hat{\boldsymbol{k}})\rangle_{\hat{\boldsymbol{k}}}}\\ &\quad\left.-\frac{\langle\lambda(\hat{\boldsymbol{k}})\cos^2\theta\rangle_{\hat{\boldsymbol{k}}}^2}{\langle\lambda(\hat{\boldsymbol{k}})\cot^2\theta\rangle_{\hat{\boldsymbol{k}}}-2(\Delta_0/\hbar\omega)^2\langle[\lambda(\hat{\boldsymbol{k}})+\Xi(\hat{\boldsymbol{k}})]\cos^2\theta\rangle_{\hat{\boldsymbol{k}}}}\right],\\ \xi_{\perp} &= 1-\frac{45}{16}\left[\tfrac{3}{2}\langle\lambda(\hat{\boldsymbol{k}})\sin^4\theta\rangle_{\hat{\boldsymbol{k}}}-\frac{\langle\lambda(\hat{\boldsymbol{k}})\sin^2\theta\rangle_{\hat{\boldsymbol{k}}}^2}{\langle\lambda(\hat{\boldsymbol{k}})\rangle_{\hat{\boldsymbol{k}}}}\right.\\ &\quad\left.-\frac{1}{4}\frac{\langle\lambda(\hat{\boldsymbol{k}})\sin^2\theta\rangle_{\hat{\boldsymbol{k}}}^2}{\langle\lambda(\hat{\boldsymbol{k}})\rangle_{\hat{\boldsymbol{k}}}-2(\Delta_0/\hbar\omega)^2\langle\lambda(\hat{\boldsymbol{k}})\sin^2\theta\rangle_{\hat{\boldsymbol{k}}}}\right]. \end{aligned}\right\} \tag{11.107}$$

As expected, the sound-propagation properties are anisotropic, depending on the angle β between the sound-propagation direction $\hat{\boldsymbol{q}}$ and the preferred direction $\hat{\boldsymbol{l}}$. The anisotropy is characterized by three functions $\xi_{\parallel}$, ξ_{c} and $\xi_{\perp}$, where $\xi_{\parallel}$ accounts for parallel and $\xi_{\perp}$ for perpendicular orientation, whereas all three functions contribute for intermediate orientation. The anisotropic sound attenuation $\alpha(\omega)$ can then be obtained from (11.91), yielding

$$\alpha(\omega) = \alpha_{\parallel} \cos^4 \beta + 2\alpha_{c} \sin^2 \beta \cos^2 \beta + \alpha_{\perp} \sin^4 \beta, \qquad (11.108)$$

where $\alpha_i = -qZ \operatorname{Im} \xi_i(\omega)$, $i = \parallel$, c, $\perp$. In general, the temperature dependence of $\alpha(\omega)$, (11.108), has to be evaluated numerically. However, in the limiting cases $T \to 0$ and $T \to T_c$ it may be calculated analytically by employing the respective results for $\lambda(\hat{\boldsymbol{k}})$, (11.62).

Thus for sound propagation parallel to $\hat{\boldsymbol{l}}$, none of the flapping and clapping modes are excited, which is easily understood from the schematic representation in Fig. 11.2. For $\hat{\boldsymbol{q}} \perp \hat{\boldsymbol{l}}$, on the other hand, only the clapping mode is excited: a compression wave perpendicular to $\hat{\boldsymbol{l}}$ will lead to a local widening and narrowing of the angle between the preferred directions $\hat{\boldsymbol{m}}$ and $\hat{\boldsymbol{n}}$, which is the clapping mode, while leaving the angle with respect to $\hat{\boldsymbol{l}}_{\mathrm{N}}$ unchanged.

In Fig. 11.6 the result of a numerical calculation of the three components of the sound attenuation, $\alpha_{\parallel}$, α_{c} and $\alpha_{\perp}$, is compared with experimental data of Paulson *et al.* (1976b,c). The theory includes both pair-breaking and collision effects (Wölfle and Koch 1978). The peaks associated with the clapping mode ($\alpha_{\perp}$) and the normal-flapping mode (α_{c}) are clearly borne out. The contribution of the super-flapping mode is masked by pair-breaking processes (Ashida and Nagai 1983, Einzel and Wölfle 1984), but a corresponding feature in the sound absorption has nevertheless been observed (Meisel *et al.* 1983b, 1985). The remaining discrepancy between theory and experiment, especially in $\alpha_{\parallel}$, is thought to be related to the incomplete orientation of the $\hat{\boldsymbol{l}}$ texture in the experiment. In fact, the strong dependence of α on the orientation of $\hat{\boldsymbol{l}}$ may be used to monitor the time dependence of the $\hat{\boldsymbol{l}}$ texture in a nonequilibrium situation (Wheatley 1978). Unfortunately, this probe is not local but samples the texture over a region of the size of the acoustic penetration depth α^{-1}. The attenuation of lower-frequency sound ($\nu = 5$ MHz) has been measured by Berg and Ihas (1985). Sound-attenuation measurements well below T_c and in the pair-breaking regime may in principle be used to determine the value of the gap parameter (Piché *et al.* 1982). The re-entrance of the normal-flapping mode at lower temperatures (see Fig. 11.3) has been observed by Ling *et al.* (1989). Analysis of the latter experiment provides evidence for weak f-wave pairing effects and nontrivial strong-coupling corrections.

Experimentally, there appears to be some excess attenuation even above T_c (Paulson and Wheatley 1978c), which might be interpreted as pair-fluctuation effects, and various contributions to the fluctuation-induced

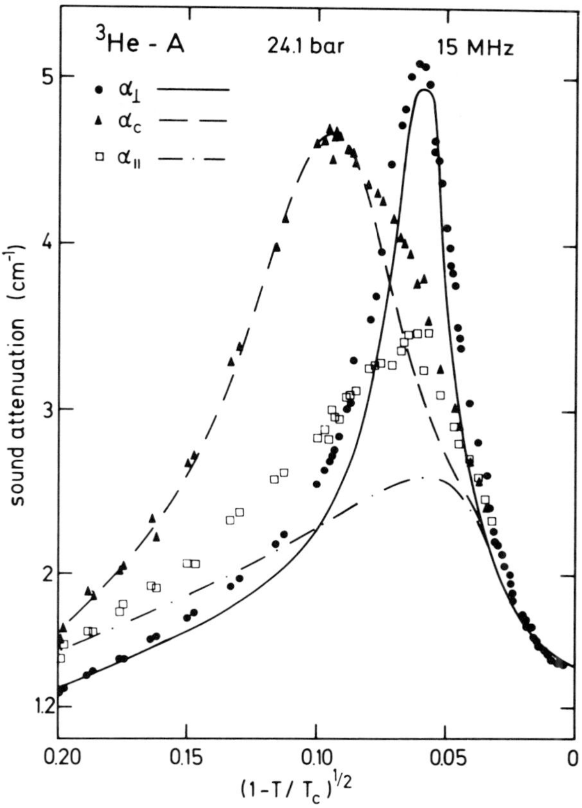

Figure 11.6 The three components of the anisotropic sound attenuation in ^{3}He-A. (After Paulson *et al.* (1976b,c).) The curves are the theoretical results of Wölfle and Koch (1978).

attenuation have been discussed (see e.g. Emery 1976, Samalam and Serene 1978).

It appears to be worth drawing explicit attention to the astonishing fact that sound absorption in ^{3}He-A is *not* isotropic, owing to the existence of the preferred direction $\hat{\boldsymbol{l}}$. As we know, this direction may be oriented by a weak (approx. 30 G) external magnetic field (via the dipolar coupling between $\hat{\boldsymbol{l}}$ and $\hat{\boldsymbol{d}}$), and will thereby change the level of sound absorption. This anisotropy is in marked contrast with the intrinsic isotropy of the ^{3}He liquid, which consists of spherical—hence structureless—atoms; it finds its origin solely in the direction of the spin of the ^{3}He nucleus. So the hadronic nuclear spin and its direction are ultimately responsible for the anisotropy of macroscopic properties like sound absorption. Here we encounter a particularly fascinating example of macroscopic quantum coherence.

11.4.4 Nonlinear effects of sound propagation

Interesting nonlinear behaviour of sound propagation has been observed in both the A and B phases. In the B phase the propagation of sound pulses near the real squashing mode at low temperatures and at correspondingly high frequencies shows the characteristics of self-induced transparency, i.e. weakly damped propagation of soliton-like excitations (Polturak *et al.* 1981a,b, de Vegvar *et al.* 1981). A simple model, based on the analogy with two-level systems in an optically resonant medium, with the Cooper-pair ground state and the pair-vibration state in question taken as the respective two levels of the system, explains some of the features observed (Sauls 1981; see also Ruoff and Varoquaux 1983, de Vegvar, 1985). However, the model does not account for the necessary strong coupling of the sound wave to the order parameter. Earlier, Paulson and Wheatley (1978b) had observed satellite peaks in the sound attenuation in the B phase at lower temperatures. As shown later by Engel *et al.* (1985), the effect is probably not an intrinsic one.

In the A phase nonlinear effects due to saturation of pair breaking have been observed (Avenel *et al.* 1981b, 1985) (see also Pushkina and Khokhlov (1976) for a discussion of nonlinear sound propagation in the hydrodynamic regime). It is likely that both effects are a consequence of resonant, i.e. coherent, pair breaking and recombination (Kopp and Wölfle 1987). The generation of higher harmonics by nonlinear coupling has been considered by Serene (1984) and McKenzie and Sauls (1989).

11.4.5 Transverse sound and spin waves

A discussion similar to the preceding one for longitudinal sound can be given for transverse zero sound and for spin waves. The salient results are that transverse zero sound continues to exist in the superfluid phases provided that it existed in the normal state, i.e. provided that the condition $F_1^s > 6$ is satisfied. In the superfluid the $l = 1$ molecular field is produced by the quasiparticles alone, in contrast with the case $l = 0$, where the rigidity of the phase takes over the role of the Landau molecular field. Consequently, the coupling parameter F_1^s is effectively replaced by the temperature-dependent quantity $F_1^s Y_0(T)$, where $Y_0(T)$ is the Yosida function. Below a temperature T_t defined by $F_1^s Y_0(T_t) \approx 6$, transverse zero sound ceases to exist (Combescot and Combescot 1976b, Maki and Ebisawa 1976b, 1977) (see Leggett (1966) for the case of s-wave pairing). By contrast, spin waves become more and more well defined as one enters the superfluid state, because here the phase rigidity helps to stabilize the mode and eventually supports it entirely (Combescot 1974a,b). The possibility of excitation of pair-vibration modes by spin waves has been investigated by Schopohl *et al.* (1982).

Transverse zero sound has been detected experimentally in the acoustic shear impedance by Roach and Ketterson (1976b), while collisionless spin waves have been studied in the form of standing waves localized by a nonuniform magnetic field (Masuhara *et al.* 1984).

FURTHER READING

Brinkman W F and Cross M C 1978 in *Progress in Low Temperature Physics,* Vol. VIIA, ed. D F Brewer (North-Holland, Amsterdam), p. 105

Brusov P N and Popov V N 1984 *Collective Excitations in Superfluid Quantum Liquids* (University of Rostov, USSR) (in Russian)

Halperin W P 1982 *Physica* **109/110B + C** 1596 (Proceedings of the 16th International Conference on Low Temperature Physics, LT-16)

Ketterson J B, Shivaram B S, Meisel M W, Sarma B K and Halperin W P *1983* in *Quantum Fluids and Solids—1983* (AIP Conf. Proc. No. 103), ed. E D Adams and G G Ihas (AIP, New York), p. 288

Kleinert H 1978 *Fortschr. Phys.* **26** 565

Serene J W 1983 in *Quantum Fluids and Solids—1983* (AIP Conf. Proc. No. 103), ed. E D Adams and G G Ihas (AIP, New York), p. 305

Wheatley J C 1978 in *Progress in Low Temperature Physics,* Vol. VIIA, ed. D F Brewer (North-Holland, Amsterdam), p. 1

Wölfle P 1977 *Physica* **90B** 96

Wölfle P 1978 in *Progress in Low Temperature Physics,* Vol. VIIA, ed. D F Brewer (North-Holland, Amsterdam), p. 191

12

Amplification of Weak Interaction Effects due to Macroscopic Quantum Coherence

The concept of macroscopic quantum coherence has been touched upon several times already: for example within the context of persistent currents (Chapter 7), in the calculation of the expectation value for a two-particle operator (Section 3.5) and when this result was applied to calculate the magnitude of the dipole–dipole interaction in the condensate (Section 6.3.1). The latter example provides a particularly striking demonstration of how a weak microscopic interaction may be amplified to macroscopic size by the specific long-range order in the system. This effect is clearly very similar to macroscopic quantum coherence in superfluid ^{4}He, where only gauge symmetry is broken. Indeed, both systems are characterized by a macroscopic wave function expressed by a complex-valued order parameter. In view of the fact that Cooper pairs are composed of two fermions with a total spin $S = 0$ or $S = 1$, these pairs may in a way be looked upon as "composite bosons". Furthermore, the pairs are all in the same quantum-mechanical state and have zero total momentum in the ground state. Hence it is tempting to view the transition to the superconducting state (or to the superfluid state in ^{3}He) as one where the formation of Cooper pairs and their Bose–Einstein condensation coincide (i.e. Cooper pairs are *automatically* Bose-condensed). In this way the macroscopic quantum coherence of the pair-correlated state and the amplification of the microscopically small dipole–dipole interaction in superfluid ^{3}He may easily be visualized. On the other hand, one has to bear in mind that a Cooper pair in superconductors or superfluid ^{3}He is neither an elementary boson nor (at least) a tightly bound composite of two fermions with an overall bosonic excitation spectrum etc. Its spatial extent ξ_0 is very *large* compared with the interparticle distance a. Consequently, a given Cooper pair overlaps with many others (Schrieffer 1964), their number in order of magnitude being given by $(\xi_0/a)^3(T_c/T_F) \approx 10^{-3}(T_F/T_c)^2$, which is about 10^5–10^6 in the case of clean conventional superconductors and about 10^2 in the case of superfluid ^{3}He. Furthermore, we already know from the discussion in Section 7.3.1 that it makes a difference whether the Bose–Einstein condensation involves overlapping or nonoverlapping diatomic molecules (Mermin and Muzikar 1980, Mermin 1980, Volovik and Mineev 1981). In spite of all these objections, a theory of superconductivity based on a Bose–Einstein condensation of pairs of particles ("Schafroth condensation") may indeed be formulated, but is mathematically very complex (see Blatt (1964) and also Baranger (1963)). Altogether, the concept of Bose–Einstein condensation is appealing since key features of superconductivity like the Meissner effect, flux quantization and superfluid mass currents in conventional superconductors are naturally implied.

Since the theory of conventional superconductivity is firmly based on BCS theory, the concept of a Bose–Einstein condensation of Cooper pairs traditionally did not receive much attention (or was even considered to be downright wrong). However, within the context of superfluid ^{3}He, this notion was taken up again by Leggett (1980a,b), who argued that tightly bound Bose–Einstein-condensed molecules on the one hand and Cooper pairs on the other may be viewed as the extreme limits of the molecule picture. This concept, which also applies to electron–hole pairs in semiconductors, has been further investigated by Nozières and Schmitt-Rink (1985) within several models. Both Leggett (1980) and the latter authors found that the transition between the two limits is *smooth*. From the point of view of critical phenomena we should therefore conclude that the formation of Cooper pairs and the Bose–Einstein condensation of tightly bound molecules is indeed not something distinctly different. This view is also implicit in the theoretical approach to high-T_c superconductivity, based on the "resonating-valence-bond" (RVB) approach, which has been advanced by Anderson (1987a,b). In this case superconductivity is viewed as the Bose–Einstein condensation of "preformed" (in the insulating regime) Cooper pairs, once the system is doped. An offspring of this idea is the concept of a gas of hard-core dimers, which may undergo Bose–Einstein condensation and thus lead to superconductivity (see e.g. Rokhsar and Kivelson 1988).

12.1 PROPERTIES OF THE BCS PAIR AMPLITUDE

Apart from the overall conceptual equality of Cooper pairs in conventional superconductors and in superfluid ^{3}He, based on BCS theory, there are also distinct qualitative differences (Leggett 1980a,b): (i) Cooper pairs in superfluid ^{3}He form in a $S=1$, $l=1$ state, i.e. they have an *internal* structure; (ii) their constituents have a strong mutual repulsion at short relative distances; (iii) these constituents are not pointlike as in a superconductor, but have a physical structure themselves. In these respects Cooper pairs in superfluid ^{3}He resemble diatomic molecules much more than do those involving electrons in a superconductor. Besides that, the coherence length ξ_0 is much shorter than in a clean conventional superconductor (see below (7.18b)).

According to Leggett (1980a,b), the picture of Cooper pairs as diatomic molecules may be helpful in gaining insight into some properties of superfluid ^{3}He. In particular, he pointed out that such a model provides guidance for a more realistic evaluation of expectation values of two-particle operators in superfluid ^{3}He than is possible in conventional BCS theory, where one only works with model wave functions.

Let us compare the results for the expectation values of a two-particle operator $W(\boldsymbol{r}, \sigma_1, \sigma_2)$ in the case of a diatomic molecule and of Cooper pairs respectively. Here $\boldsymbol{r}$ is the relative coordinate, which is supposed to be the only important coordinate. For molecules, each with a pair wave function $\psi(\boldsymbol{r}; \sigma_1, \sigma_2)$, the result is

$$\langle W \rangle = \int \mathrm{d}^3 r\, W(\boldsymbol{r})\, |\psi(\boldsymbol{r})|^2, \tag{12.1}$$

which should be taken as a schematic notation since $\langle W \rangle$ will in general be a more complicated object owing to the spin dependence. In the case of Cooper pairs the result was calculated in Section 3.5, and is given by (3.119a) and (3.120):

$$\langle W \rangle = \int \mathrm{d}^3 r\, W(\boldsymbol{r})\, |F(\boldsymbol{r})|^2, \tag{12.2}$$

where again complications due to spin have not been written explicitly. The two results are identical in form. On the other hand, (12.2) does not contain a two-particle wave function, such as the wave function $\psi(\boldsymbol{r})$ used in (3.10a) to discuss the Cooper problem for two fermions and which may equally be used to construct the total N-particle wave function Ψ_0, (3.11). Rather, it is the pair amplitude (i.e. the order parameter) F that enters. That is why F is sometimes called the "pair wave function"†. On the other hand, although F and ψ are quite different quantities (after all, F is obtained from the solution of a complex many-body problem), they are clearly related, and in fact have the same transformation and symmetry properties. This implies that F and ψ have the same dependence on spin and on the *direction* $\hat{\boldsymbol{r}}$, i.e. the orientation of the relative separation $\boldsymbol{r}$ of the atoms in the pair. However, they will differ in their *radial* dependence, i.e.

† Of course this is partly a matter of definition. But, as pointed out by Mermin (1978a), one should in fact carefully distinguish between an order parameter and a pair wave function; for this purpose, he discussed a particular two-particle wave function given by the order parameter that leads to a wrong result when used to compute the quantum-mechanical current density (7.32).

in their dependence on the magnitude of $\boldsymbol{r}$, owing to many-body effects, choice of momentum cutoff etc. entering in F.

Let us separate the angular- and spin-dependent part of F from its radial part and write

$$F(\boldsymbol{r};\sigma_1,\sigma_2)=F_\Omega(\hat{\boldsymbol{r}};\sigma_1,\sigma_2)\tilde{F}(r).$$

This is a good approximation as long as the energy eigenvalues (3.9a) of the paired state for different l components are sufficiently separated, and becomes exact for $T\to T_c$. The first part, F_Ω, reflects the structure of the order parameter in spin and orbital spaces and is known for the different superfluid phases. It can therefore be integrated over in (12.2) to yield a constant factor. The radial part, however, poses a problem because its actual form is not known. In BCS theory, i.e. mean-field theory, the pair amplitude has a particular model form, which can be obtained by transforming (3.43b) back into real space. For $r\ll\xi_0$, $\tilde{F}(r)$ is given approximately by the radial wave function of two free particles in a state of relative angular momentum l at the Fermi surface (Leggett 1975a), as discussed below (3.10b). For $r\gg\xi_0$, i.e. r larger than the maximum size of a Cooper pair, $\tilde{F}(r)\propto e^{-r/\xi_0}$, although the explicit fall-off depends on the choice of the cutoff in the momentum integral. This r dependence of $\tilde{F}(r)$ is not only unrealistic but also insufficient for two distinct reasons. (i) The operator $W(\boldsymbol{r})$ in (12.2) typically contributes mainly for small r. The dipole interaction, for example, has a r^{-3} dependence; other interactions decrease even more rapidly (see below). So the dependence of $\tilde{F}(r)$ is particularly important at small arguments. (ii) On the other hand, in the case of ^{3}He, where atoms have a finite diameter R_0, the BCS-model form for $\tilde{F}$ is especially bad at small r since it does not take into account the microscopic interaction (it even contributes for $r\leqslant R_0$). In order to improve on this point, Leggett (1977a, 1980a) suggested an ansatz for $\tilde{F}(r)$ that modifies the BCS form at distances $r\approx R_0$. It is guided by the results for the molecule picture, where the radial pair wave function at short distances is given by the relative wave function of a simple diatomic molecule. Note, however, that such an approach requires one to go beyond the regime of validity of the quasiparticle picture. Therefore it cannot yield a controlled description of the effects involved, but may only be expected to be qualitatively correct. In superfluid ^{3}He $\tilde{F}(r)$ may then be expected to be similar to $\psi_1(r)$, (3.10b), for $r\ll\xi_0$, i.e. to be given by the pair wave function for two free ^{3}He atoms in a relative $l=1$ state at the Fermi surface. A possible form is (Leggett 1977a, 1980a)

$$\tilde{F}(r)=\begin{cases}0 & (r\leqslant R_0),\\ A\dfrac{\sin k(r-R_0)}{r} & (R_0<r<\xi_0,\ k\geqslant k_F),\end{cases} \tag{12.3}$$

which cuts out the region for $r\leqslant R_0$ and assumes free-particle behaviour for

$R_0 < r < \xi_0$. (Note that this is actually a form describing an $l = 0$ rather than $l = 1$ pair state; but since $kR_0 \approx 2.3$, this choice is not expected to introduce a substantial error (Leggett 1977a,b).) This model form for $\tilde{F}$ is then used to calculate the expectation value $\langle W \rangle$ in (12.2). The prefactor A can be obtained by calculating the expectation value of the dipole interaction energy $\langle H_D \rangle$ in (6.88) or the dipole-coupling parameter

$$g_D(T) = \tfrac{1}{2}\gamma^2\hbar^2 \int d^3r\, r^{-3}\, |\tilde{F}(r)|^2, \tag{12.4}$$

which is known from experiment. This yields

$$A^2 = \frac{1}{2\pi} \frac{R_0^2}{\gamma^2\hbar^2 I(kR_0)} g_D(T), \tag{12.5}$$

where

$$I(x) = \int_0^\infty dx \frac{\sin^2 xy}{(1+y)^3}.$$

For $kR_0 \approx 2.3$, one obtains $I = 0.2$. In this way, all expectation values $\langle W \rangle$ may be expressed in terms of the dipole interaction in the superfluid.

12.2 THE PERMANENT ORBITAL MAGNETIC MOMENT OF THE A PHASE

Because of the filled 1s shell, an isolated ^{3}He atom has a perfectly spherical shape. This is why helium is chemically extremely inert. In such a case the geometric centre of the electronic charge distribution coincides with the nucleus and therefore the atom does not possess a permanent electric dipole moment. On the other hand, if two ^{3}He atoms are brought closely together, the electron shells will be distorted owing to the mutual polarization—an effect of purely chemical origin. Following Leggett (1977b), let us consider the Cooper pair in ^{3}He as a sort of diatomic molecule of two ^{3}He atoms. Since the electronic charge distributions will be slightly distorted in both atoms, their centres will no longer coincide with those of the positive nuclear charges. Since the "dimer" is rotating (it has an orbital angular momentum $l = 1$), there exists a finite magnetic moment due to the (opposite) charge currents of the circulating nuclei and electron distributions, which no longer cancel. To estimate this magnetic moment, we consider the ^{3}He nuclei in the pair to have separation R and mass M, and we imagine the atoms to rotate around an axis perpendicular to the plane of rotation (parallel to the relative orbital angular momentum $\boldsymbol{l}$). Let the centres of the electronic charge distributions be shifted by a distance $P(R)$ away from the nucleus as a result of the polarization. The orbital magnetic moments of the electrons and the nuclei for the two ^{3}He atoms are then given by

$$\left.\begin{aligned}\boldsymbol{\mu}_e &= 4\mu_B L_e \hat{\boldsymbol{l}},\\ \boldsymbol{\mu}_n &= -4\mu_B \frac{m_e}{m_n} L_n \hat{\boldsymbol{l}},\end{aligned}\right\} \tag{12.6}$$

where μ_B is the Bohr magneton and L_e and L_n are the magnitudes of the angular momenta, with $L_e = m_e |\boldsymbol{v}| (\frac{1}{2}R + P)$ and $L_n = \frac{1}{2}m_n |\boldsymbol{v}| R$ respectively. Here $\boldsymbol{v}$ is the velocity of the charges. Since the dimer is in a state of $l = 1$, it is given by $|\boldsymbol{v}| = \hbar/m_n R$ to lowest order in the small parameter m_e/m_n. The magnitude of the total magnetic moment $\boldsymbol{\mu}_{tot} = \boldsymbol{\mu}_e + \boldsymbol{\mu}_n$ along $\hat{\boldsymbol{l}}$ then follows as

$$|\boldsymbol{\mu}_{tot}| = 4\mu_B \frac{m_e}{m_n}\frac{P(R)}{R}. \tag{12.7}$$

So the polarization $P(R)$ of the electronic charge distributions leads to a finite orbital magnetic moment.

In a normal system this has no macroscopic consequences, because not only are these magnetic moments extremely small, they also average to zero because they are randomly oriented. Even a fantastically high magnetic field would not be able to align them. In the case of ^{3}He-A, however, $\hat{\boldsymbol{l}}$ is the same for every Cooper pair, so that the magnetic moments $\boldsymbol{\mu}_{tot}$ of the Cooper pairs will add up coherently. Consequently, this orbital magnetic moment is permanent and has a macroscopic magnitude. The difficulty lies in going beyond such a qualitative statement and actually calculating the size, as was done by Leggett (1977b). For this, one first of all has to know the actual polarization $P(R)$ within the Cooper pair. If one assumes that $P(R)$, measuring the distortion of the electronic wave function, depends exponentially on the separation R of the two atoms then one can write

$$P(R) = a_0 c_0 \exp\left[-\frac{\lambda(R-R_0)}{a_0}\right]. \tag{12.8}$$

Here a_0 is the Bohr radius (the characteristic length in a problem dealing with electrons), $R_0 \approx 2.5$ Å is twice the hard-core radius, i.e. the diameter, of the ^{3}He atom, and c_0 and λ are constants, which may be determined by means of an approximation for the distorted electronic wave function ($c_0 = -4.4 \times 10^{-3}$ and $\lambda \approx 1.4$). Using (12.2) to obtain the total spontaneous magnetic moment $\boldsymbol{M}(T)$, the moment for one Cooper pair, $\boldsymbol{\mu}_{tot}(R)$, has to be weighted by the probability $|F(R)|^2$ of finding the two ^{3}He nuclei at a distance R apart and then has to be integrated over all R:

$$|\boldsymbol{M}(T)| = 4\pi A^2 \mu_B \frac{m_e}{m_n} a_0 c_0 \int_{R_0}^{\infty} \mathrm{d}R \frac{\sin^2[k(R-R_0)]}{R} \exp\left[-\frac{\lambda(R-R_0)}{a_0}\right]. \tag{12.9}$$

Note that $P(R)$ only contributes for small separations $R \gtrsim R_0$ of the order of 2.5 Å and drops off exponentially for $R \gg R_0$. Now it is crucial that a Cooper pair is *not* a molecule: if it really were a molecule with a more or less fixed separation ξ_0 (the maximal spatial extent of the pair, which is several hundred ångströms) then there would be no effect at all. It is only because the position probability density of the two partners of a Cooper pair has an appreciable weight at small $R \approx R_0$ even for p-wave pairing (see (3.10b)) that the integral in (12.9) gives a non-negligible contribution.

Inserting the expression for $4\pi A^2$ in (12.5), $\boldsymbol{M}(T)$ can be calculated. Actually, it is more instructive to convert it into an equivalent magnetic field $\boldsymbol{H}_{\mathrm{eq}} = \boldsymbol{M}/\chi_{\mathrm{N}}$, where χ_{N} is the normal-state spin susceptibility. Using the fact that $5 \times 10^{-10}\mu_{\mathrm{B}}$ per atom corresponds to a field of 1 G, one finds

$$\boldsymbol{H}_{\mathrm{eq}}(T) = -H_1\left(1 - \frac{T}{T_c}\right)\hat{\boldsymbol{l}}, \tag{12.10}$$

where $H_1 \approx 0.03 f(k)$ (in gauss) and $f(k) = [1 + (\lambda/2a_0k)^2]^{-1}$. This calculation therefore predicts that there exists a spontaneous (permanent) orbital magnetic moment in the superfluid A phase, oriented opposite to the $\hat{\boldsymbol{l}}$ direction in the sample (Leggett 1977b). It corresponds to an internal magnetic field of about 10–20 mG. We note that this magnetic moment will couple to an external magnetic field. The energy $E = -\boldsymbol{M} \cdot \boldsymbol{H}$ is given by

$$E = M(T)\boldsymbol{H} \cdot \hat{\boldsymbol{l}}. \tag{12.11}$$

Hence there is a *linear* coupling between the external field and the direction of $\hat{\boldsymbol{l}}$ that prefers $\hat{\boldsymbol{l}}$ to lie antiparallel to $\boldsymbol{H}$. For the first time we encounter an orientational effect on the $\hat{\boldsymbol{l}}$ texture that distinguishes between $\hat{\boldsymbol{l}}$ and $-\hat{\boldsymbol{l}}$, or—put differently—that depends on the polarity of the external field. Consequently, the ground-state energy of the system in external magnetic fields $\boldsymbol{H}$ and $-\boldsymbol{H}$ is different, leading to a slightly different equilibrium $\hat{\boldsymbol{l}}$ texture when the magnetic field is reversed.

Indeed, Paulson and Wheatley (1978a) showed that the orientation of the $\hat{\boldsymbol{l}}$ texture does depend on the sign of the applied external magnetic field. By means of sound-attenuation measurements, they determined the prefactor of this linear coupling term $\boldsymbol{H} \cdot \hat{\boldsymbol{l}}$ and found it to be in overall agreement with the theoretical prediction. The macroscopic orbital magnetic moment makes ^{3}He-A a liquid ferromagnet.

12.3 THE PERMANENT ELECTRIC DIPOLE MOMENT OF THE B PHASE

Perhaps even more spectacular may be yet another effect, which is expected to be caused by the quantum-mechanical amplification of ultraweak interaction effects within elementary-particle physics. As is well known, the interaction between an electron and a proton (more generally, a hadron) contains, besides the usual parity-conserving Coulomb interaction, a very much weaker parity-violating part due to the exchange of a neutral Z^0 boson ("weak neutral currents"). This latter interaction is predicted to lead to a macroscopic effect in ^{3}He-B, i.e. to a permanent electric dipole moment $\boldsymbol{\mu}_c = c\langle \boldsymbol{L} \times \boldsymbol{S} \rangle$ of the Cooper pairs along the preferred direction $\hat{\boldsymbol{n}}$ (Leggett 1977a). Its existence may be understood by considering the basic properties of an electric dipole moment under time reversal T and space reflection P. Let us therefore take a look at the energy $E_\mu = -\boldsymbol{\mu}_c \cdot \boldsymbol{E}$ of an electric dipole in an external electric field $\boldsymbol{E}$. Since all electromagnetic interactions are P- *and* T-invariant, the energy must also have this invariance. To illustrate the following arguments, the symmetry behaviour of various physical quantities under the P and T operations are listed in Table 12.1.

The electric field $\boldsymbol{E}$ is invariant under T, but not under P. Consequently, an electric dipole moment must have transformation properties identical with those of $\boldsymbol{E}$, since the energy E has to be invariant under both P and T. Now the spontaneous electric dipole moment is a vector, which is something very fundamental for a single particle. In fact, in the case of a single particle at rest (e.g. an elementary particle or an atom) there is only *one* such characteristic vector: its spin $\boldsymbol{S}$ or some orbital angular momentum $\boldsymbol{J}$. Thus the electric dipole moment has to point along this one fundamental vector. However, the spin $\boldsymbol{S}$ (or $\boldsymbol{J}$) is odd under T, and therefore the energy $E = -\boldsymbol{\mu}_c \cdot \boldsymbol{E} \propto \boldsymbol{S} \cdot \boldsymbol{E}$ would change sign under time reversal. This is forbidden and hence $\boldsymbol{\mu}_c$ must be identically zero. (Demanding P invariance leads to the same result, i.e. $\boldsymbol{\mu}_c = 0$ follows from both P and T invariance separately.) This is the reason why "simple" systems, characterized by at

Table 12.1 Symmetry of various quantities under parity transformation P and time inversion T respectively.

Physical quantity	P	T
Space $\boldsymbol{r}$	$-$	$+$
Momentum $\boldsymbol{p}$ $(= \mathrm{d}\boldsymbol{r}/\mathrm{d}t)$	$-$	$-$
Angular momentum $\boldsymbol{L} = \boldsymbol{r} \times \boldsymbol{p}$	$+$	$-$
Spin $\boldsymbol{S}$	$+$	$-$
$\boldsymbol{L} \times \boldsymbol{S}$	$+$	$+$
Electric field $\boldsymbol{E} = -\mathrm{d}V/\mathrm{d}\boldsymbol{r}$	$-$	$+$
Magnetic field $\boldsymbol{H}$	$+$	$-$
Electromagnetic energy E	$+$	$+$
$\boldsymbol{p} \cdot \boldsymbol{S}$	$-$	$+$
$\boldsymbol{E} \cdot \boldsymbol{L}$	$-$	$-$
$\boldsymbol{E} \cdot (\boldsymbol{L} \times \boldsymbol{S})$	$-$	$+$
$\boldsymbol{H} \cdot \boldsymbol{L}$	$+$	$+$
$\boldsymbol{H} \cdot (\boldsymbol{L} \times \boldsymbol{S})$	$+$	$-$

most one vector, do indeed possess no electric dipole moment. (In more complicated systems like molecules (e.g. H_2O) this argument is no longer valid, since there is more than one characteristic direction; in fact, many molecules *do* have an electric dipole moment.) Let us then consider an object characterized by two independent vectors, say the spin $\boldsymbol{S}$ and angular momentum $\boldsymbol{L}$. Then one can construct a vector $\boldsymbol{L}\times\boldsymbol{S}$ that is invariant both under P and T. T invariance alone would therefore allow a dipole moment $\boldsymbol{\mu}_e \propto \boldsymbol{L}\times\boldsymbol{S}$. However, P invariance of $E=-\boldsymbol{\mu}_e\cdot\boldsymbol{E}\propto(\boldsymbol{L}\times\boldsymbol{S})\cdot\boldsymbol{E}$ is still violated and if all interactions were parity-conserving (like the electromagnetic and strong interactions), $\boldsymbol{\mu}_e$ would have to be zero after all. Here the weak interaction becomes important. Its theoretical formulation within the unified theory of the Weinberg–Salam model implies the existence of "weak neutral currents" mediated by the exchange of Z^0 bosons.

Such currents are, for example, generated by the electron–nucleon interaction within an atom, made possible by the finite probability amplitude for electrons at the nucleus. In the case of the ^{3}He atom there is a part to the interaction involving an electron and the neutron, which can be written as (Bouchiat 1977)

$$V_{e,n}(\boldsymbol{r}) = -2^{-1/2}G_F C_{An}\frac{1}{m_e}\boldsymbol{p}_e\cdot\boldsymbol{S}_n\delta(\boldsymbol{r}) + \text{Herm. conj.} \qquad (12.12)$$

Here G_F and C_{An} are the coupling constants of the usual weak interaction and of the specific lepton–neutron axial-vector interaction respectively, while m_e is the electron mass, $\boldsymbol{p}_e$ is the electron momentum and $\boldsymbol{S}_n$ is the spin of the neutron. The delta function expresses the fact that one considers a contact interaction of pointlike objects. Obviously, the above interaction is P-violating but T-conserving. Its existence removes even the last symmetry argument against a finite static dipole moment $\boldsymbol{\mu}_e = c(\boldsymbol{L}\times\boldsymbol{S})$, where c must be proportional to the coupling strength $G_F C_{An}$.

Now that we know that such a dipole moment may exist, we have to check whether the vector $\boldsymbol{L}\times\boldsymbol{S}$ makes sense at all, i.e. whether $\boldsymbol{L}$ and $\boldsymbol{S}$ are really independently conserved quantities. In a normal atom this is not the case, because $\boldsymbol{L}$ and $\boldsymbol{S}$ will couple to a total angular momentum $\boldsymbol{J}$ owing to some hyperfine interaction. Even if this were not the case, the spatial average $\langle\boldsymbol{L}\times\boldsymbol{S}\rangle$ would average to zero in a macroscopic system. In this respect ^{3}He-B is an exception since the relative orientation of $\boldsymbol{L}$ and $\boldsymbol{S}$ is the same for every Cooper pair—a consequence of the broken relative spin–orbit symmetry. (Note that in ^{3}He-A $\langle\boldsymbol{L}\times\boldsymbol{S}\rangle\propto\hat{\boldsymbol{l}}\times\langle\boldsymbol{S}\rangle$ also vanishes because only $\hat{\boldsymbol{l}}$ and $\hat{\boldsymbol{d}}$ are macroscopically defined, while $\langle\boldsymbol{S}\rangle=0$.) In the B phase one finds that $\langle\boldsymbol{L}\times\boldsymbol{S}\rangle = -\frac{4}{3}\hbar^2\sin\theta_L\,\hat{\boldsymbol{n}}$, where $\theta_L=\cos^{-1}(-\frac{1}{4})$ is the Leggett angle. We now have to calculate the magnitude of this electric dipole moment $\boldsymbol{\mu}_e\propto\langle\boldsymbol{L}\times\boldsymbol{S}\rangle$. In fact, it could still be zero for a symmetry reason not yet considered—and if the electrons in the shells of the ^{3}He atoms in the Cooper pair were in an exact 1s state, as in the isolated case, it would indeed be exactly zero. But in a Cooper pair the two ^{3}He atoms slightly polarize each other's electronic shells, giving rise to an admixture of 2p atomic orbitals to the wave function. Owing to this finite angular-momentum component, the electronic motion can couple to the rotation of the Cooper pair. The induced distortion of the electronic distribution—again a chemical effect—consequently leads to a finite value of $\boldsymbol{\mu}_e$. The electric dipole moment of a Cooper pair due to the parity-violating weak interaction can then be written as

$$\boldsymbol{\mu}_e = -2^{-3/2} G_F C_{An} \epsilon(R) \boldsymbol{L} \times \boldsymbol{S}, \qquad (12.13)$$

where $\epsilon(R)$ is called the "electrohelicity". It describes the details of the distortion of the wave function as a function of the separation of two ^{3}He atoms in the Cooper pair. Its actual calculation is quite difficult, but it appears reasonable to make an ansatz as in (12.8) with an exponential R dependence:

$$\epsilon(R) = \frac{e}{\hbar a_0} \frac{m_e}{m_n} \epsilon_0 \exp\left[-\frac{\lambda(R - R_0)}{a_0}\right], \qquad (12.14)$$

where $e/\hbar a_0$ is a characteristic unit of $\epsilon(R)$ in the case of an atom. Again using results of atomic-orbital calculations for two ^{3}He atoms at different separations, ϵ_0 and λ have been estimated by Leggett (1977a) as $\epsilon_0 = 3 \times 10^{-4}$, $\lambda \approx 0.3$–0.5. To obtain the total electric dipole moment $\boldsymbol{D}$ in a sample of ^{3}He-B, one may now proceed exactly as in the previous case of the magnetic moment: $\boldsymbol{\mu}_e(R)$ is multiplied by $|F(R)|^2$, and is then integrated over R. One obtains

$$\boldsymbol{D} = \frac{2^{1/2}}{3} G_F C_{An} Q \sin\theta_L\, \hat{\boldsymbol{n}}, \qquad (12.15)$$

where

$$Q = \int d^3r\, \epsilon(R)\, |\tilde{F}(R)|^2. \qquad (12.16)$$

To get a feeling for the size of this spontaneous electric dipole moment, we introduce a unit D_0 of the dipole moment that in an electric field of 1 V/cm leads to an orientational energy equivalent to the thermal energy $k_B T$ at 1 mK, i.e. $D_0 = 8.6 \times 10^{-8} e$ cm. Using the numerical values $G_F/\hbar a_0^2 \approx 1.6 \times 10^{-16}$, $C_{An} \approx 1$ for the coupling constants and (12.3) for $\tilde{F}(r)$, and proceeding as in Section 12.2, Leggett (1977a) obtained an order-of-magnitude estimate for $|\boldsymbol{D}|$ in terms of D_0:

$$|\boldsymbol{D}| \approx 10^{-5} D_0. \qquad (12.17)$$

The permanent parity-violating dipole moment in ^{3}He-B is hence seen to be very small. As in the case of the spontaneous orbital magnetic moment in

^{3}He-A, the dipole moment $\boldsymbol{D}$ leads to a linear coupling with an external field, this time with an external electric field $\boldsymbol{E}$. This coupling is sensitive to the polarity of $\boldsymbol{E}$. Reversing the electric field will change the energy of the system and will therefore modify the equilibrium $\hat{\boldsymbol{n}}$ texture. In principle, the change may be detected by a shift in the NMR frequencies of the texture. Although the effect will be very small—even in a strong external electric field—such an experiment does not seem to be hopelessly outside the accessible range. A positive result would show, for the first time, the consequences of parity violation of the weak interaction on a macroscopic level.

The weak interaction between electrons and hadrons not only violates parity, it also has a T- (i.e. CP-) violating part. The typical coupling constant for the T-odd interaction is about 10^{-3} times smaller than that of the P-violating part. CP violation has so far only been observed in the decay process of the neutral K meson. Clearly, it would be desirable to find evidence for this interaction by a different experimental technique. As discussed above, an electric dipole moment (EDM) of an elementary particle will violate both P and T invariance. Hence if such an EDM were indeed to be measured, it would allow one to determine the strength of the T-odd interaction. However, in spite of the long history of the search for EDMs of elementary particles (Ramsey 1982), no evidence for EDMs has emerged. There are many suggestions how to improve the sensitivity of EDM experiments, particularly using macroscopic samples (see e.g. Bialek *et al.* 1986). Based on Leggett's prediction for an EDM in the B phase, Khriplovich (1982) suggested that the A_1 phase of superfluid ^{3}He might be used to measure the EDM of the ^{3}He atom (i.e. of electrons and neutrons) by again exploiting the macroscopic amplification of this subatomic effect due to the condensate.

In this context several other similar effects may be discussed. For example, in the A phase a permanent electric dipole moment on every Cooper pair (which must be along $\hat{\boldsymbol{l}}$) would lead to an energy $E = -\boldsymbol{\mu}_e \cdot \boldsymbol{E} \propto \hat{\boldsymbol{l}} \cdot \boldsymbol{E}$ when placed into an electric field $\boldsymbol{E}$. In this case the expression is again odd under P and T; so the magnitude of $\boldsymbol{\mu}_e$ will be determined by the smaller of the P- and T-violating parts of the weak interaction, i.e. the part odd in T. Consequently, an electric dipole moment can exist in ^{3}He-A. Its magnitude is proportional to the T-odd interaction strength and may therefore be expected to be even smaller (by a few orders of magnitude) than the one predicted by Leggett for ^{3}He-B on the basis of parity violation.

Similarly, we may consider a spontaneous magnetic moment $\boldsymbol{\mu}_m \propto \langle \boldsymbol{L} \times \boldsymbol{S} \rangle$ in ^{3}He-B. The energy of such a moment in a magnetic field is given by $E_m = -\boldsymbol{\mu}_m \cdot \boldsymbol{H} \propto \langle \boldsymbol{L} \times \boldsymbol{S} \rangle \cdot \boldsymbol{H}$. Evidently, this energy is even under P but odd under T. The magnitude of any such $\boldsymbol{\mu}_m$ is therefore again determined by the strength of the T-violating part of the weak interaction.

FURTHER READING

Leggett A J 1978 *J. Physique* **39** Colloq. C-6, Vol. III, p. 1264 (Proceedings of the 15th International Conference on Low Temperature Physics, LT-15)

Leggett A J 1980 in *Proceedings of the XVI Karpacz Winter School of Theoretical Physics* (Lecture Notes in Physics, Vol. 115), ed. A Pekalski and J Przystawa (Springer, Berlin), p. 13

References

Abel W R, Anderson A C, Black W C and Wheatley J C 1965 *Physics* **1** 337

Abel W R, Anderson A C, Black W C and Wheatley J C 1966 *Phys. Rev.* **147** 111

Abragam A 1961 *The Principles of Nuclear Magnetism* (Clarendon Press, Oxford)

Abragam A and Goldman M 1982 *Nuclear Magnetism: Order and Disorder* (Clarendon Press, Oxford)

Abramowitz M and Stegun IA 1965 *Handbook of Mathematical Functions* (Dover, New York)

Abrikosov A A and Khalatnikov I M 1959 *Rep. Prog. Phys.* **22** 329

Abrikosov A A, Gorkov L P and Dzyaloshinskii I E 1963 *Methods of Quantum Field Theory in Statistical Physics* (Prentice-Hall, Englewood Cliffs, New Jersey; reprinted 1975, Dover, New York)

Abud M and Sartori G 1983 *Ann. Phys.* (*NY*) **150** 307

Adenwalla S, Zhao Z, Ketterson J B and Sarma B K 1989 *J. Low Temp. Phys.* **76** 1

Aharonov Y and Bohm D 1959 *Phys. Rev.* **115** 485

Ahonen A I, Haikala M T and Krusius M 1974a *Phys. Lett.* **47A** 215

Ahonen A I, Haikala M T, Krusius M and Lounasmaa O V 1974b *Phys. Rev. Lett.* **33** 628

Ahonen A I, Haikala M T, Krusius M and Lounasmaa O V 1974c *Phys. Rev. Lett.* **33** 1595

Ahonen A I, Alvesalo T A, Haikala M T, Krusius M and Paalanen M A 1975a *J. Phys.* **C8** L269

Ahonen A I, Haavasoja T, Haikala M T, Krusius M and Paalanen M A 1975b *Phys. Lett.* **55A** 157

Ahonen A I, Krusius M and Paalanen M A 1975c in *Proceedings of the 14th International Conference on Low Temperature Physics, LT-14,* Vol. 1, ed. M Krusius and M Vuorio (North-Holland, Amsterdam), p. 107

Ahonen A I, Kodama T, Krusius M, Paalanen M A, Richardson R C, Schoepe W and Takano Y 1976a *J. Phys.* **C9** 1665

Ahonen AI, Kokko J, Lounasmaa OV, Paalanen MA, Richardson R C, Schoepe W and Takano Y 1976b *Phys. Rev. Lett.* **37** 511

Ahonen A I, Krusius M and Paalanen M A 1976c *J. Low Temp. Phys.* **25** 421

Ahonen A I, Kokko J, Lounasmaa OV, Paalanen M A, Richardson R C, Schoepe W and Takano Y 1977a in *Quantum Fluids and Solids,* ed. S B Trickey, E D Adams and J W Dufty (Plenum, New York), p. 171

Ahonen A I, Kokko J, Lounasmaa O V, Paalanen M A, Richardson R C, Schoepe W and Takano Y 1977b in *Quantum Fluids and Solids,* ed. S B Trickey, E D Adams and J W Dufty (Plenum, New York), p. 245

Ahonen A I, Alvesalo T A, Haavasoja T and Veuro M C 1978a *J. Physique* **39** Colloq. C-6, Vol. I, p. 285 (Proceedings of the 15th International Conference on Low Temperature Physics, LT-15)

Ahonen A I, Kokko J, Paalanen M A, Richardson R C, Schoepe W and Takano Y 1978b *J. Low Temp. Phys.* **30** 205
Ainsworth T L, Bedell K S, Brown G E and Quader K F 1983 *J. Low Temp. Phys.* **50** 319
Aldrich C H and Pines D 1978 *J. Low Temp. Phys.* **32** 689
Aldrich C H, Pethick C J and Pines D 1976 *Phys. Rev. Lett.* **37** 845
Allen J F and Misener A D 1938 *Nature* **141** 75
Alonso B and Popov V N 1977 *Zh. Eksp. Teor. Fiz.* **73** 1445 [*Sov. Phys. JETP* **46** 760 (1977)]
Alvarez L W and Cornog R 1939a *Phys. Rev.* **56** 379
Alvarez L W and Cornog R 1939b *Phys. Rev.* **56** 613
Alvesalo T A, Anufriyev Yu D, Collan H K, Lounasmaa O V and Wennerström P 1973 *Phys. Rev. Lett.* **30** 962
Alvesalo T A, Collan H K, Loponen M T and Veuro M C 1974 *Phys. Rev. Lett.* **32** 981
Alvesalo T A, Collan H K, Loponen M T Lounasmaa O V and Veuro M C 1975 *J. Low Temp. Phys.* **19** 1
Alvesalo T A, Haavasoja T, Manninen M T and Soinne AT 1980 *Phys. Rev. Lett.* **44** 1076
Alvesalo T A, Haavasoja T and Manninen M T 1981 *J. Low Temp. Phys.* **45** 373
Ambegaokar V 1975 in *Quantum Statistics and the Many-Body Problem,* ed. S B Trickey, W P Kirk and J W Dufty (Plenum, New York), p. 119
Ambegaokar V and Levy M 1976 *Phys. Rev.* **B13** 1967
Ambegaokar V and Mermin N D 1973 *Phys. Rev. Lett.* **30** 81
Ambegaokar V and Rainer D 1974 Unpublished (referenced by Lee and Richardson 1978)
Ambegaokar V, de Gennes P G and Rainer D 1974 *Phys. Rev.* **A9** 2676; *ibid.* **A12** E345 (1975)
Amit D J, Kane J W and Wagner H 1968a *Phys. Rev.* **175** 313
Amit D J, Kane J W and Wagner H 1968b *Phys. Rev.* **175** 326
Anderson P W 1958a *Phys. Rev.* **110** 827
Anderson P W 1958b *Phys. Rev.* **112** 1900
Anderson P W 1959 *J. Phys. Chem. Solids* **11** 26
Anderson P W 1963 *Phys. Rev.* **130** 439
Anderson P W 1966 *Rev. Mod. Phys.* **38** 298
Anderson P W 1967 in *Progress in Low Temperature Physics,* Vol. V, ed. C J Gorter (North-Holland, Amsterdam), p. 1
Anderson P W 1973 *Phys. Rev. Lett.* **30** 368
Anderson P W 1984 *Basic Notions of Condensed Matter Physics* (Benjamin/Cummings, Menlo Park, California)
Anderson P W 1985 *Phys. Rev.* **B30** 4000
Anderson P W 1987a *Science* **235** 1196
Anderson P W 1987b in *Frontiers and Borderlines in Many Particle Physics* (*Proceedings of the Enrico Fermi International School of Physics, Varenna, July 1987*), ed. J R Schrieffer and R A Broglia (North-Holland, Amsterdam)
Anderson P W and Brinkman W F 1973 Phys. Rev. Lett. **30** 1108
Anderson P W and Brinkman W F 1975 in *The Helium Liquids* (*Proceedings of the 15th Scottish Universities Summer School, 1974*), ed. J G M Armitage and I E Farquhar (Academic Press, London), p. 315.
Anderson P W and Brinkman W F 1978 in *The Physics of Liquid and Solid Helium,* Part II, ed. K H Bennemann and J B Ketterson (Wiley, New York), p. 177.
Anderson P W and Morel P 1960 *Physica* **26** 671
Anderson P W and Morel P 1961 *Phys. Rev.* **123** 1911
Anderson P W and Palmer R G 1977 in *Quantum Fluids and Solids,* ed. S B Trickey, E D Adams and J W Dufty (Plenum, New York), p. 23

Anderson P W and Toulouse G 1977 *Phys. Rev. Lett.* **38** 508
Anderson P W and Varma C M 1973 *Nature* **241** 187
Andreev A F 1964 *Zh. Eksp. Teor. Fiz.* **46** 1823 [*Sov. Phys. JETP* **19** 1228 (1964)]
Andreev A F 1966 *Zh Eksp. Teor. Fiz.* **50** 1415 [*Sov. Phys. JETP* **23** 939 (1966)]
Andreev A F and Kagan M Yu 1987 *Zh. Eksp. Teor. Fiz.* **93** 895 [*Sov. Phys. JETP* **66** 504 (1987)]
Andres K and Bucher E 1968 *Phys. Rev. Lett.* **21** 1221
Andres K and Bucher E 1972 *J. Low Temp. Phys.* **9** 267
Andronikashvili E L and Mamaladze Yu G 1966 *Rev. Mod. Phys.* **38** 567
Anufriyev Yu D 1965 *Pis'ma Zh. Eksp. Teor. Fiz.* **1** [*JETP Lett.* **1** 155 (1965)]
Arai T and Soda T 1983 *Prog. Theor. Phys.* **69** 699
Archie C N, Alvesalo T A, Reppy J D and Richardson R C 1979 *Phys. Rev. Lett.* **43** 139
Archie C N, Alvesalo T A, Reppy J D and Richardson R C 1981 *J. Low Temp. Phys* **42** 295
Ashida M and Nagai K 1983 *Prog. Theor. Phys.* **70** 1672
Ashida M and Nagai K 1984 in *Proceedings of the 17th International Conference on Low Temperature Physics, LT-17,* Part II, ed. U Eckern, A Schmid, W Weber and H Wühl (North-Holland, Amsterdam), p. 783
Ashida M and Nagai K 1985 *Prog. Theor. Phys.* **74** 949
Ashida M and Takagi S 1981 *Prog. Theor. Phys.* **65** 1
Avenel O and Varoquaux E 1985 *Phys. Rev. Lett.* **55** 2704
Avenel O and Varoquaux E 1986 *Phys. Rev. Lett.* **57** C921
Avenel O and Varoquaux E 1987 *Jap. J. Appl. Phys.* **26** Suppl. 26-3, Part 3, p. 1798 (Proceedings of the 18th International Conference on Low Temperature Physics, LT-18)
Avenel O, Bernier M E, Varoquaux E J and Vibet C 1975 in *Proceedings of the 14th International Conference on Low Temperature Physics, LT-14,* Vol. 5, ed. M Krusius and M Vuorio (North-Holland, Amsterdam), p. 429
Avenel O, Varoquaux E and Ebisawa H 1980 *Phys. Rev. Lett.* **45** 1952
Avenel O, Piché L, Saslow W M, Varoquaux E and Combescot R 1981a *Phys. Rev. Lett.* **47** 803
Avenel O, Piché L and Varoquaux E 1981b *Physica* **107B + C** 689 (Proceedings of the 16th International Conference on Low Temperature Physics, LT-16)
Avenel O, Ruoff M, Varoquaux E and Williams G A 1983 *Phys. Rev. Lett.* **50** 1591
Avenel O, Piché L, Ruoff M, Varoquaux E, Combescot R and Maki K 1985 *Phys. Rev. Lett.* **54** 1408
Babu S and Brown G E 1973 *Ann. Phys. (NY)* **78** 1
Bagley M, Main P C, Hook J R, Sandiford D J and Hall H E 1978 *J. Phys.* **C11** L729
Bagnuls C, Dombre T and Combescot R 1979 *J. Phys.* **C12** 4399
Bailin D and Love A 1978a *J. Phys.* **A11** L219
Bailin D and Love A 1978b *J. Phys.* **C11** L909
Bailin D and Love A 1978c *J. Phys.* **A11** 821
Bailin D and Love A 1978d *J. Phys.* **C11** 1351
Bailin D and Love A 1978e *J. Phys.* **A11** 2149
Bailin D and Love A 1980 *J. Phys.* **A13** L271
Bailin D, Love A and Moore M A 1977 *J. Phys.* **C10** 1159
Balachandran A P 1986 *Nucl. Phys.* **B271** 227
Balatskii A V 1986 *Pis'ma Zh. Eksp. Teor. Fiz.* **44** 515 [*JETP Lett.* **44** 663 (1986)]
Balatskii A V 1988 *Pis'ma Zh. Eksp. Teor. Fiz.* **47** 647 [*JETP Lett.* **47** 746 (1988)]
Balatskii A V and Mineev V P 1985 *Zh. Eksp. Teor. Fiz.* **89** 2073 [*Sov. Phys. JETP* **62** 1195 (1985)]
Balatskii A V, Volovik G E and Konyshev V A 1986 *Zh. Eksp. Teor. Fiz.* **90** 2038 [*Sov. Phys. JETP* **63** 194 (1986)]

Balian R and Werthamer N R 1963 *Phys. Rev.* **131** 1553
Balinskii A A, Volovik G E and Kats E I 1984 *Zh. Eksp. Teor. Fiz.* **87** 1305 [*Sov. Phys. JETP* **60** 748 (1984)]
Baramidze G A, Gurgenishvili G E and Kharadze G A 1983 *Fiz. Nizk. Temp.* **9** 122 [*Sov. J. Low Temp. Phys.* **9** 61 (1983)]
Baranger M 1963 *Phys. Rev.* **130** 1244
Bardeen J, Cooper L N and Schrieffer J R 1957 *Phys. Rev.* **108** 1175
Bartolac T J 1980 Doctoral Thesis, University of Southern California, Los Angeles
Bartolac T J, Gould C M and Bozler H M 1981 *Phys. Rev. Lett.* **46** 126
Barton G and Moore M A 1974a *J. Phys.* **C7** 2989
Barton G and Moore M A 1974b *J. Phys.* **C7** 4220
Barton G and Moore M A 1974c *Proc. Phys. Soc.* **7** 2989
Barton G and Moore M A 1975a *J. Phys.* **C8** 970
Barton G and Moore M A 1975b *J. Low Temp. Phys.* **21** 489
Bashkin E P and Meyerovich A E 1981 *Adv. Phys.* **30** 1
Bates D M, Gould C M and Bozler H M 1983 in *Quantum Fluids and Solids—1983* (AIP Conf. Proc. No. 103), ed. E D Adams and G G Ihas (AIP, New York), p. 282
Bates D M, Ytterboe S N, Gould C M and Bozler H M 1984 *Phys. Rev. Lett.* **53** 1574
Bates D M, Ytterboe S N, Gould C M and Bozler H M 1986 *J. Low Temp. Phys.* **62** 143
Baym G and Pethick C J 1978 in *The Physics of Liquid and Solid Helium,* Part II, ed. K H Bennemann and J B Ketterson (Wiley, New York), p. 1
Baym G, Pethick C J and Salomaa M M 1977a *Phys. Rev. Lett.* **38** 845
Baym G, Pethick C J and Salomaa M M 1977b *Phys. Rev. Lett.* **39** 1290
Baym G, Pethick C J and Salomaa M M 1979 *J. Low. Temp. Phys.* **36** 431
Béal-Monod M T and Doniach S 1977 *J. Low Temp. Phys.* **28** 175
Béal-Monod M T and Mills D L 1978 *J. Low Temp. Phys.* **30** 289
Béal-Monod M T and Mills D L 1980 *J. Low Temp. Phys.* **42** 433
Béal-Monod M T, Ma S K and Fredkin D R 1968 *Phys. Rev. Lett.* **20** 929
Bedell K S 1982 *Phys. Rev.* **B26** 3747
Bedell K S and Ainsworth T L 1984 *Phys. Lett.* **102A** 49
Bedell K S and Pines D 1980a *Phys. Rev. Lett.* **45** 39
Bedell K S and Pines D 1980b *Phys. Lett.* **A78** 281
Bedell K S and Quader K F 1983 *Phys. Lett.* **96A** 91
Bedell K S and Quader K F 1984. *Phys. Rev.* **B30** 2894
Bedell K S and Sanchez-Castro C 1986 *Phys. Rev. Lett.* **57** 854
Bednorz J G and Müller K A 1986 *Z. Phys.* **B64** 189
Berezinskii V L 1970 *Zh. Eksp. Teor. Fiz.* **59** 907 [*Sov. Phys. JETP* **32** 493 (1971)]
Berezinskii V L 1974 *Pis'ma Zh. Eksp. Teor. Fiz.* **20** 628 [*JETP Lett.* **20** 287 (1975)]
Berg R F and Ihas G G 1985 *Phys. Rev.* **B32** 2851
Berk N F and Schrieffer J R 1966 *Phys. Rev. Lett.* **17** 433
Berthold J E, Giannetta R W, Smith E N and Reppy J D 1976 *Phys. Rev. Lett.* **37** 1138
Bertsch G F 1969 *Phys. Rev.* **184** 187
Betbeder-Matibet O and Nozières P 1969 *Ann. Phys. (NY)* **51** 392
Beyer J 1986 *Phys. Rev.* **B33** 3181
Bhattacharyya P, Pethick C J and Smith H 1975 *Phys. Rev. Lett.* **35** 473
Bhattacharyya P, Ho T L and Mermin N D 1977a *Phys. Rev. Lett.* **39** 1290
Bhattacharyya P, Pethick C J and Smith H 1977b *Phys. Rev.* **B15** 3367
Bialek W, Moody J and Wilczek F 1986 *Phys. Rev. Lett.* **56** 1623
Blaha S 1976 *Phys. Rev. Lett.* **36** 874
Blatt J M 1964 *Theory of Superconductivity* (Academic Press, New York)
Blount E I 1985 *Phys. Rev.* **B32** 2935

Blount E I, Brinkman W F and Anderson P W 1974 Unpublished (referenced by Anderson and Brinkman 1975, 1978)

Bloyet D, Varoquaux E, Vibet C, Avenel O, Berglund P M and Combescot R 1979 *Phys. Rev. Lett.* **42** 1158

Bogoliubov N N 1958 *Nuovo Cim.* **7** 794

Bogoliubov N N, Tolmachev V V and Shirkov D V 1959 *A New Method in the Theory of Superconductivity* (Consultant Bureau, New York)

Borovik-Romanov A S, Bun'kov Yu M, Dmitriev V V and Mukharskiy Yu M 1984a *Pis'ma Zh. Eksp. Teor. Fiz.* **39** 390 [*JETP Lett.* **39** 469 (1984)]

Borovik-Romanov A S, Bun'kov Yu M, Dmitriev V V and Mukharskiy Yu M 1984b *Pis'ma Zh. Eksp. Teor. Fiz.* **40** 256 [*JETP Lett.* **40** 1033 (1984)]

Borovik-Romanov A S, Bun'kov Yu M, Dmitriev V V, Mukharskiy Yu M and Flachbart K 1985 *Zh. Eksp. Teor. Fiz.* **88** 2025 [*Sov. Phys. JETP* **61** 1199 (1985)]

Borovik-Romanov A S, Bun'kov Yu M, Dmitriev V V and Mukharskiy Yu M 1987a *Pis'ma Zh. Eksp. Teor. Fiz.* **45** 98 [*JETP Lett.* **45** 124 (1987)]

Borovik-Romanov A S, Bun'kov Yu M, Dmitriev V V and Mukharskiy Yu M 1987b *Jap. J. Appl. Phys.* **26** Suppl. 26-3, Part 1, p. 175 (Proceedings of the 18th International Conference on Low Temperature Physics, LT-18)

Borovik-Romanov A S, Bun'kov Yu M, de Vaard A, Dmitriev V V, Makrotsieva V, Mukharskiy Yu M and Sergatskov D A 1988 *Pis'ma Zh. Eksp. Teor. Fiz.* **47** 400 [*JETP Lett.* **47** 478 (1988)]

Borovik-Romanov A S, Bun'kov Yu M, Dmitriev V V, Mukharskiy Yu M and Sergatskov D A 1989 *Phys. Rev. Lett.* **62** 1631

Bouchiat C 1977 *J. Phys.* **G3** 183

Bowley R M 1976 *J. Phys.* **C9** L151

Bowley R M 1977 *J. Phys.* **C11** 75

Bozler H M, Bartolac T, Luey K and Thomson A L 1978 *J. Physique* **39** Colloq. C-6, Vol. I, p. 283 (Proceedings of the 15th International Conference on Low Temperature Physics, LT-15)

Bozler H M, Bernier M E, Gully W J, Richardson R C and Lee D M 1974 *Phys. Rev. Lett.* **32** 875

Brand H and Cross M C 1982 *Phys. Rev. Lett.* **49** 1959

Brand H and Dörfle M 1981 *J. Low Temp. Phys.* **42** 333

Brand H and Pleiner H 1981a *Phys. Rev.* **B23** 155

Brand H and Pleiner H 1981b *J. Phys.* **C14** 97

Brand H and Pleiner H 1982 *J. Physique* **43** 369

Brand H, Dörfle M and Graham R 1979 *Ann. Phys. (NY)* **119** 434

Brenig W and Mikeska H J 1967 *Phys. Lett.* **24A** 332

Brenig W, Mikeska H J and Riedel E 1967 *Z. Phys.* **206** 439

Brewer D F 1983 in <u>Quantum Fluids and Solids—1983</u> (AIP Conf. Proc. No. 103), ed. E D Adams and G G Ihas (AIP, New York), p. 336

Brinkman W F 1974 *Phys. Lett.* **49A** 411

Brinkman W F and Anderson P W 1973 *Phys. Rev.* **A8** 2732

Brinkman W F and Cross M C 1978 in *Progress in Low Temperature Physics*, Vol. VIIA, ed. D F Brewer (North-Holland, Amsterdam), p. 105

Brinkman W F and Engelsberg S 1968 *Phys. Rev.* **169** 417

Brinkman W F and Rice T M 1970 *Phys. Rev.* **B2** 4302

Brinkman W F and Smith H 1974 *Phys. Rev.* **A10** 2325

Brinkman W F and Smith H 1975a *Phys. Lett.* **51A** 449

Brinkman W F and Smith H 1975b *Phys. Lett* **53A** 43

Brinkman W F, Serene J W and Anderson P W 1974a *Phys. Rev.* **A10** 2386

Brinkman W F, Smith H, Osheroff D D and Blount E I 1974b *Phys. Rev. Lett.* **33** 624

Bromley D J 1980 *Phys. Rev.* **B21** 2754

Bromley D J 1981 *Phys. Rev.* **B23** 4503

Brooker G A and Sykes J 1968 *Phys. Rev. Lett.* **21** 279
Brown G E, Pethick C J and Zaringhalam A 1982 *J. Low Temp. Phys.* **48** 349
Bruder C and Vollhardt D 1986 *Phys. Rev.* **B34** 131
Brueckner K A and Gammel J L 1958 *Phys. Rev.* **109** 1040
Brueckner K A, Soda T, Anderson P W and Morel P 1960 *Phys. Rev.* **118** 1442
Bruinsma R and Maki K 1978 *Phys. Rev.* **B18** 1101
Bruinsma R and Maki K 1979a *J. Low Temp. Phys.* **34** 343
Bruinsma R and Maki K 1979b *J. Low Temp. Phys.* **34** 649
Bruinsma R and Maki K 1979c *Phys. Rev.* **B20** 984
Bruinsma R and Maki K 1979d *J. Low Temp. Phys.* **37** 607
Brusov P N 1985a *Zh. Eksp Teor. Fiz.* **88** 1197 [*Sov. Phys. JETP* **61** 705 (1985)]
Brusov P N 1985b *J. Low Temp. Phys.* **58** 265
Brusov P N and Popov V N 1980a *Zh. Eksp. Teor. Fiz.* **78** 234 [*Sov. Phys. JETP* **51** 117 (1980)]
Brusov P N and Popov V N 1980b *Zh. Eksp. Teor. Fiz.* **78** 2419 [*Sov. Phys. JETP* **51** 1217 (1980)]
Brusov P N and Popov V N 1980c *Zh. Eksp. Teor. Fiz.* **79** 1879 [*Sov. Phys. JETP* **52** 945 (1980)]
Brusov P N and Popov V N 1981 *Zh. Eksp. Teor. Fiz.* **80** 1564 [*Sov. Phys. JETP* **53** 804 (1981)]
Brusov P N and Popov V N 1984a *Collective Excitations in Superfluid Quantum Liquids* (University of Rostow, USSR) (in Russian)
Brusov P N and Popov V N 1984b in *Proceedings of the 17th International Conference on Low Temperature Physics, LT-17,* Part II, ed. U Eckern, A Schmid, W Weber and H Wühl (North-Holland, Amsterdam), p. 781
Brusov P N and Popov V N 1984c *Phys. Rev.* **B30** 4060
Buchanan D S, Swift G W and Wheatley J C 1986 *Phys. Rev. Lett.* **57** 341
Buchholtz L J 1978 *Phys. Rev.* **B18** 1107
Buchholtz L J 1986 *Phys. Rev.* **B33** 1579
Buchholtz L J and Fetter A L 1976 *Phys. Lett.* **58A** 93
Buchholtz L J and Fetter A L 1977a *Phys. Rev.* **B15** 5225
Buchholtz L J and Fetter A L 1977b in *Quantum Fluids and Solids,* ed. S B Trickey, E D Adams and J W Dufty (Plenum, New York), p. 33
Buchholtz L J and Rainer D 1979 *Z. Phys.* **B35** 151
Buchholtz L J and Zwicknagl G 1981 *Phys. Rev.* **B23** 5788
Bun'kov Yu M, Hakonen P J and Krusius 1983a in *Quantum Fluids and Solids—1983* (AIP Conf. Proc. No. 103), ed. E D Adams and G G Ihas (AIP, New York), p. 194
Bun'kov Yu M, Krusius M and Hakonen P J 1983b *Pis'ma Zh. Eksp. Teor. Fiz.* **37** 395 [*JETP Lett.* **37** 468 (1983)]
Bun'kov Yu M, Dmitriev V V and Mukharskii Yu M 1984a *Phys. Lett.* **102A** 194
Bun'kov Yu M, Gurgenishvili G E, Krusius M and Kharadze G A 1984b *Usp. Fiz. Nauk.* **144** 141 [*Sov. Phys. Usp.* **27** 731 (1984)]
Bun'kov Yu M, Dmitriev V V and Mukharskii Yu M 1985 *Zh. Eksp. Teor. Fiz.* **88** 1218 [*Sov. Phys. JETP* **61** 719 (1985)]
Burckhardt T W 1968 *Ann. Phys. (NY)* **47** 516
Calder I D, Mast D B, Sarma B K, Owers-Bradley J R, Ketterson J B and Halperin W P 1980 *Phys. Rev. Lett.* **45** 1866
Campbell C E 1978 in *Progress in Liquid Physics,* ed. C A Croxton (Wiley, New York), p. 213
Capel H W, Nijhoff F W and den Breems A 1986 *Physica* **139A** 256
Carless D C, Hall H E and Hook J R 1983a *J. Low Temp. Phys.* **50** 583
Carless D C, Hall H E and Hook J R 1983b *J. Low Temp. Phys.* **50** 605
Carneiro G M and Pethick C J 1977 *Phys. Rev.* **B16** 1933

Carney J P, Guénault A M, Pickett G R and Spencer G F 1989 *Phys. Rev. Lett.* **62** 3042
Carton J P 1975 *J. Physique* **36** 213
Castaing B and Nozières P 1979 *J. Physique* **40** 257
Castelijns C A M, Coates K F, Guénault A M, Mussett S G and Pickett G R 1985 *Phys. Rev. Lett.* **55** 2021
Castelijns C A M, Coates K F, Guénault A M, Mussett S G and Pickett G R 1986 *Phys. Rev. Lett.* **56** 69
Ceperley D M and Kalos M H 1979 in *Monte Carlo Methods in Statistical Physics*, ed. K Binder (Springer, Berlin), p. 145
Chainer T, Morii Y and Kojima H 1980 *Phys. Rev.* **B21** 3941
Chainer T, Morii Y and Kojima H 1984 *J. Low Temp. Phys.* **55** 353
Chechetkin V R 1976a *Zh. Eksp. Teor. Fiz.* **71** 1463 [*Sov. Phys. JETP* **44** 766 (1976)]
Chechetkin V R 1976b *Fiz. Nizk. Temp.* **2** 1365 [*Sov. J. Low Temp. Phys.* **2** 665 (1976)]
Chechetkin V R 1976c *Fiz. Nizk. Temp.* **2** 434 [*Sov. J. Low Temp. Phys.* **2** 215 (1976)]
Chechetkin V R 1976d *Fiz. Nizk. Temp.* **2** 443 [*Sov. J. Low Temp. Phys.* **2** 220 (1976)]
Cheng T P and Li L F 1984 *Gauge Theory of Elementary Particle Physics* (Clarendon Press, Oxford)
Chervonko E 1976 *Zh. Eksp. Teor. Fiz.* **71** 1099 [*Sov. Phys. JETP* **44** 202 (1976)]
Choi C H and Muzikar P 1986 *Phys. Rev.* **B33** 2033
Coates K F, Guénault A M, Mussett S G and Pickett G R 1986 *Europhys. Lett.* **2** 523
Coffey D and Pethick C J 1988 *Phys. Rev.* **B37** 1647
Coleman S 1985 *Nucl. Phys.* **B262** 263
Combescot M and Combescot R 1976a *Phys. Rev. Lett.* **37** 388
Combescot M and Combescot R 1976b *Phys. Lett.* **58A** 181
Combescot R 1974a *Phys. Rev.* **A10** 1700
Combescot R 1974b *Phys. Rev. Lett.* **33** 946
Combescot R 1975a *Phys. Rev. Lett.* **35** 471
Combescot R 1975b *Phys. Rev.* **B12** 4839
Combescot R 1975c *Phys. Rev. Lett.* **34** 8
Combescot R 1975d *Phys. Rev. Lett.* **35** 1646
Combescot R 1975e *J. Low Temp. Phys.* **18** 537
Combescot R 1976 *Phys. Rev.* **B13** 126
Combescot R 1978a *Phys. Rev.* **B18** 3139
Combescot R 1978b *Phys. Rev.* **B18** 6071
Combescot R 1980a *Phys. Lett.* **78A** 85
Combescot R 1980b *J. Physique Lett.* **41** 207
Combescot R 1981a *J. Phys.* **C14** 1619
Combescot R 1981b *J. Phys.* **C14** 4765
Combescot R 1982a *Phys. Rev.* **B25** 3392
Combescot R 1982b *J. Low Temp. Phys.* **49** 295
Combescot R and Dombre T 1980 *Phys. Lett.* **76A** 293
Combescot R and Dombre T 1983a *Phys. Rev.* **B28** 5140
Combescot R and Dombre T 1983b in *Quantum Fluids and Solids—1983* (AIP Conf. Proc. No. 103), ed. E D Adams and G G Ihas (AIP, New York), p. 261
Combescot R and Dombre T 1985a *Phys. Rev. Lett.* **54** 1461
Combescot R and Dombre T 1985b *Phys. Rev.* **B32** 2960
Combescot R and Dombre T 1986 *Phys. Rev.* **B33** 79
Combescot R and Ebisawa H 1974 *Phys. Rev. Lett.* **33** 810
Combescot R and Saslow WM 1977 *J. Low Temp. Phys.* **26** 519

Cooper L N 1956 *Phys. Rev.* **104** 1189
Cooper L N, Mills R L and Sessler A M 1959 *Phys. Rev.* **114** 1377
Corruccini L R and Osheroff D D 1975a *Phys. Rev. Lett.* **34** 564
Corruccini L R and Osheroff D D 1975b *Phys. Rev. Lett.* **34** 695
Corruccini L R and Osheroff D D 1978 *Phys. Rev.* **B17** 126
Corruccini L R and Osheroff D D 1980 *Phys. Rev. Lett.* **45** 2029
Corruccini L R, Osheroff D D, Lee D M and Richardson R C 1971 *Phys. Rev. Lett.* **27** 650
Crooker B C 1983 Doctoral Thesis, Cornell University
Crooker B C, Hebral B and Reppy J D 1981 *Physica* **108B + C** 795 (Proceedings of the 16th International Conference on Low Temperature Physics, LT-16)
Cross M C 1975 *J. Low Temp. Phys.* **21** 525, *ibid.* **24** E261
Cross M C 1977a *J. Low Temp. Phys.* **26** 165
Cross M C 1977b in *Quantum Fluids and Solids,* ed. S B Trickey, E D Adams and J W Dufty (Plenum, New York), p. 183
Cross M C 1977c *Sci. Prog.* **64** 157
Cross M C 1978 *J. Low Temp. Phys.* **30** 481
Cross M C 1983 in *Quantum Fluids and Solids—1983* (AIP Conf. Proc. No. 103), ed. E D Adams and G G Ihas (AIP, New York), p. 325
Cross M C and Anderson P W 1975 in *Proceedings of the 14th International Conference on Low Temperature Physics LT-14,* Part I, ed. M Krusius and M Vuorio (North-Holland, Amsterdam), p. 29
Cross M C and Brinkman W F 1977 *J. Low Temp. Phys.* **27** 683
Cross M C and Liu M 1978 *J. Phys.* **C11** 1795
Czerwonko J 1967 *Acta Phys. Polon.* **32** 335
Czerwonko J 1980a *Physica* **100A** 277
Czerwonko J 1980b *Physica* **100A** 291
Czerwonko J 1982 *Physica* **115A** 531
Czerwonko J 1985 *Acta Phys. Polon.* **A67** 1111
Dahl D A 1977 *J. Low Temp. Phys.* **27** 139
Dahm A J, Betts D S, Brewer D F, Hutchins J D, Saunders J and Truscott W S 1980 *Phys. Rev. Lett.* **45** 1411
Daniels M E, Dobbs E R, Saunders J and Ward P L 1983 *Phys. Rev.* **B27** 6988
Das P, de Bruyn Ouboter R and Taconis K W 1965 in *Proceedings of the 9th International Conference on Low Temperature Physics, LT-9,* Part B, ed. J G Daunt, D O Edwards, F J Milford and M Yaqub (Plenum, New York), p. 1253
Davis J C, Amar A, Pekola J P and Packard R E 1988 *Phys. Rev. Lett.* **60** 302
de Gennes P G 1966 *Superconductivity of Metals and Alloys* (Benjamin, New York)
de Gennes P G 1973 *Phys. Lett.* **44A** 271
de Gennes P G 1974 *The Physics of Liquid Crystals* (Clarendon Press, Oxford)
de Gennes P G and Rainer D 1974 *Phys. Lett.* **46A** 429
Delrieu, J M 1974 *J. Physique Lett.* **35** 189; *ibid.* **36** E22
Delrieu, J M 1977 *J. Physique Lett.* **38** 127
de Vegvar P G N 1984 *Phys. Rev.* **B30** 6349
de Vegvar P G N 1985 *J. Stat. Phys.* **39** 675
de Vegvar P G N, Yan S S, Sagan D, Polturak E and Lee D M 1981, *Physica* **107B + C** 683 (Proceedings of the 16th International Conference on Low Temperature Physics, LT-16)
Dobbs E R 1983 *Contemp. Phys.* **24** 389
Dombre T and Combescot R 1981 *Physica* **107B + C** 51 (Proceedings of the 16th International Coference on Low Temperature Physics, LT-16)
Dombre T and Combescot R 1982a *J. Phys.* **C15** 6925
Dombre T and Combescot R 1982b *Phys. Rev. Lett.* **48** 495
Dombre T and Combescot R 1984 *Phys. Rev.* **B30** 3765

Dombre T and Combescot R 1985 *Phys. Rev.* **B32** 1751
Dombre T, Combescot R and Bagnuls C 1980 *J. Phys.* **C13** 4297
Doniach S and Engelsberg S 1966 *Phys. Rev. Lett.* **17** 750
Donnelly R J 1965 *Phys. Rev. Lett.* **14** 39
Dörfle M, Brand H and Graham R 1980 *J. Phys.* **C13** 3337
Dow R C M 1984 Doctoral Thesis, Lancester University
Dow R C M and Hook J R 1984 in *Proceedings of the 17th International Conference on Low Temperature Physics, LT-17,* Part I, ed. U Eckern, A Schmid, W Weber and H Wühl (North-Holland, Amsterdam), p. 53
Dow R C M and Hook J R 1985 *Phys. Rev. Lett.* **55** 2305
Duru I H and Ünal N 1982 *J. Low Temp. Phys.* **49** 559
Dy K S and Pethick C J 1969 *Phys. Rev.* **185** 373
Dyugaev A M 1984 *Zh. Eksp. Teor. Fiz.* **87** 1232 [*Sov. Phys. JETP* **60** 704 (1984)]
Ebisawa H 1980 *J. Phys.* **C13** 6453
Ebisawa H and Maki K 1974 *Prog. Theor. Phys.* **51** 337
Eckern U 1981 *Ann. Phys. (NY)* **133** 390
Edwards D O, Feder J D and Nayak V S 1977 in *Quantum Fluids and Solids,* ed. S B Trickey, E D Adams and J W Dufty (Plenum, New York), p. 375
Efetov K B and Salomaa M M 1981 *J. Low Temp. Phys.* **42** 35
Eilenberger G 1965 *Z. Phys.* **182** 427
Eilenberger G and Schopohl N 1988 *Europhys. Lett.* **7** 139
Einzel D 1980 Doctoral Thesis, Technische Universität München
Einzel D 1981 *Physica* **108B + C** 1143 (Proceedings of the 16th International Conference on Low Temperature Physics, LT-16)
Einzel D 1984 *J. Low Temp. Phys.* **54** 427
Einzel D and Parpia J M 1987a *Phys. Rev. Lett.* **58** 1937
Einzel D and Parpia J M 1987b *Jap. J. Appl. Phys.* **26** Suppl. 26-3, Part 1, p. 161 (Proceedings of the 18th International Conference on Low Temperature Physics, LT-18)
Einzel D and Wölfle P 1978 *J. Low Temp. Phys.* **32** 19
Einzel D and Wölfle P 1984 in *Proceedings of the 17th International Conference on Low Temperature Physics, LT-17,* Part II, ed. U Eckern, A Schmid, W Weber and H Wühl (North-Holland, Amsterdam), p. 773
Einzel D, Højgaard Jensen H, Smith H and Wölfle P 1983 *J. Low Temp. Phys.* **53** 695
Einzel D, Eska G, Hirayoshi Y, Kopp T and Wölfle P 1984a *Phys. Rev. Lett.* **53** 2312
Einzel D, Wölfle P, Højgaard Jensen H and Smith H 1984b *Phys. Rev. Lett.* **52** 1705
Einzel D, Hirschfeld P J and Wölfle P 1987 *Jap. J. Appl. Phys.* **26** Suppl. 26-3, Part 1, p. 129 (Proceedings of the 18th International Conference on Low Temperature Physics, LT-18)
Eisenstein J P 1983 in *Quantum Fluids and Solids—1983* (AIP Conf. Proc. No. 103), ed. E D Adams and G G Ihas (AIP, New York), p. 344
Eisenstein J P and Packard R E 1982 *Phys. Rev. Lett.* **49** 564
Eisenstein J P, Swift G W and Packard R E 1979 *Phys. Rev. Lett.* **43** 1676
Eisenstein J P, Swift G W and Packard R E 1980a *Phys. Rev. Lett.* **45** 1199
Eisenstein J P, Swift G W and Packard R E 1980b *Phys. Rev. Lett.* **45** 1569
Emery V J 1964 *Ann. Phys. (NY)* **28** 1
Emery V J 1976 *J. Low Temp. Phys.* **22** 467
Emery V J and Sessler A M 1960 *Phys. Rev.* **119** 43
Engel B N and Ihas G G 1985 *Phys. Rev. Lett.* **55** 955
Engel B N, Berg R F and Ihas G G 1985 *Phys. Rev. Lett.* **54** 1331
Engelsberg S 1977 *Phys. Lett.* **62A** 223
Engelsberg S, Brinkman W F and Anderson P W 1974 *Phys. Rev.* **A9** 2592
Engelsberg S, Brinkman W F and Osheroff D D 1977 *J. Low Temp. Phys.* **29** 29

Eska G, Neumaier K, Schoepe W, Uhlig K, Wiedemann W and Wölfle P 1980 *Phys. Rev. Lett.* **44** 1337
Eska G, Willers H G, Amend B and Wiedemann W 1981 *Physica* **108B + C** 1155 (Proceedings of the 16th International Conference on Low Temperature Physics, LT-16)
Eska G, Neumaier K, Schoepe W, Uhlig K and Wiedemann W 1982 *Phys. Lett.* **87A** 311
Eska G, Neumaier K, Schoepe W, Uhlig K and Wiedemann W 1983 *Phys. Rev.* **B27** 5534
Essmann U and Träuble H 1967 *Phys. Lett* **24A** 526
Fal'ko V I 1985 *Pis'ma Zh. Eksp. Teor. Fiz.* **42** 213 [*JETP Lett.* **42** 264 (1985)]
Fay D 1975 *Phys. Lett.* **51A** 365
Feder J D, Edwards D O, Gully W J, Muething K A and Scholz H N 1981 *Phys. Rev. Lett.* **47** 428
Feenberg E 1969 *Theory of Quantum Fluids* (Academic Press, New York)
Fetter A L 1975a in *Quantum Statistics and the Many Body Problem,* ed. S B Trickey, W P Kirk and J W Dufty (Plenum, New York), p. 127
Fetter A L 1975b *Phys. Lett.* **54A** 63
Fetter A L 1976a *J. Low Temp. Phys.* **23** 245
Fetter A L 1976b *Phys. Rev.* **B14** 2801
Fetter A L 1977 *Phys. Rev.* **B15** 1350
Fetter A L 1978a *Phys. Rev. Lett.* **40** 1656
Fetter A L 1978b *J. Physique* **39** Colloq. C-6, Vol. I, p. 46 (Proceedings of the 15th International Conference on Low Temperature Physics, LT-15)
Fetter A L 1978c *Phys. Rev.* **B17** 1152
Fetter A L 1979 *Phys. Rev.* **B20** 303
Fetter A L 1981 *Phys. Rev.* **B24** 1181
Fetter A L 1982 *J. Low Temp. Phys.* **48** 435
Fetter A L 1983 in *Quantum Fluids and Solids—1983* (AIP Conf. Proc. No. 103), ed. E D Adams and G G Ihas (AIP, New York), p. 229
Fetter A L 1985a *J. Low Temp. Phys.* **58** 545
Fetter A L 1985b *Phys. Rev.* **B31** 7012
Fetter A L 1986 in *Progress in Low Temperature Physics,* Vol. X, ed. D F Brewer (North-Holland, Amsterdam), p. 1
Fetter A L 1987 *J. Low Temp. Phys.* **67** 17
Fetter A L 1988 *J. Low Temp. Phys.* **70** 499
Fetter A L 1989 Private communication
Fetter A L and Kurkijärvi J 1977a in *Quantum Fluids and Solids,* ed. S B Trickey, E D Adams and J W Dufty (Plenum, New York), p. 233
Fetter A L and Kurkijärvi J 1977b *Phys. Rev.* **B15** 4272
Fetter A L and Theodorakis S 1984 *Phys. Rev. Lett.* **52** 2007
Fetter A L and Ullah S 1988 *J. Low Temp. Phys.* **70** 515
Fetter A L and Walecka J D 1971 *Quantum Theory of Many-Particle Systems* (McGraw-Hill, New York)
Fetter A L and Williams M R 1979 *Phys. Rev. Lett.* **43** 1601
Fetter A L and Williams M R 1981 *Phys. Rev.* **B23** 2186
Fetter A L, Sauls J A and Stein D L 1983 *Phys. Rev.* **B28** 5061
Feynman R P 1955 in *Progress in Low Temperature Physics,* Vol. I, ed. C J Gorter (North-Holland, Amsterdam), p. 17
Feynman R P 1972 *Statistical Physics* (Benjamin, Reading, Massachusetts)
Fishman F 1978 *J. Low Temp. Phys.* **33** 331
Fishman F 1979 *J. Low Temp. Phys.* **36** 357
Fishman F and Folk R 1984a *Phys. Rev.* **B29** 6367
Fishman F and Folk R 1984b *Phys. Rev.* **B30** 1547
Fishman F and Folk R 1985 *Phys. Rev.* **B32** 98

Fishman F and Privorotskii I A 1976 *J. Low Temp. Phys.* **25** 225
Fishman R S and Sauls J A 1985 *Phys. Rev.* **B31** 251
Fishman R S and Sauls J A 1986 *Phys. Rev.* **B33** 6068
Fishman R S and Sauls J A 1988a *Phys. Rev. Lett.* **61** 2871
Fishman R S and Sauls J A 1988b *Phys. Rev.* **B38** 2526
Flint E B, Mueller R M and Adams E D 1978 *J. Low Temp. Phys.* **33** 43
Fomin I A 1976 *Zh. Eksp. Teor. Fiz.* **71** 791 [*Sov. Phys. JETP* **44** 416 (1976)]
Fomin I A 1978a *J. Low Temp. Phys.* **31** 509
Fomin I A 1978b *Pis'ma Zh. Eksp. Teor. Fiz.* **28** 362 [*JETP Lett.* **28** 334 (1978)]
Fomin I A 1978c *Pis'ma Zh. Eksp. Teor. Fiz.* **28** 679 [*JETP Lett.* **28** 631 (1978)]
Fomin I A 1978d *Phys. Lett* **66A** 47
Fomin I A 1979a *Zh. Eksp. Teor. Fiz.* **77** 279 [*Sov. Phys. JETP* **50** 144 (1979)]
Fomin I A 1979b *Pis'ma Zh. Eksp. Teor. Fiz.* **30** 179 [*JETP Lett.* **30** 164 (1979)]
Fomin I A 1980 *Zh. Eksp. Teor. Fiz.* **78** 2392 [*Sov. Phys. JETP* **51** 1203 (1980)]
Fomin I A 1981a *Pis'ma Zh. Eksp. Teor. Fiz.* **33** 317 [*JETP Lett.* **33** 301 (1981)]
Fomin I A 1981b in *Soviet Scientific Reviews,* Section A, *Physics Reviews,* Vol. 3 (Harwood Academic Publishers, Chur), p. 275
Fomin I A 1983 *Zh. Eksp. Teor. Fiz.* **84** 2109 [*Sov. Phys. JETP* **57** 1227 (1983)]
Fomin I A 1984a *Pis'ma Zh. Eksp. Teor. Fiz.* **39** 387 [*JETP Lett.* **39** 466 (1984)
Fomin I A 1984b *Pis'ma Zh. Eksp. Teor. Fiz.* **40** 260 [*JETP Lett.* **40** 1037 (1984)]
Fomin I A 1985a *Pis'ma Zh. Eksp. Teor. Fiz.* **43** 134 [*JETP Lett.* **43** 170 (1985)]
Fomin I A 1985b *Zh. Eksp. Teor. Fiz.* **88** 2039 [*Sov. Phys. JETP* **61** 1207 (1985)]
Fomin I A 1986 *Pis'ma Zh. Eksp. Teor. Fiz.* **43** 134 [*JETP Lett.* **43** 171 (1986)]
Fomin I A 1987a *Pis'ma Zh. Eksp. Teor. Fiz.* **45** 106 [*JETP Lett.* **45** 135 (1987)]
Fomin I A 1987b *Zh. Eksp. Teor. Fiz.* **93** 2002 [*Sov. Phys. JETP* **66** 1142 (1987)]
Fomin I A 1988 *Zh. Eksp. Teor. Fiz.* **94** 112 [*Sov. Phys. JETP* **67** 1148 (1988)]
Fomin I A and Kamenskii V G, 1982 *Pis'ma Zh. Eksp. Teor. Fiz.* **35** 241 [*JETP Lett.* **35** 302 (1982)]
Fomin I A and Shopova D V 1989 *Pis'ma Zh. Eksp. Teor. Fiz.* **50** 143 [*JETP Lett.* **50** 160 (1989)]
Fomin I A and Vuorio M 1975 *J. Low Temp. Phys.* **21** 271
Fomin I A, Pethick C J and Serene J W 1978 *Phys. Rev. Lett.* **40** 1144
Fomin I A, Schertler R and Schoepe W 1982 *Phys. Lett.* **92A** 408
Fomin I A, Eska G and Wölfle P 1984 *J. Low Temp. Phys.* **56** 315
Forster D 1975 *Hydrodynamic Fluctuations, Broken Symmetry and Correlation Functions* (Benjamin, Reading, Massachusetts)
Frautschi S and Kim J S 1982 *Nucl. Phys.* **B196** 301
Freeman M R, Germain R S, Thuneberg E V and Richardson R C 1988 *Phys. Rev. Lett.* **60** 596
Fujita T 1975 *Prog. Theor. Phys.* **53** 1219
Fujita T and Tsuneto T 1975 *Prog. Theor. Phys.* **53** 289
Fujita T, Nakahara M, Ohmi T, Tsuneto T 1978a *Prog. Theor. Phys.* **60** 671
Fujita T, Ohmi T and Tsuneto T 1978b *Prog. Theor. Phys.* **59** 664
Fujita T, Ohmi T and Tsuneto T 1978c *Prog. Theor. Phys.* **60** 661
Fujita T, Nakahara M, Ohmi T and Tsuneto T 1980 *Prog. Theor. Phys.* **64** 396
Fulde P, Keller J and Zwicknagl G 1988 in *Solid State Physics* (*Advances in Research and Applications*), Vol. 41, ed. H Ehrenreich and D Turnbull (Academic Press, New York), p. 1
Galasiewicz Z 1960 *Acta Phys. Polon.* **19** 467
Galasiewicz Z 1969 *Superconductivity and Quantum Fluids* (Pergamon, Oxford)
Gammel P L and Reppy J D 1984 in *Proceedings of the 17th International Conference on Low Temperature Physics, LT-17,* Part I, ed. U Eckern, A Schmid, W Weber and H Wühl (North-Holland, Amsterdam), p. 1
Gammel P L, Hall H E and Reppy J D 1984 *Phys. Rev. Lett.* **52** 121

Gammel P L, Ho T L and Reppy J D 1985 *Phys. Rev. Lett.* **55** 2708
Garg A 1987 *Phys. Rev.* **B36** 6794
Garg A, Nair V P and Stone M 1987 *Ann. Phys. (NY)* **173** 149
Gay R, Hall H E, Hook J R and Sandiford D J 1981 *Physica* **108B + C** 797 (Proceedings of the 16th International Conference on Low Temperature Physics, LT-16)
Gay R, Bagley M, Hook J R, Sandiford D J and Hall H E 1983 *J. Low Temp. Phys.* **51** 227
Geilikman B T and Chechetkin V R 1975 *Zh. Eksp. Teor. Fiz.* **69** 286 [*Sov. Phys. JETP* **42** 148 (1975)]
Giannetta R W, Ahonen A I, Polturak E, Saunders J, Zeise E K, Richardson R C and Lee D M 1980 *Phys. Rev. Lett.* **45** 262
Giannetta R W, Smith E N and Lee D M 1981 *J. Low Temp. Phys.* **45** 295
Ginzburg V L 1960 *Fiz. Tverd. Tela* **2** 2031 [*Sov. Phys. Solid State* **2** 1824 (1961)]
Glaberson W I and Donnelly R J 1986 in *Progress in Low Temperature Physics,* Vol. IX, ed. D F Brewer (North-Holland, Amsterdam), p. 1
Glassgold A E and Sessler A M 1961 *Nuovo Cim.* **19** 723
Glyde H R and Hernadi S I 1983 *Phys. Rev.* **B28** 141
Glyde H R and Hernadi S I 1984a *Phys. Rev.* **B29** 3873
Glyde H R and Hernadi S I 1984b *Phys. Rev.* **B29** 4926
Glyde H R and Khanna F C 1977 *Can. J. Phys.* **55** 1906
Glyde H R and Khanna F C 1980 *Can. J. Phys.* **58** 343
Godfrin H, Frossati G, Thoulouze D, Chapellier M and Clark W G 1978 *J. Physique* **39** Colloq. C-6, Vol. I, p. 287 (Proceedings of the 15th International Conference on Low Temperature Physics, LT-15)
Goldstein L 1954 *Phys. Rev.* **96** 1445
Goldstone J 1961 *Nuovo Cim.* **19** 154
Golo V L 1981a *Zh. Eksp. Teor. Fiz.* **81** 942 [*Sov. Phys. JETP* **54** 501 (1982)]
Golo V L 1981b *Lett. Math. Phys.* **5** 155
Golo V L 1984 *Zh. Eksp. Teor. Fiz.* **86** 2100 [*Sov. Phys. JETP* **59** 1221 (1984)]
Golo V L and Kats E I 1984 *Pis'ma Zh. Eksp. Teor. Fiz.* **40** 418 [*JETP Lett.* **40** 1238 (1984)]
Golo V L and Kats E I 1986 *Zh. Eksp. Teor. Fiz.* **90** 952 [*Sov. Phys. JETP* **63** 555 (1986)]
Golo V L and Leman A A 1982a *Pis'ma Zh. Eksp. Teor. Fiz.* **35** 227 [*JETP Lett.* **35** 284 (1982)]
Golo V L and Leman A A 1982b *Zh. Eksp. Teor. Fiz.* **83** 1546 [*Sov. Phys. JETP* **56** 891 (1982)]
Golo V L and Leman A A 1983 *Zh. Eksp. Teor. Fiz.* **85** 932 [*Sov. Phys. JETP* **58** 541 (1983)]
Golo V L and Leman A A 1989 in *Helium 3,* ed. W P Halperin and L P Pitaevskii (North-Holland, Amsterdam)
Golo V L and Monastyrsky M I 1977 *Pis'ma Zh. Eksp. Teor. Fiz.* **25** 272 [*JETP Lett.* **25** 251 (1977)]
Golo V L and Monastyrsky M I 1978a *J. Physique* **39** Colloq. C-6, Vol. I, p. 48 (Proceedings of the 15th International Conference on Low Temperature Physics, LT-15)
Golo V L and Monastyrsky M I 1978b *J. Physique* **39** Colloq. C-6, Vol. I, p. 50 (Proceedings of the 15th International Conference on Low Temperature Physics, LT-15)
Golo V L and Monastyrsky M I 1978c *Lett. Math. Phys.* **2** 373
Golo V L and Monastyrsky M I 1978d *Ann. Inst. Henri Poincaré* **A28** 75
Golo V L, Monastyrsky M I and Novikov S P 1979 *Commun. Math. Phys.* **69** 237

Golo V L, Leman A A and Fomin I A 1983 *Pis'ma Zh. Eksp. Teor. Fiz.* **38** 123 [*JETP Lett.* **38** 146 (1983)]
Gongadze A D, Gurgenishvili G E and Kharadze G A 1976 *Pis'ma Zh. Eksp. Teor. Fiz.* **23** 667 [*JETP Lett.* **23** 622 (1976)]
Gongadze A D, Gurgenishvili G E and Kharadze G A 1977 *Zh. Eksp. Teor. Fiz.* **72** 1534 [*Sov. Phys. JETP* **45** 805 (1977)]
Gongadze A D, Gurgenishvili G E and Kharadze G A 1978a *Zh. Eksp. Teor. Fiz.* **75** 1504 [*Sov. Phys. JETP* **48** 759 (1978)]
Gongadze A D, Gurgenishvili G E and Kharadze G A 1978b *Fiz. Nizk. Temp.* **4** 582 [*Sov. J. Low Temp. Phys.* **4** 290 (1978)]
Gongadze A D, Gurgenishvili G E and Kharadze G A 1980 *Zh. Eksp. Teor. Fiz.* **78** 615 [*Sov. Phys. JETP* **51** 310 (1980)]
Gongadze A D, Gurgenishvili G E and Kharadze G A 1981 *Fiz. Nizk. Temp.* **7** 821 [*Sov. J. Low Temp. Phys.* **7** 397 (1984)]
Gorkov L P 1959 *Zh. Eksp. Teor. Fiz.* **36** 1918 [*Sov. Phys. JETP* **9** 1364 (1959)]
Gorkov L P 1987 in *Soviet Scientific Reviews,* Section A, *Physics Reviews,* Vol. 9 (Harwood Academic Publishers, Chur), p. 1
Gould C M 1979 Doctoral Thesis, Cornell University
Gould C M and Lee D M 1976 *Phys. Rev. Lett.* **37** 1223
Gould C M and Lee D M 1978 *Phys. Rev. Lett.* **41** 967
Gould C M, Bartolac T J and Bozler H M 1980 *J. Low Temp. Phys.* **39** 291
Grabinski M 1989 *Phys. Rev. Lett.* **63** C814
Grabinski M and Liu M 1987 *Phys. Rev. Lett.* **58** 800
Graham R 1974 *Phys. Rev. Lett.* **33** 1431
Graham R and Pleiner H 1975 *Phys. Rev. Lett.* **34** 792
Graham R and Pleiner H 1976 *J. Phys.* **C9** 279
Greaves N A 1976 *J. Phys.* **C9** L181
Greaves N A and Leggett A J 1983a *J. Phys.* **C16** 4383
Greaves N A and Leggett A J 1983b in *Quantum Fluids and Solids—1983* (AIP Conf. Proc. No. 103), ed. E D Adams and G G Ihas, (AIP, New York), p. 254
Greytak T J, Johnson R T, Paulson D N and Wheatley J C 1973 *Phys. Rev. Lett.* **31** 452
Greywall D S 1983 *Phys. Rev.* **B27** 2747
Greywall D S 1984 *Phys. Rev.* **B29** 4933
Greywall D S 1986 *Phys. Rev.* **B33** 7520
Gu G-Q 1982a *Commun. Theor. Phys. (China)* **1** 319
Gu G-Q 1982b *Acta Phys. Temp. Humilis Sin. (China)* **4** 16
Guénault A M, Keith V, Kennedy C J and Pickett G R 1983a *Phys. Rev. Lett.* **50** 522
Guénault A M, Keith V, Kennedy C J and Pickett G R 1983b *Phys. Rev. Lett.* **51** 589
Guénault A M, Keith V, Kennedy C J, Mussett S G and Pickett G R 1986 *J. Low Temp. Phys.* **62** 511
Guernsey R W, McCoy R J, Steinback M and Lyden J K 1976 *Phys. Lett.* **58A** 26
Gully W J, Osheroff D D, Lawson D T, Richardson R C and Lee D M 1973 *Phys. Rev.* **A8** 1633
Gully W J, Gould C M, Richardson R C and Lee D M 1975 *Phys. Lett.* **55A** 27
Gully W J, Gould C M, Richardson R C and Lee D M 1976 *J. Low Temp. Phys.* **24** 563
Gurgenishvili G E and Kharadze G A 1980 *Pis'ma Zh. Eksp. Teor. Fiz.* **31** 593 [*JETP Lett.* **31** 557 (1980)]
Gutzwiller M C 1963 *Phys. Rev. Lett.* **10** 159
Gutzwiller M C 1964 *Phys. Rev.* **134** A923
Gutzwiller M C 1965 *Phys. Rev.* **137** A1726

Hakonen P and Lounasmaa O V 1987 *Phys. Today* **40** 70
Hakonen P J and Mineev V P 1987 *J. Low Temp. Phys.* **67** 313
Hakonen P J and Volovik G E 1982 *J. Phys.* **C15** L1277
Hakonen P J, Ikkala O T and Islander S T 1982a *Phys. Rev. Lett.* **49** 1258
Hakonen P J, Ikkala O T, Islander S T, Lounasmaa O V, Markkula T K, Roubeau P, Saloheimo K M, Volovik G E, Andronikashvili E L, Garibashvili D I and Tsakadze J S, 1982b *Phys. Rev. Lett.* **48** 1838
Hakonen P J, Ikkala O T, Islander S T, Lounasmaa O V and Volovik G E 1983a *J. Low Temp. Phys.* **53** 425
Hakonen P J, Ikkala O T, Islander S T, Markkula T K, Roubeau P M, Saloheimo K M, Garibashvili D I and Tsakadze J S 1983b *Cryogenics* **23** 243
Hakonen P J, Krusius M, Salomaa M M, Simola J T, Bun'kov Yu M, Mineev V P and Volovik G E 1983c *Phys. Rev. Lett.* **51** 1362
Hakonen P J, Krusius M, Salomaa M M and Simola J T 1985a *Phys. Rev. Lett.* **54** 245
Hakonen P J, Krusius M and Seppälä H K 1985b *J. Low Temp. Phys.* **60** 187
Hall H E 1975 in *Proceedings of the 14th International Conference on Low Temperature Physics, LT-14,* Part I, ed. M Krusius and M Vuorio (North-Holland, Amsterdam), p. 33
Hall H E 1976 *J. Phys.* **C9** L443
Hall H E 1977 *Physica* **90B + C** 68
Hall H E 1978 *J. Physique* **39** Colloq. C-6, Vol. I, p. 15 (Proceedings of the 15th International Conference on Low Temperature Physics, LT-15)
Hall H E 1983 in *Quantum Fluids and Solids—1983* (AIP Conf. Proc. No. 103), ed. E D Adams and G G Ihas (AIP, New York) p. 265
Hall H E 1985 *Phys. Rev. Lett.* **54** 205
Hall H E and Hook J R 1977 *J. Phys.* **C10** L91
Hall H E and Hook J R 1986 in *Progress in Low Temperature Physics,* Vol. IX, ed. D F Brewer (North-Holland, Amsterdam), p. 143
Hall H E, Ford P J and Thompson K 1966 *Cryogenics* **6** 80
Hall H E, Gammel P L and Reppy J D 1984 *Phys. Rev. Lett.* **52** 1701
Halperin W P 1982 *Physica* **109/110B + C** 1596 (Proceedings of the 16th International Conference on Low Temperature Physics, LT-16)
Halperin W P, Buhrman R A, Lee D M and Richardson R C 1973 *Phys. Lett.* **45A** 233
Halperin W P, Archie C N, Rasmussen F B, Buhrmann R A and Richardson R C 1974 *Phys. Rev. Lett.* **32** 927
Halperin W P, Archie C N, Rasmussen F B and Richardson R C 1975 *Phys. Rev. Lett.* **34** 718
Halperin W P, Archie C W, Rasmussen F B, Alvesalo T A and Richardson R C 1976 *Phys. Rev.* **B13** 2124
Ham T E and Hu C R 1980 *J. Low Temp. Phys.* **40** 373
Hara J 1981 *J. Low Temp. Phys.* **43** 533
Hara J and Nagai K 1979 *J. Low Temp. Phys.* **34** 351
Hara J, Ono Y A, Nagai K and Kawamura K 1980 *J. Low Temp. Phys.* **39** 603
Harris C G 1980 *J. Phys.* **C13** L1061
Hasegawa Y 1982 *Prog. Theor. Phys.* **67** 1232
Hasegawa Y 1983 *Prog. Theor. Phys.* **70** 1141
Hasegawa Y 1985 *Prog. Theor. Phys.* **73** 1258
Hasegawa Y and Namaizawa H 1982 *Prog. Theor. Phys.* **67** 389
Herring C 1966 *Magnetism,* Vol. IV, ed. G Rado and H Suhl (Academic Press, New York)
Higgs P W 1964 *Phys. Rev. Lett.* **12** 132
Hirashima D S 1987 *Phys. Rev. Lett.* **59** C2386

Hirschfeld P and Stein D L 1983 *J. Low Temp. Phys.* **51** 257
Ho T L 1977 in *Quantum Fluids and Solids,* ed. S B Trickey, E D Adams and J W Dufty (Plenum, New York), p. 97
Ho T L 1978a *Phys. Rev.* **B18** 1144
Ho T L 1978b *Phys. Rev. Lett.* **41** 1473
Ho T L 1978c Doctoral Thesis, Cornell University
Ho T L and Mermin N D 1980 *Phys. Rev. Lett.* **44** 330
Ho T L, Fulco J R, Schrieffer J R and Wilczek F 1984 *Phys. Rev. Lett* **52** 1524
Hoffberg M, Glassgold A E, Richardson R W and Ruderman M 1970 *Phys. Rev. Lett.* **24** 175
Højgaard Jensen H, Smith H and Wilkins J W 1968 *Phys. Lett.* **27A** 532
Højgaard Jensen H, Smith H and Wilkins J W 1969 *Phys. Rev.* **185** 323
Højgaard Jensen H, Smith H, Wölfle P, Nagai K and Maack Bisgaard T 1980 *J. Low Temp. Phys.* **41** 473
Højgaard Jensen H, Smith H and Wölfle P 1983 *J. Low Temp. Phys.* **51** 81
Hong D K 1988 *J. Low Temp. Phys.* **71** 483
Hook J R 1978 *J. Physique* **39** Colloq. C-6, Vol. 1, p. 17 (Prooceedings of the 15th International Conference on Low Temperature Physics, LT-15)
Hook J R 1989 *J. Low Temp. Phys.* **74** 19
Hook J R and Hall H E 1979 *J. Phys.* **C12** 783
Hook J R and Zimmermann W 1981 *Physica* **108B + C** 1145 (Proceedings of the 16th International Conference on Low Temperature Physics, LT-16)
Hook J R, Eastop A D, Faraj E, Gould S G and Hall H E 1986 *Phys. Rev. Lett.* **57** 1749
Hook J R, Eastop A D, Faraj E, Gould S G and Hall H E 1987 *Can. J. Phys.* **65** 1486
Hook J R, Faraj E, Gould S G and Hall H E 1989 *J. Low Temp. Phys.* **74** 45
Houghton A and Maki K 1977 *Phys. Lett.* **61A** 317
Hoyt R F, Scholz H N and Edwards D O 1981 *Physica* **107B + C** 287 (Proceedings of the 16th International Conference on Low Temperature Physics, LT-16)
Hsu W 1984 Doctoral Thesis, University of Illinois at Urbana–Champaign
Hsu W and Pines D 1985 *J. Stat. Phys.* **38** 273
Hu C R 1978 *J. Low Temp. Phys.* **30** 267
Hu C R 1979a *Phys. Rev.* **B20** 276
Hu C R 1979b *Phys. Rev. Lett.* **43** 1811; *ibid.* **44** E299
Hu C R 1979c *Phys. Rev.* **B20** 1938
Hu C R and Ham T E 1978 *J. Physique* **39** Colloq. C-6, Vol I, p. 55 (Proceedings of the 15th International Conference on Low Temperature, LT-15)
Hu C R and Saslow W M 1977 *Phys. Rev. Lett.* **38** 605
Hu C R, Kumar P and Maki K 1977 Unpublished (referenced by Maki 1977a)
Hu C R, Ham T E and Saslow W M 1978 *J. Low Temp. Phys.* **32** 301
Huang K 1982 *Quarks, Leptons and Gauge Fields* (World Scientific, Singapore)
Huang K 1987 *Statistical Mechanics,* 2nd edn (Wiley, New York)
Hubbard J 1963 *Proc. R. Soc. Lond.* **A276** 238
Hutchins J D, Betts D S, Brewer D F, Dahm A J and Truscott W S 1981a *Physica* **108B + C** 1159 (Proceedings of the 16th International Conference on Low Temperature Physics, LT-16)
Hutchins J D, Betts D S, Brewer D F and Truscott W S 1981b *Physica* **108B + C** 1157 (Proceedings of the 16th International Conference on Low Temperature Physics, LT-16)
Hutchins J D, Brewer D F and Kruppa D 1985 *Phys. Rev. Lett.* **55** 1410
Ichikawa K, Yamasaki H, Akimoto T, Kodama T, Shigi T and Kojima H 1987 *Phys. Rev. Lett.* **58** 1949
Ihas G G, Engel B N and Berg R F 1985 *Phys. Lett.* **109A** 105

Ikkala O T, Volovik G E, Hakonen P J, Bun'kov Yu M, Islander S T and Kharadze G A 1982 *Pis'ma Zh. Eksp. Teor. Fiz.* **35** 338 [*JETP Lett.* **35** 416 (1982)]
Ishikawa M 1976 *Prog. Theor. Phys.* **55** 2014
Ishikawa M 1977a *Prog. Theor. Phys.* **57** 1
Ishikawa M 1977b *Prog. Theor. Phys.* **57** 1836
Ishikawa M 1980 *Prog. Theor. Phys.* **63** 338
Ishikawa M, Miyake K and Usui T 1980 *Prog. Theor. Phys.* **63** 1083
Ishikawa O, Sasaki Y, Mizusaki T, Hirai and Tsubota M 1989 *J. Low Temp. Phys.* **75** 35
Israelsson U E, Crooker B C, Bozler H M and Gould C M 1984 *Phys. Rev. Lett.* **53** 1943
Jacak L 1977 *J. Phys.* **C10** 4701
Jacak L 1981 *Acta Phys. Polon.* **A60** 205
Jacobsen K W and Smith H 1983 *J. Low Temp. Phys.* **52** 527
Jacobsen K W and Smith H 1987 *J. Low Temp. Phys.* **67** 83
Janke W and Kleinert H 1980 *Phys. Lett.* **78A** 363
Jarić M V 1980 in *Group Theoretical Methods in Physics* (Lecture Notes in Physics, Vol. 135), ed. K B Wolf (Springer, Berlin), p. 12
Jarić M V 1982 *Phys. Rev. Lett.* **48** 1641
Johnson D L 1982 *Phys. Rev. Lett.* **49** 1361
Johnson R T, Kleinberg R L, Webb R A, and Wheatley J C 1975 *J. Low Temp. Phys.* **18** 501
Jones D R T, Love A and Moore M A 1976 *J. Phys.* **C9** 743
Jones R B 1977 *J. Phys.* **C10** 657
Josephson B D 1962 *Phys. Lett.* **1** 251
Josephson B D 1965 *Adv. Phys.* **14** 419
Josephson B D and Lekner J 1969 *Phys. Rev. Lett.* **23** 111
Kamerlingh Onnes H 1911a *Leiden Commun.* **108**; *Proc. R. Acad. Amsterdam* **11** 168
Kamerlingh Onnes H 1911b *Proc. R. Acad. Amsterdam* **13** 1903
Kamerlingh Onnes H and Boks J D A 1924 *Rep. Comm. 4th International Congress on Refrigeration, London*; *Leiden Commun.* **170a**
Kanamori J 1963 *Prog. Theor. Phys.* **30** 275
Kapitza P L 1938 *Nature* **141** 74
Katayama T 1981 *Prog. Theor. Phys.* **65** 1158
Katayama T, Ooiwa K and Ishii C 1980a *J. Low Temp. Phys.* **40** 1
Katayama T, Ooiwa K and Ishii C 1980b *Phys. Lett.* **76A** 300
Katayama T, Ooiwa K and Ishii C 1980c *Phys. Lett.* **79A**91
Katayama T, Yamaguchi Y, Furukawa F and Ishii C 1983a *Phys. Rev.* **B27** 3096
Katayama T, Yamaguchi Y and Ishii C 1983b *Phys. Rev.* **B27** 3096
Kaul R and Kleinert H 1980 *J. Low Temp. Phys.* **38** 539
Keen B E, Mathews P W and Wilks J 1963 *Phys. Lett.* **5** 5
Keesom W H 1942 *Helium* (Elsevier, Amsterdam)
Keesom W H and Wolfke M 1928 *Proc. R. Acad. Amsterdam* **31** 90; *Leiden Commun.* **190b**
Keller W E 1969 *Helium-3 and Helium-4* (Plenum, New York)
Kesaev V I and Ugulava A I 1984 *Zh. Eksp. Teor. Fiz.* **87** 1058 [*Sov. Phys. JETP* **60** 605 (1984)]
Ketterson J B 1983 *Phys. Rev. Lett.* **50** 259
Ketterson J B, Roach P R, Abraham B M and Roach P D 1975 in *Quantum Statistics and the Many-Body Problem,* ed. S B Trickey, W P Kirk and J W Dufty (Plenum, New York), p. 35
Ketterson J B, Meisel M W, Shivaram B S, Sarma B K and Halperin W P 1983a in *1983 Ultrasonics Symposium Proceedings* (IEEE, New York), p. 1074

Ketterson J B, Shivaram B S, Meisel M W, Sarma B K and Halperin W P 1983b in *Quantum Fluids and Solids—1983* (AIP Conf. Proc. No. 103), ed. E D Adams and G G Ihas (AIP, New York), p. 288
Khalatnikov I M 1965 *An Introduction to the Theory of Superfluidity* (Benjamin, New York)
Khalatnikov I M and Lebedev V V 1977a *Pis'ma Zh. Eksp. Teor. Fiz.* **25** 377 [*JETP Lett.* **25** 351 (1977)]
Khalatnikov I M and Lebedev V V 1977b *Phys. Lett.* **61A** 319
Khalatnikov I M and Lebedev V V 1980 *Prog. Theor. Phys. Suppl.* **69** 269
Kharadze G A and Maki K 1982 *Phys. Rev.* **B26** 1182
Kharadze G A and Smith H 1978 *Phys. Rev.* **B18** 3185
Khazan M V 1985 *Pis'ma Zh. Eksp Teor. Fiz.* **41** 396 [*JETP Lett.* **41** 486 (1985)]
Khriplovich I B 1982 *Pis'ma Zh. Eksp. Teor. Fiz.* **35** 392 [*JETP Lett.* **35** 485 (1982)]
Kieselmann G and Rainer D 1983 *Z. Phys.* **B52** 267
Kiewiet C W, Main P C and Hall H E 1975 in *Proceedings of the 14th International Conference on Low Temperature Physics, LT-14,* Vol. 5, ed. M Krusius and M Vuorio (North-Holland, Amsterdam), p. 413
Kim J S 1982 *Nucl. Phys.* **B196** 285
Kim J S 1985 *Phys. Rev.* **B31** 1433
Kitchenside P W, Bullough R K and Caudrey P J 1978 in *Solitons and Condensed Matter Physics,* ed. A R Bishop and T Schneider (Springer, Berlin), p. 291
Kitchenside P W, Bullough R K and Caudrey P J 1979a in *Structural Stability in Physics,* ed. W Güttinger and H Eikemeier (Springer, Berlin), p. 254
Kitchenside P W, Caudrey P J and Bullough R K 1979b *Physica Scripta* **20** 673
Kjäldman L H, Kurkijärvi J and Rainer D 1978 *J. Low Temp. Phys.* **33** 577
Kleinberg R L 1979a *Phys. Rev. Lett.* **42** 182
Kleinberg R L 1979b *J. Low Temp. Phys.* **35** 489
Kleinberg R L, Paulson D N, Webb R A and Wheatley J C 1974 *J. Low Temp. Phys.* **17** 521
Kleinert H 1978 *Fortschr. Phys.* **26** 565
Kleinert H 1979a *Phys. Lett.* **71A** 66
Kleinert H 1979b Lecture notes presented at the 17th Erice Summer School (unpublished)
Kleinert H 1980a *Phys. Lett.* **78A** 155
Kleinert H 1980b *J. Low Temp. Phys.* **39** 451
Kleinert H, Lin-Liu Y R and Maki K 1978 *J. Physique* **39** Colloq. C-6, Vol. I, p. 59 (Proceedings of the 15th International Conference on Low Temperature Physics, LT-15)
Kleinert H, Lin-Liu Y R and Maki K 1979 *Phys. Lett.* **70A** 27
Koch V E 1980 Doctoral Thesis, Technische Universität München
Koch V E and Wölfle P 1981 *Phys. Rev. Lett.* **46** 486
Kodama T and Kojima H 1981 *Phys. Lett.* **87A** 103
Kojima H, Paulson D N and Wheatley J C 1974 *Phys. Rev. Lett.* **32** 141
Kojima H, Paulson D N and Wheatley J C 1975 *J. Low Temp. Phys.* **21** 283
Kokko J, Paalanen M A, Richardson R C and Takano Y 1978a *J. Phys.* **C11** L125
Kokko J, Paalanen M A, Schoepe W and Takano Y 1978b *J. Low Temp. Phys.* **33** 69
Kondo J and Soda T 1983 *J. Low Temp. Phys.* **50** 21
Kopnin N B 1978 *Zh. Eksp. Teor. Fiz.* **74** 1538 [*Sov. Phys. JETP* **47** 804 (1978)]
Kopnin N B 1986a *Pis'ma Zh. Eksp. Teor. Fiz.* **43** 541 [*JETP Lett.* **43** 700 (1986)]
Kopnin N B 1986b *J. Low Temp. Phys.* **65** 433
Kopnin N B 1987 *Zh. Eksp. Teor. Fiz.* **92** 2106 [*Sov. Phys. JETP* **65** 1187 (1987)]
Kopp T and Wölfle P 1987 *Phys. Rev. Lett.* **59** 2979
Korshunov S E 1985 *Zh. Eksp. Teor. Fiz.* **89** 531 [*Sov. Phys. JETP* **62** 301 (1985)]

Kosterlitz J M 1974 *J. Phys.* **C7** 1046

Kosterlitz J M and Thouless D J 1972 *J. Phys.* **C5** L124

Kosterlitz J M and Thouless D J 1973 *J. Phys.* **C6** 1181

Kotsubo V, Hahn K D and Parpia J M 1987 *Phys. Rev. Lett.* **58** 804

Krotscheck E 1983 in *Quantum Fluids and Solids—1983* (AIP Conf. Proc. No. 103), ed. E D Adams and G G Ihas (AIP, New York), p. 132

Krusius M, Paulson D N and Wheatley J C 1978 *J. Low Temp. Phys.* **33** 255

Kumar P 1977 in *Quantum Fluids and Solids*, ed. S B Trickey, E D Adams and J W Dufty (Plenum, New York), p. 45

Kumar P 1978 *Phys. Lett.* **66A** 487

Kurkijärvi J 1978 *J. Physique* **39** Colloq. C-6, Vol. I, p. 63 (Proceedings of the 15th International Conference on Low Temperature Physics, LT-15)

Kurkijärvi J and Rainer D 1989 in *Helium 3,* ed. W P Halperin and L P Pitaevskii (North-Holland, Amsterdam)

Kuroda Y 1974 *Prog. Theor. Phys.* **51** 1269

Kuroda Y 1975 *Prog. Theor. Phys.* **53** 349

Kuroda Y and Nagi A D S 1975 *Prog. Theor. Phys.* **53** 1206

Kuroda Y and Nagi A D S 1976a *J. Low Temp. Phys.* **23** 751

Kuroda Y and Nagi A D S 1976b *Prog. Theor. Phys.* **55** 2018

Kuroda Y and Nagi A D S 1976c *J. Low Temp. Phys.* **25** 569

Kuroda Y and Nagi A D S 1977 *Physica* **85B + C** 131

Kuroda Y and Nagi A D S 1978 *J. Low Temp. Phys.* **30** 755

Kurten K E and Campbell C E 1981 *J. Low Temp. Phys.* **44** 149

Kurti N, Robinson F N, Simon F and Spohr D A 1956 *Nature* **178** 450

Kyynäräinen J M, Pekola J P, Manninen A J and Torizuka K 1989 Superfluid ^{3}He in strong magnetic fields: anomalous sound attenuation in the B phase and evidence for splitting of the A–B transition. Preprint, Helsinki University of Technology

Landau L D 1941 *Zh. Eksp. Teor. Fiz.* **11** 592

Landau L D 1956 *Zh. Eksp. Teor. Fiz.* **30** 1058 [*Sov. Phys. JETP* **3** 920 (1957)]

Landau L D 1957 *Zh. Eksp. Teor. Fiz.* **32** 59 [*Sov. Phys. JETP* **5** 101 (1957)]

Landau L D 1958 *Zh. Eksp. Teor. Fiz.* **35** 97 [*Sov. Phys. JETP* **8** 70 (1959)]

Landau L D and Lifshitz E M 1958 *Quantum Mechanics* (Pergamon, London)

Landau L D and Lifshitz E M 1959a *Statistical Physics* (Pergamon, London)

Landau L D 1959b *Fluid Mechanics* (Pergamon, London)

Langer J S and Fisher M E 1967 *Phys. Rev. Lett.* **19** 560

Langer J S and Reppy J D 1970 in *Progress in Low Temperature Physics,* Vol. VI, ed. C J Gorter (North-Holland, Amsterdam), p. 1

Lawson D T, Gully W J, Goldstein S, Richardson R C and Lee D M 1973 *Phys. Rev. Lett.* **30** 541

Lawson D T, Gully W J, Goldstein S, Richardson R C and Lee D M 1974 *J. Low Temp. Phys.* **15** 169

Lawson D T, Bozler H M and Lee D M 1975a *Phys. Rev. Lett.* **34** 121

Lawson D T, Bozler H M and Lee D M 1975b in *Quantum Statistics and the Many-Body Problem,* ed. S B Trickey, W P Kirk and J W Dufty (Plenum, New York), p. 19

Layzer A and Fay D 1968 in *Proceedings of the 11th International Conference on Low Temperature Physics, LT-11,* Vol. 1, ed. J F Allen, D M Finlayson and D M McCall (University of St Andrews Printing Dept), p. 404

Layzer A and Fay D 1971 *Int. J. Magn.* **1** 135

Layzer A and Fay D 1974 *Solid State Commun.* **15** 599

Lea M J, Butcher K J and Dobbs E R 1977 *Commun. Phys.* **2** 59

Lebedev V V and Khalatnikov I M 1977 *Zh. Eksp. Teor. Fiz.* **73** 4537 [*Sov. Phys. JETP* **46** 808 (1977)]

Lee D M and Richardson R C 1978 in *Physics of Liquid and Solid Helium,* Part II, ed. K H Bennemann and J B Ketterson (Wiley, New York), p. 287
Lee M A, Schmidt K E, Kalos M H and Chester G V 1981 *Phys. Rev. Lett.* **46** 728
Lee P A, Rice T M, Serene J W, Sham L J and Wilkins J W 1986 *Comments Cond. Matter Phys.* **12** 99
Leggett A J 1965a *Phys. Rev.* **140A** 1869
Leggett A J 1965b *Phys. Rev. Lett.* **14** 536
Leggett A J 1966 *Phys. Rev.* **147** 119
Leggett A J 1972 *Phys. Rev. Lett.* **29** 1227
Leggett A J 1973a *Phys. Rev. Lett.* **31** 352
Leggett A J 1973b *J. Phys.* **C6** 3187
Leggett A J 1974a *Ann. Phys. (NY)* **85** 11
Leggett A J 1974b *Prog. Theor. Phys.* **51** 1275
Leggett A J 1975a *Rev. Mod. Phys.* **47** 331
Leggett A J 1975b *Phys. Rev. Lett.* **35** 1178
Leggett A J 1976 *Endeavour* **35** 83
Leggett A J 1977a *Phys. Rev. Lett.* **39** 587
Leggett A J 1977b *Nature* **270** 585
Leggett A J 1978a *J. Physique* **39** Colloq. C-6, Vol. III, p. 1264 (Proceedings of the 15th International Conference on Low Temperature Physics, LT-15)
Leggett A J 1978b in *Quantum Liquids,* ed. J Ruvalds and T Regge (North-Holland, Amsterdam), p. 167
Leggett A J 1980a in *Proceedings of the XVI Karpacz Winter School of Theoretical Physics* (Lecture Notes in Physics, Vol. 115), ed. A Pekalski and J Przyslawa (Springer, Berlin), p. 13
Leggett A J 1980b *J. Physique* **41** Colloq. C-7, p. 19
Leggett A J 1984 *Phys. Rev. Lett.* **53** 1096
Leggett A J 1985 *Phys. Rev. Lett.* **54** 246
Leggett A J 1987 *J. Magn. Magn. Mater.* **63&64** 406
Leggett A J 1989 Private communication
Leggett A J and Takagi S 1975 *Phys. Rev. Lett.* **34** 1424
Leggett A J and Takagi S 1976 *Phys. Rev. Lett.* **36** 1379
Leggett A J and Takagi S 1977 *Ann. Phys. (NY)* **106** 79
Leggett A J and Takagi S 1978 *Ann. Phys. (NY)* **110** 353
Leggett A J and Yip S 1989 in *Helium 3,* ed W P Halperin and L P Pitaevskii (North-Holland, Amsterdam)
Levin K 1975 *Phys. Rev. Lett.* **34** 1002
Levin K and Valls O T 1977 *Phys. Rev.* **B15** 4256
Levin K and Valls O T 1979a *Phys. Rev.* **B20** 105
Levin K and Valls O T 1979b *Phys. Rev.* **B20** 120
Levin K and Valls O T 1981 *Phys. Rev.* **B23** 6154
Levin K and Valls O T 1983 *Phys. Rep.* **98** 1
Lhuillier D 1977 *J. Physique Lett.* **38** 121
Lifshitz E M and Pitaevskii L P 1980 *Statistical Physics,* Part 2 (Vol. 9 of the Landau–Lifshitz *Course of Theoretical Physics*) (Pergamon, Oxford)
Ling R-Z, Betts D S and Brewer D F 1984a *Phys. Rev. Lett.* **53** 930
Ling R-Z, Betts D S and Brewer D F 1984b in *Proceedings of the 17th International Conference on Low Temperature Physics, LT-17,* Part I, ed. U Eckern, A Schmid, W Weber and H Wühl (North-Holland, Amsterdam), p. 3
Ling R-Z, Betts D S and Brewer D F 1984c in *Proceedings of the 17th International Conference on Low Temperature Physics, LT-17,* Part I, ed. U Eckern, A Schmid, W Weber and H Wühl (North-Holland, Amsterdam), p. 33
Ling R-Z, Saunders J and Dobbs E R 1987 *Phys. Rev. Lett.* **59** 461

Ling R-Z, Saunders J, Wojtanowski W and Dobbs E R 1989 *Europhys. Lett.* **10** 323
Lin-Liu Y R and Maki K 1978 *Phys. Rev.* **B18** 4724
Lin-Liu Y R and Maki K 1980a *Phys. Rev.* **B21** 990
Lin-Liu Y R and Maki K 1980b *Phys. Rev.* **B21** 5153
Lin-Liu Y R and Maki K 1981 *Phys. Rev.* **B24** 131
Lin-Liu Y R, Maki K and Vollhardt D 1978 *J. Physique Lett.* **39** 381
Lin-Liu Y R, Vollhardt D and Maki K 1979 *Phys. Rev.* **B20** 159
Liu M 1975 *Phys. Rev. Lett.* **35** 1577
Liu M 1976 *Phys. Rev.* **B13** 4174
Liu M 1977 *Physica* **90B** 78
Liu M 1978 *Phys. Rev.* **B18** 1165
Liu M 1979a *Phys. Rev. Lett.* **43** 1740
Liu M 1979b *Phys. Lett.* **73A** 29
Liu M 1980 *Z. Phys.* **B40** 175
Liu M 1982 *Physica* **109/110B + C** 1615 (Proceedings of the 16th International Conference on Low Temperature Physics, LT-16)
Liu M 1985 *Phys. Rev. Lett.* **55** 441
Liu M and Brinkman W F 1978 *J. Low Temp. Phys.* **30** 551
Liu M and Cross M C 1978 *Phys. Rev. Lett.* **41** 250
Liu M and Cross M C 1979 *Phys. Rev. Lett.* **43** 296
Liu M and Stern M 1982 *Phys. Rev. Lett.* **48** 1842
London F 1950 *Superfluids,* Vol. I (Wiley, New York)
London F 1954 *Superfluids,* Vol. II (Wiley, New York)
London H 1951 in *Proceedings of the 3rd International Conference on Low Temperature Physics, Oxford,* p. 157
London H, Clarke G R and Mendoza E 1962 *Phys. Rev.* **128** 1992
Lounasmaa O V 1974 *Contemp. Phys.* **15** 353
Lounasmaa O V 1984 *Physica* **126B + C** (Proceedings of the 17th International Conference on Low Temperature Physics, LT-17)
Lounasmaa O V, Manninen M T, Nenonen S A, Pekola J P, Sharma R G and Tagirov M S 1983 *Phys. Rev.* **B28** 6536
Love A and Zakrzewski S 1979 *J. Phys.* **A12** L151
Lu S T and Kojima H 1985 *Phys. Rev. Lett.* **55** 1677
Lu S T, Jiang Q and Kojima H 1989 *Phys. Rev. Lett.* **62** 1639
McClure M G and Takagi S 1979 *Phys. Rev. Lett.* **43** 596
McInerney M F 1980 *J. Low Temp. Phys.* **41** 563
McInerney M F 1981 *J. Low Temp. Phys.* **45** 249
McKenzie R H and Sauls J A 1989 Acoustic-order parameter three-wave resonance in superfluid ^{3}He-B. Preprint, Northwestern University, Evanston, Illinois.
Main P C, Kiewiet C W, Band W T, Hook J R, Sandiford D J and Hall H E 1976 *J. Phys.* **C9** 397
Main P C, Band W T, Hook J R, Hall H E and Sandiford D J 1977 in *Quantum Fluids and Solids,* ed. S B Trickey, E D Adams and J W Dufty (Plenum, New York), p. 117
Maki K 1973 *Phys. Lett.* **46A** 173
Maki K 1974 *J. Low Temp. Phys.* **16** 465
Maki K 1975a *Phys. Lett.* **51A** 337
Maki K 1975b *Phys. Rev.* **B11** 4264
Maki K 1975c in *Quantum Statistics and the Many-Body Problem,* ed. S B Trickey, W P Kirk and J W Dufty (Plenum, New York), p. 101
Maki K 1976a *Phys. Lett.* **56A** 101
Maki K 1976b *J. Low Temp. Phys.* **24** 755
Maki K 1977a *Physica* **90B** 84

Maki K 1977b in *Quantum Fluids and Solids,* ed. S B Trickey, E D Adams and J W Dufty (Plenum, New York), p. 65
Maki K 1977c in *Physics at Ultralow Temperatures,* ed. T Sugawara, S Nakajima, T Ohtsuka and T Usui (The Physical Society of Japan, Tokyo), p. 66
Maki K 1978a *J. Low Temp. Phys.* **32** 1
Maki K 1978b in *Solitons and Condensed Matter Physics,* ed. A R Bishop and F. Schneider (Springer, Berlin), p. 278
Maki K 1980 *Phys. Rev.* **B22** 3493
Maki K 1982 in *Progress in Low Temperature Physics,* Vol. VIII, ed. D F Brewer (North-Holland, Amsterdam), p. 1
Maki K 1983a in *Quantum Fluids and Solids—1983* (AIP Conf. Proc. No. 103), ed. E D Adams and G G Ihas (AIP, New York), p. 244
Maki K 1983b *Phys. Rev.* **B27** 4173
Maki K 1983c *Phys. Rev.* **B28** 2452
Maki K 1986a in *Solitons,* ed. S E Trullinger, V E Zakharov and V L Pokrovskii (North-Holland, Amsterdam), p. 435
Maki K 1986b *Phys. Rev. Lett.* **56** 1312
Maki K and Bruinsma R 1980 *Phys. Rev.* **B21** 148
Maki K and Combescot R 1985 *Phys. Rev. Lett.* **54** 2257
Maki K and Ebisawa H 1973 *Prog. Theor. Phys.* **50** 1452
Maki K and Ebisawa H 1974a *J. Low Temp. Phys.* **15** 213
Maki K and Ebisawa H 1974b *Prog. Theor. Phys.* **51** 690
Maki K and Ebisawa H 1974c *Phys. Rev. Lett.* **32** 520
Maki K and Ebisawa H 1975a *J. Low Temp. Phys.* **21** 475
Maki K and Ebisawa H 1975b *Phys. Rev.* **B12** 2667
Maki K and Ebisawa H 1976a *Phys. Rev.* **B13** 2924
Maki K and Ebisawa H 1976b *J. Low Temp. Phys.* **22** 285
Maki K and Ebisawa H 1976c *J. Low Temp. Phys.* **23** 351
Maki K and Ebisawa H 1976d *Phys. Rev.* **B13** 4845
Maki K and Ebisawa H 1977 *J. Low Temp. Phys.* **26** 627
Maki K and Hu C R 1975a *J. Low Temp. Phys.* **18** 377
Maki K and Hu C R 1975b *J. Low Temp. Phys.* **19** 259
Maki K and Kumar P 1976a *Phys. Rev.* **B14** 118
Maki K and Kumar P 1976b *Phys. Rev.* **B14** 3920
Maki K and Kumar P 1977a *Phys. Rev. Lett.* **38** 557
Maki K and Kumar P 1977b *Phys. Rev.* **B16** 182
Maki K and Kumar P 1977c *Phys. Rev.* **B16** 4805
Maki K and Kumar P 1977d *Phys. Rev.* **B16** 174
Maki K and Kumar P 1978 *Phys. Rev.* **B17** 1088
Maki K and Lin-Liu Y R 1978 *Phys. Rev.* **B17** 3558
Maki K and Nakahara M 1983 *Phys. Rev.* **B27** 4181
Maki K and Tsuneto T 1974a *Prog. Theor. Phys.* **52** 617
Maki K and Tsuneto T 1974b *Prog. Theor. Phys.* **52** 773
Maki K and Tsuneto T 1975 *Phys. Rev.* **B11** 2539
Maki K and Tsuneto T 1977a *J. Low Temp. Phys.* **27** 537
Maki K and Tsuneto T 1977b *J. Low Temp. Phys.* **27** 635
Maki K and Zotos X 1985a *Phys. Rev.* **B31** 177
Maki K and Zotos X 1985b *Phys. Rev.* **B31** 3116
Manninen M T and Pekola J P 1982 *Phys. Rev. Lett.* **48** 812; *ibid.* **48** E1369
Manninen M T and Pekola J P 1983 *J. Low Temp. Phys.* **52** 497
Manninen M T, Pekola J P, Sharma R G and Tagirov M S 1982 *Phys. Rev.* **B26** 5233
Manousakis E, Fantoni S, Pandharipande V R and Usmani Q N 1983 *Phys. Rev.* **B28** 3770
Markelov A V 1987 *Zh. Eksp. Teor. Fiz.* **92** 1714 [*Sov. Phys. JETP* **65** 962 (1987)]

Markelov A V 1988 *Zh. Eksp. Teor. Fiz.* **94** 156 [*Sov. Phys. JETP* **67** 520 (1988)]
Martin P C, Parodi P and Pershan P S 1972 *Phys. Rev.* **A6** 2401
Mast D B, Sarma B K, Owers-Bradley J R, Calder I D, Ketterson J B and Halperin W P 1980 *Phys. Rev. Lett.* **45** 266
Masuhara M, Candela D, Edwards D O, Hoyt R F, Scholz H N and Sherill D S 1984 *Phys. Rev. Lett.* **53** 1168
Mattuck, R D and Johansson B 1968 *Adv. Phys.* **17** 509
Meisel M W, Shivaram B S, Sarma B K, Ketterson J B and Halperin W P 1983a *Phys. Lett.* **98A** 437
Meisel M W, Shivaram B S, Sarma B K, Ketterson J B and Halperin W P 1983b *Phys. Rev. Lett.* **50** 361
Meisel M W, Shivaram B S, Sarma B K, Mast D B, Ketterson J B and Halperin W P 1983c *Phys. Rev.* **B27** 6982
Meisel M W, Shivaram B S, Sarma B K, Ketterson J B and Halperin W P 1985 *Phys. Lett.* **110A** 49
Meisel M W Avenel O and Varoquaux E 1987 *Jap. J. Appl. Phys.* **26** Suppl 26-3, Vol. I, p. 183 (Proceedings of the 18th International Conference on Low Temperature Physics, LT-18)
Mendelssohn K 1977 *The Quest for Absolute Zero,* 2nd edn (Taylor & Francis, London)
Mermin N D 1974 *Phys. Rev.* **A9** 868
Mermin N D 1975 *Phys. Rev. Lett.* **34** 1651
Mermin N D 1976 *Phys. Rev.* **B13** 112
Mermin N D 1977a *Physica* **90B + C** 1
Mermin N D 1977b in *Quantum Fluids and Solids,* ed. S B Trickey, E D Adams and J W Dufty (Plenum, New York), p. 3
Mermin N D 1978a in *Quantum Liquids,* ed. J Ruvalds and T Regge (North-Holland, Amsterdam), p. 195
Mermin N D 1978b *J. Physique* **39** Colloq. C-6, Vol. III, p. 1283 (Proceedings of the 15th International Conference on Low Temperature Physics, LT-15)
Mermin N D 1979 *Rev. Mod. Phys.* **51** 591
Mermin N D 1980 in *Proceedings of the XIV Karpacz Winter School of Theoretical Physics* (Lecture Notes in Physics, Vol. 115), ed. A Pekalski and J Przystawa (Springer, Berlin), p. 28
Mermin N D 1981 *Phys. Today* **34** 46
Mermin N D and Ho T L 1976 *Phys. Rev. Lett.* **36** 594
Mermin N D and Lee D M 1976 *Scientific American* **235** (December) 56
Mermin N D and Muzikar P. 1980 *Phys. Rev.* **B21** 980
Mermin N D and Stare C 1973 *Phys. Rev. Lett.* **30** 1135
Mermin N D and Stare C 1974 Unpublished (referenced in Barton and Moore 1974b)
Mermin N D, Mineev V P and Volovik G E 1978 *J. Low Temp. Phys.* **33** 117
Metzner W and Vollhardt D 1988 *Phys. Rev.* **B37** 7382
Metzner W and Vollhardt D 1989 *Phys. Rev. Lett.* **62** 324
Michel L 1971 *C. R. Acad. Sci. Paris* **A272** 433
Michel L 1980 *Rev. Mod. Phys.* **52** 617
Millis A J, Sachdev S and Varma C M 1988 *Phys Rev.* **B37** 4975
Mineev V P 1980 in *Soviet Scientific Reviews,* Section A, *Physics Reviews,* Vol. 2 (Harwood Academic Publishers, Chur), p. 173
Mineev V P 1983 *Usp. Fiz. Nauk* **139** 303 [*Sov. Phys. Usp.* **26** 160 (1983)]
Mineev V P 1985 *Zh. Eksp. Teor. Fiz.* **88** 507 [*Sov. Phys. JETP* **61** 297 (1985)]
Mineev V P 1986 *Zh. Eksp. Teor. Fiz.* **90** 1236 [*Sov. Phys. JETP* **63** 721 (1986)]
Mineev V P and Salomaa M M 1984 *J. Phys.* **C17** L181
Mineev V P and Volovik G E 1978 *Phys. Rev.* **B18** 3197

Mineev V P and Volovik G E 1984 in *Proceedings of the 17th International Conference on Low Temperature Physics, LT-17,* Part I, ed. U Eckern, A Schmid, W Weber and H Wühl (North-Holland, Amsterdam), p. 39
Mineev V P, Salomaa M M and Lounasmaa O V 1986 *Nature* **324** 333
Mishra S G and Ramakrishnan T V 1985 *Phys. Rev.* **B31** 2825
Miyake K 1983 *Prog. Theor. Phys.* **69** 1794
Miyake K and Usui T 1980a *Prog. Theor. Phys.* **63** 711
Miyake K and Usui T 1980b *Prog. Theor. Phys.* **64** 1119
Miyake K, Takagi H and Usui T 1981 *Prog. Theor. Phys.* **65** 1115
Monien H and Tewordt L 1985 *J. Low Temp. Phys.* **60** 323
Monien H and Tewordt L 1986 *J. Low Temp. Phys.* **62** 277
Montgomery R 1985 *J. Low Temp. Phys.* **58** 417
Movshovich R, Varoquaux E, Kim N and Lee D M 1988 *Phys. Rev. Lett.* **61** 1732
Mühlschlegel B 1959 *Z. Phys.* **155** 313
Muzikar P 1978 *J. Physique* **39** Colloq. C-6, Vol. I, p. 53 (Proceedings of the 15th International Conference on Low Temperature Physics, LT-15)
Muzikar P 1979 *J. Low Temp. Phys.* **36** 225
Muzikar P 1980 *Phys. Rev.* **B22** 3200
Muzikar P 1982 *J. Low Temp. Phys.* **46** 533
Muzikar P 1984 in *Proceedings of the 17th International Conference on Low Temperature Physics, LT-17,* Part I, ed. U Eckern, A Schmid, W Weber and H Wühl (North Holland, Amsterdam), p. 45
Muzikar P and Rainer D 1983 *Phys. Rev.* **B27** 4243
Muzikar P and Rainer D 1984 *Phys. Rev.* **B29** 6360
Muzikar P, Sauls J A and Serene J W 1980 *Phys. Rev.* **D21** 1494
Nagai K 1975 *Prog. Theor. Phys.* **54** 1
Nagai K 1977 *J. Low Temp. Phys.* **28** 139
Nagai K 1979 *J. Low Temp. Phys.* **36** 485
Nagai K 1980 *J. Low Temp. Phys.* **38** 677
Nagai K 1981a *Prog. Theor. Phys.* **65** 793
Nagai K 1981b *Physica* **108B + C** 1151 (Proceedings of the 16th International Conference on Low Temperature Physics, LT-16)
Nagai K 1984 *J. Low Temp. Phys.* **55** 233
Nagai K and Wölfle P 1981 *J. Low Temp. Phys.* **42** 227
Nakahara M 1986 *J. Phys.* **C19** L195
Nakahara M and Maki K 1983 *Phys. Rev.* **B27** 4456
Nakahara M and Ohmi T 1979 *Prog. Theor. Phys.* **61** 709
Nakahara M, Ohmi T, Tsuneto T and Fujita T 1979 *Prog. Theor. Phys.* **62** 874
Nakajima S 1973 *Prog. Theor. Phys.* **50** 1101
Nambu Y 1960 *Phys. Rev.* **117** 648
Nambu Y 1985 *Physica* **15D** 147
Neganov B S, Borisov N and Liburg M 1966 *Zh. Eksp. Teor. Fiz.* **50** 1445 [*Sov. Phys. JETP* **23** 959 (1966)]
Nijhoff F W and Capel H W 1981 *Physica* **106A** 369
Nijhoff F W and Capel H W 1982 *Physica* **111A** 371
Nijhoff F W, Capel H W and den Breems A 1985 *Physica* **130A** 375
Nozières P 1964 *Theory of Interacting Fermi Systems* (Benjamin, New York)
Nozières P and Schmitt-Rink S 1985 *J. Low Temp. Phys.* **59** 195
Nummila K K, Simola J T and Korhonen J S 1989 *J. Low Temp. Phys.* **75** 111
Ohmi T 1984 *J. Low Temp. Phys.* **56** 183
Ohmi T, Tsuneto T and Fujita T 1981 *Physica* **107B + C** 49 (Proceedings of the 16th International Conference on Low Temperature Physics, LT-16)
Ohmi T, Tsuneto T and Fujita T 1983 *Prog. Theor. Phys.* **70** 647
Oliphant M L E, Kinsey B B and Rutherford E 1933 *Proc. R. Soc. Lond.* **A141** 722

Ono Y A and Hara J 1981 *J. Phys.* **C14** 2093
Ono Y A, Hara J, Nagai K and Kawamura K 1977 *J. Low Temp. Phys.* **27** 513
Ono Y A, Hara J and Nagai K 1982 *J. Low Temp. Phys.* **48** 167
Ooiwa K 1982 *J. Low Temp. Phys.* **47** 137
Ooiwa K, Katayama T and Ishii C 1981 *J. Low Temp. Phys.* **42** 187
Osgood E B and Goodkind J M 1967 *Phys. Rev. Lett.* **18** 894
Osheroff D D 1974 *Phys. Rev. Lett.* **33** 1009
Osheroff D D 1975 *Phys. Lett.* **51A** 447
Osheroff D D 1977 *Physica* **90B + C** 20
Osheroff D D 1979 *J. Physique* **39** Colloq. C-6, Vol. III, p. 1270 (Proceedings of the 15th International Conference on Low Temperature Physics, LT-15)
Osheroff D D and Anderson P W 1974 *Phys. Rev. Lett.* **33** 686
Osheroff D D and Brinkman W F 1974 *Phys. Rev. Lett.* **32** 584
Osheroff D D and Brinkman W F 1977 Unpublished (referenced in Brinkman and Cross 1978)
Osheroff D D and Corruccini L R 1975a *Phys. Lett.* **51A** 447
Osheroff D D and Corruccini L R 1975b in *Proceedings of the 14th International Conference on Low Temperature Physics, LT-14,* Vol. 1, ed. M Krusius and M Vuorio (North-Holland, Amsterdam), p. 100
Osheroff D D and Cross M C 1979 *Phys. Rev. Lett.* **38** 905
Osheroff D D, Richardson R C and Lee D M 1972a *Phys. Rev. Lett.* **28** 885
Osheroff D D, Gully W J, Richardson R C and Lee D M 1972b *Phys. Rev. Lett.* **29** 920
Osheroff D D, Engelsberg S, Brinkman W F and Corruccini L R 1975 *Phys. Rev. Lett.* **34** 190
Osheroff D D, van Roosbroeck W, Smith H and Brinkman W F 1977 *Phys. Rev. Lett.* **38** 134
Østgard E 1968a *Phys. Rev.* **170** 257
Østgard E 1968b *Phys. Rev.* **171** 248
Østgard E 1968c *Phys. Rev.* **176** 351
Østgard E 1969a *Phys. Rev.* **180** 263
Østgard E 1969b *Phys. Rev.* **187** 371
Østgard E 1973 *Phys. Lett.* **46A** 103
Østgard E 1974 *Physica* **71** 415
Østgard E 1975 *Phys. Lett.* **54A** 39
Ott H R 1987 in *Progress in Low Temperature Physics,* Vol. XI, ed. D F Brewer (North-Holland, Amsterdam), p. 215
Owen J C and Ripka G 1983 *Riv. Nuovo Cim.* **6** 1
Owers-Bradley J R, Chocolacs H, Mueller R M, Buchal C, Kubota M and Pobell F 1983 *Phys. Rev. Lett.* **51** 2120
Paalanen M A and Osheroff D D 1980 *Phys. Rev. Lett.* **45** 362; *ibid.* **45** E1895
Paalanen M A, Cross M C, Sprenger W O, van Roosbroeck W and Osheroff D D 1979 *J. Low Temp. Phys.* **34** 607
Palmeri J 1989 *Phys. Rev. Lett.* **62** 1872
Pandharipande V R and Wiringa R B 1979 *Rev. Mod. Phys.* **51** 821
Panoff R M and Carlson J 1989 *Phys. Rev. Lett.* **62** 1130
Parpia J M 1985 *Phys. Rev.* **B32** 7564
Parpia J M and Reppy J D 1979 *Phys. Rev. Lett.* **43** 1332
Parpia J M and Rhodes T L 1983 *Phys. Rev. Lett.* **51** 805
Parpia J M, Yanof J M and Reppy J D 1976 *Bull. Am. Phys. Soc.* **21** 617
Parpia J M, Sandiford D J, Berthold J E and Reppy J D 1978 *Phys. Rev. Lett.* **28** 885
Passvogel T, Schopohl N, Warnke M and Tewordt L 1982 *J. Low Temp. Phys.* **46** 161

Passvogel T, Schopohl N and Tewordt L 1983 *J. Low Temp. Phys.* **50** 509
Passvogel T, Tewordt L and Schopohl N 1984 *J. Low Temp. Phys.* **56** 383
Pati J C and Salam A 1974 *Phys. Rev.* **D10** 275
Patton B R 1974 *Phys. Lett.* **47A** 459
Patton B R and Zaringhalam A 1975 *Phys. Lett.* **55A** 95
Paulson D N and Wheatley J C 1978a *Phys. Rev. Lett.* **40** 557
Paulson D N and Wheatley J C 1978b *Phys. Rev. Lett.* **41** 254
Paulson D N and Wheatley J C 1978c *Phys. Rev. Lett.* **41** 561
Paulson D N and Wheatley J C 1978d *J. Low Temp. Phys.* **33** 277
Paulson D N, Johnson R T and Wheatley J C 1973a *Phys. Rev. Lett.* **30** 829
Paulson D N, Johnson R T and Wheatley J C 1973b *Phys. Rev. Lett.* **31** 746
Paulson D N, Kojima H and Wheatley J C 1974a *Phys. Rev. Lett.* **32** 1098
Paulson D N, Kojima H and Wheatley J C 1974b *Phys. Lett.* **47A** 457
Paulson D N, Kleinberg R L and Wheatley J C 1976a *J. Low Temp. Phys.* **23** 725
Paulson D N, Krusius M and Wheatley J C 1976b *J. Low Temp. Phys.* **25** 695
Paulson D N, Krusius M and Wheatley J C 1976c *Phys. Rev. Lett.* **36** 1322
Paulson D N, Krusius M and Wheatley J C 1976d *Phys. Rev. Lett.* **37** 599
Paulson D N, Krusius M and Wheatley J C 1977 *J. Low Temp. Phys.* **26** 73
Pekola J P and Simola J T 1985 *J. Low Temp. Phys.* **58** 555
Pekola J P, Simola J T, Hakonen P J, Krusius M, Lounasmaa O V, Nummila K K, Mammiashvili G, Packard R E and Volovik G E 1984a *Phys. Rev. Lett.* **53** 584
Pekola J P, Simola J T, Nummila K K, Lounasmaa O V and Packard R E 1984b *Phys. Rev. Lett.* **53** 70
Pekola J P, Simola J T, Nummila K K, Lounasmaa O V and Packard R E 1984c *Phys. Rev. Lett.* **53** 584
Pekola J P, Davis J C, Zhu Y-Q, Spohr R N R, Price P B and Packard R E 1987 *J. Low Temp. Phys.* **67** 47
Pekola J P, Davis J C and Packard R E 1988 *J. Low Temp. Phys.* **71** 141
Penrose O and Onsager L 1956 *Phys. Rev.* **104** 576
Pethick C J and Carneiro G M 1973 *Phys. Rev.* **A7** 304
Pethick C J and Smith H 1976 *Phys. Rev. Lett.* **37** 226
Pethick C J and Smith H 1977 *Physica* **90B + C** 107
Pethick C J, Smith H and Bhattacharyya P 1975 *Phys. Rev. Lett.* **34** 643
Pethick C J, Smith H and Bhattacharyya P 1976 *J. Low Temp. Phys.* **23** 225
Pethick C J, Bhattacharyya P and Smith H 1977 *Phys. Rev.* **B15** 3384
Pfitzner M 1984 Doctoral Thesis, Technische Universität München
Pfitzner M 1985a *J. Low Temp. Phys.* **61** 141
Pfitzner M 1985b *J. Low Temp. Phys.* **61** 433
Pfitzner M and Wölfle P 1983a *J. Low Temp. Phys.* **51** 535
Pfitzner M and Wölfle P 1983b in *Quantum Fluids and Solids—1983* (AIP Conf. Proc. No. 103), ed. E D Adams and G G Ihas (AIP, New York), p. 148
Pfitzner M and Wölfle P 1984 in *Proceedings of the 17th International Conference on Low Temperature Physics, LT-17,* Part II, ed. U Eckern, A Schmid, W Weber and H Wühl (North-Holland, Amsterdam), p. 1253
Pfitzner M and Wölfle P 1986 *Phys. Rev.* **B33** 2003
Pfitzner M and Wölfle P 1987 *Phys. Rev.* **B35** 4699
Piché L, Ruoff M, Varoquaux E and Avenel O 1982 *Phys. Rev. Lett.* **49** 744
Pines D 1966 in *Quantum Fluids,* ed. D F Brewer (North-Holland, Amsterdam), p. 166
Pines D 1985 in *Highlights of Condensed Matter Theory*: *Course 89 of the Varenna Summer School* (Societa Italiana di Fisica, Bologna, Italy), p. 580
Pines D 1987 *Can. J. Phys.* **65** 1357
Pines D and Alpar A 1985 *Nature* **316** 27
Pines D and Nozières P 1966 *The Theory of Quantum Liquids* (Benjamin, New York)

Pitaevskii L P 1959 *Zh. Eksp. Teor. Fiz.* **37** 1794 [*Sov. Phys. JETP* **10** 1267 (1960)]
Pleiner H 1977a *J. Phys.* **C10** 2337
Pleiner H 1977b *J. Phys.* **C10** 4241
Pleiner H and Brand H 1981 *Phys. Rev.* **B24** 6430
Pleiner H and Brand H 1983 *Phys. Rev.* **B28** 3782
Pleiner H and Graham R 1976 *J. Phys.* **C9** 4109
Podd'yakova E V and Fomin I A 1988 *Pis'ma Zh. Eksp. Teor. Fiz.* **47** 519 [*JETP Lett.* **47** 606 (1988)]
Polturak E, de Vegvar P G N, Zeise E K and Lee D M 1981a *Phys. Rev. Lett.* **46** 1588
Polturak E, de Vegvar P G N, Zeise E K and Lee D M 1981b *Physica* **107B + C** 687 (Proceedings of the 16th International Conference on Low Temperature Physics, LT-16)
Poluéktov Yu M 1982 *Fiz. Nizk. Temp.* **8** 1125 [*Sov. J. Low Temp. Phys.* **8** 587 (1982)]
Poluéktov Yu M 1983 *Fiz. Nizk. Temp.* **9** 469 [*Sov. J. Low Temp. Phys.* **9** 221 (1983)]
Poluéktov Yu M 1984 *Fiz. Nizk. Temp.* **10** 1013 [*Sov. J. Low Temp. Phys.* **10** 527 (1984)]
Pomeranchuk I 1950 *Zh. Eksp. Teor. Fiz.* **20** 919
Privorotskii I A 1977 *J. Low Temp. Phys.* **26** 379
Pushkina N I and Khokhlov R V 1976 *Pis'ma Zh. Tekh. Fiz.* **2** 1135 [*Sov. Tech. Phys. Lett.* **2** 581 (1976)]
Putterman S J 1974 *Superfluid Hydrodynamics* (North-Holland, Amsterdam)
Quader K F and Bedell K S 1985 *J. Low Temp. Phys.* **58** 89
Rainer D 1986 in *Progress in Low Temperature Physics,* Vol. X, ed. D F Brewer (North-Holland, Amsterdam), p. 371
Rainer D and Serene J W 1976 *Phys. Rev.* **B13** 4745
Rainer D and Serene J W 1980 *J. Low Temp. Phys.* **38** 601
Rainer D and Vuorio M 1977 *J. Phys.* **C10** 3093
Ramsey N 1982 *Ann. Rev. Nucl. Part. Sci.* **32** 211
Rice M J 1967a *Phys. Rev.* **159** 153
Rice M J 1967b *Phys. Rev.* **162** 189
Richardson R C 1975 in *Proceedings of the 14th International Conference on Low Temperature Physics, LT-14,* Part V, ed. M Krusius and M Vuorio (North-Holland, Amsterdam), p. 69
Rickayzen G 1965 *Theory of Superconductivity* (Wiley, New York)
Riedel E 1968 *Z. Phys.* **210** 403
Roach P D, Ketterson J B and Roach P R 1977 *Phys. Rev. Lett.* **39** 626
Roach P R and Ketterson J B 1976a *Phys. Rev. Lett.* **36** 736
Roach P R and Ketterson J B 1976b *J. Low Temp. Phys.* **25** 637
Roach P R, Abraham B M, Kuchnir M and Ketterson J B 1975a *Phys. Rev. Lett.* **34** 711
Roach P R, Abraham B M, Roach P D and Ketterson J B 1975b *Phys. Rev. Lett.* **34** 715
Roach P R, Meisel M W and Eckstein Y 1982 *Phys. Rev. Lett.* **48** 330
Roach P R, Eckstein Y, Meisel MW and Aniola-Jedrzejek L 1983 *J. Low Temp. Phys.* **52** 433
Rokhsar D S and Kivelson S A 1988 *Phys. Rev. Lett.* **61** 2376
Rosen G 1982 *Phys. Rev. Lett.* **49** 1885
Rozhkov S S 1986 *Pis'ma Zh. Eksp. Teor Fiz.* **44** 32 [*JETP Lett.* **44** 38 (1986)]
Ruel R and Kojima H 1985 *Phys. Rev. Lett.* **54** 2238
Ruel R and Kojima H 1986 *Phys. Rev.* **B34** 6511

Ruoff M and Varoquaux E 1983 *Phys. Rev. Lett.* **51** 1107
Sachrajda A S, Brewer D F and Truscott W S 1984 *J. Low Temp. Phys.* **56** 617
Sagan D C, de Vegvar P G N, Polturak E, Friedman L, Yan S S, Ziercher E L and Lee D M 1984 *Phys. Rev. Lett.* **53** 1939
Sager R E, Kleinberg R L, Warkentin P A and Wheatley J C 1977 *Phys. Rev. Lett.* **39** 1343
Sager R E, Kleinberg R L, Warkentin P A and Wheatley J C 1978a *J. Low Temp. Phys.* **31** 409
Sager R E, Kleinberg R L, Warkentin P A and Wheatley J C 1978b *J. Low Temp. Phys.* **32** 263
Salmelin R H and Salomaa M M 1987a *J. Phys.* **C20** L681
Salmelin R H and Salomaa M M 1987b *J. Phys.* **C20** L689
Salmelin R H, Salomaa M M and Mineev V P 1989a *Phys. Rev. Lett.* **63** 868
Salmelin R H, Pekola J P, Manninen A J, Torizuka K, Berglund P M, Kyynäräinen J M, Lounasmaa O V, Tvalashvili G K, Magradze O V, Varoquaux E, Avenel O and Mineev V P 1989b *Phys. Rev. Lett.* **63** 620
Salomaa M 1982 *Contemp. Phys.* **23** 169
Salomaa M 1988 *J. Phys.* **C21** 4425
Salomaa M M and Volovik G E 1983a in *Quantum Fluids and Solids—1983* (AIP Conf. Proc. No. 103), ed. E D Adams and G G Ihas (AIP, New York), p. 210
Salomaa M M and Volovik G E 1983b *Phys. Rev. Lett.* **51** 2040
Salomaa M M and Volovik G E 1984 *Phys. Rev. Lett.* **52** 2008
Salomaa M M and Volovik G E 1985a *Phys. Rev.* **B31** 203
Salomaa M M and Volovik G E 1985b *Phys. Rev. Lett.* **54** 2127
Salomaa M M and Volovik G E 1985c *Phys. Rev. Lett.* **55** 1184
Salomaa M M and Volovik G E 1986a *Phys. Rev. Lett.* **56** 363
Salomaa M M and Volovik G E 1986b *Europhys. Lett.* **2** 781
Salomaa M M and Volovik G E 1987 *Rev. Mod. Phys.* **59** 533
Salomaa M M and Volovik G E 1988 *Phys. Rev.* **B37** 9298
Salomaa M M and Volovik G E 1989 *J. Low Temp. Phys.* **77** 17
Salomaa M, Pethick C J and Baym G 1980a *Phys. Rev. Lett.* **44** 998
Salomaa M, Pethick C J and Baym G 1980b *J. Low Temp. Phys.* **40** 297
Samalam V K and Serene J W 1978 *Phys. Rev. Lett.* **41** 497
Saslow W M 1971 *Phys. Lett.* **35A** 241
Saslow W M 1973 *Phys. Rev. Lett.* **31** 870
Saslow W M 1976a *J. Low Temp. Phys.* **23** 495
Saslow W M 1976b *J. Low Temp. Phys.* **26** 583
Saslow W M and Hu C R 1978 *J. Physique Lett.* **39** 379
Saslow W M and Hu C R 1981 *Phys. Rev.* **B23** 4523
Sauls J A 1981 *Phys. Rev. Lett.* **47** 530
Sauls J A 1986 *Phys. Rev.* **B34** 4861
Sauls J A and Serene J W 1978 *Phys. Rev.* **D17** 1524
Sauls J A and Serene J W 1981a *Phys. Rev.* **B24** 183
Sauls J A and Serene J W 1981b *Physica* **108B + C** 1137 (Proceedings of the 16th International Conference on Low Temperature Physics, LT-16)
Sauls J A and Serene J W 1981c *Phys. Rev.* **B23** 4798
Sauls J A and Serene J W 1982 *Phys. Rev. Lett.* **49** 1183
Sauls J A and Serene J W 1984 in *Proceedings of the 17th International Conference on Low Temperature Physics, LT-17,* Part II, ed. U Eckern, A Schmid, W Weber and H Wühl (North-Holland, Amsterdam), p. 775
Sauls J A and Serene J W 1985 *Phys Rev.* **B32** 4782
Sauls J A, Stein D L and Serene J W 1982 *Phys. Rev.* **D25** 967
Saunders J, Betts D S, Brewer D F, Swithenby S J and Truscott W S 1978 *Phys. Rev. Lett.* **40** 1278

Saunders J, Daniels M E, Dobbs E R and Ward P L 1983 in *Quantum Fluids and Solids—1983* (AIP Conf. Proc. No. 103), ed. E D Adams and G G Ihas (AIP, New York), p. 314
Schakel A M J 1989 Doctoral Thesis, University of Amsterdam
Schakel A M J and Bais F A 1989 *J. Phys.: Condens. Matter* **1** 1743
Schakel A M J and Batenburg P 1989 *Ann. Phys. (NY)*, in press
Scherm R, Stirling W G, Woods A D B, Cowley R A and Coombs G J 1974 *J. Phys.* **C7** L341
Scherm R, Guckelsberger K, Fak B, Sköld K, Dianoux A J, Godfrin H and Stirling W G 1987 *Phys. Rev. Lett.* **57** 217
Schmidt K E and Kalos M H 1984 in *Applications of the Monte Carlo Method in Statistical Physics*, ed. K Binder (Springer, Berlin), p. 125
Schopohl N 1980 *J. Low Temp. Phys.* **41** 409
Schopohl N 1982 *J. Low Temp. Phys.* **49** 347
Schopohl N 1987 *Phys. Rev. Lett.* **58** 1664
Schopohl N and Tewordt L 1980 *J. Low Temp. Phys.* **41** 305
Schopohl N and Tewordt L 1981 *J. Low Temp. Phys.* **45** 67
Schopohl N and Tewordt L 1984 *J. Low Temp. Phys.* **57** 601
Schopohl N, Streckwall H and Tewordt L 1982 *J. Low Temp. Phys.* **49** 319
Schopohl N, Warnke M and Tewordt L 1983 *Phys. Rev. Lett.* **50** 1066
Schopohl N, Marquardt W and Tewordt L 1985 *J. Low Temp. Phys.* **59** 469
Schrieffer J R 1964 *Theory of Superconductivity* (Benjamin, Reading, Massachusetts)
Schwarz K W 1978 *Phys. Rev.* **B18** 245
Seiler K, Gros C, Rice T M, Ueda K and Vollhardt D 1986 *J. Low Temp. Phys.* **64** 195
Seppälä H K and Volovik G E 1983 *J. Low Temp. Phys.* **51** 279
Seppälä H K, Hakonen P J, Krusius M, Ohmi T, Salomaa M M, Simola J T and Volovik G E 1984 *Phys. Rev. Lett.* **52** 1802
Serene J W 1974 Doctoral Thesis, Cornell University
Serene J W 1983 in *Quantum Fluids and Solids—1983* (AIP Conf. Proc. No. 103), ed. E D Adams and G G Ihas (AIP, New York), p. 305
Serene J W 1984 *Phys. Rev.* **B30** 5373
Serene J W and Rainer D 1977 in *Quantum Fluids and Solids*, ed. S B Trickey, E D Adams and J W Dufty (Plenum, New York), p. 111
Serene J W and Rainer D 1978 *Phys. Rev.* **B17** 2901
Serene J W and Rainer D 1979 *J. Low Temp. Phys.* **34** 589
Serene J W and Rainer D 1983 *Phys. Rep.* **101** 221
Shahzamanian M A 1975 *J. Low Temp. Phys.* **21** 589
Shields S E and Goodkind J M 1977 *J. Low Temp. Phys.* **27** 259
Shivaram B S, Meisel M W, Sarma B K, Mast D B, Halperin W P and Ketterson J B 1982 *Phys. Rev. Lett.* **49** 1646
Shivaram B S, Meisel M W, Sarma B K, Halperin W P and Ketterson J B 1983 *Phys. Rev. Lett.* **50** 1070
Shivaram B S, Meisel M W, Sarma B K, Halperin W P and Ketterson J B 1986 *J. Low Temp. Phys.* **63** 57
Shumeiko V S 1972 *Zh. Eksp. Teor. Fiz.* **63** 621 [*Sov. Phys. JETP* **36** 330 (1973)]
Simola J T, Nummila K K, Hirai A, Korhonen J S, Schoepe W and Skrbek L 1986 *Phys. Rev. Lett.* **57** 1923
Simola J T, Skrbek L, Nummila K K and Korhonen J S 1987 *Phys. Rev. Lett.* **58** 904
Sköld K and Pelizzari C A 1976 *Phys. Rev. Lett.* **37** 842
Sköld K and Pelizzari C A 1980 in *Recent Developments in Condensed Matter Physics*, Vol. 1, ed. J R Devreese (Plenum, New York), p. 111
Sköld K, Pelizzari C A, Kleb R and Ostrowski G E 1976 *Phys. Rev. Lett.* **37** 842

Smith H 1984 *Physica* **126B + C** 267 (Proceedings of the 17th International Conference on Low Temperature Physics, LT-17)
Smith H 1987 In *Progress in Low Temperature Physics,* Vol. XI, ed. D F Brewer (North-Holland, Amsterdam), p. 75
Smith H, Brinkman W F and Engelsberg S 1977 *Phys. Rev.* **B15** 199
Soda T 1975 *Prog. Theor. Phys.* **53** 903
Soda T 1977a *Prog. Theor. Phys.* **58** 395
Soda T 1977b *Prog. Theor. Phys.* **58** 1096
Soda T 1984 *Prog. Theor. Phys.* **72** 387
Soda T and Arai T 1981 *J. Phys. Soc. Jpn* **50** 3222
Sokolov A I 1979 *Pis'ma Zh. Eksp. Teor. Fiz.* **29** 618 [*JETP Lett.* **29** 590 (1979)]
Sokolov A I 1980 *Zh. Eksp. Teor. Fiz.* **78** 1985 [*Sov. Phys. JETP* **51** 983 (1980)]
Sokolov A I 1983 *Zh. Eksp. Teor. Fiz.* **84** 1373 [*Sov. Phys. JETP* **57** 798 (1983)]
Solomon A I 1981 *J. Phys.* **A14** 2177
Soni V 1978 *Phys. Rev.* **B18** 6076
Sonin E B 1979 *Pis'ma Zh. Eksp. Teor. Fiz.* **30** 697 [*JETP Lett.* **30** 681 (1979)]
Sonin E B 1981 *Pis'ma Zh. Eksp. Teor. Fiz.* **34** 97 [*JETP Lett.* **34** 91 (1982)]
Sonin E B 1983a *Pis'ma Zh. Eksp. Teor. Fiz.* **38** 11 [*JETP Lett.* **38** 11 (1983)]
Sonin E B 1983b *Phys. Lett.* **93A** 308
Sonin E B 1984a *J. Low Temp. Phys.* **55** 533
Sonin E B 1984b *Zh. Eksp. Teor. Fiz.* **87** 1670 [*Sov. Phys. JETP* **60** 959 (1985)]
Sonin E B 1987 *Rev. Mod. Phys.* **59** 87
Sonin E B and Fomin N V 1985 *Pis'ma Zh. Eksp. Teor. Fiz.* **42** 185 [*JETP Lett.* **42** 229 (1985)]
Spencer G F and Ihas G G 1982 *Phys. Rev. Lett.* **48** 1118
Stamp P C E 1985 *J. Phys.* **F15** 1829
Steglich F, Aarts J, Bredl C D, Lieke W, Meschede D, Franz W and Schäfer J 1979 *Phys. Rev. Lett.* **43** 1892
Stein D L and Cross M C 1979 *Phys. Rev. Lett.* **42** 504
Stern M R and Liu M 1982 *Physica* **109/110B + C** 2099 (Proceedings of the 16th International Conference on Low Temperature Physics, LT-16)
Stewart G R 1984 *Rev. Mod. Phys.* **56** 755
Stirling W G 1978 *J. Physique* **39** Colloq. C-6, Vol. III, p. 1334 (Proceedings of the 15th International Conference on Low Temperature Physics, LT-15)
Stirling W G, Scherm R, Hilton O A and Cowley R A 1976 *J. Phys.* **C9** 1643
Stone M, Garg A and Muzikar P 1985 *Phys. Rev. Lett.* **55** 2328
Struik D J 1961 *Lectures on Classical Differential Geometry* (Addison-Wesley, London)
Sun Y, Wölfle P and Yip S 1989 *Phys. Rev. Lett.* **63** 1613
Swift G W and Packard R E 1981 *J. Low Temp. Phys.* **43** 517
Swift G W, Eisenstein J P and Packard R E 1980 *Phys. Rev. Lett.* **45** 1955
Sydoriak S G, Grilly E R and Hammel E F 1949a *Phys. Rev.* **75** 303
Sydoriak S G, Grilly E R and Hammel E F 1949b *Phys. Rev.* **75** 1103
Sykes J and Brooker G A 1970 *Ann. Phys.* (*NY*) **56** 1
Takagi H 1981a *Prog. Theor. Phys.* **65** 1145
Takagi H 1981b *J. Phys. Soc. Jpn* **50** 723
Takagi S 1973 Doctoral Thesis, University of Tokyo
Takagi S 1974a *Prog. Theor. Phys.* **51** 1998
Takagi S 1974b *Prog. Theor. Phys.* **51** 674
Takagi S 1974c *Prog. Theor. Phys.* **51** 635
Takagi S 1975a *J. Phys.* **C8** 1507
Takagi S 1975b *J. Low Temp. Phys.* **18** 309
Takagi S 1977 *J. Phys.* **C10** 4045
Takagi S 1978 *Prog. Theor. Phys.* **60** 934

Tankersley L L 1979 *J. Low Temp. Phys.* **37** 141
Tesanovich Z 1984 *Phys. Lett.* **100A** 158
Tesanovich Z and Valls O T 1986a *Phys. Rev.* **B31** 1374
Tesanovich Z and Valls O T 1986b *Phys. Rev.* **B33** 3139
Tesanovich Z and Valls O T 1986c *Phys. Rev.* **B34** 7610
Tewordt L 1974a *J. Low Temp. Phys.* **15** 349
Tewordt L 1974b *Z. Phys.* **268** 207
Tewordt L and Einzel D 1976 *Phys. Lett.* **56A** 97
Tewordt L and Schopohl N 1979a *J. Low Temp. Phys.* **34** 489
Tewordt L and Schopohl N 1979b *J. Low Temp. Phys.* **37** 421
Tewordt L, Fay D, Dörre P and Einzel D 1975a *Phys. Rev.* **B11** 1914
Tewordt L, Fay D, Dörre P and Einzel D 1975b *J. Low Temp. Phys.* **21** 645
Tewordt L, Schopohl N and Vollhardt D 1977 *J. Low Temp. Phys.* **29** 119
Tewordt L, Fay D and Schopohl N 1978 *J. Low Temp. Phys.* **30** 177
Theodorakis S and Fetter A L 1983 *J. Low Temp. Phys.* **52** 559
Thouless D J 1960 *Ann. Phys. (NY)* **10** 553
Thuneberg E V 1986a *Phys. Rev. Lett.* **56** 359
Thuneberg E V 1986b *Phys. Rev.* **B33** 5124
Thuneberg E V 1987a *Phys. Rev.* **B36** 3583
Thuneberg E V 1987b *Europhys. Lett.* **3** 711
Thuneberg E V and Kurkijärvi J 1981 *Phys. Lett.* **86A** 35
Thuneberg E V, Kurkijärvi J and Rainer D 1981a *Physica* **107B** + **C** 43 (Proceedings of the 16th International Conference on Low Temperature Physics, LT-16)
Thuneberg E V, Kurkijärvi J and Rainer D 1981b *J. Phys.* **C14** 5615
Thuneberg E V, Kurkijärvi J and Rainer D 1982 *Phys. Rev. Lett.* **48** 1853
Tilley D R and Tilley J 1986 *Superfluidity and Superconductivity,* 2nd edn (Adam Hilger, Bristol)
Tinkham M 1975 *Introduction to Superconductivity* (McGraw-Hill, New York)
Topsoe F and Højgaard Jensen H 1984 *J. Low Temp. Phys.* **55** 469
Tkachenko V K 1965 *Zh. Eksp. Teor. Fiz.* **49** 1875 [*Sov. Phys. JETP* **22** 1282 (1966)]
Tough J T 1982 in *Progress in Low Temperature Physics,* Vol. VIII, ed. D F Brewer (North-Holland, Amsterdam), p. 133
Toulouse G and Kléman M 1976 *J. Physique Lett.* **37** 149
Truscott W S 1979 *Phys. Lett* **74A** 80
Tsuneto T 1960 *Phys. Rev.* **118** 1029
Tsuneto T 1977 *Prog. Theor. Phys.* **58** 719
Tsuneto T, Ohmi T and Fujita T 1978 in *Physics at Ultralow Temperatures,* ed. T Suguwara, S Nakajima, T Ohtsuka and T Usui (The Physical Society of Japan, Tokyo), p. 153
Ueda K and Rice T M 1985 *Phys. Rev.* **B31** 7114
Ugulava A I 1984 *Zh. Eksp. Teor. Fiz.* **86** 497 [*Sov. Phys. JETP* **59** 290 (1984)]
Ullah S 1988 *Phys. Rev.* **B37** 5010
Valatin J G 1959 *Nuovo Cim.* **7** 843
Varma C M and Werthamer N R 1974 *Phys. Rev.* **A9** 1465
Varoquaux E, Williams G A and Avenel O 1986 *Phys. Rev.* **B34** 7617
Vdovin Y A 1963 in *Application of Methods of Quantum Field Theory to Problems of Many Particles,* ed. A I Alekseyeva (Moscow: GOS ATOM ISDAT), p. 94 (in Russian)
Vinen W F 1957a *Proc. R. Soc. Lond.* **A240** 114
Vinen W F 1957b *Proc. R. Soc. Lond.* **A240** 128
Vinen W F 1957c *Proc. R. Soc. Lond.* **A242** 493
Vinen W F 1961 in *Progress in Low Temperature Physics,* Vol. III, ed. C J Gorter (North-Holland, Amsterdam), p. 1

Vollhardt D 1979 Doctoral Thesis, Universität Hamburg
Vollhardt D 1982a *Phys. Blätter* **39** 41
Vollhardt D 1982b *Phys. Blätter* **39** 120
Vollhardt D 1982c *Phys. Blätter* **39** 151
Vollhardt D 1984a *Rev. Mod. Phys.* **56** 99
Vollhardt D 1984b in *Proceedings of the 17th International Conference on Low Temperature Physics, LT-17,* Part I ed. U Eckern, A Schmid, W Weber and H Wühl (North-Holland, Amsterdam), p. 5
Vollhardt D and Kumar P 1978 *J. Physique* **39** Colloq. C-6, Vol. I, p. 43 (Proceedings of the 15th International Conference on Low Temperature Physics, LT-15)
Vollhardt D and Maki K 1978 *J. Low Temp. Phys.* **31** 457
Vollhardt D and Maki K 1979a *Phys. Lett.* **72A** 21
Vollhardt D and Maki K 1979b *Phys. Rev.* **B20** 963; *ibid.* **B23** E1489
Vollhardt D and Wölfle P 1981a *Phys. Rev. Lett.* **47** 190
Vollhardt D and Wölfle P 1981b *Physica* **108B + C** 1055 (Proceedings of the 16th International Conference on Low Temperature Physics, LT-16)
Vollhardt D, Lin-Liu Y R and Maki K 1979 *J. Low Temp. Phys.* **37** 627
Vollhardt D, Maki K and Schopohl N 1980 *J. Low Temp. Phys.* **39** 79
Vollhardt D, Lin-Liu Y R and Maki K 1981 *J. Low Temp. Phys.* **43** 189
Vollhardt D, Wölfle P and Anderson P W 1987 *Phys. Rev.* **B35** 6703
Volovik G E 1975a *Pis'ma Zh. Eksp. Teor. Fiz.* **22** 234 [*JETP Lett.* **22** 108 (1975)]
Volovik G E 1975b *Pis'ma Zh. Eksp. Teor. Fiz.* **22** 412 [*JETP Lett.* **22** 198 (1975)]
Volovik G E 1978a *Pis'ma Zh. Eksp. Teor Fiz.* **27** 605 [*JETP Lett.* **27** 573 (1978)]
Volovik G E 1978b *Pis'ma Zh. Eksp. Teor. Fiz.* **28** 65 [*JETP Lett.* **28** 59 (1978)]
Volovik G E 1979 in *Soviet Scientific Reviews,* Section A, *Physics Reviews,* Vol. 1 (Harwood Academic Publishers, Chur), p. 23
Volovik G E 1980 *Zh. Eksp. Teor. Fiz.* **79** 309 [*Sov. Phys. JETP* **52** 156 (1980)]
Volovik G E 1984a *Pis'ma Zh. Eksp. Teor. Fiz.* **39** 169 [*JETP Lett.* **39** 200 (1984)]
Volovik G E 1984b *Usp. Fiz. Nauk.* **143** 73 [*Sov. Phys. Usp.* **27** 363 (1984)]
Volovik G E 1984c *Pis'ma Zh. Eksp. Teor. Fiz.* **39** 304 [*JETP Lett.* **39** 365 (1984)]
Volovik G E 1985 *Pis'ma Zh. Eksp. Teor. Fiz.* **42** 294 [*JETP Lett.* **42** 363 (1985)]
Volovik G E 1986a *Pis'ma Zh. Eksp. Teor. Fiz.* **43** 428 [*JETP Lett.* **43** 551 (1986)]
Volovik G E 1986b *Pis'ma Zh. Eksp. Teor. Fiz.* **43** 535 [*JETP Lett.* **43** 693 (1986)]
Volovik G E 1986c *Pis'ma Zh. Eksp. Teor. Fiz.* **44** 144 [*JETP Lett.* **44** 185 (1986)]
Volovik G E 1986d *Pis'ma Zh. Eksp. Teor. Fiz.* **44** 388 [*JETP Lett.* **44** 498 (1986)]
Volovik G E 1987 *J. Low Temp. Phys.* **67** 301
Volovik G E and Balatskii A V 1985 *J. Low Temp. Phys.* **58** 1
Volovik G E and Gorkov L P 1984 *Pis'ma Zh. Eksp. Teor. Fiz.* **39** 550 [*JETP Lett.* **39** 674 (1984)]
Volovik G E and Gorkov L P 1985 *Zh. Eksp. Teor. Fiz.* **88** 1412 [*Sov. Phys. JETP* **61** 843 (1985)]
Volovik G E and Hakonen P J 1981 *J. Low Temp. Phys.* **42** 503
Volovik G E and Khazan M V 1982 *Zh. Eksp. Teor. Fiz.* **82** 1498 [*Sov. Phys. JETP* **55** 898 (1982)]
Volovik G E and Khazan M V 1983 *Zh. Eksp. Teor. Fiz.* **85** 948 [*Sov. Phys. JETP* **58** 551 (1983)]
Volovik G E and Khazan M V 1984 *Zh. Eksp. Teor. Fiz.* **87** 583 [*Sov. Phys. JETP* **60** 276 (1984)]
Volovik G E and Konyshev V A 1988 *Pis'ma Zh. Eksp. Teor. Fiz.* **47** 207 [*JETP Lett.* **47** 250 (1988)]
Volovik G E and Kopnin N B 1977 *Pis'ma Zh. Eksp. Teor. Fiz.* **25** 26 [*JETP Lett.* **25** 22 (1977)]

Volovik G E and Mineev V P 1976a *Pis'ma Zh. Eksp. Teor. Fiz.* **23** 647 [*JETP Lett.* **23** 593 (1976)]
Volovik G E and Mineev V P 1976b *Pis'ma Zh. Eksp. Teor. Fiz.* **24** 605 [*JETP Lett.* **24** 561 (1976)]
Volovik G E and Mineev V P 1976c *Zh. Eksp. Teor. Fiz.* **71** 1129 [*Sov. Phys. JETP* **44** 591 (1976)]
Volovik G E and Mineev V P 1977a *Zh. Eksp. Teor. Fiz.* **72** 2256 [*Sov. Phys. JETP* **45** 1186 (1977)]
Volovik G E and Mineev V P 1977b *Zh. Eksp. Teor. Fiz.* **73** 767 [*Sov. Phys. JETP* **46** 401 (1977)]
Volovik G E and Mineev V P 1981 *Zh. Eksp. Teor. Fiz.* **81** 989 [*Sov. Phys. JETP* **54** 524 (1981)]
Volovik G E and Mineev V P 1982 *Zh. Eksp. Teor. Fiz.* **83** 1025 [*Sov. Phys. JETP* **56** 579 (1982)]
Volovik G E and Mineev V P 1983 *Pis'ma Zh. Eksp. Teor. Fiz.* **37** 103 [*JETP Lett.* **37** 127 (1983)]
Volovik G E and Mineev V P 1984 *Zh. Eksp. Teor. Fiz.* **86** 1667 [*Sov. Phys. JETP* **59** 972 (1984)]
Volovik G E and Salomaa M M 1985a *Pis'ma Zh. Eksp. Teor. Fiz.* **42** 421 [*JETP Lett.* **42** 521 (1985)]
Volovik G E and Salomaa M M 1985b *Zh. Eksp. Teor. Fiz.* **88** 1656 [*Sov. Phys. JETP* **61** 986 (1985)]
Volovik G E and Salomaa M M 1987 *Pis'ma Zh. Eksp. Teor. Fiz.* **45** 44 [*JETP Lett.* **45** 56 (1987)]
Volovik G E and Yakovenko V M 1989 *J. Phys.: Condens. Matter* **1** 5263
Vulovic V Z, Stein D L and Fetter A L 1984 *Phys. Rev.* **B29** 6090
Vuorio M 1974 *J. Phys.* **C7** L5
Vuorio M 1976 *J. Phys.* **C9** L267
Wartak M S 1978 *J. Physique* **39** Colloq. C-6, Vol. I, p. 11 (Proceedings of the 15th International Conference on Low Temperature Physics, LT-15)
Wartak M S 1980a *Phys. Lett.* **79A** 435
Wartak M S 1980b *Acta Phys. Polon.* **A57** 99
Wartak M S 1981 *J. Phys.* **C14** L155
Wartak M S 1982 *Phys. Lett.* **88A** 84
Webb R A 1977a *Phys. Rev. Lett.* **38** 1151
Webb R A 1977b *Phys. Rev. Lett.* **39** 1008
Webb R A 1978a *Phys. Rev. Lett.* **40** 883
Webb R A 1978b *J. Physique* **39** Colloq. C-6, Vol. I, p. 31 (Proceedings of the 15th International Conference on Low Temperature Physics, LT-15)
Webb R A 1978c *Phys. Lett.* **67A** 197
Webb R A, Greytak T J, Johnson R T and Wheatley J C 1973 *Phys. Rev. Lett.* **30** 210
Webb R A, Kleinberg R L and Wheatley J C 1974a *Phys. Lett.* **48A** 421
Webb R A, Kleinberg R L and Wheatley J C 1974b *Phys. Rev. Lett.* **33** 145
Webb R A, Sager R E and Wheatley J C 1975a *Phys. Rev. Lett.* **35** 615
Webb R A, Sager R E and Wheatley J C 1975b *Phys. Rev. Lett.* **35** 1010
Webb R A, Sager R E and Wheatley J C 1975c *Phys. Rev. Lett.* **35** 1164
Webb R A, Sager R E and Wheatley J C 1977 *J. Low Temp. Phys.* **26** 439
Wellard N V 1982 Doctoral Thesis, University of Manchester
Wellard N V, Alexander P W, Hall H E and Hook J R 1982 *Physica* **109/110B + C** 2096
Wheatley J C 1966 in *Quantum Fluids,* ed. D F Brewer (North-Holland, Amsterdam), p. 183
Wheatley J C 1970 in *Progress in Low Temperature Physics,* Vol. VI, ed. J C Gorter (North-Holland, Amsterdam), p. 77

Wheatley J C 1973 *Physica* **69** 218
Wheatley J C 1975 *Rev. Mod. Phys.* **47** 415
Wheatley J C 1976 *Phys. Today* **29** 32
Wheatley J C 1978 in *Progress in Low Temperature Physics,* Vol. VIIA, ed. D F Brewer (North-Holland, Amsterdam), p. 77
Wheatley J C, Rapp R E and Johnson R T 1971 *J. Low Temp. Phys.* **4** 1
Whitham G B 1974 *Linear and Nonlinear Waves* (Wiley, New York)
Wilczek F and Zee A 1983 *Phys. Rev. Lett.* **51** 2250
Wilks J 1967 *The Properties of Liquid and Solid Helium* (Clarendon Press, Oxford)
Wilks J and Betts D S 1987 *An Introduction to Liquid Helium,* 2nd edn (Clarendon Press, Oxford)
Williams G A and Packard R E 1980 *J. Low Temp. Phys.* **39** 553
Williams M R and Fetter A L 1979 *Phys. Rev.* **B20** 169
Witten E 1979 *Nucl. Phys.* **B156** 269
Wojtanowski W 1986 *J. Low Temp. Phys.* **64** 309
Wojtanowski W and Wölfle P 1986a *Phys. Rev. Lett.* **56** 488
Wojtanowski W and Wölfle P 1986b *Phys. Lett.* **115A** 49
Wojtanowski W and Wölfle P 1987 *Phys. Rev. Lett.* **59** C2387
Wölfle P 1973a *Phys. Rev. Lett.* **30** 1169
Wölfle P 1973b *Phys. Rev. Lett.* **31** 1437
Wölfle P 1974 *Phys. Lett.* **47A** 224
Wölfle P 1975a *Phys. Rev. Lett.* **34** 1210
Wölfle P 1975b in *Quantum Statistics and the Many-Body Problem,* ed. S B Trickey, W P Kirk and J W Dufty (Plenum, New York), p. 9
Wölfle P 1976a *J. Low Temp. Phys.* **22** 152
Wölfle P 1976b *Phys. Rev.* **B14** 89
Wölfle P 1976c *Phys. Rev. Lett.* **37** 1279
Wölfle P 1977a *Physica* **90B** 96
Wölfle P 1977b *J. Low Temp. Phys.* **26** 659
Wölfle P 1978a in *Progress in Low Temperature Physics,* Vol. VIIA, ed. D F Brewer (North-Holland, Amsterdam), p. 191
Wölfle P 1978b in *Physics at Ultralow Temperatures,* ed. T Sugawara, S Nakajima, T Ohtsuka and T Usui (The Physical Society of Japan, Tokyo), p. 87
Wölfle P 1978c *J. Physique* **39** Colloq. C-6, Vol. III, p. 1278 (Proceedings of the 15th International Conference on Low Temperature Physics, L T-15)
Wölfle P 1979 *Rep. Prog. Phys.* **42** 269
Wölfle P 1980 in *Proceedings of the XVI Karpacz Winter School of Theoretical Physics* (Lecture Notes in Physics, Vol. 115), ed. A Pekalski and J Przyslawa (Springer, Berlin), p. 47
Wölfle P and Einzel D 1978 *J. Low Temp. Phys.* **32** 39
Wölfle P and Koch V E 1978 *J. Low Temp. Phys.* **30** 61
Wölfle P, Götze W, Kurkijärvi J and Smith H 1980 *J. Phys.* **C13** 2461
Woo C W 1976 in *The Physics of Liquid and Solid Helium,* Part I, ed. K H Bennemann and J B Ketterson (Wiley, New York), p. 349
Wu M K, Ashburn J R, Torng C J, Hor P H, Meng R L, Gao L, Huang Z J, Wang Y Q and Chu C W 1987 *Phys. Rev. Lett.* **58** 908
Yamaguchi Y 1983 *Prog. Theor. Phys.* **69** 1377
Yamaguchi Y, Katayama T and Ishii C 1982 *Phys. Lett.* **91A** 299
Yang C N 1962 *Rev. Mod. Phys.* **34** 694
Yanof A W and Reppy J D 1974 *Phys. Rev. Lett.* **33** 631
Yarmchuk E J and Packard R E 1982 *J. Low Temp. Phys.* **46** 479
Yip S 1985 *Phys. Rev.* **B32** 2915
Yip S 1986 *J. Phys.* **C19** 1491
Yip S and Leggett A J 1986 *Phys. Rev. Lett.* **57** 345
Yosida K 1958 *Phys. Rev.* **110** 769

Zhang W and Kurkijärvi J 1988 *J. Low Temp. Phys.* **73** 483
Zhang W, Kurkijärvi J and Thuneberg E V 1985 *Phys. Lett.* **109A** 238
Zhang W, Kurkijärvi J and Thuneberg E V 1987 *Phys. Rev.* **B36** 1987
Zhang W, Kurkijärvi J, Rainer D and Thuneberg E V 1988 *Phys. Rev.* **B37** 3336
Zinov'eva K N and Boldarev T S 1969 *Zh. Eksp. Teor. Fiz.* **56** 1089 [*Sov. Phys. JETP* **29** 585 (1969)]
Zotos X and Maki K 1984 *Phys. Rev.* **B30** 145
Zotos X and Maki K 1985 *Phys. Rev.* **B31** 7120

Author Index

Note: page numbers in bold refer to reference section

Subject Index